Materials Science

for Engineers

5th Edition

PERIODIC TABLE OF THE ELEMENTS

1	2	3	4	5	6	7	8	9	10	11	12	13	14	15	16	17	18
1·008 **H** 1																	4·003 **He** 2
6·940 **Li** 3	9·012 **Be** 4											10·811 **B** 5	12·011 **C** 6	14·007 **N** 7	15·999 **O** 8	18·998 **F** 9	20·183 **Ne** 10
22·990 **Na** 11	24·312 **Mg** 12											26·982 **Al** 13	28·086 **Si** 14	30·974 **P** 15	32·064 **S** 16	35·453 **Cl** 17	39·948 **A** 18
39·102 **K** 19	40·080 **Ca** 20	44·956 **Sc** 21	47·900 **Ti** 22	50·942 **V** 23	51·996 **Cr** 24	54·938 **Mn** 25	55·847 **Fe** 26	58·993 **Co** 27	58·710 **Ni** 28	63·540 **Cu** 29	65·370 **Zn** 30	69·720 **Ga** 31	72·590 **Ge** 32	74·922 **As** 33	78·960 **Se** 34	79·909 **Br** 35	83·800 **Kr** 36
85·470 **Rb** 37	87·620 **Sr** 38	88·905 **Y** 39	91·220 **Zr** 40	92·906 **Nb** 41	95·940 **Mo** 42	99 **Tc** 43	101·070 **Ru** 44	102·905 **Rh** 45	106·400 **Pd** 46	107·870 **Ag** 47	112·400 **Cd** 48	114·820 **In** 49	118·690 **Sn** 50	121·750 **Sb** 51	127·600 **Te** 52	126·904 **I** 53	131·300 **Xe** 54
132·905 **Cs** 55	137·340 **Ba** 56	RARE EARTHS **La-Lu**	178·490 **Hf** 72	180·948 **Ta** 73	183·85 **W** 74	186·20 **Re** 75	190·2 **Os** 76	192·2 **Ir** 77	195·09 **Pt** 78	196·967 **Au** 79	200·590 **Hg** 80	204·370 **Tl** 81	207·190 **Pb** 82	208·980 **Bi** 83	210 **Po** 84	210 **At** 85	222 **Rn** 86
223 **Fr** 87	227 **Ra** 88	ACTINIDE SERIES **Ac-Lw**															

RARE EARTHS	138·92 **La** 57	140·13 **Ce** 58	140·92 **Pr** 59	144·24 **Nd** 60	145 **Pm** 61	150·35 **Sm** 62	152·0 **Eu** 63	157·25 **Gd** 64	158·92 **Tb** 65	162·50 **Dy** 66	164·93 **Ho** 67	167·3 **Er** 68	169 **Tm** 69	173·04 **Yb** 70	174·98 **Lu** 71
ACTINIDE SERIES	227 **Ac** 89	232·038 **Th** 90	231 **Pa** 91	238·030 **U** 92	237 **Np** 93	242 **Pu** 94	243 **Am** 95	247 **Cm** 96	249 **Bk** 97	251 **Cf** 98	254 **E** 99	255 **Fm** 100	256 **Mv** 101	255 **No** 102	257 **Lw** 103

The atomic number of an element appears below its symbol, while its atomic weight is above.

Materials Science
for Engineers

5th Edition

J.C. Anderson
K.D. Leaver
P. Leevers
R.D. Rawlings

First edition published in 1969:
by Chapman & Hall
Second edition 1974
Third edition 1985
Fourth edition 1990

Fifth edition published in 2003 by:
Nelson Thornes Ltd
Delta Place
27 Bath Road
CHELTENHAM
GL53 7TH
United Kingdom

03 04 05 06 07 / 10 9 8 7 6 5 4 3 2 1

A catalogue record for this book is available from the British Library

ISBN 0 7487 6365 1

Page make-up by Aarontype Ltd, Easton, Bristol

Printed and bound in Croatia by Zrinski

Contents

Preface to the fifth edition

A scientific textbook whose first edition appeared in 1969 has inevitably seen many changes. That is certainly true of this edition, for the pace of development of new materials and new uses has accelerated, necessitating many small additions as well as major revisions of whole chapters. Since the fourth edition, high temperature superconductors have matured into practical materials, in use in mobile phone base stations around the world and in MRI (magnetic resonance imaging) machines. Semiconductor light-emitting diodes (LEDs) are replacing tungsten lamps in traffic lights, while the range of polymer products in common use has increased almost beyond measure.

For the new reader's guidance we have added an overview of its contents in Chapter 1, and for the same reason we have divided the book into three parts discussing respectively the structure, the mechanical properties and the electromagnetic properties of materials. Each chapter now includes an outline of the relevance of its topics, and a sample of the many applications for the materials mentioned therein.

Polymers receive an appropriately increased share of space in this edition, and the treatment of their structural and mechanical properties is now fully integrated into the book.

The team of authors welcomes Dr Patrick Leevers as a full co-author of this edition. We all wish to pay tribute to Professor 'Andy' Anderson, who died during its preparation, and for whom this fifth edition is an appropriate memorial, for the first edition was his inspiration. We are, as always, indebted to many colleagues at Imperial College and elsewhere who have given us advice and assistance in preparing this edition.

J.C.A.
K.D.L.
P.L.
R.D.R.

Preface to the first edition

The study of the science of materials has become in recent years an integral part of virtually all university courses in engineering. The physicist, the chemist and the metallurgist may, rightly, claim that they study materials scientifically, but the reason for the emergence of the 'new' subject of materials science is that it encompasses all these disciplines. It was with this in mind that the present book was written. We hope that, in addition to providing for the engineer an introductory text on the structure and properties of engineering materials, the book will assist the student of physics, chemistry or metallurgy to comprehend the essential unity of these subjects under the all-embracing, though ill-defined, title 'materials science'.

The text is based on the introductory materials course given to all engineers at Imperial College, London. One of the problems in teaching an introductory course arises from the varying amounts of background knowledge possessed by the students. We have, therefore, assumed only an elementary knowledge of chemistry and a reasonable grounding in physics, because this is the combination most frequently encountered in engineering faculties. On the other hand, the student with a good grasp of more advanced chemistry will not find the treatment familiar and therefore dull. This is because of the novel approach to the teaching of basic atomic structure, in which the ideas of wave mechanics are used, in a simplified form, from the outset. We believe that this method has several virtues: not only does it provide for a smooth development of the electronic properties of materials, but it inculcates a feeling for the uncertainty principle and for thinking in terms of probability, which are more fundamental than the deterministic picture of a particle electron moving along a specific orbit about the nucleus. We recognize that this approach is conceptually difficult, but no more so than the conventional one if one remembers the 'act of faith' which is necessary to accept the quantization condition in the Bohr theory. The success of this approach with our own students reinforces the belief that this is the right way to begin.

In view of the differences which are bound to exist between courses given in different universities and colleges, some of the more advanced material has been separated from the main body of the text and placed at the end of the appropriate chapter. These sections may, therefore, be omitted by the reader without impairing comprehension of later chapters.

In writing a book of this kind, one accumulates indebtedness to a wide range of people, not least to the authors of earlier books in the field. We particularly wish to acknowledge the help and encouragement given by our academic colleagues.

Our students have given us much welcome stimulation and the direct help of many of our graduate students is gratefully acknowledged. Finally, we wish to express our thanks to the publishers, who have been a constant source of encouragement and assistance.

J.C.A.
K.D.L.

Preface to the second edition

The goals of the first edition remain unchanged, but the need was felt to provide in the second edition a wider and more detailed coverage of the mechanical and metallurgical aspects of materials. Accordingly we have extensively rewritten the relevant chapters, which now cover mechanical properties on the basis of continuum theory as well as explaining the microscopic atomic mechanisms which underlie the macroscopic behaviour. The opportunity has also been taken to revise other chapters in the light of the many helpful comments on the first edition which we have received from colleagues around the world. We would like to thank here all who have taken the trouble to point out errors and inconsistencies, and especially our colleagues who have read the manuscript of this and the first edition. We are also indebted to Dr D. L. Thomas and Dr F. A. A. Crane for several micrographs which appear here for the first time.

J.C.A.
K.D.L.
J.M.A.
R.D.R.

Preface to the third edition

This edition represents a general updating and revision of the text of the second edition to take account of continuing developments in the field of materials science and the helpful comments and criticisms that we have received from our colleagues around the world.

Chapter 7, on thermodynamics, has been revised and extended to include the basic ideas of lattice waves and this is applied, in the chapter on electrical conductivity, to phonon scattering of electrons.

Chapters 8 and 9 on mechanical properties have been revised and extended and the fundamental principles of the relatively new subject of fracture toughness have been introduced; the microstructural aspects of fracture, creep and fatigue, together with the interpretation of creep and fatigue data, are dealt with in more detail.

The section on steel in Chapter 10 has been considerably extended. A completely new chapter on ceramics and composites has been added to take account of the increasing development and importance of these materials.

The chapter on semiconductors has been revised and updated by including the field effect transistor. An innovation is the introduction of a complete new chapter on semiconductor processing which is an important area of application of materials science in modern technology.

The addition of a chapter on optical properties has given us the opportunity of discussing the principles of spectroscopy, absorption and scattering, optical fibre materials and laser materials.

In general, all chapters have been reviewed and minor revision, correction and updating has been carried out where required.

J.C.A.
K.D.L.
R.D.R.
J.M.A.

Preface to the fourth edition

Since 1985, when the third edition was published, there have been rapid and significant developments in materials science and this new edition has been revised to include the most important of these.

Chapter 11 has been subject to major revision and updating to take account of the newly developed ceramics and high-strength composites.

A completely new chapter on plastics has replaced the original Chapter 12 on organic polymers. This has been written by Dr P. S. Leevers of Imperial College and includes the basic science and technology relating to the plethora of modern plastic materials that have had such a profound influence on everyday life. The authors are most grateful to Dr Leevers for his valuable contribution.

Chapter 13 now includes a section on superconductors and covers the new 'high-temperature' superconducting materials, discovered in 1986, for which the Nobel Prize in Physics was awarded in 1987.

Chapter 15 has been extensively revised to cover the most modern technologies adopted by the semiconductor industry, including molecular beam epitaxy and ion implantation.

Chapter 16 now includes coverage of modern permanent magnet materials and of the new 'glassy' amorphous magnetic materials.

Chapter 17 has been extended to give more details of piezo-, pyro- and ferroelectric materials, which have found increasingly wide applications in medical diagnostics, infrared imaging and communications.

Chapter 18 has been revised and extended to cover materials and techniques used in opto-electronic applications, which are increasingly important in information technology.

Finally, the authors would once again like to express their appreciation of the helpful comments and suggestions received from colleagues and of the publisher's continued enthusiam for our book.   J.C.A.
K.D.L.
R.D.R.
J.M.A.

Self-assessment questions

A series of self-assessment questions, with answers, will be found at the end of each chapter. By using these the student can easily test his understanding of the text and identify sections that he or she needs to re-read. The questions are framed so that the answer is a choice between two or more alternatives and the answer is simply a letter (a), (b), (c), etc. The correct answers are given at the end of each set of questions and it should be noted that, where a question involves more than two possibilities, there may be more than one right answer.

part I

Physics, chemistry and structure

The basis of materials science | 1

1.1 Introduction

Science is very much concerned with the identification of patterns, and the recognition of these is the first step in a process that leads to identification of the building bricks used to construct the patterns. This process has both the challenge and excitement of exploration and the fascination of a good detective story, and it lies at the heart of materials science.

At the end of the nineteenth century, a pattern had begun to emerge in the chemical properties of elements – this was fully recognized by Mendeléev when he constructed his periodic table. It was immediately apparent that there must be common properties and similar types of behaviour among the atoms of the different elements – the long process of understanding atomic structure had begun. There were many wonders along the way. For instance, was it not remarkable that among the elements *only* iron, nickel and cobalt showed the property of ferromagnetism? (Gadolinium, a fourth ferromagnetic element, was discovered much later.) A satisfactory theory of the atom must be able to explain this apparent oddity. Not only is magnetism exclusive to these elements, but also actual pieces of the materials sometimes appear magnetized and sometimes do not, depending on their history. Thus a theory merely stating that the atoms of the element are magnetic is not enough – we must consider what happens when the atoms come together to form a solid.

Similarly, we may wonder at the extraordinary range – and beauty – of the shapes of crystals. Here we have clear patterns – how can they be explained? Why are metals ductile while rocks are brittle? What rules determine the strength of a material, and is there a theoretical limit to the strength? Why do metals conduct electricity while ceramics and many plastics do not? All these questions are related to the properties of aggregations of atoms. Thus an understanding of the atom must be followed by an understanding of how atoms interact when they form a solid, because this must be the foundation on which explanations of the properties of materials are based.

This book attempts to describe the modern theories through which many of the above questions have been answered. We are not interested in tracing the history of their development but prefer to present, from the beginning, the quantum mechanical concepts that have been so successful in modern atomic theory. The pattern of the book parallels the pattern of understanding outlined above. We should start with a thorough grasp of fundamental atomic theory before going on to the theories of solid aggregates of atoms. As each theoretical concept emerges it is used to explain relevant observed properties. With such a foundation the many electrical, mechanical, thermal and other properties of materials can be described, discussed and explained.

1.1.1 The role of materials science in engineering

The study of materials for their own sake is not enough. Materials are of interest because we use them in an enormous variety of ways – they are a necessary part of all engineering activity. Even software products rely for their delivery on materials – the polymers used in compact disks; the optical fibres, copper and insulators in the interconnecting cables of the Internet; the glasses, polymers and liquid crystals used to make screens for displaying both images and text. Although new materials do not always lead to new engineering products, they frequently have done – semiconductors, liquid crystals, aluminium, titanium and polymers are only a few examples. Thus an engineer needs materials science, because the materials scientist can frequently supply him or her with novel properties that can only be effectively exploited if the engineer understands the language of materials science. This enables them to understand the limitations and opportunities that a new material presents, and also helps to optimize the use of both new and old materials. The properties we shall discuss are therefore primarily those of engineering relevance, although in the early chapters this will not always seem obvious.

Why, for example, is it useful to study the symmetry of crystals? In later chapters it will become clear that most materials are composed of an aggregation of many tiny crystals, called crystallites. Mechanical properties such as strength and ductility are influenced by the size and distribution of these crystallites. Heat treatments can be used to control crystallite size and thus improve a material's properties. Similar control of crystallite size is necessary to optimize the magnetic properties of materials destined for use in electrical voltage transformers, in magnetic tape video recorders and in computer disk memories. The size of the crystallites in the microscopic metallic conductors fabricated on computer chips even influences the lifetime before failure of the conductor.

Electronic 'chips', LEDs (light-emitting diodes), lasers and mobile phone electronic components are all made from highly perfect crystals of silicon, gallium arsenide, or similar materials. The nature of irregularities, collectively called defects, in the way that the atoms are stacked in crystals of these materials is critical to ensuring perfect operation of such devices. Indeed, without the efforts of many thousands of materials scientists, there is no doubt that computers would never have reached their present stage of development. Crystal defects, in particular those called *dislocations*, also determine how the dimensions of structural metals 'creep' in response to a constant, large force, and how the same materials 'fatigue' when repeatedly subjected to rather low forces.

Another apparently esoteric topic, discussed in Chapter 2, is the peculiar wavelike properties of electrons, which mean that they are capable of possessing only a limited number of values of energy. But the same features lead to the development of 'quantum' devices that make use of these very properties, which become noticeable when materials are made with dimensions of less than about 10^{-7} m.

1.2 Outline of the book

We summarize below the contents of the book, chapter by chapter, in an attempt to clarify the interrelationship between a material's structure and its properties – the basic aim of materials science. At the same time we will point out the relevance of the topics examined in each chapter. The book is divided into three main parts, covering successively the structure, mechanical properties and electromagnetic properties of materials, all of which are closely interrelated, as we shall see.

Part I: Physics, chemistry and structure

Chapters 2–5 begin by considering first isolated electrons, then electrons in atoms, leading on to how electrons join atoms together by forming bonds. This is followed in Chapter 6 by a discussion of the arrangements, or patterns, which atoms form when bonded together in solids, after which we explain how these patterns are disturbed by heat energy. We end our discussion of structure with a description of the various irregularities that result when atoms are displaced from their usual positions.

Chapter 2

The wavelike properties of electrons, which underlie quantum mechanics, are fully described in Chapter 2, because they are needed in subsequent chapters to explain atomic structure, the chemical behaviour of elements, the way that atoms bond together, and how these bonds stretch or bend when forces are exerted on them. The behaviour of electrons thus underpins much more than electronic engineering. The reader who already has a good grasp of the quantum nature of electrons and the structure of the hydrogen atom may wish to move straight to Chapter 4, where we use these ideas to explain the periodic table of the elements, a table that is frequently referred to in later chapters.

Chapter 3

Chapter 3 covers the basic features of atomic structure, using atomic hydrogen as a simple example. The quantum nature of the electron leads to behaviour that appears strange to someone familiar only with the physical properties of large bodies. The characteristic feature is that the total energy of an electron (kinetic + potential) cannot have just any value; there is a set of precisely defined values, or 'energy levels', which it may possess. This, when coupled with the peculiarity that only two electrons in any atom may share exactly the same motion around the nucleus, determines both the shape of the periodic table of the elements and the way in which atoms bond together to form molecules and solids.

Chapter 4

We show next just how atomic structure determines the periodic table of the elements. The periodic table, which forms the frontispiece of this book, is like a map on which elements with similar properties are clustered in a clearly recognizable way. Thus the simple metals on the left-hand side are well separated from the non-metals on the extreme right, while the middle ground is occupied by the so-called *transition metals*, which exhibit multiple oxidation states (multiple valencies).

The table is an invaluable guide when searching for alternatives to a known compound or alloy, as the substitution of one element by another from the same column of the table leads to more or less predictable but limited changes in properties.

Chapter 5

The bonds that join atoms together in a solid are formed by electrons. Their nature determines most of the properties discussed in Parts II and III of this book. We show how the behaviour of the outermost electrons in atoms results in bonds that fall into a small number of types: metallic, ionic, covalent, van der Waals and hydrogen bonds being the principal ones. The type depends on the electronic structure of the atom concerned, which means in turn the atom's location in the periodic table. This is one reason why so many engineering properties are related to position in this table.

Chapter 6

We next concentrate on the regularity and symmetry of the ways in which atoms form up in ranks in the solid state. Regular crystalline shapes formed by gemstones and in the naturally occurring forms of many crystalline salts are the external evidence of internal regularity. The types of bonding discussed in Chapter 5 affect the pattern. For example, metallic bonds result in structures that are tightly packed, while the bonds formed in carbon, silicon and germanium all lead to a more rigid, but less tightly packed, arrangement (see Fig. 6.4). The number of patterns that geometry allows is limited, so that classification into types enables us to predict whether properties depend, for example, on direction, or whether such properties as Young's modulus or refractive index are independent of the direction in which the measurement is made. We have already remarked on the need for near-perfect crystals in modern electronics.

Chapter 7

The regularity of crystals is disturbed by the heat energy that is present in all solids at normal temperatures. A solid may appear rigid, but internally its atoms are constantly vibrating back and forth at high frequency. We show how some atoms vibrate more energetically than others, which has implications, outlined in Chapter 9, for the way in which solids change their crystalline structure as the temperature increases. This becomes crucial, for example, when metallic alloys are discussed, because they depend for their mechanical strength on the control of their structure by heating and cooling in prescribed ways (see Chapter 9).

The movement of atoms by thermal diffusion, while easy to understand in gases and liquids, also occurs in solids, thanks to the presence of both the thermal energy possessed by atoms and the defects discussed in Chapter 8. Diffusion is explained in this chapter, in preparation for later use in connection with the mechanical behaviour of polymers and metals. An explanation of diffusion in solids, however, requires an understanding of the defects of structure discussed in the next chapter.

Chapter 8

To understand the strength of all but glassy materials, we need to study defects in the regular structure of crystals and crystallites, especially the kind called *dislocations*. The reason is that, although the strength of materials depends ultimately on the strength of interatomic bonds, nature has many ingenious ways of concentrating the forces applied to a solid onto just a few of these bonds, so that movement begins, and eventually fracture occurs, when the external forces are vastly lower than those needed to rupture all these bonds simultaneously. Other defects of structure affect many properties: electrical, magnetic and optical, as well as mechanical. Many defects are created by thermal energy, and can become frozen into place when a solid is cooled from just below its melting point.

Part II: Mechanical properties and applications

Chapter 9

We introduce in Chapter 9 the main tests used to characterize the mechanical properties of materials intended for structural use. These properties include stiffness, strength and toughness; we look at how they are measured under creep (constant loading over long periods), fatigue (many repeated loads) and impact (a single, suddenly applied load) conditions. If the test results are to be used to predict future behaviour under different conditions, they must be represented using more general 'material models', based on our understanding of mechanisms occurring at atomic and molecular scale. We look at simple models (linear elastic, viscoelastic, etc.) suitable for representing the three very different classes of materials represented in the following three chapters.

In each of Chapters 10–12 we then discuss the mechanical properties of one of the three major classes of materials – metals, ceramics and polymers.

Chapter 10

Metals were once the prime engineering materials – a car such as the one illustrated in cut-away in Fig. 1.1 would (before 1950) have been made almost entirely of metal parts. Even now, metals are used for the largest, hardest wearing and highest precision components, and for the electrical wiring.

Metals are characterized not just by their strength and by being good conductors, but also because they donate electrons when they combine chemically with non-metallic elements. The reason for this is readily understood from a study of their atomic structure (Chapter 4). The feature that makes them good structural engineering materials is that mechanical strength is coupled with *ductility* (Chapter 9), properties that derive from the way electrons bond the atoms together (Chapter 5) into structures (Chapter 6) that are much less rigid than materials such as silicon or diamond. Thus the different types of chemical bonds that join atoms together play a major role in determining the variety of mechanical properties exhibited by materials.

Metals deform under mechanical stress in unique ways, which depend upon the defects in the regular structure of the crystallites referred to earlier. We saw before that the crystallite size often affects their strength as well as various other properties.

Fig. 1.1

A car contains thousands of components made of many types of materials – metals, plastics, ceramics, semiconductors, magnets – each optimized for a particular function (drawing by kind permission of the Ford Motor Company Ltd).

Metallic elements mix readily to form alloys, often referred to as *solid solutions*. Alloys change their crystal structure as the temperature changes, as well as when the starting composition is altered, in such a way that many engineering alloys after heat treatment end up as a finely divided mixture of two or more different crystal structures, known as *phases*. This leads to changes in properties, which can be exploited to make the material either stronger, or more ductile, or alternatively, to alter both its electrical and thermal conductivities. Alloying two or more metallic elements together therefore creates an enormously enriched range of available properties. Thus there are literally hundreds of different stainless steels, each having a slightly different combination of properties. Because of this, we can only cover basic principles in this book, which is not intended as a guide to the selection of specific materials, except in very general terms. A good grasp of these basic principles will help an engineer to understand how to narrow down the field for detailed study when he or she turns to specialized books on materials selection, where a bewilderingly wide choice will often be presented.

Chapter 11

Ceramics and glasses broadly include all non-metallic solids that are not built up from organic substances. They thus include a vast range of compounds, and many are composed either of naturally occurring mixtures of compounds, or of artificial mixtures referred to as composite materials. Materials such as brick, cement, earthenware and glass are all ceramics whose uses need no introduction to readers, but there are many other ceramics of engineering importance, especially for use as high-temperature components in aero engines, rockets and, most famously, as cladding for NASA's space shuttle. The basic principles underlying these are explained in this chapter. Many ceramics are oxides of metals, or of semiconducting elements (e.g. SiO_2 and GeO_2).

Ceramics differ from most metals in being much more brittle: they fracture in a quite distinct way, and are susceptible to thermal shock, which metals are not. Their extreme mechanical hardness and high melting points are discussed in terms of chemical bonding – the connection between these two properties and bond strength will become clear in earlier chapters. Their brittle nature means that ceramic goods must be manufactured by methods that are quite different from those suitable for most metals and alloys. Hence Chapter 11 covers manufacturing methods for the more common ceramics – modern materials incorporating alumina (Al_2O_3), zirconia (ZrO_2) and silicon nitride (Si_3N_4), as well as cement and concrete, receive particular attention.

The properties of glasses, epitomized by the simplest, pure SiO_2 (quartz glass), are also explained in terms of their random network structure. Both the structure and properties are often modified by the addition of other oxides such as Na_2O, K_2O or CaO. Like ceramics they are both brittle and are processed at high temperature, so their thermal properties have to be discussed, as well as their mechanical strength.

Chapter 12

Polymer molecules have a unique shape: they are extremely long compared with atomic spacings, being constructed from a simple unit that is repeated many times (see, for example, Fig. 6.20). Chapter 12 begins by illustrating how different polymer types result from combining the simple units in various ways. For example, because they are so long, the molecules can be either simply intertwined, or bonded chemically to their neighbours, leading to two distinct types of polymer, thermoplastic and thermosetting, which are discussed separately.

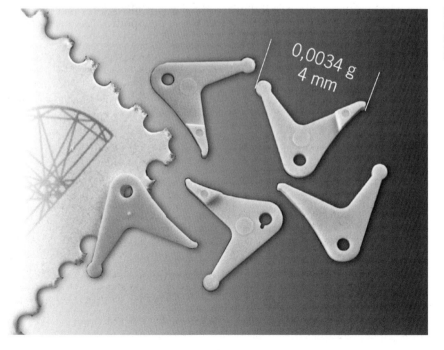

Fig. 1.2

Some polyacetal watch components (courtesy of Battenfield Injection Moulding Technology).

The mechanics of polymers, including types that are rubbery, are then explained. Having understood Chapter 4's explanation of how carbon or silicon atoms bond to each other, you will appreciate the strength and flexibility that this lends to such molecules.

Following a description of several important thermosetting polymers, we explain how shaped products are made by extrusion through a shaped aperture or by injection into a mould.

Amorphous polymers, with a glass-like structure, form the next important class to be covered, after which those polymers that crystallize are shown to possess rather different properties. Then we show how cross-linking the chain-like molecules affects the properties. Liquid crystal polymers are discussed briefly, before we survey the important properties of typical polymers: their mechanical, chemical, optical and electrical properties are treated in turn. Finally, some specialized products such as gels and adhesives are introduced.

The use of polymers to replace many of the former functions of metals such as steel, brass and bronze (including many parts of the car in Fig. 1.1), is a result of the efforts of materials scientists to improve polymer properties, as well as engineers, who have learnt how best to employ these properties without suffering from the accompanying disadvantages, such as higher thermal expansion, and mechanical failure at temperatures that for metals would be no problem.

The low price of oil, from which polymers are made, and the ease with which they can be formed into complex shapes, have together led to their use for nearly all types of small containers, as well as for fasteners, fabrics, furniture, paints, small machinery, and as a direct replacement for all but the largest wooden building components, including doors and window frames. Very large polymer products are still uncommon, for a variety of reasons, and mechanical reinforcement is usually necessary. In such cases a polymer's inherent resistance to atmospheric corrosion becomes an important factor, especially where metal reinforcement is used. Polymers with appropriate properties can also compete with metals for use in many

Fig. 1.3

A hard disk drive from a personal computer uses at least five different magnetic materials.

fine precision parts. Figure 1.2 shows one example: some watch components weighing just a few thousandths of a gram, moulded in polyacetal. Moulded into the final shape in a single operation, these materials hold their dimensions with great precision over time.

Chapter 13

Many materials developed for mechanical strength are *composites*, made of strong ceramics together with either polymers or metals chosen for their toughness. A simple, well-known example is fibreglass, which combines to great effect the flexibility of polymers with the tensile strength of glass made in the form of long hair-like fibres (Fig. 1.4). Fibreglass is widely used as thin panels in building construction, in all types of vehicles (especially small boats) and for strong casings around large equipment, and is discussed in depth in this chapter.

The principles underlying the optimum combination of the strength of the one with the toughness of the other are first introduced using simple examples of polymers combined with powdered ceramic 'fillers'. More elaborate examples using rubbers and fibres are then discussed, before comparing the mechanical performance of composites using polymers, metals or ceramics as the main constituent. Examples of several commercial composites are also included.

Part III: Electromagnetic properties and applications

This section covers all electrical, electronic, magnetic, optical and dielectric properties, which are dealt with in turn, a chapter at a time.

Chapter 14

Metallic conduction is not difficult to understand in the light of the metallic bonds introduced in Chapter 5. The fact that electrons cannot move without some hindrance (giving rise to electrical resistance) is explained in terms of crystal defects (Chapter 8) and thermal vibration of atoms (Chapter 7). Otherwise, metals would all be superconductors, with zero resistance. The unusual aspect of superconductors, discussed next, is that their electrons move in pairs, which, strangely, enables them to avoid the resistance presented by defects in the regular spacing of atoms. Superconductors have recently found new uses in mobile telephone base stations, where they help to reduce energy losses in critical circuit components, even though the need for them to operate at very low temperatures ($\sim$20 K) greatly increases costs.

Chapter 15

Our current ability to make microscopic electronic switches simultaneously by the billion depends wholly upon the study of materials we call semiconductors, which in their pure state do not conduct electricity particularly well. Materials science has played an enormous role here.

In semiconductors, as in metals, electrons move quite easily, but there are only a few that can do so. We explain how the number of 'conduction electrons' in a semiconductor depends greatly on temperature, unless the material is 'doped' with impurities chosen from particular columns of the periodic table. The construction and operation of transistors is also explained briefly. As mentioned in the Introduction, defects such as dislocations can cause transistors to malfunction.

Chapter 16

The manufacture of silicon 'chips' (an example is shown in Fig. 16.19) depends so heavily upon materials science and technology that we have included a chapter describing the methods used. As a result we introduce the reader to some scientific tools, like ion implantation, which have spawned new developments in many different materials and many applications.

Chapter 17

Magnetism in materials depends entirely upon the fact that an electron spins about its own axis. The combination of many electrons spinning about the *same* axis makes for strong magnetic properties. The various kinds of magnetic behaviour observed in different materials are related to their position in the periodic table. The useful materials, described either as ferromagnetic or ferrimagnetic, are optimized by intelligent selection from amongst those elements whose atoms contain the largest number of electrons spinning together, although we will show that other factors also come into play.

The computer disk drive shown in Fig. 1.4, for example, uses at least five different types of magnetic material, each to perform a different function, and we explain how the different types arise.

Chapter 18

Insulators and dielectrics can be ceramics, plastics or glasses, some of which possess interesting and unexpected properties. They include *piezoelectrics*, materials that can generate an electrical voltage when compressed or bent. Conversely, these materials will bend or change length when a voltage is

Fig. 1.4

Optical fibres enable the Internet to carry vastly more information than copper cables. Bundles of similar but much cheaper glass fibres are also used in combination with plastics to make strong flexible materials.

applied to electrodes in contact with them. Piezoelectric quartz crystals, for example, are used to control the rate of almost every modern watch or clock, whilst piezoelectric ceramics and plastics are used in a variety of motors and actuators for making fine adjustments to the position of items to accuracies of better than $1\,\mu$m. Pyroelectric materials, which respond to heat and are used as heat sensors, are also described, and we explain the underlying factors making for good insulating properties and high dielectric constants.

Chapter 19

The optical properties of materials are governed by the electromagnetic nature of light, as well as its quantum nature. Both of these determine the interaction of light with electrons in the solid. Thus transparency and absorption depend upon the wavelength of the light. Together with optical properties we include here the behaviour of materials in the infrared and ultraviolet portions of the electromagnetic frequency spectrum.

Quantum properties become even more apparent in materials in which laser action – light amplification – is possible. The electronic energy levels in atoms are seen to determine the way a material responds to light, just as they determine so many other properties. This brings us back full circle to the subject we shall begin in the next section: the properties of electrons, and particularly those properties that they lend to the atoms in which they endlessly circulate.

1.3 Atoms as planetary systems

All solids are composed of atoms, and so we begin by surveying some of the simpler facts about atoms that we will need to use, and which will be familiar to most readers, but perhaps not to those with less background in basic science. For example, how do we know that they consist of a heavy, charged nucleus around which circulate electrons carrying negative charge? Readers whose memories need no refreshing may wish to move straight to Chapter 2. We will take a more or less historical approach, as it clarifies the arguments that force us to make use of quantum theory. We shall see that the 'planetary' model of an atom, which was proposed in 1911 by Rutherford and then worked out in detail by Bohr and Sommerfeld, gives a reasonably good account of the size of atoms, but contains various arbitrary assumptions, and is inconsistent with other known facts, unlike modern atomic theory, which itself will be properly explained in Chapters 2 and 3. We shall attempt to build simple mathematical models of behaviour wherever possible, because such models, when tested against experiment, allow us to explain and predict all kinds of material properties.

1.3.1 Avogadro's number and the mass of an atom

We begin with Avogadro's number and the mole, and we use them first to calculate the mass of a single hydrogen atom, and compare it with an electron. Experiments on the electrolysis of water show that it takes 95,720 coulombs of electricity to liberate 1 gram of hydrogen, indicating that the hydrogen ion (the name for a charged hydrogen atom) carries a positive charge. Knowing how many atoms of hydrogen make up 1 gram it is possible to calculate the charge on each hydrogen ion. This can be done using Avogadro's number, which is defined as the number of atoms in a mole of a substance, i.e. the amount of substance whose mass in grams equals its atomic weight. Similarly a mole of a compound is the amount whose mass in grams equals its *molecular* weight, which, in turn, is the sum of the atomic weights of the atoms making up the molecule. In either case, the number of atoms or molecules making up a mole is equal to Avogadro's number. Atomic weights are all defined relative to the carbon-12 isotope: a mole of carbon-12 weighs exactly 12 grams[†], while a mole of hydrogen weighs 1.0080 grams.

Returning to our electrolysis experiment, the amount of electricity required to liberate a gram-atom of hydrogen will be $95,720 \times 1.0080 = 96,486$ coulombs. Now, *assuming* that the charge on the hydrogen ion is equal to that on the electron (measured by other means as 1.602×10^{-19} C, then we may calculate Avogadro's number, N:

$$N = \frac{96,486}{1.602 \times 10^{-19}} = 6.023 \times 10^{23} \, \text{mol}^{-1}$$

Experimentally, the accurate value has been determined as $6.023 \times 10^{23} \, \text{mol}^{-1}$, and so it may be concluded that the charge on the hydrogen ion is indeed equal in magnitude and opposite in sign to that on the electron.

Now, using Avogadro's number we may also calculate the mass of a hydrogen atom as

$$\frac{\text{atomic weight}}{\text{Avogadro's number}} = \frac{1.0080}{6.023 \times 10^{23}} = 1.672 \times 10^{-24} \, \text{g}$$

This is just 1840 times the mass of the electron.

Such a calculation suggests that there is another constituent of atoms, apart from the electron, which is relatively much more massive and positively charged.

1.3.2 The planetary atom

In 1911 Rutherford made use of the α-particle emission from a radioactive source to make the first exploration of the structure of the atom. On passing a stream of α-articles through a thin gold foil and measuring the angles through which the beam of α-articles was scattered, he noticed that a large proportion were deflected through only very small angles: the gold atoms must contain a lot of empty space! He concluded that most of the mass of the gold atom (atomic weight 187) is concentrated in a small volume called the nucleus, which carried a positive charge. He was also able to show that the radius of the gold nucleus is not greater than 3.2×10^{-11} mm.

Because the number of gold atoms in a mole of gold is given by Avogadro's number and, from the density of gold we may calculate the volume occupied by a mole, then we may estimate the volume of each

---

[†] In 1815 Prout suggested that if the atomic weight of hydrogen were taken as unity, the atomic weights of all other elements should be whole numbers. This turned out to be a simplification that was not subsequently supported by accurate measurements, largely because an element can have different isotopes, i.e. atoms of the same atomic *number* could have differing atomic *weights*: this will be discussed in a later chapter. Mixtures of naturally occurring isotopes were the principal cause of the atomic weights not being exact whole numbers. In 1962 it was internationally agreed to use the carbon isotope ^{12}C, with an exact atomic weight of 12, as the basis of all atomic weights, which gives hydrogen an atomic weight of 1.0080.

atom. If we assume them to be spheres packed together as closely as possible we can deduce that the radius of the gold atom is in the region of 10^{-7} mm, which is 10,000 times larger than the radius obtained by Rutherford. Thus the atom seems to comprise mainly empty space.

To incorporate these findings with the discovery of the electron, Rutherford proposed a 'planetary' model of the atom – this postulated a small dense nucleus carrying a positive charge about which the negative electrons orbited like planets round the Sun. The positive charge on the nucleus was taken to be equal to the sum of the electron charges so that the atom was electrically neutral.

This proposal has an attractive simplicity. Referring to Fig. 1.5, if an electron of mass m moves in a circular orbit of radius r with a constant linear velocity v then it will be subject to an inward acceleration v^2/r, maintained by the force of attraction between the electrical charges. This force F of electrostatic attraction is described by Coulomb's law:

$$F = \frac{q_1 q_2}{4\pi\varepsilon_0 r^2} \tag{1.1}$$

where q_1 and q_2 are the positive and negative charges (in coulombs), and ε_0 is the permittivity of free space, equal to $10^{-9}/36\pi\,\mathrm{F\,m^{-1}}$. In the atom of hydrogen, q_1 and q_2 each equal the value of the charge, e, on the electron. The force is related to the electron's inward radial acceleration by Newton's second law. The orbit of the electron thus settles down to a stable value when the force just balances the mass times the acceleration, that is, when

$$\frac{e^2}{4\pi\varepsilon_0 r^2} = \frac{mv^2}{r} \tag{1.2}$$

Unfortunately the model has a basic flaw. Maxwell's equations, which describe the laws of electromagnetic radiation, can be used to show that an electron undergoing a change in velocity (i.e. acceleration or deceleration) will radiate energy in the form of electromagnetic waves. This can be seen by analogy: if the electron in a circular orbit were viewed from the side it would appear to be oscillating rapidly backwards and forwards. Now it carries a charge and a moving charge represents an electric current. Thus its alternating motion corresponds to an alternating electric current at a very high frequency. Just such a high-frequency alternating current is supplied to an aerial by a radio transmitter and the electromagnetic radiation from the aerial is readily detectable. Thus we must expect the electron in Rutherford's model of the atom to continuously radiate energy.

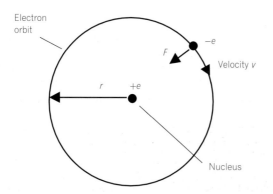

Electron orbit

$-e$

F

Velocity v

r $+e$

Nucleus

Fig. 1.5

Rutherford's planetary atomic model: an electron moving at velocity v in a circular orbit of radius r around a heavy nucleus carrying a charge $+e$ experiences an inward acceleration of v^2/r, sustained by the force F attracting it to the nucleus.

Expressed formally, an electron that is subject to a continual radial acceleration of magnitude v^2/r must, by the laws of electromagnetic radiation, continuously radiate energy.

The kinetic energy of the electron in its orbit is proportional to the square of its velocity, and if it loses energy its velocity must diminish. The outward, radial component of acceleration v^2/r will therefore also diminish, so that Eq. (1.2) is no longer satisfied unless r also diminishes, and so the electrostatic attraction between the positive nucleus and the negative electron will pull the electron closer to the nucleus. It is not difficult to see that the electron would ultimately spiral into the nucleus.

We shall see in Chapters 2 and 3 how wave mechanics removes this and various other difficulties that arise when this model is explored in more detail, as we shall do next.

1.3.3 Bohr's theory of the planetary atom

Following the suggestion that the atom may be like a planetary system, as outlined above, Bohr calculated the energy levels of the electron using such a model. He assumed that the electron was a particle that orbited the nucleus along a circular path, as in Fig. 1.5.

As mentioned above, the accelerating electron is expected to radiate energy. In order to get around this, Bohr had to postulate that certain orbits were stable, non-radiating states. In order to obtain agreement with the wavelengths of the spectral lines emitted he proposed an arbitrary quantization condition: the angular momentum of the electron should be an integral multiple of $(h/2\pi)$, where h, Planck's constant, had already been introduced to explain the quantum of energy hf carried by each photon of frequency f.

Thus the product of the linear momentum, mv, and the radius, r, is given by

$$mvr = \frac{nh}{2\pi} \tag{1.3}$$

where $n = 1, 2, 3, \ldots, \infty$ is an integer, called a *quantum number*, each value of which is associated with a different orbit. There is no justification for the *quantization condition* given in Eq. (1.3), but it is identical to Eq. (3.2) in Chapter 3, which results, as we shall see, from the assumption that the electron is wavelike. Ignorance of the wave properties gives incorrect results, in spite of the apparently correct choice made by Bohr for the quantization condition.

Bohr then proceeded to calculate the total energy of the electron. The kinetic energy E_k is

$$E_k = \tfrac{1}{2}mv^2$$

At a distance r from nuclear charge $+e$, where the electric potential is $V(r)$, the potential energy E_p of the electron is just

$$E_p = -eV(r) = -e\left(\frac{e}{4\pi\varepsilon_0 r}\right) = \frac{-e^2}{4\pi\varepsilon_0 r}$$

The total energy, E, is the sum of these:

$$E = E_k + E_p = \tfrac{1}{2}mv^2 - \frac{e^2}{4\pi\varepsilon_0 r}$$

With the help of Eqs (1.2) and (1.3), the radius r and velocity v can be eliminated to give the result

$$E = -\frac{me^4}{8h^2\varepsilon_0^2 n^2} \qquad (1.4)$$

which is exactly the result we shall obtain in Chapter 3 [Eq. (3.9)], and gives the correct energy levels for hydrogen atoms, as we shall see there.

However, this model ignores the possibility of non-circular orbits. Sommerfeld therefore extended the theory to allow for elliptical orbits, which are also possible according to Newtonian mechanics. To do this he had to introduce arbitrarily a second quantum number. Unfortunately the lack of the wave model leads to an error in the value of the angular momentum. A major difficulty lies in the fact that the zero angular momentum must be impossible if the electron is a particle. This is because the 'orbit' must then be a straight line passing through the nucleus, which should be impossible for a particle-like electron. But we shall see from the wave theory that the angular momentum may indeed be zero, and yet does not imply a collision between the electron and the nucleus.

In spite of these shortcomings the Bohr–Sommerfeld theory of the atom successfully explained many previously inexplicable facts. Corrections were made that brought the theory into very close agreement with the energy levels the electron may occupy, as deduced from spectroscopic measurements.

However, it could not deal with the criticisms above or with the fact that many of the expected wavelengths either do not appear in the spectrum of radiation, or are extremely weak. In all these respects the wave theory has proved entirely satisfactory.

The resolution of these difficulties was only made possible by bringing together a variety of observations and theories, and we shall consider these briefly in turn. To begin at the beginning we consider next the building bricks of the atoms themselves, starting with the electron.

Self-assessment questions

1   The mole is

   (a) the molecular weight of a substance in grams

   (b) the amount of a substance whose mass is numerically equal to its molecular weight

   (c) the amount of a substance whose volume is equal to that of 1 gram of hydrogen gas at standard temperature and pressure

2   Avogadro's number is

   (a) the number of atoms in a gram-atom

   (b) the number of molecules in a gram-molecule

   (c) both (a) and (b)

3   What is the mass of a helium atom, whose atomic weight is 4.003?

   (a) 1.672×10^{-26} g   b) 6.64×10^{-24} g

   (c) 2.41×10^{24} g

4   Rutherford's α-particle scattering experiment showed that

   (a) the nuclear charge is proportional to atomic number

   (b) electrons are small compared with the atom

   (c) the nucleus is small compared with the atom

Answers

1   (b)   2   (c)   3   (b)   4   (c)

2 | The smallest building blocks: electrons, photons and their behaviour

2.1 Introduction

In this book we will need to make frequent use of the properties of electrons. They control the way atoms combine to form molecules and solids, and so affect the strength of materials, and they determine the selection of materials for electrical, electronic and optical engineering. Their quantum mechanical properties underlie the 20th century revolutions involving silicon electronics, in lasers and superconductors. Proposals for quantum computing 'machines' could revolutionize computers later this century.

The wavelike, quantum properties of electrons show up most clearly in the smallest man-made objects – in so-called quantum dots, quantum wires and quantum pillars (Fig. 2.1), and in the primary building blocks of materials, namely atoms. Wavelike electron behaviour underlies the

Fig. 2.1

'Quantum pillars' of metals made for research into 'smart' material properties, viewed under an electron microscope (courtesy Mino Green, Imperial College).

structure of the periodic table of the elements, so our quest for an understanding of materials begins with electrons.

We begin in Section 2.2 by simply stating the properties of the electron as they were known at the start of the 20th century, before outlining the wave-mechanical aspects of electron behaviour, which we will need to discuss atoms in the next chapter. We then compare the electron with the photon, the 'particle' of light that shows its particulate nature when interacting with electrons. The universal character of the dual wave-particle nature of all forms of energy and matter thus becomes apparent. Readers wanting to know more about the experimental evidence for our knowledge of fundamental electronic properties will find further details in Sections 2.4 and 2.5.

2.2 Fundamental properties of electrons and photons

The electron was first clearly identified as an elementary particle in 1897 by J. J. Thomson. A description of his experiment is given in Section 2.5; here we simply note that he proved the electron to be a constituent of all matter. It has a fixed, negative charge, e, of 1.6022×10^{-19} coulomb and a mass of 9.1095×10^{-31} kg when at rest[†].

As the smallest unit of matter known at the beginning of the 20th century, electrons were soon found to exhibit wavelike behaviour, like light, as well as having fixed charge and mass. The reconciliation of these apparently conflicting aspects will be discussed in Section 2.3, and the experimental evidence for it in Sections 2.4 and 2.5. Here we simply give Louis de Broglie's equation for the measured wavelength λ of an electron, which is directly related to the momentum p that the particle possesses when in motion:

$$\lambda = \frac{h}{p} \tag{2.1}$$

where h, Planck's constant, has the value 6.625×10^{-34} J s. This is one of the most fundamental relationships in science and applies to all matter, from electrons to billiard balls. It implies that all moving matter will exhibit wavelike behaviour.

Consider for example an electron that has been accelerated to a high velocity v by a voltage V of, say, 1.0 kV, applied between an electrode and a source of electrons such as a hot wire. In losing the potential energy eV, the electron has acquired kinetic energy E_k given by

$$E_k = \tfrac{1}{2}mv^2 = eV \tag{2.2}$$

$$= 1.602 \times 10^{-16} \text{ J}$$

Now an electron's kinetic energy can also be expressed in terms of its momentum $p = mv$, thus

$$E_k = p^2/2m \tag{2.3}$$

[†] In case the reader compares these values with those in an earlier edition of this book, we should point out that refinements in the accuracy of measurement result in small changes over time.

so, using Eq. (2.1), the above electron has a wavelength equal to

$$\lambda = \frac{h}{p} = \frac{h}{\sqrt{2mE_k}} = \frac{6.625 \times 10^{-34}}{\sqrt{2 \times 9.11 \times 10^{-31} \times 1.602 \times 10^{-16}}}$$

$$= 1.248 \times 10^{-11} \text{ m}$$

The wavelike property results in electrons being *diffracted* and showing *interference* patterns. A crystalline material like carbon or silicon has atoms regularly spaced apart by a distance of about 0.3 nm (1 nm $= 10^{-9}$ m), which is small enough to cause a diffraction pattern like that shown in Fig. 2.7 when an electron beam passes through a thin layer of such a material.

Another consequence of the charge that the electron carries is that it experiences a force when it is moving in a magnetic field. The strength F_B of this force is proportional to both the flux density B and the electron velocity v. When the electron velocity v is directed at right angles to the direction of the magnetic field, the force is given by the equation

$$F_B = Bev \tag{2.4}$$

The behaviour of free electrons in electric and magnetic fields is reviewed in Section 2.5, where we describe the experimental methods for determining the basic properties of electrons.

Another 'particle' with which we shall frequently be concerned is the photon, the smallest packet of light energy. Consider light of a single wavelength λ and frequency f, related to one another by the velocity c of light in the equation

$$\lambda f = c$$

The smallest packet of energy, E, called a quantum, is directly related to the frequency by Planck's equation:

$$E = hf \tag{2.5}$$

where h is again Planck's constant, and has the value given earlier. The evidence underlying these strange properties of light is also reviewed in Section 2.5.

2.3 Particles or waves?

Possibly the most challenging puzzle that gradually emerged from experimental physics as more and more physical phenomena were explored was this conflict between the particle and wave views of electrons and photons. Certain aspects remain a puzzle at the beginning of the 21st century.

How can we reconcile these views? One way is to think about the many possible forms a wave may take. The first is a simple plane wave. If we plot amplitude vertically and distance horizontally we can draw such a wave in one dimension, as in Fig. 2.2(a). This wave has a single wavelength and frequency. If we add together two plane waves with slightly different wavelengths the result is a single wave of different 'shape', as shown in Fig. 2.2(b). Its amplitude varies with position so that there is a maximum at regular intervals. Adding a third plane wave with yet another wavelength makes some of these maxima bigger than others. If enough plane waves are added together just *one* of the maxima is bigger than all the others, as

Fig. 2.2

Combination of waves of different
frequency to form a wave packet.

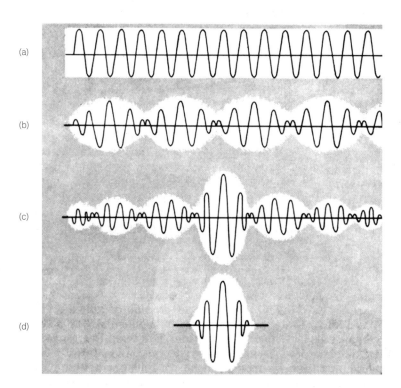

in Fig. 2.2(c): if they are added together in the correct phases all the other maxima disappear and the wave so formed has a noticeable amplitude only in one (relatively) small region of space, as shown in Fig. 2.2(d). This type of wave is often called a *wave packet* and it clearly has some resemblance to what we call a particle, in that it is localized. The distance over which a wave packet such as that in Fig. 2.2(d) has more than about half of its maximum amplitude is often used as a rough measure of the uncertainty in the particle's position. The mathematics underlying this description of wave packets is given in Appendix 5.

So we can regard an electron, or a light wave, when it acts in the manner of what we call a 'particle', as being a wave packet. It is still useful at such times to use the language that applies to particles, just as a matter of convenience.

2.3.1 Wave nature of larger 'particles'

To treat the electron purely as a 'particle', then, is an approximate way of describing it. Its wavelike behaviour is expected to show up when the electron wavelength is comparable, for example, to the diameter of an atom or the distance between atoms. This is the case, too, with light – its wave nature is apparent when it passes through a slit whose width is comparable to the wavelength of the light.

Now an atom, having mass like an electron, also has a characteristic wavelength given by the relationship $\lambda = h/p$. It thus shows wavelike properties when it interacts with something of the right size – roughly equal to its wavelength.

However the lightest atom (hydrogen) is about 2000 times heavier than the electron, so for the same velocity its momentum, p, is about 2000 times greater and so the wavelength, λ, is correspondingly 2000 times smaller than that of the electron. Its wavelike properties will rarely be observable, because its wavelength is so small. The same applies, even more so, as we consider bigger and bigger 'particles'. So a dust particle consisting of several million atoms virtually never displays wavelike behaviour and can always be treated as a particle. This means, in terms of physics, that Newton's laws will always describe the motion of a dust particle while they cannot be expected to apply, at any rate in the same form, to electrons.

Thus we see that treating an electron that is orbiting the nucleus of an atom as if it were a particle subject to Newtonian mechanics (see Section 1.3) is bound to be inadequate. It is necessary to develop a new mechanics – called either quantum mechanics or wave mechanics – to deal with the problem. This is the subject of a later section, where a mathematical description of the electron wave is developed.

If you want to learn about the physical evidence that underpins the dual wave-particle picture of matter outlined above, you should now read Sections 2.4 and 2.5. Otherwise, go to Section 2.6.

2.4 Waves and particles – the evidence

How do we confirm that both light and electrons behave in some ways like particles, in others like waves? In this section we examine some of the relevant experiments.

2.4.1 Electron waves

The properties that convince us that light is a wave motion are *diffraction* and *interference*. For instance, a beam of light impinging on a very narrow slit is diffracted (i.e. spreads out behind the slit). A beam of light shining on two narrow adjacent slits, as shown in Fig. 2.3, is diffracted, and if a screen is placed beyond the slits interference occurs, giving the characteristic intensity distribution in the figure. Yet another pattern is produced if we use many parallel slits, the device so produced being called a *diffraction grating*. If we take a linear source of light of one colour and project its image through a diffraction grating, as sketched in Fig. 2.4, we see a set of lines whose separation depends on the wavelength of the light and the distance between the slits in the diffraction grating. If two identical diffraction gratings are superimposed with their lines intersecting at right angles, a spot pattern is produced at the intersections of the two sets of linear images on the screen. The distances between the spots are a

Fig. 2.3

Diffraction of light through two slits.

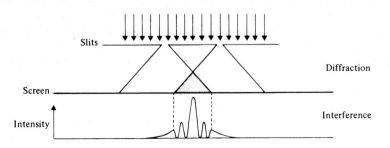

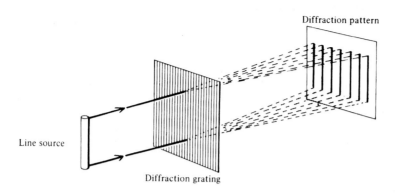

Fig. 2.4

Diffraction grating and diffraction pattern.

function of the wavelength of the light for a given pair of gratings. For the experiment to work the dimensions of the slits in the grating *must be comparable to the wavelength of the light*, i.e. the slits must be about 10^{-4} cm wide and separated by a similar distance.

If we are to observe the wavelike behaviour of electrons, we can expect to do so with a diffraction grating of suitable dimensions. It so happens that the size required is comparable to the mean distance between atoms in a typical crystal, i.e. in the region of 10^{-7} mm. Thus if we 'shine' a beam of electrons on a crystal that is thin enough for them to pass through, the crystal will act as a diffraction grating if the electrons behave as waves.

Referring to Fig. 2.5, a source of electrons, such as a hot filament, is mounted in an evacuated enclosure with suitable electrodes such that the electrons can be accelerated and collimated into a narrow beam. The beam is allowed to fall upon a thin carbon layer, on the other side of which is placed a fluorescent screen of the type commonly used in television tubes (a description of suitable fluorescent materials is given in Section 19.10.2). Electrons arriving at the screen produce a glow of light proportional to their number and energy.

The energy imparted to the electron beam is determined by the potential, V, applied to the accelerating electrode. If we are regarding it as a charged particle, the potential energy lost by an electron on reaching an electrode with an applied voltage of V volts is, by definition of potential, just eV, where e is its charge in coulombs. The accelerating electron increases its kinetic energy at the expense of its potential energy and, if we treat it as a particle, it will reach a velocity, v, given by Eq. (2.2):

$$\tfrac{1}{2}mv^2 = eV$$

Although it is not clear how we can attribute a kinetic energy of $\tfrac{1}{2}mv^2$ to a wave, we can nevertheless say that it has gained energy, and it is convenient to define this energy as the change in potential energy of the wave packet in passing through a potential difference of V volts. This energy will be eV joules and, for convenience, we define a new unit of energy, the *electron volt*. This is defined as the energy an electron acquires when it passes through a potential drop of 1 volt and is equal to 1.602×10^{-19} J.

Returning to Fig. 2.5, the electrons thus impinge upon the carbon layer with high energy. If they were particles we would expect them to be scattered randomly in all directions when they hit atoms in the carbon crystals. The intensity distribution beyond the carbon layer would then be

Fig. 2.5

Schematic arrangement to demonstrate electron diffraction.

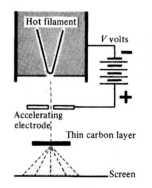

Fig. 2.6

Intensity distribution produced by randomly scattered light.

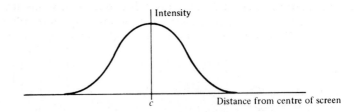

greatest at the centre, falling off uniformly towards the edges, as shown in Fig. 2.6, and the pattern of light on the screen would be a diffuse one, shading off gradually in brightness from the centre outwards.

In fact the experiment produces a quite different result. The pattern seen on the screen consists of rings of intense light separated by regions in which no electrons are detected, as shown by the photograph in Fig. 2.7. This can only be explained by assuming that the electron is a wave. If layers of atoms in a crystal act as a set of diffraction gratings we can regard them as line gratings superimposed with their slits at right angles and the diffraction pattern should be a set of spots, as described earlier. The pattern seen is a series of rings rather than spots, because the carbon layer is made up of many tiny crystals, all oriented in different directions; the rings correspond to many superimposed spot patterns, each of which is rotated by a different angle around the incident beam. Some of the rings correspond to diffraction from different sets of atomic planes in the crystal. Diffraction of electrons by crystals is a powerful tool in the determination of crystal structures and is dealt with in detail in more advanced textbooks.

If in this experiment the accelerating voltage, V, is varied, the diameters of the rings change, and it is not difficult to establish that the radius, R, of any ring is inversely proportional to the square root of the accelerating voltage, i.e. $R \propto 1/V^{1/2}$.

To understand this we must introduce yet another branch of physics, that of X-ray crystallography. X-rays are electromagnetic waves having the same nature as light but with much shorter wavelengths. 'Hard' X-rays have wavelengths in the range of 0.1 to 1 nm and are therefore diffracted by the planes of atoms in a crystal, in the same way as light is diffracted by

Fig. 2.7

Electron diffraction rings.

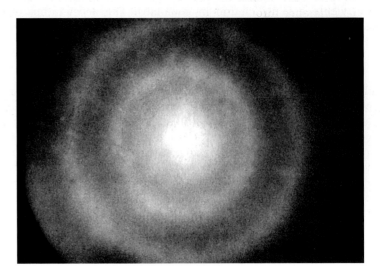

a diffraction grating. The study of X-ray diffraction by crystals led to the well-known *Bragg's law*, relating wavelength to the angle through which a ray is diffracted. This is derived in Chapter 6 and the law can be written as

$$n\lambda = 2d \sin \theta \qquad (2.6)$$

where n is an integer (the 'order' of the diffracted ray), λ is the wavelength, d the distance between the planes of atoms, and 2θ is the angle between the direction of the diffracted ray and that of the emergent ray, as shown in Fig. 2.8. Because the electrons appear to be behaving as waves it is reasonable to assume that we may also apply Bragg's law to them.

Referring again to Fig. 2.5, in our experiment the electron beam fell normally on the carbon film and, on emergence, had been split into diffracted rays forming a cone with an angle measuring 4θ at its apex. From the diagram, the radius R of the diffraction ring is given by

$$R = \ell \tan 2\theta$$

where ℓ is the distance from the carbon film to the screen. Because θ is commonly only a few degrees we may approximate

$$\tan 2\theta \approx 2\theta \approx 2 \sin \theta, \text{ so that}$$
$$R \approx 2\ell \sin \theta$$

and, using Bragg's law from Eq. (2.6),

$$R \approx \frac{n\ell\lambda}{d} \qquad (2.7)$$

Now n has integer values, giving several rings, the innermost corresponding to $n = 1$. Because R and ℓ can be measured for this ring, we may determine λ from this experiment, given a knowledge of the spacing, d, between the atomic planes in carbon, which is known by measuring the angle through which they diffract X-rays of known wavelength.

The importance of this result for our present purposes is that it shows that R is proportional to λ, the other quantities being constants. But we have already observed that $R \propto 1/V^{1/2}$ and so we conclude that $\lambda \propto 1/V^{1/2}$. Now we earlier defined the energy of the wave, E, in terms of the electric potential difference through which it moved, that is, $E = eV$. Because e is a constant it follows that

$$\lambda \propto \frac{1}{E^{1/2}} \qquad (2.8)$$

and we have the important result that *the electron wavelength is inversely proportional to the square root of the kinetic energy of the wave*. Although the electron might be behaving like a wave we do know that its kinetic energy is also related to its mass and velocity by Eq. (2.2):

$$E = \tfrac{1}{2} mv^2$$

so that it must have a momentum, p, given by $p = mv$. Relating the kinetic energy, E, to the momentum we have

$$E = \tfrac{1}{2} mv^2 = \frac{(mv)^2}{2m} = \frac{p^2}{2m} \qquad (2.9)$$

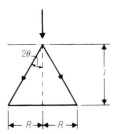

Thus, because m is a constant, $E^{1/2} \propto p$ and using the experimental result of Eq. (2.8), we may write

$$\lambda \propto \frac{1}{p}$$

The experiment suggests that the electron wave carries a momentum that is inversely proportional to its wavelength. The wave, in some way, represents a mass m moving with a velocity v and we refer to it as a 'matter wave'.

The relationship between λ and p was, in fact, proposed by de Broglie before the phenomenon of electron diffraction was discovered. Putting in a constant of proportionality, h, we obtain de Broglie's relationship:

$$\lambda = \frac{h}{p} \tag{2.10}$$

in which h is the universal constant called *Planck's constant* and has the value given earlier, 6.625×10^{-34} J s. Further experiments with atoms and molecules in place of electrons confirm that all moving matter exhibits wavelike behaviour that is governed by de Broglie's relation.

2.4.2 Particles of light

We have already mentioned that the properties of diffraction and interference show that light is a wave motion. But now consider the experiment shown in Fig. 2.9. Some materials (e.g. caesium) when irradiated with light will emit electrons that can be detected by attracting them to a positively-charged electrode and observing the flow of current.

Other experiments show that to remove an electron from a solid it is necessary to give the electron enough kinetic energy to surmount a 'potential energy barrier' (often called just the 'potential barrier') that otherwise traps the electron within the solid. Any energy supplied by the light beam that is in excess of a minimum amount, W, needed to surmount the barrier, must be carried away by the electron as its kinetic energy, E_k, i.e.

$$\text{total energy supplied by light} = E_L = W + E_k \tag{2.11}$$

If we reduce E_L by reducing the amount of light supplied (i.e. its intensity), then E_k should also decrease measurably. But this does not happen! What we observe is that the *number* of electrons released by the

Fig. 2.9

Schematic arrangement to demonstrate photoemission.

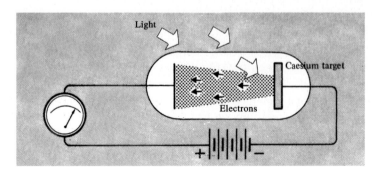

light reduces proportionally with the light intensity but the *energy* of the electrons remains the same. This was first explained by Einstein, by supposing the light beam to consist of particles of fixed energy. Reducing the intensity would then reduce the number of particles flowing per unit time. If we suppose that when a particle 'collides' with an electron it can give the electron its whole energy, then reducing the rate of flow of particles would reduce the number of collisions, but the energy transferred at each collision would remain the same. Thus we would reduce the rate of emission of electrons without changing their kinetic energy.

We call the light particles 'photons'. How do we know how much energy a photon has? We can get some idea by changing the *frequency* of the light. By trying several different wavelengths (i.e. different frequencies) it is possible to show that the energy, E, of each photon is proportional to the frequency, f, that is, $E \propto f$. This is clearly different from the case of the electron, for which $E^{1/2} \propto 1/\lambda$ [see Eq. (2.8)]. Nevertheless, the constant of proportionality is again h, Planck's constant, so that the photon's energy is $E = hf$.

Thus the energy carried by the wave is apparently broken up into discrete lumps or 'quanta'. The energy of each quantum depends only on the frequency (or wavelength) of the light while the intensity (or brightness) of the light is determined by the number of quanta arriving on the illuminated surface in each second. This idea emerged around the time the electron was discovered. It was in 1900 that Planck suggested that heat radiation, which is infrared light, is emitted or absorbed in multiples of a definite amount. Just as wavelike behaviour can be assumed to apply to all moving matter, so all energy is *quantized*, whatever its form.

In most practical applications in the real world, such as in the oscillations of a pendulum, the frequency is so low that a single quantum is too small to be observable. However, at high frequencies of light, e.g. X-rays and electron waves, quantum effects will be readily apparent. This general truth is the basis of quantum theory.

2.5 Finding the values of e/m and e for electrons

2.5.1 J. J. Thomson's experiment: measurement of e/m

The following descriptions are not historically accurate but are intended to highlight the basic behaviour of electrons in the presence of electric and magnetic fields.

Beginning with J. J. Thomson's type of experiment, suppose that we have the apparatus illustrated in Fig. 2.10. Electrons are emitted from a hot filament, are focused into a beam, and accelerated by a potential V to a velocity v, finally impinging on a fluorescent screen. The beam is arranged so that it passes between two plane parallel electrodes between which a voltage can be applied to produce a uniform electrostatic field, E V m^{-1}, across the path of the beam. Treating the electrons as particles with a charge $-e$, they will be deflected sideways by the field by virtue of the force it exerts, F_{E}. This force is given, from the definition of an electric field, by $F_{\mathrm{E}} = -eE$, and the resulting deflection of the beam moves its arrival point at the screen through a distance d.

Fig. 2.10

Schematic diagram of Thomson's experiment.

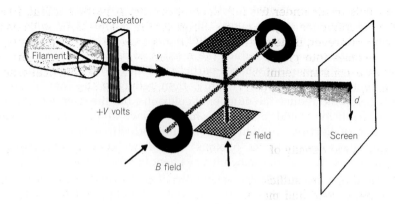

In the same position as the electrodes we also have a pair of magnetic coils that, when supplied with current, produce a uniform magnetic field of flux density B (Teslas) perpendicular to the path of the electron beam. Now the moving electrons, because they are charged, are equivalent to a current of value ev/l over a path of length l. One of the fundamental laws of electromagnetism is that a wire carrying a current, i, in a uniform magnetic flux density, B, experiences a force perpendicular to both the direction of current and of the magnetic field given by BiL, where L is the length of the wire. In the present case, the electrons will therefore experience a force given by

$$F_B = Bev$$

If we now arrange that the forces due to the electric and magnetic fields oppose each other and we adjust the fields so that the spot on the screen is returned to its original straight-through position we have

$$eE = Bev \qquad (2.12)$$

The velocity of the beam is found from the relationship $E = \frac{1}{2}mv^2$ so that

$$v = \left(\frac{2eV}{m}\right)^{1/2} \qquad (2.13)$$

Substituting this expression for v into Eq. (2.12) gives the result

$$E = B\left(\frac{2eV}{m}\right)^{1/2} \qquad \text{or} \qquad \frac{e}{m} = \frac{E^2}{2VB^2} \qquad (2.14)$$

As E, V and B are known, we can obtain a value of e/m for the electron, which is $1.759 \times 10^{11}\,\text{C}\,\text{kg}^{-1}$.

By using different materials for the filament wire we can establish that the same value for e/m is always obtained, supporting the statement that the electron is a constituent of all matter.

2.5.2 Millikan's experiment: measurement of e

Because the experiment described above is only able to yield the charge-to-mass ratio of the electron we need a means of measuring either the charge or the mass separately. Thomson designed a suitable method for measuring e, which was improved in Millikan's well-known oil-drop experiment. Both depended upon the speed with which a small drop of liquid falls in air under the influence of gravity. A particle falling freely under the influence of gravity accelerates continuously – nothing limits its velocity. However, if the particle is falling in air, the friction, or viscosity, of the air comes into play as the velocity increases and the particle finally reaches a constant terminal velocity. If the oil drop is spherical with a radius a, its velocity v is given by a law, derived by Stokes:

$$v = \frac{2ga^2\rho}{9\eta} \tag{2.15}$$

where ρ is the density of the sphere and η is the coefficient of viscosity of the air.

If the drops are sufficiently small, v may be only a fraction of a milli-metre per second and may be measured by direct observation. It is then possible to deduce the radius, a, and hence the mass of the drop. This is the prerequisite for the oil-drop experiment. A mist of oil is produced by spraying through a fine nozzle into a glass chamber. The drops may be observed through the side of the chamber using a suitable microscope with a graticule in the eyepiece. The free fall of a suitable drop is observed in order to calculate its mass from Eq. (2.15). At the top and bottom of the chamber are plane parallel electrodes to which a potential V can be applied, producing a uniform electric field in the vertical direction. The drop of oil is charged by irradiating it with ultraviolet light, which causes the loss of one or more electrons. Let us concentrate on a particular drop that has acquired a positive charge, $+e$. The potential V is adjusted so that the drop appears to be stationary when the upward force due to the field just balances the downward force due to the weight of the drop. Thus,

$$\frac{eV}{d} = mg \tag{2.16}$$

where d is the distance between the electrodes, m is the mass of the drop, and g the acceleration due to gravity.

The charging of the drop by ejecting an electron from it will be better understood when we have dealt with atomic structure in more detail. It is impossible to be sure that the drop has not lost two, three, or even more, electrons. However, by observing a large number of drops the charge always works out to be an integral multiple of 1.6×10^{-19} C. Thus the *smallest* charge the drop can acquire has this value, which we therefore assume to be the charge carried by a single electron. Because there is no theoretical proof that the drop cannot charge except by losing an electron we are, in fact, jumping to conclusions. However, once we have studied the detailed atomic model in Chapter 3, no other explanation will be conceivable.

2.6 Wave mechanics – matter, waves and probability

In the previous sections we saw how the properties of the electron as a particle (which is localized) and a wave (which extends over a region of space) may be brought together by treating an electron as a wave with wavelength $\lambda = h/p$. We need a mathematical means of representing the wave so that we can perform calculations of the kind for which we normally use Newtonian mechanics.

Wave function Ψ as a function of (a) ωt for $x = 0$ and (b) position x for $\omega t = 0$.

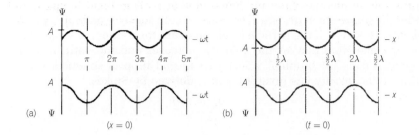

A plane wave travelling in the x-direction and having a wavelength λ and frequency f can be represented by expressions of the type

$$\Psi = A \cos\left(\omega t - \frac{2\pi x}{\lambda}\right) \qquad (2.17)$$

where $t =$ time and $\omega = 2\pi f$, and A, the peak amplitude, is in the simplest case just a constant. We may plot the time variation of Ψ, the wave's instantaneous amplitude, at a given point on the x-axis, e.g. at $x = 0$, as shown in Fig. 2.11(a). This is what would be observed by standing at the point $x = 0$ and observing the variation in amplitude of the wave with time as it passes. Alternatively, we may choose to look at a length of space at a given time, for example at time $t = 0$, and plot the variation of amplitude of the wave with position along the length chosen, as in Fig. 2.11(b).

If we describe an electron by such an equation an immediate difficulty arises, namely, what does the amplitude of the wave represent? In all our experiments we can detect only that an electron is present or is not present – we can never detect only a part of it. Thus the electron must be represented by the *whole* wave and we must somehow relate the point-to-point variation in amplitude to this fact. It was the physicist Max Born who suggested that the wave should represent the probability of finding the electron at a given point in space and time, that is, Ψ is related to this probability. But Ψ itself cannot actually equal this probability because the sine and cosine functions can be negative as well as positive, and there can be no such thing as a negative probability. Thus we use the intensity of the wave, which is the square of its amplitude, to define the probability. However, at this point our ideas and our mathematical representation are out of step. We can see that Ψ has no 'real' meaning and yet in Eq. (2.17) Ψ is a mathematically real quantity. As described in Section A5.1 of Appendix 5, Ψ is normally a mathematically complex quantity with real and imaginary parts, i.e. we can represent Ψ by a complex number of the type $(A + jB)$, where $j = \sqrt{-1}$.

Now the square of the modulus of a complex number is always a real, positive number, and so we can use this to represent the probability defined above. Thus we will have the wave function Ψ in the form:

$$\Psi = (A + jB)$$

and the probability given by

$$|\Psi|^2 = A^2 + B^2$$

Traditionally – and in most physics textbooks – this is put in a different way that, nevertheless, is precisely the same thing. If $\Psi = (A + jB)$, its complex conjugate is $\Psi^* = (A - jB)$. The probability of finding an electron that is represented by the wave function Ψ at a given point in space and time is defined by the product of Ψ and its complex conjugate Ψ^*, that is:

$$\begin{aligned}
\text{probability} &= \Psi\Psi^* \\
&= (A + jB)(A - jB) \\
&= A^2 + B^2 \\
&= |\Psi|^2 \qquad\qquad (2.18)
\end{aligned}$$

The entire wave motion can therefore be regarded as a shimmering distribution of charge density of a certain shape. This nebulous solution is in contrast to the sharp solutions of classical mechanics, the definite predictions of the latter being replaced by probabilities.

2.7 Wave vectors, momentum and energy

The electron can thus be described as a wave by means of a mathematical expression. It must be realized that Fig. 2.11 and Eq. (2.17) both describe a wave moving in one direction only (the +x-direction). It is perfectly possible, however, to have a wave whose amplitude and wavelength are different in different directions at the same time. For example, we can have a two-dimensional wave as shown in Fig. 2.12. In this, the variation of amplitude with distance and the wavelengths are different as we proceed along the x- and along the z-directions, and thus the mathematical expression for the amplitude must be a function of x and z as well as of time. Equally, although it is almost impossible to draw, we could clearly have a three-dimensional wave described by a mathematical expression involving three space coordinates. However, even in three dimensions, there will always be a resultant direction of travel for the wave along which its wavelength can be measured.

When we describe a particle in motion we must specify not only its speed but also the direction in which it is moving and this is done by treating its velocity, v, as a vector. A vector is defined as a quantity having magnitude and direction and the rules for adding vectors, multiplying them, and so on are described by *vector algebra*.

The reader is referred to mathematical textbooks for the details of vector algebra. It is sufficient for our purposes here to remember that a quantity printed in bold type, for example $\mathbf{v}$, is a vector. This will have components of v_x, v_y and v_z, in a Cartesian coordinate system, as shown in Fig. 2.13.

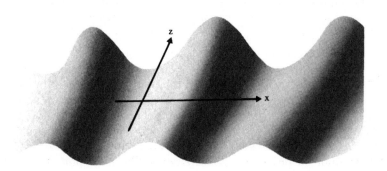

Fig. 2.12

A two-dimensional wave.

Fig. 2.13

Showing the Cartesian
components of the velocity
vector **v**.

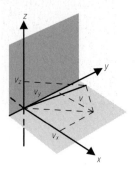

The length of the vector is called its scalar magnitude and is printed in
ordinary type, v, or in italic, v. The relationship between the vector's length
v and the magnitude of its components in a Cartesian coordinate system is
always given by

$$v = (v_x^2 + v_y^2 + v_z^2)^{1/2}$$

Although the electron is not simply a particle, its momentum has a
direction, and this is the direction in which the wave itself is travelling.
Now the momentum is directly proportional to $1/\lambda$, the inverse of the
wavelength, and its direction is that along which the wavelength is
measured. So we shall define a vector, **k**, which has a magnitude $2\pi/\lambda$ and
the direction defined above and which we call the *wave vector*. This is often
a more convenient quantity to use than wavelength.

In particular, we can relate the wave vector directly to the momentum, p,
carried by the wave through the de Broglie relationship, $p = h/\lambda$. Now for a
particle, $p = mv$ and because v is a vector, p must be a vector also. Similarly
with the wave: because $k = 2\pi/\lambda$ we have

$$\mathrm{p} = \frac{h}{2\pi}\,\mathrm{k} \tag{2.19}$$

Writing this as an equation relating two vectors implies that the wave
vector defines the direction in which momentum is carried by the wave.

In many of our calculations the actual direction of travel of the
wave may not be important and k is simply used as an ordinary scalar
quantity. This will be the case when the wave is assumed to be a plane
wave travelling in a single direction: however, in two-dimensional or
three-dimensional problems we must remember the vector nature of k.

Energy is, of course, a scalar quantity because it can never be said to
have a direction. As usual, the kinetic energy of a particle can be expressed
either in terms of its velocity v or its momentum p:

$$\frac{1}{2}\,mv^2 = \frac{p^2}{2m}$$

where both the total velocity v and the total momentum p are used without
regard for direction. Similarly, a wave with a constant value of k has an
associated kinetic energy, E_k, given by

$$E_k = \frac{p^2}{2m} = \frac{h^2 k^2}{8\pi^2 m} \tag{2.20}$$

2.8 Potential energy for an electron

By means of the wave vector **k** we can describe the direction of travel of a
wave and the energy that it carries. However, we must also be able to
describe the environment in which it moves because the electron will, in
general, interact with its environment. Because the electron carries a charge
a convenient method of doing this is to ascribe to the electron a potential
energy, E_p, which may be a function of position. Thus, for example, if the
electron, which is negatively charged, is in the vicinity of a positive charge
it will have a potential energy in accordance with Coulomb's law of electrostatics. This is proved in textbooks covering elementary electrostatics – it is
sufficient for our purposes to define the potential energy of an electron at

some point as the work done in bringing an electron from infinity to that point. For example, an electron at a point distance r from a positive charge of strength $+e$ has a potential energy given by

Fig. 2.14

Illustrating potential energy.

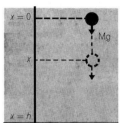

$$E_p = \frac{-e^2}{4\pi\varepsilon_0 r} \tag{2.21}$$

where ε_0 is the electrical permittivity of free space.

There will, of course, be other ways in which E_p could vary with position, different from the $1/r$ form in Eq. (2.21). For example, an electron wave moving through a solid such as a metal encounters positively-charged atoms (ions) at regular intervals and in such a case the potential energy is a periodic function of position.

To take a simple mechanical analogy, suppose we have a mass M (kg), at a height h (m) above the floor, as illustrated in Fig. 2.14, and we define the potential energy as zero at floor level. The potential energy of the mass at the initial position will be given by the work done in raising it to that position from the floor, that is, by force $\times$ distance.

If the mass is at rest in this position its kinetic energy is zero and its total energy is equal to its potential energy. When it is released it falls, converting its potential energy to kinetic energy so that, at some intermediate position x, its total energy E is given by

$$E = E_k + E_p$$
$$= \tfrac{1}{2}mv^2 + Mg(h - x)$$

where v is the velocity acquired at the point x and g is the acceleration due to gravity. When it reaches the floor, all its energy is kinetic energy and the potential energy is zero. Thus we see that the potential energy, being a function of position, describes accurately the effects of the environment.

The total energy E of an electron, in order to include the effect of its environment, must thus be expressed as

$$E = E_k + E_p \tag{2.22}$$

By rearranging Eq. (2.20) we find that

$$k = \frac{2\pi}{h}\sqrt{2mE_k}$$

and by using Eq. (2.22) to express E_k in terms of the total energy E, this becomes

$$k = \frac{2\pi}{h}\sqrt{2m(E - E_p)} \tag{2.23}$$

So the inclusion of potential energy changes the magnitude of the wave vector $\mathbf{k}$ and the corresponding wavelength $\lambda = 2\pi/k$. In any situation in which E_p varies with position, $\mathbf{k}$ must therefore depend on position, and the wavelength varies from place to place.

2.9 Schrödinger's wave equation

All of the above ideas are brought together in a general equation describing the wave function, Ψ, which is known as Schrödinger's wave equation.

The wave described in Eq. (2.17) is just one of many solutions to the *differential equation*

$$\frac{\partial^2 \psi}{\partial x^2} = k^2 \Psi$$

This is easily proved by differentiating Eq. (2.17) twice.

Because the electron wave function Ψ differs in being a complex function, it obeys a slightly different equation. Ψ can be written as a product of a space-dependent complex function that we shall write either as $\psi(x)$ or just ψ, with a purely time-dependent complex function of the form $\exp(j\omega t)$, thus[†]

$$\Psi = \psi(x)\exp(j\omega t)$$

The space-dependent part $\psi(x)$ by itself obeys a differential equation that is similar, but not identical, to the one above, namely

$$\frac{d^2 \psi}{dx^2} = -k^2 \psi$$

where k is the wave vector as before, equal to $2\pi/\lambda$. The mathematics behind this is explained more fully in Sections A5.1 and A5.2 of Appendix 5. Substituting for k^2 using Eq. (2.23), we may express the right-hand side in terms of the electron's total energy E and its potential energy E_p, leading to the result:

$$\frac{d^2 \psi}{dx^2} + \frac{8\pi^2 m}{h^2}(E - E_p)\psi = 0 \qquad (2.24)$$

and this is the differential wave equation known as Schrödinger's equation, for an electron wave moving along the x-axis. Its solution will yield information about the position of the electron in an environment described by the potential energy, E_p, which itself may be a function of position, x.

Equation (2.24) cannot be constructed without making assumptions (for example that Ψ is a complex quantity): the equation was put forward by Schrödinger as an *assumption*, to be tested by its ability to predict the behaviour of real electrons, which it does with great precision.

2.10 Electron confined in a 'box': quantized energy and wavelength

Let us now examine an example of the wave function ψ for a particular environment in which an electron is confined to a specific region of empty space. We shall find that confinement results in the energy of the electron being restricted to one of a set of values, called *energy levels*.

In three dimensions the electron might be confined in a rectangular 'box', but we shall treat the problem as one-dimensional, with the electron confined to a line of length L. This situation can be modelled by letting the potential energy, E_p, of the electron have the value zero within the length L, and infinity beyond it, as illustrated in Fig. 2.15. Because the probability

Fig. 2.15

Potential energy plotted against x, for the problem discussed in the text.

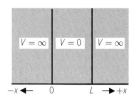

[†] Throughout this book we will use the function exp(x) to represent e^x, where e is the base of natural logarithms, in order to avoid confusion with the electronic charge $-e$.

of finding the electron is zero wherever $E_p = \infty$, the value of ψ is then zero *except* within the length L, as is shown in Appendix 5.3 in more detail.

The stable solution of Eq. (2.24) for the wave function ψ within the length L is given below and illustrated in Fig. 2.16:

$$\psi = j\left(\frac{2}{L}\right)^{1/2} \sin\left(\frac{n\pi x}{L}\right) \qquad (0 \leq x \leq L) \tag{2.25}$$

where n is an arbitrary positive integer. This result is justified mathematically in Appendix 5. It is easy to show by substituting $x = 0$ or $x = L$ into Eq. (2.25) that, at those particular points, $\psi = 0$, as required. If x lies outside the range 0 to L, then Eq. (2.25) is not valid. The magnitude, $|\psi|$, found by dividing Eq. (2.25) by j, is plotted against x in Fig. 2.16(a) for the three cases $n = 1$, 2 and 3, while the function $|\psi|^2$, which gives the dependence on x of the probability (defined in Section 2.6), is drawn in Fig. 2.16(b).

The shapes of $|\psi|$ in Fig. 2.16(a) look exactly like those of a wave on a stretched string. The motion of the electron is almost entirely analogous to that of a disturbance travelling along such a string. When the electron wave reaches the boundary at $x = L$ it is totally reflected and travels back to $x = 0$ where it is again reflected. The result is a *standing electron wave* with a probability distribution that is stationary in space. The analogy with waves on a string will prove useful in Chapter 3.

The fact that only certain shapes of wave function ψ describe states of the electron that are stable means that only certain values of the wavelength give stability. It will be seen from Fig. 2.16 that the wavelength of the electron for the case $n = 1$ is such that $L = \lambda/2$, while for $n = 2$, $L = \lambda$ and for $n = 3$, $L = 3\lambda/2$, and so on. To have a stable wave, then, there must be an integral number n of wavelengths in the length $2L$ of the wave's round trip from end to end and back. The wavelength is constrained to one of a particular set of values – we say it is *quantized*. It is this feature that makes the results of wave mechanics quite different from those of classical mechanics.

The stable wavelengths λ are given by the general equation

$$\lambda = 2L/n \qquad \text{where } n = 1, 2, 3, \ldots \infty \tag{2.26}$$

Each stable standing wave has a different energy E from the others, which we may calculate as follows. Because the potential energy is always zero, the energy is all kinetic, and we may use Eq. (2.9) to relate it to the wavelength, by remembering that momentum $p = h/\lambda$. Thus

$$E = \frac{p^2}{2m}$$

$$= \frac{h^2/\lambda^2}{2m}$$

Using Eq. (2.26) for the wavelength λ gives

$$E = \frac{h^2 n^2}{8mL^2} \tag{2.27}$$

The energy therefore increases in proportion to n^2. Thus, when $n = 2$, the electron has four times the energy that it has when $n = 1$, and so on; the electron's energy can have only certain specific values, and it is thus

Fig. 2.16

The wave function ψ for an electron in a 'box' of length L. Plotted as a function of position are (a) the wave function ψ, and (b) the probability $|\psi|^2$ for three values of the quantum number n.

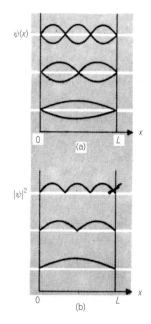

quantized. These *energy levels* are separated by gaps, i.e. by a range of energies that the electron can never have as long as it is confined to the length L. The number n, the value of which determines both the number of wavelengths making up the length $2L$, and the electron's total energy E, is called a *quantum number*. In subsequent chapters we shall see that in practice more than one quantum number is needed for each electron, and that these quantum numbers have great significance when explaining the properties of atoms, as well for as the properties of semiconductors and metals.

Problems

2.1 What reasons are there for believing that matter consists of atoms?

2.2 Outline two experiments, one showing the electron to be a particle, the other showing it to be a wave.

2.3 The proton has a positive charge equal in magnitude to that of the electron and a mass 1840 times that of the electron. Calculate the energy, in electron volts, of a proton with a wavelength of 0.001 nm. What voltage difference would be needed to give it that much energy?

2.4 A dust particle of mass $1\,\mu g$ travels in outer space at a velocity of 25,000 mph. Calculate its wavelength.

2.5 In the electron diffraction experiment described in the text the distance between the carbon film and the screen is 100 mm. The radius of the innermost diffraction ring is 3.5 mm and the accelerating voltage for the electron beam is 10 kV. Calculate the spacing between the atomic planes of carbon.

2.6 X-rays are produced when high-energy electrons are suddenly brought to rest by collision with a solid. If all the energy of each electron is transferred to a photon, calculate the wavelength of the X-rays when 25 keV electrons are used.

2.7 A static charge of 0.1 C is placed by friction on a spherical piece of ebonite of radius 200 mm. How many electronic charges are there per square centimetre of the surface?

2.8 A parallel beam of light of wavelength $0.5\,\mu m$ has an intensity of $1.0\,\mu W\,mm^{-2}$ and is reflected at normal incidence from a mirror. Calculate how many photons are incident on a square centimetre of the mirror in each second.

2.9 Calculate the magnitude of the wave vector of an electron moving with a velocity of $10^6\,m\,s^{-1}$.

2.10 If an electron having total energy of $10^{-21}\,J$ moves in a circle at a constant distance of 10 nm from a static proton, what will be its velocity and its value of wave vector?

The following questions require an understanding of Appendix 5.

2.11 Explain why it is necessary to use the *intensity* of a matter wave in order to provide a physical interpretation of the mathematical expression

$$\psi = A\exp[j(\omega t + kx)]$$

2.12 If an electron is in a region of constant electrical potential V, show that $\psi = A\exp(jkx)$ (where $j = \sqrt{-1}$) is a solution to Schrödinger's equation:

$$\frac{\partial^2 \psi}{\partial x^2} + \frac{8\pi^2 m}{h^2}(E + eV)\psi = 0$$

provided that we set

$$k^2 = \frac{8\pi^2 m}{h^2}(E + eV)$$

2.13 If $\psi = A\exp(jkx)$ and the value of ψ at any point x is equal to the value of ψ at a point $(x + a)$, where a is a constant, show that $k = 2n\pi/a$, where n is an integer.

2.14 Why is the function $\psi = \text{constant}$ not a permissible solution to Problem 2.12? *(Hint: consider normalization.)*

2.15 Assume that, at very large distances from the origin of the axes, a wave function may be written in the form

$$\psi = \frac{C}{r^n}$$

where C is a constant and r is the distance from the origin. Show by considering normalization of ψ that this expression is only a permissible solution of Schrödinger's equation if $n > 2$.

Self-assessment questions

1 The electron is a constituent of all materials

   (a) true   (b) false

2 The ratio of charge to mass for an electron is

   (a) $1.759 \times 10^{-11}\,\mathrm{C\,kg^{-1}}$ (b) $1.759\,\mathrm{C\,kg^{-1}}$

   (c) $1.759 \times 10^{11}\,\mathrm{C\,kg^{-1}}$

3 We know that an electron shows wavelike behaviour because

   (a) it is diffracted by a crystal

   (b) it is scattered by collision with a lattice of atoms

   (c) it can penetrate solid objects

4 The distance between slits in a diffraction grating must be

   (a) much greater than

   (b) comparable to

   (c) much less than

   the wavelength of light

5 An electron volt is

   (a) the voltage required to accelerate an electron to a velocity of $1\,\mathrm{m\,s^{-1}}$

   (b) the kinetic energy acquired by an electron in falling through a potential difference of 1 volt

   (c) the voltage required to give an electron 1 joule of energy

6 A pattern of concentric rings on a fluorescent screen, produced by projecting a beam of electrons through a thin carbon film, is evidence for the wave nature of electrons

   (a) true   (b) false

7 The diameters of the rings described in the previous question are

   (a) inversely proportional to the electron velocity

   (b) inversely proportional to the square root of the electron velocity

   (c) proportional to the electron velocity

8 The kinetic energy of an electron accelerated from rest through a potential difference of $2\,\mathrm{kV}$ will be

   (a) $2000\,\mathrm{J}$ (b) $3.2 \times 10^{-16}\,\mathrm{J}$ (c) $4.1 \times 10^{-10}\,\mathrm{J}$

9 What will be the velocity of the electron in Question 8?

   (a) $2.65 \times 10^{7}\,\mathrm{m\,s^{-1}}$ (b) $3 \times 10^{10}\,\mathrm{m\,s^{-1}}$

   (c) $6.63 \times 10^{16}\,\mathrm{m\,s^{-1}}$

10 De Broglie's relationship gives the wavelength of a hydrogen atom moving with a velocity of $10^{3}\,\mathrm{m\,s^{-1}}$ to be

   (a) $1.46 \times 10^{-9}\,\mathrm{mm}$ (b) $3.96 \times 10^{-10}\,\mathrm{m}$

   (c) $7.27 \times 10^{-7}\,\mathrm{m}$

11 In photoemission of electrons the *number* of electrons emitted depends on

   (a) intensity   (b) wavelength

   (c) velocity of the light

12 In photoemission of electrons the *energies* of electrons emitted depend on the

   (a) intensity   (b) wavelength

   (c) velocity of the light

13 The velocity of light is $3 \times 10^{10}\,\mathrm{m\,s^{-1}}$. The photon energy of light of $6 \times 10^{-7}\mathrm{m}$ wavelength is

(a) $3.97 \times 10^{-40}\,\mathrm{J}$ (b) $3.31 \times 10^{-17}\,\mathrm{J}$

(c) $1.99 \times 10^{-23}\,\mathrm{J}$

14 A wave packet is the result of adding together a large number of waves of the same frequency but different amplitudes

(a) true   (b) false

15 Newton's laws do not apply without modification to a wave packet

(a) true   (b) false

16 A wave packet has no mass

(a) true   (b) false

17 A wave packet cannot carry momentum

(a) true   (b) false

18 A wave packet cannot be formed by a single sine wave

(a) true   (b) false

19 If the expression

$$\Psi = A \sin\left(\omega t - \frac{2\pi x}{\lambda}\right)$$

represents an electron as a wave, the quantity A is the magnitude of the electron wave function at

(a) any time t for a fixed position x

(b) any position x at a fixed time t

(c) a time and position such that

$$\left(\omega t - \frac{2\pi x}{\lambda}\right) = \frac{n\pi}{2} \quad \text{where } n \text{ is an integer}$$

20 The intensity of any electron wave function Ψ at any time t and position x is given by

(a) A^2 (b) $|\Psi|^2$

(c) the condition that $\left(\omega t - \frac{2\pi x}{\lambda}\right) = n\pi$

21 The probability of finding an electron at a point x at a time t is given by

(a) the amplitude of the wave

(b) the intensity of the wave

(c) the peak amplitude of the wave

22 The charge density due to any electron wave function Ψ at a point x at a time t is given by

(a) eA^2 (b) $e|\Psi|^2$ (c) $e\Psi$

where e is the charge on an electron.

23 De Broglie's relationship between momentum and wavelength for an electron is

(a) $\lambda = \dfrac{h}{p}$ (b) $p = h\lambda$ (c) $p = \dfrac{\lambda}{h}$

24 The magnitude of a wave vector is proportional to

(a) the wavelength

(b) the reciprocal of the wavelength

(c) the reciprocal of the frequency

25 The wave vector changes direction if a wave is deflected from its original path

(a) true   (b) false

26 The kinetic energy associated with a plane electron wave having $k = 2\pi/\lambda$ is given by

(a) hk (b) $\frac{1}{2}mk^2$ (c) $\dfrac{h^2 k^2}{8\pi^2 m}$

where h is Planck's constant.

27 The environment of an electron may be described by attributing a potential energy to the electron

(a) true   (b) false

28 The total energy of the electron is

(a) the difference between its kinetic and potential energies

(b) the sum of its kinetic and potential energies

(c) the product of its kinetic and potential energies

29 The wave vector is

(a) independent of

(b) proportional to

(c) proportional to the square root of

the total energy of the electron.

30 The total energy of an electron having potential energy depends on its position if the potential energy is

(a) constant   (b) zero

(c) a function of position

31 Schrödinger's equation relates the probability of an electron being at a particular point to its total energy when at that point

(a) true   (b) false

32 The solution of Schrödinger's equation allows the position of an electron to be fixed exactly

(a) true   (b) false

33 If the potential energy of an electron is zero, Schrödinger's equation does not apply

(a) true   (b) false

34 If the wavelength of an electron wave is infinite, the electron must be stationary

(a) true   (b) false

35 The energy of an electron in a deep potential well (i.e. in a 'box') is quantized

(a) true   (b) false

36 The solutions for Schrödinger's equation for an electron in a deep potential well are standing waves

(a) true   (b) false

The following questions test your understanding of Appendix 5.

37 If $\Psi = (A + jB)$, then $\Psi\Psi^*$ is equal to

(a) $(A + jB)^2$ (b) $(A - jB)^2$ (c) $(A^2 + B^2)$

38 A complex wave, given by $\Psi = A \exp[j(\omega t + kx)]$ represents the product of a space-dependent wave function with a time-dependent wave function

(a) true   (b) false

Answers

1	(a)	**2**	(c)	**3**	(a)	**4**	(b)
5	(b)	**6**	(a)	**7**	(a)	**8**	(b)
9	(a)	**10**	(b)	**11**	(a)	**12**	(b)
13	(b)	**14**	(b)	**15**	(a)	**16**	(a)
17	(b)	**18**	(a)	**19**	(c)	**20**	(b)
21	(b)	**22**	(b)	**23**	(a)	**24**	(b)
25	(a)	**26**	(c)	**27**	(a)	**28**	(b)
29	(c)	**30**	(c)	**31**	(a)	**32**	(b)
33	(b)	**34**	(a)	**35**	(a)	**36**	(a)
37	(c)	**38**	(a)				

3 | The simplest atom: hydrogen

3.1 Introduction

Now that we have formed a clear picture of how an electron behaves in a rather simple environment we are able to approach the more complicated problem of an electron bound to a single positively-charged 'particle'. Such a combination forms a simple atom – an atom of hydrogen is exactly like this. In hydrogen a single electron moves in the vicinity of a particle, a *proton*, which carries a positive charge equal in magnitude to that of the electron. The proton's mass, however, is 1840 times the mass of the electron so that at the same velocity its wavelength is 1840 times smaller, and it behaves much more like a particle than does the electron. We shall treat it therefore as a fixed point charge around which the electron wave travels. The shape of this wave can be calculated by solving Schrödinger's equation, using an appropriate expression for its potential energy E_p. This is mathematically rather difficult, except in the case of spherically symmetrical solutions, so we will restrict ourselves to studying just two of these, and merely describe the results for other cases.

Although the electron wave clearly must be three-dimensional, we will build up our atomic model gradually, and begin in Section 3.2 by seeing how far it is possible to go while limiting consideration to one dimension only. In Chapter 2 we saw that confining an electron in a limited space leads to its energy being quantized, and the same will be found here, but the energy levels are differently spaced, dependent on a single integer, the *principal quantum number*. To describe non-spherical wave functions, we must introduce the second and third dimensions in space, when it is necessary to introduce two more quantum numbers, the *angular momentum quantum number*, and the *magnetic quantum number*. The interpretation of these integral numbers is explained in Sections 3.2.2 to 3.2.4. In Section 3.3, when we consider the *internal* motion of an electron, spinning about its centre of mass, the fourth and final quantum number is introduced, the *spin quantum number*. The shapes of the electron wave functions are discussed further in Section 3.4, for they affect the degree of overlap of the electrons from neighbouring atoms, which, as later chapters will explain, determine all kinds of interactions between atoms in a solid.

Section 3.5 relates all this theory to reality by showing how it correctly predicts the wavelengths present in light emitted by hydrogen atoms in an electrical discharge.

3.2 Electron wave functions in the atom

Chemical bonds between atoms have properties that are strongly influenced by the geometric shapes of the probability density 'cloud' given by the solutions of Schrödinger's equation for the wave function ψ. We shall explore here only the simplest shapes, without explaining how the mathematical solutions may be found, because that is covered in many textbooks on the quantum mechanics of atoms and molecules, some of which are to be found in the Further reading section on page 633. We will explain the basic assumptions and also the results of calculating the wave function for a hydrogen atom.

When an electron is at a radial distance r from a nucleus carrying a charge $+e$, its potential energy E_p is the product of the electronic charge $-e$ and the electrostatic potential, $V(r)$, at a distance r from a nucleus whose charge is $+e$, i.e.

$$E_p = -eV(r)$$

where

$$V(r) = \frac{+e}{4\pi\varepsilon_0 r}$$

Thus

$$E_p = -\frac{e^2}{4\pi\varepsilon_0 r} \tag{3.1}$$

Note that the electron's potential energy is zero when at an infinite distance ($r \to \infty$), and elsewhere it has negative values. The energy E_p is independent of the two angular coordinates θ and ϕ that denote the many points around the nucleus at the *same* radius r. Hence we shall assume that the wave function ψ does not depend on θ or ϕ, i.e. $\partial\psi/\partial\theta = 0$ and $\partial\psi/\partial\phi = 0$. So we insert the value of E_p from Eq. (3.1) into Schrödinger's equation, choosing the form appropriate to radial coordinates, which is given in Appendix 5, Eq. (A5.8), as:

$$\frac{1}{r^2}\frac{\partial}{\partial r}\left(r^2\frac{\partial\psi}{\partial r}\right) + \frac{8\pi^2 m}{h^2}\left(E + \frac{e^2}{4\pi\varepsilon_0 r}\right)\psi = 0 \tag{3.2}$$

The simplest solution of this equation has the form

$$\psi = A\,e^{-r/r_0} \tag{3.3}$$

where A and r_0 are constants whose values need to be found. This solution is plotted against r in Fig. 3.1(a), and is known as the *ground state*, because it is the solution with the *lowest* energy E.

We shall now show that this expression for ψ satisfies Eq. (3.3), and we shall find values for the energy E and the length r_0. Differentiating Eq. (3.3) once with respect to r gives:

$$\frac{\partial\psi}{\partial r} = -\frac{1}{r_0}A\,e^{-r/r_0} = -\frac{\psi}{r_0}$$

Differentiating again with respect to r enables us to find the first term in Eq. (3.2):

$$\frac{1}{r^2}\frac{\partial}{\partial r}\left(r^2\frac{\partial\psi}{\partial r}\right) = \frac{\partial^2\psi}{\partial r^2} + \frac{2}{r}\frac{\partial\psi}{\partial r} = \frac{A\,e^{-r/r_0}}{r_0^2} - \frac{2}{r}\frac{A\,e^{-r/r_0}}{r_0} \tag{3.4}$$

Fig. 3.1

Wave functions ψ, and probability densities $|r\psi|^2$ plotted against radial distance r from the nucleus, for two of the states of hydrogen: (a) $n=1$, $\ell=0$, and (b) $n=2$, $\ell=0$.

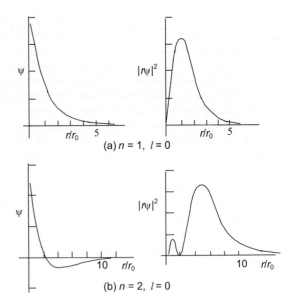

(a) $n = 1$, $l = 0$

(b) $n = 2$, $l = 0$

When Eq. (3.3) is used for ψ, the remaining terms in Eq. (3.2) become

$$\frac{8\pi^2 m}{h^2}\left(E + \frac{e^2}{4\pi\varepsilon_0 r}\right)A\,e^{-r/r_0} \tag{3.5}$$

Adding Eqs (3.4) and (3.5) to form the left-hand side of Eq. (3.2), we find two terms that vary as e^{-r/r_0}, while the other terms have the form $(1/r)\,e^{-r/r_0}$. The coefficients of the terms of the first type must sum to zero, and the coefficients of the latter type must also separately add up to zero. This is because the addition of any two terms with differing dependence on r can only result in zero at one particular value of r, while Eq. (3.2) must hold for all values of r. So, setting each sum of coefficients separately to zero gives us two equations, which together are the equivalent of Eq. (3.2), as follows:

$$\frac{8\pi^2 mE}{h^2} + \frac{1}{r_0^2} = 0 \tag{3.6a}$$

and

$$\frac{8\pi^2 m}{h^2}\left(\frac{e^2}{4\pi\varepsilon_0}\right) - \frac{2}{r_0} = 0 \tag{3.6b}$$

Now, Eq. (3.6b) contains no unknown quantities except r_0, so that we can solve for r_0 to find that:

$$r_0 = \frac{h^2 \varepsilon_0}{\pi e^2 m} \tag{3.7}$$

Putting this result into Eq. (3.6a) gives us the value of the energy E:

$$E = -\frac{me^4}{8h^2\varepsilon_0^2} \tag{3.8}$$

These two equations give us useful information about the hydrogen atom. First, when Eq. (3.8) is evaluated, its energy is found to equal -13.6 eV (electron volts), when compared with an electron and proton separated by an infinitely large distance. This agrees well with the measured energy needed to extract an electron from a hydrogen atom.

Box 3.1   The most probable distance of the electron from the nucleus

The most probable value of r in the ground state of the hydrogen atom is the distance r from the nucleus at which the probability of finding the electron is a maximum. The wave function for the ground state is given by Eq. (3.4):

$$\psi = A\,e^{-r/r_0}$$

The probability that the electron lies somewhere between radii r and $r + dr$, anywhere within a thin spherical shell of radius r and thickness dr, is just $|\psi|^2$ times the volume of the thin shell, a volume that increases as r^2. So we should find where the maximum value of the product $r^2|\psi|^2$ lies by setting the derivative of $(r\psi)^2$ to zero. Thus

$$\frac{d}{dr}\,(r\psi)^2 = A^2\,\frac{d}{dr}\left[r^2\exp\left(\frac{-2r}{r_0}\right)\right]$$

$$= A^2\left[2r - 2\,\frac{r^2}{r_0}\right]\exp\left(\frac{-2r}{r_0}\right)$$

This is clearly zero when $2r = 2r^2/r_0$, i.e. when $r = r_0$. Thus r_0 is the radius at which the probability is a maximum.

Second, the radius r_0 at which the wave amplitude falls by $1/e$ is found by evaluating Eq. (3.7), and is 0.053 nm. This is both the radius that Bohr's theory predicted for the orbit of the electron, and, as Box 3.1 shows, it is the most probable distance of the electron from the nucleus according to the measure of probability $\psi\psi^*$ introduced in Section 2.6. It is also quite a good fit to the measured size of a hydrogen atom.

We have not calculated the amplitude A of the wave function given in Eq. (3.3). Its value should be chosen to make the function $\psi\psi^*$ (which is the probability per unit volume of finding the electron at any point) have the correct value. We expect the probability of finding the electron somewhere within the whole atom to be 1.0 exactly, so that the amplitude A should be adjusted to ensure that, when $\psi\psi^*$ is integrated over the volume of the atom, it has the value unity. In this way A can be found to equal $2/r_0^{3/2}$.

3.2.1 Solutions for ψ with higher energy

Other solutions of Schrödinger's equation involve differentials of ψ with respect to angular coordinates θ and ϕ as well as r [see Eq. (A5.7) in Appendix 5], and they are thus more complicated. It turns out that the only valid solutions contain three distinct integers, called *quantum numbers*, given the symbols n, ℓ and m_ℓ. For example, Schrödinger's equation requires that the dependence of the wave function ψ on the angular coordinate ϕ must satisfy the equation

$$\frac{\partial^2\psi}{\partial\phi^2} = m_\ell^2\psi \tag{3.9}$$

and it happens that solutions of Eq. (3.9) are only possible when m_l has an integral value. Instead of presenting the detailed mathematics here, we shall try to explain in the next few sections how the three integral quantum numbers n, ℓ and m_ℓ arise from three different conditions that the wave function must satisfy.

Each quantum number is related to a particular physical property of the electron's motion, as follows.

n is related to the electron's total energy by the important equation

$$E = -\frac{E_0}{n^2} \qquad (3.10)$$

where E_0 is the lowest possible energy, given in Eq. (3.8), corresponding to the case $n = 1$. n is called the *principal quantum number*, and may have values $1, 2, 3, \ldots \infty$.

ℓ is related to the total angular momentum of the electron, which equals $\sqrt{\ell(\ell+1)}\,(h/2\pi)$. ℓ is called the *angular momentum quantum number*, and may take only positive values up to a maximum of $(n-1)$, i.e. $1, 2, 3, \ldots (n-1)$.

m_l is proportional to the angular momentum of the electron about any specially selected axis such as the direction of a magnetic field that is applied. This angular momentum equals $m_\ell\,(h/2\pi)$. m_ℓ is called the *magnetic quantum number*, and may have values of either sign, with a maximum of $\pm\ell$, i.e. $-\ell, -(\ell-1), -(\ell-2), \ldots, 0, 1, 2, 3, \ldots \ell$.

Summarising these values:

$$n = 1, 2, 3, \ldots \infty$$

$$\ell = 1, 2, 3, \ldots (n-1)$$

$$m_\ell = -\ell, -(\ell-1), -(\ell-2), \ldots, 0, 1, 2, 3, \ldots (\ell-1), \ell$$

These rules show that the mathematical analysis above treated the case in which $n = 1$, and both ℓ and m_ℓ are zero. They also mean that no other solution has the same total energy of $-13.6\,\text{eV}$. For $n = 2$ or more, however, there are several solutions, called *degenerate* solutions, with the same energy.

The two quantum numbers ℓ and m_ℓ relate to simple conditions satisfied by the *wavelength* of the electron (remember the wavelength and momentum are intimately related). We shall now show how this comes about.

3.2.2 *The angular momentum quantum number ℓ*

The angular momentum quantum number, ℓ, is the simplest quantum number to explain. It equals the number of wavelengths that fit into a suitable circular path centred on the nucleus, one that lies on the plane in which the electron's mass circulates. This plane sometimes lies in a moving frame of reference, for the circulating electron behaves rather like a gyroscope that *precesses*, as shown in Fig. 3.2, about a vertical axis. We will return to this idea when we consider the quantum number m_ℓ.

The three-dimensional electron wave is difficult to picture. A two-dimensional wave is illustrated in Fig. 3.3, showing that the wavelength depends on the direction in which we measure it. Along the direction in which the wave moves (the normal to the wavefront), the wavelength is a minimum, while at other angles it has various values that can even be as large as infinity when measured along a wavefront.

The angular momentum is defined as the radial distance r multiplied by the linear momentum mv_t at right angles to r. Thus it is $mv_t r$, or $p_t r$, where p_t is the momentum component tangential to any circular path that has a radius of r [Fig. 3.4(a)]. The wave's amplitude varies from point to point around such a path as in Fig. 3.4(a), completing exactly ℓ wavelengths in the circle. If ℓ were not integral, a discontinuity would exist such as that shown

Fig. 3.2

The motion of a gyroscope under a gravitational couple is called *precession*.

Path followed by tip of spindle

Anticlockwise couple due to weight of gyroscope and reaction of support

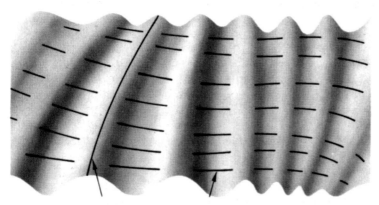

Fig. 3.3
Wavefronts and normals pictured in a two-dimensional wave.

Wavefront   Normals to wavefronts

in Fig. 3.4(b) or (c), giving locally an infinite mathematical derivative, and failing to satisfy Schrödinger's equation [Eq. (A5.7) in Appendix 5].

Now the momentum p_t around the circle is found using de Broglie's relation $\lambda_t = h/p_t$ in terms of the wavelength λ_t measured around the circle. The average wavelength has a value $2\pi r/\ell$, so that

$$p_t = \frac{h}{\lambda_t} = \ell \frac{h}{2\pi r}$$

This makes the angular momentum, $p_t r$, equal to $\ell h/2\pi$, a relation that is close to the correct value of $\sqrt{\ell(\ell+1)}\,(h/2\pi)$ given earlier, and differs only because a diagram cannot depict a wave function with both real and imaginary components, so can give only an approximate result.

As a simple example, consider the case of the wave given by Eq. (3.3), repeated here:

$$\psi = A\,e^{-r/r_0}$$

Because r is constant around any circle centred on the nucleus, ψ is also constant. On such a path the number of wavelengths is therefore zero, so that $\ell = 0$. This wave thus possesses *no* angular momentum.

This would be a quite impossible result if we treated the electron simply as a large particle orbiting like a moon around a planet! The same is true of *all* states for which $\ell = 0$, in each of which the momentum is everywhere in the radial direction.

3.2.3 *The principal quantum number* n

The principal quantum number n has a particularly simple interpretation in the states of the atom in which $\ell = 0$. When $\ell = 0$, there are exactly n wavelengths in a complete 'orbit' consisting of a long, infinitesimally thin ellipse surrounding the nucleus. In states having non-zero *angular* momentum, there are fewer than n wavelengths, but in all cases, the quantum number n governs only the shape of the wave in the *radial* direction, and not its angular variation.

Consider, for example, the case $n = 2$, $\ell = 0$, illustrated in Fig. 3.1(b) as a function of r, and described by the equation:

$$\psi = \frac{1}{\sqrt{2}r_0^{3/2}} \left(1 - \frac{r}{2r_0}\right) e^{-r/2r_0} \tag{3.11}$$

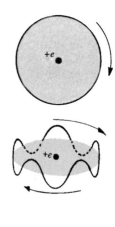

(a)

(b)

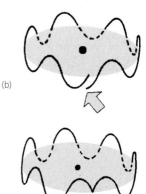

(c)

The 'orbit' is simply from $r = +\infty$ to $r = -\infty$ and back again. Starting at $r = +\infty$, the wave function ψ first peaks at a negative value where $r = r_0$, then crosses zero and climbs to a positive peak at $r = 0$, before passing zero again at $r = -r_0$, thus completing a single cycle of variation. It remains negative in value all the way to $r = -\infty$ and back to the same zero point, and then only completes another full cycle when we reach $r = +\infty$ again. ψ thus goes through two complete periods in the orbit, determined, as we stated, by the fact that $n = 2$.

3.2.4 The magnetic quantum number m_l

The magnetic quantum number m_l is closely related to the angular momentum quantum number ℓ. When ℓ is non-zero, the electron wave possesses angular momentum about the nucleus, which implies that the electron mass is circulating with the wave. The electron charge is intimately associated with its mass, so it, too, must be circulating as if there were an electric current flowing in a loop around the nucleus (Fig. 3.5). Indeed, the equivalent current is directly proportional to the angular momentum.

Such a current loop behaves exactly like a small permanent magnet. It has the same distribution of magnetic flux around it [Fig. 3.5(b)], and both behave in a similar way when a magnetic field is applied from an external source. Thus the two-dimensional electron wave experiences a couple on it when a magnetic field with flux density B is applied as shown in Fig. 3.5(c). The couple acts in such a way as to try to turn the plane of the wave perpendicular to the flux lines. But the wave is not free to turn in that direction, because its circulating mass makes it behave like a gyroscope. A gyroscope under the action of the couple due to gravity does not fall over but *precesses* about the direction of the gravitational field, as Fig. 3.2 illustrated.

In the same way, the electron wave precesses about the direction of the magnetic flux density B, as illustrated in Fig. 3.6. At all times a fixed number of wavelengths must fit into any fixed circular path – in particular, one in a plane normal to the magnetic field direction. Such a path is exactly of length m_l times the average wavelength around the path, which is of course related to the angular momentum about an axis parallel to the magnetic field. An identical argument to that used earlier in connection

Fig. 3.5

A loop carrying a current i is equivalent to a bar magnet (a). They are accompanied by similar magnetic flux distributions (b), and experience similar couples (c) when in an externally applied magnetic flux density B.

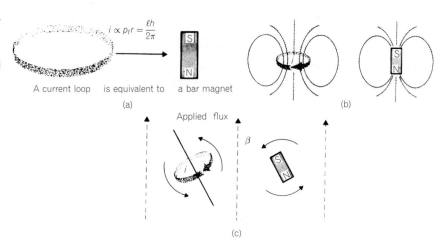

$i \propto p_l r = \dfrac{\ell h}{2\pi}$

A current loop   is equivalent to   a bar magnet

(a)

(b)

Applied flux

(c)

with ℓ leads to the conclusion that this angular momentum is $m_\ell h/2\pi$. This condition results directly from the solution of Eq. (3.9) for the angular dependence of the wave function ψ, and is not an approximation.

Because the total angular momentum is $\sqrt{l(l+1)}\,(h/2\pi)$, the momentum about the flux axis must be a component of it – in fact we can represent angular momentum in vectorial form as we did linear momentum. The vector representing the total angular momentum is drawn along the axis of rotation, pointing in the direction about which the rotation is clockwise. Its length is made proportional to the magnitude of the angular momentum. Figure 3.7(a) shows this, and we see that this vector precesses about the flux axis with a *constant component* along that axis. It is this component that is equal to $m_\ell h/2\pi$. The remaining component rotates about that axis and is not quantized in any simple way.

From this diagram it is easy to see that the magnetic quantum number m_l has a maximum value equal to ℓ, when the vector $\sqrt{\ell(\ell+1)}(h/2\pi)$ points very nearly along B. Figure 3.7(b) shows how the component $m_\ell h/2\pi$ can sometimes point in the *opposite* direction to B. This is described mathematically by giving m_l a negative value. The range of values permitted for m_ℓ thus extends from $-\ell$ through zero to $+\ell$. If, for example, $\ell = 3$, then m_ℓ can have one of the values $-3, -2, -1, 0, 1, 2, 3$, i.e. seven possibilities in all.

3.3 *Spin of the electron*

It is worth noting that each spatial dimension r, θ and ϕ in the above problem is associated with a quantum number, and that this was also true in the cases discussed in Chapter 2. It is indeed generally true that three quantum numbers are necessary to describe a three-dimensional wave function.

So far, however, we have not considered the possibility of motion within the electron itself. The wavelike model of the electron does not include this possibility, but it has been found necessary to assume that the electron is spinning internally about an axis in order to explain many aspects of the magnetic properties of atoms. Pictorially we may represent the spinning electron as a fuzzy distribution of mass rotating about its own centre of gravity, rather as a planet spins on its axis while orbiting the Sun. It is difficult to couple this image with that of the wave motion around the nucleus and the reader is not advised to try to do so! However, a spinning wave packet is not too difficult to imagine, although more difficult to draw.

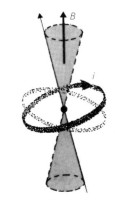

Fig. 3.6

The circulating electron wave (represented here as a current loop) precesses around the magnetic field direction labelled B.

Cone swept out by axis normal to plane of 2D waves

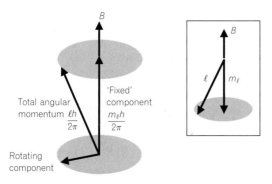

Fig. 3.7

The components of angular momentum for the precessing wave. (b) shows that the component $m_\ell h/2\pi$ can sometimes point in the opposite direction to a magnetic flux density B, when m_ℓ has a negative value.

Fig. 3.8

A pictorial spinning electron (a) and equivalent motion of a 'particle electron' (b).

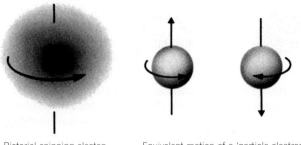

Pictorial spinning electon   Equivalent motion of a 'particle electron'

As might be expected, the angular momentum of this spinning motion is quantized and in units of $h/2\pi$. To distinguish it from the *orbital* angular momentum, $\ell h/2\pi$, we call it the *spin angular momentum* and its magnitude is $m_s h/2\pi$, where m_s is the spin quantum number. Unlike ℓ and m_ℓ, however, m_s does not have integral values, but can only take one of two values, $+\frac{1}{2}$ and $-\frac{1}{2}$. Thus, when a magnetic field is present, the motion can be either clockwise or anticlockwise about the magnetic field direction, the spin angular momentum being equal to $\pm m_s h/2\pi$. Because the vector representing this points along the axis of rotation, the two states are often referred to as 'spin up' and 'spin down' (Fig. 3.8). An explanation of the reason for the non-integral values of m_s is beyond the scope of this book.

3.4 Electron clouds in the hydrogen atom

Because the shape of the wave function ψ is defined by the quantum numbers n, ℓ and m_ℓ, so too is the probability density distribution for the electron. Thus we may picture an 'electron cloud' that represents the charge distribution in a hydrogen atom using the value of $e|\psi|^2$ as a measure, just as described in Chapter 2. Such a cloud can be illustrated as in Fig. 3.9, where some examples are given that represent the many different states of the hydrogen atom. The quantum numbers corresponding to each state are given below the illustrations. Each cloud is shown in cross-section, and the density of the image rises with the charge density, i.e. with $|\psi|^2$. These clouds all have rotational symmetry about a vertical axis, which coincides with the axis along which m_ℓ is quantized. Note that because the electron is not a standing but a running wave, the probability density does not vary along a circle described around that axis, because it is a time-averaged probability. The $\ell = 0$ states are all perfectly spherical in shape, while the others have quite different angular symmetry. These shapes will be referred to again when we consider atomic bonding in Chapter 5. In view of the variety of shapes, it is surprising that all states having the same value of n have the same total energy E, although their angular momentum may differ, as it changes with the value of ℓ. States with the same energy are said to be *degenerate*.

3.5 Energy levels and atomic spectra

One of the most direct checks on the accuracy of the expression for electron energy is to measure the energy released when the electron 'jumps' from any energy level to a lower one. Such a representation

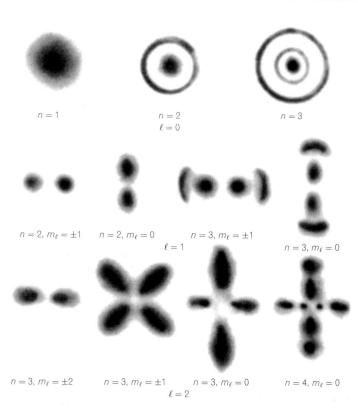

Fig. 3.9

Electron clouds for the hydrogen
atom, illustrating the distribution
of charge density in states with the
quantum numbers given.

between energy levels is represented in Fig. 3.10 by a vertical arrow whose
length is proportional to the energy released. Each horizontal line
represents one of the energy levels predicted by Eq. (3.10).

Because no matter is expelled from the atom during the transition the
only form that the emitted energy can take is an electromagnetic wave –
a photon. The photon energy must be exactly equal to the difference in
energy between the two electron energy levels concerned[†] so it can be
calculated from Eq. (3.10), given again here:

$$E = -\frac{E_0}{n^2}$$

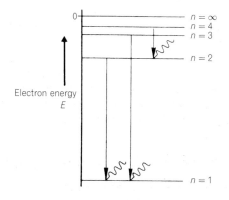

Fig. 3.10

Transitions between energy levels
involved in the emission of
photons.

[†] There is the possibility that two photons might be emitted but the chances of this are so small that they are
negligible.

Fig. 3.11

A discharge tube and its power
supply.

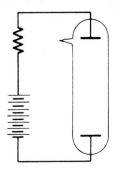

where $-E_0$ is the energy of the ground state (equal to $-13.6\,\text{eV}$), and n is the principal quantum number. If the electron makes a transition from an energy level with quantum number n_1 to one having the quantum number n_2 the photon frequency f may thus be found from the equation

$$E_{\text{photon}} = hf = -\frac{E_0}{n_1^2} + \frac{E_0}{n_2^2} \tag{3.12}$$

Using the value of E_0 from Eq. (3.8), we find that

$$f = \frac{e^4 m}{8h^2 \varepsilon_0^2}\left(\frac{1}{n_2^2} - \frac{1}{n_1^2}\right) \tag{3.13}$$

so the frequency may have any one of a discrete set of values obtained by substituting different values for n_1 and n_2.

How can we arrange to observe these? The first requirement is a source of hydrogen containing many atoms in *excited* states, that is, states other than the one with lowest energy (the ground state). Fortunately this is easily achieved by setting up an electric discharge in gaseous hydrogen. In a discharge tube (closely related to the fluorescent light tube) a large potential difference is applied between two electrodes (Fig. 3.11). Once the discharge has been started the gas glows brightly and a large current flows. The current is carried across the gas by ions – atoms that have lost or gained an electron and so are charged. The potential gradient across the tube accelerates electrons to high velocities whereupon they collide with neutral atoms and excite the latter into states of high energy. As these excited atoms return to the ground state they emit photons, as described above.

The emitted light contains several wavelengths and the corresponding frequencies are given by Eq. (3.12). The light from a discharge tube is usually coloured as a result. The wavelengths present in the light may be

Fig. 3.12

Apparatus for recording the
emission spectrum of a gas, and a
spectrum of hydrogen obtained in
this way.

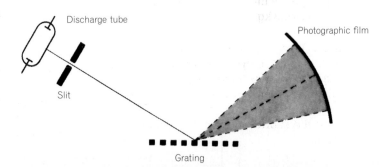

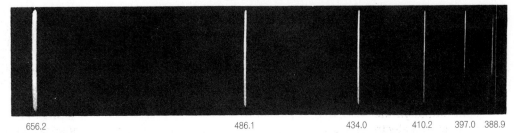

656.2   486.1   434.0   410.2   397.0   388.9

Wavelength (nm)

separated and measured using a spectrometer. In its simplest form, this has a glass prism to split the light into its components, but more often a diffraction grating is used. Figure 3.12 shows how the beam transmitted by a narrow slit is diffracted onto a photographic film.

If we know the spacing of the grating lines and Bragg's law, we can measure the wavelengths. In Fig. 3.12 we also show a photographic record of the spectrum of hydrogen obtained in this way. It is found that Eq. (3.13) accurately predicts the measured wavelengths. Even more accurate agreement is obtained when corrections are made for the finite mass of the nucleus and for relativistic effects.

If the discharge tube is placed in a large magnetic field during the measurement, the quantum states with different values of the magnetic and spin quantum numbers m_l and m_s have different energies. The highest energy occurs when the total angular momentum vector points along the field direction and the lowest when it points in the opposite direction. The resulting shifts in the energy levels shown in Fig. 3.10 may be observed experimentally, and confirm the predictions of the theory. It was through such measurements of spectral lines that the theory was originally built up, and the existence of the spin of the electron was first established.

Problems

3.1 Show that the wave function given in Eq. (3.3) satisfies Schrödinger's equation.

3.2 Compare the average diameter of the wave function of the hydrogen atom in its ground state with the distance between the atoms in solid hydrogen, which may be calculated from the density, $76.3 \, \text{kg m}^{-3}$.

3.3 Calculate the energies of the first three energy levels of the hydrogen atom. What are the frequencies of radiation emitted in transitions between those levels? Find the first transition for which the radiation is visible, if the shortest visible wavelength is 400 nm.

3.4 How many different values may the quantum number m_ℓ have for an electron in each of the three states $n = 3$, $\ell = 2$; $n = 4$, $\ell = 2$; $n = 4$, $\ell = 3$?

3.5 Deduce a formula for the energies of the helium ion, He^+, in which a single electron moves around a nucleus whose charge is $+2e$. [*Hint:* although the nuclear charge is doubled, the electronic charge is not, so that it is not correct to replace e by $2e$ in Eq. (3.8).]

3.6 Discuss the consequences if Planck's constant were to have the value $10^{-3} \, \text{J s}$.

3.7 Draw a rough plot of the amplitude of the electron wave versus distance from the nucleus for the state $n = 3$, $\ell = 0$ illustrated in Fig. 3.9. Do the same for the state $n = 3$, $\ell = 2$, $m_\ell = 0$, along the vertical axis. How many maxima are there in each case?

3.8 Explain why, in Fig. 3.9, the state $n = 2$, $\ell = 1$, $m_\ell = 0$ has *two* maxima around a circumference, while the fact that $\ell = 1$ indicates that there is only *one* wavelength in the same distance.

3.9 What do you expect might happen if an electron with kinetic energy greater than 13.6 eV were to collide with a hydrogen atom?

3.10 If the space around the nucleus of a hydrogen atom were filled with a dielectric medium with relative permittivity ε_r, calculate the new expression for the energy levels. Hence find the energy of the ground state ($n = 1$) when $\varepsilon_r = 11.7$. (The relevance of this question to semiconductor theory will be discussed in Chapter 15.)

Self-assessment questions

1 How many quantum numbers are needed to define the wave function of an electron moving in two dimensions (excluding spin)?

(a) one   (b) two   (c) three   (d) four

2 A quantization condition for the electron wave requires that

(a) the value of ψ alone must not be discontinuous

(b) the value of $d\psi/dx$ alone must not be discontinuous

(c) the values of both ψ and $d\psi/dx$ must not be discontinuous

3 The quantization condition for a closed path leads to the result that

(a) the wavelength of the electron is an integral multiple of the circumference of the path

(b) there is a half-integral number of wavelengths in the circumference of the path

(c) there is an integral number of wavelengths in the circumference of the path

4 The quantization condition applied to a circular path of radius r leads to an equation of the form:

(a) $n\lambda = 2\pi r$

(b) $\lambda = 2\pi rn$

(c) $r\lambda = 2\pi n$

(d) $n\lambda = 2\pi/r$

5 The equation in Question 4 leads to quantization of the angular momentum in units of

(a) $2\pi/h$ (b) $2\pi h$ (c) $h/2\pi$ (d) $\pi h/2$

6 The kinetic energy of an electron of mass m and momentum p is

(a) $\frac{1}{2}mp^2$ (b) $\frac{1}{2}p^2/m$ (c) $2m/p^2$

7 The spherically symmetrical wave functions in a hydrogen atom all have the following values for their quantum numbers

(a) $n = 0$ (b) $\ell = 0$ (c) $m_\ell = 0$

8 The total energy E of the electron in equilibrium is proportional to which of the following expressions involving the principal quantum number n:

(a) $-n$ (b) $-1/n$ (c) $-n^2$ (d) $-1/n^2$

9 The total energy E is also inversely proportional to the angular momentum quantum number ℓ

(a) true   (b) false

10 The energy differences between adjacent energy levels of the hydrogen atom

(a) decrease with increasing energy

(b) increase with increasing energy

(c) are independent of energy

11 The gyroscopic nature of the electron motion about the nucleus leads to

(a) quantization of the orbital angular momentum

(b) magnetic properties of the electron

(c) the spin of the electron

(d) precession of the electron orbit (wave function)

12 The spin quantum number of the electron determines

(a) the angular momentum about the nucleus

(b) the total angular momentum of the electron

(c) the angular momentum of the electron about its own centre of mass

13 The principal quantum number n may have only the values

(a) $0, 1, 2, \ldots$ (b) $0, \pm1, \pm2, \pm3, \ldots$

(c) $1, 2, 3, \ldots$

14 The angular momentum quantum number ℓ may take only the values

(a) $0, 1, 2, 3, \ldots (n-1)$ (b) $0, 1, 2, 3, \ldots n$

(c) $1, 2, 3, \ldots n$ (d) $1, 2, 3, \ldots (n-1)$

15 The magnetic quantum number m_l may have only the values

(a) $0, \pm 1, \pm 2, \ldots \pm \ell$

(b) $0, \pm 1, \pm 2, \ldots \pm n$

(c) $0, \pm 1, \pm 2, \ldots \pm (\ell - 1)$

(d) $0, \pm 1, \pm 2, \ldots \pm (n - 1)$

16 The spin quantum number m_s may have only the values

(a) $0, \pm \frac{1}{2}$ (b) $0, \pm \frac{1}{2}, \pm 1, \pm \frac{3}{2} \ldots \pm (\ell - \frac{1}{2})$ (c) $\pm \frac{1}{2}$

17 When an electron 'jumps' from an energy level to a lower one, the energy released is usually

(a) absorbed by the nucleus

(b) emitted as heat

(c) emitted as light

(d) emitted as a continuous electromagnetic wave

(e) emitted as a photon

18 The frequency and wavelength of the emitted radiation can be found from which two equations?

(a) $\lambda = h/p$ (b) $E = hf$

(c) $E = \frac{1}{2}mc^2$ (d) $c = f\lambda$

19 In a magnetic field the energy of the electron depends additionally on the value of the quantum number

(a) ℓ (b) m_ℓ (c) m_s

20 The radiation emitted by a heated gas of hydrogen atoms contains

(a) all wavelengths

(b) one specific wavelength

(c) a set of discrete values of wavelength

21 The emission of radiation from a gas of atoms occurs when

(a) an electron is spiralling towards the nucleus

(b) an electron jumps between two energy levels

(c) the wavelength of an electron changes

22 In the emission spectrum of hydrogen the effect of a magnetic field will be

(a) to increase the number of spectral lines

(b) to decrease the number of spectral lines

(c) to change the wavelength of the spectral lines without increasing their number

Each of the sentences in Questions 23–29 consists of an assertion followed by a reason.
Answer:

(a) If both assertion and reason are true statements and the reason is a correct explanation of the assertion.

(b) If both assertion and reason are true statements but the reason is not a correct explanation of the assertion.

(c) If the assertion is true but the reason contains a false statement.

(d) If the assertion is false but the reason contains a true statement.

(e) If both the assertion and reason are false statements.

23 The value of the quantum number m_l is always less than or equal to l *because* the absolute value of the orbital angular momentum must always be greater than one of its components.

24 The hydrogen atom in its ground state can emit radiation *because* the electron can make a transition to a higher energy level.

25 Hydrogen gas at normal temperatures does not emit light *because* the radiation that is emitted has wavelengths too short to be visible.

26 In a magnetic field the electron's angular momentum precesses like a gyroscope *because* the field exerts a couple on it.

27 The electron wave in motion around an atom must take a closed path *because* the electron is confined within the atom.

28 The electron behaves like a magnet *because* it is charged.

29 The hydrogen atom is the simplest atom *because* it is spherical.

Answers

1 (b)	**2** (c)	**3** (c)	**4** (a)
5 (c)	**6** (b)	**7** (b), (c)	**8** (d)
9 (b)	**10** (a)	**11** (d)	**12** (c)
13 (c)	**14** (a)	**15** (a)	**16** (c)
17 (e)	**18** (b), (d)	**19** (b), (c)	**20** (c)
21 (b)	**22** (a)	**23** (a)	**24** (d)
25 (c)	**26** (a)	**27** (a)	**28** (b)
29 (b)			

Atoms with many electrons: the periodic table | 4

4.1 Introduction – the nuclear atom

In the early years of the last century the structure of atoms had to be laboriously deduced from a varied collection of facts and experimental results. It was soon discovered that atoms of most elements contained several electrons, because they acquired electrical charge in multiples of $\pm e$ when electrons were either knocked out of atoms or added to them by collision. An atom so charged is called an *ion*. Because an isolated atom is normally electrically neutral, it must also contain a number of positive charges, to balance the negative charges on the electrons.

From the experiment by Rutherford mentioned in Chapter 1 it was also possible to deduce the actual charge on the atomic nucleus, by analysing the paths of the deflected alpha particles. In this way Rutherford showed that the number of positive charges of magnitude e on the nucleus just equalled the atomic number – the number assigned to an element when placed with the other elements in sequence in the periodic table. As the reader who has studied more advanced chemistry will know, this sequence is nearly identical to that obtained by placing the elements in order of increasing atomic weight, and has the merit that the relationships between elements of similar chemical behaviour are clearly displayed. This is a topic that we will cover later in this chapter.

The behaviour of electrons in motion around the nucleus underlies the structure of the periodic table. So, following a brief discussion of nuclear structure, we will concentrate on the way electrons interact with one another in an atom. This leads to a basic principle, Pauli's exclusion principle, which we then use to construct the periodic table. There then emerges quite naturally a connection between electron quantum numbers and chemical properties, and this leads eventually (in Chapter 5) to an explanation of how chemical bonds between atoms are formed by electrons from each atom having particular quantum numbers, called the *valence* electrons.

Now if the nucleus of an atom contains Z positive charges, each equal in magnitude to the electronic charge, it follows that the neutral atom must contain Z electrons. By analogy with the case of hydrogen we therefore anticipate that these electrons are in motion around the nucleus, bound to it by the mutual attraction of opposite charges. It remains only to assign appropriate quantum numbers to each electron and we shall then have a model of the atom that we can use to explain many of its properties, including chemical combination. This is the aim of this and the next few chapters.

But before proceeding, note one point that is as yet unexplained. We have already identified the proton as a stable particle and we expect an atomic nucleus to contain Z such protons. However, the mass of an atom is much greater than the mass of Z protons so there must be some other constituent of the nucleus. In any case, it is unreasonable to expect a group of protons, all having the same charge, to form a stable arrangement without some assistance. The extra ingredient has been identified as an electrically neutral particle of nearly the same mass as the proton, called the *neutron*. Several neutrons are found in each nucleus – the number may vary without changing the chemical properties of the atom. The only quantity that changes is the atomic weight, and this explains the existence of different isotopes of an element, as mentioned in Chapter 1.

Because the proton and neutron have nearly identical masses and the electron masses may be neglected in comparison, the atomic weight A must be nearly equal to the total number of protons and neutrons. So the number of neutrons is just the integer nearest to $(A - Z)$. The force that binds the uncharged neutrons to the protons is a new kind of force, called *nuclear force*, which is very strong compared to the electrical repulsion between the protons.

As the structure of the nucleus has little or no bearing on the chemical and physical properties of an element (except for the radioactive elements, e.g. radium and uranium) we shall not study it further.

4.2 Pauli's exclusion principle

In assigning quantum numbers to the Z electrons in an atom the first consideration must be that the atom should have the minimum possible energy because, if one electron could make a transition to a lower energy level, it would do so, emitting radiation on the way. At first sight this implies that the electrons must all be in the lowest energy level, each having the same quantum numbers, i.e. $n = 1$, $\ell = 0$, $m_\ell = 0$ and $m_s = \frac{1}{2}$. But this is not the case and, in fact, the truth is almost exactly the opposite. It is found that each electron has its own set of quantum numbers, which is different from the set belonging to every other electron in the atom. This result has become enshrined in a universal principle named after Wolfgang Pauli, who first deduced it. In its simplest form, *Pauli's exclusion principle* states that:

> No more than one electron in a given atom can have a given set of the four quantum numbers.

To emphasize this important principle we state it again in a different way – no two electrons in an atom may have all four quantum numbers the same.

What is surprising about this principle is that it is possible for *two* electrons having opposite spin, one with $m_s = \frac{1}{2}$ and one with $m_s = -\frac{1}{2}$, to have the same spatial wave function, defined by a particular set of values of the three quantum numbers n, l and m_l. Opposite spins seem to prevent the two charge distributions from repelling one another, so that they can occupy the same region of space! It is less surprising that in other respects, electrons in the same atom avoid one another by occupying different regions of space, and hence have different spatial wave functions. We shall discover later that Pauli's exclusion principle is responsible for the wide variety of different ways in which electrons bond atoms together.

4.3 Electron states in multi-electron atoms

We may now use Pauli's exclusion principle to assign quantum numbers for the first few atoms in the table of elements and we will find that the sequence of values repeats in such a way that we can readily use them to interpret the periodic table, which was originally constructed to reflect the chemical similarities and differences between the elements.

The element with an atomic number, Z, of 2 is helium. It contains two electrons, and at least one of the quantum numbers of the second electron must differ from those of the first. On the other hand both electrons must have the lowest possible energy. Both these requirements are met if the electrons have the following quantum numbers:

1st electron: $n = 1$ $\ell = 0$ $m_\ell = 0$ $m_s = +\frac{1}{2}$

2nd electron: $n = 1$ $\ell = 0$ $m_\ell = 0$ $m_s = -\frac{1}{2}$

This places both electrons in states with the same wave function, but with opposite spins. Atomic number $Z = 3$ corresponds to lithium. Two of the three electrons may have the sets of quantum numbers given above for helium. The third electron must have

$$n = 2 \qquad \ell = 0 \qquad m_\ell = 0 \qquad m_s = +\frac{1}{2}$$

Here we note that the lowest energy for this electron is the level $n = 2$, which is higher than the energy level with $n = 1$. This should mean that this electron can be removed more readily from the atom than either of the electrons in the helium atom, because less energy is needed to remove the electron to an infinite distance. This is indeed the case, for while helium is a noble gas, lithium is metallic, and we know that metals readily emit electrons when heated in a vacuum while helium certainly does not.

In assigning the value $\ell = 0$ to this third electron we have made use of another rule – in a multi-electron atom the levels with lowest ℓ values fill up first. In other words, the energy of an electron increases with ℓ as well as with n. Remember that this was not so in the hydrogen atom – the expression for the energy [Eq. (3.9)] contained only the quantum number n. Now, however, the electrical repulsion between the electrons alters the energy of each level as indicated in Fig. 4.1, splitting what was a single level in the hydrogen atom into a series of energy levels, each with a different value for the angular momentum quantum number, ℓ. In the present case, we can understand this by noting that Fig. 3.1(b) shows that there is zero probability that the 2s electron can be found at the radius r_0, i.e. exactly where the probability of finding the 1s electron is a maximum (see Box 3.1). On the other hand, the wave function for a 2p electron (not shown) has quite a large value of ψ at the same radius r_0. As a result of its closer proximity to the 1s electron, the 2p electron has higher potential energy than the 2s electron.

Fig. 4.1

A comparison of the energy levels
in hydrogen and in atoms with
more than one electron.

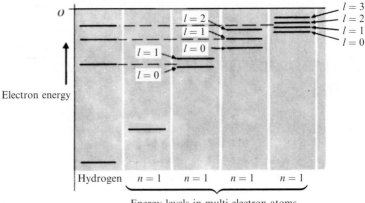

4.4 Notation for quantum states

Before going further we introduce a shorthand notation for the quantum
numbers and their values. The principal quantum number, n, defines a
series of energy levels (often called 'shells' because each level corresponds
to a different average radius, $\bar{r}$, of the wave function). Each shell corresponds to one value of n, and each of these shells is assigned a letter
according to the scheme:

n:	1	2	3	4
Letter:	K	L	M	N

Similarly, the values of l are characterized by another series of letters,
which, like the above, derive from the early days of spectroscopy:

ℓ:	0	1	2	3
Letter:	s	p	d	f

All energy levels belonging to given values of n and ℓ are said to form
a *subshell* and subshells are labelled by the number corresponding to
the value of n and the appropriate letter for the value of ℓ. Thus the subshell with $n=3$, $\ell=2$ is denoted by 3d, that with $n=2$, $\ell=0$ is denoted
by 2s.

We now continue to build up the electronic structure of the elements,
assuming that the levels are filled sequentially with increasing values of n
and ℓ. Table 4.1 shows the number of electrons in each subshell for each of
the first 18 elements.

Note that there is a maximum number of electrons that can be put into
each subshell, and by looking at the details of quantum numbers we can
see why this is. For instance in the K (or 1s) shell, we have $n=1$, $\ell=0$,
$m_\ell = 0$. There are two possibilities for m_s, $+\frac{1}{2}$ and $-\frac{1}{2}$. Hence only two
electrons are permitted in the K shell.

In the L shell, where $n=2$, the possible values for ℓ, m_ℓ and m_s are
$\ell = 0$, 1; $m_\ell = -1$, 0, +1; and $m_s = \pm\frac{1}{2}$.

Table 4.1

Electrons in each subshell for the first 18 elements

Atomic weight (A)	Atomic number (Z)	Element	K 1s	L		M		
				2s	2p	3s	3p	3d
1.008	1	H	1					
4.003	2	He	2					
6.94	3	Li	2	1				
9.01	4	Be	2	2				
10.81	5	B	2	2	1			
12.01	6	C	2	2	2			
14.01	7	N	2	2	3			
16.00	8	O	2	2	4			
19.00	9	F	2	2	5			
20.18	10	Ne	2	2	6			
22.99	11	Na	2	2	6	1		
24.31	12	Mg	2	2	6	2		
26.98	13	Al	2	2	6	2	1	
28.09	14	Si	2	2	6	2	2	
30.97	15	P	2	2	6	2	3	
32.06	16	S	2	2	6	2	4	
35.43	17	Cl	2	2	6	2	5	
39.95	18	Ar	2	2	6	2	6	

So the possible combinations of these numbers label the various 'states' into which electrons may go:

$$\left. \begin{array}{lll} \ell = 0 & m_\ell = 0 & m_s = +\frac{1}{2} \\ \ell = 0 & m_\ell = 0 & m_s = -\frac{1}{2} \end{array} \right\} \quad \text{2 states in the 2s subshell}$$

$$\left. \begin{array}{lll} \ell = 0 & m_\ell = 0 & m_s = \pm\frac{1}{2} \\ \ell = 0 & m_\ell = 1 & m_s = \pm\frac{1}{2} \\ \ell = 0 & m_\ell = -1 & m_s = \pm\frac{1}{2} \end{array} \right\} \quad \text{6 states in the 2p subshell}$$

making a grand total of eight in the L shell.

It is easy to derive the corresponding numbers for subsequent shells from the rules:

- for each value of n there are n possible values of ℓ
- for each value of ℓ there are $(2\ell + 1)$ possible values of m_ℓ
- for each value of m_ℓ there are two possible values of m_s

In this way the total numbers of electrons that can be accommodated in each succeeding shell are found to be

K: 2   L: 8   M: 18   N: 32

4.5 The periodic table

Certain interesting features arise in Table 4.1. The elements with atomic numbers 2, 10 and 18 are the rare gases, which are not merely stable but are extremely inert chemically. Thus we may equate inert characteristics with completely filled s and p subshells of electrons. It will be noted that the outermost occupied shell (that with the largest value of n) contains 2 electrons in helium and 8 in both neon and argon. This is the first indication, as we move down the list of elements, that similar arrangements of outer electrons appear periodically and that this periodic variation is reflected in the chemical properties. In fact the periodicity of chemical properties was noted before the electronic structure of the elements was known, and as long ago as 1870 the chemist Mendeléev devised a way of tabulating the elements that demonstrated it very effectively. This is now known as the periodic table of the elements. To construct the periodic table we take the elements listed in Table 4.1 and place them in order in horizontal rows so that the outer electron structure changes stepwise as we proceed along the row. We begin a new row whenever a p subshell becomes filled with electrons. Ignoring for the moment elements 1 and 2, this means that each row finishes with a rare gas, for example neon with its full L shell or argon with full 2s and 2p subshells. In this way elements with identical numbers of electrons in their outermost shells appear directly beneath one another. Thus lithium, sodium and potassium, each with one electron outside full s and p subshells, appear in the first column. More remarkably, these elements all display very similar chemical behaviour – they are all very reactive, they are all metals that become singly-charged negative ions when they combine chemically with a variety of other elements, and they form similar compounds with, for example, fluorine or chlorine (the *halides* from Group 17). We may amplify this last point by noting the properties of the chlorides: they all form transparent, insulating crystals, which are readily cleaved to form regular and similar shapes. They all dissolve, to a greater or lesser extent, in water and they all have fairly high melting points (about 700°C).

We observe corresponding similarities between the elements in the second column: beryllium, magnesium and calcium are all light metals that always combine with *two* halogen atoms; they form very stable oxides that have even higher melting points than the alkali halides mentioned above and tend to be reactive, although not to the degree shown by the alkali metals.

It would be possible to fill a book by listing all the properties shared by elements in the same column but by now the reader should be able to recognize that this unity of chemical behaviour is common to all the columns (or *groups,* as they are called) of the periodic table. Moreover, it is reasonable to associate similar chemical behaviour with a similarity in the occupancy of states in the outermost electron shells. The importance of the outer electrons in forming chemical bonds to other atoms was quickly apparent to 20th century chemists, who noted that the number of electrons outside filled s and p subshells bore a simple relation to the number of neighbouring atoms to which an atom could simultaneously bond. A rather loosely defined number, termed the 'valency', equalled the group number in Groups 1–14, while in Groups 15, 16 and 17 it was equal to the number of electrons that would have to be added to completely fill the outermost s and p subshells. It is clear from the discussion above that the combination of two s electrons and six p electrons, commonly denoted as s^2p^6, has special stability.

A better understanding of chemical bonding, to be explained in the next chapter, has led to the imprecise concept of valency being replaced by an improved measure: the *oxidation state*, described by its corresponding *oxidation number*. The latter is defined as the charge on an ion in electron units, when the ion is bonded in a compound to ions of opposite charge. Examples will be given in Section 4.7.

We shall return to these topics later, as our first glimpse of the role of electrons in chemical combination shows them to be of prime importance.

4.6 Transition elements

The reader will observe that we have so far limited discussion to the first three rows of the periodic table plus potassium and calcium. The reason is that while the periodicity of properties continues beyond this point it is not exemplified in such a simple fashion.

Let us study the filling of energy levels in the elements of the fourth row. These are shown in Table 4.2 (note the new notation here), and immediately an anomaly is apparent. The rule concerning the order of filling the various levels has been broken! Instead of the outermost electrons of potassium (K) and calcium (Ca) entering the 3d subshell, they go into the 4s subshell. Only when the 4s subshell is full do electrons begin to enter the 3d subshell – there is one 3d electron in scandium, two in titanium, and so on.

The reason for this oddity is that, in potassium (K) and calcium (Ca), the 4s levels have slightly *lower* energy than the 3d levels, so that they fill first, in keeping with the minimum energy principle. This is illustrated in the energy diagram in Fig. 4.1, where the positions of the energy levels in the hydrogen atom are shown for comparison. The shift in the relative energies in a multi-electron atom is another example of the way the interactions between electrons modify the wave functions and their energies. The more electrons there are competing for space around the nucleus, the more important these interactions become. So, while the effects of such interactions have been small up to this point, from now on we will find more and more examples of interchanged energy levels.

Returning to the filling of the 3d subshell in elements 21 to 29 we note the irregularities at 24 chromium (Cr) and 29 copper (Cu), each of which

Z	Element	Electron configuration[†]				
19	K	(filled K and L shells)	$3s^2$	$3p^6$	$3d^0$	$4s^1$
20	Ca	(filled K and L shells)	$3s^2$	$3p^6$	$3d^0$	$4s^2$
21	Sc	(filled K and L shells)	$3s^2$	$3p^6$	$3d^1$	$4s^2$
22	Ti	(filled K and L shells)	$3s^2$	$3p^6$	$3d^2$	$4s^2$
23	V	(filled K and L shells)	$3s^2$	$3p^6$	$3d^3$	$4s^2$
24	Cr	(filled K and L shells)	$3s^2$	$3p^6$	$3d^5$	$4s^1$
25	Mn	(filled K and L shells)	$3s^2$	$3p^6$	$3d^5$	$4s^2$
26	Fe	(filled K and L shells)	$3s^2$	$3p^6$	$3d^6$	$4s^2$
27	Co	(filled K and L shells)	$3s^2$	$3p^6$	$3d^7$	$4s^2$
28	Ni	(filled K and L shells)	$3s^2$	$3p^6$	$3d^8$	$4s^2$
29	Cu	(filled K and L shells)	$3s^2$	$3p^6$	$3d^{10}$	$4s^1$

Table 4.2

Arrangement of electrons for elements 19 to 29

[†] In the standard notation used here the superscript indicates the number of electrons that occupy the subshell.

contains one 4s electron instead of two. This is because the exactly half-filled 3d subshell and the filled 3d subshell are more stable configurations (i.e. they have lower energy) than the neighbouring occupancies of four and nine electrons respectively. Thus in Cu, the energy level of the 4s state has become higher than the energy of the 3d states, now that the latter subshell contains 10 electrons – another example of energy levels changing places as they continue to be filled up.

How do we assign the elements in Table 4.2 to their correct groups in the periodic table? In spite of the complications, it is clear that the first two elements, potassium (K) and calcium (Ca), fall respectively into Groups 1 and 2, both on grounds of chemical similarity and because of the similarity in the 'core' of electrons remaining after removal of the valence electrons. Assignment of subsequent elements to their groups is easier if we first discuss Cu, and then succeeding elements up to number 36 where we arrive at krypton, another inert gas, with a stable s^2p^6 octet of outer electrons (Table 4.3). The intervening elements, 29 to 35, all have complete s^2p^6 subshells and part-filled outer subshells; they might fall naturally into Groups 11 to 17 in sequence. However, Cu does not fit quite so naturally with the very reactive lithium (Li) and sodium (Na), because removal of a single electron leaves not a highly stable core but the configuration $3d^{10}$. As we remarked above, this arrangement has lower energy than the 4s electron states. Cu is less reactive than Na and Li, and, unlike them, it can donate up to three electrons when in combination with non-metals. The atom also has a much smaller radius than the lighter Group 12 metals. It is clear that a full d subshell is not as stable as the 'stable octet' of s and p subshells. We therefore put Cu and Zn into two new Groups, 11 and 12.

Let us now return to the elements 21 scandium (Sc) to 28 nickel (Ni). This set is called the first series of *transition elements*, which, because of the presence of the 4s electrons, all have similar properties (they are all metals). Although the first five are similar to Groups 13–17, again we find it appropriate to create five new Groups, numbered 3–7. But for the elements Fe, Co and Ni there are no precedents in the table, and these we assign to new Groups 8–10.

Having dealt with the first series of transition elements, we are not surprised to find another series in the fifth row of the table, and also in the sixth. The latter, however, is more complicated, owing to the filling of two inner subshells (4f and 5d) before the transition is complete and the 6p subshell begins to fill. Here, the series of elements 57 to 71, in which the 4f

---

Table 4.3

Arrangement of electrons for elements 29 to 36

Z	Element	Electron configuration[†]		
29	Cu	(filled K, L, M shells)	$4s^1$	
30	Zn	(filled K and L shells)	$4s^2$	
31	Ga	(filled K and L shells)	$4s^2$	$4p^1$
32	Ge	(filled K and L shells)	$4s^2$	$4p^2$
33	As	(filled K and L shells)	$4s^2$	$4p^3$
34	Se	(filled K and L shells)	$4s^2$	$4p^4$
35	Br	(filled K and L shells)	$4s^2$	$4p^5$
36	Kr	(filled K and L shells)	$4s^2$	$4p^6$

[†] As before, the superscript indicates the number of electrons occupying the subshell.

subshell is being filled, have almost identical chemical properties – as a group they are called the *rare earth metals*. The reason for their chemical similarity is that these elements differ only in the number of electrons in a subshell that is well inside the atom, far removed from the outermost electrons, while it is the latter that determine chemical behaviour. Their almost indistinguishable qualities place all these elements into one pigeonhole in the periodic table, in Group 3.

The subsequent filling of the 5d subshell beyond its solitary occupancy in 71 lutetium (Lu) marks the continuation of the periodic table, for the 5d electrons, like the 3d electrons in the first transition series, are of chemical importance. Thus the elements from 72 hafnium (Hf) to 78 platinum (Pt) appear in Groups 4–10.

Another group of elements, similar to the rare earths, is found in the seventh periodic row, although the majority of these do not occur in nature because their nuclei are unstable – they have, however, been manufactured artificially. The elements in this last series in Group 3 are often referred to as the *actinide elements* (they all behave like actinium, the first of the series), just as the rare earths are sometimes called *lanthanide elements*.

4.7 Group number and chemical combination

We can now clarify some points concerning chemical combination, particularly in the transition elements. We have already mentioned that, in an ideal *ionic* compound, elements combine as ions carrying charges that are multiples of $\pm e$. The ease with which an atom parts with one of its electrons is measured by the energy, in eV, that is required to ionize it, a quantity called the *first ionization energy*. It is lowest for the alkali metals, lying between 4 eV and 6 eV, while much the highest values are found in the rare gases, peaking at 24 eV in He and 22 eV in Ne, and reducing in periodic fashion as the atomic number Z rises (Fig. 4.2). This graph gives particularly clear confirmation of the structure of the periodic table, and of the exceptional stability of the s^2p^6 electron grouping.

Because materials are normally uncharged, the atomic composition, as expressed by a chemical formula such as $FeCl_3$, must carry zero charge – the sum of the oxidation numbers in $FeCl_3$ is thus zero. Three chlorine ions, each with oxidation number -1, combine with Fe, which therefore has an oxidation number $+3$.

Many elements, like Cl, have the same oxidation state when combined in different compounds. Thus magnesium (Mg) and calcium (Ca) readily lose the two s electrons, known as the *valence electrons*, from the neutral atoms. Removal of a third electron would involve destruction of the complete and very stable octet of inner electrons. The stability of closed s and p subshells is thus displayed in all atoms, not just in the rare gases that happen to possess closed s and p subshells in the electrically neutral state.

By contrast, a transition element deprived of its outermost s electrons (for example, the two 4s electrons in iron) is not nearly as stable. It may easily lose yet more electrons, and can consequently have more than one value of ionic charge. Thus we find ions such as Fe^{3+} and Fe^{2+} (the superscript gives the resultant charge on the ion, i.e. the oxidation state). Fe^{2+} ions are found in compounds such as ferrous oxide (FeO) or ferrous bromide ($FeBr_2$), while Fe^{3+} occurs in ferric oxide (Fe_2O_3) and ferric bromide ($FeBr_3$).

Fig. 4.2

First ionization energies of the elements, plotted against their atomic number Z.

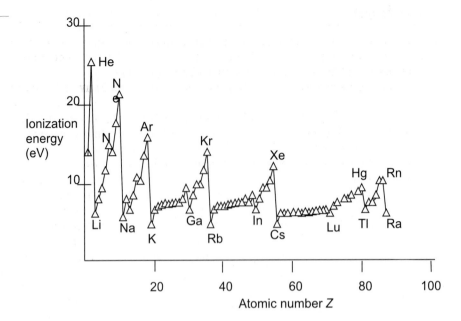

In these examples the oxygen and bromine atoms have oxidation numbers of −2 and −1 respectively, because they require the addition of this number of electrons to attain a stable s^2p^6 configuration. Because, in this stable configuration, the s and p subshells together contain eight electrons the number of electrons to be added is $(8 − N)$ or $(18 − N)$, where N is the group number. Thus the principal oxidation state of the elements in Groups 4–7 is given by the so-called $8 − N$ rule.

Hydrogen is a unique case because the nearest stable arrangement contains either two electrons (a filled K shell), or alternatively no electrons. Hydrogen atoms can either donate electrons to, or attract electrons from, a neighbouring atom, permitting combination with halogens to form acids (HF, HCl, HBr, HI) and also with metals, giving hydrides (NaH, ZrH, CaH_2, etc.). Hydrogen's oxidation number is either +1 or −1.

The noble gases are assigned to Group 18: their oxidation number is nominally zero, and they were for a long time thought to be completely inert. It is now known that, with the exception of helium and neon, they form a few simple compounds, and that their atoms can also combine to form solids. Because the solids all vaporize at very low temperatures, the attractive forces must be very weak.

It is interesting to note at this point that, although the individual wave functions of electrons in a subshell have quite complex shapes (see Fig. 3.12), in a completed subshell the total charge density is quite simply spherically symmetrical. The electron cloud of a filled subshell is thus like that of one of the s states in Fig. 3.12, and a noble gas atom may therefore be pictured roughly as a small sphere.

When we come to discuss compounds of Group 14 elements (C, Si, Ge) we will find that, while the nature of the chemical bonding differs dramatically, the number of other atoms to which each one can bond is still determined by the number of outer (valence) electrons available. The subject of bonding will be discussed in detail next.

Problems

4.1 Discuss as many chemical similarities as you can find that might lead us to place the elements of Groups 2 and 12 together. Discuss also the differences between these Groups.

4.2 Plot the melting points and densities against atomic number for the first three rows of the periodic table. Notice how the changes reflect the periodicity represented by the table. (Data in Appendix 3.)

4.3 Look up the densities of the elements in Group 2 (Appendix 3) and explain why they depend on atomic number.

4.4 What differences would there be in the periodic table if (a) the 5p levels had lower energy than the 4d levels; (b) the 5d levels had lower energy than the 4f levels?

4.5 In which elements do the nuclei contain the following numbers of neutrons?

10   14   22

Suggest how the existence of several isotopes might explain why chlorine apparently has the non-integral atomic weight 35.4.

4.6 Write down the quantum numbers of each electron in the M shell, in the order in which they are filled. Hence show that this shell may accommodate only 18 electrons.

4.7 Why do the oxides of the Group 2 metals have higher melting points than the alkali halides? Why are the Group 2 metals less reactive than those in Group 1?

4.8 Without consulting Table 4.1, write down the electronic configuration of the elements with atomic numbers 4, 7, 10 and 15. By counting the valence electrons, decide to which group each element belongs.

4.9 Copper has a single valence electron in the 4s subshell and belongs to Group 11. What is the electronic structure of the Cu ion? Why does copper not behave chemically in the same way as sodium or potassium?

4.10 The outermost shells of all the inert gases are filled s and p subshells. Silicon and germanium have four electrons outside closed s and p subshells, and so have titanium and zirconium. Why, then, do the chemical characteristics of the first pair of elements differ from those of the second pair?

Self-assessment questions

1 If the atomic number of an element is Z and its atomic weight is A, the number of protons in the nucleus is

(a) Z (b) $A - Z$ (c) A (d) $Z - A$

2 The difference between A and Z is a result of the presence in the nucleus of

(a) electrons   (b) protons

(c) photons   (d) neutrons

3 Pauli's exclusion principle states that, within one atom

(a) no more than two electrons may have the same energy

(b) the spins of the electrons interact so as to become parallel if possible

(c) no two electrons may have the same four quantum numbers

(d) there are only two values for the quantum number m_s

4 In atoms containing many electrons the subshells are filled in order of

(a) increasing n and ℓ

(b) decreasing n and ℓ

(c) increasing energy

5 The maximum number of electrons in the L shell ($n = 2$) is

(a) 4   (b) 6   (c) 8   (d) 14

6 In the notation $2p^6$, $3s^2$, etc. the meanings of the symbols are

(a) the first number is the value of ℓ, the letter gives the value of n, and the superscript is the number of electrons in the subshell

(b) the first number is the number of electrons in the subshell, the letter gives the value of ℓ and the superscript the value of n

(c) the first number is the value of n, the letter gives the value of ℓ and the superscript is the number of electrons

7 The maximum number of electrons allowed in the 4d subshell is

(a) 14   (b) 10   (c) 8   (d) 4

8 $3f^6$ denotes a subshell containing 6 electrons for which

(a) $n = 3$ and $\ell = 3$ (b) $n = 3$ and $\ell = 4$

(c) neither a) nor b)

9 The lithium atom, which contains three electrons, has the structure

(a) $1s^22s^1$ (b) $1s^22p^1$ (c) $1s^12p^2$ (d) $2s^22p^1$

10 The atomic number of the element whose outermost electron fills the 3s subshell exactly is

(a) 13   (b) 8   (c) 10   (d) 12

11 The characteristic feature of the transition elements is

(a) a partly filled valence subshell

(b) an empty inner subshell

(c) an unfilled outer subshell

(d) a partly filled inner subshell

12 The first series of transition elements, in which the 3d subshell is gradually filled, begins at atomic number

(a) 19   (b) 21   (c) 11   (d) 13

13 The element with electronic structure $1s^22s^22p^63s^23p^63d^84s^2$ is a transition element

(a) true   (b) false

14 In the periodic table, the elements are arranged in order of increasing

(a) atomic weight

(b) chemical equivalent weight

(c) molecular weight

(d) atomic number

15 The polonium atom has the electronic structure
$$1s^22s^22p^63s^23p^63d^{10}4s^24p^64d^{10}4f^{14}$$
$$5s^25p^65d^{10}6s^26p^4$$

Its group number is

(a) 2   (b) 4   (c) 16   (d) 8

16 The manganese atom has the electronic structure $1s^22s^22p^63s^23p^63d^54s^2$

Its group number is therefore

(a) 2   (b) 7   (c) 12   (d) 17

17 The outer electron configuration that gives the noble gases their extreme inertness is

(a) s^2p^6 (b) s^2p^4 (c) $s^2p^6d^{10}$ (d) s^2p^2

18 Copper, atomic number 29, has the structure $1s^22s^22p^63s^23p^63d^{10}4s^1$. Its chemical properties are unlike those of the Group 1 metals Na, Li because

(a) removal of an electron leaves a particularly stable structure in the outermost shell

(b) removal of two electrons requires only a little more energy than removing one

(c) the electronic structure of Cu^+ is less stable than that of Group 1 ions

19 Transition elements can have one of several oxidation numbers because

(a) they contain several electrons in the outermost subshell

(b) removal of the electrons in the outermost subshell does not leave a stable octet

(c) the electrons in the valence subshell can readily be removed singly from the atom

20 The principal oxidation number of the elements in Groups 1–4 of the periodic table is equal to

(a) 6 minus the group number

(b) the group number

(c) 8 minus the group number

21 The oxidation number of an ion of the element with structure $1s^2 2s^2$ is

(a) 2   (b) 4   (c) 6   (d) 0

22 The principal oxidation number of the elements in Groups 5–7 of the periodic table is

(a) 9 minus the group number

(b) the group number

(c) 8 minus the group number

(d) the group number minus 4

23 Boron (Group 13) and fluorine (Group 17) may form a compound with the formula

(a) B_2F_3 (b) B_3F (c) BF_3 (d) B_3F_2

Each of the sentences in Questions 24–31 consists of an assertion followed by a reason.

Answer:

(a) If both assertion and reason are true statements and the reason is a correct explanation of the assertion.

(b) If both assertion and reason are true statements but the reason is *not* a correct explanation of the assertion.

(c) If the assertion is true but the reason contains a false statement.

(d) If the assertion is false but the reason contains a true statement.

(e) If both the assertion and reason contain false statements.

24 The ground state of the helium atom is $2s^2$ *because* this is the lowest energy state.

25 The electronic structure

$$1s^2 2s^2 2p^6 3s^2 3p^6 3d^{10} 4s^2 4p^5 4d^2$$

does not normally occur in a real atom *because* the subshells have been filled in the wrong order.

26 The rare gases do not form compounds *because* they are in Group 18 of the periodic table.

27 The elements in any one group of the periodic table are chemically similar *because* they all contain the same number of electrons in the outermost subshell.

28 There are 18 groups in the periodic table *because* there are nine stable arrangements of electrons in the s and p subshells.

29 Copper (Group 11) is chemically not as similar to lithium (Group 1) as is sodium (Group 1) *because* its atoms do not contain the same number of electrons in the outermost subshell.

30 The first series of transition elements are all placed together in Group 3 *because* of their chemical similarity.

31 Iron and oxygen can combine to form more than one oxide, e.g. FeO and Fe_2O_3, *because* iron and oxygen may have more than one oxidation number.

Answers

1	(a)	**2**	(d)	**3**	(c)	**4**	(c)
5	(c)	**6**	(c)	**7**	(b)	**8**	(b) (no such subshell exists!)
9	(a)	**10**	(d)	**11**	(d)	**12**	(b)
13	(a)	**14**	(d)	**15**	(c)	**16**	(b)
17	(a)	**18**	(b), (c)	**19**	(b)	**20**	(b)
21	(a)	**22**	(c)	**23**	(c)	**24**	(e)
25	(a)	**26**	(b)	**27**	(c)	**28**	(a)
29	(c)	**30**	(e)	**31**	(c)		

5 | Molecules and interatomic bonding

5.1 Introduction – classification of bonding mechanisms

When a gas of neutral atoms condenses to form a solid the atoms are held almost rigidly together by mutual attraction. The pull between them is much stronger than a mere gravitational force, and we say that atoms *bond* to one another. This chapter does not attempt to cover all aspects of bonding in materials, but concentrates on illustrating some of the basic mechanisms, at the same time warning the reader that, in most individual cases, several mechanisms are found in combination.

Bonding is a result of the way that electronic wave functions of two or more atoms merge together when their nuclei approach closely enough that the outer (valence) electrons are attracted to *both* nuclei.

In a solid or a large molecule, each atom bonds with two or more of its nearest neighbours, while in a diatomic molecule a single bond may be studied in isolation. We therefore begin our discussion with a study of a single diatomic molecule (hydrogen) in Section 5.2, before moving on to the mechanisms that operate in solids (Section 5.3). In Sections 5.4 to 5.9 we describe in turn the principal bonding mechanisms in solids. Classification into different mechanisms is only approximate because the boundaries between types are imprecise, but it is particularly useful to distinguish between *primary* or *interatomic* bonding mechanisms acting between atoms, from weaker *secondary* or *intermolecular* bonding that acts between molecules. We describe first four mechanisms of interatomic bonding – *ionic*, *covalent*, *metallic* and *hydrogen bonding* (the first and last of which can also be intermolecular), before showing how in particular cases two or more mechanisms commonly act together. We also discuss delocalized, or conjugated, bonding, which is important in larger molecules such as benzene, as well as in polymers and living materials. We then discuss the major intermolecular bonding mechanisms, van der Waals and hydrogen bonding. The interrelationship between these mechanisms in molecular solids is summarized in Box 5.1.

Some bonding mechanisms act in particular directions, while others are non-directional. This fact becomes important in Chapter 6 when we discuss the way atoms pack together in solids, and later still when we come to discuss how real materials fracture. The mechanism of fracture of solids depends on the strength of bonds, which are studied in Sections 5.10 and 5.11 by comparing the heats of vaporization and the melting points of various materials with different bonding mechanisms.

5.2 Electron pairing in a diatomic molecule

Just as the simplest atom to study is hydrogen, so the simplest molecule is formed by the combination of two hydrogen atoms – the diatomic hydrogen molecule.

This combination of two protons and two electrons is not as easy to treat mathematically as the single atom, unless a computer is used. However, a qualitative picture may easily be obtained by imagining two atoms, H_a and H_b, approaching one another as depicted in Fig. 5.1. As the separation between them decreases, each electron feels two new competing forces: an attraction towards the other nucleus and a repulsion from the other electron. These forces both vary inversely with the mean square of the separation r between the charges concerned. In practice the attraction is the stronger, because the repulsion is reduced when the two electrons have opposite spins. This can be appreciated in the light of Pauli's exclusion principle, which allows electrons to have the same spatial wave functions, i.e. to occupy the same region of space, if their spins are antiparallel. Each electron thus moves closer to the 'other' nucleus than the distance d between the two nuclei. As a result, the attractive forces between electron a (or b) and nucleus b (or a) overcomes the repulsion between the two nuclei.

The atoms are pulled together until each electron cloud surrounds both nuclei, and the two electrons, with opposite spins, share the same spatial wave function [Fig. 5.1(c)]. Closer approach is impossible because the repulsion between the two nuclei would be too great. At the equilibrium separation there is an exact balance between the attraction of the electrons and the repulsion of the nuclei. This is shown in Fig. 5.2, where each force is plotted against the internuclear distance d. As the atoms approach one

Fig. 5.1

(a) When atoms are well separated there is a weak attractive force. (b) At closer approach the electron clouds are distorted by the attractive forces, and the two electron spins are antiparallel. (c) At equilibrium the electrons share the same wave function.

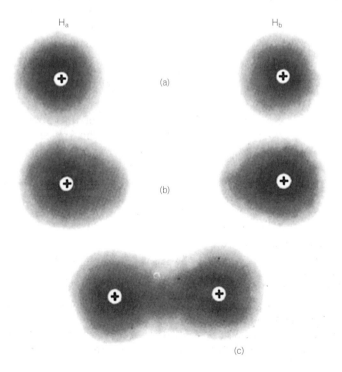

H_a H_b

(a)

(b)

(c)

Fig. 5.2

Forces between two atoms plotted against interatomic distance.

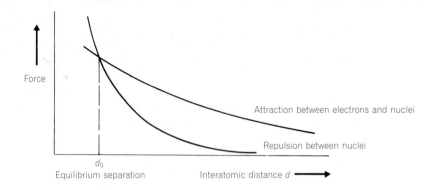

another, both attractive and repulsive forces build up, the latter more slowly at first. On close approach the repulsive force increases rapidly until it equals the attractive force, and the molecule is in equilibrium. In hydrogen this occurs at a nuclear separation d_0 of 0.074 nm. Thus it is possible to calculate the equilibrium separation if we know how the forces depend on separation.

Note that bonding results in a marked overlap of the two electron clouds. This is a universal characteristic, and it means that a pair of atoms bond together only when they approach close enough for the valence electron clouds to overlap one another. This allows the valence electrons to pair up, with opposite spins in a single wave function that may be situated largely between the atoms, as here, or concentrated near one of the atoms, as we describe in the next section.

When two identical but heavier atoms combine there is one significant difference from the case just discussed. Consider, then, the approach of two sodium atoms. We may picture each as a nucleus surrounded by a core of negative charge (the completed K and L shells), outside which moves the valence electron cloud (Fig. 5.3). We have already seen that the closed

Fig. 5.3

The formation of a 'molecule' of Na. This structure is not stable in practice, but the addition of further atoms produces a stable crystal of sodium (see later).

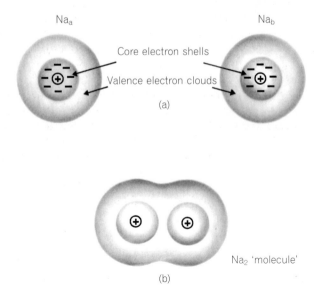

shells of the core are spherical in shape, so they are represented as spheres in the figure. The bonding force arises exactly as in the hydrogen molecule, but the repulsive force is now augmented at short separations by the reluctance of the closed inner shells to overlap. They could overlap only if the electron spins were opposed, but because each shell contains electrons with spins already opposed, overlap is not allowed. Hence the equilibrium separation is determined largely by the radius of the inner core, and the molecule is like a pair of nearly rigid spheres glued together. This simple model is very useful when we consider how atoms pack together in crystals in Chapter 6. Nevertheless, electron clouds do not have sharp boundaries, so the radius of the core of electrons is not as precisely defined as the above argument suggests, and differs slightly in different crystals.

These molecules demonstrate a key idea – that when bonding two atoms, valence electrons form pairs with opposing spins, each pair occupying a single spatial wave function. An s^2 pair is formed in the cases above, but more often the pairing results in completion of a stable *octet*, like the s^2p^6 core of each Na atom, or, as we shall see later, like an equivalent set of four spatial wave functions having a non-spherical shape. The pairing of electrons to form completed subshells or octets pervades much of the subsequent discussion of interatomic bonding mechanisms.

5.3 Bonding in solids

We have seen how the overlapping of the valence electron clouds of two atoms can lead to an overall attraction and hence to the formation of a molecule. The formation of a solid body may be described in similar terms, but involves attraction between each atom and several of its neighbours simultaneously, instead of with just one other atom. The question of how many neighbours surround each atom and in what arrangement is left to the next chapter. Here we will look primarily at bonding between one pair of atoms in a solid, but we must remember that both the atoms concerned may also be bonded to others.

Bonding between atoms varies in nature according to the electronic structure of the atoms involved. We would not expect the closed-shell structure of the inert gases to behave in the same way as the alkali metals, each of which can so easily lose its outer electron. Indeed, we know from experiment that, while these metals readily form compounds with other elements, the inert gases do not. Because the pairing of valence electrons can happen in a variety of ways, depending on the available number of valence electrons, we discuss these different ways in separate sections under the headings of *ionic*, *covalent* and *metallic solids*. However, the boundaries between some of these categories are not well defined, so that we will also need to discuss cases in which more than one mechanism operates.

In the hydrogen molecule, the electrons responsible for bonding are heavily concentrated in the space between two atoms. In such cases it is useful to think of the pair of atoms being linked by a *localized bond*, as suggested by the lines drawn between atoms in conventional diagrams of molecular structure. But electron wave functions are spread out, as we have seen, and often this concept of individual bonds is not an accurate description. The idea of a localized bond between particular atoms is, however, very useful in dealing with physical properties like the strength and rigidity of solid materials, so we will use it quite frequently.

5.4 Ionic solids

As remarked earlier, the valence electron is fairly easily removed from an alkali metal, leaving behind a very stable structure resembling an inert gas, but with an extra positive nuclear charge. The ionization potentials in Fig. 4.2 showed this very clearly. In contrast to the alkali metals, an element from Group 17 (fluorine, chlorine, bromine or iodine – the halides) is only one electron short of an inert gas structure. Because the electronic structure of an inert gas is so stable, we might expect that a halide atom would readily accept an extra electron and might even be reluctant to lose it, given suitable conditions.

Taking these two tendencies together, we can understand what happens when, for example, sodium and chlorine atoms are brought together in equal numbers. It costs little energy to transfer the valence electron of each sodium atom to a chlorine atom, giving a rather stable s^2p^6 grouping in each. But now there must be an electrostatic attraction between the ions so formed, for each sodium ion carries a positive charge (Fig. 5.4) and each chlorine ion a negative one. The attractive forces pull them together until the inner electron clouds begin to overlap. At this point a strong repulsive force grows rapidly, exactly as for the case of two sodium atoms described earlier, and the two forces just balance one another. The way in which these forces depend upon the separation, d, is shown in Fig. 5.5, where we can see the similarity to the case of the hydrogen molecule in Fig. 5.2.

Naturally there is also an electrostatic repulsion between ions of the same charge, so that in solid sodium chloride (NaCl) we do not expect to find like ions side by side but rather alternating, as shown in Fig. 5.6. In this way the attraction of unlike ions overcomes the repulsion of like ions and a stable structure is formed. We will meet this and other kinds of atomic packing in Chapter 6.

In Fig. 5.6 the ions are represented by spheres, for we have already seen (Fig. 5.3) that this is a realistic way of picturing the inner core of electrons.

It is clear from Fig. 5.6 that the concept of an individual molecule is of no use in such a solid. Indeed, a single pair of ions does not necessarily

Fig. 5.4

Schematic representation of the formation of an ionic 'molecule' of sodium chloride.

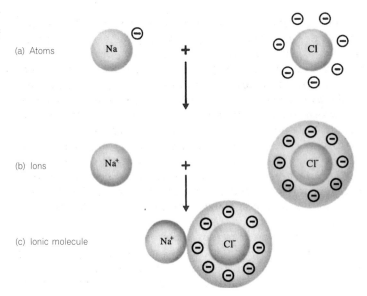

(a) Atoms

(b) Ions

(c) Ionic molecule

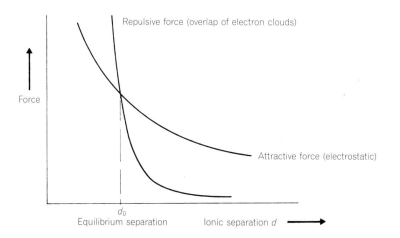

Fig. 5.5

Dependence of interionic forces on ionic separation.

Repulsive force (overlap of electron clouds)

Force

Attractive force (electrostatic)

d_0
Equilibrium separation

Ionic separation d ⟶

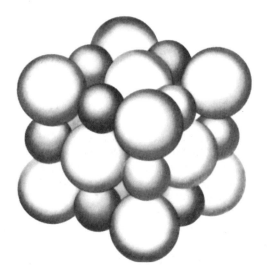

Fig. 5.6

Part of the crystal of NaCl modelled as a close packed arrangement of spheres.

form a stable molecule. Although the molecule of sodium chloride may exist in the gaseous state, it is less stable than the assembly of many ions in Fig. 5.6, where each Na^+ ion has six Cl^- ions as neighbours.

A crystal of magnesium oxide (MgO) has a similar internal structure to that of NaCl. In this case, a modest expenditure of energy will allow *two* electrons to transfer from each magnesium atom to an oxygen atom, again leaving each ion with a stable octet of outer electrons (Fig. 5.7). Because these ions are doubly ionized and hence carry a charge of $-2e$, it is understandable that the interatomic cohesion should be much stronger than in NaCl. This accounts for the much higher melting point of magnesia (2800°C compared with 800°C for NaCl). The relationship between melting point and bond strength will be discussed later in this chapter. For now it is sufficient to note that the stronger a bond, the higher is the temperature needed to break it.

Further examples of ionic bonding occur in the compounds cupric oxide, chromous oxide and manganese difluoride (CuO, CrO, MnF_2), showing that the metallic element need not be from Groups 1 or 2, but that any metal may become ionized by losing some or all of its valence electrons.

Fig. 5.7

Representation of the formation of an ionic 'molecule' of MgO. Like NaCl, it is not stable in isolation.

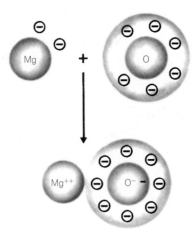

However, a metal ion with a high positive charge will often modify the wave functions of the bonding electrons, resulting in bonds that are no longer purely ionic in nature, as Section 5.7 will explain. Ionized molecular groups such as NH_4^+ and SO_4^{2-} can also combine with counterions in ionic solids, some of which have a structure similar to NaCl (Fig. 5.6).

5.4.1 The special case of hydrogen

The hydrogen atom has unique properties that it does not share with Group 1 elements: it forms bonds in its own way. The first distinctive feature is the absence of an inner core of electrons. Unlike an Li^+ ion, for example, which has a K electron shell of radius 0.06 nm, the bare nucleus of hydrogen has nothing to prevent it from being closely surrounded by the electron wave functions of a neighbouring atom such as F or Cl. For example, the nuclear separation in the HF molecule is 0.0917 nm, smaller than the radius of the F^- ion, which is 0.136 nm. This results in a strong influence of the H^+ ion on the valence electrons of the atom with which it bonds, so strong that their motion is modified, as we saw in the H_2 molecule studied in Section 5.2. Because the wave functions envelop both nuclei, a very strong bond is formed that cannot be described as ionic, but is more akin in both character and strength to the *covalent* bond discussed in the next section. Thus the bonding mechanism for hydrogen is usually termed covalent, although a more accurate description for HF will be given in Section 5.7.

The second peculiarity of hydrogen is that it can form another unique type of bond, called a *hydrogen bond*, which is rather weaker than the bonding mechanisms we have so far encountered. Hydrogen's small size permits it to share electrons simultaneously with two neighbouring atoms, when it is situated between them. For example in solid HF, there are two fluorine atoms flanking every hydrogen atom, which are drawn closer together than their ionic radii would lead us to expect. This is because the electron wave functions are modified by the H^+ nucleus, which thereby shares all the valence electrons that are available in the three atoms taking part in this hydrogen bond.

The hydrogen bond plays an important role in crystals like potassium dihydrogen phosphate, KH_2PO_4, in which regularly spaced PO_4 groups are joined to one another by O—H—O bonds. Because several mechanisms contribute to the strength of the hydrogen bond, we will leave further discussion of it to Section 5.9.2.

Finally, hydrogen actually becomes a *negative* ion in the salt-like metal hydrides NaH, CaH_2, etc. mentioned in Chapter 4, attracting to itself one of the metal's valence electrons.

5.5 Covalent solids

Elements from the central Groups 13, 14 and 15 of the periodic table – notably Group 14 – are not readily ionized. The energy required to remove all the valence electrons can be too large for ionic bonding to be possible. It is still possible for each atom to complete its outer s and p subshells, however, by sharing one or more pairs of electrons with its neighbours.

We saw earlier how the hydrogen molecule is formed by sharing electrons in this way but we can now show that this bond is more universal and may be present in solids.

We take carbon as an example. It has a filled K shell and four electrons in the L shell, i.e. $1s^2 2s^2 2p^2$. Four more electrons are required to fill the L shell, and these may be acquired by sharing an electron with each of four neighbours when carbon is in its solid form. One way in which this could be done is shown in Fig. 5.8, although this precise arrangement does not occur in practice because of the directional nature of the bonds.

To understand this we must look at the shape of the electron clouds in each carbon atom. The four electrons in the L shell interact rather strongly with one another. Each electron moves in the electric potential due to the nuclear charge $+4e$ and three separate electron charges $-e$. The shapes of the s and p wave functions become modified until they are identical in shape, each becoming what is called an sp^3 *hybrid wave function*. The electrostatic repulsion between these is so strong that each electron's 'cloud' concentrates itself away from those of the other three, as illustrated in Fig. 5.9. This means that each cloud is sausage-shaped and points away from the nucleus, the four arranging themselves with the largest possible angle, 109.5°, between each pair. It is easy to see that they point towards the corners of an imaginary tetrahedron.

Because of the strong repulsion this arrangement is difficult to distort and the carbon atoms join up as shown in Fig. 5.9, with the electron clouds of neighbouring atoms pointing towards one another, each pair of atoms effectively sharing two electrons of opposite spins at any moment. In this way each carbon atom is surrounded by eight electrons that, like the $s^2 p^6$ case met earlier, is a stable arrangement. The structure so formed is very strong and rigid – it is the structure of diamond, the crystalline form of carbon. The term crystal (from the Greek *kristallos* – clear ice) originally

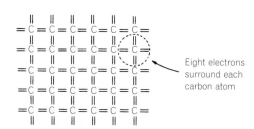

Eight electrons surround each carbon atom

Fig. 5.8

By sharing electrons (shown by dashes) with its neighbours, a carbon atom forms four bonds, and a repetitive structure is created.

Fig. 5.9

(a) The lobe-shaped valence
electron clouds in the carbon
atom. (b) The carbon atom in its
crystalline surroundings has a
similarly shaped electron cloud.

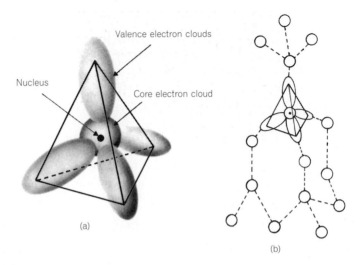

referred to a solid that can be cleaved to form a regular geometrical shape,
or which occurs naturally in such a shape. The modern definition of a
crystal will be given in Chapter 6.

Just as in the case of sodium chloride, no molecule can be distinguished
here, but the solid is rather like one huge molecule (a so-called *macromolecule*) because it is a never-ending structure – it is always possible to
add more carbon atoms to it. The other Group 14 elements, silicon and
germanium, also crystallize in the same structure.

It is useful to note here how to deduce the number of neighbours with
which each atom may bond in a covalent solid of an element. In the
covalent bonding mechanism the atom acquires electrons by sharing, until
it has stable outer subshells. The number of valence electrons needed thus
equals the number of electrons lacking from the outer s and p subshells; if
there are N electrons present in these outer subshells of the neutral atom
then bonding with $(8 - N)$ neighbours results.

It is interesting that compounds of an element of Group 13 with one
of Group 15 also form a structure related to that of diamond, and which
we will meet in Chapter 6. In these compounds each atom is again tetra-
hedrally coordinated with four others, even though the isolated Group 13
and Group 15 atoms have three and five valence electrons respectively.
The $(8 - N)$ rule clearly does not apply within compounds.

Covalent bonding sometimes results in two, or occasionally three,
electron pairs being shared by two atoms. We refer to them as double and
triple covalent bonds, and show them as two or three parallel lines in a
structural diagram. Thus O=O and N≡N represent two famil-
iar molecules, while calcium dicarbide, CaC_2, is an ionic solid comprised
of the ions Ca^+ and $(C≡C)^{2-}$.

5.6 Metallic solids

As the name implies, these are confined to metals and near-metals, many
of which are found in Groups 1–3 and 11–13 of the periodic table. If we
take copper as an example, we see that shells K, L and M are full, while
there is just one 4s electron in the N shell. In solid copper the outer

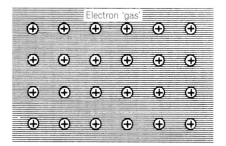

Fig. 5.10
A metallic crystal, pictured as a lattice of positive ions embedded in a 'sea' of electrons. The charge distribution of electrons is here shown uniform, although in practice it is not quite so.

electron is readily released from the parent atom and all the valence electrons can move about freely between the copper ions. The positively charged ions are held together by their attraction to the cloud of negative electrons in which they are embedded (Fig. 5.10), rather like ball bearings in a liquid glue.

In some respects this arrangement is like an ionic solid, but instead of the negative 'ions' being like rigid spheres, they fill all the space between the positive ions. We could say that a metallic solid is a sort of ionic solid in which the free electron is donated to all the other atoms in the solid. Although that is the easiest way to understand metallic bonding, it is often useful and legitimate to view it as a kind of covalent solid in which the electrons form a temporary covalent bond between a pair of atoms and then move on to another pair. Thus the bonds are not directional and their most important characteristic is the freedom of the valence electrons to move. In later chapters we will see how this mobility of the electrons is responsible for the high electrical and thermal conductivities of metals.

Because the ions are bonded to the valence electrons and only indirectly to each other, it is possible to form a metallic solid from a mixture of two or more metallic elements, for example copper and gold. Moreover, it is not necessary for the two constituents to be present in any fixed ratio in order that the solid be stable. The composition may thus be anywhere between that of gold with a small proportion of copper added, and that of copper containing a small proportion of gold. It is rather as if one metal were soluble in the other and, indeed, we often refer to this type of material as a *solid solution*, although the term *alloy* is used more commonly. We will learn more about alloys, some of which are very important because of their great strength, in Chapter 10.

5.7 Combinations of bonding mechanisms

Although at first sight the interatomic bonding mechanisms seem quite distinct from one another, the bonding in many substances does not fit easily into just one or other of these categories. Here we concentrate on covalent and ionic mechanisms, and show how they can act cooperatively. To illustrate this, consider a series of molecules containing the same total number of electrons – known as *isoelectronic* – methane, ammonia, water and hydrogen fluoride.

In methane (CH_4) the carbon atom forms four symmetrically placed covalent bonds, one with each hydrogen atom (Fig. 5.11). Hydrogen fluoride (HF), at the other extreme, is naturally and permanently dipolar

Fig. 5.11

Electron clouds for covalent, ionic
and mixed ionic/covalent bonding.

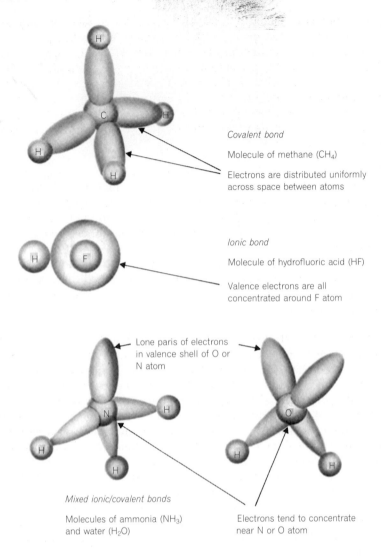

Covalent bond

Molecule of methane (CH_4)

Electrons are distributed uniformly
across space between atoms

Ionic bond

Molecule of hydrofluoric acid (HF)

Valence electrons are all
concentrated around F atom

Lone paris of electrons
in valence shell of O or
N atom

Mixed ionic/covalent bonds

Molecules of ammonia (NH_3)
and water (H_2O)

Electrons tend to concentrate
near N or O atom

(chemists refer to it as *polar*), because the effective centre of the positive
charge is not located centrally within the electron charge distribution.
The H and F atoms can be treated as *partially* ionized: a charge of about
$-0.4e$ can be regarded as having transferred from the hydrogen atom
to the fluorine atom, leading to strong electrostatic attraction and giving
the bonding a partly ionic nature. In reality this means merely that the
valence electron clouds are distorted roughly as shown in Fig. 5.11, so that
there is a heavy predominance of electronic charge density around the
fluorine nucleus.

In the intervening compounds, ammonia and water (NH_3 and H_2O), a
smaller proportion of an electronic charge may be considered to be
transferred, while the rest of it is shared equally in a covalent fashion. This
shift of the 'centre of charge' increases as we go through the series from the
C—H bond to the F—H bond. The carbon, nitrogen, oxygen and fluorine
atoms are arranged in order of increasing *electronegativity*, which is
defined in terms of the degree to which an atom can attract electrons to
itself. It is a semi-quantitative concept based on an analysis of the energy

stored in bonds of mixed nature. Various scales of electronegativity are used[†], but we will just give the electronegativities worked out by Nobel prize winner Linus Pauling for the above elements. They are H: 2.1, C: 2.5, N: 3.0, O: 3.5 and F: 4.0. In general, the electronegativity increases steadily across the periodic table. It also varies within a group, decreasing with increasing atomic number, except in the case of transition elements.

A difference in electronegativity is a measure of the tendency of a pair of elements to form ions when bonding, so that we can formulate rules about bonding as follows:

- Two elements of similar electronegativity form either a *metallic* solid or *covalent* solid, according to whether they can release electrons (low electronegativity) or accept electrons (high electronegativity).
- Where the electronegativities differ, the bonding mechanism is partially ionic, the ionic character increasing with the difference in electronegativity.
- Because covalent bonding is directional, while ionic bonding is not, the degree of directionality changes with the bonding character. Such changes have a marked influence on the crystal structure, as we shall see in Chapter 6.

A further example of mixed ionic/covalent bonding occurs in an important class of substances based upon the dioxide of silicon (SiO_2). This compound can form both a regular, ordered structure, i.e. a crystal – in the commonest form this is called quartz – and an irregular structure called quartz glass, vitreous silica, or simply silica. This and many other types of glass are *amorphous* structures containing silicon dioxide and other compounds. *Amorphous* – having no form or structure – derives from the Greek *morphe* meaning form. Amorphous solids do not form crystals with regular shapes, because there is no regularity in the way in which their atoms are packed together. (Note that not all solids that do not form regular geometrical crystals are amorphous.) Various ceramic materials, for example porcelain, are made from SiO_2 and other oxides, but are in the form of crystalline regions embedded in an amorphous matrix. A variety of amorphous materials will be discussed in more detail in Chapters 11, 12, 13 and 17.

Just as bonds occur with mixed ionic/covalent natures, so we find that there is a continuous change in bonding character in a series of alloys of metals such as Cu-Ni, Cu-Zn, Cu-Ga, Cu-As and Cu-Se. In fact, the last few of these tend to form definite compounds (i.e. the elements combine in simple ratios), indicating the presence of a bonding mechanism other than metallic. However, at the same time they retain certain characteristically metallic properties.

5.8 Conjugated bonding: delocalized electrons in molecular orbitals

Organic materials, commonly called hydrocarbons, are compounds of carbon with hydrogen and other (usually minor) constituents. Particular

---

[†] Caution: do not compare electronegativity values obtained from different sources unless you are sure that they are quoted on the same scale (the Pauling, Goldschmidt, Allred, Mulliken or spectroscopic scale).

combinations of the covalent carbon–carbon bonding in some hydro-carbons show distinctive properties, as we now explain.

In all covalent molecules and solids discussed above, the electrons were *localized* in the space between the two atoms that they join. Wave functions in molecules are conventionally termed *molecular orbitals*, a term we will use here. Sometimes electrons can be shared between more than two atoms within a molecule – they become *delocalized*, but the charge they carry nevertheless results in stabilizing the molecular shape. A classic case is the ring-shaped benzene molecule C_6H_6, illustrated in Fig. 5.12. One of the four outer electrons from each atom bonds covalently with the nearby hydrogen atom, choosing the opposite spin to the hydrogen's electron. Two of the remaining three electrons of each carbon atom are paired with those from a neighbouring carbon atom, and form so-called *sigma* orbitals that, as in any single covalent bond, are localized between the carbon nuclei and labelled σ in Fig. 5.12(b). The fourth bonding electron of each C atom cannot be accommodated in a σ orbital, and it might be thought that pairing six such electrons would bond only three pairs of the six nuclei, giving three single and three double bonds as in Fig. 5.12(a). But this does not happen. These electrons must avoid overlapping with the σ orbitals lying within the plane occupied by the six C atoms, but are forced to concentrate their charge alongside that ring, both above and below it. Each such electron then experiences attraction, not only to the parent nucleus, but also that of neighbouring nuclei. As a result, the fourth bonding electron of each carbon atom spreads itself around the ring in a *molecular orbital*. The six electrons occupy three such orbitals, each with a different angular momentum. Each of the six electron clouds consists of two roughly doughnut-shaped rings, one above and one below the plane of the nuclear ring, as in Fig. 5.13(b); each wave function ψ has zero amplitude in the plane occupied by the C nuclei, and peak values inside each doughnut.

In this way all the carbon–carbon bonds in benzene end up with identical strength, each composed of *three* pairs of electrons. The energy needed to break all six of these bonds in benzene is higher than that needed to break three single and three double bonds, and this fact led

---

Fig. 5.12

(a) A conventional structural diagram of the benzene molecule. As seen in (b), the molecule is indeed flat. The π molecular orbital in benzene has a doughnut shape, while the σ orbitals are each lozenge-shaped. (c) A structural diagram of a cyclohexane molecule, which is not planar, as the bond angles in (d) indicate, because all bonds are σ orbitals.

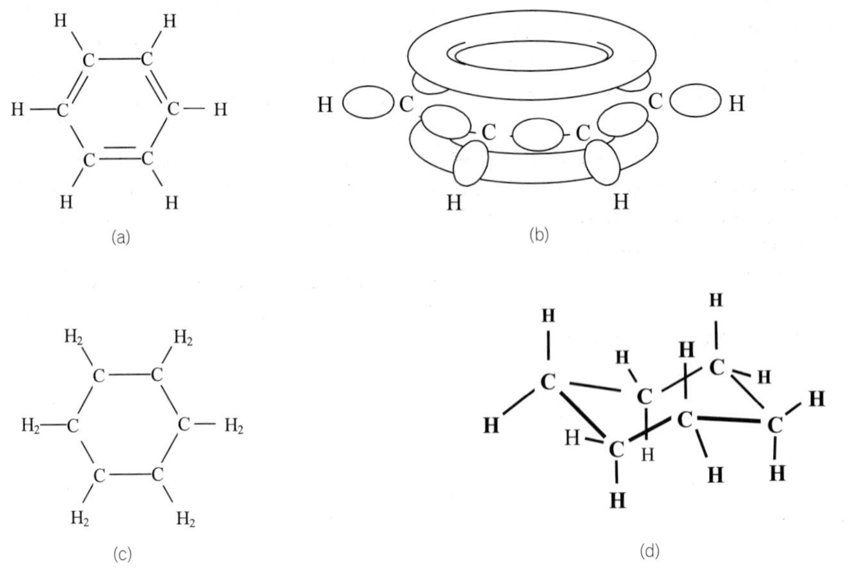

chemists to recognize a distinct type of bonding, long before quantum mechanics was understood. Such bonds are referred to as *conjugated,* and are often drawn by convention in structural chemical diagrams with alternate single and double bonds, as in Fig. 5.12(a). But measurements with X-rays of the separation between the atoms show that *all six* C—C bonds have an identical length of 0.139 nm, rather than alternating between 0.133 nm and 0.154 nm, as would be expected if single and double covalent bonds were involved.

The *delocalized* nature of electron orbitals within these π bonds gives such molecules as benzene distinctive properties. It allows freedom of electron motion around the whole molecule (or along it, in the case of long-chain molecules held together by conjugated bonding). The special electrical and optical properties conferred on molecules by delocalized conjugated bonding will be discussed in Chapters 16 and 20.

Conjugated bonds give benzene and its derivatives (derived molecules are formed by substituting the hydrogen atoms by other atoms or groups of atoms) different properties from cyclohexane (C_6H_{12}) and its derivatives. Cyclohexane's six carbon atoms are also joined in a ring-shaped molecule [Fig. 5.12(c) and (d)], but by *single* covalent bonds, each having two electrons. The ring is not flat, but puckered, because of the need to accommodate four such bonding orbitals symmetrically around each carbon atom, as in Fig. 5.9. In molecules such as cyclohexane, the bonding is said to be *saturated* – no more interatomic bonds can be formed, as there are no spare electrons such as benzene possesses in its *unsaturated* bonding system. Note, however, that many unsaturated molecules are not conjugated. A conjugated system may be recognized in the conventional structural diagram by the alternate single and double bonds.

5.9 Intermolecular bonding in molecular solids

The inert gases condense to form solids at sufficiently low temperatures, although we would not expect them to bond together in any of the ways described so far. The requirements for filled s and p subshells of electrons are fulfilled by the combination of atoms into molecules of the gases methane, carbon dioxide, hydrogen, etc. (CH_4, CO_2, H_2, etc.), leaving no spare electrons available for bonding by metallic, covalent or ionic means. Yet these gases, too, can solidify. Heavier molecules like benzene (C_2H_6) are liquids at room temperature, while many molecular solids, as they are called, remain solid to temperatures of several hundred degrees Celsius. It is clear that there must be other kinds of mechanisms for *intermolecular* bonding of uncharged molecules, in which no major modification occurs of the electronic structure of the atoms or molecules concerned, in contrast with the *interatomic* bonds already discussed. We shall discuss just two of the more important mechanisms.

5.9.1 van der Waals forces

A major group of intermolecular bonding forces is named after van der Waals, winner of the Nobel Prize for physics in 1910. Electrical forces of the van der Waals type are a result of *electrical dipoles* within each molecule or atom.

Fig. 5.13

Illustrating the effect of dipole
orientation on the strength of van
der Waals forces. All three pairs at
(a) mutually repel, unlike the pairs
at (b), which attract.

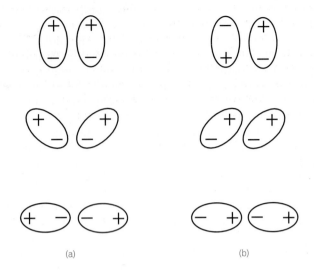

(a)   (b)

Consider the six pairs of molecules in Fig. 5.13, in which each molecule
is shown as an electric dipole. Between the members of each of the three
pairs in Fig. 5.13(a) there is mutual repulsion. Figure 5.13(b), on the other
hand, illustrates three orientations that result in a marked attractive force,
because the unlike charges lie closer than the like charges. You may test
these conclusions with two identical bar magnets, each of which behaves
as a dipole.

Although all the van der Waals mechanisms involve forces between
electric dipoles, they nevertheless fall into three types, as we will now
explain. Molecules like HCl and HF are naturally *polar* – as explained in
Section 5.7, they contain electric dipoles. A polar molecule can attract a
non-polar molecule, because the electric field set up by the permanent
dipole within the non-polar molecule can *induce* the electrons in the non-
polar molecule to move, creating an *induced* dipole. We say that the
second molecule is *polarizable*, and its *polarizability* (defined in Chapter
18) determines the strength of the induced dipole. Because the induced
displacement of electrons is towards the positively charged end of the
permanent dipole and away from its negatively charged end, the results
always resemble those in Fig. 5.13(b) and the force is always attractive.

But even non-polar molecules (i.e. those with no permanent dipoles)
and inert gas atoms may experience attractive forces. We know that an
atom with closed electron shells consists essentially of a positive nucleus
surrounded by a cloud of negative charge. If the electron clouds of two
such atoms were static and not deformable, no force would exist between
them. In fact, however, the electron clouds result from the motion of the
various electrons with their accompanying charge. Thus whilst an atom or
molecule may on average have no electrical dipole moment, at any instant
the centre of the negative charge distribution does not coincide with the
nucleus, but rapidly fluctuates about it. Thus an atom has a rapidly fluctuating dipole moment.

When, therefore, two such atoms approach one another, the rapidly
fluctuating dipole moment of each affects the motion of the electrons in
the other atom and a lower energy (i.e. an attraction) is produced if the
fluctuations occur in sympathy with one another. The fluctuations occur in

such a way that situations like those in Fig. 5.13(b) occur most frequently and so a net attractive force exists.

All of these forces can act over distances that are greater than molecular dimensions, for the electron clouds do not need to overlap one another. As a result they makes themselves felt even in the gaseous state, and cause small deviations from the perfect gas laws. The improved gas equation suggested by van der Waals incorporates these effects.

Van der Waals forces thus fall into the following three types:

1. Forces between polar molecules, i.e. permanent dipole–permanent dipole forces.
2. Forces between a permanent dipole and an induced dipole – sometimes called an *induction* force.
3. Forces between temporary, fluctuating dipoles within each molecule or atom, sometimes referred to for historical reasons as *dispersion* forces.

Because atoms can possess induced or temporary dipoles, mechanisms 2 and 3 can add strength to interatomic bonding.

Simple electrostatic calculations can be used to predict how these forces depend on the separation r between the centres of the dipoles in a pair of molecules. Thus type 1 varies as $1/r^3$, while types 2 and 3 vary as $1/r^6$. So all three types become very weak if r increases by, say, a factor of two or more. This fact means that, when comparing different materials, a change in the smallest dimension of a molecule has a large effect – usually greater than those caused by differences in the polarizability of different molecules. It is worth noting that, contrary to intuition, all three types of force can have comparable strengths. The strengths of these intermolecular forces in various materials may be gauged from the melting points of some molecular crystals, which will be discussed in Section 5.11 and are listed there in Table 5.3.

We will come across many solid substances that are bonded primarily by van der Waals forces, including plastics, graphite and paraffins.

5.9.2 Hydrogen bonding revisited

In a group of atoms with no electrons to spare but containing hydrogen, an important addition to intermolecular bonding strength may result from hydrogen bonding, introduced in Section 5.4.1. In a molecular solid, a proton on the periphery of one molecule bonds to a negative ion on a neighbouring molecule. The strength of hydrogen bonding derives from several contributing factors of roughly equal importance:

• The sharing of electrons between the three atoms concerned, as described in Section 5.4.1, i.e. a covalent mechanism, but reduced in strength owing to a greater separation.
• Electrostatic forces between permanent dipoles and ions, as explained below.
• Fluctuating dipoles, as described in Section 5.9.1.

The resemblance to van der Waals bonding, which, as we saw, is due to the attraction between dipoles, is only superficial. The small size of the hydrogen nucleus is crucial. Consider a polar molecule formed by an atom of hydrogen and some other atom or group of atoms that we shall call X.

We may describe the structure by the symbol X^-—H^+. This molecule is a small dipole. The positive charge on the hydrogen ion can attract the negative electron on a second negative ion Y^-, which may or may not be part of another molecule. So far we have a weak dipole–dipole attraction like the van der Waals force. But the H^+ nucleus, being very small, allows the third ionic group, Y^-, to approach very closely, forming a stronger bond than is possible with any other positive ion. We may denote this structure symbolically as:

$$X^-\text{——}H^+\text{——}Y^-$$

The resulting bond is noticeably stronger than van der Waals forces (see Table 5.1, Section 5.10).

Hydrogen bonds frequently play a major role in organic materials, including DNA. We will come across them again in Chapter 12, as they can make an important contribution to the strength of plastics (polymers). Nylon, for example, consists of long chain-like molecules containing both CO groups and NH groups that are linked to neighbouring chains by O—H—N bonds. As might be expected, hydrogen bonds also appear in ice and water – raising the surface tension, among other things – and they explain the occurrence of such hydrated inorganic salts as nickel sulphate ($NiSO_4.6H_2O$).

Both van der Waals bonding and hydrogen bonding differ in quite distinct ways from the stronger interatomic bonding mechanisms, and they give quite different properties to a solid. In Sections 5.10 and 5.11 we illustrate this by comparing the strengths of the main bonding mechanisms.

5.9.3 Bonding between charged molecules

We have already remarked that ionic solids may be formed from *molecular* ions. This is often referred to as *charge-transfer bonding*. The charge on large molecular ions is often shared around the whole molecule, so that the effective ion size is large, and the separation between the neighbouring centres is therefore much greater than typical atomic diameters. The electrostatic attractive force is therefore expected to be weakened when compared with the ionic bonding between atoms, but is much stronger than bonding between uncharged molecules. This is confirmed if such a solid vaporizes into molecules rather than atoms.

An example of a charge-transfer solid will appear in Section 6.13, where its crystal structure is described.

5.10 Bonding strengths measured by heat of vaporization

Because there is a smooth transition between covalent and ionic and metallic bonding, there is also a smooth variation in bonding strength. However, some characteristic distinctions emerge. The strength of bonding is best measured by the energy required to break up the solid, i.e. the amount of heat that must be supplied to vaporize it and hence separate the constituent atoms. If we measure the heat of vaporization of 1 mole, this is the energy used in separating N_0 individual atoms or molecules, where N_0 is Avogadro's number.

Table 5.1

Heats of vaporization

Bond type	Material	Heat of vaporization, H (kJ mol^{-1})
van der Waals	He	0.08
	N$_2$	7.79
	CH$_4$	10.05
Hydrogen:		
O—H—O	Phenol	18.4
F—H—F	HF	28.5
Metallic	Na	109
	Fe	394
	Zn	117
Ionic	NaCl	641[†]
	LiI	544[†]
	NaI	507[†]
	MgO	1013[†]
Covalent	C (diamond)	712
	SiC	1185
	SiO$_2$	1696
	Si	356

[†] Not strictly the heat of vaporization, because the solid does not break down into its constituent monatomic gases when vaporized. These figures are thus obtained by adding those for a sequence of several changes of state with the required end result.

Figures derived in this way are given in Table 5.1, alongside the dominant type of bonding for the solid concerned. As might be expected, the weakest mechanisms are van der Waals and hydrogen bonding. Next comes metallic bonding, followed by ionic and covalent bonding, whose strengths are nearly comparable.

Comparison of *intermolecular* bonding strengths can only be made directly in this way if the constituent molecules are stable in the vapour phase.

5.11 Bonding strength and melting point

Melting points are also indicative of bonding strengths. The atoms in a solid are in constant vibration at normal temperatures. In fact the heat stored in the material when its temperature is raised is stored in the vibrational energy of the atoms, which is proportional to the absolute temperature (these ideas are explained more fully in Chapter 7). When melting occurs, the thermal vibration has become great enough to shear the bonds, and the atoms, now mobile, slide past one another. Note that in most cases the volume change on melting is small, so that the atoms or molecules are not pulled much apart. In fact their average separation is little changed, and in some cases it decreases.

By comparing melting points it is possible to illustrate some interesting variations in bonding strength. For instance we have already remarked

Table 5.2

Melting points of Group IV elements

Material	Atomic number	Melting point (K)
Carbon (diamond)	6	4023
Silicon	14	1694
Germanium	32	1110
Tin	50	505

upon the extra strength conferred on ionic bonding by transfer of more electrons (i.e. by higher oxidation number), and how this results in the high melting point of magnesium oxide compared to sodium chloride. It is also clear from Table 5.2, showing melting points of elements in Group IV, that bond strength decreases with increasing atomic number. We conclude from these and other data that electrons with higher principal quantum numbers form weaker bonds.

Bonding strength is clearly important in determining the strength of materials, although the form of the crystal structure has at least as much, if not more, influence on strength. Thus we will leave a discussion of this to a later chapter.

In molecular solids two kinds of bonding are separately present: the solid forms of the gases are a case in point. Nitrogen (N_2), carbon dioxide (CO_2), oxygen (O_2), nitrous oxide (NO_2) and hexane (C_6H_{14}) all have covalent or mixed ionic–covalent bonds within the molecule. In solid and in liquid form the molecules are held together by van der Waals forces. This explains how it is that they solidify only at very low temperatures, because the van der Waals bond is rather easily broken by thermal agitation. On the other hand these gases break down into their constituent atoms (we say they *dissociate*) only at very high temperatures, because of the strong intramolecular bonds. Similar behaviour is found in polymers (see Chapter 12) and in many organic materials.

The strengths of the intermolecular forces in various molecular solids may be gauged from the melting points of some molecular crystals listed in Table 5.3. Comparing materials with similar molecular size and shape like benzene and cyclohexane, differences are affected mostly by molecular polarizability. Toluene has a polarizability close to that of benzene, but the bulky CH_3 side group prevents it from packing as tightly in the solid as the flat benzene molecule. Hence it has weaker bonds and melts at a lower temperature.

Table 5.3

Melting points of some molecular solids

Substance	Formula	Melting point (K)
Chlorine	Cl_2	172
Hexane	C_6H_{14}	178
Toluene (methylbenzene)	$C_6H_5(CH_3)$	178
Benzene	C_6H_6	279
Iodine	I_2	303
Naphthalene	$C_{10}H_8$	353

Problems

5.1 Using the data in Table 5.1, calculate the energy in eV required to break each individual bond in MgO.

5.2 Calculate the angles between the tetrahedral bonding directions in solid carbon.

5.3 Given that the distance between the H and F nuclei in the HF molecule is 0.0917 nm, calculate the maximum possible electrostatic force between two HF molecules, when their centres are separated by a distance of 0.27 nm (the molecular diameter). [*Hint*: the effective charges are $\pm 0.4e$].

Compare your result with the attractive electrostatic force between the partially ionized H and F atoms in the molecule. What points from this chapter do your results illustrate?

5.4 What kind of bonding do you expect in the following materials: SO_2, RbI, FeC, C_6H_6, InAs, AgCl, UH_3, GaSb, CaS, BN, and Cu-Fe, GdO, GdTe?

5.5 The attractive force between ions with unlike charges is $e^2/4\pi\varepsilon_0 r^2$ while the repulsive force may be written Ce^2/r^n, where the exponent n is about 10 and C is a constant.

Obtain an expression for the equilibrium distance r_0 between the ions in terms of C and n. Hence deduce an expression for the energy E required to separate the ions to an infinite distance apart, and show that it may be written in the form

$$E = \frac{e^2}{4\pi\varepsilon_0 r_0}\left(1 - \frac{1}{n-1}\right)$$

(*Hint*: Energy = integral of force with respect to distance.)

5.6 To remove an electron to infinity from a sodium atom, 5.13 eV of energy must be used, while adding an electron from infinity to a Cl atom to form the Cl ion liberates 3.8 eV of energy. What is the energy expenditure in transferring an electron from an isolated sodium atom to an isolated chlorine atom?

Using the expression given in Problem 5.5, calculate also the energy change when the two isolated atoms are brought together to form an NaCl molecule, given that the ionic separation is 0.27 nm at equilibrium, and that the exponent n in the expression for the energy is equal to 10. Compare your result with the figure given in Table 5.1 for the heat of sublimation. Comment on the discrepancy between the two figures.

5.7 The elements in Group 14 display a decrease in melting point with increasing atomic number. Does this occur in all groups in the periodic table? (Data in Appendix 3.)

Self-assessment questions

1 When the atoms in a solid are separated by their equilibrium distance

(a) the potential energy of the solid is lowest

(b) the force of attraction between the atoms is a maximum

(c) the force of repulsion is zero

2 The repulsive component in the force between two atoms that bond together is due primarily to

(a) electrostatic repulsion between their nuclei

(b) electrostatic repulsion between the ion cores

(c) repulsion between the overlapping electron clouds of inner filled shells

(d) repulsion between the overlapping electron clouds of the valence electrons

3 Ionic bonding is a result of the sharing of electrons of two atoms

(a) true   (b) false

4 The electrostatic nature of ionic bonding makes it

(a) non-directional   (b) weak

(c) applicable only to Group 1 and 2 elements

5   Atoms in the hydrogen molecule are held
    together by hydrogen bonding

    (a) true   (b) false

6   The figure illustrates which of the following
    bonding mechanisms?

    (a) covalent   (b) ionic

    (c) metallic   (d) van der Waals

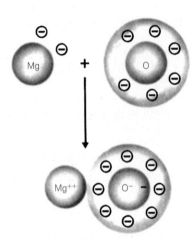

7   Two Group 2 atoms may be covalently
    bonded by

    (a) one electron from each atom

    (b) two electrons from each atom

    (c) a completed octet of electrons from the two
        atoms

    (d) a double bond

8   Covalent bonding occurs only between
    elements in Group 14 of the periodic table

    (a) true   (b) false

9   Metallic solids are held together by

    (a) the attraction between the ion cores

    (b) the attraction between ion cores and the
        electrons

    (c) electrons shared between adjacent pairs of
        atoms

10 Metallic bonding is directional

    (a) true   (b) false

11 van der Waals bonding is prominent

    (a) between molecular ions

    (b) between covalent molecules

    (c) when there are no valence electrons
        available to form primary bonds

12 van der Waals bonding is a result of

    (a) attraction between magnetic dipoles

    (b) attraction between saturated covalent
        bonds

    (c) attraction between electrostatic dipoles

13 When a hydrogen atom shares its electron with
    two chlorine atoms, a hydrogen bond is formed

    (a) true   (b) false

14 Hydrogen bonds play a part in bonding

    (a) water of hydration to a salt

    (b) solid methane (CH_4)

    (c) ice

15 In germanium the bonding is

    (a) covalent   (b) tetravalent

    (c) semi-metallic   (d) ionic

16 The bonding between boron (Group 13) and
    oxygen (Group 16) is

    (a) ionic   (b) metallic

    (c) mixed ionic–covalent   (d) covalent

    (e) van der Waals

17 Which of the answers given for Question 16
    describes the bond between silicon
    (Group 14) and carbon (Group 14)?

18 Which of the answers given for Question 16
    describes the bond between caesium
    (Group 1) and chlorine (Group 17)?

19 Solid CO_2 has a low sublimation point because

    (a) the carbon–oxygen bonds are covalent

    (b) the intermolecular bonding is due to van
        der Waals' forces

    (c) there are no spare electrons for bonding the
        molecules to one another

20 (i) Chlorine solidifies at a temperature in the range

    (a) 0–300 K   (b) 300–1000 K
    (c) above 1000 K

  (ii) because of the action of

    (a) ionic bonding

    (b) covalent bonding

    (c) metallic bonding

    (d) mixed ionic–covalent bonding

    (e) van der Waals bonding

21 The electronegativity of an element is a measure of

  (a) the excess of electrons over protons

  (b) the number of electrons in the valence shell

  (c) the strength with which electrons are attracted to the atom

Each of the sentences in Questions 22–27 consists of an assertion followed by a reason. Answer:

(a) If both assertion and reason are true statements and the reason is a correct explanation of the assertion.

(b) If both assertion and reason are true statements but the reason is *not* a correct explanation of the assertion.

(c) If the assertion is true but the reason contains a false statement.

(d) If the assertion is false but the reason contains a true statement.

(e) If both the assertion and reason are false statements.

22 Magnesium oxide is bonded ionically *because* magnesium and oxygen are both divalent.

23 Bonding in diamond is covalent *because* carbon lies in Group 14 of the periodic table.

24 van der Waals bonding is weak *because* it bonds only molecules together.

25 Liquid carbon tetrachloride cannot conduct electrically *because* it contains neither C^+ ions nor Cl^- ions.

26 A substance whose melting point is −154°C must consist of small covalent molecules *because* only van der Waals forces could give rise to such a low melting point.

27 CO_2 and SiO_2, whose melting points are respectively −56°C and 1700°C, both form small covalent molecules *because* carbon and silicon are both in Group 14 of the periodic table.

Answers

1	(a)	2	(c)	3	(b)	4	(a)
5	(b)	6	(b)	7	(b), (d)	8	(b)
9	(b)	10	(b)	11	(b), (c)	12	(c)
13	(b)	14	(a), (c)	15	(a)	16	(c)
17	(b)	18	(a)	19	(b), (c)	20	(i) (a) (ii) (e)
21	(c)	22	(b)	23	(a)	24	(c)
25	(a)	26	(d)	27	(d)		

6 | The internal structure of crystals

6.1 Introduction

It was mentioned in the last chapter that the properties of a solid depend as much upon the arrangement of atoms as on the strength of the bonds between them. We met there some examples of such arrangements, diamond and rock salt (NaCl), and also an irregular solid, glass. We know that regular internal structure often leads to regular crystalline external shapes, as in many natural precious stones, and that amorphous solids do not naturally form regular external shapes. Between these extremes lie other categories such as the metals, which show regularity in the way their atoms pack together, but which do not normally take on crystalline shapes.

Materials in which perfectly aligned rows of atoms extend throughout their volume are commonly seen in use only as precious stones, but artificially grown crystals containing such aligned rows of silicon atoms are used inside every desktop computer, television and electronic controller of equipment. Mobile phones use crystals of both silicon and gallium arsenide, while the internal crystalline nature of metals, ceramics and plastics is crucial to their strength and ductility (or lack of it). The role of structure in controlling material properties will become much clearer in later chapters.

In this chapter, we introduce the concept of atomic *order* within crystals. Beginning with examples of simple inorganic elements and compounds, we then describe the different degrees and classes of atomic ordering, and how they depend on the interatomic bonding. Along the way, we explain the standard ways of describing directions and planes within periodic lattices. A description of the 14 possible crystal lattice patterns in which atoms or molecules may be arranged is followed by some examples of crystals formed from molecular compounds, and by polymers. A brief introduction to the advanced topic of quasicrystals illustrates the restrictions imposed on crystal symmetry. Finally we outline some methods of measuring crystalline structure. Liquid crystals, which are not solids, but are of engineering importance, are left until Chapter 20.

6.2 Crystalline order, polycrystals and grain boundaries

The concept we use to distinguish the regularity or otherwise of the atomic packing is that of *order*. Thus in a perfect diamond crystal [Fig. 5.9(b)] there is perfect order, because each atom is in the same position relative to

Fig. 6.1

A polycrystalline solid is composed of many grains oriented at random and separated by grain boundaries.

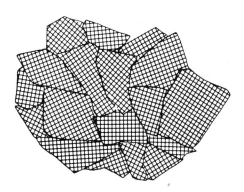

its neighbours, irrespective of where in the crystal it lies. We say that the crystal has *long-range order*. By contrast, an amorphous solid displays almost no order – the surroundings of each atom are different from those of all others.

In yet other cases the long-range order exists only over distances up to some maximum, perhaps 0.1 mm or even smaller. In all cases where this maximum is a very long distance compared to the atomic spacing of about 0.2 nm, the term 'long range' still applies. We illustrate this kind of structure in two dimensions in Fig. 6.1. Each region of long-range order is called a crystallite or grain and the solid is termed polycrystalline. Metals, some ceramics, and many ionic and covalent solids are often found in this form.

The boundary between crystallites is called a *grain boundary*, and it is at this interface that the regularity 'changes direction'. Grain boundaries can be shown up by etching a polished metal with a suitable acid – the metal dissolves more readily at the boundaries, forming grooves[†] – and a micrograph of an etched polycrystalline specimen of copper is shown in Fig. 6.2. The dimensions of the grains are commonly anything upwards of about 10 nm across.

To distinguish those solids in which the order extends right through we call them either *single crystals* or *monocrystalline* solids.

A different situation arises in materials like window glass, in which order extends over distances embracing a few atoms only. This we call *short-range* order – it is also found in liquids, and in many solids that are classified as glassy. The structure is *amorphous*, as defined in Section 5.9. Many solids that display short-range order are indeed supercooled liquids and are not strictly in equilibrium in this state. Some types of glass may actually crystallize if given sufficient time (several hundred years) or if heat treated (see Chapter 11), although modern, well-stabilized glasses will not do so.

Polymers can also display order, but in a unique way, so that we leave discussion of their structure to Section 6.14.

6.3 Single crystals and unit cells

Now that we have met the three principal kinds of solid – monocrystalline, polycrystalline, and amorphous – we shall study next the different forms of crystalline order.

[†] Sometimes the etch rate differs sufficiently with orientation of the crystallite axes that height variations occur between crystallites.

Fig. 6.2

A photomicrograph of an etched polycrystalline specimen of copper showing grain boundaries. Twins (see Chapter 8) are also present.

30 μm

Historically, *crystallography* began in about 1700 with the study of the external form or morphology of crystals, but it was only in the first years of the nineteenth century that the concept of a crystal as a regular array of atoms was established. The discovery in 1912 of the diffraction of X-rays by crystals led to rapid advances. X-rays are like light waves of very short wavelength, about a few tenths of a nanometre. Because this is comparable to the interatomic spacing in a crystal, the X-ray is readily diffracted by the crystal, and this property can be used to study the arrangement of atoms and their spacing. The wavelength of the X-rays can be independently measured by recording their diffraction using a calibrated ruled grating.

Figure 6.3 shows a diffraction photograph, taken using the set-up illustrated in Fig. 6.23, of a single crystal of an oxide of chromium, Cr_2O_3. The exposed dots on the photographic plate are at the intersections of the

Fig. 6.3

An X-ray diffraction pattern obtained photographically from a single crystal of Cr_2O_3. (The dark lines and the central disc are not produced by the X-rays.)

diffracted beams with the plate. The analysis of diffraction patterns to determine the crystal structure is complex but the techniques for doing this are well established. Later in this chapter we describe the experimental method and prove Bragg's law. In the meantime we consider only the results, and will show atomic arrangements that have been deduced from much patient analysis of X-ray photographs. We will emphasize the way in which the crystal structure follows from consideration of the bonding requirements for the material concerned.

First we must define more carefully what is meant by crystalline order. In a crystal it is possible to choose a small group of atoms that we can imagine to be contained in a regular sided box. The smallest such unit is called a *unit cell* [Fig. 6.4(a)]. If many such boxes are stacked together, all in the same orientation like bricks [Fig. 6.4(b)], the atoms are automatically placed in their correct positions in the crystal. The example shown in the figure is diamond, which may look unfamiliar as it has been drawn in a different orientation. This unit cell contains 8 atoms, but in other solids the unit cell may contain more or fewer atoms, and the cell may be other than

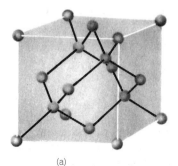

(a)

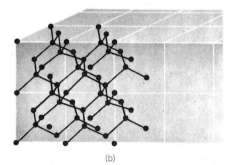

(b)

Fig. 6.4

(a) The unit cell of the diamond structure; (b) the crystal of diamond is built by stacking many unit cells together.

cubic in shape. The formal definition of the unit cell is the smallest group of atoms that, by repeated translation in three dimensions, builds up the whole crystal[†].

6.4 Interatomic distances and ionic radii

It was mentioned in Chapter 5 that the closed electron shell that is created when an ionic or metallic bond is formed can be regarded roughly as a rigid sphere, so that adjacent atoms in a solid pack together like solid balls. The radius of the equivalent rigid sphere is called the *ionic radius* of the element, and some values are given in Table 6.1. These have been deduced from the interatomic distances in ionic and metallic solids, measured using X-ray diffraction. It is also possible to calculate them from the diameter of the electronic wave functions obtained by solving Schrödinger's equation, as described in Chapter 3. The agreement with measured values is good, in spite of the approximations that must be made when performing the complex calculations.

An obvious starting point for the study of crystal structures is therefore the consideration of the kinds of structures that can be built by packing identical spheres together as closely as possible. Later on, structures composed of spheres of two different sizes will be discussed.

6.5 Close-packed structures of identical spheres

Figure 6.5(a) shows balls packed together on a plane surface. Each is surrounded by six neighbours that just touch it. If we place two such planes in contact, each ball in the second layer rests on three from the lower layer. In Fig. 6.5(b) the centres of the spheres in the lower layer are labelled A while those of the second layer are labelled B. When we come to add a third layer, the centres of the spheres can be placed either over the gaps in the second layer marked C or over those above the A's. In the latter case they are directly above the spheres in the first layer. We show a side view of this arrangement in Fig. 6.5(c): the sequence of positioning the layers may be represented as A-B-A-B-A ..., etc. However, in the other case the third layer does not sit over the first so it is positioned differently from either

Table 6.1

Ionic radii

Ion	Radius (nm)	Ion	Radius (nm)
Li^+	0.078	U^{4+}	0.105
Na^+	0.098	Ba^{2+}	0.135
K^+	0.133	Cl^-	0.181
Rb^+	0.149	Br^-	0.196
Cu^+	0.096	I^-	0.220
Be^{2+}	0.034	O^{2-}	0.132
Mg^{2+}	0.078	S^{2-}	0.174

[†] A few materials are composed of extended lattices of atoms that are built with units *not* in the same orientation or *not* precisely identical, called *quasicrystals*. They will be introduced later.

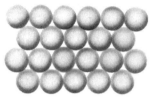

(a) Plan view of spheres resting on a plane surface

(b) A second layer of spheres (shown dashed) is placed on the first (plan view)

Fig. 6.5

The formation, (a) and (b), of the two close-packed structures of spheres from layers: close-packed hexagonal (c) and face-centred cubic (d).

(c) Side elevation of several layers stacked A-B-A-B-A

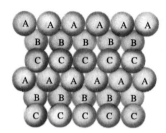

(d) Side elevation of several layers stacked A-B-C-A-B-C-A-

layer A or B. Because the fourth layer must begin to repeat the sequence we obtain the pattern A-B-C-A-B-C . . . , etc. as shown in Fig. 6.5(d).

These two closest-packed arrangements are shown again in Fig. 6.6(a) and (b), where the relative atomic positions are seen more clearly and the unit cell is indicated in each case. The two structures are known as *close-packed hexagonal* (c.p.h.) and *face-centred cubic* (f.c.c.) respectively. In the latter structure the cube shown is not the smallest possible 'building brick' but it is nevertheless usually referred to as the unit cell.

These close-packed structures are expected to be preferred by solids having predominantly either van der Waals or metallic bonding, because the atoms or ions have closed shells and therefore may be treated as hard spheres. Many metals do indeed crystallize in the face-centred cubic structure, as do all the noble gases apart from helium, which chooses the close-packed hexagonal structure. Among the metals the c.p.h. structure is exemplified by cobalt, zinc and several others. Some molecular solids having roughly spherical molecules also solidify as close-packed structures, for example methane, in the NaCl structure.

Because ionic solids do not normally consist of atoms of identical size they do not form these close-packed structures (see, however, Section 6.5). Similarly, covalent substances do not normally seek closest packing: the main consideration is that the direction of their bonds be maintained because these are rather rigid.

These are the simplest ways in which identical spheres can be closely packed together. Some arrangements with looser packing also appear quite often in nature. One of these is shown in Fig. 6.7, and is called body-centred cubic (b.c.c.). As can be seen by inspecting the atom in the centre of the cube, each atom has only eight nearest neighbours, compared to twelve in both of the close-packed structures. This is indicative of the loose packing, in spite of which many metals crystallize in the b.c.c. form. They include all

Fig. 6.6

Isometric views of the c.p.h. and f.c.c. structures, showing the unit cells.

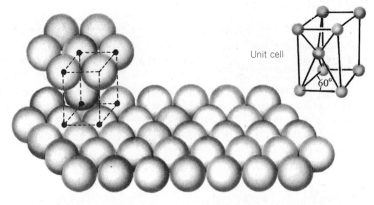

(a) Close-packed hexagonal structure

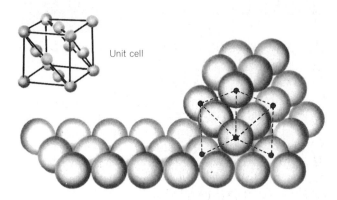

(b) Face-centred cubic structure

Fig. 6.7

The unit cells of the body-centred cubic and simple cubic structures.

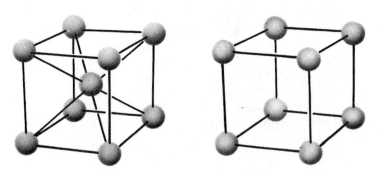

the alkali metals and, most important, iron. The number of nearest neighbours, eight in this case, is called the *coordination number* of each atom in this structure.

The simplest cubic arrangement possible, also shown in Fig. 6.7, is called *simple cubic*. This is a very loosely packed structure and is not of much importance because it is exhibited only by the metal polonium. Each atom has only six nearest neighbours.

6.6 Ionic crystals

It was mentioned earlier that the ratio of the ionic radii is important in determining the structure of an ionic solid. To understand this question it helps to study the problem in two dimensions, for example by arranging coins on a table. With identical coins, it is possible to make each touch six neighbours. Now take coins the ratio of whose radii is 0.73, e.g. in the UK a 5 pence piece and some 10 pence coins, or in the USA, 10 and 25 cent coins. It is possible to place five 10 p (25 c) coins in contact with one 5 p (10 c) coin, but one cannot make a periodically repetitive lattice in this way. However, an arrangement with four nearest neighbours, making a square lattice, can be repeated indefinitely in every direction.

The three-dimensional case is illustrated in Fig. 6.8(a), where we show how a positive ion (a *cation)* placed in the body-centred position in a simple cubic structure of *anions* (negative ions) does not touch the anions if it is too small compared to them. Because of this, the forces of attraction

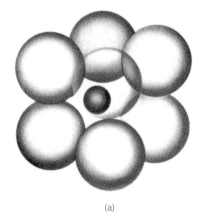

(a)

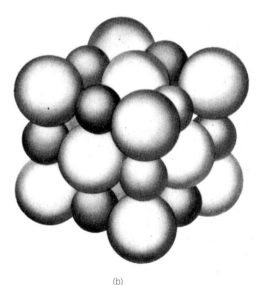

(b)

Fig. 6.8

(a) The CsCl unit cell, showing that if the central cation is too small it cannot touch the anions; (b) the rock salt unit cell.

and repulsion between the ions do not balance and the structure is not stable. It is easy to show (Problem 6.1) that the critical ratio of radii for which the b.c.c.-like structure is expected to be just stable is $r^+/r^- = 0.732$. If the ratio is smaller than this, it is possible that the arrangement shown in Fig. 6.8(b) will be stable. With the same radius ratio as in Fig. 6.8(a), the cation, when placed in the body-centred position in an f.c.c. arrangement of anions, just touches the six face-centred anions, giving a stable arrangement. Figure 6.9(a) and (b) shows the extended lattices corresponding to those in Fig. 6.8. We see that the lattice in Fig. 6.9(a) can be regarded as two interpenetrating simple cubic lattices, one for each kind of ion. This structure is named after the ionic compound caesium chloride (CsCl), which crystallizes in this way. Eight nearest neighbours surround each ion and eight is hence the coordination number of this structure.

We have met the structure in Fig. 6.9(b) before in NaCl, and it takes the name of the natural crystalline form of this material, *rock salt*. It can be represented as two interpenetrating f.c.c. lattices, one of anions and the other of cations, shifted with respect to one another by half a cube edge. Because of this, the structure is basically f.c.c., as it may be regarded as a

Fig. 6.9

(a) Part of the extended lattice of CsCl, showing how it is built of unit cells; (b) the rock salt structure – a single unit cell is shown here; (c) the unit cell of the zinc blende structure.

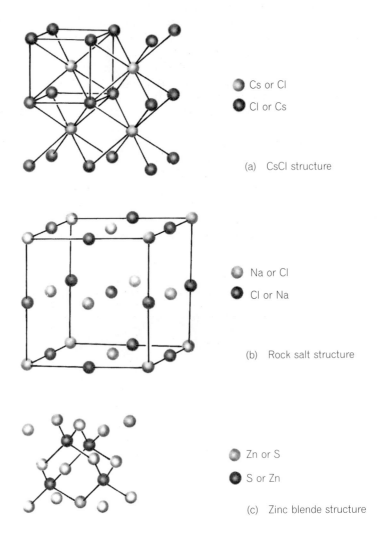

Cs or Cl

Cl or Cs

(a)   CsCl structure

Na or Cl

Cl or Na

(b)   Rock salt structure

Zn or S

S or Zn

(c)   Zinc blende structure

single f.c.c. lattice in which each lattice point is associated with an NaCl atom pair. Note, however, that it is not close-packed in the sense of the previous section. Because each ion touches six neighbours the coordination number is six and the packing is even looser than in the body-centred arrangement. The coordination number is actually the same as that of the simple cubic lattice, and indeed the two lattices become identical if we ignore the difference between the positive and negative ions.

It is clear that in the rock salt structure also there is a limit to how small the cation may be. This is easily found to be 0.414 times the anion size (see Problem 6.2).

For radius ratios below this value we might expect the coordination number to reduce yet further. Because arrangements with five neighbours cannot fit together to give an extended space lattice, four is the next lowest coordination number. Once again we have met this already when we considered carbon – the diamond structure (Fig. 6.4). It should be emphasized that carbon itself has this structure because of the directional nature of its four covalent bonds and not because of its radius ratio (its radius ratio is unity).

The ionic relative of the diamond structure is shown in Fig. 6.9(c). Like rock salt and caesium chloride it can be represented as two interpenetrating lattices, both f.c.c. in this case. It is named after zinc blende (ZnS), a mineral that itself is not wholly ionic but which is the commonest compound that crystallizes in this arrangement.

As an example of a predominantly ionic solid with the zinc blende structure we may take beryllium oxide (BeO), although the bonds here are also partly covalent. However, this structure is favoured for ionic bonding because the radii of the two ions are 0.031 nm and 0.170 nm, i.e. the radius ratio is 0.222.

It should be noted, however, that the radius ratio acts only as a rough guide to the structure of an ionic compound. There are many crystals that do not follow the rule, like lithium iodide (LiI), which, with a radius ratio of 0.28, adopts the rock salt structure in spite of the calculated radius ratio limit of 0.14. At the other end of the scale, potassium fluoride (KF), radius ratio 0.98, also crystallizes in the rock salt structure. This is probably because the difference between the stabilities of the caesium chloride and

Compound	Radius ratio	Predicted structure	Observed structure	Coordination number
CsCl	0.93	CsCl	CsCl	8
AgF	0.93	CsCl	NaCl	6
CsBr	0.87	CsCl	CsCl	8
RbCl	0.82	CsCl	CsCl*	8
RbCl	0.82	CsCl	NaCl	6
KBr	0.68	NaCl	NaCl	6
NaCl	0.52	NaCl	NaCl	6
BeO[†]	0.22	ZnS	ZnS	4
BeS[†]	0.17	ZnS	ZnS	4
CuBr[†]	0.49	NaCl	ZnS	4
LiI[†]	0.28	ZnS	NaCl	6

Table 6.2

Ionic structures

* Takes this structure when under high pressure.
† Bonding in this compound is partially covalent.

sodium chloride structures is rather small, and minor effects such as van der Waals forces can just tip the balance.

To summarize, we give below the rules that govern the crystal structures of ionic solids, and finally in Table 6.2 we show how well (or how badly) they are followed.

- The non-directional ionic bond favours closest possible packing consistent with geometrical considerations.
- If anion and cation cannot touch one another in a given arrangement, it is generally not stable.
- The radius ratio of the ions commonly determines the most likely crystal structure.

6.7 Covalent crystals

It may be thought that the previous section deals with solids of little practical interest. Unfortunately, because of the mixed ionic–covalent bonding in so many solids, the underlying principles of ionic bonding are only clearly exemplified in the alkali halides, which have few applications. Among the covalent solids, on the other hand, we have a number of practical importance in the electronics industry. Carbon, silicon and germanium all crystallize in the diamond structure, as we have already seen. The main requirement in these covalent structures is that the direction of the bonds be appropriate to those of the wave functions of the bonding electrons. The compounds formed between Group 13 and Group 15 atoms (e.g. GaAs and InSb) have the related zinc blende structure. This retains the four-fold coordination of diamond with the nearest neighbours at the corners of an imaginary tetrahedron – an arrangement called *tetrahedral* bonding. The preference for this structure has nothing to do with atomic radii but is determined by the directional nature of the four covalent bonds.

Several compounds between Group 12 and Group 16 elements also choose this structure, e.g. ZnS (obviously) and BeS, while others like CdS and CdSe are normally found in a modified form of the zinc blende structure called *wurtzite*. Naturally, there is more than an element of ionic bonding present in these compounds. In other sulphides, notably PbS and MgS, the ionic nature of the bond predominates and the rock salt structure results. This is also true of many oxides of Group 12 elements such as MgO and CaO, although BeO, as we saw, takes the zinc blende structure on account of the smallness of the Be ion. Among the covalent oxides is ZnO, which has the wurtzite structure, and CuO, PdO and PtO, which all form yet another kind of structure with four-fold coordination.

Because oxygen has a lower atomic number than sulphur it is more electronegative and hence tends to form ionic bonds more readily. Thus the covalent oxides mentioned above are in the minority.

6.8 Crystals with mixed bonding

We also find some covalently bonded solids with a marked metallic character. Although not really in this category, graphite (an allotrope of carbon) conducts electricity and is a soft material quite unlike diamond. The structure (Fig. 6.10) explains this: it consists of sheets of covalently bonded carbon atoms with three-fold coordination, loosely bonded by van der Waals

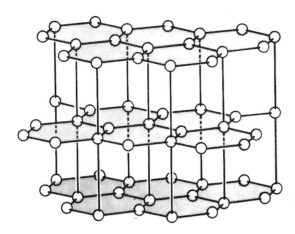

Fig. 6.10

The structure of graphite.

forces to neighbouring sheets. The large distance between sheets is a reflection of the weakness of van der Waals bonding, which allows the sheets to slide over one another rather readily. This gives graphite its lubricating properties, which is of such great importance because it is one of the few lubricants capable of withstanding high temperatures without vaporizing. In addition there is one spare electron per atom, because only three covalent bonds are formed. At temperatures above 0 K, these become freely mobile electrons, but they cannot easily cross from sheet to sheet because of the large distance involved, so the conducting property is directional, and the resistance is low only along the sheets of atoms.

Because thermal energy is needed to free the conduction electrons, graphite is classed as semi-metallic, and is not strictly metallic. However antimony, Sb, and bismuth, Bi, certainly are, in spite of their position in Group 15 of the periodic table. Their structure is related to that of graphite. The sheets of atoms are puckered, not flat, but the important difference is that the distance between sheets is much less. This is because the inter-sheet bonding is predominantly metallic and therefore stronger. As a result the conductivity is much higher and less directional than in graphite. Note that the coordination number is much lower than is expected of a metal, because of the covalent bonds. However, in liquid form this influence of the covalent bonds is lost and the atoms pack more closely together, so that Sb and Bi both contract on melting. It is this property that made them important in typecasting for printing, where the expansion as they solidify enables them to fill all the interstices in a mould.

6.9 Polymorphism

The impression given so far in this chapter is that, with the exception of carbon, each substance has but one structure that it possesses under all conditions. This is not the case, as can be seen from a variety of examples – sulphur, tin, caesium chloride, zinc sulphide, iron and many others all display allotropic forms that are generally stable over different ranges of temperature and pressure. Thus iron at a pressure of 1 atmosphere changes reversibly from a body-centred cubic form to a face-centred structure at a temperature of about 910°C, and again to the body-centred structure at 1400°C. We say that such solids are *polymorphic*, i.e. they may have several different forms.

Aluminium oxide, Al_2O_3, is another polymorphic example, and has considerable engineering importance as a component of many ceramics, including cements (discussed in Chapter 11), and it forms the hard gemstones sapphire and ruby when 'doped' with coloured ionic impurities. The bonding is mixed ionic–covalent, because of the high charge and small size of the ions. In two of its crystalline forms, called α-Al_2O_3 and γ-Al_2O_3, the oxygen ions form a close-packed structure, while the Al^{3+} ions fill a fraction of the interstices. In the less stable γ-form, the oxygens take an f.c.c. lattice, while in the highly stable α-form, they form an h.c.p. lattice, the Al ions filling two-thirds of only those interstices that have *octahedral* symmetry (having six-fold coordination), and not the four-fold coordinated tetrahedral sites that in the ZnS structure are occupied by Zn ions. Unlike the case of Fe, heating Al_2O_3 causes *irreversible* changes, for example it permanently converts the γ-form to the α-form, called corundum, an extremely hard crystal.

Changes that come about as a result of the action of heat cannot be discussed simply in terms of bond types and strengths as in this chapter, and we shall not be considering them until the nature of heat energy in solids has been explained in Chapter 7.

6.10 Miller indices of atomic planes

Sometimes we wish to refer to planes of atoms in a lattice, such as the cubic faces in the f.c.c. structure, or the sheets of atoms in graphite. A shorthand notation has been devised for this purpose, comprising a set of three numbers, called *Miller indices*, which identify a given group of parallel planes. Because there are many parallel planes, all containing a similar arrangement of atoms, it is clearly unnecessary to distinguish between them. The Miller indices for a plane may be derived by reference to Fig. 6.11.

We first set up three axes along three adjacent edges of the unit cell, which in the case illustrated is cubic. We choose to measure distances along x, y and z as multiples of (or fractions of) the length of the corresponding edge of the unit cell. Thus the three atoms labelled A, B and C are all at unit distance from the origin, and this would apply even if the cell were not cubic.

Suppose that we wish to describe a plane, such as the one shown shaded in the figure, which passes through the centres of particular atoms in the

Fig. 6.11

Construction to determine the Miller indices of a plane (shown shaded) in a cubic lattice.

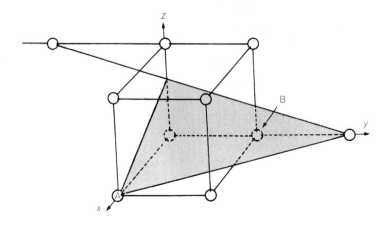

lattice. If it intersects the axes at distances x_1, y_1 and z_1 from the origin, then these three intercepts are characteristic of this one plane. In the case shown their magnitudes are respectively 1, 2 and 2/3. As these numbers are unique to the plane shown and are not the same for all parallel planes, we take their reciprocals, and express them as fractions with a common denominator; thus the reciprocals 1, 1/2 and 3/2 may be expressed as 2/2, 1/2 and 3/2. The numerators of these fractions are the *Miller indices* of the plane. The indices are normally enclosed in round brackets, so that the plane in question is denoted a (213) plane (pronounced 'two one three plane').

To show that a parallel plane to the one chosen has identical Miller indices, we will consider a plane that is parallel to the one shaded in Fig. 6.11, and which passes through the atom marked B. The intercepts this plane makes on the x, y and z axes are 1/2, 1 and 1/3 respectively. Their reciprocals are 2, 1 and 3 which, because none are fractions, are the Miller indices of this plane also. A little thought will show that *all* such parallel planes are (213) planes.

In a finite crystal it is useful to be able to distinguish opposite faces, so we allow any intercept and its corresponding Miller index to be negative. A negative sign is denoted by placing a bar above the index. A (21$\bar{3}$) plane may thus make intercepts +1/2, +1 and −1/3 on the axes and is called a 'two one three-bar plane'.

If a plane is parallel to an axis, the corresponding intercept is of infinite length and the Miller index is therefore zero.

As an example, the cube face parallel to both the x and z axes has intercepts ∞, 1, ∞ [see Fig. 6.12(a)] and hence is denoted by (010). The plane having equal positive intercepts on all axes is denoted by (111) and that containing two face diagonals is denoted (110). These are all illustrated in Fig. 6.12.

In cubic crystals, there are families of planes of the same character with different indices, for example (110), (101), (011), ($\bar{1}$10), ($\bar{1}$0$\bar{1}$), etc. We often

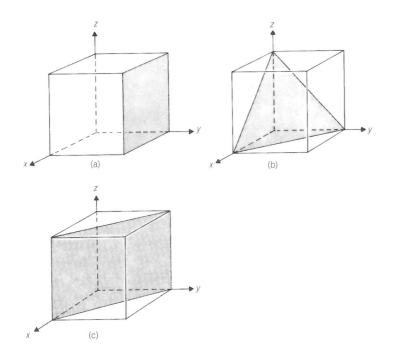

Fig. 6.12

Planes in a cubic lattice: (a) (010), (b) (111), (c) (110).

need to refer to such a set of planes, so the following notation has been adopted: {110} denotes a family of planes of the same basic type as the (110) planes, and would include all those listed above; similarly, there are families of {100} planes and {111} planes. These three are the types you will meet most frequently in this book.

6.11 Crystallographic directions and the zone law

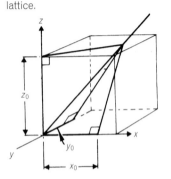

The direction of a line of atoms may be described using a similar notation. Here we take a parallel line that passes through the origin (Fig. 6.13) and note the length of the projections of this line on the x, y and z axes; say they are x_0, y_0 and z_0. Then we reduce these to the smallest integers having the same ratios. So if the intercepts were $0.75a$, $0.25a$ and a, the indices of the direction would be 3, 1 and 4 because the relation $3:1:4$ is the same as $0.75:0.25:1$. This direction is denoted the [314] direction: note the square brackets to distinguish it from the (314) plane.

Just like the principal planes of importance, the directions with which we shall often be primarily concerned are [110], [100] and [111]. These are, respectively, a cube face diagonal, a cube edge, and a body diagonal. Note that the fact that a plane and its normal have the same indices is a peculiarity of the cubic lattice alone, and does not apply to other lattice types. Families of directions are labelled by special brackets as are families of planes. Thus $\langle 100 \rangle$ denotes a family of directions that includes [100], [010], [001], [$\bar{1}$00], [0$\bar{1}$0], and [00$\bar{1}$].

A useful relationship exists between the Miller indices (hkl) of any plane that lies parallel to a crystallographic direction whose indices are $[u\,v\,w]$. This relationship, called the *zone law*, is expressed by the equation

$$uh + vk + wl = 0$$

This equation makes it easy to determine the unknown Miller indices of a plane in which two directions are already known, e.g. $[u\,v\,w]$ and $[u'\,v'\,w']$. Then the unknown Miller indices $(h\,k\,l)$ must satisfy the two equations

$$uh + vk + wl = 0$$

$$u'h + v'k + w'l = 0$$

The ratios h/l and k/l are readily found by dividing these simultaneous equations by l and solving by any convenient method.

6.12 Classification of crystal structures: Bravais lattices

In this chapter atomic arrangements have been discussed only in a qualitative way. An alternative approach is to study them in a formal, abstract way, just as if they are purely geometrical structures. If we do this, we find that only a limited number of arrangements is possible. We have already mentioned this fact when considering five-fold coordination around an atom. It is just not possible to build a regular, extended lattice with such an arrangement. Similarly, many other polyhedra (i.e. arrangements with different coordination numbers) cannot be stacked to form periodic space lattices. There is another example of the packing of coins that demonstrates this: the reader may care to try building a two-dimensional lattice with a seven-sided coin, e.g. the UK's 20-pence piece (you are advised not to spend too long trying!). In this analogy the seven-sided coin would represent the unit cell, if the lattice were not impossible.

As a result of such considerations Bravais discovered that there are only 14 different kinds of unit cell that can form an extended periodic three-dimensional lattice. They are known after him as *Bravais lattices*, and we show them in Fig. 6.14. Each lattice point (represented by a circle in the figure) may accommodate more than a single atom and in many crystals it does so – Section 6.13 gives some examples. Consequently there are many more than just 14 kinds of crystal structure. These variations may

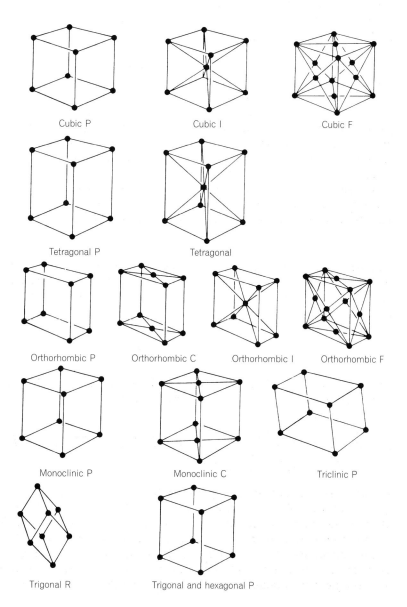

Fig. 6.14

The 14 Bravais lattices.

Cubic P   Cubic I   Cubic F

Tetragonal P   Tetragonal

Orthorhombic P   Orthorhombic C   Orthorhombic I   Orthorhombic F

Monoclinic P   Monoclinic C   Triclinic P

Trigonal R   Trigonal and hexagonal P

be classified according to the degree of symmetry exhibited by the group of atoms situated at each lattice point. Because there is a limited number of kinds of molecular symmetry the total number of true crystal symmetries is restricted to 230.

It is important to note that each of these 230 arrangements may be realized in many different ways using different atoms or molecules, so that there is virtually no restriction on the range of different materials that may be made.

6.13 Advanced topic: Crystal symmetry and quasicrystals

The Bravais lattices are limited in number because only certain symmetries are possible in a repetitive, periodic structure that fills all space. The easiest symmetry operations to consider are those of rotation about an axis. Thus an equilateral triangle and a similar triangular prism display three-fold rotational symmetry, while a square or a box with square cross-section displays four-fold symmetry. Only two-fold, three-fold, four-fold and six-fold rotational symmetries are found in cells that can be stacked repetitively without leaving space between them. Thus unit cells of the Bravais lattices contain groups of lattice points with rectangular, triangular, square and hexagonal cross-sections, but *not pentagonal or heptagonal*. Similarly, pentagonal floor tiles cannot be placed in a regular floor-covering pattern. Because of this, everyone believed until 1982 that, in the words of the eminent American mathematics professor Hermann Weyl, 'While pentagonal symmetry is found in the organic world[†], . . . in crystals . . . no other rotational symmetries are possible.' A description of all possible combinations of the symmetries that occur in crystals can be found in any good book on crystallography, but is beyond the scope of this book.

But in 1984, scientist Dany Schectman published a diffraction pattern (see Section 6.1) showing five-fold symmetry in an Al–Mn alloy. Since then, many similar alloys, mostly containing 60–70% aluminium, have been discovered with this symmetry. It has been convincingly demonstrated that, in these materials, cells with pentagonal or octagonal symmetry are repeated not at regular intervals, but in a non-periodic fashion. These materials also have unusual properties, some of which are useful, and are called *quasicrystals*. In 1991, by international agreement, the term 'crystal' was redefined to include all solids that display essentially a sharp diffraction pattern. Periodic and aperiodic (non-periodic) crystals are thus two distinct classes.

Table 6.3 lists a few of the alloys discovered with aperiodic lattices and symmetry. They fall into two main classes. In one class the unit cell is an *icosahedron* (a polyhedron having 20 identical equilateral triangular faces), and the lattice is aperiodic along all three coordinate axes. The other class has cell shapes with eight-fold or ten-fold symmetry and shows periodicity like a crystal only along the symmetry axis, while in the plane normal to that axis, the structure is aperiodic.

The rather obscure symmetry of quasicrystals would be simply of academic interest, were it not for the unusual properties it appears to give to these alloys. Because they are metallic alloys, it is remarkable that they

---

[†] He was referring to the many sea shells and plants with pentagonal patterns.

Three-dimensional non-periodic packing Icosahedral cell	One-dimensional periodic, two-dimensional non-periodic packing Decagonal or octahedral cell
Al + an early transition metal, e.g. Al–Cr, Al–V, Al–Mn, Al–Mo Mg–Al–Zn Al 65%–Cu 20%–Fe 15% Al–Pd–Mn	Al + a late transition metal, e.g. Al–Ni, Al–Pd, Al–Rh Al 65%–Cu 15%–Co 20% Al 70%–Ni 15%–Co 15% Cr–Ni–Si Mn–Si, Mn–Si–Al

Table 6.3

Some alloys[†] that form quasicrystals

[†] Where no composition is quoted, the elements are in decreasing order of atomic per cent.

have very low thermal and electrical conductivities. They are also poorly reactive, have high surface hardness, low friction, and display poor adhesion. As a result, Al–Pd–Mn alloys are already in commercial use as a replacement for coatings of polytetrafluorethylene (PTFE, or Teflon[TM]) on non-stick cookware, for they have superior hardness and temperature resistance. Other applications are developing rapidly.

These uncommon symmetries occur only when the composition of an alloy has specific values, e.g. 65% Al–20% Cu–15% Fe (atomic per cent). The earliest known quasicrystals were produced by rapid solidification of binary alloys of aluminium and a transition metal such as Mn, and were metastable, like a glass. Addition of a small amount of a third element is found to stabilize the five-fold lattice symmetry and to assist in producing structures with few imperfections. These stable quasicrystals can be grown slowly, like conventional crystals. Their stability is determined in part by their electronic properties – for example many of the icosahedral alloys possess exactly 1.75 valence electrons per atom.

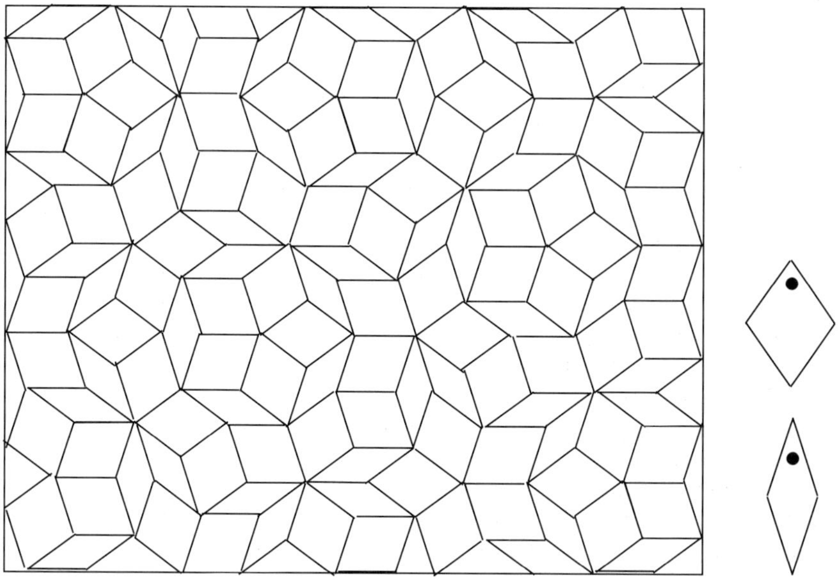

(a)   (b)

Fig. 6.15

Illustrating quasicrystal symmetry: (a) a Penrose tiling made with the two tiles shown at (b). Each tile is a rhombus, the angles marked ● being $\pi/5$ and $2\pi/5$.

6.14 Molecular crystals

The crystal structures we have discussed in detail so far have almost all had just one atom present at each lattice point. This is a rather restricted set of materials, because it excludes *molecular* solids. These are solids in which molecules experience relatively weak intermolecular attraction, rather than being held together by primary bonds. Their crystalline forms are known as molecular crystals.

Crystalline iodine is a simple example: below 30°C it has a face-centred structure like NaCl, but is rectangular rather than square in cross-section. Each corner site and face centre site is occupied by a covalent molecule I_2, as shown in Fig. 6.16 (for clarity the molecular shape is not drawn to scale). Van der Waals forces that are nearly non-directional pull the molecules together, making it close-packed. Bromine and chlorine form similar crystals, which melt below room temperature.

In the general case, each lattice point in any of the Bravais lattice shapes shown in Fig. 6.14 could be occupied by an identical molecule instead of an atom – provided, of course, that the intermolecular bonds are such as to stabilize the lattice. Each molecule may be in an identical orientation, or, as in crystalline iodine, they may have one of a set of different orientations. The requirement of long-range order means, however, that the structure must repeat itself at regular intervals. In crystals of simple molecules, the repetition lengths (the *lattice constants*) are at most a few times the smallest intermolecular separation. However when large molecules form crystals the lattice constant may be anything from a few intermolecular distances to thousands of them. In the latter case, it may be difficult to distinguish the structure from one that is not ordered.

The enormous variety of molecular solids, which includes most organic substances, means that only a tiny selection of simple crystal structures can be discussed here. We will concentrate on explaining a few of the principles governing them, and discuss three main types of molecular crystal, divided according to the types of attractive force discussed in Section 5.9.1. For convenience we summarize these again now.

Fig. 6.16

The structure of solid iodine, a molecular crystal. It is like a face-centred rectangular lattice, (a), in which a diatomic molecule, I_2, is situated at each of the lattice points. The unit cell nevertheless has a *rhombic* shape, (b).

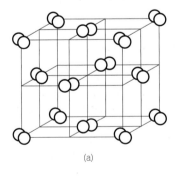

(a)

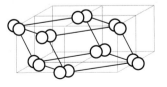

(b)

In the absence of charge transfer between molecules, electrical forces are dipolar in origin. These forces were classified in Section 5.9.1 as follows:

- Permanent dipole–permanent dipole attractions, varying as $1/r^3$.
- Induction forces, between a permanent dipole and an induced dipole, varying as $1/r^6$.
- Dispersion forces between fluctuating dipoles within each molecule, varying as $1/r^6$.

Charge transfer creates molecular ions, and results in somewhat stronger bonds with a longer range, falling off as $1/r^2$. Because the intermolecular distance r is often related to molecular size, the melting points of charge-transfer molecular solids are usually well below those of ionic solids.

The net force in all cases is *attractive*, pulling the molecules together until the electronic orbitals on neighbouring molecules just begin to overlap, at which point there arises a strong repulsive force, as a result of Pauli's exclusion principle, which in equilibrium just balances all the attractive forces described above.

We shall give just one example of each of three crystal types. The first type contains non-polar molecules in which only induction and dispersion forces operate. The second consists of polar molecules in which permanent dipoles play a role, and the third involves charge transfer between the molecules. Hydrogen-bonded molecular solids fall into the polar type, but behave slightly differently.

The common features that are found are as follows:

- The packing of molecules is in all cases as tight as the molecular shape allows, the only exceptions to this being hydrogen-bonded crystals, which tend to have more open space in them.
- Non-polar molecules that are large and also planar (i.e. flat) tend to pack face-to-face in layered stacks.
- *If* charge transfer occurs between molecules, it is made possible by an overlap of specific part-filled electron orbitals in the two molecules. This requires tight molecular packing, as well as molecular orientations that permit the two orbitals concerned to overlap one another.

Thus in all of these cases the molecules fill most of the space available in the unit cell, taking up between 70% and 80% of the volume in most crystals. In spite of this the density is not as high as in a covalent solid, because the outer dimensions of each molecule include all of the valence electrons, rather than just the diameter of the inner core.

Figure 6.17(a) and (b) illustrate two views of a non-polar crystal of hexamethyl benzene. The internal molecular structure, shown in (c), has the planar symmetry of the benzene ring. Molecules occupy the corners of a so-called *triclinic* lattice, which can be regarded as a stack of planes, each plane rather like a garden trellis [view (a)], and consisting of coplanar molecules at the corners of adjacent cells. Alternatively, and usefully, it can be envisaged as a close-packed set of parallel columns, three of which can be seen side-on in view (b). Each column is formed by stacking molecules one on another at a fixed angle to the axis of the column. This angle allows each pair of molecules to fit slightly closer together than they would if the molecules all lay at 90° to the axis.

Fig. 6.17

(a), (b) Two views of the crystal structure of solid hexamethyl benzene. Only the foremost plane of molecules is shown in each view. The non-polar molecules are shown reduced in size relative to their spacing, to clarify the lattice. (c) The molecular structure.

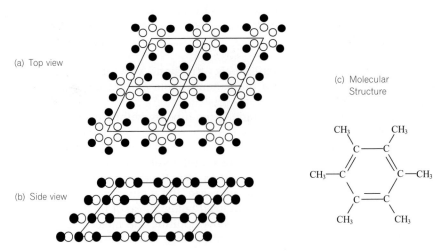

(a) Top view

(b) Side view

(c) Molecular Structure

In other crystals of this type, alternate columns may each be rotated by a different angle about their own axes, destroying the uniform planes of molecules referred to in the first interpretation above. We will meet one of these in Fig. 6.19. Such columnar structures are often encountered in the case of planar molecules.

Many of the properties of these materials are determined by the properties of the isolated molecules themselves, but others depend on the crystal structure, for example on whether the molecules all have the same orientation in space. It is easy to see that a crystal containing polar molecules, all of which have *identical* orientations, would make the crystal *as a whole* into a permanent electric dipole, while other arrangements, such as that adopted by 4-nitroaniline (Fig. 6.18), usually result in one molecule cancelling the dipole moment of another. We shall discuss such properties in Chapter 18.

As our last example, Fig. 6.19 shows schematically the structure of an organic charge-transfer salt. This salt of the organic molecule TCNQ (tetracyanoquinodimethane) is made by combining it with the electron donor TTF (tetrathiafulvalene). This material has recently become important in the development of organic semiconducting electronic devices, for it is a reasonably good conductor. Both donor and acceptor are flat, planar molecules. The salt crystallizes in a monoclinic lattice in which the donor molecules (TTF$^+$) are stacked in columns alongside corresponding stacks of acceptor molecules (TCNQ$^-$).

Fig. 6.18

(a) The molecule of 4-nitroaniline, showing the direction of its dipole moment. The arrowed hydrogen atom is referred to in the text. (b) The way that dipoles pair up in crystalline 4-nitroaniline.

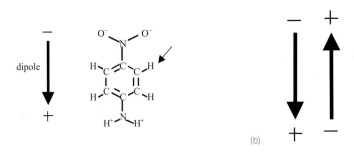

(a)

(b)

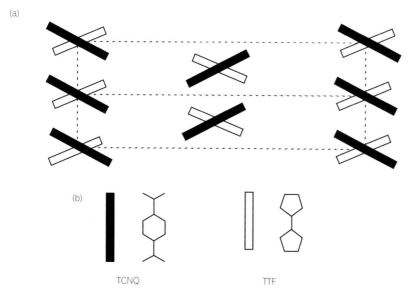

(a)

Fig. 6.19

(a) The lattice structure of the charge-transfer salt TTF/TCNQ, viewed along a direction parallel to one axis of the unit cell. In this view the columns of TCNQ⁻ (shown as dark rods) lie in a plane in front of the columns of TTF⁺. (b) Stick diagrams illustrate the shapes of the two hydrocarbon molecules.

(b)

TCNQ   TTF

6.15 Crystallinity in polymers

Polymers have a distinctive molecular structure whose principal feature is their sheer size. They consist of *macromolecules* with molecular weights of at least several thousand. The basic elements of these giant molecules are long, thread-like, covalently-bonded chains of repeated chemical units. In some polymers these chains are branched, and may be interconnected at points along their length by long branch chains or by short crosslinks, to form a *network*.

The simplest polymers consist of separate linear chains of identical *repeat units*, for example polyethylene $(C_2H_4)_n$, a chain of n C_2H_4 units, is shown in Fig. 6.20 in both a two-dimensional and a three-dimensional representation. The tetrahedral arrangement of the four covalent bonds of each carbon atom dictates a zigzag shape for the pair of bonds in each repeat unit. The chain

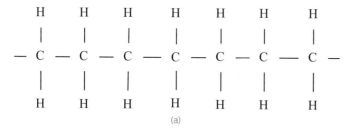

(a)

Fig. 6.20

Polyethylene: (a) a two-dimensional representation of the chemical formula, (b) its three-dimensional structure.

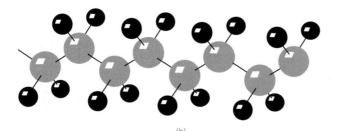

(b)

can be anything from a few hundred to several tens of thousands of C_2H_4 units long. If its *backbone* of carbon bonds is stretched out as straight as possible, as in Fig. 6.20(b), each zigzag pair of bonds is 254 pm in length and the entire chain might be 1 mm long. In other linear polymers the hydrogen atoms are replaced by other atoms or larger chemical groups: in polytetra-fluorethylene or PTFE, for example, all four hydrogen atoms have been substituted by fluorine.

Although the chain molecules of linear polyethylene are extremely long, they can be separated by heating to slide against each other in a semi-fluid melt: such a polymer is a *thermoplastic*. In the same way, it can be dis-solved in certain solvents. On cooling, a polyethylene melt might be expected to crystallize readily, because it displays long-range order along its length. The intermolecular forces to drive crystallization fall into the same types as were discussed in connection with molecular crystals. But the molecule is not rigid: rotation can occur about any of the covalent bonds between each pair of carbon atoms. If a single kink occurs by rotating about one bond in a zigzag chain, such as the right-most bond drawn in Fig. 6.20(b), we can calculate and plot the potential energy of the molecule as a function of the angle of rotation. It changes because there are forces of attraction or repulsion between the atoms attached on opposite ends of that bond. As we would expect, the *trans* position, where the kink rotation is 0° and the chain continues its previous zigzag course, has the lowest stored energy. However, there are two other stable positions in which the energy is a minimum with respect to *small* rotations, at which points the repulsive forces between atoms just balance one another. These are the gauche states, called *gauche*$^+$ and *gauche*$^-$, in which the next carbon atom rotates into the position that (before rotation) was occupied by one of the hydrogen atoms, for example, either of the two hydrogen atoms on the far right of Fig. 6.20(b). Successive *gauche* states would wind the chain into a tight coil in which it collided with itself and are therefore impossible, but the sequence *trans–gauche*$^+$*–gauche*$^-$*–trans* takes the chain round a hairpin bend, folded back on itself! The ability of a chain to do this is crucial to the crystallization of polyethylene.

As the temperature increases, the energy required to change between *trans* and *gauche* states becomes available more and more frequently in each molecule, so that random changes in conformation (i.e. shape) take place. In the melt state, then, the chain can be thought of as a long wriggling string, which readily becomes entangled with its neighbours, making crystallization difficult. In Chapter 12 we will see that polyethylene in fact freezes slowly and with difficulty from this state into a *semicrystal-line* polymer, in which the entanglements allow only partial crystallization, the crystalline regions being interspersed with amorphous, glassy regions. Surprisingly, polyethylene and all solid polymers like it are much stronger and tougher as a result.

Polyethylene shows off its ability to crystallize much more easily from a warm, dilute solution, in which separate polymer chains float and wriggle freely without interacting with one another. Under these conditions almost perfect crystals do form as the temperature is reduced. This is to be expected, because linear polyethylene is really no more or less than a high molecular weight alkane – one of the chemical series $C_{2i}H_{4i+2}$ (see Fig. 12.1). Starting from ethane, C_2H_6 (i.e. $i = 1$), the alkanes show steadily changing physical properties as i increases and the chain becomes longer. Ethane itself and butane ($i = 2$) are gases, while n-octane ($i = 4$) is a liquid; paraffins are liquids such as $C_{60}H_{122}$ ($i = 30$) and heavier molecules form greases and waxes. As the chain molecule gets longer and longer, the extra hydrogen atom at each end has less and less effect on properties.

All of these alkanes will crystallize at a sufficiently low temperature; the melting point and boiling point increase with the size of the molecule. As we would expect, the molecules crystallize by straightening into their zigzag conformations and lining up side by side. The hydrogen atom 'end caps' lie neatly in a plane at the top and bottom surfaces. Similarly, polyethylene forms plate-like crystals known as *lamellae*. These are extremely thin (about 10 nm), but 100 or more times larger in the other two directions. It is clear that to form a lamella of these dimensions many polyethylene chains must have participated in a very orderly construction process without entangling, and that each chain must have undergone many hairpin bend *chain folds* at the top and bottom surfaces.

In general, the degree of crystallinity achievable in a solid polymer depends on the size of its molecules and the complexity of their basic shape, as will become clear in Chapter 12, where polymer structures are discussed in more detail.

6.16 Measurements on crystals and Bragg's law

Earlier in this chapter it was mentioned that the distances between atomic planes can be measured by using the diffraction of X-rays. The basic requirements are (i) a narrow beam of X-rays of known wavelength, (ii) a means of recording the direction of the diffracted beam (often a photographic film or plate), and (iii) a turntable on which the crystal is mounted so that it may be suitably rotated with respect to the beam. These elements are shown in Fig. 6.21 but the details are beyond the scope of this book and the reader is referred to the list of further reading at the end of the book. In this way the dimensions and angles in the unit cell of every common material have been measured. The dimensions quoted in reference books are usually the lengths of the edges of the unit cell, and these are called the *lattice constants*. Naturally, in a cubic material, only one such constant is required to specify the size of the cell.

We conclude with a derivation of the law that relates the wavelength of an X-ray to the spacing of the atomic planes – *Bragg's law*. We begin by assuming that each plane of atoms partially reflects the wave as might a part-silvered mirror. Thus the diffracted wave shown in Fig. 6.22 is

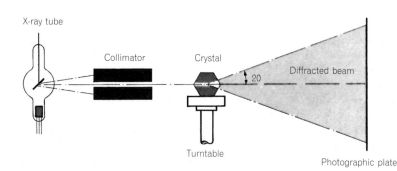

Fig. 6.21

Apparatus for the determination of crystal structures by X-ray diffraction.

Fig. 6.22

Diagram for the proof of Bragg's law.

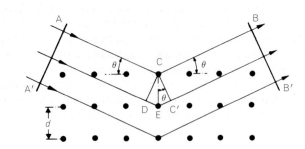

Fig. 6.23

Illustrating the reflection of X-rays from a plane of atoms.

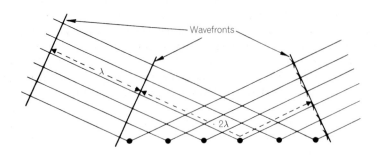

assumed to make the same angle, θ, with the atomic planes as does the incident wave.

Now the criterion for the existence of the diffracted wave is that the reflected rays should all be in phase across a wavefront such as BB'. For this to be so, the path lengths between AA' and BB' for the two rays shown must differ by exactly an integral number n of wavelengths λ.

Thus δ, the path difference, is given by $\delta = n\lambda$ (where $n = 1, 2, 3, \ldots$). Now because lines CC' and CD are also wavefronts we have

$$\delta = 2EC'$$
$$= 2CE \sin \theta$$
$$= 2d \sin \theta$$

Thus

$$2d \sin \theta = n\lambda \qquad (n = 1, 2, 3, \ldots)$$

This is known as Bragg's law.

The assumption that each atomic plane reflects like a mirror may be justified by a similar argument to that above, also assuming that each atom in the plane is the source of a secondary Huyghen's wavelet (Fig. 6.23). The scattered waves all add up in phase as shown to form the reflected ray.

Problems

6.1 In the CsCl structure shown in Fig. 6.9(a) let the ionic radii be r_1 and r_2 ($r_2 > r_1$). Assuming that the anions just touch the cations, calculate the length of the *body* diagonal, and hence derive the value of the radius ratio at which the structure just becomes unstable.

Using the ionic radii listed in Table 6.1, find the size of the largest impurity ion that can be accommodated interstitially (i.e. between the host ions) in the CsCl lattice with its centre at the point $\frac{1}{2}, \frac{1}{2}, 0$ in the unit cell.

6.2 If the radii of the ions in the NaCl structure are r_1 and r_2 $(r_2 > r_1)$, calculate the length of the face diagonal in the unit cell and hence show that the structure is stable only when $r_1 > 0.414r_2$.

6.3 What crystal structure do you expect to find in the following solids: BaO, RbBr, SiC, GaAs, Cu, NH_3, BN, SnTe, Ni? (Use the ionic radii in Table 6.1.)

6.4 Calculate the angles for first-order diffraction (i.e. the case $n = 1$) from the (100) and (110) planes of a simple cubic lattice of side 0.3 nm when the wavelength is 0.1 nm.

   If the lattice were b.c.c., would you expect to find diffracted beams at the same angles?

6.5 How many atoms are there in the unit cell of (a) the b.c.c. lattice, (b) the f.c.c. lattice?

6.6 Determine the Miller indices of a plane containing three atoms that are nearest neighbours in the diamond lattice. [Use Fig. 6.4(a).]

6.7 Calculate the density of graphite if the interplanar distance is 0.34 nm and the interatomic distance within the plane is 0.14 nm. (Note: density = mass of unit cell ÷ volume of unit cell.)

6.8 The density of solid copper is 8.9×10^3 kg m^{-3}. Calculate the number of atoms per cubic metre.

   In an X-ray diffraction experiment the unit cell of copper is found to be face-centred cubic, and the lattice constant (the length of an edge of the unit cell cube) is 0.361 nm. Deduce another figure for the number of atoms per cubic metre and compare the two results. What factors might give rise to a discrepancy between them?

6.9 Place the materials listed in Problem 6.3 in the expected order of increasing melting points. Compare with the melting points given in a standard reference book.

6.10 The lattice constants of the metals Na, K, Cu and Ag are respectively 0.424, 0.462, 0.361 and 0.408 nm. Both Na and K have b.c.c. structures, while Cu and Ag are f.c.c. Calculate in each case the distance between the centres of neighbouring atoms, and hence deduce the atomic radius of each metal.

6.11 What are the Miller indices of the close-packed planes of atoms in the f.c.c. lattice (see Fig. 6.6)?

Self-assessment questions

1 A solid in which all similar atoms are in similar positions relative to their neighbours is said to

   (a) have long-range order   (b) be crystalline

   (c) be amorphous

2 A polycrystalline material always contains

   (a) crystals of different chemical composition

   (b) crystallites of the same composition but different structures

   (c) crystallites with different orientations

3 A solid with long-range order is ordered over distances large compared with the size of

   (a) an electron   (b) an atom   (c) a crystallite

4 A material that is

   (a) amorphous   (b) glassy   (c) polycrystalline

consists of ordered regions containing only a few atoms.

5 With isolated exceptions, a unit cell is

   (a) the smallest group of atoms that when regularly repeated forms the crystal

   (b) a group of atoms that form a cubic arrangement

   (c) a unit cube containing the smallest number of atoms

   (d) the smallest group of atoms that will diffract X-rays

6 Which of the following are close-packed structures?

(a) c.p.h.   (b) tetragonal   (c) b.c.c.

(d) f.c.c.   (e) diamond

7 Close-packed structures are chosen by elements in which the bonding is

(a) directional   (b) non-directional

(c) metallic

8 The structure of an ionic crystal is determined primarily by

(a) the relative diameters of the constituent ions

(b) the nature of the chemical bond

(c) the charges on the ions

(d) the coordination number of each ion

9 The number of atoms per unit cell in the f.c.c. structure is

(a) 4   (b) 2   (c) 14

10 The number of atoms per unit cell in the b.c.c. structure is

(a) 6   (b) 9   (c) 2

11 From the diagram it can be seen that the coordination number of atoms in the CsCl structure is

(a) 4   (b) 6   (c) 8   (d) 10   (e) 12

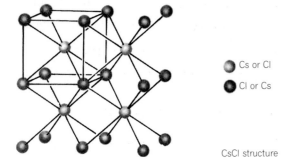

Cs or Cl
Cl or Cs

CsCl structure

12 The coordination number of atoms in the rock salt structure shown next is

(a) 4   (b) 6   (c) 8   (d) 10   (e) 12

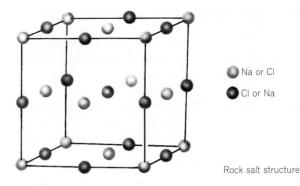

Na or Cl
Cl or Na

Rock salt structure

13 The coordination number of atoms in the zinc blende structure is

(a) 4   (b) 6   (c) 8   (d) 10   (e) 12

Zn or S
S or Zn

Zinc blende structure

14 Tetrahedral bonding is characteristic of

(a) ionic bonds

(b) covalent bonds

(c) metallic bonds

(d) van der Waals bonds

15 A covalently bonded solid always chooses the diamond structure

(a) true   (b) false

16 In graphite the bonding is primarily

(a) ionic   (b) covalent

(c) van der Waals   (d) metallic

17 Graphite is a good lubricant because

(a) sheets of atoms are bonded together covalently

(b) the atoms in the sheets are bonded covalently to one another

(c) the sheets are bonded to one another by van der Waals forces

18 A polymorphic material is one that

(a) is found naturally in many different shapes

(b) has more than one kind of crystal structure

(c) displays allotropic forms

19 In defining Miller indices we set up coordinate axes

(a) along the edges of the unit cell

(b) along the $\langle 100 \rangle$ directions

(c) perpendicular to the faces of the unit cell

20 The Miller indices of a plane are proportional to

(a) the intercepts of the plane on the coordinate axes

(b) the reciprocals of the intercepts

21 In a cubic structure the [111] direction and the (111) plane are parallel

(a) true   (b) false

22 The figure shows a cubic unit cell; the planes shown shaded in (a), (b) and (c) are

(a) (110)   (b) (111)   (c) (102)   (d) (010)

(e) (100)   (f) (001)

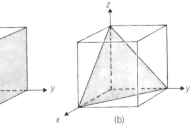

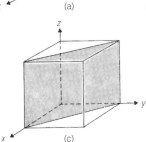

23 The indices of all directions that are parallel to one another are identical.

(a) true   (b) false

In the next two questions, R is the atomic radius and the material is an element.

24 The area of the (100) plane in an f.c.c. unit cell is

(a) $8R^2$ (b) $\sqrt{5R^2}$ (c) $(16/3)R^2$

25 The area of the (101) plane in a b.c.c. unit cell is

(a) $4\sqrt{5R^2}$

(b) $(16/3)R^2$

(c) $(16\sqrt{2}/3)R^2$

26 The [110] direction in a unit cell is parallel to

(a) the diagonal of one face of the cell

(b) the body diagonal of the cell

(c) one edge of the cube

27 X-ray diffraction can be used to deduce

(a) the regularity of atomic stacking in a crystal

(b) the magnitude of the interatomic spacing

(c) the orientation of particular atomic planes relative to the crystal faces

(d) the electronic structure of the atoms

28 In the equation $n\lambda = 2d \sin \theta$, known as Bragg's law, the angle θ is

(a) the angle between the incident and diffracted X-ray beams

(b) the angle between the incident beam and the normal to the diffracting planes

(c) the angle between the incident beam and the diffracting planes

29 Polymer molecules differ from other types of molecules because

(a) they are long

(b) they contain many similar atoms

(c) they contain many repetitions of the same chemical unit

(d) they do not crystallize readily

30 A polymer often contains regions that are

(a) quasicrystalline

(b) amorphous

(c) glassy

(d) crystalline

31 A material containing grain boundaries is

(a) metallic

(b) grainy

(c) a polymer

(d) polycrystalline

Each of the sentences in Questions 32–40 consists of an assertion followed by a reason. Answer:

(a) If both assertion and reason are true statements and the reason is a correct explanation of the assertion.

(b) If both assertion and reason are true statements but the reason is *not* a correct explanation of the assertion.

(c) If the assertion is true but the reason contains a false statement.

(d) If the assertion is false but the reason contains a true statement.

(e) If both the assertion and reason are false statements.

32 The coordination number in a diamond crystal equals 4, the oxidation number of carbon, *because* 4 is the number of bonds that can be formed by a carbon atom.

33 Graphite remains a good lubricant up to high temperatures *because* planes of atoms are bonded together with van der Waals bonds.

34 Graphite does not conduct electricity as well as does a metal *because* electrons cannot move freely from one sheet of atoms to the next.

35 Polymers do not crystallize easily *because* the molecules are bonded to one another by van der Waals forces and/or hydrogen bonding.

36 An ion in a solid can be assigned a fairly well-defined diameter *because* the interionic distance is independent of the kind of ion with which it bonds.

37 Nickel has a close-packed structure *because* its bonding is metallic.

38 Caesium chloride does not exhibit the rock salt structure *because* its radius ratio is unsuitable for closest packing.

39 Grain boundaries exist in polycrystalline solids *because* no long-range order is present.

40 Polyethylene is semicrystalline *because* it crystallizes readily from the molten state.

Answers

1 (a), (b)	**2** (c)	**3** (b)	**4** (b)
5 (a)	**6** (a), (d)	**7** (b), (c)	**8** (a)
9 (a)	**10** (c)	**11** (c)	**12** (b)
13 (a)	**14** (b)	**15** (b)	**16** (b), (c)
17 (c)	**18** (b), (c)	**19** (a)	**20** (b)
21 (b)	**22** (a) d (b) b (c) a	**23** (a)	**24** (a)
25 (c)	**26** (a)	**27** (a), (b), (c)	**28** (c)
29 (c)	**30** (b), (c), (d)	**31** (d)	**32** (c)
33 (b)	**34** (a)	**35** (b)	**36** (c)
37 (a)	**38** (c)	**39** (c)	**40** (c)

Thermal properties: kinetic theory, phonons and phase changes | 7

7.1 Introduction

The perfect crystal, which we met in Chapter 6, is unfortunately an ideal that is rarely attained. In subsequent chapters we will learn of the many ways in which real materials differ from the ideal, regular structure proposed for the perfect crystal. In this chapter we concentrate on the internal vibrational motion associated with the storage of heat energy in a solid. By considering this motion we can learn something of the relationship between thermal energy and temperature, and hence of specific heat and thermal expansion.

More importantly for the purposes of this book, we will pave the way for an understanding of solid-state phase changes, such as the transformation of b.c.c. iron into f.c.c. iron at 910°C, which was mentioned in Chapter 6. Such changes are basic to a study of the microstructure of alloys like steel and brass, which in turn leads to an understanding of many of their properties.

In Chapter 8 you will learn about irregularities in atomic packing in crystals, and how the action of heat energy inside a material ensures that there are always a few defects present. Heat energy, usually known as thermal energy, has a strong influence on many properties of real materials. You may have experienced how many plastics change in stiffness, becoming flexible if immersed in hot water, and quite brittle at temperatures around and below 0°C. Similar changes, less easily observed, occur in metals – blacksmiths relied on this to shape iron, and steel is still shaped into strip and sheet forms at temperatures that are high enough for them to be malleable, i.e. soft but still solid. Thermal energy also enables atoms to change their location by diffusing – surprisingly, perhaps, this is possible even in a solid, thanks to the presence of the microscopic crystal defects that we will discuss in Chapter 8. This normally happens quite slowly, unless the temperature is raised, so that such processes do not cause failures in heat engines, nuclear reactors and furnaces, to mention just a few of the more obvious cases.

Electrical properties, too, depend on temperature – very strongly in the case of semiconductors, so that the performance of most silicon 'chips' degrades markedly if the temperature rises much above 70°C, although some semiconducting materials are much less affected. Magnetic materials, too, whether they are metals or ceramics, lose their magnetic strength entirely above a critical temperature, as Chapter 18 explains.

Table 7.1 lists a small sample of applications of materials that require particular thermal properties, together with some examples of materials that possess them, and some of the applications in which they are used. The linear expansion coefficients and thermal conductivities of the materials mentioned will be listed in Tables 7.2 and 7.4. It is notable that materials such as glasses and ceramics, which we think of as insulators, have thermal conductivities as much as 10^5 times that of dry air, and only about 100 times less than those of typical metals. Thermal conductivity in insulating solids is one of the more difficult properties to explain, and is left until the end of this chapter.

Figure 7.1 illustrates a high-temperature application using a variety of materials that must be chosen for their compatibility with one another and for operation at the high temperature of a jet engine combustion chamber. The ignition plug, which serves the same function as the petrol engine's spark plug, uses several metals, a glass, a ceramic insulator and semi-conducting silicon carbide, SiC, all bonded together into a solid tubular shape that must withstand repeated cycles of electrical and thermal stress.

We begin in Section 7.2 of this chapter with a justification for treating atoms hereafter as if they are hard spheres. In order to help you understand and anticipate the effects of thermal energy in engineering situations, we then apply this concept in Sections 7.3 and 7.4 to the simplest case to understand: the properties of gases. Then we shall need to discuss the importance of the very wide spread in the thermal energies of individual

Table 7.1

Thermal applications of some materials

Application requirement	Property required	Examples of materials	Some applications of materials in column 3	More detail in Chapter or Section
Heat removal (efficient cooling)	High thermal conductivity	Diamond Sapphire (corundum) Silicon Copper, aluminium Al–Cu alloy	Electronic chip carrier Electronic chips Cooking pans Lightweight petrol engines	15, 16 10.8
Heat insulation	Low conductivity	SiO_2 (glassy) Air (dry) Glass wool Glass	 Building insulation Building insulation Building material	11
High-temperature strength		Silica Silicon carbide Alumina, Al_2O_3 Diamond Various ceramics Nickel alloys	Space flight, jet and rocket engines High-temperature manufacturing, e.g. Si chips, cutting edges of machine tools	11 16
Thermal shock resistance	Resistance to fracture on rapid cooling or heating	Silicon nitride, Si_3N_4 Alumina	Robust containers Spark plugs	11

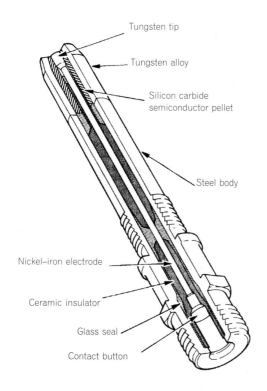

Tungsten tip

Tungsten alloy

Silicon carbide
semiconductor pellet

Steel body

Nickel–iron electrode

Ceramic insulator

Glass seal

Contact button

Fig. 7.1

Construction of an igniter plug for a jet engine's combustion chamber. The widely different materials used must bond together to form a solid cylinder capable of withstanding the thermal, mechanical and electrical stresses applied. The plug initiates an electrical discharge in the fuel–air mixture, which is in contact with the surface of the silicon carbide semiconductor at its upper end, when a voltage of about 2 kV is applied between the central electrode and the outer tungsten alloy body. The silicon carbide pellet then passes a small current that causes ionization of the gas at its surface. This pellet reduces the voltage needed, which would otherwise be 25 kV. (Reproduced with kind permission of Rolls Royce plc.)

atoms (Sections 7.5 and 7.6), which enables us, in Section 7.7, to define thermal equilibrium in a precise way.

In Section 7.8 we introduce heat as vibrational motion in solids, enabling us to describe how thermal expansion occurs. Sections 7.10 and 7.11 show that the specific heat of a solid depends on the frequencies of the atomic vibrations, and hence is linked to the subject of elasticity. Section 7.13 describes how thermal energy causes structural changes such as melting, and introduces the idea of metastable (i.e. temporarily stable) structural phases, which survive because their *rate of change* to a stable form can be slowed by rapid cooling. We then survey the influence of heat on chemical reactions (important in the corrosion of materials) in Sections 7.20 and 7.21, and we introduce the subject of thermodynamics.

The important topic of diffusion of atoms is introduced in Section 7.22, and the chapter concludes with an explanation of thermal conductivity by the diffusion of heat energy in both metals and non-metals.

7.2 'Hard sphere' model of a crystal

In the earlier chapters of this book we emphasized the quantum-mechanical aspects of the electrons in an atom. In particular, the idea of the particulate electron confined to a specific orbit was shown to be an inadequate approximation for some purposes, and the electron distribution around the atom was best thought of in terms of charge clouds. Nevertheless, when considering crystal structures, especially of metallic and ionic solids, we have seen that the arrangements of the atoms are consistent with a picture of them as 'hard' spheres, closely packed together. This is justified quantum-mechanically by (a) the spherical symmetry of closed electron shells and (b) the repulsive force between overlapping electron clouds, which rises very rapidly as the overlap increases. In this and subsequent chapters on thermal and mechanical properties we shall make the tentative assumption that no serious error is involved in treating atoms and ions in the solid state as slightly elastic spheres.

It must be borne in mind that this approximation may break down, in which case the observed behaviour of solids will differ slightly from what we might expect. Such deviations are, in fact, quite small and are rarely important.

7.3 The nature of thermal energy

How do we know that heat energy is associated with the motion of atoms, and what meaning can be attached to our mental picture of vibrating atoms?

The first direct piece of evidence for the nature of heat applies not to solids but to gases, and is the so-called 'Brownian motion' named after the botanist Brown. In 1828 he observed that, when still air containing specks of pollen dust is viewed in a microscope, the dust can be seen in continuous but irregular motion. He suggested that this is caused by the particles being bombarded from random directions by gas atoms which are themselves continuously moving about because they have thermal energy. Confirmation of this view comes from the success of the kinetic theory of gases, which is based on this idea of continuous motion of the atoms, and which explains in considerable detail the gas laws of Boyle and Charles and also many of the deviations from them.

7.4 Summary of the kinetic theory of monatomic gases

The starting point for the kinetic theory of gases is the idea that the pressure exerted by a gas on the walls of the containing vessel arises from continual bombardment by the gas atoms. The atoms of a monatomic gas are regarded for the purpose of calculation as small elastic spheres – a viewpoint discussed above. The rate of change of the momentum of the gas atoms as they bounce off the wall can be calculated, and is equal to the force that they exert on the wall. The latter is just the pressure, P, times the area. The momentum change depends on the speed, c, of the gas atoms (we use the term speed to represent the magnitude of the velocity, independently of its direction). The calculation, which will be familiar to many readers, can be found in any book describing the kinetic theory of gases (see the list of further reading on page 633). The result is

$$P = \frac{1}{3}\frac{N}{V}\,m\overline{c^2} \tag{7.1}$$

where $\overline{c^2}$ is the mean square speed of the gas atoms, N is the number of atoms in the volume V and m the mass of an atom.

Now the total kinetic energy, E, of the atoms is just

$$E = \tfrac{1}{2}Nm\overline{c^2} \tag{7.2}$$

so by combining Eqs (7.1) and (7.2) we find that

$$PV = \tfrac{2}{3}E \tag{7.3}$$

By comparison with the gas laws we see that the kinetic energy of the gas atoms is proportional to temperature. In fact, absolute temperature on the ideal gas scale of temperature may be *defined* by the relationship

$$\tfrac{1}{2}m\overline{c^2} = \tfrac{3}{2}kT \qquad (7.4)$$

where k is called Boltzmann's constant and has the value 1.380×10^{-23} $J\,K^{-1}$. It is a *universal* constant, being the same for all ideal gases.

Substituting from Eq. (7.4) into Eq. (7.3) we obtain

$$PV = NkT$$

which is the *equation of state* of an ideal gas. If the volume V contains N_0 atoms, where N_0 is Avogadro's number, the equation of state is usually written as $PV = RT$, where R is the gas constant for a gramme molecule.

We may obtain some idea of the magnitude of the molecular speed from Eq. (7.1) if we note that the expression Nm/V is just the total mass divided by the volume – that is, the density ρ of the gas. So the mean square velocity is given by

$$\overline{c^2} = 3P/\rho \qquad (7.5)$$

As an example we may take hydrogen, whose density at standard temperature and pressure is $0.09\,kg\,m^{-3}$. Because atmospheric pressure is approximately $10^5\,N\,m^{-2}$, this gives a root mean square velocity of $1.84\,km\,s^{-1}$ (about 4000 miles per hour!). A similar calculation for carbon dioxide gives a velocity of $0.393\,km\,s^{-1}$ at 0°C.

The average *energy* is also of interest. From Eq. (7.4) and the value of k we find that the average energy per molecule at 0°C is $5.65 \times 10^{-21}\,J$, or $3.53 \times 10^{-2}\,eV$. This energy is of course very small compared with the energy required to excite electrons to higher states within the atoms. The number of atoms containing electrons in excited states is therefore entirely negligible and we conclude that the energy $E = \tfrac{1}{2}Nm\overline{c^2}$ given in Eq. (7.2) represents the total heat energy of the gas atoms. The specific heat of a monatomic gas must, therefore, be a constant, independent of temperature, because the amount of heat that must be added to raise the temperature by one degree is the same whatever the temperature. The specific heat can be defined as the differential of heat energy with respect to temperature and so is constant when the internal energy, E, is proportional to temperature.

7.5 Energy distributions

The definition of temperature quoted above, Eq. (7.4), might lead us to expect the heat energy of all matter to be proportional to temperature. But this is manifestly not so, for the specific heats of solids, and polyatomic gases, are not by any means constant and independent of temperature. Even monatomic gases deviate from the ideal behaviour in this respect as they approach the point of condensation to the liquid or solid state.

We naturally ask, then, what is the same in two substances which are at the same temperature? The answer lies not in the equality of total energy, but in the way in which that energy is distributed among all the atoms in the substance, as we shall now explain.

The picture of a gas in the kinetic theory is one of molecules that are moving in all possible directions at very high speeds. Naturally the molecules will frequently collide with each other and may gain or lose velocity at each collision, although the sum of the energies of the two colliding atoms must remain constant. Thus their speeds and their kinetic energies may vary wildly about an average value but, provided that the gas is in a condition of thermal equilibrium, it is possible to calculate the distribution of the molecular population among the various possible velocities by determining the number having each particular value of velocity.

Details of the calculation can be found elsewhere; here we simply quote the results.

Because we wish to express the distribution of the particles among all possible velocities from zero to infinity we can do this by stating the number of particles with a particular velocity. However, if all velocities are possible we must decide on a minimum separation between the velocities at which we, as it were, count the number of particles. This can be expressed mathematically by letting $f(c)\,dc$ be the number of particles having velocities between c and $(c + dc)$, where dc is a small increment of velocity. The quantity $f(c)$ is then the number of particles per unit velocity range centred about the value c. It is argued in Section 7.7 that this is given by

$$f(c) = A\exp(-\tfrac{1}{2}\beta mc^2) \tag{7.6}$$

where A and β are constants, and m is the mass of each particle.

It will be remembered that a velocity has a magnitude and a direction and Eq. (7.6) gives the number of particles having a particular velocity in a given direction. In calculating quantities like the total kinetic energy of the particles, we need to know the total number of particles per unit volume having a velocity in the range c to $(c + dc)$ irrespective of direction. We define this as $N(c)\,dc$ and, by a process of integration, we can show it to be given by

$$N(c)\,dc = 4\pi c^2 N \left(\frac{m}{2\pi kT}\right)^{3/2} \exp\left(-\frac{mc^2}{2kT}\right) dc \tag{7.7}$$

where m is the mass of the particle, N is the number of particles per unit volume, T is absolute temperature, and k is Boltzmann's constant. This is the famous Maxwell–Boltzmann distribution law and describes, in mathematical terms, how an assembly of fast-moving particles distribute themselves over a range of velocities when they are constantly colliding with each other. We may express the result in the form of graphs of $N(c)$ against velocity, c, as in Fig. 7.2. This shows, for the case of hydrogen, how $N(c)$ varies with velocity at different temperatures and it will be seen that, as temperature rises, the velocities 'spread out' over a wider range. This is to be expected because, if on average they move faster, they will collide more often. By definition of $N(c)$ it follows that $\int_0^\infty N(c)\,dc$ is equal to N, the total number of particles per unit volume. But $\int_0^\infty N(c)\,dc$ is the area under the graph of $N(c)$ against c, so that the area under each of the curves of Fig. 7.2 is the same.

In a monatomic gas the expression $mc^2/2$ is just equal to the thermal energy, E, of an atom, so that Eq. (7.7) can be expressed in terms of E. In doing this we must remember that the range of energies, dE, corresponding to the range, dc, of speed is the differential of $mc^2/2$ with respect to c. Accordingly, Eq. (7.7) can be rewritten

$$N(E)\,dE = \frac{2\pi N E^{1/2}}{(\pi kT)^{3/2}} \exp(-E/kT)\,dE \tag{7.8}$$

where $N(E)\,dE$ is defined as the number of atoms having an energy between E and $(E + dE)$.

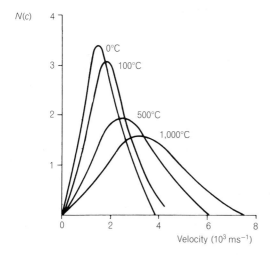

Although Eq. (7.8) applies only to a monatomic gas, a corresponding expression can be derived for polyatomic gases in which some of the molecular energy is a result of rotation of the molecule or vibration of its atoms with respect to one another. In each case the expression for $N(E)$ contains the factor $\exp(-E/kT)$, where E is now the total molecular energy.

In general we may write

$$N(E) = C \exp(-E/kT) \qquad (7.9)$$

where C is a factor that depends on the particular system to which the equation applies. In most cases, C is a function of energy E and temperature T which varies only slowly with E and T when compared to the exponential.

Equation (7.9) applies to any assembly of atoms in which the effects of energy quantization are negligible (in its derivation quantum effects are entirely ignored). The exponential factor is therefore a general one – it is known as the Boltzmann factor, and it arises from the random nature of all thermal energy distributions, a feature common to solids, gases and liquids. The Boltzmann factor can, in fact, be deduced from much more general statistical considerations than those of gas kinetics alone. On the other hand, factor C in Eq. (7.9) is dependent on the detailed distribution of available energy levels and, unlike the exponential, is not governed by the randomness of thermal processes.

It is possible to interpret Eq. (7.9) in a very simple way if we first remember that in all real situations the energy is quantized, even though the available quantum states may often be so closely spaced that for practical purposes energy may be treated as a continuous variable. In this case, the factor C may be written as the product of two quantities. One of these may be written dS/dE, where dS is the number of quantum states (i.e. states with distinguishable sets of quantum numbers) available in the energy interval dE. This factor, usually called the *density of states*, is therefore the number of states per unit energy interval at the energy E, and in general it depends on E itself. The other factor we shall call A.

We can thus rewrite Eq. (7.9) in the form

$$N(E)\,dE = \frac{dS}{dE} A \exp(-E/kT)\,dE \qquad (7.10)$$

where $N(E)\,dE$ is, of course, the number of *atoms* having energies due to thermal vibrations within the interval dE. The factor $A\exp(-E/kT)$ therefore represents the *fraction* of quantum states that are actually occupied. It consequently also represents the *probability*, $p(E)$, that an individual quantum state of energy E is occupied. Furthermore, because the number of occupied states at energy E is equal to the number of atoms having energy E, the probability that an individual *atom* has an energy E is proportional to the same factor.

If this probability is small compared to unity, then A does not depend upon the energy E and we can say that the relative numbers of atoms in individual quantum states of differing energies E_1, E_2, E_3, etc. are in the ratios $\exp(-E_1/kT):\exp(-E_2/kT):\exp(-E_3/kT)$, etc. This very simple result will prove useful in discussing many atomic processes in this book.

It is worth remarking that these ratios do not *necessarily* give the relative numbers of atoms with energy E_1, E_2, E_3, etc., because there may be different numbers of quantum states at these energies. In this case recourse to Eq. (7.10) gives the correct result.

7.6 Some other energy distributions

We have implied above that, when the fraction of atoms in a quantum state becomes comparable to unity, the 'constant' A in Eq. (7.10) depends on the energy E. We can see that this must be so because, if there is a high probability of a particle being in a particular state, then the probability of its existing in other states is correspondingly low. This cannot be the case unless A depends upon energy.

The correct expression for the probability $p(E)$ can be deduced with the aid of statistical theory, with the result

$$p(E) = \frac{1}{(1/A)\exp(E/kT) - 1} \tag{7.11}$$

where the quantity A is now *independent* of energy.

The reason for retaining the symbol A can be seen by noticing that, when it is very small, the 1 in the denominator may be neglected. Equation (7.11) then becomes approximately the same as Eq. (7.10); Eq. (7.11) is often referred to as the Bose–Einstein distribution.

At this point it is worth remarking that Eqs (7.10) and (7.11) need not apply to assemblies of atoms alone, but to any system of 'particles' that possess thermal energy and are in thermal equilibrium. The only exceptions are those particles, like electrons, to which Pauli's principle applies, for then only particles with different spin are allowed in each quantum state. This leads to the Fermi–Dirac probability $f(E)$, the derivation of which will be found in more advanced texts on solid state physics (see list of further reading on page 633). The probability that a quantum state, E, will be occupied by an *electron* is given by

$$f(E) = \frac{1}{1 + \exp[(E - E_F)/kT]} \tag{7.12}$$

where E_F is a constant called the Fermi energy and is discussed in Chapter 15. Unless $(E - E_F)$ is of the order of, say, two times kT or less, the exponential term will be much greater than 1 so the expression can be approximated to $f(E) \simeq \exp[-(E - E_F)/kT] = \exp(E_F/kT)\exp(-E/kT)$ and we again have the Boltzmann distribution of the type of Eq. (7.10).

In this book we are interested mainly in situations in which the occupation probability p of each quantum state is fairly small compared to unity.

In summary, we may say that:

(1) In most cases of interest in which atoms are involved, the Boltzmann distribution, Eq. (7.9) or (7.10), is applicable. A particularly useful form of it states that the numbers of atoms in quantum states of energies E_1, E_2, E_3, etc. are in the ratios $\exp(-E_1/kT)$: $\exp(-E_2/kT)$: $\exp(-E_3/kT)$, etc.
(2) In cases where the probability that a state is occupied is not small, the Bose–Einstein distribution Eq. (7.11) is applicable.
(3) For electrons, yet another distribution must be used, because they obey Pauli's principle. This is the Fermi–Dirac probability distribution, Eq. (7.12).

7.7 Thermal equilibrium

We are now in a better position to understand the meaning of temperature. Because the Boltzmann factor is applicable to assemblies of atoms at almost all temperatures (except those situations mentioned in (2) above), it can be seen that this is the common factor shared by two bodies which are in thermal equilibrium. Hence, the probability that a quantum state of a given energy is filled is *the same in both bodies* when they are at the same temperature.

The significance of this fact can be better understood by considering how heat is exchanged between two bodies A and B when they are placed in contact. The atoms at the common surface can 'collide' with one another because they are in thermal motion, and these collisions involve the transfer of energy.

Let us suppose that both A and B are at the same temperature and consider the atoms in body A in a quantum state of energy E_1 and those in body B in a quantum state E_2. The probability that an atom in A with energy E_1 can collide with a B atom having energy E_2 is proportional to the probability that both states will be simultaneously occupied; it is therefore the product $f_1 F_2$ of the individual probabilities f_1 and F_2 that E_1 and E_2 be separately occupied. For the moment we assume that f (which refers to body A) and F (body B) depend upon energy E in different ways, as illustrated in Fig. 7.3.

After the collision, the two atoms may have any energy provided that the sum of their energies is $E_1 + E_2$. One atom can therefore have an energy E_3 anywhere in the range 0 to $E_1 + E_2$ while the other atom must have the remainder, say E_4. This is illustrated schematically by the arrows on the graph in Fig. 7.3.

The probability that the A atom moves to a state with energy E_3 depends on the number of states it has to choose from – in fact it is equal to the reciprocal of that number. The total number of available states is the number lying between the energies 0 and $E_1 + E_2$, and we shall denote it by N_a. The collision probability is then proportional to

$$\frac{f_1 F_2}{N_a}$$

If the two bodies are in equilibrium, this expression must equal the collision probability for the reverse process: that is, when an A atom with

Fig. 7.3

Effect of an atomic collision on the Boltzmann distribution $f(E)$ and $F(E)$ for two bodies A and B.

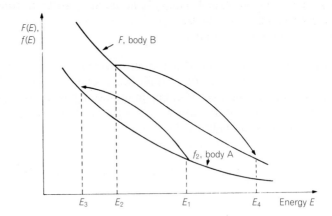

energy E_3 and a B atom with energy E_4 collide and end up with the respective energies E_1 and E_2. This probability is proportional to

$$\frac{f_3 F_4}{N_a'}$$

where N_a' is the number of states lying between the energies 0 and $E_3 + E_4$.

    The reason why the two probabilities must be equal is that, were they not, atoms would accumulate in, say, levels E_3 and E_4, while the population in levels E_1 and E_2 diminished. In this situation the probability distributions would change when a body were placed in contact with another at the same temperature, which is clearly unreasonable.

    It follows that

$$\frac{f_1 F_2}{N_a} = \frac{f_3 F_4}{N_a'}$$

But by the laws of conservation of energy, $E_1 + E_2 = E_3 + E_4$, so that the numbers N_a and N_a' must, according to their definitions, be equal. They therefore cancel in the above equation, leaving the result

$$\frac{f_1}{f_3} = \frac{F_4}{F_2}$$

This is a statement of the condition of equilibrium between the two bodies. It can only be satisfied for all energies if f and F depend on energy in the same way, and have the same value for the same energy. Thus if $f = \exp(-E/kT_a)$ and $F = \exp(-E/kT_b)$, where E is the energy and T_a and T_b are the temperatures of A and B respectively, then for equilibrium we require that $T_a = T_b$, i.e. that the bodies should be at the same temperature.

    It is interesting to note that, if A and B are just two halves of the same body, the condition of equilibrium is

$$f_1 f_2 = f_3 f_4$$

Taking logarithms, we have

$$\log f_1 + \log f_2 = \log f_3 + \log f_4$$

which may be compared with the equation

$$E_1 + E_2 = E_3 + E_4$$

The comparison suggests that $E_1 \propto \log f_1$, $E_2 \propto \log f_2$, etc. The comparison also suggests a solution for f of the type

$$\log f \propto E$$

or

$$f(E) = A \exp \beta E \tag{7.13}$$

This is the nearest we can get to a simple justification of the Boltzmann probability factor, in which of course $\beta = -1/kT$. It leads, in particular, to the result quoted in Eq. (7.6).

7.8 Kinetic theory of solids – interatomic forces

We now return to the problem of heat energy in solids. It has already been noted that the specific heat of a solid varies considerably with temperature, and in fact it tends to zero at the absolute zero of temperature. Figure 7.4 gives an example of the form of the temperature dependence for a pure solid that undergoes no change in crystal structure in the temperature range of interest.

In view of this marked difference from the behaviour displayed by monatomic gases whose specific heat is constant, we must investigate more closely the way in which heat energy is stored in solids.

With our picture of the solid as a lattice of atoms held together by bonds that are not perfectly rigid, we can imagine that the atoms can be displaced from their equilibrium positions as if they were mounted on springs and, because they have mass, they can oscillate about a mean position with an amplitude that depends upon the amount of heat energy possessed by the solid (see Fig. 7.6).

Let us make this picture more precise. In Chapter 5 we saw that a bond beween two atoms or ions provides a net attractive force that depends upon their separation and which is balanced at the equilibrium distance by a repulsion caused by overlap of the electronic charge clouds.

We can express this behaviour for a pair of atoms by assigning to them a potential energy, V, which is a function of the distance, r, between them. Thus we write

$$V = -\frac{C_1}{r^n} + \frac{C_2}{r^m} \tag{7.14}$$

where, following the usual convention for potential energy, the term arising from an *attractive* force is given a negative sign and that from a repulsive force a positive sign. In this expression C_1 and C_2 are constants and n and m are to be determined; in the case of electrostatic attraction between unlike charges (the case of the ionic bond) for example, n would be unity.

Again, by definition, the net force, F, between the atoms will be given by

$$F = -\frac{dV}{dr} = \frac{-nC_1}{r^{n+1}} + \frac{mC_2}{r^{m+1}} \tag{7.15}$$

and, if the atoms are in equilibrium, this force will be zero at some critical separation r_0, so that

$$0 = -\frac{nC_1}{r_0^{n+1}} + \frac{mC_2}{r_0^{m+1}}$$

Fig. 7.4

The dependence on temperature of the specific heat C_p of aluminium and copper, measured at a constant pressure of 1 atmosphere.

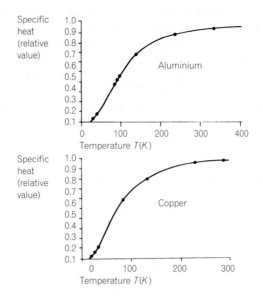

These equations are illustrated graphically for arbitrary values of n and m in Fig. 7.5; they are known as the Condon–Morse curves after their originators.

While Fig. 7.5 is constructed for a pair of atoms, similar behaviour will obviously occur between each pair of atoms in a crystal lattice. The actual values of n, m, C_1 and C_2 will depend upon the nature of the crystal bonding forces and on the crystal structure itself. They have, for example, been calculated for ionic crystals such as rock salt, in which n is unity, as would be expected for electrostatic attraction, and m is in the region of ten.

Now imagine it is possible to push one atom of the crystal aside from its equilibrium position, from point a to point b in the curve of Fig. 7.5, while keeping the rest of the lattice undisturbed. Figure 7.5(b) shows that there is an attractive force, ΔF, which is proportional to the displacement, Δr, if the latter is very small (it is exaggerated in the figure). An equal and opposite displacement, to point c, produces an (approximately) equal and opposite force that, remembering the finite atomic mass, we perceive that the conditions for simple harmonic motion are met.

Fig. 7.5

The Condon–Morse curves: (a) potential energy V and (b) force F as a function of interatomic spacing r.

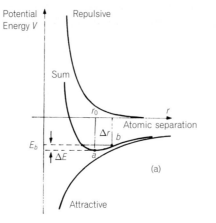

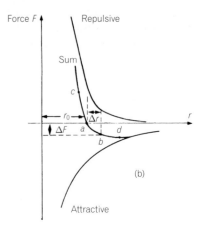

During the atom's motion its total energy remains constant, for the law of conservation of energy must be obeyed. This energy is made up partly of kinetic, partly of potential energy. At the turning points of its motion (points *a* and *b* in Fig. 7.5) the atom is momentarily stationary and its kinetic energy is necessarily zero. At these points, then, its potential energy – given by the curve – is equal to its total energy. The horizontal line *bc* in Fig. 7.5(a) therefore represents a plot of the total energy against position, and the vertical distance between this line and the potential energy curve gives the kinetic energy of the atom. The maximum value of this, shown as ΔE in Fig. 7.5(a), is the energy that would be given up if the atom were to come permanently to rest.

Now from the kinetic theory of gases, all the atoms in a gas would be at rest in the limit of zero temperature. This can be seen from Eq. (7.4). It is reasonable to suppose that the same applies to solids and therefore the energy ΔE in the above discussion represents the heat energy possessed by the atom in question.

Naturally, this hypothetical situation does not occur in nature; a single atom cannot vibrate in isolation, for in doing so it exerts forces on its neighbours and sets them moving also. We may imagine the crystal to be rather like the model depicted in Fig. 7.6, where the 'atoms' are joined by springs and the motion will obviously be complicated. We shall consider this, more realistic, model in Section 7.10.

Fig. 7.6

Mechanical model of an elastic crystal lattice.

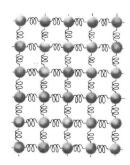

7.9 Thermal expansion and the kinetic theory

Before pursuing the subject, it is interesting to note that the asymmetry of the Condon–Morse curves of Fig. 7.5(a) means that expansion will occur when a solid is heated. Figure 7.7 shows the same curve with two horizontal lines corresponding to different total energies superimposed on it. It is clear that the average separation between the atoms (point *a*) increases with the total energy and hence the temperature, because of the asymmetry of the potential energy curve.

This is the first direct evidence of the correctness of the model we have chosen. In addition to the fact of expansion itself, it is possible to demonstrate experimentally that the length of a solid bar fluctuates randomly by a very small amount simply because of the randomness of the thermal vibrations of its atoms. The relative motion of two ends of a bar can be magnified until it can be visually observed, in a manner analogous to the observation of Brownian motion.

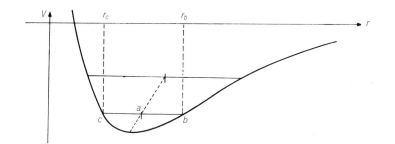

Fig. 7.7

Relationship of thermal expansion to the Condon–Morse curve for a solid.

Table 7.2

Linear thermal expansion coefficients for a selection of solids (data from various sources)

Table 7.2

Linear thermal expansion coefficients for a selection of solids (data from various sources)

Material	Expansion coefficient (K^{-1})	Melting temperature (K)
NaCl	40×10^{-6}	1074
MgO	14×10^{-6}	3070
Si	3×10^{-6}	1683
Silica, SiO_2	0.55×10^{-6}	1983
Alumina, Al_2O_3	8.8×10^{-6}	3253
Diamond	1×10^{-6}	~4600
Aluminium, Al	25×10^{-6}	933
Lead, Pb	29×10^{-6}	600
Tungsten, W	4.5×10^{-6}	3683
Polyethylene	~100×10^{-6}	
Polystyrene	~70×10^{-6}	

The expansion coefficient must increase with the width of the trough in the Condon–Morse curve in Fig. 7.7. This width, which varies with the material and its structure (see, for example, Fig. 7.4), is also related to the melting temperature, discussed in Section 7.15, and to the ease with which the bonds stretch, i.e. to Young's modulus of elasticity. Linear thermal expansion coefficients for various materials are compared in Table 7.2 with their melting temperatures, to demonstrate the connection. Polymers, included for comparison in the table, will be discussed separately in Chapter 12.

7.10 Lattice waves and phonons

The model of the motion of atoms about their idealized equilibrium positions in a solid is that each atom vibrates in a potential well of the force fields of its neighbours. Applying the kinetic theory of gases, the vibrations of all atoms are treated as independent (referred to as the Einstein model) and this is a good enough approximation in some cases, especially at high temperatures. However, if one or more atoms move in unison, the forces between them tending to restore them to equilibrium positions are reduced, and this must be taken into account.

A direct consequence of this cooperative behaviour of the atoms is that only certain elastic vibrations can exist in a crystal. In order to specify these elastic vibrations of the atoms it is convenient to describe them as waves that may be designated by wavelength, direction of propagation, frequency of vibration, and direction of polarization. There are two distinct types of waves – acoustic and optical. Acoustic waves occur in all crystalline materials and are produced by neighbouring atoms (or ions) moving in the same direction, but by slightly differing amounts, so that the spacing between the atoms remains relatively constant. This is shown graphically in Fig. 7.8(a), in which the displacements of the atoms in a linear chain from their ideal rest positions are plotted. However, the neighbouring ions in an ionically bonded crystal may move in opposite directions; this leads to a change in the spacing between the ions, so producing electric dipole moments as shown in Fig. 7.8(b). The resulting waves are called optical waves and are of higher frequency than the acoustic waves. These are the

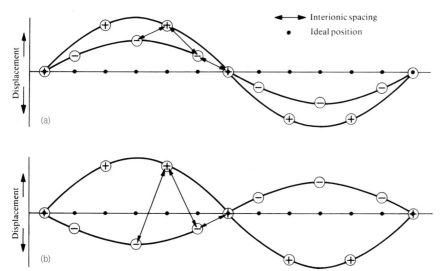

Fig. 7.8

The displacement of ions, on an exaggerated scale, in an ionic crystal giving (a) acoustic and (b) optical waves.

modes responsible for optical scattering in a crystal and are further discussed in Chapter 19 in connection with the optical properties of solids. There are in general three modes, or polarizations, of the atomic vibrations. In the simplest case these correspond to one longitudinal mode (atoms moving along the line of atoms), and two transverse modes (atoms moving perpendicular to the line of atoms) (see Fig. 7.9).

If classical mechanics were valid, the energy of the vibrational waves could have any value. However, quantum mechanics impose restrictions that limit the energy of the waves to discrete values, i.e. the oscillations are quantized. The discrete energy values that are allowed are the same as those of any simple harmonic oscillator vibrating at a frequency ν and are given by

$$E_n = (n + \tfrac{1}{2})h\nu \qquad (7.16)$$

where h is Planck's constant, and the quantum number n is any positive integer including zero (n is zero for all modes at absolute zero of temperature). (Note that frequency is denoted by ν rather than f in this chapter.)

The energy change accompanying a transition from one energy state with the quantum number n_1, to the next highest energy state with the quantum number n_2, is

$$\Delta E = E_{n_2} - E_{n_1} = (n_2 + \tfrac{1}{2})h\nu - (n_1 + \tfrac{1}{2})h\nu = (n_2 - n_1)h\nu \qquad (7.17)$$

Because n_1 and n_2 differ by unity $\Delta E = h\nu$, i.e. the transition is accompanied by the absorption of one quantum of thermal energy $h\nu$. The reverse transition from a higher energy state to one of lower energy is accompanied by the emission of $h\nu$ of thermal energy. The spectrum of energy levels is illustrated in Fig. 7.10.

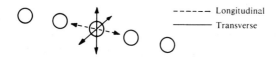

Fig. 7.9

The three modes of atomic vibrations in a chain of atoms.

Fig. 7.10

Energy levels of atoms vibrating at a single frequency ν.

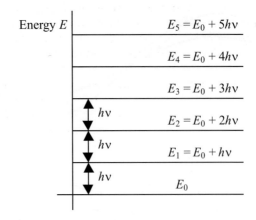

The expression for the average energy of a quantum oscillator at frequency ν is different from the average energy, kT, of a vibrating atom in classical theory. Thus the average energy of a quantum oscillator of frequency ν is shown in Box 7.1 to be

$$\bar{E} = \frac{h\nu}{\exp(h\nu/kT) - 1} \qquad (7.18)$$

Translating this in terms of lattice waves we can say that the average energy in a lattice wave of frequency ν is given by Eq. (7.18).

If the wave is regarded as a stream of 'particles', each of energy $h\nu$, then the average energy would be $h\nu$ times the number of particles. Thus, from Eq. (7.18), the number of particles in a mode of frequency ν is given by

$$n_q = \frac{1}{\exp(h\nu/kT) - 1} \qquad (7.19)$$

Box 7.1 Average energy of a quantum oscillation of frequency ν

In Section 7.6, we saw that the numbers of atoms in quantum states E_1 and E_2 are in the ratios $\exp(-E_1/kT)$ to $\exp(-E_2/kT)$, i.e.

$$\frac{N_1}{N_2} = \frac{\exp(-E_1/kT)}{\exp(-E_2/kT)} = \exp\left(-\frac{(E_1 - E_2)}{kT}\right)$$

But, from Eq. 7.17, $E_1 - E_2 = h\nu$ for adjacent energy levels and so the ratio of populations in the n and $(n + 1)$th levels is $\exp(-h\nu/kT)$.

The *average* energy of the oscillators will be given by

$$\bar{E} = \frac{\text{Sum of energies of all oscillators}}{\text{Number of oscillators}}$$

$$= \frac{\displaystyle\sum_{n=0}^{\infty} nh\nu \exp(-nh\nu/kT)}{\displaystyle\sum_{n=0}^{\infty} \exp(-nh\nu/kT)}$$

$$= \frac{h\nu(e^{-h\nu/kT} + 2e^{-2h\nu/kT} + 3e^{-3h\nu/kT} + \cdots)}{(1 + e^{-h\nu/kT} + e^{-2h\nu/kT} + e^{-3h\nu/kT} + \cdots)}$$

Writing $x = -h\nu/kT$, the above can be written as

$$\bar{E} = h\nu \frac{d}{dx} \ln(1 + e^x + e^{2x} + \cdots)$$

$$= h\nu \frac{d}{dx} \ln\left(\frac{1}{1 - e^x}\right)$$

$$= \frac{h\nu}{e^{-x} - 1}$$

The name phonon has been given to these 'particles', each of which is a quantum of thermal energy, by analogy with the transitions involving the absorption or emission of photons (quanta of electromagnetic energy). The phonon is the most universal of the imperfections found in crystalline materials.

At any temperature a crystal will contain a wide spectrum of phonons as the atoms will be vibrating at many frequencies, which may range from low-frequency fundamental acoustic to frequencies of 10^{13} Hz. As the atomic vibrations are not independent of each other the frequencies are quantized according to

$$\nu = \frac{u}{2}\sqrt{\left(\frac{n_x}{X}\right)^2 + \left(\frac{n_y}{Y}\right)^2 + \left(\frac{n_z}{Z}\right)^2} \tag{7.20}$$

where u is the velocity of propagation, and n_x, n_y and n_z are integers in the direction of the three orthogonal axes, x, y and z. The quantities X, Y and Z are the dimensions of the crystal in the directions x, y and z. Thus, only those phonons whose energies, $h\nu$, or frequencies, ν, satisfy this equation can be absorbed by the crystal. The total number of allowed frequencies is $3N$ where N is the number of atoms in the crystal. The number of allowed frequencies dN in a frequency range $d\nu$ varies markedly with frequency. For a crystal with a primitive cubic structure dN is given by

$$dN = 4\pi\nu\,d\nu\left(\frac{1}{u_l^3} + \frac{2}{u_t^3}\right)V_c \tag{7.21}$$

where u_l and u_t are the velocities of propagation of the longitudinal and transverse vibrations respectively, and V_c is the volume of the crystal, i.e. $V_c = XYZ$. In general, u_l will be the velocity of propagation of a longitudinal sound wave through the crystal. Equation (7.21) shows that the frequency density of allowed states $dN/d\nu$ is directly proportional to the crystal volume. Thus, increasing the volume of a crystal increases the number of allowed frequencies, but does not change the distribution of these frequencies. Figure 7.11 shows the frequency density of allowed states according to Eq. (7.21), assuming $u_l = u_t$. Figure 7.11 is for a primitive cubic structure, but similar complex density of allowed states distributions are found for all structures. Indeed, in certain cases it is even possible to obtain a forbidden frequency range within the spectrum as is the case, shown in Fig. 7.11(b), in lithium chloride.

Figure 7.11 shows examples of the distribution of allowed frequencies, but it must be appreciated that not all these frequencies need exist in a crystal in a given situation. In fact, the occupation of the allowed states varies markedly with temperature, as illustrated in Fig. 7.12. At very low temperatures kT, which is a measure of the available thermal energy, is so

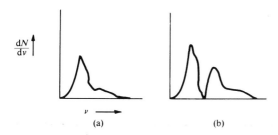

$\frac{dN}{d\nu}$

$\nu \longrightarrow$

(a)   (b)

Fig. 7.11

The frequency density of allowed states for (a) primitive cubic structure and (b) lithium chloride [After A.M. Karo and I.B. Hardy, *Phys. Rev.*, **129**, pp. 2024–2036 (1963)].

Fig. 7.12

The occupation (shaded) of the allowed states at (a) a low temperature, (b) an intermediate temperature and (c) an elevated temperature.

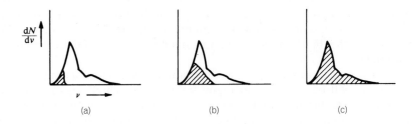

small compared to $h\nu$ of most of the allowed waves that very few phonons exist and these will be of low frequency. At somewhat higher temperatures there are many more low-frequency phonons and also some of higher frequencies. At elevated temperature kT becomes greater than $h\nu$ for even the highest frequency waves, and so a large number of phonons with a spectrum of frequencies is found.

7.11 Specific heats of solids

We consider now the heat capacity of a solid, for which we shall use the ideas about phonons described above.

For simplicity, let us suppose that the atoms all vibrate at a frequency ν. Because each atom can vibrate in three orthogonal directions, each atom can be in one of three quantum states for each energy and so the total number of states for N atoms is $3N$. Then the total heat energy, Q, which the lattice can hold, i.e. the heat capacity of the solid, will be given by the number of states times the mean energy of the phonons associated with them. Thus, using Eq. (7.18)

$$Q = \frac{3Nh\nu}{\exp(h\nu/kT) - 1} \tag{7.22}$$

The specific heat at constant volume, C_{v}, can now be found by differentiating this with respect to T, giving

$$C_{\mathrm{v}} = \frac{\mathrm{d}Q}{\mathrm{d}T} = 3Nk\left(\frac{h\nu}{kT}\right)^2 \frac{\exp(h\nu/kT)}{[\exp(h\nu/kT) - 1]^2} \tag{7.23}$$

where N is now to be taken as the number of atoms in unit mass of the solid.

This result is illustrated graphically in Fig. 7.13 and, except at very low temperatures, it agrees with measured data for many materials fairly well if an appropriate value of ν is taken. To obtain better agreement with experiment at low temperatures it is necessary to take account of the fact that more than one frequency of oscillation is allowed for the atoms.

At very low temperatures the atoms are mostly at rest, and few can be excited into the first energy level, because of the small amount of heat energy available. As the temperature rises, the number of excited atoms rises exponentially, so that the specific heat increases rapidly. At higher temperatures still, the mean thermal energy becomes large compared to the spacing of the energy levels, and the latter can be regarded as nearly continuous. It is therefore not surprising that the solid behaves much like a gas, so that the specific heat tends to a constant value of $3N_0k$ per mole (N_0 is Avogadro's number) (see Problem 7.5). This result is an expression

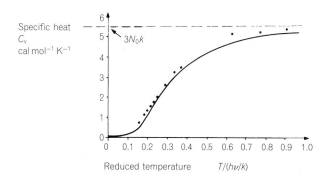

Specific heat
C_v
cal mol⁻¹ K⁻¹

Reduced temperature $T/(h\nu/k)$

Fig. 7.13

Dependence of C_v on temperature according to Eq. (7.23), compared with measurements (dots) for diamond.

of the empirical law discovered in 1819 by Dulong and Petit which stated that, at ordinary temperatures, the specific heats of all solids per mole are approximately the same.

Equation (7.23) predicts that this law holds accurately only if the temperature is high enough, when $\exp(h\nu/kT) \ll 1$ (see Problem 7.5). In fact, Fig. 7.13 shows that it is already quite a good approximation when $T = k/h\nu$, a temperature known as the *Debye temperature* of a material. Some values are quoted in Table 7.3. The fact that some materials are exceptions to Dulong and Petit's law is simply because, for such materials, 'normal' temperatures are not high enough.

Note that, except in the limit of high temperature, C_v depends upon the frequency ν. Hence the specific heat is expected to vary from solid to solid. In particular, from Eq. (7.23) it can be seen that a higher value of ν results in a *lower* specific heat. This is because, as we add the same quantity of heat to two solids, thus raising their temperatures, the atoms must be excited to higher energies in the solid with higher vibrational frequencies, because the spacing of the levels is greater.

This dependence of specific heat on frequency is of utmost importance, because it leads to the possibility of observing allotropic states of solids, as we shall see in Section 7.13. Further, the existence of such allotropes is fundamental to the microstructure and the mechanical behaviour of many structural materials such as steel.

The final point to note in this context is that the specific heat of allotropic forms of a substance must differ. Because their structures differ, the Condon–Morse curves must also differ in shape, and this affects the atomic vibration frequency. Inspection of Fig. 7.5 shows that if the force constant is increased, then the potential energy minimum displays more curvature. This is simply an expression of the mathematical relation

$$\frac{\mathrm{d}F}{\mathrm{d}r} = -\frac{\mathrm{d}^2 V}{\mathrm{d}r^2}$$

A higher force constant gives a higher vibrational frequency (it is as if we made the springs joining the atoms stiffer) and this in turn gives rise to a

Table 7.3

Debye temperatures for some materials

Material	Pb	Au	NaCl	Fe	Se	C (diamond)
Deybe temperature (K)	95	170	280	360	650	1850

lower specific heat. Hitherto, we have discussed only the specific heats at constant volume, C_v, whereas it will shortly be necessary to discuss the specific heat at constant pressure, C_p. The latter is the more easily measured quantity, so that C_v is usually obtained from C_p by making a theoretically derived correction. The correction for most solids at normal temperatures is quite small, so that the general temperature dependence of C_p is similar to that of C_v, except at very low temperatures. As might be anticipated, though, C_p tends to zero as does C_v when the temperature tends to the absolute zero.

7.12 Advanced topic: Specific heats of polyatomic gases

Before moving on to other topics, it is appropriate to add a brief postscript on specific heats in polyatomic gases, because these form the majority of gases in nature, including those commonly used in heat exchangers and whose thermal properties are therefore of practical interest.

A polyatomic molecule has means of motion other than purely translational. For instance, in a diatomic molecule the atomic bond can stretch and each atom can vibrate back and forth with respect to the other, without influencing its flight. In addition, the molecule may rotate about an axis pendicular to the axis of the bond. The rotational velocity has two orthogonal components, both of which can be associated with a mean thermal energy of $\frac{1}{2}kT$, and the vibrational energy can also be $\frac{1}{2}kT$ per molecule.

But at low temperatures the vibrational and rotational motions may not be excited, as their energies are quantized – just as in a solid at low temperatures. Hence at low temperatures a diatomic gas (provided it does not liquefy) behaves like a monatomic one. As the temperature rises the rotational states begin to participate and the specific heat rises accordingly. Because the interatomic bond is normally very stiff, the vibrational mode is the last to be excited. However, the full range of behaviour is not shown by any one gas, as liquefaction and dissociation of the molecules limit the temperature range in which measurements can be made.

7.13 Allotropic phase changes

We have already noted that some solids change their structure when their temperature is raised or lowered. For example, iron changes from b.c.c. to f.c.c. at 910°C, and tin from the diamond structure to a related, tetragonal, structure at 13°C. These allotropic forms are often referred to as different *phases* of the solid material, and the transformation from one to another is very similar to the phase changes of melting and vaporization. Latent heat is usually evolved when the high-temperature phase is cooled through the transition point, and the specific heat changes abruptly at the same temperature. We may well ask why it is that the change occurs at all? For if latent heat is evolved when iron cools through 910°C, then the b.c.c. form (commonly called α-iron) to which it transforms has lower energy and it should, according to earlier precepts, be the more stable of the two forms. This energy difference must be reflected in the Condon–Morse curves of atomic potential energy vs. atomic separation for the two structures, and we might imagine them to be as illustrated in Fig. 7.14. The deeper

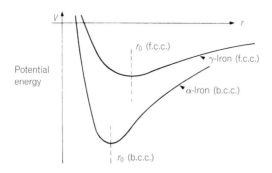

Fig. 7.14

Schematic Condon–Morse curves for α-iron and γ-iron.

minimum of the α-iron curve represents a lower energy in the equilibrium state, when the atomic separation is exactly r_0.

But in fact at normal temperatures the average energy per atom is higher than the minimum of the Condon–Morse curve, because the solid contains a quantity of heat energy. The amount of energy added to a solid that is heated from 0 K to a temperature T is just

$$H = \int_0^T C_p \, dT$$

where C_p is the specific heat at constant pressure.

We note immediately that solids with different values of C_p contain different amounts of heat energy at the same temperature. This, however, does not explain the phase change in iron because at and above 910°C extra heat must be given to α-iron before it will change to γ-iron (the f.c.c. form). The answer to the problem lies in the way the heat energy is distributed among the atoms.

Let us consider the energy diagrams for the atoms in the two structures of iron. We saw in the previous section that the vibrational energy levels of the atoms in a solid may be reasonably assumed to be equally spaced by an energy $h\nu$, and that the frequency ν is related to the degree of curvature of the Condon–Morse curve at its minimum. The higher the curvature, the higher is ν, so that from Fig. 7.14 we see that the deeper minimum, which has the higher curvature, gives a higher frequency. Thus, the b.c.c. iron lattice has the higher frequency and its vibrational energy levels are therefore more widely spaced than those of f.c.c. iron.

To see more clearly what effect this has on the relative stabilities of the two phases let us consider an extreme case. We suppose that the two phases of the same material have such widely differing frequencies ν that one, phase B, has a nearly continuous band of energy levels where the other, phase A, has but two. This is illustrated in Fig. 7.15. We further suppose for simplicity that only these two lowest levels of phase A are populated at the temperature of interest. Because the higher frequency phase is the stable one at low temperatures, its lowest energy level must lie below that of phase B. Superimposed on each energy diagram is the Boltzmann probability distribution. It has the identical shape for both phases because they share the same temperature.

Now let the two phases be in contact with one another, so that at the interface the atoms may 'choose' one structure or the other.

In 'collision' with one another, the atoms will exchange energy, as explained earlier. Consider a 'collision' at the interface between a phase A

Fig. 7.15

Energy levels and distribution functions for two phases A and B of a solid held at the same temperature.

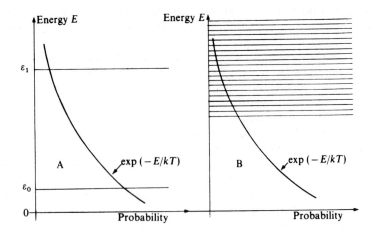

atom of energy ε_1 and a phase B atom of any energy E_2. The amount of energy exchanged between them, ΔE, is a purely random quantity owing to the random nature of such collisions. The energy of the A atom subsequent to the collision is $\varepsilon_1 \pm \Delta E$, which in most cases is different from both ε_0 and ε_1. It is consequently impossible for this atom to remain in phase A, because there is no suitable energy level available to it. However, it may move to a phase B energy level as these are plentiful. The atom therefore has a very high chance of making the transition from phase A to phase B.

At the same time the atom that began life in phase B is similarly unlikely to acquire exactly the correct amount of energy to allow it to make the transition to phase A, so that it has a very small chance of making that transition. There is thus a net transfer of atoms from phase A to phase B.

It can be seen that the basic cause of the phase change is the difference in the number of energy levels available in the two phases. For all the atoms in phase B to get into phase A, their purely random 'collisions' must simultanously put them each into one of the two energy levels available to phase A. The chance of this happening is roughly like that of putting all the snooker balls on a snooker table into two pockets, by hitting them randomly with the cue. Because there are many other places in which the balls may settle, the probability is low, like that of the phase transition B to A.

On the other hand the probability of getting the snooker balls *out* of their pockets by hitting them with the cue (although against the rules of snooker) is high – as before, there are plenty of alternative resting places for them.

In summary we see that, because of the random nature of heat energy, the criterion for stability of one phase *vis-à-vis* a second is not simply that it must have lower energy, but also that a transition to the second phase must be less probable than the reverse transition.

From the above description it may be thought that the transition from phase B to phase A will never occur, but this is not so. If the temperature is lowered so that fewer atoms are present in the excited state of phase A, then the rate of transfer of atoms from phase A to phase B is lowered and may eventually fall to equality with the reverse transition. This temperature is then what is normally called the transition temperature and, below it, transitions to the lower energy phase become the more probable.

In summary of this description, the effect of *rates* of transition on thermal equilibrium may be put into two general statements.

(1) When two phases of a substance are in equilibrium at the transition temperature, there are equal and opposite rates of exchange of atoms between the two phases.

(2) If the two opposite rates become unequal through a change in temperature for example, one or other of the two phases is the more stable and hence predominates.

We have deliberately avoided quantifying the concept of stability, as it is a rather complicated one. It is possible, though, to define a quantity called the *thermodynamic probability*, which measures the tendency of materials to choose a structure with a larger number of energy levels rather than one with lower thermal energy. Thermodynamic probability is a measure of the number of different ways in which the atoms can be distributed among their quantum states without changing the energy distribution – thus the larger the number of populated levels, the larger the thermodynamic probability.

7.14 Latent heat and specific heat

The two imaginary phases of the previous section must have different specific heats, according to the arguments of Section 7.11. In fact it is apparent that, were the two phases to have the same specific heat, the phase change would not occur, because the distribution of the energy levels would be the same in both phases.

The normal transition temperature, T_t, between two phases is governed by the distribution of the energy levels. The latent heat at this temperature is governed by the specific heats of the two phases. The latent heat is equal to the total energy difference between the phases at T_t. If the energy difference is H_0 at the absolute zero of temperature (the difference between the minima of the respective Condon–Morse curves), then the energy difference ΔH at the temperature T_t is

$$\Delta H = H_0 + \int_0^{T_t} C_p^a \, dT - \int_0^{T_t} C_p^b \, dT$$

where C_p^a and C_p^b are the specific heats at temperature T of the two phases. Normally the C_p of the high-temperature phase (b) cannot be measured below T_t, but it can be calculated theoretically as described earlier (Section 7.11) and values of latent heat calculated in this way confirm the theories we have been discussing in this chapter.

7.15 Melting

Allotropes will be observable where the two possible structures have Condon–Morse curves of the kind illustrated in Fig. 7.14, provided, of course, that the low-temperature phase does not melt first! But melting is also a phase transition, and can be discussed in exactly the same way. The only difference is that the high-temperature phase does not have a regular structure, a fact which can be understood if we note that at some temperature the amplitude of vibration of the atoms in a solid must become comparable to the atomic size. Thus the lattice can no longer be regarded as rigid, for as a pair of atoms move apart, another may insert itself between them, destroying the basic structure of the unit cell. Such a structure must be liquid-like, for the *mean* atomic separation will still be close to that of the solid, but rigidity is totally lost.

It is instructive to note that melting a crystal is not a gradual process, as might be expected from the previous paragraph. A crystalline solid does not become progressively less rigid as the temperature rises until complete fluidity exists: in fact the transition occurs abruptly at a well-defined temperature. This is because the liquid state has quite a distinct *structure*, and therefore has quite a different specific heat from the solid. As the temperature is raised through the melting point, the stability of the liquid structure first becomes equal to and then greater than the stability of the solid state, exactly as for two solid structures. The transition is therefore abrupt, and there are *no stable intermediate states* between the solid and liquid phases.

In contrast, an amorphous solid such as glass undergoes no structural change, and melting is not abrupt, but involves a gradual fall in viscosity as the temperature rises.

7.16 Thermodynamics

In this chapter, the thermal behaviour of solids has been discussed in relation to their structure, with the emphasis on the use of atomic models to explain various phenomena. This is not the only way of treating thermal properties; historically, the measurement of heat and work and the changes they effect in matter came first, and the subject of *thermodynamics* was then built up to coordinate all the observed phenomena. Its basic tenets are that heat and work are interchangeable and that energy cannot be created or destroyed. These are enshrined in the *first law* of thermodynamics, which states that if heat dQ is put into a body and work dw is done on it, their sum is equal to the change in internal energy dU. By convention, the work dw is given a negative sign while work done *by* the body on its surroundings is given a positive sign. Thus

$$dU = dQ - dw$$

The total heat content of the body at a temperature T is thus $\int_0^T (dU + dw)$, which might be thought to be the same for all bodies in thermal equilibrium. We have seen, however, that equilibrium is also governed by *thermodynamic probability*, a concept that is alien to thermodynamics and can only be appreciated by recourse to atomic models. It was therefore found necessary to 'invent' another energy term which substitutes for the concept of probability, so that phase changes could be ascribed to differences in energy of the phases concerned. The invented quantity S, called *entropy*, when multiplied by the temperature T, gives an energy term that increases as a body is heated. Differences in thermodynamic probability are equivalent to differences in the product TS. The change in 'free energy' ΔG when a phase change occurs is then the total heat change minus $T\Delta S$, i.e.

$$\Delta G = \Delta U + \Delta w - T\Delta S \tag{7.24}$$

where ΔU, Δw and ΔS are the changes in the respective quantities. Two phases of an element are in equilibrium at constant pressure if $\Delta G = 0$, so that their free energies are exactly the same. Otherwise the phase with lower free energy is the equilibrium one. In this way the minimum-energy concept was retained in the scheme. The condition for equilibrium in a two-component system like an *alloy* is slightly different and is discussed in Chapter 10. It is frequently convenient to lump ΔU and Δw into one term, ΔH, called the *enthalpy* or *total heat*, in which case Eq. (7.24) reads

$$\Delta G = \Delta H - T\Delta S$$

Not surprisingly, a connection between entropy, S, and thermodynamic probability, W, was established at a later date – the one increases with the other according to the equation

$$S = k \ln W \tag{7.25}$$

where k is Boltzmann's constant. This means that the higher the probability, the lower is the free energy, because the entropy term subtracts from the 'true' energy (the enthalpy) [Eq. (7.24)]. Thus the subject of thermodynamics was made self-consistent, and provides a powerful tool for discussing all kinds of changes that occur in materials. Some examples will occur in subsequent chapters.

For a more thorough treatment of thermodynamics the reader is referred to the list of further reading at the end of this book.

A limitation of thermodynamics is that in spite of its name it cannot deal with *rates* of phase changes and this makes it unable to explain non-equilibrium or metastable structures, which constitute a number of engineering materials. It is with such questions that the next three sections deal, in the same terms as those used in earlier sections.

7.17 Multiphase solids

In Chapter 10 materials that have many phases will be discussed. In some of these the phases exist in equilibrium at room temperature, but we shall find others exhibiting phases that are not in true equilibrium with one another. That is, the two or more phases exist side by side at room temperature, although the temperature at which they are truly in equilibrium is much higher. Although we can now understand why different phases may exist, our description so far makes their simultaneous co-existence seem impossible at temperatures other than the one at which they are in equilibrium. For if one is more stable than the other at all temperatures bar one, then the more stable one should exist alone. To explain the apparent anomaly it must be supposed that either there is some barrier to the establishment of the more stable phase, or that the transition occurs so slowly that it cannot be observed in times of less than hundreds of years. In fact both of these explanations are involved, as we shall now discover by investigating what controls the *rate* of a phase change.

7.18 Rate theory of phase changes

Until now we have discussed only differences in energy between the initial and final states of a system undergoing a phase change. However, during the change the structure must go *through* intermediate stages, though obviously none of them are stable or they would be observable over a finite temperature range. Indeed, it is important to understand that the only stable structures lie at the two end-points of the process. This implies that the structure goes through a stage (the *activated* phase) having *higher energy* than both the low-temperature and the high-temperature structures, for stable structures must have a minimum energy with respect to small changes of structure. Confirmation of this comes from the fact that in most phase changes the high-temperature state can be supercooled – it can exist slightly *below* the transition temperature if mechanical disturbances are removed – and the low-temperature phase can similarly exist at slightly higher temperatures. It is not surprising that mechanical vibration can, for instance, cause supercooled f.c.c. iron to revert to the b.c.c. form, for the lattice actually has to change shape during the process.

We can make this idea more concrete by plotting the potential energy of the structure against some convenient dimension z of the lattice that increases monotonically during the transition. A hypothetical plot is shown in Fig. 7.16, where the values z_A and z_B of the lattice dimension are the equilibrium values in the initial and final states at the absolute zero of temperature. Figure 7.16 is therefore simply a Condon–Morse curve of a different kind to those we have considered hitherto.

Because the atoms have thermal energies, their total energy is higher than E_1 or E_2, and we shall suppose that the mean total energy per atom is E_A in one phase and E_B in the other, when the temperature of both phases is the same. The differences $(E_A - E_1)$ and $(E_B - E_2)$ therefore represent the average thermal energies of atoms in either state. The points x, x' and y, y' on the potential energy curves at the energies E_A and E_B correspond to the turning points of vibrating atoms with the energies E_A and E_B respectively.

How, then, can the phase change occur, if a state of higher energy (up to E_C) exists between states A and B? The clue lies in the reminder, above, that E_A and E_B represent only *mean* energies and that, if the barrier E_C is not too high, there may be many atoms with enough energy to surmount it and hence make the transition to the state B. Once there, they may lose energy by collision with neighbours so that they are trapped there in the potential energy 'well'. Note that this picture does not rule out the possibility of atoms making the reverse transition – as we have discussed earlier (Section 7.13), the *rate* of the latter change is lower, for reasons unconnected with Fig. 7.16.

We describe such a change as this as an *activated process* – the idea being that the solid in phase A must be *activated* before the change can occur, and that the activated state (state C of Fig. 7.16) has a higher energy E_C. The energy difference $E_C - E_1$ in the figure is referred to as the *activation energy* of the change. The reason why this energy difference rather than $E_C - E_A$ is used to characterize the activated state will be clear

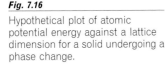

Fig. 7.16

Hypothetical plot of atomic potential energy against a lattice dimension for a solid undergoing a phase change.

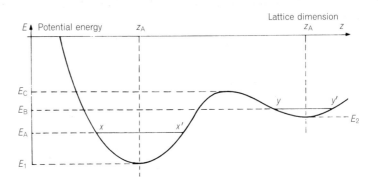

when we consider how it affects the *rate* at which the change occurs. This rate can be roughly estimated as follows. The net transition rate is the difference between two flows in opposite directions – one of atoms changing from state A to state B and the other making the reverse transition. The former rate r_A may be defined as the number of atoms that make the A–B transition in unit time, and this number must be proportional to the following factors:

(a) the number n_A of atoms in phase A with energies higher than E_C;
(b) the frequency, f, with which the atoms approach the barrier;
(c) the relative thermodynamic probability of phase B compared to phase A.

We shall now discuss each of these in turn. The number n_A is given by the Boltzmann expression, Eq. (7.9), integrated over the energy range above E_C. Thus

$$n_A = \int_{E_C}^{\infty} C \exp - \left(\frac{E - E_1}{kT} \right) dE$$

where E_1 appears because it is the energy at which atoms possess no kinetic energy in phase A. The factor C in the above equation depends weakly upon E, so for our present purposes we shall assume it is a constant. Performing the integration thus gives the result

$$n_A = \frac{C}{kT} \exp - \left(\frac{E_C - E_1}{kT} \right)$$

The factor C can be found by extending the range of integration down to E_1, for then the number obtained is equal to the total number of atoms N_A in state A. Thus

$$N_A = \frac{C}{kT} \quad \text{and hence} \quad n_A = N_A \exp - \left(\frac{E_C - E_1}{kT} \right)$$

The second quantity, f, the frequency of approaching the barrier, may be taken to be the frequency of vibration ν_A of atoms in phase A. In most solids this is around 10^{13} Hz. The final term to be evaluated is the thermodynamic probability of phase B *relative* to phase A. This quantity has only briefly been discussed in this book, and here it will simply be assumed that the thermodynamic probabilities W_A and W_B of phases A and B can be defined, and that their ratio W_B/W_A gives the required relative probability. This ratio has the right form, because it is still finite when $W_A = W_B$, while the difference $(W_B - W_A)$ would not satisfy this requirement.

The rate, r_A, being proportional to each of the above factors, is therefore proportional to their product. Thus

$$r_A = K\nu_A \frac{W_B}{W_A} N_A \exp - \left(\frac{E_C - E_1}{kT} \right) \tag{7.26}$$

where K is a constant dependent only on the geometry of the interface between phases A and B.

But Eq. (7.26) does not give the net transition rate because the reverse transition also occurs. Its rate, r_B, is given by an exactly similar expression:

$$r_B = K\nu_B \frac{W_A}{W_B} N_B \exp - \left(\frac{E_C - E_2}{kT} \right) \tag{7.27}$$

The net transition rate r is just $r_A - r_B$, so that

$$r = K\nu_A \frac{W_B}{W_A} N_A \exp - \left(\frac{E_C - E_1}{kT} \right) - K\nu_B \frac{W_A}{W_B} N_B \exp - \left(\frac{E_C - E_2}{kT} \right) \quad (7.28)$$

Because only the temperature dependence of r is of interest at present it is convenient to replace the coefficients of the two exponential terms by the symbols F and G, giving

$$r = F \exp - \left(\frac{E_C - E_1}{kT} \right) - G \exp - \left(\frac{E_C - E_2}{kT} \right) \quad (7.29)$$

This equation may be used to discuss the temperature dependence of r if we concede that all factors other than the exponential depend rather weakly on temperature, at least for small temperature changes. Let us therefore investigate how the rate of transition varies according to this theory for temperatures close to the normal transition temperature.

At the transition temperature T_t the two phases can exist side by side – they are in equilibrium and the net rate is zero. Thus

$$F \exp - \left(\frac{E_C - E_1}{kT_t} \right) - G \exp - \left(\frac{E_C - E_2}{kT_t} \right) = 0 \quad (7.30)$$

The dependence of r on T at temperatures close to T_t may be found by differentiation of Eq. (7.29) as follows

$$\frac{dr}{dT} = F \frac{(E_C - E_1)}{kT^2} \exp - \left(\frac{E_C - E_1}{kT} \right) - G \frac{(E_C - E_2)}{kT^2} \exp - \left(\frac{E_C - E_2}{kT} \right)$$

$$= \frac{(E_C - E_1)}{kT^2} \left\{ F \exp - \left(\frac{E_C - E_1}{kT} \right) - G \left(\frac{E_C - E_2}{E_C - E_1} \right) \exp - \left(\frac{E_C - E_2}{kT} \right) \right\}$$

Because $(E_C - E_2) < (E_C - E_1)$, a comparison of this with Eq. (7.30) shows immediately that the second term is smaller than the first when $T = T_t$, and hence dr/dT is positive. An increase in temperature therefore causes the transition from phase A to phase B to proceed, while a reduction has the reverse effect.

7.19 Metastable phases

Suppose that the solid discussed above is cooled instantaneously from above T_t to a very low temperature. Initially, the number of atoms in phase A will be negligible, so that transition from this phase may be neglected. The net transition rate is then equal to r_B, given by Eq. (7.27). That is,

$$r = r_B = K\nu_B \frac{W_A}{W_B} N_B \exp - \left(\frac{E_C - E_2}{kT} \right) \quad (7.31)$$

If the new, low temperature is small enough so that $E_C - E_2 \gg kT$, then this rate is so small that for practical purposes it can often be regarded as zero. Thus, by supercooling a body very rapidly to a low temperature before the transition can begin, it can effectively be halted. The number of atoms able to surmount the energy barrier is reduced to the extent that the (normal) high-temperature phase appears to be stable.

We shall encounter several examples of such metastable phases in later chapters. The most widely quoted instance is that of glass, which is a super-cooled liquid. Its structure will be discussed in more detail in Chapter 11.

7.20 Other applications of the rate theory

The above equations describe the rate of *any* process of change between two states separated by an energy barrier. One very large class of such changes is the class of chemical reactions. The rates of chemical reactions are of fundamental importance to a discussion of the corrosion in engineering materials so let us consider now the application of the ideas of the previous section to this problem.

In general, corrosion of a solid is a two-stage process. The formation, for instance, of an oxide layer on a metal obviously constitutes a chemical reaction between the metal and oxygen (also often involving water), but once the oxide is formed it covers the surface of the metal so that the latter is no longer in direct contact with the atmosphere. For the chemical reaction to proceed further it is necessary for the atoms either of the metal or of the atmosphere to migrate through the oxide layer until they meet one another. The process of migration is called *diffusion* and, because it occurs via thermal atomic motion, it is affected by temperature.

Diffusion, therefore, controls the rate at which atoms in a chemical reaction can become close enough to one another for the reaction to proceed. In some cases, diffusion may occur fast enough for the overall rate to be controlled by the rate of the reaction itself, while in others, diffusion may be slow enough to be important in determining the overall rate of the process. We shall consider diffusion later.

7.21 Chemical reactions

Let us take as a simple example the reaction of hydrogen and chlorine to form hydrochloric acid, usually written

$$H_2 + Cl_2 \rightleftharpoons 2HCl$$

This equation can, in fact, proceed in either direction, but for the present we shall regard the two acid molecules as the final state of the reaction and the two molecules of hydrogen and chlorine as the initial state. [According to the discussion in Section 7.16 on classical thermodynamics, the final state ($2HCl$) must have lower free energy than the initial state ($H_2 + Cl_2$).]

Now the reaction of hydrogen with chlorine is exothermic at normal temperatures – that is, heat is evolved in the production of hydrochloric acid. We may therefore conclude that the energy of the final state (the acid molecules) is lower than that of the initial state, unlike the case of the phase change we discussed in Section 7.18. [In the language of thermodynamics, this energy should be called *enthalpy* (see Section 7.16).] There is, however, an energy barrier between them, for otherwise hydrogen and chlorine would react spontaneously whenever they came into contact with one another. A mixture of the two gases does in fact appear quite stable – the reaction rate is so low that it can be ignored for practical purposes, except in the presence of sunlight.

An energy diagram can now be drawn (Fig. 7.17), analogous to Fig. 7.16, in which energy is plotted against the separation of the molecules taking part in the reaction. The rate of the reaction is just the number of molecules crossing the barrier in Fig. 7.17 in unit time. The process is almost exactly like that of a phase change as discussed in Section 7.18, and it will therefore obey the same equations. There will thus be a temperature at which the reaction is in equilibrium – the rate of dissociation of HCl

Fig. 7.17

Hypothetical plot of potential energy vs separation of the H_2 and Cl_2 molecules in the reaction $H_2 + Cl_2 \rightleftharpoons 2HCl$.

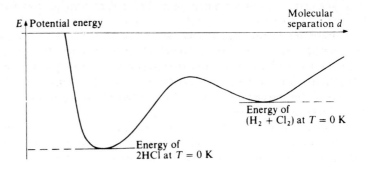

into H_2 and Cl_2 is equal to the rate of combination at this temperature. At temperatures well removed from the equilibrium value, one or other of the two rates will be negligible and the net reaction rate, r, is therefore given by an equation similar to Eq. (7.31) having the form

$$r = A \exp - \Delta E/kT$$

where ΔE is the activation energy, as before, and A is a constant.

The rate of a chemical reaction should therefore increase rapidly with the temperature. This is indeed the case: a flame applied to a mixture of H_2 and Cl_2 gases raises the local temperature, so that molecules are more rapidly excited over the energy barrier in Fig. 7.17. The reaction goes to completion because the energy released is transmitted to other molecules, exciting them over the energy barrier, and so on. (Incidentally, the action of sunlight can also be understood if we assume that the atoms absorb photons and so acquire enough energy to surmount the barrier.)

In this instance, it is quite clear that the constituents of the reaction, being gases, can mix readily so that the rate of diffusion is high and has a negligible effect on the overall reaction rate.

7.22 *Diffusion*

What is diffusion, and why is it important? It is the movement of atoms from point to point within a material, resulting from the ever-present thermal energy. In the case of solids, it is also due to the fact that a small proportion of atoms has more energy than is needed to displace an atom by temporarily breaking the bonds that hold it in place. In fact, diffusion is one of the clearest demonstrations that thermal energy is present in all materials. It occurs at different rates in solids, liquids and gases.

The example of diffusion that is easiest to envisage involves two gases, initially separated as in Fig. 7.18, which diffuse into one another when the separating wall is removed (of course, no bond breaking occurs in gaseous diffusion). However, it was remarked earlier in connection with corrosion that diffusion can occur through solids, unlikely as this may seem. If this were not so, then only the surface layer of atoms on a metal could oxidize, while in practice the layer of oxide that forms is many molecules thick. A further example is the diffusion of gases through glass, which causes vacuum flasks and cathode ray tubes to go 'soft' after many years of service, even though the glass envelope remains intact. Diffusion can also be a useful means of inserting atoms *into* a material from its surface, enabling the engineer to modify and control its mechanical and/or its electronic

Fig. 7.18

Two gases, initially separated by a wall, will diffuse into one another when the separating wall is removed.

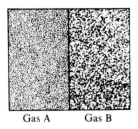

Gas A   Gas B

properties, e.g. surface hardening of steel by in-diffusing carbon (which almost certainly happened accidentally before the 19th century whenever iron was worked by a smith in a carbon-rich fire), and in the fabrication of transistors within silicon, to form integrated circuits. It is also exploited usefully in solid-state fuel cells, where for example the rather fast diffusion rate of Na^+ ions through sodium β-alumina, a so-called super-ionic conductor, enables this material to be used in rechargeable batteries that are about four times lighter than their lead-acid equivalents.

In the above examples, atoms move from a region of high concentration of the diffusing atoms to one of low concentration. We call this process *chemical diffusion*. In contrast, *self-diffusion*, which is the random wandering through a material of *its own* atoms does not need such a concentration difference. Self-diffusion is what enables us to change some properties of materials by heat treatments, called *annealing*, and (more slowly) can result at 300 K in the gradual relief of stresses, sometimes resulting in mechanical failure. We shall concentrate here on chemical diffusion, leaving self-diffusion until Chapter 8.

The characteristic features are that

- there is a net transfer of atoms across a material;
- diffusion is a result of a gradient in the number of atoms per unit volume (a *concentration gradient*).

It is easiest to make measurements of diffusion when the concentration gradient is both uniform and controllable. Thus, measurements on gases diffusing through porous solids show that the flow or flux, F, of atoms is simply proportional to the concentration gradient, i.e.

$$F = -D \frac{dn}{dx} \tag{7.32}$$

where F is the number of atoms or molecules flowing across a unit cross-sectional area in the y–z plane in 1 s, and dn/dx is the concentration gradient along the x-direction, in units of m^{-4}. If the flow F is towards the positive x-direction, as in Fig. 7.19, the gradient in the concentration, n, must be *negative*. The constant D is known as the *diffusion coefficient*, which depends on the nature both of the diffusing species and of the medium in which diffusion occurs. Equation (7.32) is known as Fick's first law, and it applies in many situations concerning atomic diffusion in solids.

It is important to realize that the motion of individual atoms is entirely random: this is the characteristic that distinguishes diffusion from all other types of flow, such as convection. The reason that a net flow occurs towards the right in Fig. 7.19 is that there are *more atoms* on the left. Half of them are moving to the left, and half to the right at any moment, while half of the smaller number on the right-hand side of the diagram are moving to the left at the same instant.

Because an atom in a crystal lattice is strongly bound to its neighbours, it must gain enough energy to move. Thus the diffusing atoms must be thermally 'activated', i.e. they must overcome an energy barrier. The mathematics developed in Section 7.20 for reaction rates is applicable here, and the outcome is that the diffusion coefficient D is proportional to an exponential factor. Thus

$$D = D_0 \exp(-E_0/kT) \tag{7.33}$$

where D_0 and E_0 are constants.

Fig. 7.19

Illustrating Fick's first law of diffusion.

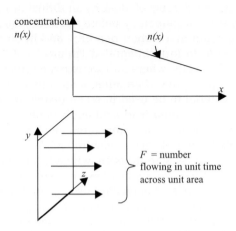

Diffusion in solids is therefore strongly temperature-dependent, while in gases and liquids, where no energy barrier exists, diffusion coefficients are generally independent of temperature. At normal temperatures, diffusion in solids is relatively slow and may have a controlling effect on the rate of a chemical reaction. Thus, if the surface of a metal normally oxidizes fairly rapidly it may be because diffusion of oxygen through the surface oxide layer is easy, though other factors, discussed in Chapter 10, are equally important.

Diffusion in solids is dealt with in more detail in the next chapter, because it depends on some of the many different kinds of defects, or irregularities in structure, that are found in crystals. These defects cannot be entirely eliminated because of thermal energy, and at the same time, some defects actually facilitate atomic diffusion, as we shall explain.

7.23 Thermal conductivity in solids

The thermal conductivity κ of all solids is defined in terms of the rate of flow of heat Q across unit area per unit time of a sample in which there is a uniform temperature gradient dT/dx, thus

$$Q = \kappa\, dT/dx$$

The similarity of this equation to Fick's first law of diffusion suggests strongly that diffusion of something is the mechanism of heat transfer. In metals, the diffusing particles are electrons, while in non-metals they are quantum 'energy packets' rather than particles having mass: they are the phonons that we introduced in Section 7.10. Let us consider metals first.

Metals are among the best conductors, as Table 7.4 shows, and are also the best electrical conductors. Indeed, many groups of metallic alloys display a simple relationship between κ and the electrical conductivity σ, described by the equation:

$$\kappa = L\sigma T + C \tag{7.34}$$

where T is the absolute temperature, and both C and L are constants for each alloy group. Their values are tabulated in the book *Tables of Physical and Chemical Constants* by Kaye and Laby in the list of further reading on page 633.

Equation (7.34) is an extension of the 'law' of Wiedemann and Franz, which states that for many metallic *elements*, the ratio $\kappa\sigma/T$ is approximately constant, within a factor of about 2. The variation of this ratio can be

Table 7.4

Thermal conductivities of some materials

Material	Thermal conductivity κ $(\text{W m}^{-1}\text{K}^{-1})$ @ 300 K
Diamond*	400–2000
Copper	403
Aluminium	237
Al–Cu alloy	160–180
Silicon	150
Sapphire (corundum)	34
SiO_2 (glassy)	2
ZrO_2	1.5
Glass	~1.0
Glass	4.2×10^{-2}
Air (dry)[†]	1.4×10^{-5}

* Varies widely in synthetic diamond with the degree of crystalline perfection and the proportions of the isotopes.
[†] Conduction only, i.e. when convection is prevented by design.

explained by taking account of the quantized energy levels available to free electrons, but the theory is too complex to introduce here. These relations do not apply to non-metals, confirming that different particles must be involved in heat transfer.

We shall treat the valence electron model of electrical conduction mathematically in Chapter 14. Here we shall simply note that the conduction electrons move about rather like gas particles: their average kinetic energy, and hence the heat they carry, rises linearly with the absolute temperature. They collide frequently with lattice imperfections as they fly around, with a well-defined mean time between collisions that is calculated in Chapter 14. Because of these collisions, they carry heat energy by diffusing, rather than by direct flow between hot and cold regions. We shall see in Chapter 11 that the conductivity is directly proportional to the mean time between collisions, a quantity that for historical reasons is called the *relaxation time* of electrons.

Now we turn to non-metals, which of course contain no freely moving particles such as electrons. As we have seen, atomic vibrations travel as waves of frequency ν, and carry heat energy in multiples of the *quantum* of energy $h\nu$. The velocity u of the waves however is not the velocity at which heat travels along a temperature gradient, for the phonons 'collide' randomly with one another and so are scattered like gas atoms. This is why they carry heat by diffusion in a similar way to electrons in metals. The reason they scatter one another is because, where two waves coincide as they cross paths, the resulting displacement of atoms is *not* simply the sum of the two wave amplitudes, for force and displacement are not exactly proportional, as we have seen. The resulting thermal conductivity, κ, can be shown to be proportional to the heat capacity of the phonons, i.e. to the specific heat C, and to the phonon velocity u and their *mean free path*, Λ, the average distance between collisions:

$$\kappa \propto Cu\Lambda$$

Thus a material that has high heat capacity has a high phonon velocity, due to high stiffness, and a long mean free path, which results when the material is most nearly perfectly elastic. Thus hard materials like diamond display particularly high values of thermal conductivity amongst the insulating materials.

The thermal conductivities of some materials, including those in Table 7.1, are compared in Table 7.4. The wide range of values of κ amongst synthetic diamonds in the table is due to different mean free paths, which vary with the concentration of lattice defects in these materials.

Problems

7.1 List as many pieces of experimental evidence as you can which support the view that atoms are in constant motion.

7.2 Using Eq. (7.7) find the most probable speed for hydrogen gas molecules at 0°C. Explain with the aid of Fig. 7.2 why your result is lower than the root mean square speed quoted in Section 7.4.

7.3 A gas of hydrogen atoms is heated to 10 000 K so that collisions between them result in electrons being excited into higher energy levels. Calculate the number of atoms in the 2s and 2p states as a proportion of the number in the ground state.
(*Hint*: There are six 2p electron states having the same energy, but only one 1s state.)

7.4 Show that in Eq. (7.15) F is approximately proportional to $(r - r_0)$, if the extension of the interatomic distance is kept small enough. What implication does this have for the change in potential energy of the pair of atoms with $(r - r_0)$?

7.5 Show that, in the limit of very high temperature, i.e. $h\nu/kT \ll 1$, Eq. (7.23) becomes approximately

$$C_{\mathrm{v}} = 3Nk$$

Show also that this result is identical to that for a monatomic gas.

7.6 Find the value of C_{v} from Eq. (7.23) at the temperature $T_{\mathrm{D}} = h\nu/k$, as a fraction of its value at $T \rightarrow \infty$.

The value of this temperature is given for some materials below. Find some of the corresponding values of ν, and explain why high values of this temperature occur in the stiffer and lighter materials.

Material	Pb	Au	NaCl	Fe	Se	Diamond
T_{D} (K)	95	170	280	360	650	1850

7.7 How many degrees of translational, rotational and vibrational freedom has a triatomic molecule? Hence determine the maximum possible specific heat of a gas of such molecules at constant volume.

7.8 Describe the process of vaporization of a liquid from the point of view of the kinetic theory. Explain the reason for the existence of a latent heat of vaporization.

7.9 Explain what is meant by an *activated process*. The logarithm of the rate of an activated process is plotted against $1/T$ (T = absolute temperature). How can the *activation energy* be found from the graph? (The rate of the reverse process can be neglected.)

7.10 The diffusion coefficient of copper atoms in aluminium is found to be $1.28 \times 10^{-22} \, \mathrm{m^2 \, s^{-1}}$ at 400 K and $5.75 \times 10^{-19} \, \mathrm{m^2 \, s^{-1}}$ at $T = 500$ K. Find the temperature at which its value is $10^{-16} \, \mathrm{m^2 \, s^{-1}}$.

Copper atoms diffuse under steady state conditions at this temperature through 0.1 mm thick aluminium foil. If the concentration of copper atoms is maintained at $10^{29} \, \mathrm{m^{-3}}$ on one side of the foil and negligible on the other, what is the mass rate of flow of copper through the foil?

Self-assessment questions

1 The kinetic theory of gases shows that

   (a) the mean kinetic energy of a gas molecule is proportional to temperature

   (b) Boyle's law is a natural consequence of the theory

   (c) the pressure exerted by a gas is proportional to the mean square speed of its atoms

2 The amount of energy possessed by a monatomic gas molecule for *each* cartesian direction of motion is

   (a) $\frac{3}{2}kT$ (b) $\frac{1}{2}kT$ (c) $2kT^2$

3 If the molecules in a gas may rotate and or vibrate then

   (a) the average thermal energy for each component of the motion is $\frac{1}{2}kT$

   (b) the specific heat in the high-temperature limit is increased

   (c) the material vaporizes more readily

   (d) the mean square speed of the molecules at a given temperature is raised

4 The amount of heat energy in all matter is proportional to the absolute temperature, only the constant of proportionality differing from substance to substance

   (a) true   (b) false

5 The differential of heat content with respect to temperature is called

   (a) enthalpy   (b) internal energy

   (c) specific heat of the substance

6 The Maxwell–Boltzmann distribution law shows that

   (a) the mean square speed of the molecules in a gas is proportional to temperature

   (b) there is an upper limit to the velocity of a molecule at any temperature

   (c) the most probable velocity increases with temperature

7 The energy distribution law for atoms with quantized energy levels that are closely spaced

can be written in the form

$$N(E) = \frac{dS}{dE} A \exp\left(-\frac{E}{kT}\right)$$

In this equation

   (a) $N(E)$ is the fraction of atoms having an energy E

   (b) dS/dE is the number of quantum states per unit energy interval at the energy E

   (c) $A \exp(-E/kT)$ is the probability that an atom lies in a quantum state of energy E

8 In an atomic system in which the population of each quantum state is *small*, the numbers of atoms in states of differing energies E_1, E_2, E_3 are in the ratios

   (a) $E_1 : E_2 : E_3$

   (b) $\exp(E_1/kT) : \exp(E_2/kT) : \exp(E_3/kT)$

   (c) $\exp(-E_1/kT) : \exp(-E_2/kT) : \exp(-E_3/kT)$

   (d) $\dfrac{1}{\exp(E_1/kT) - 1} : \dfrac{1}{\exp(E_2/kT) - 1}$

   $: \dfrac{1}{\exp(E_3/kT) - 1}$

9 Which of the answers to Question 8 is correct when the average population of some states approaches unity? If none, answer (e).

10 Two bodies in contact are in thermal equilibrium when:

   (a) there is no net flow of heat between them

   (b) there is no exchange of energy between their atoms

   (c) the probabilities of atomic collision at the interface are the same in both bodies

   (d) the distribution of energy among the atoms is the same in both bodies

11 A lattice wave can be represented as a stream of particles called phonons each of energy proportional to

   (a) the velocity of propagation of the wave

   (b) the heat content of the body

   (c) the volume of the crystal

12 Figure (a) below is a graph of the forces between two atoms separated by a distance r. The attractive and repulsive forces can each be closely approximated by an expression of the form

(a) $\pm CR^n$ (b) $\pm CR^{1/n}$ (c) $\pm Cr^{-n}$ (d) $\pm C e^{-nr}$

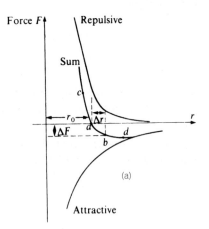

(a)

13 Figure (b) above is a graph of the potential energy of an atom versus displacement from its rest position.

(i)   The rest position is at

(a) a (b) b (c) c

(ii) If the turning points of the motion of the atom are at b and c, the straight line bc represents a plot of

(a) the kinetic energy

(b) the potential energy

(c) the total energy of the atom versus position

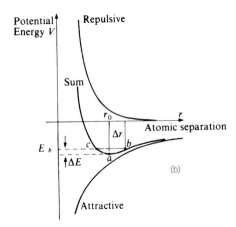

(b)

(iii) The energy ΔE represents

(a) the value of the kinetic energy of the atom

(b) the energy given to the atom to set it in motion

(c) the excess potential energy of the atom in motion

(iv) The mean position of the atom when in motion is the point

(a) a (b) b (c) c

(d) midway between b and c

(v) The change in this mean position with ΔE provides an explanation of

(a) thermal expansion

(b) the change of specific heat with temperature

(c) the thermal fluctuations in the length of a solid bar

14 To calculate the specific heat of a body we can begin by assuming that

(i)   all the atoms vibrate at a frequency ν

(a) true   (b) false

(ii) the energy in a lattice wave is quantized in units of

(a) hc/λ (b) $h\nu$ (c) $\exp(-h\nu/kT)$

(iii) the number of phonons in a mode of frequency ν is proportional to

(a) $h\nu/kT$ (b) $\dfrac{1}{1 - \exp(h\nu/kT)}$

(c) $1/[\exp(h\nu/kT) - 1]$

15 As a result of the calculation referred to in Question 14, it is found that the total heat content Q of a body containing N atoms is

$$Q = \frac{3Nh\nu}{\exp(h\nu/kT) - 1}$$

From this equation it can easily be shown that the specific heat C_v at constant volume is equal to

(a) $\dfrac{3Nh\nu}{T[\exp(h\nu/kT) - 1]}$

(b) $\dfrac{3Nh^2\nu^2 \exp(h\nu/kT)}{kT^2[\exp(h\nu/kT) - 1]^2}$

(c) $\dfrac{3N[\exp(h\nu/kT)(1-h\nu/kT)-1]}{[\exp(h\nu/kT)-1]^2}$

16 Dulong and Petit's Law states that the specific heats of all substances have nearly the same value when expressed per

(a) mol   (b) unit volume   (c) gramme

of the substance.

17 Dulong and Petit's Law is confirmed by the expression in the answer to Question 15, in the limit of

(a) high   (b) low

temperatures.

18 The term 'phase change' is used to denote changes such as

(a) changes of lattice structure of a solid as the temperature is changed

(b) the melting of a solid

(c) the change of specific heat with temperature

(d) the change of the amplitude and vibration of atoms with temperature

(e) the change in volume of a gas with pressure

(f) the vaporization of a liquid

(g) a chemical reaction

19 A phase change *usually* occurs

(a) at a specific temperature and pressure

(b) with the evolution or absorption of heat

(c) over a range of temperature at a given pressure

(d) spontaneously, without the absorption or evolution of heat

(e) without passing through stable intermediate states

20 When two phases of a single substance are in equilibrium with one another at a given pressure

(a) the heat content (enthalpy) in each phase is the same

(b) their temperatures are the same

(c) their thermodynamic probabilities are the same

(d) their free energies are the same

21 The latent heat associated with a change of phase is equal to the difference in

(a) enthalpy (heat content)

(b) thermodynamic probability

(c) free energy

between two phases.

22 The thermodynamic probability of a phase is a measure of

(a) the tendency to change to another phase

(b) the number of ways its atoms can be distributed among the available quantum states

(c) the tendency of its atoms to acquire more than the average amount of thermal energy.

23 The quantity $\int_0^{T_1} C_p(\alpha)\,dT$ is equal to

(a) the free energy

(b) the heat content, or enthalpy

(c) the internal energy

of a phase α at a temperature T_1

24 The first law of thermodynamics relates the amount of heat ΔQ put into a body, the work ΔW done by it on its surroundings and the change in its internal energy ΔU, according to the equation

(a) $\Delta U = \Delta Q + \Delta W$

(b) $\Delta Q = \Delta U + \Delta W$

(c) $\Delta W = \Delta U + \Delta Q$

25 Two phases that coexist but are not in equilibrium at a given temperature

(a) must eventually revert to equilibrium, albeit slowly

(b) can remain indefinitely in the non-equilibirum state if there is a barrier to change

26 The barrier to change referred to in Question 25 may be due to

(a) an intermediate state of higher energy

(b) insufficient energy per atom to promote the change

(c) insufficient time for the atoms to rearrange themselves

27 In the figure below, the potential energy of atoms in a two-phase material is plotted against some convenient dimension z that changes with the phase. The change between states with average energies E_A and E_B is possible because

(a) atoms can 'tunnel' through the barrier

(b) the height of the barrier fluctuates with time

(c) atoms can 'hop' over the barrier by acquiring extra thermal energy

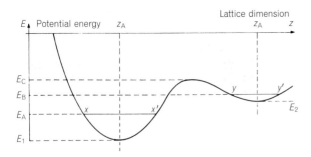

28 An activated state is

(a) one that results from a phase change over an energy barrier

(b) the unstable state of higher energy that constitutes the barrier

(c) any state with more than average thermal energy

29 The activation energy of the phase change illustrated in the figure in Question 27 is equal to

(a) $E_C - E_1$ (b) $E_A - E_1$

(c) $E_B - E_A$ (d) $E_C - E_A$

30 The rate of an activated phase change depends upon

(a) the number of atoms with sufficient energy to cross the barrier

(b) the frequency with which the atoms approach the barrier

(c) the relative thermodynamic probabilities of the two phases

(d) the height of the barrier

31 A metastable phase is

(a) fully stable against all disturbances

(b) unstable to any disturbance

(c) stable only against small disturbances

32 The following are examples of metastable phases in air at room temperature

(a) glass   (b) uncoated steel

(c) water vapour   (d) b.c.c. iron

33 Aluminium does not corrode in a normal atmosphere because

(a) no reaction with oxygen can occur

(b) the reaction with oxygen has too high an activation energy

(c) atmospheric oxygen can only diffuse very slowly through the oxide layer that is formed

34 Chemical reaction rates obey the same rate equations as phase changes

(a) true   (b) false

35 The slow passage of air through the walls of a sealed vacuum vessel is called

(a) thermal activation

(b) self-diffusion

(c) chemical diffusion

(d) thermal transpiration

36 In chemical diffusion the flux of atoms is directly proportional to

(a) the temperature

(b) the concentration gradient

(c) the number of atoms per unit volume

37 If activated diffusion occurs via an energy barrier of height E, the diffusion rate

(a) increases with E

(b) decreases with increasing E

(c) is independent of E

38 When diffusion is activated, the dependence of the diffusion coefficient on temperature is

(a) $\exp - E/kT$ (b) $\exp E/kT$ (c) T

(d) $1/T$ (e) zero

where E is a constant.

39 When diffusion is not activated the temperature dependence of the diffusion coefficient is

(a) T (b) $1/T$ (c) $\exp E/kT$

(d) $\exp - E/kT$ (e) zero

40 Diffusion is normally an activated process in

(a) solids   (b) liquids   (c) gases

Answers

1 (b), (c)	**2** (b)	**3** (a), (b)	**4** (b)
5 (c)	**6** (c)	**7** (b), (c)	**8** (c), (d)
9 (d)	**10** (a), (d)	**11** (a)	**12** (c)
13 (i) (a)	**14** (i) (a)	**15** (b)	**16** (a)
(ii) (c)	(ii) (a), (b)		
(iii) (b)	(iii) (c)		
(iv) (d)			
(v) (a)			
17 (a)	**18** (a), (b), (f)	**19** (a), (b), (e)	**20** (b), (d)
21 (a)	**22** (b)	**23** (b)	**24** (b)
25 (a)	**26** (a)	**27** (c)	**28** (b)
29 (a)	**30** (a), (b), (c), (d)	**31** (c)	**32** (a), (b)
33 (c)	**34** (a)	**35** (c)	**36** (b)
37 (b)	**38** (a)	**39** (e)	**40** (a)

8 | Crystal defects

8.1 Introduction

If a perfect crystal could be obtained, every unit cell would be occupied by an identical group of atoms, atoms would only exist in their expected sites, each atom would have its full quota of electrons in the lowest energy levels, and the atoms would be stationary. We have already seen that in practice the atoms vibrate about their lattice sites, and later, in Chapter 15, we shall see that thermal energy excites electrons into higher energy levels. In terms of the ideal perfect crystal, atomic vibrations and excited electrons can be regarded as defects. There are also many other types of defect in crystals, and all defects are of great importance to the materials scientist, because their presence affects many of the bulk properties of materials. The types of defects covered in this chapter are listed in Table 8.1, which also shows some of the applications in which they are important, and their relevance to some material properties.

Table 8.1

Crystal defects and their engineering relevance

Type of defect	Properties affected	Application or use
Impurity atoms:		
Substitutional	Electrical conductivity	Semiconductor diodes and transistors
	Mechanical strength	Solute hardening of alloys
Interstitial	Magnetic coercivity	
	Dielectric strength	
	Optical transparency	Colouring of glasses and plastics
Vacancies	Solid-state diffusion, mechanical creep	Annealing
Dislocations, twin boundaries	Plastic deformation (ductility)	Strain hardening
Grain boundaries	Dislocation movement	Mechanical hardening
	Electrical resistance	
	Magnetic coercivity	Strong permanent magnets
	Optical transparency	
Stacking faults	Mechanical strength	
	Electrical resistance	

Mechanical strength, discussed first in Chapter 9, is particularly affected by the presence of defects, sometimes to advantage, sometimes not. Semiconductor devices require extraordinarily low concentrations of many kinds of defect in order to work at all.

Defects may be classified according to their size. Atomic vibrations are small compared with atomic dimensions and are termed *subatomic defects*. The defects with which we are concerned in this chapter are, in ascending order of size, *point defects*, *line defects* and *planar defects*. Several kinds of point defects are described in Section 8.2. In Section 8.3, we explain how point defects enable atoms to diffuse readily in their presence. In Section 8.4 a description of line defects (called *dislocations*) is followed by an explanation of their role in the plastic flow of metals. Section 8.5 describes the common planar defects.

8.2 Point defects

Figure 8.1 illustrates the various point defects that can exist in a crystal. All of these are characterized by having atomic dimensions in all crystallographic directions. We see from this diagram that a foreign atom (the solute), whether it is an impurity atom or a deliberate alloying addition, can occupy one of two distinct positions in a crystal of the parent material (the solvent). If the solute substitutes for an atom of the solvent it is said to be a *substitutional atom,* while if it occupies a hole or interstice in the parent lattice it is called an *interstitial atom*. Which of these two alternatives a solute atom takes depends on its size relative to the solvent atoms – the smaller the solute atom the more likely it is to exist as an interstitial. As far as metals are concerned the most important interstitials are carbon, nitrogen and oxygen, which are small atoms of less than 0.08 nm radius.

The parent atoms surrounding a solute atom are usually displaced slightly to accommodate it. The lattice is therefore locally strained, and is

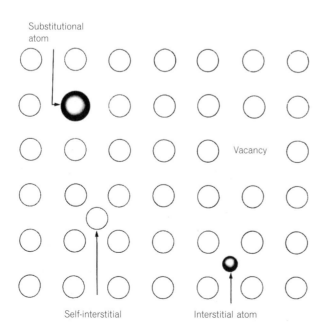

Substitutional atom

Vacancy

Self-interstitial

Interstitial atom

Fig. 8.1

The various point defects that can exist in a crystal.

Fig. 8.2

The formulation of (a) a Schottky defect and (b) a Frenkel defect.

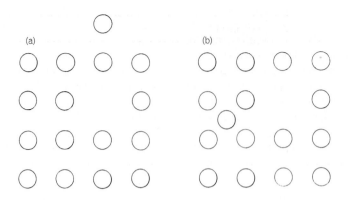

either in tension or compression in a radial direction, depending on the size of the solute atom. The mechanical energy stored in this distortion is a factor in determining the effect of the solute on the mechanical strength of a material. We will come back to interstitial and substitutional atoms in later chapters, where such factors as the amount of solute that can be accommodated by a solvent and the effect these defects have on the properties of the solvent will be discussed.

Now let us turn our attention to the vacant lattice site or *vacancy*. It is important to remember that, at all temperatures, the atoms in a solid are subject to thermal vibrations. This means that they continuously vibrate about their equilibrium positions in the lattice with an average amplitude of vibration that increases with increasing temperature.

At a given temperature there is always a wide spectrum of vibration amplitudes, and so occasionally in a localized region the vibrations may be so intense that an atom is displaced from its lattice site and a vacancy is formed. The displaced atom can either move into an interstice, in which case it is called a *self-interstitial*, or on to a surface lattice site. The vacancy self-interstitial is known as a *Frenkel defect* and the simple vacancy itself as a *Schottky defect* (Fig. 8.2).

Experimentally it has been found that at a given temperature there is an equilibrium concentration of Frenkel and Schottky defects. This means that the imperfect crystal must have a lower free energy than the perfect crystal. Reference to the previous chapter reminds us that a change in the free energy ΔG of a system is related to the changes in enthalpy ΔH and entropy ΔS by

$$\Delta G = \Delta H - T\Delta S$$

The energy E_D to form a defect, i.e. the thermal energy required to displace an atom from its lattice site and into an interstice (Frenkel defect) or on to a surface site (Schottky defect), is identified with ΔH in this equation. E_D is positive and therefore corresponds to an increase in the free energy of the crystal. On the other hand, the presence of defects increases the degree of disorder of the crystal and so raises the entropy, which favours a decrease in free energy. The relative contributions of these two terms to the free energy of the crystal are given as a function of the number of defects, n, in Fig. 8.3. We can see that as the number of defects increases the free energy falls to a minimum and then increases; the equilibrium number of defects n_e corresponds to the minimum free energy condition. By calculating the

Fig. 8.3

The concentration dependence of the enthalpy E_D and the entropy ΔS leads to a minimum in the free energy curve. The equilibrium number of defects n_e corresponds to the minimum free energy condition.

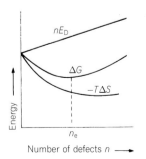

Table 8.2

The equilibrium concentration of defects as a function of temperature according to $C_e = A \exp(-E_D/kT)$ with $E_D = 1\,eV/atom$ $(1.6 \times 10^{-19}$ J/atom) and $A = 1$

Temperature K	c_e
0	$10^{-\infty}$
200	6.5×10^{-26}
400	2.5×10^{-13}
600	4.2×10^{-9}
800	5.0×10^{-7}
1000	9.2×10^{-6}
1200	6.1×10^{-5}

minimum free energy condition as a function of temperature, Boltzmann obtained the following expression for the equilibrium number, n_e, and thus the equilibrium concentration, c_e, of defects

$$\frac{n_e}{N} = c_e = A \exp\left(\frac{-E_D}{kT}\right) \qquad (8.1)$$

where N is the number of atoms in the crystal, A is a constant often taken as unity, T is the absolute temperature, and k is Boltzmann's constant. Values of c_e calculated from this equation with $A = 1$ and $E_D = 1\,eV/atom$ $(1.6 \times 10^{-19}\,J/atom)$ are listed in Table 8.2.

The energies of formation, and hence the equilibrium concentration at a given temperature, of Frenkel and Schottky defects differ from material to material. If we look at metals we find that E_D for Schottky defects is relatively low. The value of 1 eV/atom used to calculate the concentrations given in Table 8.2 is typical of silver, gold and copper, which have melting points of 1234 K, 1336 K and 1356 K respectively. With this information we conclude that, although the equilibrium Schottky defect concentration increases rapidly as the temperature is raised, even at temperatures approaching the melting point only about one lattice site in $10^5 - 10^4$ is vacant. The energy of formation of Frenkel defects in a metal is much greater than 1 eV/atom and therefore the concentration of these defects under equilibrium conditions is even less than for Schottky defects and is generally negligible.

In contrast to metals, in ionic crystals there may be a preponderance of either Frenkel or Schottky defects depending on which has the smaller energy of formation. Frenkel defects are more likely to be important in crystals with open lattice structures that can accommodate interstitials without much distortion. These crystals have structures with low coordination numbers, for example the zinc blende and zinc oxide structures.

Table 8.3

Point defects in ionic crystals

Crystal	Structure	Dominant defect	Energy of formation	
			(eV atom^{-1})	(kJ mol^{-1})
CdTe	ZnS	Frenkel	1.04	100
AgI	ZnO	Frenkel	0.69	67
NaCl	NaCl	Schottky	2.08	201
NaBr	NaCl	Schottky	1.69	163

Fig. 8.4

Self-diffusion taking place by the
vacancy mechanism.

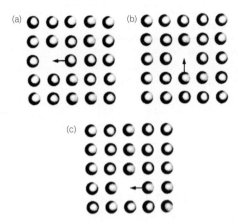

In contrast we would expect to find Schottky defects in crystals with high
coordination numbers such as the sodium chloride structure (Table 8.3).
Where there is a large difference in size between the cations (the positive
ions) and the anions (the negative ions), Frenkel defects are usually formed
by the displacement of the smaller ion, which is most often the cation.

Many important solid-state processes to be discussed in later chapters,
such as recovery, recrystallization and precipitation, proceed by atoms
diffusing from one site to another in the crystal lattice. One of the reasons
for our interest in the equilibrium concentration of vacancies is that
diffusion, both of atoms of the host material and of substitutional atoms,
occurs by a *vacancy mechanism*, i.e. atoms diffuse by jumping into vacant
sites in the lattice (Fig. 8.4). Thus self-diffusion of the host atoms results in
vacancies apparently moving continually from site to site. In the next
section we will see that the activation energy for diffusion, E_0 [see Eq.
(7.33)], consists of two contributions, the energy, E_v, of formation of a
vacancy, and the energy of its motion E_j.

8.3 Vacancy-assisted diffusion of impurities and self-diffusion

In Chapter 7 we explained that the presence of defects in a solid enhances
the rate at which chemical diffusion takes place. In Chapters 9 and 10 we
will find how the distribution of chemical species in a solid affects its
mechanical properties. Here we explore diffusion further.

The driving force for a chemical diffusion process, as was discussed in
Section 7.22, is a gradient in the concentration of the diffusing atoms. The
diffusion flux, F, according to Fick's first law [Eq. (7.32)] is

$$F = -D\frac{dN}{dx} \tag{8.2}$$

where dN/dx is the concentration gradient and D is the diffusion coefficient, measured in $m^2\,s^{-1}$.

We can construct a simple model from which Fick's first law may be
deduced (see Box 8.1), if some straightforward assumptions are made
about the diffusion process within a crystalline solid. These are that the
individual molecules move in a series of individual 'jumps', which occur at

Box 8.1 Fick's law of diffusion in a crystal

Suppose that a species of impurity within a crystal has a concentration $N(x)$ that decreases uniformly in the x-direction. The impurity atoms must therefore be diffusing along the x-direction.

Following any instant in time, one-half of the diffusing atoms in a y–z plane at a location x will jump in the +x-direction, and one-half in the −x-direction. This is because of the random nature of thermal energy. If we wish to calculate the flow of atoms across unit area from the plane x to the next atomic plane at $(x + a)$, we must multiply the number $\frac{1}{2}N(x)$ by the frequency of a jump of size a. It is reasonable to suppose that this frequency, which we shall call J, is independent of location x.

Then $JN(x)/2$ is the flow of atoms in the +x-direction. But at the location $(x + a)$, there are $N(x + a)$ atoms, half of which are jumping with the same frequency J in the −x-direction. The net flow is the difference between the flow from x to $(x + a)$ and that in the reverse direction, which is just

$$\frac{J}{2}[N(x) - N(x + a)] = \frac{J}{2}\left[a\frac{dN(x)}{dx}\right]$$

Because a and J are independent of x, this is just proportional to $dN(x)/dx$, as Fick's empirical law states. The diffusion coefficient D is proportional to both J and a.

a fixed average frequency, and in random directions – Fig. 8.4 illustrates the last point. (In gaseous diffusion, the 'jumps' would be the periods of free flight between collisions.) In a solid the jumps require the expenditure of some energy, for, in order to move, the atom must briefly compress or stretch the bonds it makes to its neighbours (Fig 8.5), raising the energy stored in those bonds. In this case, the frequency with which the jumps occur must depend upon the minimum energy E_0 that the atom must acquire in order to move, and which is therefore referred to as an 'energy barrier' over which the atom must jump. The mathematics developed in Chapter 7 for reaction rates is applicable here, for the diffusing atoms must be 'activated', i.e. they must gain enough energy to cross the energy barrier. When the temperature is T, at any one time only a fraction of them very nearly equal to $\exp(-E_0/kT)$ have this much energy or more, so that the average frequency of performing the jumps is proportional to this factor. The outcome is that the diffusion coefficient D is proportional to the same factor. Thus

$$D = D_0 \exp(-E_0/kT) \tag{8.3}$$

where the factor D_0 is normally a constant. This equation, introduced in the previous chapter, is almost universally applied to diffusion mechanisms in solids.

8.3.1 Substitutional impurity diffusion: a mechanism involving vacancies

The value of the diffusion coefficient D of any atomic species depends on the detailed mechanism by which the diffusion occurs. If we were to imagine an ideal crystal in which there were no vacancies or defects, it would obviously be difficult for foreign atoms to penetrate the crystal, and they could only do so into interstitial positions. Thus diffusion of substitutional impurities normally differs significantly from the diffusion of interstitials.

Fig. 8.5

The diffusion of an impurity atom through a crystal: (a) initial position with adjacent vacancy; (b) the activated step; (c) new position exchanged with vacancy; (d) potential energy variation of impurity atom during the diffusion process.

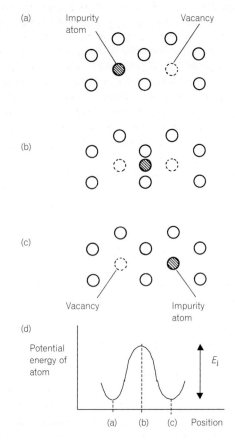

There will always be a few vacancies in the lattice because of thermal agitation, where an atom has moved from its regular site in the lattice into an interstitial position (a *Frenkel defect*). If this interstitial atom diffuses away from the vacancy, the latter becomes a *Schottky defect*. Thus a substitutional impurity can diffuse most readily through the crystal if a neighbouring site is occupied by a vacancy. In diffusing, therefore, the impurity atom changes places with a neighbouring vacancy, effectively jumping from vacancy to vacancy as illustrated in Fig. 8.5. Whether it is able to move from one vacancy to the next will depend on the availability of a vacant neighbouring site, and the probability of this being available will be the probability of an adjacent atom jumping out of its lattice site to form a Frenkel defect. We therefore have two *activated* processes involved sequentially in the diffusion mechanism. The first excites a solvent atom out of its normal position to create a vacancy, requiring an activation energy E_v, and the second activates the jump of the impurity atom from its initial site, over the energy barrier, and into the vacant site, requiring an energy E_j, as illustrated in Fig. 8.5(d). The joint probability of both steps occurring is the product of their individual probabilities, which are respectively proportional to $\exp(-E_v/kT)$ and $\exp(-E_j/kT)$. So the diffusion coefficient, D, has a strong temperature dependence, given by

$$D = D_0 \exp\left[-\frac{(E_v + E_j)}{kT}\right] = D_0 \exp\left[-\frac{E_0}{kT}\right] \quad (8.4)$$

where D_0 is a constant, and $E_0 = E_v + E_j$ is therefore the activation energy for diffusion. In practice, the value of the measured activation energy is modified by the effects of electrostatic attraction between the impurity atom and charged defects. This leads to a small lowering of the energy necessary to promote the diffusion. Because of such interactions between defects, the diffusion coefficient may change its value somewhat whenever the concentration of impurities rises to the point where individual impurity atoms are separated on average by only a few lattice constants.

Diffusion of impurity atoms in semiconductor crystals to create transistors has become one important application of diffusion in solids, and is dealt with in Chapter 16. Impurities from Groups 13 and 15 of the periodic table are used, and they diffuse readily at temperatures near 1000°C in silicon, with approximately the same average total activation energy $E_0 \approx 3.5\,\mathrm{eV}$ and with about the same value of $D_0 = 10^{-4}\,\mathrm{m}^2\,\mathrm{s}^{-1}$.

8.3.2 Defect-assisted diffusion in solids

The vacancy-assisted mechanism for diffusion that is described above is by far the most important of all diffusion mechanisms for substitutional impurities in all types of crystalline materials, because it has a low activation energy E_0. If the concentration of vacancies in a material changes for any reason, the diffusion coefficient of the impurity is affected. For example, impurities in ionic crystals often affect D, because the introduction of, say, a doubly charged positive ion such as Ca^{2+} in place of a monovalent one, in, for example, NaCl, means that two Na^+ ions must be expelled to maintain charge neutrality – thus creating an extra vacancy, which assists in diffusion. These effects are much greater in ionic solids than they are in alloyed metals, where the freely mobile electrons modify the activation energy.

Other defects that can assist diffusion are, in approximate order of importance, interstitials, divacancies (i.e. vacancy pairs), grain boundaries and surfaces. It is important to note that quite small differences between the values of the total activation energy E_0 for different mechanisms cause only slight differences in the exponent in Eq. (8.4), but make a very large difference to the diffusion coefficient, while the constant D_0 varies much less. Hence rates of diffusion vary widely for different mechanisms. For example, at a temperature of 300 K, where the value of kT is 0.026 eV, the addition of 0.06 eV to the activation energy E_0 results in a *multiplication* of D by a factor

$$e^{\frac{0.06}{0.026}} = e^{2.3} = 10.05$$

while an increase of 0.12 eV in E_0 results in multiplication of D by a factor of 100, for the same D_0!

Unlike substitutional impurities, interstitials, being small, can often diffuse readily without the aid of vacancies, and hence they usually diffuse faster than substitutional impurities. Carbon in iron is a classic example – at a temperature of 1000 K it diffuses as much as 10^6 times faster than the self-diffusion rate of iron atoms.

Diffusion along *grain boundaries*, which will be introduced in Section 8.5, forms a special case worthy of attention, and is dealt with there.

8.3.3 Prediction of depth profiles for diffused dopant concentrations

We now show how to predetermine the depth of penetration of an impurity into the surface of a material by diffusion, and to predict the profile of the impurity concentration plotted versus depth.

Referring to Fig. 8.6, consider a rectangular elementary volume of unit cross-sectional area, and thickness dx. Let the flux of atoms entering it from the left-hand side be $F(x)$ and that leaving from the right-hand side be $F(x + dx)$. The rate of change in the number of atoms contained in the elementary volume must equal the difference between the flux in and the flux out, which is just $F(x) - F(x + dx)$. Now if N is the average concentration, then the number of atoms in the elementary volume will be $N\,dx$. The increase in this number with time can be written as

$$\frac{\partial(N\,dx)}{\partial t} = \frac{\partial N}{\partial t}\,dx = F(x) - F(x + dx)$$

But $F(x + dx) - F(x) = (dF/dx)\,dx$, so that

$$\frac{\partial N}{\partial t} = -\frac{\partial F}{\partial x} \tag{8.5}$$

where the negative sign reflects the fact that if the flux $F(x)$ into the volume is greater than flux $F(x + dx)$ leaving it, the gradient of flux in the positive x-direction must be negative.

Substituting for F from Eq. (8.2) we have

$$\frac{\partial N}{\partial t} = -D\frac{\partial^2 N}{\partial x^2} \tag{8.6}$$

which is the *diffusion transport equation.*

The appropriate solution of this differential equation will depend upon the *boundary conditions*, e.g. the values of $N(x, t)$ at particular positions x at time $t = 0$.

One possible solution of this equation is

$$N(x, t) = At^{-1/2} \exp(-x^2/4Dt) \tag{8.7}$$

Fig. 8.6

An elementary slice for derivation of the diffusion transport equation.

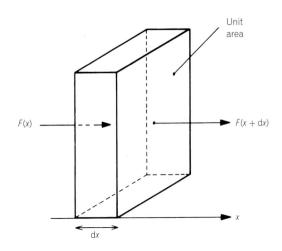

as may be confirmed by differentiating Eq. (8.7) and substituting into Eq. (8.6). Although this is not the most general solution, it is useful for many purposes, and will prove useful in Chapter 16.

A more general solution will be the sum, or integral, over many such solutions. This may be written as

$$N(x, t) = \frac{A}{2\sqrt{\pi Dt}} \int_{-\infty}^{+\infty} f(x') \exp[-(x - x')^2/4Dt]\, dx' \qquad (8.8)$$

where x' is a dummy variable and $f(x')$ is the distribution of the concentration of diffusing atoms at the starting time, at $t = 0$. Clearly the precise form of $f(x')$ must be known before the integral in Eq. (8.8) can be evaluated. The simplest case to consider is a solid of unlimited thickness containing no impurities at time $t = 0$, into which an impurity is to be diffused from a flat surface at the plane $x = 0$. The concentration at the surface can be maintained constant, for example by a vapour of the diffusing atoms. If the surface concentration $N(0, t)$ is held at the value N_s, then that found at a depth x after a time t is given by

$$N(x, t) = N_s \operatorname{erfc}(x^2/4Dt) \qquad (8.9)$$

where the function erfc is called the *complementary error function* whose value is tabulated in mathematical tables.

The Gaussian function in Eq. (8.7) and the error function in Eq. (8.9) are compared in Fig. 8.7, where it can be seen that the latter drops more rapidly with increasing depth x, which is proportional to the argument $u = x/2(Dt)^{1/2}$.

The Gaussian function gives a close approximation to what results after a fixed amount of impurity is deposited as a thin layer on the surface of the solvent material, following which the temperature is raised to speed

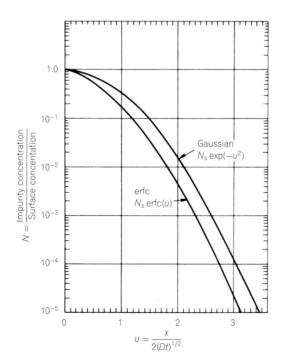

Fig. 8.7

The erfc(u) error function and exp($-u^2$) Gaussian diffusion profiles plotted after a time t against u, which is proportional to depth $x/t^{1/2}$.

the diffusion process. The way in which the distribution of the impurity concentration changes with time is described in connection with a practical situation in Section 16.10.

8.4 Line defects

Line defects are long in one direction, while measuring only a few atomic diameters at right angles to their length. The only defect of this class is the *dislocation*. Knowledge of dislocations has allowed us to reconcile the rate of crystal growth from a vapour predicted by classical theory with the significantly faster growth rates obtained experimentally. It also enables us to account for the discrepancy between the stress theoretically needed to deform a perfect crystal plastically and the actual stresses measured on 'ordinary', i.e. imperfect crystals, which are much lower. In fact dislocations were first postulated in order to account for the latter; we will follow a similar path in our study of dislocations, so first we must look at the macroscopic aspects of plastic deformation.

8.4.1 Slip planes and slip directions

Consider Fig. 8.8(a) which shows a crystal loaded as in a tensile test. It must be appreciated that the applied tensile force can be resolved into components normal and parallel to any plane, such as XX', in the specimen. This is illustrated in Fig. 8.8(b) where the applied force, P, is resolved into a normal, tensile, component P_T and parallel component P_S (i.e. a shear force) on the plane XX'. This means that even a simple tensile force produces *shear forces*, and hence *shear stresses*, in a specimen. The value of the shear stress varies with the orientation of the plane and is a maximum for the planes at 45° to the tensile axis.

If we deform a single crystal plastically, that is, we increase the tensile force until there is a permanent shape change, and then closely examine the surface under a microscope, we see a series of parallel black lines such as those shown in Fig. 8.9. These lines are small steps on the surface of the crystal, called *slip lines* or slip steps. On increasing the tensile force still further, so causing more plastic strain, the number and or height of the slip

Fig. 8.8

Block slip mechanism for plastic deformation: (a) crystal before testing; (b) resolution of the applied force; (c) crystal after testing, i.e. plastically deformed; (d) a slip line (step) magnified showing the slip direction and the slip plane.

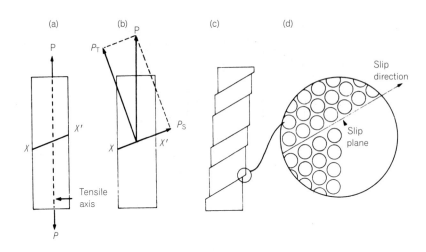

Fig. 8.9

Slip lines on the surface of a deformed specimen.

lines increases. It is as if whole blocks of the crystal were slipping over one another under the action of the resolved shear stresses as shown in Fig. 8.8(c). The fact that the slip lines are parallel and that they do not necessarily correspond to slip along the planes of maximum shear indicates that the slip process is closely connected with the crystallography of the crystal. This is indeed the case, as shown in the magnified diagram of Fig. 8.8(d); slip only occurs on preferred crystallographic planes, *the slip planes*, and in certain directions, *the slip directions*. The combination of a slip plane and a slip direction constitutes a *slip system*. As we will see in the following discussion on the slip systems in the f.c.c., c.p.h. and b.c.c. crystal structures, the slip direction is usually the direction in which the atoms are most closely packed and the slip plane the most densely packed plane containing that direction.

(a) *FCC.* The ⟨110⟩ directions are the closest packed directions in the face-centred cubic structure and are therefore the slip directions. The ⟨110⟩

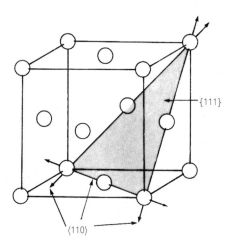

Fig. 8.10

One of the four {111} slip planes with its three ⟨110⟩ slip directions in the f.c.c. structure.

Fig. 8.11

The single slip plane (the basal plane) with its three slip directions giving three slip systems in the c.p.h. structure.

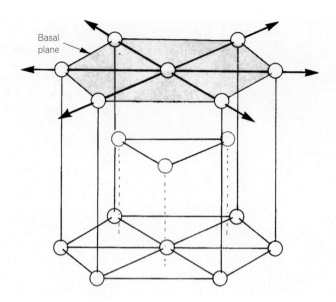

directions lie in the most densely packed planes, which are the close-packed {111} planes (Fig. 8.10). Thus, the four {111} planes are the slip planes, and as each plane contains three of the family of ⟨110⟩ directions there are altogether $3 \times 4 = 12$ slip systems.

(b) *CPH*. The slip directions are the closest packed directions. Three such directions lie in the close-packed basal plane, giving only $3 \times 1 = 3$ slip systems. The three slip directions are shown in Fig. 8.11.

(c) *BCC*. The most closely packed directions in the b.c.c. structure are the ⟨111⟩, which are therefore the slip directions. However, unlike the f.c.c. and c.p.h. structures, the b.c.c. structure is not constructed from close-packed planes. Nevertheless, there are three planes in the b.c.c. structure with relatively high packing densities. These are, in descending order of packing density, the {110}, {100} and {112} planes. However, the {100} planes do not contain the ⟨111⟩ directions, so only the {110} and the {112} are slip planes. There are six {110} planes with two of the ⟨111⟩ directions in each, giving $6 \times 2 = 12$ slip systems of the type {110} ⟨111⟩ (Fig. 8.12).

Fig. 8.12

One example from each of the three families of planes with high packing density, namely {100} {110} and {112} in the b.c.c. structure. The two slip directions of type ⟨111⟩ for the [110] slip plane are also marked.

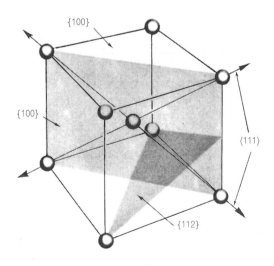

There are also twelve systems of the type {112} ⟨111⟩ made up from one of the ⟨111⟩ directions in each of the twelve {112} planes.

So far we have considered plastic deformation to take place by the shearing of complete blocks of a crystal on the slip planes and in the slip directions. From a knowledge of the bonding energies between atoms it is possible to estimate the magnitude of the shear stress necessary to cause plastic flow by this mechanism. These estimates give stresses of the order of $G/30$, where G is the shear modulus. In practice, shear stresses of only about $G/1000$ are required to produce plastic flow in ordinary crystalline materials so the mechanism must be different from that of simple block shearing of complete planes over one another. In the next section we will see how plastic deformation really occurs by the movement of dislocations.

8.4.2 Geometry and Burgers vector of dislocations

An extra half-plane of atoms inserted between the planes of atoms in a crystal can be accommodated as shown in Fig. 8.13. If the top of the crystal, above the line XX', is viewed in isolation it appears perfect. Similarly, the bottom of the crystal below the line XX' is perfect. Only when the two halves of the crystal are together is there a defect, the defect running in a line perpendicular to the page through the symbol ⊥. This line defect is an *edge dislocation* and by convention is represented by the symbol ⊥. The edge dislocation is able to move through a crystal causing slip at much lower stresses than those required for the block slip mechanism. We can see this in a general way by referring to Fig. 8.14. We are going to concentrate on the atoms marked 1, 2, 3, 4 and 5. These atoms are displaced from their equilibrium positions and therefore their bond energies are altered (see Fig. 7.5). In the unstressed condition the atom pairs 1–2 and

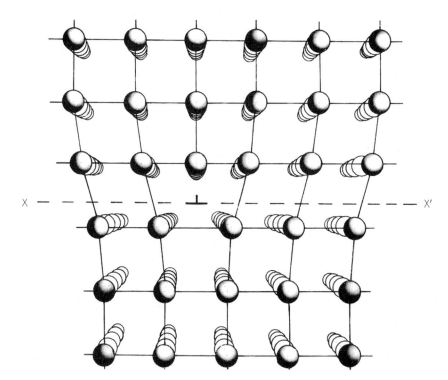

Fig. 8.13
An edge dislocation.

Fig. 8.14

Movement of an edge dislocation
under the influence of a shear
stress: (a) unstressed; (b) stress τ
applied and the dislocation moves
slightly to the left; (c) stress τ
increased and the dislocation
moves one atomic spacing to the
left; (d) this procedure is repeated
until the dislocation leaves the
crystal forming a step of height b.

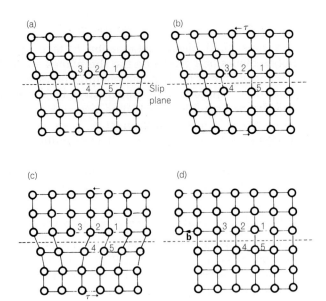

Fig. 8.14

Movement of an edge dislocation under the influence of a shear stress: (a) unstressed; (b) stress τ applied and the dislocation moves slightly to the left; (c) stress τ increased and the dislocation moves one atomic spacing to the left; (d) this procedure is repeated until the dislocation leaves the crystal forming a step of height b.

2–3 are closer together and the pairs 1–5 and 3–4 are further apart than the normal equilibrium spacing [Fig. 8.14(a)]. On applying a shear stress τ [Fig. 8.14(b)] the distances apart of the atoms change; atom 2 moves a little to the left so that it is nearer atom 3 and further from atom 1, at the same time the distance between atoms 1–5 is reduced while between atoms 3–4 it is increased. The result is that there is an increase in energy of the 2–3 and 3–4 bonds (the interatomic distances between these atoms have become further removed from the equilibrium value) and a decrease in the 1–2 and 1–5 bond energies (the interatomic distances have become nearer the equilibrium value). The increase in energy from the former is greater than the decrease in energy because of the latter effect, the difference being supplied by the shear stress. On increasing the stress further, atom 2 continues to move to the left until eventually it bonds more strongly with atom 4 than does atom 3 and the dislocation has effectively moved one interatomic distance to the left [Fig. 8.14(c)]. This procedure is repeated until the dislocation leaves the crystal giving a step of height b [Fig. 8.14(d)]. If a number of dislocations move along the same slip plane the step height increases to give a configuration such as that shown in Fig. 8.8(d), i.e. a visible slip line. From this discussion it is clear that the movement of dislocations is similar in many ways to block slip (e.g. dislocations move on the slip plane and give slip lines) but requires smaller stresses as only a few bonds are being altered or broken at any one time.

The magnitude and direction of the slip resulting from the motion of a single dislocation, **b** in Fig. 8.14, is called the *Burgers vector*. We see that for the edge dislocation of Fig. 8.14 the Burgers vector is perpendicular to the line of the dislocation. This is a special case and dislocations exist with the Burgers vector at all angles to the dislocation. Another special case is when b is parallel to the dislocation; this is the *screw dislocation* and is illustrated in Fig. 8.15. As for the edge dislocation, under the influence of a shear stress the screw dislocation moves in the slip plane but this time the direction of motion of the dislocation is perpendicular to the Burgers vector. The motions of an edge, screw, and an arbitrary dislocation loop are

Fig. 8.15

A screw dislocation. Note that the Burgers vector is parallel to the dislocation.

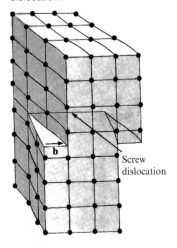

Screw dislocation

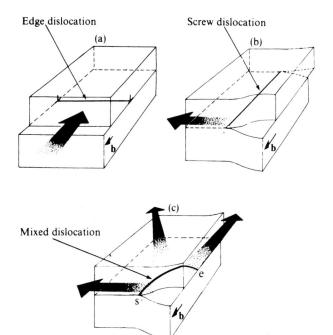

Fig. 8.16

The motion of (a) an edge
dislocation; (b) a screw dislocation;
(c) a mixed dislocation; s and e are
screw and edge components
respectively.

compared in Fig. 8.16. The arbitrary dislocation loop has an edge component (**b** perpendicular to the dislocation), a screw component (**b** parallel to the dislocation), and a large mixed component (angle between **b** and the dislocation greater than 0° but less than 90°).

So far we have defined the Burgers vector of a dislocation in terms of the atomic displacement caused during slip. This is rather restrictive because, as pointed out earlier, dislocations play an important role in processes other than plastic deformation. A general method for obtaining the magnitude and direction of the Burgers vector is to determine the discontinuity due to the presence of the dislocation by drawing what is called a Burgers circuit. First a closed circuit is drawn around the dislocation by jumping from atom to atom in a clockwise direction [Fig. 8.17(a) and (b)]. The same number of jumps is then made in the same directions in a perfect crystal. This time the circuit does not close, i.e. it does not return to the atom at which it started, and the vector needed to complete the circuit is the Burgers vector of the dislocation [Fig. 8.17(c) and (d)]. It will be noted, of course, that this gives the magnitude and direction of the Burgers vector but, like all vectors, it has no fixed position. The position of the arrow in the diagram is not significant.

All dislocations are line defects and therefore only measure a few atomic diameters at right angles to their length. We define the *width*, w, of a dislocation as the distance over which the displacement of atoms is greater than $b/4$. When w is one or two atomic spacings the dislocation is said to be narrow. On the other hand a wide dislocation has a width of several atomic spacings (Fig. 8.18). Typically metals have wide dislocations, whereas covalently bonded materials have narrow dislocations.

We have already noted that the stress to move a dislocation is several orders of magnitude less than that for block slip. The reader will learn in the next chapter how other defects and second phase particles hinder

Determination of the Burgers
vector by means of a Burgers
circuit. (a) and (b) show the
circuits around an edge and a
screw dislocation respectively.
(c) and (d) are the corresponding
circuits in a perfect crystal.

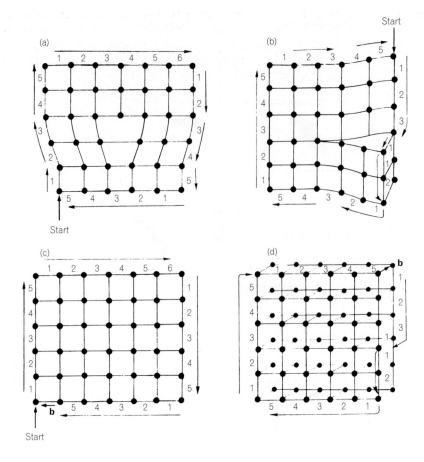

The width of a dislocation: (a) a
wide dislocation; (b) a narrow
dislocation.

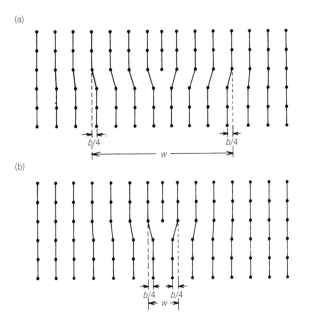

dislocation motion. It follows that the minimum stress for the type of motion shown in Fig. 8.16 is the stress to move a dislocation in an otherwise perfect crystal. This intrinsic lattice friction stress is known as the *Peierls–Nabarro* stress. The significance of the width of a dislocation is that it determines the magnitude of the Peierls–Nabarro stress, τ_{PN}, as can be seen from:

$$\tau_{PN} \sim G\exp(-2\pi w/b) \tag{8.10}$$

τ_{PN} is very sensitive to w as w appears in an exponential term. The wider the dislocation the lower is τ_{PN} and the easier it is to move the dislocation. We can also see from this equation that a small Burgers vector, b, leads to a low τ_{PN} value. This is one of the reasons why the dislocations responsible for plastic deformation often have the smallest b compatible with the crystal structure, or in terms of macroscopic slip, why the slip direction is generally the closest packed direction.

8.4.3 Cross-slip and climb

The dislocation motion described in the previous section and illustrated in Figs 8.14 and 8.16 is usually referred to as *glide*. Glide may be defined as the motion of a dislocation along a slip plane containing both the dislocation and its Burgers vector. In the case of an edge dislocation the Burgers vector is perpendicular to the dislocation, thereby specifying a single plane. It follows that an edge dislocation is only able to glide on this specified slip plane. In contrast, the Burgers vector is parallel to a screw

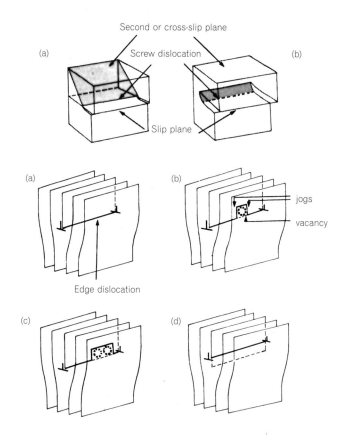

Second or cross-slip plane
Screw dislocation
Slip plane
Edge dislocation
jogs
vacancy

Fig. 8.19

Cross-slip of a screw dislocation: (a) screw dislocation gliding on the first plane; (b) screw dislocation has cross-slipped and is now gliding on the second, or cross-slip, plane.

Fig. 8.20

Climb of an edge dislocation: (a) the edge dislocation before climb; (b) a vacancy diffuses to the dislocation producing jogs; (c) the jogs move apart as more vacancies arrive at the dislocation; (d) the dislocation has climbed one atomic spacing.

dislocation and so a unique plane is not defined. This means that a screw dislocation can glide in any slip planes that have a common slip direction. This gliding from one slip plane to another, which is known as *cross-slip*, enables screw dislocations to bypass obstacles to their motion (Fig 8.19).

Although edge dislocations cannot avoid obstacles by cross-slip they can avoid them by moving normal to their slip planes by the process of *climb*. The most common climb mechanism is the diffusion of vacancies to an edge dislocation and this is illustrated in Fig. 8.20. The first vacancy to reach the dislocation produces a step, the sides of which are termed *jogs* [Fig. 8.20(b)]. As more vacancies diffuse to the dislocation the jogs move apart [Fig. 8.20(c)] until eventually the whole dislocation has climbed one atomic spacing [Fig. 8.20(d)]. This process can be repeated many times so the edge dislocation can climb a number of atomic spacings. Climb is assisted by elevated temperatures, which speed up the rate of vacancy diffusion.

8.4.4 Dislocation energy

The regularity of a crystal lattice results from each atom taking up a position that minimizes its potential energy; it follows that a dislocation must represent a raising of the potential energies of all the atoms whose positions are affected by its presence. Thus an energy may be ascribed to a dislocation that, physically, is the strain energy built into the crystal structure by displacement of the atoms from their regular positions.

We may estimate this energy by considering a screw dislocation in a cylinder of crystal of length l, as shown in Fig. 8.21(a). At a radius, r, the deformation in a thin annulus of thickness dr is given by the magnitude, b, of the Burgers vector so that the shear strain, γ, is $b/2\pi r$ [Fig. 8.21(b)], and using Hooke's law for shear, the average stress, τ, will be given by $Gb/2\pi r$, where G is the elastic shear modulus.

It is interesting to note from this that the stress field of a dislocation is inversely proportional to distance from the dislocation and may, therefore, be described as long range, compared with the interatomic forces. Thus dislocations will interact with each other at quite large distances apart on the atomic scale. In fact at a distance of 10,000 atoms from the core the stress field may still have a value of the order of $10^{-5} G$.

The elastic strain energy dw of a small volume element dv is given by $\frac{1}{2}\tau\gamma\,dv$ so we have

$$dw = \frac{1}{2}\tau\gamma\,dv = \frac{1}{2}\frac{Gb}{2\pi r}\cdot\frac{b}{2\pi r}\,dv$$

$$= \frac{1}{2}G\left(\frac{b}{2\pi r}\right)^2 dv$$

Fig. 8.21

(a) Model for the calculation of the energy of a screw dislocation.
(b) Opening out the cylinder we can see that the shear strain $\gamma = \tan\theta = b/2\pi r$.

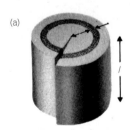

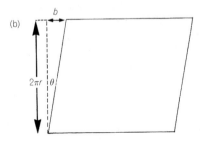

The volume of the annular element is $2\pi r/dr$, hence

$$dw = \frac{Gb^2 l}{4\pi}\frac{dr}{r}$$

The elastic strain energy contained within a cylinder of radius R around the dislocation is obtained by integrating this equation up to the limit R. However, the lower limit of integration is not taken as zero for two reasons: (1) the calculation assumes the material is an isotropic continuum, but in the region near to the dislocation this assumption is unrealistic and it is necessary to consider the displacements of, and the forces between, individual atoms; and (2) the strains near the dislocation are large and Hooke's law, on which the calculation is based, fails at large strains. Thus the lower limit of integration is taken as some small value, r_0 (which is often taken to be equal to b). The region inside r_0 is referred to as the *dislocation core*. We therefore have

$$E = \frac{Gb^2 l}{4\pi}\int_{r_0}^{R}\frac{dr}{r} + E_c l$$

where $E_c l$ is the strain energy within the radius 0 to r_0, i.e. the core energy. The calculation for the core energy is complex but a reasonable estimate is $Gb^2 l/10$. Thus, on integrating to obtain the elastic strain energy we get for the total strain energy

$$E = \frac{Gb^2 l}{4\pi}\ln\frac{R}{r_0} + \frac{Gb^2 l}{10}$$

Because generally $R \gg r_0$, the logarithmic term in this expression only varies slowly with R/r_0. As an approximation, $\ln(R/r_0)$ may be taken as 4π, and so

$$E \approx Gb^2 l + \frac{Gb^2 l}{10} \tag{8.11}$$

Two important features of this formula are: (1) the total energy is proportional to b^2; and (2) the core energy is only one-tenth of the elastic strain energy.

For typical metals such as chromium or copper $G \approx 5 \times 10^{10}\,\mathrm{N\,m^{-2}}$, and taking $b = 2.5 \times 10^{-10}$ m we obtain for the energy of a screw dislocation a value in the region of $5.5\,\mathrm{eV\,atom^{-1}}$. An equation similar to Eq. (8.11) can be derived for an edge dislocation and would give for the energy of an edge dislocation about $8\,\mathrm{eV\,atom^{-1}}$. These energies are considerably greater than the formation energies of Schottky defects in metals. We have seen how Schottky defects are an equilibrium feature of a crystal because the increase in entropy associated with them outweighs the increase in enthalpy and leads to a decrease in free energy of the crystal (Fig. 8.3). Dislocations also increase the entropy, but in this case the entropy change does not outweigh the very large enthalpy increase of 5.5–$8\,\mathrm{eV\,atom^{-1}}$, and consequently the free energy of the crystal is increased. Therefore, dislocations are non-equilibrium defects and if possible a crystal will decrease its dislocation density in order that its free energy may be reduced.

8.4.5 Super and partial dislocations

An established nomenclature for describing dislocations has been developed in terms of the Burgers vector. The dislocations considered so far have had Burgers vectors equal to one lattice spacing, see for example Fig. 8.17, and are called *unit* dislocations. A unit dislocation is a specific case of a *perfect* dislocation, which is a dislocation whose Burgers vector is an integral number of lattice spacings. Conversely, a dislocation is *imperfect* if its Burgers vector is not an integral number of lattice spacings. A *partial* dislocation is an imperfect dislocation with its Burgers vector less than unity, while a *super* dislocation can be perfect or imperfect with a Burgers vector greater than unity.

Sometimes a dislocation of Burgers vector $\mathbf{b}_1$ splits into two, or more, dislocations of Burgers vectors $\mathbf{b}_2$, $\mathbf{b}_3$, $\mathbf{b}_4$, etc. The reverse can also occur, namely dislocations of Burgers vectors $\mathbf{b}_2$, $\mathbf{b}_3$, $\mathbf{b}_4$, etc., can combine to form a single dislocation of Burgers vector $\mathbf{b}_1$. This behaviour can be explained in terms of the energies of the dislocations involved. We know that the energy of a dislocation is proportional to the Burgers vector squared [Eq. (8.11)], therefore a dislocation will split, if it is geometrically possible, when

$$\mathbf{b}_1^2 > \mathbf{b}_2^2 + \mathbf{b}_3^2 + \cdots \tag{8.12}$$

Alternatively, a number of dislocations will combine if

$$\mathbf{b}_2^2 > \mathbf{b}_3^2 + \cdots > \mathbf{b}_1^2 \tag{8.13}$$

Relationships (8.12) and (8.13) are known as *Frank's rule*.

Let us use Frank's rule to examine some examples of super and partial dislocations in more detail. First consider a super perfect dislocation of Burgers vector $n\mathbf{b}$ where $n = 2, 3, 4 \ldots$. Generally it is geometrically possible for such a dislocation to split into n unit dislocations each of Burgers vector $\mathbf{b}$, but is this splitting energetically favourable? Applying Frank's rule we see that it is, because

$$n^2 b^2 > n b^2$$

Therefore super dislocations are rarely found in crystals as they split into unit dislocations.

Partial dislocations, however, and their associated planar defects, *stacking faults*, are not at all uncommon. We shall use the f.c.c. structure for the following example of partial dislocations. The stacking sequence of the close-packed {111} planes, which are also the slip planes, is ABCABC. This is illustrated in Fig. 8.22(a) where the lower layer comprises atoms marked A, the next layer has atoms marked B, and the top layer is in the C positions. Figure 8.22(b) shows that the unit dislocation that would cause a displacement in the slip direction would move atom B_1 to position B_2. In order for B_1 to move to B_2 it would have to rise over atom A in the plane below. This is a difficult process, and it is geometrically easier for slip to take place in two stages. The first stage is to move B_1 down the 'valley' between the A atoms to a C position, C_1, and the second stage is to move from C_1, again down the 'valley' between the A atoms, to B_2. This zigzag motion could be produced by two partial dislocations, so let us see if it is energetically possible for suitable partial dislocations to exist in the f.c.c. structure. The Burgers vectors of the unit dislocation and the possible partial dislocations are given in Fig. 8.22(c). From the vector diagram we have for the Burgers vectors:

$$b_2 = \frac{b_1}{2 \cos 30°} = \frac{b_1}{\sqrt{3}}$$

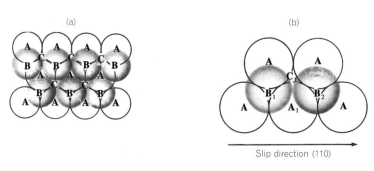

Fig. 8.22

Partial dislocations in the f.c.c. structure: (a) the packing of the close-packed {111} planes; (b) comparison of slip by a unit dislocation ($B_1 \rightarrow B_2$) to that by partial dislocations ($B_1 \rightarrow C_1 \rightarrow B_2$); (c) vector diagram for a unit dislocation b_1 and the two partial dislocations b_2 and b_3.

Slip direction (110)

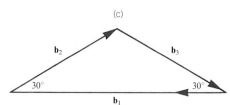

Similarly $b_3 = b_1/\sqrt{3}$. Therefore the partial dislocations have lower energy and so are favoured over a unit dislocation, because the lengths of the Burgers vectors obey the relation

$$b_1^2 > (b_2^2 + b_3^2) = \left(\frac{b_1}{\sqrt{3}}\right)^2 + \left(\frac{b_1}{\sqrt{3}}\right)^2 = \frac{2b_1^2}{3}$$

Partial dislocations are therefore responsible for slip in an f.c.c. structure. In this combination of two partial dislocations, the leading partial dislocation produces a *stacking fault*, while the following partial dislocation re-establishes the normal f.c.c. stacking sequence. The stacking of atoms within the stacking fault in this example has become BCBC ..., exactly like a close-packed hexagonal lattice, and this repeating pattern continues until the second partial dislocation is reached, when the ABC sequence is resumed. The pair of dislocations is known as an *extended dislocation*.

The separation between a pair of partial dislocations depends on the amount of energy per unit area stored in the planar stacking fault lying between them. This can range from a measured value of $16\,\mathrm{mJ\,m^{-2}}$ in f.c.c. silver, to an estimated $940\,\mathrm{mJ\,m^{-2}}$ in b.c.c. iron.

The complicated structures created by splitting unit dislocations into partials appears to add mechanical strength to a solid, an important example being the case of alumina, Al_2O_3. A variety of stacking faults is sometimes found in layers of pure crystalline silicon grown during the preparation of 'integrated circuits' (see Chapter 16). Their presence usually reveals irregularities in the silicon crystal onto which they are grown, and is invariably damaging to the operation of transistors made in the faulted region.

8.4.6 Forces on dislocations

The force on a dislocation caused by an applied stress is simply related to the Burgers vector of the dislocation. We refer to Fig. 8.23, which shows a crystal of unit thickness that has been sheared by the movement of a single edge dislocation from face A to face B. If F is the force per unit length of dislocation, then the work done when a dislocation moves from face A to face B is FL. Now, the force on the slip plane caused by the applied shear

Fig. 8.23

Model for the calculation of the force on a dislocation; an edge dislocation has moved from face A to face B under the action of a shear stress τ.

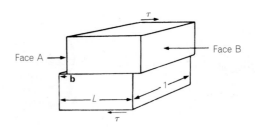

Face A

Face B

$\mathbf{b}$

L

τ

stress τ is $\tau \times$ area $= \tau \times L \times 1 = \tau L$, and this force does work in producing a displacement b. The amount of work is given by $\tau L b$ and this must equal the work done in moving the dislocation, so we have

$$FL = \tau L b$$

Therefore,

$$F = \tau b \qquad (8.14)$$

that is, the force on a dislocation is simply the product of the shear stress and the Burgers vector.

Dislocations are not always straight, see for example Fig. 8.16(c), and using Eq. (8.14) we can calculate the shear stress required to bow a dislocation. To do this the concept of *line tension* has to be introduced. Just as a soap bubble has a surface energy and a related surface tension, so a dislocation has a strain energy and an associated line tension. It can be shown that the line tension, T, per unit length is approximately equal to the elastic strain energy per unit length, that is

$$T \simeq Gb^2 \qquad (8.15)$$

Figure 8.24 shows a dislocation bowed to a radius of curvature R. The line tension will produce a force tending to straighten the dislocation and it will only remain bowed if a shear stress τ_0 produces an equal and opposite force. Resolving the forces acting on the dislocation segment dS horizontally,

$$\tau_0 b \, dS = 2T \sin\left(\frac{d\theta}{2}\right)$$

For small values of $d\theta$ we can take $\sin(d\theta/2)$ to be equal to $d\theta/2$, and as $d\theta/dS = 1/R$ we obtain, for the stress required to keep the dislocation bowed,

$$\tau_0 = \frac{T}{bR} = \frac{Gb}{R} \qquad (8.16)$$

Fig. 8.24

The line tension of a dislocation tends to straighten the dislocation.

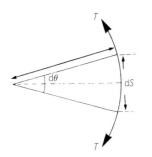

T

$d\theta$

dS

T

The significance of this equation will become apparent in the following section on dislocation multiplication and in the later discussions on precipitation hardening.

8.4.7 Dislocation multiplication

The density, ρ, of dislocations in a crystal is strictly expressed as length of dislocation line per unit volume, i.e. as metres per cubic metre. This is equivalent (although not exactly) to the number of dislocations piercing a unit area anywhere in the crystal, so that ρ is usually quoted as the number per unit area and, for convenience, is usually taken as the number per square centimetre. The most perfect bulk single crystal of metal that can be

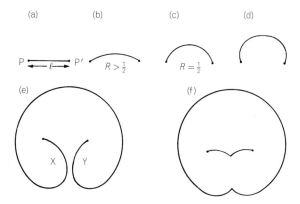

Fig. 8.25
The Frank–Read source. The stress on the dislocation increases to a maximum corresponding to minimum radius R [configuration (c)]. The sequence of events represented by (d), (e) and (f) can occur at lower stresses.

produced will usually have a dislocation density of 10^2 to 10^3 cm^{-2} while a normal polycrystal has a density in the region 10^7 to 10^8 cm^{-2}. These dislocations are built in 'naturally' in the fabrication processes, but when the specimen has undergone severe plastic deformation the density is found to increase to 10^{11} to 10^{12} cm^{-2}. We must conclude, therefore, that dislocations somehow multiply in the process of deformation. A possible multiplication mechanism is called the *Frank–Read* source after its originators.

Suppose that a dislocation is pinned at two points P and P' a distance ℓ apart, Fig. 8.25(a). Under the action of an applied stress the dislocation will bow out, the radius of curvature R being related to the stress by Eq. (8.16) [Fig. 8.25(b)]. Increasing the stress further causes more bowing, i.e. R decreases, until a minimum value of R is reached that is equal to $\ell/2$, Fig. 8.25(c). As the stress is inversely proportional to R, this minimum R configuration must correspond to the maximum stress. Consequently, the dislocation configurations shown in Fig. 8.25(d), (e) and (f) occur spontaneously at stresses less than that required for $R = \ell/2$. The sequence of events is that the dislocation expands into a loop, Fig. 8.25(d), the segments X and Y meet, Fig. 8.25(e), and annihilate each other forming a complete loop and reforming the pinned dislocation. The loop expands and moves away from the source PP', and the pinned dislocation can continue to multiply.

8.5 Planar defects

Planar defects are atomic in one dimension and large in the others. They may be classified as either external or internal defects. External planar defects are the solid–gas, solid–liquid and liquid–gas interfaces. As the name suggests, internal planar defects are found in the interior of crystalline materials and the most common are grain boundaries and twin interfaces. Stacking faults have already been covered in Section 8.3.

The structure and energy of the various planar defects differ and it is particularly important to have an appreciation of the relative magnitudes of the energies. Approximate energies per unit area, as a fraction of the solid–gas interface energy, E(S–G), are:

$$E(\text{solid–liquid}) \sim \tfrac{1}{12} E(\text{S–G}) \qquad E(\text{liquid–gas}) \sim \tfrac{5}{6} E(\text{S–G})$$

$$E(\text{twin}) \sim \tfrac{1}{20} E(\text{S–G}) \qquad E(\text{high-angle grain boundary}) \sim \tfrac{1}{3} E(\text{S–G})$$

8.5.1 Grain boundaries

Grain boundaries may be defined as the interface between crystals that differ in crystallographic orientation, composition or dimensions of the crystal lattice (in some cases in two or all of these). Grain boundaries between crystals of different orientation have been described in an earlier chapter and are schematically represented in Fig. 6.1, and may be seen in the micrograph of Fig. 6.2. The energy E of these boundaries varies with the misorientation θ across the boundary according to:

$$E = E_0\theta(A - \ln\theta)$$

where E_0 and A are constants. The variation in energy with misorientation is considerable at low misorientations of less than about 12°, but at high misorientations the energy is nearly constant (Fig. 8.26). The grain boundaries seen in micrographs of single-phase materials, such as those of Fig. 6.2, are normally associated with large misorientation and are called high-angle grain boundaries. The other two types of grain boundaries need no explaining, and many examples of the boundaries between crystals of

Fig. 8.26

Variation of grain boundary energy with misorientation for copper [the data are from N.A. Gjostein and F.N. Rhines, *Acta Met.*, **7**, 319 (1959) and for a particular type of grain boundary called a simple twist boundary].

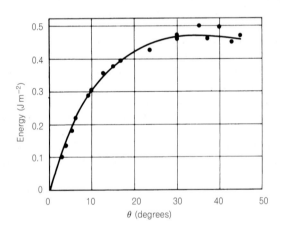

Fig. 8.27

Three processes that occur because they lead to a reduction in the total grain boundary energy: (a) grain growth in a single-phase material; (b) spheroidization in a two-phase material; (c) particle coarsening in a two-phase material.

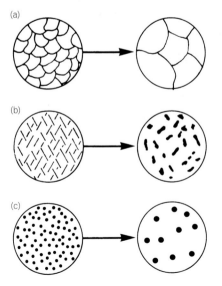

different compositions may be found in the micrographs in the following two chapters. The energy of these boundaries depends on the mismatch of the crystal lattices and the compositions of the two crystals.

All grain boundaries have a relatively high energy, typically 0.2–$1.0\,\mathrm{J\,m^{-2}}$, and as a result are non-equilibrium defects. Thus, a crystal will always try to reduce the total grain boundary area in order to reduce its free energy. This is why the processes illustrated in Fig. 8.27, namely grain growth, spheroidization and particle coarsening occur in materials held at elevated temperatures.

8.5.2 Diffusion in grain boundaries

Diffusion of atoms along grain boundaries is often much faster than in the surrounding crystallites. It is sometimes called 'pipe' diffusion, because the relatively open structure of a grain boundary, like that in a dislocation, allows fast diffusion to occur along it, so it acts like a pipe. Because grain boundaries normally extend to a surface, this enables the 'injection' of a raised impurity concentration into the interior of a polycrystalline solid along the easy pathways of the grain boundaries, rather like the use of a hypodermic syringe to inject painkiller before dental treatment. Figure 8.28 illustrates the progression by diffusion of a source of impurity (such as a gas) placed in contact with a flat surface, into a polycrystalline solid when the diffusion coefficient within the grain boundaries is very much higher than it is in the bulk of a (crystalline) grain. In an extreme case, the grain boundaries fill first with the impurity atoms, before any significant depth of penetration into the individual grains is seen.

In such a case, a manufacturing process that is aimed at uniformly 'doping' a surface layer in a component with an impurity can be enormously speeded up by using a material with reduced grain size, so that the penetration depth of the impurity becomes comparable to the average grain diameter after a relatively short diffusion time.

On the other hand, if we were to consider a corrosion process, for example the oxidation of steel, then raising the grain size at the surface may be beneficial (although adding a protective coating would clearly be even better). In practice, other factors often dominate corrosion mechanisms, such as the presence of precipitates within grain boundaries (see Chapter 10).

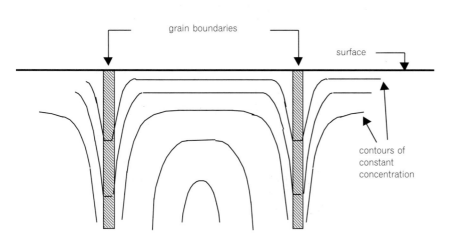

Fig. 8.28

Pipe diffusion of impurity atoms along grain boundaries. Each successive contour corresponds to a change in concentration by a constant multiplying factor. The diffusion rate along the grain boundaries is assumed to be many times that within the grains.

8.5.3 Twin interfaces

There are other internal planar defects than grain boundaries, the most familiar being the *twin* interface. A crystal is twinned when one portion of the lattice is a mirror image of the neighbouring portion, the mirror being the twinning plane as shown in Fig. 8.29. This may be formed during the growth of the crystal or may be the result of dislocation motion caused by an applied stress. The straight lines traversing the grains in Fig. 6.2 are where the former type of twin interface intersects the specimen surface; these are called *annealing* twins and are formed in many f.c.c. metals and alloys during recrystallization (cf. Chapter 9).

The other type of twin, the *deformation* twin, is produced by the movement of specific dislocations. It can be seen from Fig. 8.29 that the atomic displacements required to form a twin are not integral numbers of lattice spacings, therefore we conclude that the dislocations involved are imperfect dislocations. They are in fact partial dislocations and have to move in a certain complex manner in order to produce twins.

The stress to twin a crystal tends to be higher than that required for slip, hence except under certain conditions slip is the normal deformation mechanism. For example, the twinning stress is less sensitive to temperature than the stress for slip, consequently twinning becomes more favourable as the deformation temperature becomes lower. This is the case when b.c.c. iron and its alloys are rapidly loaded at low temperatures where thin, lamellar twins appear, called *Neumann bands*. Twinning may also play a significant role in the deformation of c.p.h. materials, especially when in

Fig. 8.29

The arrangement of the atoms on either side of a twin interface.

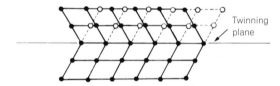

Twinning plane

Fig. 8.30

Deformation twins in this electron micrograph of a thin slice of polycrystalline magnesium appear dark in contrast with the light untwinned regions. Closely spaced slip lines that are nearly horizontal are also visible.

25 μm

the polycrystalline condition. This is because the c.p.h. structure only has three slip systems, therefore slip is restricted. The twins in Fig. 8.30 were produced by deforming polycrystalline magnesium; the presence of slip lines indicates that slip has also occurred.

Problems

8.1 The energies of formation of Schottky and Frenkel defects in a material are $1\,eV\,atom^{-1}$ ($1.6 \times 10^{-19}\,J\,atom^{-1}$) and $4\,eV\,atom^{-1}$ ($6.4 \times 10^{-19}\,J\,atom^{-1}$) respectively. What are the equilibrium concentrations of these defects at 1000 K? Assume that the pre-exponential constant, A, is equal to unity.

8.2 The equilibrium vacancy concentrations of Schottky defects in copper are 4×10^{-6} and 15×10^{-5} at temperatures of 1000 K and 1325 K respectively. What is the energy of formation of these defects?

8.3 Distinguish between edge and screw dislocations. With the aid of sketches, show that the normal glide motion of both types of dislocation produces slip lines on the surface of a crystal.

8.4 Draw a Burgers circuit for an edge dislocation in a primitive cubic crystal.
  In a certain crystal the Burgers vector for an edge dislocation is $2.5 \times 10^{-10}\,m$. Calculate the force per unit length on the dislocation when a shear stress of $350\,N\,m^{-2}$ is applied.

8.5 A two-phase material has precipitate particles of radius $1.25 \times 10^{-8}\,m$ and of concentration $10^{20}\,m^{-3}$. The energy of the grain boundary between the particle and the surrounding material (the matrix) is $0.5\,J\,m^{-2}$. After heat treatment at an elevated temperature the particles were found to have coarsened to $3 \times 10^{-8}\,m$ radius and their concentration decreased to $7.2 \times 10^{18}\,m^{-3}$. What is the reduction in the total grain boundary energy as a result of the coarsening?

Self-assessment questions

1 Point defects have atomic dimensions in all directions

(a) true   (b) false

2 Substitutional atoms occupy interstices in the parent lattice

(a) true   (b) false

3 Many different point defects exist in crystals, for example,

(a) excited electrons   (b) self-interstitials

(c) slip lines   (d) vacancies

(e) precipitate particles   (f) substitutional atoms

4 The following figure shows the formation of a

(a) Schottky defect   (b) grain boundary

(c) Frenkel defect   (d) substitutional atom

e) line defect   f) point defect

5 The following graph refers to the formation of Schottky or Frenkel defects. It shows that

(a) the entropy term favours an increase in free energy

(b) the enthalpy term favours an increase in free energy

(c) a crystal will always try to minimize the point defect concentration

(d) the free energy of a crystal is a minimum when the equilibrium number of point defects is present

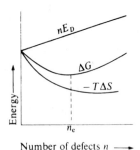

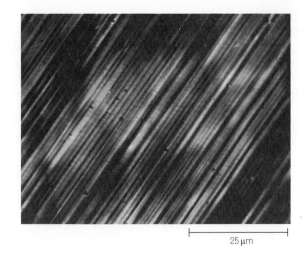

|— 25 μm —|

6   The equilibrium concentration of Schottky defects c_e, at any given temperature T, may be obtained using the expression

(a) $c_e = AT^Q$

(b) $c_e = A[\ln T - \ln Q]$

(c) $c_e = A\exp(-Q/kT)$

(d) $c_e = kT\ln Q$

where A is a constant, Q is the energy of formation and k is Boltzmann's constant.

7   In metals the energy of formation of a Schottky defect is of the order of

(a) 10^{-8} eV   (b) 10^{-3} eV   c) 1 eV

(d) 10^2 eV   (e) 10^8 eV

8   In ionic crystals the Frenkel defect concentration is always greater than the Schottky defect concentration

(a) true   (b) false

9   Vacancies play an important role in

(a) deformation twinning   (b) self-diffusion

(c) climb   (d) cross-slip

10   The following is a micrograph of

(a) grain boundaries

(b) annealing twins

(c) climb of edge dislocations

(d) slip lines

(e) a plastically deformed crystal

(f) an undeformed crystal

11   The slip plane and the slip direction in the f.c.c. structure are the {111} and ⟨110⟩ respectively. Consequently the number of slip systems in the f.c.c. structure is

(a) 12   (b) 6   (c) 24   (d) 3

12   The Burgers vector of an edge dislocation is perpendicular to the dislocation

(a) true   (b) false

13   To obtain the Burgers vector of a dislocation, Burgers circuits are drawn

(a) first around the dislocation, and then, with the same number of jumps, in the perfect crystal

(b) first around the dislocation and then, with one fewer jump, in the perfect crystal

(c) first in the perfect crystal, and then with the same number of jumps, around the dislocation

(d) first in the perfect crystal, and then with one fewer jump, around the dislocation

(e) in an anticlockwise direction

(f) in a clockwise direction

14   Both edge and screw dislocations can climb.

(a) true   (b) false

15   The total energy of a dislocation is proportional to

(a) G   (b) G^2 (c) b   (d) b^2 (e) b^{-1}

where G and $\mathbf{b}$ are the shear modulus and Burgers vector respectively.

16 The core energy of a dislocation is only about one-tenth of the elastic strain energy

   (a) true   (b) false

17 The energy of an edge dislocation is greater than that of a screw dislocation in the same material

   (a) true   (b) false

18 The Burgers vector of an imperfect dislocation is

   (a) always less than one lattice spacing

   (b) always more than one lattice spacing

   (c) an integral number of lattice spacings

   (d) not an integral number of lattice spacings

19 A dislocation of Burgers vector $\mathbf{b}_1$ will be split into two dislocations of Burgers vectors $\mathbf{b}_2$ and $\mathbf{b}_3$ when

   (a) it is geometrically possible

   (b) $\mathbf{b}_1 > \mathbf{b}_2 + \mathbf{b}_3$

   (c) $\mathbf{b}_1^2 > \mathbf{b}_2^2 + \mathbf{b}_3^2$

   (d) $\mathbf{b}_1^2 = \mathbf{b}_2 + \mathbf{b}_3$

   (e) $\mathbf{b}_1$ is parallel to $\mathbf{b}_2$ and $\mathbf{b}_3$

20 Dislocation motion in f.c.c. materials is represented in the diagram below (i) $\mathbf{b}_1$ is the Burgers vector of a ... (ii) $\mathbf{b}_2$ is the Burgers vector of a ...

   (a) perfect dislocation

   (b) imperfect dislocation

   (c) screw dislocation

   (d) partial dislocation

   (e) unit dislocation

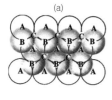

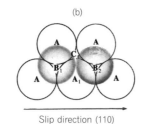

Slip direction (110)

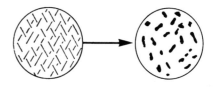

21 The sequence of events in the operation of a Frank–Read source is shown in (a), (b), (c), (d), (e) and (f) below. The maximum stress to operate the source

   (a) corresponds to event (b)

   (b) corresponds to event (c)

   (c) corresponds to event (d)

   (d) corresponds to event (e)

   (e) is proportional to ℓ

   (f) is proportional to ℓ^{-1}

   (g) is much greater for a screw dislocation than an edge dislocation

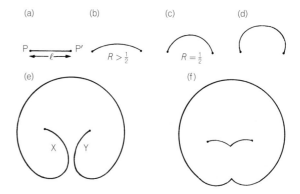

22 The diagram below illustrates

   (a) grain growth

   (b) deformation twinning

   (c) spheroidization

   (d) vacancy formation

   (e) a process that reduces the total grain boundary energy

   (f) clustering of self-interstitials

23 Annealing twins are formed in many f.c.c. metals and alloys during

   (a) casting

   (b) plastic deformation

   (c) rapid quenching from a high temperature

   (d) recrystallization

24 Deformation twins are only formed in b.c.c. metals

    (a) true   (b) false

Each of the sentences in Questions 25–34 consists of an assertion followed by a reason. Answer:

(a) If both assertion and reason are true statements and the reason is a correct explanation of the assertion

(b) If both assertion and reason are true statements but the reason is *not* a correct explanation of the assertion

(c) If the assertion is true but the reason contains a false statement

(d) If the assertion is false but the reason contains a true statement

(e) If both the assertion and reason are false statements

25 Nitrogen and oxygen generally occupy interstitial sites in metals *because* they are gases.

26 When carbon is alloyed with a metal it normally substitutes for the parent metal atoms, i.e. forms substitutional atoms, *because* it is a small atom.

27 The slip planes in the b.c.c. structure are the {100} planes *because* they are some of the more densely packed planes in the b.c.c. structure.

28 Plastic deformation does not take place by the shearing of complete blocks of a crystal (block shear mechanism) *because* the stress for this process is of the order of G/30 (where G is the shear modulus), which is much higher than the stresses required to produce plastic deformation.

29 Dislocations can move at stresses as low as $G/10^3$ (where G is the shear modulus) because they are non-equilibrium defects.

30 An edge dislocation cannot cross-slip *because* it has a uniquely defined slip plane.

31 The glide motion of a screw dislocation is not so restricted as that of an edge dislocation *because* a screw dislocation can climb.

32 Dislocations are non-equilibrium defects *because* the increase in enthalpy associated with them outweighs the entropy contribution to the free energy.

33 When a crystal is plastically deformed the dislocation density decreases *because* some of the dislocations reach the surface of the crystal.

34 Grain boundaries are line defects *because* they are formed during plastic deformation.

Answers

1 (a)	**2** (b)	**3** (b), (d), (f)	**4** (a), (f)
5 (b), (d)	**6** (c)	**7** (c)	**8** (b)
9 (b), (c)	**10** (d), (e)	**11** (a)	**12** (a)
13 (a), (f)	**14** (b)	**15** (a), (d)	**16** (a)
17 (a)	**18** (d)	**19** (a), (c)	**20** (i) (a), (e) (ii) (b), (d)
21 (b), (f)	**22** (c), (e)	**23** (d)	**24** (b)
25 (b)	**26** (d)	**27** (d)	**28** (a)
29 (b)	**30** (a)	**31** (c)	**32** (a)
33 (d)	**34** (e)		

part II

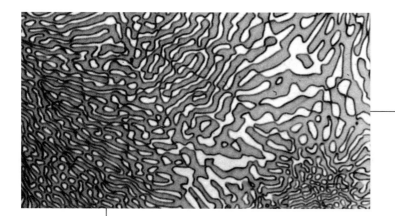

Mechanical
properties and
applications

Mechanical properties | 9

9.1 Introduction

The mechanical properties of a material determine how applied forces affect its shape and size, and sometimes its movement in space. It is usually structural solids – the metals, ceramics and plastics that we study in later chapters – whose mechanical properties are of most interest. These must contain pressure or withstand forces without suffering excessive deflection or deformation. For example, gas turbine blades must maintain a precisely defined shape under enormous stresses, at high temperatures, for long periods, because it is this shape that contains and guides the fast-moving fluid. Many important materials, however, lie at the frontiers of what we think of as solids. The mechanical properties of gel inside the earpiece seals of an aviation headset are of primary importance to their function of conforming to the contours of the head. Even the mechanical properties of foods, such as chocolate, biscuits and cheese, influence their function of providing aesthetic pleasure as well as nourishment.

This chapter will introduce the many ways in which different classes of materials respond to forces, for example elastic (recoverable), visco-elastic (temporary) and plastic (permanent) deformation, flow and fracture. Unless the material is to be used in a very simple component such as a beam or rod, its behaviour will probably be predicted using stress analysis, usually by computer. The details of stress analysis theory are of little concern to us here, but we will look at some of the basic *material models* (e.g. Hookeian, linearly viscoelastic, von Mises), which scientists and engineers employ to represent materials. We will also look at the test methods used to provide specific property data for these models (e.g. Young's modulus, yield strength, fracture toughness), and at some of the general micromechanisms that determine these properties. Micromechanisms specific to particular classes of materials will be dealt with in the chapters that follow.

The task of materials science is usually to explain these different ways of behaviour so that new materials can be designed to perform better. This task must begin by subjecting materials to tests.

9.1.1 The stress–strain curve

The simplest mechanical test to visualize is the uniaxial tensile test, in which a specimen is extended on one axis, usually at a constant speed, while the corresponding tensile force is monitored, as in Fig. 9.1(a). The *stress*, σ, is defined as the tensile force per unit cross-sectional area of

Fig. 9.1

The tensile test: (a) an increasing
stress $\sigma = F/A$ is applied to the
specimen; (b) the resulting stress–
strain curve.

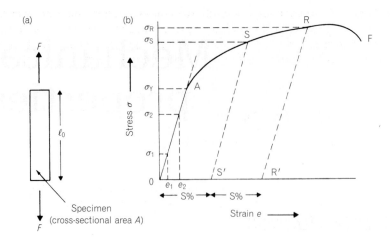

the specimen. Thus if the force is F newtons and the cross-sectional area
is A square metres, then the stress is given by

$$\sigma = \frac{F}{A} \text{N} \text{m}^{-2}$$

In practice the $\text{N} \text{m}^{-2}$ (also called a pascal, Pa) is such a small quantity that
$\text{MN} \text{m}^{-2}$ (meganewtons per square metre) or MPa are often used.

One $\text{MN} \text{m}^{-2}$ is roughly the stress needed to double the length of an
ordinary elastic band. In response to tensile stress, every material will
extend, though few, of course, by this much. Because the extension is
initially proportional to the original length ℓ_0 of the specimen we define the
engineering strain, *e*, as the fractional change in length, i.e.

$$e = \frac{\Delta \ell}{\ell_0}$$

where $\Delta \ell$ is the increase in length. Few structural materials can
accommodate tensile strains of more than a few percent without sustaining
permanent damage or fracture.

A typical stress–strain curve, obtained from a tensile test on a metal
such as mild steel, is shown in Fig. 9.1(b). As extension begins the area A
hardly changes from its initial value A_0 and the *engineering stress* or
nominal stress F/A_0 increases with load. At this stage the specimen extends
elastically, returning to its original length as soon as the stress is removed.

Beyond a certain strain – the *elastic limit* – part of the strain remains after
removal of the applied stress. This permanent change in shape is known as
plastic deformation. If the specimen is extended beyond this limit the load,
and with it the engineering stress, continues to increase until the *ultimate
tensile stress* or the *tensile strength* (maximum load/original cross-sectional
area) is reached. This maximum stress point is of great practical importance because if the specimen is subjected to an equivalent 'dead' load,
e.g. by a suspended weight, it will extend uncontrollably.

We can only continue with the test by carefully controlling extension.
This can best be achieved using a rigid testing machine in which linear
movement is generated by a nut running along a rotating screw. A *neck*
now begins to develop somewhere along the length of the specimen, and
further deformation is concentrated within this neck while the stress in the

remainder of the specimen actually decreases. Finally, fracture occurs at the minimum cross-section of the neck. Because the state of uniform deformation that existed throughout the specimen has been lost, it is very difficult to obtain useful information from the latter part of the test.

9.2 Elastic deformation

9.2.1 Characteristics of small-strain elastic deformation

We begin our detailed study with the elastic region 0–A of Fig. 9.1(b). In many materials the engineering stress is directly proportional to strain in this region, so that

$$\sigma = Ee \qquad (9.1)$$

where E is a material constant called *Young's modulus*. This equation is known as *Hooke's law*, and materials that conform to it as Hookeian. Young's modulus has the units of stress ($N\,m^{-2}$). In principle, a stress E would cause an engineering strain of unity, doubling the length of the specimen, so for most structural materials its value is extremely high and is often measured in giganewtons (thousands of meganewtons) per square metre, $GN\,m^{-2}$.

A linear relationship between stress and strain while a material is being deformed elastically is also found under other kinds of stress. Stress or relative displacement can be applied parallel rather than normal to

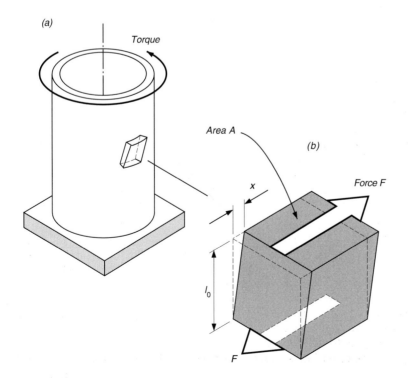

Fig. 9.2

(a) The wall of a thin tube loaded by torsion is subjected to pure shear stress; (b) shear stress and strain on an elemental cube of material.

the loaded surfaces; if a thin tube is held at one end and a torque is applied at the other, the material of its wall is subjected to pure shear stress (Fig. 9.2). The shear stress $\tau = F/A_0$ and the shear strain $\gamma = \Delta x/l_0$ are linearly related by

$$\tau = G\gamma \tag{9.2}$$

where G is the *shear modulus*. A third modulus is defined by the response of the material to a hydrostatic pressure p so that the specimen volume V is reduced elastically by ΔV. In this case the *bulk modulus*, K, is defined by the equation

$$p = K\frac{\Delta V}{V} \tag{9.3}$$

Elastic deformation is instantaneous and reversible. If a load is changed suddenly, the inertia of the material may delay the stress reaching a particular point, but the reaction to that stress when it arrives will be instantaneous. When the stress is removed, the specimen immediately reverts to its original size and shape.

9.2.2 Atomic mechanism of elastic deformation

The important features of elastic deformation in rigid solids like metals and ceramics are that it is instantaneous and linear (stress and strain are proportional). In order to understand why, we must look at what is happening on the atomic scale during elastic deformation.

When an external tensile force is applied to a crystal of length l_0, the atoms are immediately pulled further apart and this is seen as an elastic elongation Δl of the crystal. We have already used a plot of force against interatomic distance [Fig. 7.4(b)] and the earlier diagram is reproduced in Fig. 9.3. The atoms are pulled from their equilibrium positions, say from a to b in Fig. 9.3, until the applied force is balanced by the increase, ΔF, in the attractive force between atoms. The elastic strain of the crystal is $\Delta l/l_0$ and this is equal to $\Delta r/r_0$, where Δr is the equilibrium interatomic distance. On removal of the external force the atoms instantaneously return to their equilibrium positions under the action of the interatomic forces. This is why elastic deformation is reversible.

Fig. 9.3

Interatomic force plotted against interatomic distance.

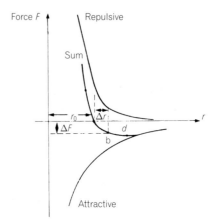

In practice, deformation in such materials remains elastic only up to strains of about 0.005 (i.e. 0.5%), so that we are concerned only with the region of the force–displacement curve around the equilibrium spacing r_0. Figure 9.3 shows that the force–displacement curve can be regarded as linear near to r_0 for small changes of r; it follows that under these circumstances the external force required to elastically deform a material will be linearly proportional to the displacement, and that stress will be linearly proportional to strain, i.e. Hooke's law.

An approximate estimate of Young's modulus for a solid of this kind can be obtained from the interatomic forces. A *tensile* (positive) stress is balanced by an interatomic *attractive* force $-F$, so that for a single atom in a crystal during elastic deformation,

$$\sigma \approx -\frac{\Delta F}{r^2}$$

$$e \approx \frac{\Delta r}{r_0}$$

Now Young's modulus is equal to σ/e, and as the displacements are small during elastic deformation, $r \approx r_0$, and so

$$E \approx -\frac{\Delta F}{r_0 \Delta r} = -\frac{1}{r_0}\left(\frac{\partial F}{\partial r}\right)_{r=r_0} \tag{9.4}$$

In Chapter 7 the relationship between F and r was described mathematically by Eq. (7.14):

$$F = -\frac{nC_1}{r^{n+1}} + \frac{mC_2}{r^{m+1}}$$

The value of r_0 was found by putting $F = 0$ so that

$$-\frac{nC_1}{r^{n+1}} + \frac{mC_2}{r^{m+1}} = 0$$

The differential $(\partial F/\partial r)_{r=r_0}$ is obtained from these two equations. Thus

$$\left(\frac{\partial F}{\partial r}\right)_{r=r_0} = \frac{n(n+1)C_1}{r_0^{n+2}} - \frac{m(m+1)C_2}{r_0^{m+2}} = \frac{nC_1(n-m)}{r_0^{n+2}}$$

and substituted into Eq. (9.4) it gives for Young's modulus

$$E \approx \frac{nC_1(m-n)}{r_0^{n+3}} \tag{9.5}$$

This equation is only approximate because it relates to two atoms only, ignoring the influence of the other neighbouring atoms. Nevertheless, we will see later that it gives a rough guide to the dependence of Young's modulus on atomic parameters, particularly interatomic spacing.

In a crystalline solid the interatomic spacing, and in some cases the bonding, varies with direction within a single crystal. Young's modulus therefore depends on the direction of the stress in relation to the crystal axes, i.e. single crystals are elastically *anisotropic*. A typical degree of variation of E with direction is shown in Fig. 9.4. Although single crystals are elastically anisotropic, a polycrystalline material in which the grains, or crystals, are randomly orientated behaves *isotropically*, i.e. its properties are the same in every direction. Amorphous materials such as glass and

Fig. 9.4

The anisotropy of Young's modulus for Fe–3%Si single crystals.

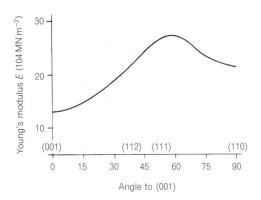

non-crystallizing polymers, except when produced in such a way as to cause some alignment of the molecular structure, are also isotropic.

Now, it is perfectly feasible to carry out a uniaxial tensile test on ice – another isotropic polycrystalline material – and to determine its Young's modulus (which is about $9.1\,\mathrm{MN\,m^{-2}}$). On the other hand a tensile test on liquid water, however it was carried out, would yield a Young's modulus of zero. How can this be so, because the interatomic spacing in water does not change much when it freezes? To answer this question, we must consider elastic deformation in more depth. We will see that the Young's modulus result from a tensile test really depends on two more basic properties of the material – its elastic recovery from a change of *volume* (which is indeed determined principally by the stiffness of interatomic forces) and its elastic recovery from a change of *shape* (which is only partly so). This diversion will later help us to understand the elastic properties of many less 'solid' materials – particularly the most 'elastic' of all, rubber.

9.2.3 Shear stress

First of all we need to extend our idea of stress, by considering the forces acting on a small cubic element within a material under load (Fig. 9.5). Because here we are concerned only with uniform states of stress, the size of the cube is unimportant. The material surrounding the cube has been removed and replaced by the forces that it exerted on each face. In a solid, the force and therefore the stress acting on each face of a cube does not generally act in the surface-normal direction. It can be resolved into three components acting as shown, the two new components being *shear* stresses τ.

Now in Fig. 9.6(a) the cube direction within a bar under uniform tension has been chosen so that one pair of faces cuts across the axis. Only *direct* stresses σ_{11}, σ_{22} and σ_{33} exist and here two of these are zero (i.e. the stress is *uniaxial*). The values of these stresses could be changed by, for example, removing the specimen from the testing machine and placing it in a chamber under pressure p. This would induce compressive (i.e. negative) direct stresses $\sigma_{11} = \sigma_{22} = \sigma_{33} = -p$. In general the three direct stresses are not equal and we define the state of 'negative pressure' in the material – the *hydrostatic* stress – by their average $\sigma_{\mathrm{h}} = (\frac{1}{3}\sigma_{11} + \sigma_{22} + \sigma_{33})$. Clearly $\sigma_{\mathrm{h}} = -p$ for the case of hydrostatic pressure.

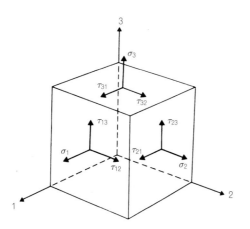

Fig. 9.5

The tensile (σ) and shear (τ) stresses on a cubical element.

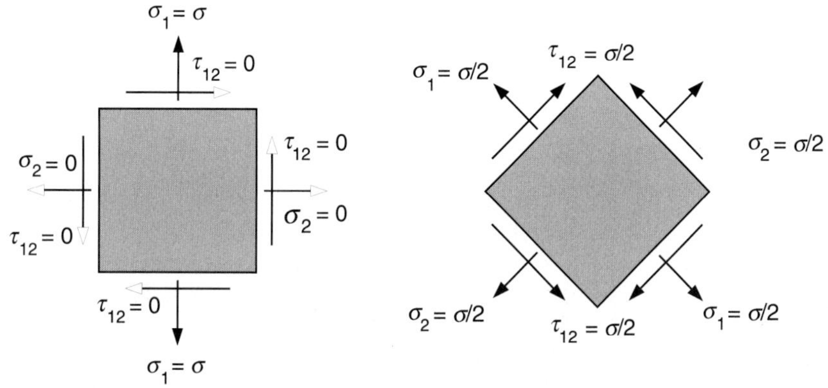

Fig. 9.6

Stresses acting within a uniaxially stressed bar (a) in the principal stress directions and (b) at 45° to one of these directions.

Within a solid body under load we can always find at least one cube orientation for which all shear stresses are zero and only direct stresses act, as in Fig. 9.6(a). The surface-normal directions of this special cube are the *principal axes*. The direct stresses vary with orientation but take their maximum or minimum values, the *principal stresses*, on the planes normal to these axes. The paths traced out by the principal axes show very clearly how applied forces 'flow' through a body. Often, as in the case of the uniaxially stressed bar, it is easy to identify the principal axes by inspection. Usually neither the surfaces nor the boundaries of a thin plate support a shear stress, so these must be principal planes. Figure 9.7 shows the way in which principal stress directions flow around a circular hole in such a plate. An interesting observation is that the uniaxial tensile stress applied at the top and bottom surfaces is, as we might expect, concentrated at the equator of the hole where the force-carrying principal stress directions are squeezed together. The general equations relating strain to stress, and used to analyse situations like that of Fig. 9.7, are given in textbooks on elasticity and will not be discussed here.

For the tensile bar one principal axis is the bar axis. If we now tilt the cube so that two surface normals lie at ±45° to the principal axis, as in Fig. 9.6(b), shear stresses have appeared and in fact have reached maximum values. This is easily verified by choosing a new cube within the original one and resolving forces on one of the wedge-shaped regions that

Fig. 9.7

Stress concentration. Principal axis directions indicate the 'flow' of force through a body. Where they are concentrated near a hole in a thin, uniaxially stressed sheet, the stress is increased.

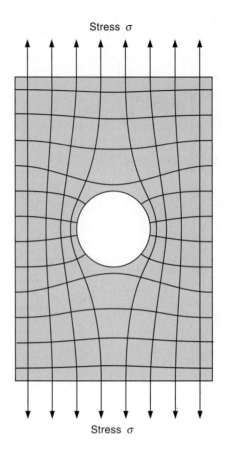

Stress σ

Stress σ

lie between them. The direct stress is now shared between the two faces tilted in its direction, which bear direct stresses of $\sigma_1 = \sigma_2 = \sigma/\sqrt{2}$, and the shear stress is $\tau = \frac{1}{2}\sigma$. In an elemental cube under pure hydrostatic stress $\sigma_1 = \sigma_2 = \sigma_3$, however, all directions are equivalent and there are no shear stresses in any of them: *shear stresses always correspond to differences between principal stresses.*

Materials are much more distinctive in their response to shear stress than to direct stress. Indeed, as we hinted when distinguishing water from ice, we recognize a solid by its ability to recover elastically from shear stress whereas both solids and liquids recover elastically from pressure. It is shear rather than direct stress that causes plastic flow by making planes of atoms or molecules slide against each other – although, as we will see in Section 9.6.2, it is direct tensile stresses that cause materials to separate by brittle fracture.

9.2.4 Elastic deformation of an isotropic material

As we saw earlier, shear stresses lower than those needed to cause permanent flow set up elastic *shear strains*, whose magnitude is determined by the shear modulus G. In a simple tensile test these shear strains acting on ±45° elements cause the well-known effect of an elastic contraction transverse to the bar axis. If the transverse dimension of the specimen is d, then this transverse strain is $e_t = -\Delta d/d$ (negative because Δd is negative) and another elastic constant, Poisson's ratio, is defined as

$$\nu = -e_t/e \qquad (9.6)$$

For many metals and ceramics, Poisson's ratio ν is about 0.3.

For an isotropic Hookeian material the total strain under a combination of stresses is given simply by adding the contributions from each stress. Thus a cube under uniaxial stress σ_1 will extend by a strain σ_1/E in the load direction and by strains $-\nu\sigma_1/E$ in the other two directions. If tensile stresses σ_2 and σ_3 are now applied as well and the material is Hookeian, the strain e_1 will be *reduced* from σ_1/E to

$$e_1 = \frac{1}{E}[\sigma_1 - \nu(\sigma_2 + \sigma_3)]$$

Corresponding equations can be written down for the strains in the other two directions.

It is not difficult to see that if a pressure p is applied on the cube, i.e. $\sigma_1 = \sigma_2 = \sigma_3 = -p$, there will be a strain $-(1-2\nu)p/E$ in all three directions. Now the volumetric strain is always simply the sum of the three direct strains: this leads quickly to the result that the bulk modulus of Eq. (9.3) is directly related to E by the equation

$$K = \frac{E}{3(1-2\nu)} \qquad (9.7)$$

If a material has a Poisson's ratio equal to 0.5 and is subjected to uniaxial tension, the increase in volume caused by elastic extension in that direction is exactly balanced by the reduction in volume caused by lateral contraction in the other two directions. Because an average *hydrostatic stress* (a negative pressure) of $\frac{1}{3}\sigma$ has resulted in no net volume expansion, this material is *incompressible*, with an infinite bulk modulus. On the other hand if, as we assumed in the atomic force calculation of the previous section, another material can be extended in one direction without any lateral contraction, then $\nu = 0$ and we find from Eq. (9.7) that $K = \frac{1}{3}E$.

A similar argument can be used to show that Young's modulus E is related to the shear modulus G:

$$G = \frac{E}{2(1+\nu)} \qquad (9.8)$$

E, G and Poisson's ratio ν are widely used by engineers for stress analysis: they allow the elastic strain resulting from the application of a given load to be calculated. This is essential when, for example, elastic deformations in the moving parts of an engine need to be known because close clearances are to be maintained. Furthermore, when more than one material is used in a structure we can make sure that the strains in the different materials are compatible and that the materials share the load in the way that we want them to. However, it is in many ways useful to regard the bulk modulus K and the shear modulus G as more fundamental properties of a material. Rearranging Eqs (9.7) and (9.8) to express Young's modulus and Poisson's ratio in terms of K and G we find that

$$E = \frac{3}{1 + \dfrac{G}{K}}\, G \qquad (9.9)$$

and

$$\nu = \frac{1}{2}\frac{3 - 2\dfrac{G}{K}}{3 + \dfrac{G}{K}} \tag{9.10}$$

9.2.5 Bulk modulus for an atomic solid

We can now return to our investigation of the elastic deformation of atomic solids with a new insight: in considering extension of the solid without lateral contraction, we are really estimating the bulk modulus, K. This interpretation makes it much easier to understand the bulk modulus of liquids and rubbery solids for which Eq. (9.5) fails, but a new version of it given by $K = \frac{1}{3}E$,

$$K \approx \frac{nC_1(m - n)}{3r_0^{n+3}} \tag{9.11}$$

still applies.

Accurate calculations have been done with a range of ionically bonded materials, such as lithium fluoride and sodium chloride, and the relation shown to be approximately correct, with $n = 1$ (i.e. the value used to determine the Coulomb forces of electrostatic attraction between unlike charges). This low value of n also applies to the low melting point metals such as lithium, sodium and potassium, which have metallic bonding. Higher melting point metals have higher values of modulus, consistent with the atomic bond strength increasing with a decrease in r_0. For the metals tungsten, molybdenum, nickel, iron and aluminium $n = 4$, showing that the simple Coulombic attraction no longer applies. For covalent bonded materials such as carbon, silicon, silicon carbide and germanium, $n = 2$.

Liquids, as well as solids, have a resistance to pressure that depends on the strength of bonding and the interatomic spacing. The bulk modulus of an organic liquid such as oil is typically $1\,\text{GN}\,\text{m}^{-2}$, so that a pressure of $10\,\text{MN}\,\text{m}^{-2}$ is needed to decrease the volume of a given mass by 1%. This compressibility is important in many engineering applications, particularly where oil is used as a lubricant or as a medium to transmit force in hydraulic systems. Liquids show little or no elastic recovery from shear, while in most metals and ceramics the shear modulus G is about 25–50% of the bulk modulus K. We shall see in Chapter 12 that in elastomers or rubbers – and many natural materials that resemble them – the elastic forces that restore shape are very weak and their shear moduli are low, but their bulk moduli are much the same as those of other organic liquids and solids.

9.2.6 Strain energy

At the beginning of Section 9.2 we pointed out that elastic deformation was by definition recoverable, in contrast to viscous deformation, which dissipates energy as heat. We have seen that at an atomic level a stress applied to a crystalline material increases the energy of bonds within it. In such an ordered solid there can be little accompanying change in entropy, so that this internal energy is stored as a form of free energy known as *strain energy*. We will now calculate the magnitude of the strain energy density U_E in terms of the applied stresses or strains acting on a material.

Consider first a cube of side ℓ_0 strained in tension by a direct force that increases from zero to a value F, so that opposite sides move apart by a distance $\delta\ell$. The total work done will be equal to the mean force multiplied by the distance through which the force acts: $\frac{1}{2}F\delta\ell$. As the volume of the cube is ℓ_0^3, the strain energy density is given by $\frac{1}{2}F\delta\ell/\ell_0^3$. Now by definition the tensile stress is $\sigma = F/\ell_0^2$ and the tensile strain is $e = \delta\ell/\ell_0$, therefore the strain energy density is given by

$$\frac{1}{2}\frac{F}{\ell_0^3}\delta\ell = \frac{1}{2}\sigma\ell_0^2\frac{\delta\ell}{\ell_0^3} = \frac{1}{2}\sigma e \qquad (9.12)$$

Because $\sigma = Ee$, we can introduce Young's modulus into this expression, giving

$$U_E = \frac{1}{2}\sigma e = \frac{1}{2}\frac{\sigma^2}{E} = \frac{1}{2}e^2 E \qquad (9.13)$$

The same argument can now be applied for a shear force, giving

$$U_E = \frac{1}{2}\tau\gamma = \frac{1}{2}\frac{\tau^2}{G} = \frac{1}{2}\gamma^2 G \qquad (9.14)$$

It is interesting to calculate the maximum strain energy density that can be attained by a material. Substituting the strength and Young's modulus typical of a strong steel into Eq. (9.12), we arrive at about $2.5 \times 10^6\,\mathrm{J\,m^{-3}}$, which works out at about $2.5 \times 10^{-4}\,\mathrm{eV}$ per atomic bond. Compared with bond energies that are of the order of 2–4 eV it can be seen that the strain energy stored is extremely small on an atomic scale.

9.2.7 Large strain deformation

We cannot leave elastic deformation without mentioning a class of materials that illustrate it more perfectly than any other – the rubbers or *elastomers* (elastic polymers). These are characterized by very low Young's modulus but an ability to recover from enormously high strains. An understanding of this large strain elasticity is important not only for elastomers but also for other polymers – which may recover from large strains only when heated or after a long period of time – and for many other 'soft solids' such as biological materials. We will see in Chapter 12 that these materials gain their spectacular elasticity not from the stretching of interatomic bonds but from the extension of much larger coiled 'molecular springs'. The local stretching of interatomic bonds is almost negligible; in a similar way, the local stretching of the steel used in a coil spring remains very small even when the spring is highly extended.

To exploit these materials fully, we must deal with the fact that they are non-Hookeian, or *non-linearly elastic*: at the large strains from which they are capable of recovering, the stress–strain curve is not even approximately straight. Much of this non-linearity arises directly from large changes in the overall shape of the body during stretching.

Large elastic deformations are usually analysed in terms of the *extension ratio*, which is just the ratio of the current length ℓ to the original length ℓ_0:

$$\lambda = \frac{\ell}{\ell_0} = 1 + e \qquad (9.15)$$

To express the idea that compressing a material to one-half of its original length ($\lambda = \frac{1}{2}$) is in some sense the inverse of extending it to twice

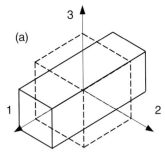

(a)

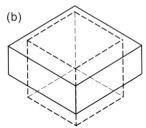

(b)

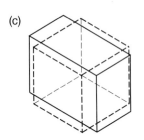

(c)

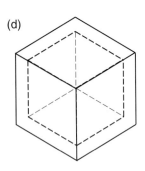

(d)

its original length ($\lambda = 2$), the *compression ratio*, equal to ℓ_0/ℓ, is sometimes used. We can represent any state of strain, even shear, by three extension ratios λ_1, λ_2 and λ_3, which transform a cube into a rectangular parallelopiped of volume $\lambda_1\lambda_2\lambda_3$.

Some of these states of strain are illustrated in Fig. 9.8. Figures 9.8(a) to (c) represent deformations in which there has been negligible change in volume (as is the case for an elastomer), so that:

- *uniaxial extension* to $\lambda_1 = 2$ is accompanied by contraction to $\lambda_2 = \lambda_3 = 1/\sqrt{2}$
- *biaxial extension* to $\lambda_2 = \lambda_3 = \sqrt{2}$ is accompanied by contraction to $\lambda_1 = \frac{1}{2}$ and so is equivalent to a uniaxial compression ratio of 2
- *pure shear* leaves $\lambda_3 = 1$ while $\lambda_1 = 2$ and $\lambda_2 = \frac{1}{2}$

Figure 9.8(d) represents an increase of volume without shape change by a factor of 2, so that $\lambda_1 = \lambda_2 = \lambda_3 = 2^{1/3}$.

For such large extensions we must find a new definition for stress that accounts for the change in area. If the current (instantaneous) load is F and the instantaneous cross-sectional area is A then the *true stress*, σ_T, is

$$\sigma_T = \frac{F}{A} \tag{9.16}$$

which at the beginning of deformation is of course the same as the engineering stress F/A_0.

9.3 Viscous deformation

Although under pressure a liquid behaves like a solid, it flows steadily when a shear stress is applied. This *viscous flow*, which ceases when the stress is removed, also occurs slowly in some solids. It is common in the melts of glasses and polymers, although in the latter it is usually also accompanied by viscoelastic deformation (viscous flow in glass and polymer melts is discussed further in Chapters 11 and 12 respectively). In contrast, viscous flow in metals is found only under specific conditions of high temperatures and low stresses. When a material flows viscously the resulting shear strain γ is in general a function of both the shear stress τ and time, t. Ideal viscous behaviour is represented by Newton's law of viscous flow:

$$\tau = \eta\frac{d\gamma}{dt} \tag{9.17}$$

where η is the viscosity; η is independent of the rate of shear, as shown in Fig. 9.9(a). The units of η are $\mathrm{N\,s\,m^{-2}}$ (or pascal seconds, Pa s) and typical values for fluids near room temperature are air, 0.18×10^{-4}; water, 10^{-3}; treacle, 10^2; pitch, $10^9\,\mathrm{N\,s\,m^{-2}}$.

In many situations of practical importance, viscous liquids flow at a constant rate determined by a constant applied stress, or are sheared at a constant rate and resist with a constant drag stress. This is the situation, for example, when glass or polymer melts flow within channels or pipes. An infinitesimally thin layer adjacent to each fixed surface sticks to it and has zero velocity. This leads to a shearing action between layers of liquid progressively further into the moving bulk. The stress that a liquid exerts in response to this shear is the origin of viscous drag.

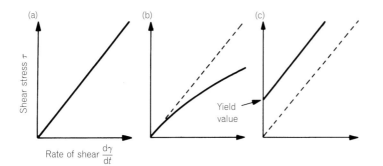

Fig. 9.9

Viscous flow: (a) ideal Newtonian behaviour; (b) non-ideal behaviour, shear thinning; (c) non-ideal behaviour, yield value.

Pure shear flow occurs between two flat parallel plates that move with a relative velocity dx/dt. At each instant a cube of height l_0 is sheared to a strain dx/l_0, so that the rate of shear is just

$$\frac{d\gamma}{dt} = \frac{1}{l_0}\frac{dx}{dt} \qquad (9.18)$$

A simple and widely used method of measuring viscosity based on this expression is the *cone-and-plate viscometer* shown in Fig. 9.10. A sample is held between a stationary flat disc and a cone of small angle α that rotates at a constant speed ω rad s^{-1} around the same axis. The relative velocity of the two surfaces is therefore given by $r\omega$ and the distance between the surfaces is $l_0 = r\alpha$. Now the rate at which the material is sheared is given by

$$\frac{d\gamma}{dt} = \frac{\omega}{\alpha} \qquad (9.19)$$

which is independent of radius, and is therefore uniform throughout the sample of viscous material. This is an important feature of the test for *non-Newtonian* liquids in which viscosity varies with shear rate.

The shear stress that the material in a cone-and-plate viscometer exerts on the discs gives rise to a drag torque, from which the viscosity can be calculated. An annular element of liquid at radius r and of radial thickness δr exerts a drag stress τ over an area $2\pi r\delta r$, giving rise to a torque

$$\delta T = 2\pi r^2 \tau \delta r$$

Integrating this from the centre ($r = 0$) to an outside radius R gives a total torque

$$T = \tfrac{2}{3}\pi R^3 \tau \qquad (9.20)$$

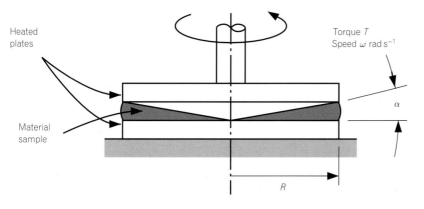

Fig. 9.10

Schematic diagram of the cone-and-plate viscometer. The rate of circumferential shear is uniform throughout the material sample.

Graphs such as Fig. 9.9 can be plotted by calculating the shear strain rate from the cone angle and speed using Eq. (9.19), and the shear stress from the drag torque using Eq. (9.20).

An alternative method of examining the response of a material to shear is to apply an oscillating displacement. This could be done using the cone-and-plate viscometer, although other methods are more convenient. Oscillatory straining is a useful method of investigating viscoelastic (combined viscous and elastic) behaviour, as we will see in Chapter 12 on polymers. Consider a cube of material of side l_0 sheared by a small displacement

$$x = x_{max} \sin 2\pi ft$$

Fig. 9.11

Force, displacement and work for shear by sinusoidal oscillation of (a) a Hookeian solid and (b) a Newtonian liquid.

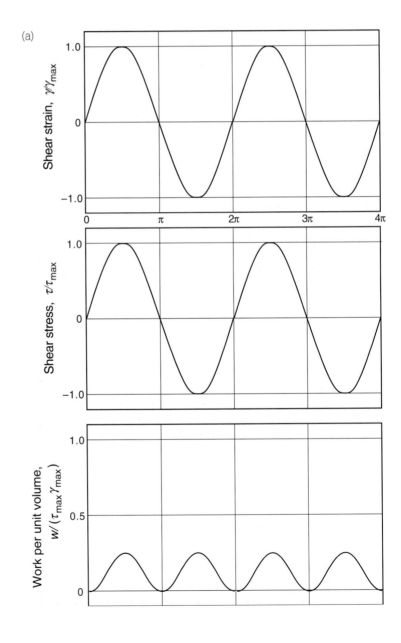

i.e. with f displacement cycles per second of rotation between $+x_{max}$ and $-x_{max}$, as shown in Fig. 9.11. The shear strain is x/ℓ_0 and if the material is linearly elastic it will simply react at any instant with a shear stress determined through Eq. (9.2) by the shear modulus:

$$\tau = G\gamma_{max} \sin 2\pi ft \qquad (9.21)$$

where $\gamma_{max} = x_{max}/\ell_0$. The force remains *in phase* with the displacement, as Fig. 9.11 shows. Plotted beneath the force curve is the integral of force and speed – the total work done by the force – per unit volume. It is easily seen that the total work done during a cycle is zero. The material stores the work done on it during either extension or compression and returns it when the displacement is removed.

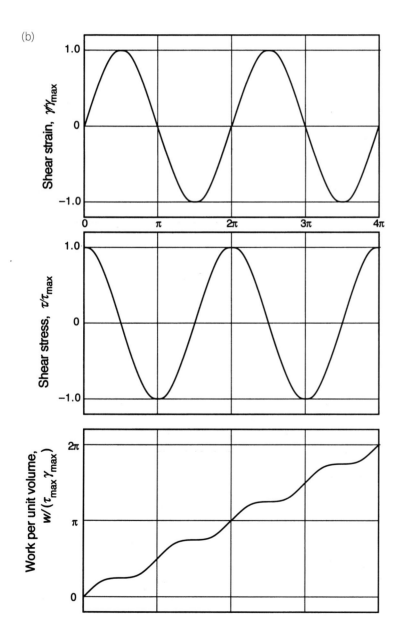

(b)

Fig. 9.11

Continued.

However, if the same oscillating displacement is applied to a Newtonian liquid, we have

$$\tau = \eta \frac{\mathrm{d}}{\mathrm{d}t}\left(\frac{x}{\ell_0}\right) = \eta\gamma_{max}\frac{\mathrm{d}}{\mathrm{d}t}(\sin 2\pi ft)$$

$$= \eta 2\pi f\gamma_{max}\cos 2\pi ft \tag{9.22}$$

$$= \eta 2\pi f\gamma_{max}\sin\left(2\pi ft + \tfrac{1}{2}\pi\right)$$

The stress reaches a maximum not when the *displacement* reaches a maximum but when the displacement *rate* reaches a maximum, which is one-quarter of a cycle earlier. We say that there is a phase shift of 90°, or $\tfrac{1}{2}\pi$ radians, between stress and strain. This time the work done on unit volume of the material per cycle is not zero (in fact it is $\pi\eta\gamma_{max}^2$), because the force always resists the imposed displacement. This work appears as heat within the material.

9.3.1 Spring and dashpot models

The ability of viscous fluids to absorb work is exploited in a *dashpot* (Fig. 9.12), a mechanical device used to damp oscillations in, for example, automotive suspension systems. Relative displacement of its ends forces a piston through a viscous fluid, usually oil, which fills a closed cylinder. The oil exerts a resisting force proportional to the displacement rate.

We now introduce the idea of representing stress/strain properties using imaginary 'mechanisms' of springs and dashpots. For simplicity, we will consider only small *shear* stresses and strains. As shown in Fig. 9.12, the stress needed to shear a unit cube of the material is thought of as arising not from processes within a volume of material but from forces exerted by

Fig. 9.12

(a) The dashpot: a mechanical element with a constant ratio η of force to displacement rate. (b) Hookeian elastic and Newtonian viscous behaviour of an elemental cube of material represented using a spring and a dashpot.

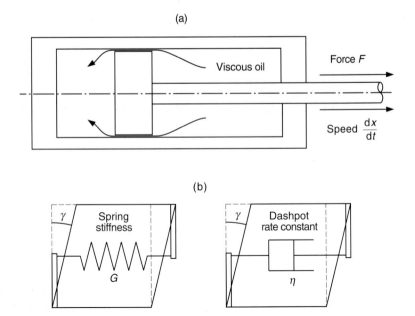

mechanical elements connected inside an empty, deformable box. If a Hookeian elastic material can be represented as a spring, a Newtonian viscous material can be represented as a dashpot, using the symbol shown in Fig. 9.12. For a unit cube, shearing of the top surface by a distance γ causes a shear strain γ, which an internal spring of stiffness G resists with a force $G\gamma$. Shearing of the top surface at a velocity $d\gamma/dt$ causes a shear strain rate $d\gamma/dt$ which an internal dashpot of constant η resists with a force $\eta\, d\gamma/dt$. Both these forces act on unit area and appear as stresses.

All this seems an unnecessarily complicated way to represent elastic or viscous materials, and it is. However in the next section we will encounter *viscoelastic* materials, which lie between these two simple extremes, have phase shifts between 0 and $\frac{1}{2}\pi$ and show intermediate rates of energy dissipation. These materials can be represented in a particularly intuitive way using models consisting of interconnected springs and dashpots. The exact way in which the springs and dashpots are connected represents the relationship between the internal mechanisms that they represent, and is revealed in the behaviour of the material.

Before we leave viscous behaviour, we must consider two deviations that are often encountered and are illustrated in Figs 9.10(b) and (c). The non-linear relationship between τ and $d\tau/dt$ of Fig. 9.10(b) is termed *pseudoplasticity* or *shear thinning* and is a result of a reversible decrease in viscosity with increasing shear rate. Shear thinning is a desirable characteristic for some paints. During brushing, which imposes high strain rates, it is preferable for the viscosity to be not too high. However, after application, the paint should not run extensively under the action of gravitational forces and therefore the viscosity must not be too low. The opposite effect, called shear thickening, exists but is rarely observed. The second deviation from Newtonian flow is a yield value, i.e. a critical stress below which apparently no flow occurs and above which flow is viscous. Above the yield value flow may either be Newtonian, as shown in Fig. 9.10(c), or non-ideal.

Viscous flow in solids normally involves the thermally activated movement of atoms or molecules within the material. The rate of a thermally activated process such as this is controlled by the Boltzmann exponential factor, as discussed in Chapter 7. Consequently, over limited ranges of temperatures the viscosity varies with temperature according to

$$\frac{1}{\eta} = A\exp\left(\frac{-Q}{RT}\right) \qquad (9.23)$$

where A is a constant and Q is the activation energy for viscous flow.

9.4 Anelastic and viscoelastic deformation

Unlike elastic deformation, which is instantaneous on the application of a stress, anelastic or viscoelastic deformation is time dependent, i.e. the strain response lags behind the stress (Fig. 9.13). Viscoelastic deformation is the result of processes within the material that depend on time, such as the movement of point crystalline defects or the straightening out of kinked or coiled molecular chains, in response to a constant stress. The viscous and elastic processes are coupled in such a way that on removal

Fig. 9.13

The response of ideal Hookeian and Newtonian materials to (a) creep and (b) relaxation loading.

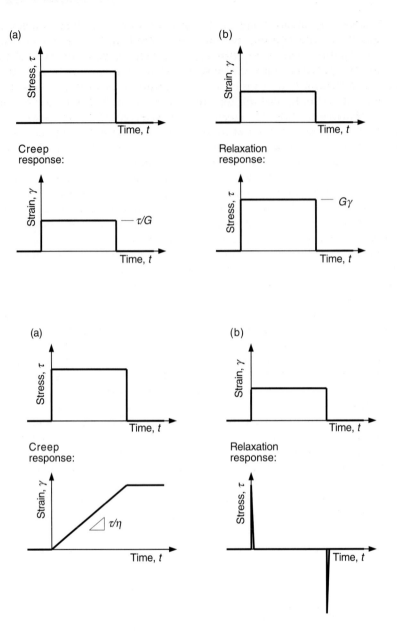

of the applied stress, these processes will reverse and the material will gradually revert to its original size and shape. Thus viscoelastic deformation is elastic in the sense that it is reversible, but viscous in the sense that the reverse process takes time. Although viscoelastic processes occur in all materials, the proportion of deformation that is 'delayed' by them may not be significant.

The phenomena of viscoelasticity are most clearly illustrated in creep and relaxation tests – in fact one test of each type can provide enough information to fully describe a viscoelastic material. An alternative is to apply oscillating stress over a wide range of frequencies, and a method based on this idea is widely used for plastics, as we will see in Chapter 12.

A creep test involves applying a stress (i.e. a load) instantaneously at time zero, and monitoring the increase in strain with time as that stress is held constant. Creep is observed in components such as the wall of a pipe under constant pressure. A relaxation test, on the other hand, involves applying a strain at time zero, and monitoring the reduction in stress while that strain is held constant. In other words, the specimen is stretched to and held at a new length. Relaxation is observed in components such as a bolt, holding two very rigid components together and pre-tensioned by a fixed number of turns on a nut. The loaded length of the bolt cannot change, but its state of internal stress can and does.

To see the 'elastic' component of viscoelastic response, we can reverse the creep test by complete unloading. Only part of the extension is immediately recovered. Similarly, a relaxation test can be reversed by returning the specimen to its original length, but because some of the extension is not immediately reversible a compressive stress will be needed.

For an elastic material, of course, neither the creep test nor the relaxation test reveals any new behaviour [Fig. 9.13(a)] because elastic deformation is by definition instantaneous and fully reversible. A viscous material subjected to constant stress simply flows at a constant rate, as defined by Eq. (9.17). If we attempt to carry out a relaxation test on a viscous material, however, the same equation tells us that the material will resist the 'instantaneous' deformation needed at the beginning of the test with an infinite stress – but will present no resistance to the constant displacement held thereafter. In effect, a viscous liquid relaxes instantaneously.

The results of applying these tests to viscoelastic materials are much more interesting. For the first moments of each test, a viscoelastic material shows the response of an elastic one. Under constant load, a viscoelastic material then continues to extend, usually at a decreasing rate: this is *creep*, a phenomenon of great importance in structural metals at high temperature, which will be discussed in Section 9.7.7. In metals, removal of the stress results in recovery only of the instantaneous strain. In polymers and biomaterials, however, given enough time, virtually all of the creep deformation will be recovered and the specimen will have returned

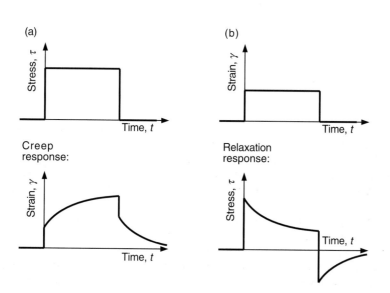

Creep
response:

Relaxation
response:

Fig. 9.14

The response of viscoelastic materials to (a) creep and (b) relaxation loading.

to its original length. Corresponding phenomena are seen if a constant strain is applied in a relaxation test. Again, all materials react instantaneously by developing a component of stress, but in viscoelastic materials the stress decays with time. Reversal of the initial strain will induce a compressive stress that in polymers will eventually relax nearly to zero (in metals, too, this *recovery* process takes place but it is much less complete).

Under sinusoidal displacement, we saw that the load in an elastic material showed no phase shift whereas a viscous material shows a phase lag of $\delta = 90°$. For viscoelastic materials δ lies between these extremes and for viscoelastic solids it is usually small – they are much more elastic than viscous. The phase lag δ (universally quoted as the tangent of δ, 'tan δ') determines the proportion of energy dissipated and therefore the material's capacity to damp out vibrations. Metals, having low damping capacities, are the materials of bells and piano strings. Polymers and other materials that show rubber elasticity generally provide very good damping – this is one of many reasons for their widespread use in automotive interiors.

9.4.1 Viscoelastic models

A *material model* replaces the complexity of real behaviour by a simplified representation that captures the main features, usually in mathematical form. We have already met basic Hookeian and Newtonian models of mechanical behaviour, represented by simple equations. The best models for more complicated behaviour are based on some understanding of the underlying internal processes – the *micromechanisms*. If the model then represents the material well, we can usually conclude that this understanding was correct. The linear viscoelastic models that we are about to meet illustrate this point rather well.

For small strains polymers exhibit *linear* viscoelasticity, strains and strain rates varying linearly with stress. Models for this behaviour can be constructed by interconnecting the spring and dashpot units of Fig. 9.12. For the time being we will consider the two simplest, each of which contains just one spring and one dashpot.

In the Maxwell model (Fig. 9.15) a spring and a dashpot are connected in *series*. We will see in Chapter 12 that Maxwell behaviour arises when spring-like polymer chains extend between attachment points that are not secure, but slip at a rate proportional to load. On the basis that each element carries the same load, and their extensions add, we can derive the differential equation

$$\frac{\mathrm{d}\gamma}{\mathrm{d}t} = \frac{1}{G}\left(\frac{\mathrm{d}\tau}{\mathrm{d}t} + \frac{G}{\eta}\tau\right) \tag{9.24}$$

Equation (9.24) is a complete representation of the model, and can be solved to find the behaviour of the material under different conditions. Under constant stress the spring extension will remain fixed, the dashpot will extend indefinitely at a constant rate, and when the stress is removed only the spring extension will be recovered [Fig. 9.15(b)]. Under constant strain, the dashpot will extend at a rate that decreases exponentially as the initially stretched spring relaxes [Fig. 9.15(c)]. The ratio η/G, which

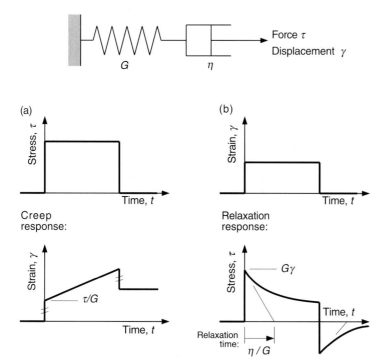

Fig. 9.15

Response of a Maxwell material to (a) creep and (b) relaxation loading.

is the time after which the stress has relaxed to 1/e of its initial value, is referred to as the *relaxation time*.

In the Voigt model (Fig. 9.16) a spring and a dashpot are connected in parallel; whatever extends elastically in the real material, must do so against some kind of viscous resistance. On the basis that spring and dashpot experience the same extension but share the load, we can derive the differential equation

$$\gamma + \frac{\eta}{G}\frac{d\gamma}{dt} = \frac{\tau}{G} \tag{9.25}$$

Under constant stress, the solid will initially flow in a viscous manner at an almost constant rate, the spring hardly being extended, but this rate will fall as the spring bears more and more of the load [Fig. 9.16(b)]. This is quite a good account of the behaviour shown in Fig. 9.14. However the Voigt model cannot be subjected to a relaxation test because it cannot be displaced instantaneously – the dashpot would resist with an infinite force. Even if the zero-strain rate condition for a relaxation test could be applied, the dashpot would thereafter support none of the applied load and the spring would bear all of it [Fig. 9.16(c)].

Having laid the groundwork for understanding how they behave, we will now put viscoelastic 'model' materials aside. In Chapter 12 we will meet polymers, the class of material of which viscoelastic behaviour is most typical; we will then consider the underlying mechanisms in much more detail. For the time being, we turn to another kind of model material whose behaviour is in many ways simpler because it is not, in general, time-dependent.

Fig. 9.16

Response of a Voigt material to (a) creep and (b) relaxation loading.

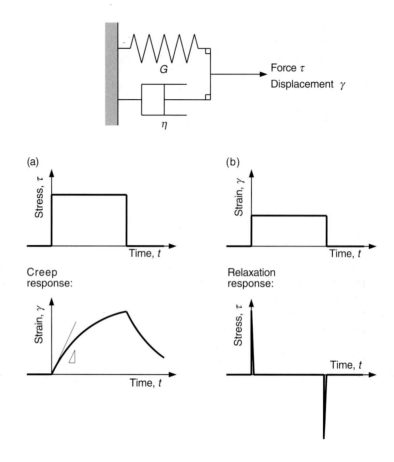

9.5 Plastic deformation

Remember that a tensile test on a metal is characterized by linear elastic behaviour up to a specific stress, above which part of the extension becomes irreversible. Elastic–plastic materials, which are almost exclusively metals and alloys, differ from viscoelastic materials firstly in that the stress needed to drive flow depends relatively little on the rate of extension, and secondly in that the deformation resulting from flow is never recovered.

9.5.1 Dislocations and stress–strain curves

In Chapter 8 we saw how plastic deformation in crystalline materials, whether in single crystal or polycrystalline form, occurs by the movement of dislocations. In this section we will concentrate on the stress–strain curves obtained from tensile tests and will correlate these curves with the movement and interactions of the dislocations responsible for plastic flow.

Figure 9.17 is a generalized stress–strain curve for a single crystal. The details of the stress–strain curve, for example the length of stage I, depend on many factors such as the crystal structure, orientation and purity of the specimen. However, we will not concern ourselves with these finer points but will concentrate on the general features of the curve. The initial linear

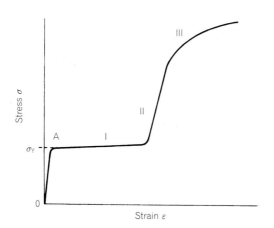

Fig. 9.17
Generalized stress–strain curve for a single crystal.

region, O–A, of the curve is the elastic region and was discussed in Section 9.2. We are now interested in the plastic deformation that takes place in stages I, II and III of the curve.

First let us consider the tensile stress at the beginning of stage I, i.e. the tensile stress σ_y required for the onset of plastic deformation. This stress σ_y, which is called the yield stress, depends on the orientation of the crystal with respect to the tensile axis; in particular, as shown in Fig. 9.18, it is a function of the angles the tensile axis makes with the normal to the slip plane and the

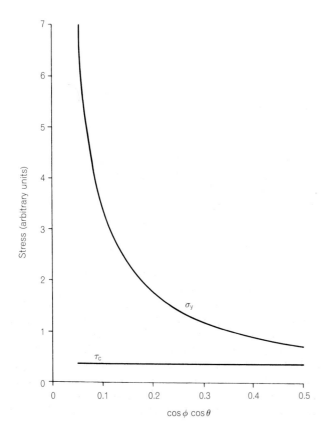

Fig. 9.18
The variation of yield stress σ_y of a single crystal with orientation. In contrast, the critical resolved shear stress τ_c is constant.

Fig. 9.19

Determination of the critical resolved shear stress on the slip plane and in the slip direction for a single crystal.

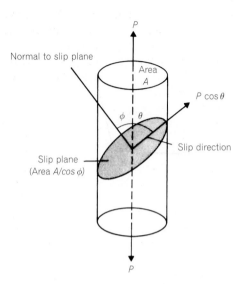

slip direction (angles ϕ and θ respectively in Fig. 9.19). This is not surprising as it is the shear stress on the slip plane and in the slip direction that causes dislocations to glide, hence this is the stress that has to be determined. Referring to Fig. 9.9, the cross-sectional area of the slip plane is $A/\cos\phi$, where A is the cross-sectional area of the crystal, and the tensile force F resolved in the slip direction is $F\cos\theta$. Therefore, the resolved shear stress, τ, on the slip plane and in the slip direction is given by

$$\tau = \frac{(F\cos\theta)}{(A/\cos\phi)} = \frac{F}{A}\cos\theta\cos\phi \qquad (9.26)$$

$$= \sigma\cos\theta\cos\phi$$

This equation is known as Schmid's law and the term $\cos\theta\cos\phi$ as the Schmid factor. If the different tensile yield stresses from tests on single crystals of varying orientations are inserted into Eq. (9.26), a constant value of τ, called the critical resolved shear stress, τ_c, is obtained (Fig. 9.18).

We know that crystals have many slip systems, e.g. c.p.h. and f.c.c. have three and twelve systems respectively, and for a given stress direction each system will have different values for ϕ and θ and therefore different values for the Schmid factor. The system that operates at the onset of stage I, referred to as the primary slip system, is that on which the critical resolved shear stress is first reached. Reference to Eq. (9.26) and Fig. 9.18 indicates that this is the system with the highest Schmid factor.

As shown in Fig. 9.17, stage I, which is termed the *easy glide* region, is approximately linear. In this region the rate at which the material is becoming harder to deform is low. A measure of this rate is the slope dσ/de, which is called the *work-hardening rate*. This is low because only the primary slip system is active, i.e. dislocations are only moving on one set of parallel planes. This means that the dislocations do not interfere with each other's motion, consequently they move unhindered for large distances and many reach the surface of the crystal. In stage II, the *linear hardening* region, dislocations are generated and move on intersecting slip planes: more than one slip system is operating. The dislocations become entangled within the crystal, thereby forming obstacles to further dislocation motion.

The number of such obstacles increases with increasing strain, thus making plastic deformation more and more difficult – this is reflected in the high work-hardening rate in stage II. When the stress is sufficiently high, cross-slip commences, which enables screw dislocations to bypass obstacles to their motion. This leads to the *parabolic* curve of stage III with its reduced work-hardening rate.

The stress–strain curve shown in Fig. 9.1(b) is typical of a polycrystalline material. The plastic deformation of a polycrystal is complex as, if the material is not going to fall apart at the grain boundaries, the strain in each grain must be accommodated by all the neighbouring grains. Because of this constraint, many slip systems have to operate in each grain before plastic deformation can commence. Not all of the required slip systems can be favourably orientated for slip, i.e. some must have low Schmid factors, and hence the tensile yield stress of a polycrystalline material is always high. In addition, the grain boundaries act as barriers to dislocations and this further hinders the onset of plastic deformation.

A consequence of the many slip systems operating, and the high stresses involved, is that the stress–strain curves of polycrystals do not exhibit stages I and II shown by single crystals. The work-hardening rate of a polycrystal up to about 10% strain ($e = 0.1$) is usually greater than that found in stage III for a single crystal; at higher strains however, the two work-hardening rates are very similar.

9.5.2 Plasticity theory for an isotropic material

The previous sections have dealt in detail with the microscopic aspects of plastic deformation in crystalline materials. We will see in Chapter 12 that plastic deformation in polymers shares some of these micromechanisms while adding others of its own. Next we will try to relate these micromechanisms to the behaviour of a material under the many different states of stress, strain and strain rate that it might experience in service. It is often convenient when analysing plastic flow to consider the material to behave as an isotropic continuum. This approach has led to the development of a mathematical theory of plasticity, which allows relationships between stress and strain to be established in the same way as they were for elasticity (see Section 9.2.4).

Now for elastic deformation we saw that each material has a property that defines its 'volumetric stiffness' – the bulk modulus K – and a property that defines its 'shape stiffness' – the shear modulus G. These properties determine the Young's modulus E and the Poisson's ratio ν measured in a tensile test. Once E and ν have been measured, elasticity theory allows the deformation of a Hookeian material to be computed for any shape of component under any geometry of loading. One basic feature of elasticity theory is *superposition* – the principle that if several loads act simultaneously, the resulting strains are simply the sum of those arising from each. For plasticity theory, this is not the case.

Fortunately there is one way in which plasticity theory is often simpler. Plastic flow proceeds, in metals at least, by the movement of dislocations, which results in the slipping of planes of atoms over one another. It is therefore reasonable to assume that no volume change takes place during plastic straining. Thus, the effective bulk modulus for plastic flow is infinite, corresponding to a value for Poisson's ratio of 0.5. Metals under pure hydrostatic pressure cannot be 'crushed' by isotropic flow – but it should also be remembered that some very important materials *can*. These are often materials whose ability to absorb energy by crushing is their most important feature, e.g. polystyrene foam (Styrofoam). Similarly many solid polymers, particularly those toughened by small rubber inclusions, expand under hydrostatic *tensile* stress.

These thoughts emphasize that the uniaxial stress–strain curve from a tensile test will not be sufficient to describe plastic flow under general, multiaxial states of stress, because if a tensile stress is applied in all three directions a metal will not yield at σ_y of Fig. 9.1(b) or, indeed, at *any* realistic stress. However, it is still possible to predict the deformation of an element of material under multiaxial stress from a uniaxial stress–strain curve. This is achieved using the concept of an *effective* stress, σ_{eff}, which expresses the combined 'effectiveness' of the three principal direct stresses in producing plastic flow. Because hydrostatic tension or compression, i.e. $\sigma_1 = \sigma_2 = \sigma_3$, cannot cause plastic flow its effectiveness, and therefore the corresponding σ_{eff}, must be zero.

One widely used definition of the effective stress is

$$\sigma_{eff} = \frac{1}{\sqrt{2}} \sqrt{[(\sigma_1 - \sigma_2)^2 + (\sigma_2 - \sigma_3)^2 + (\sigma_3 - \sigma_1)^2]} \qquad (9.27)$$

which is referred to as the von Mises effective stress. It can be seen from this equation that, when $\sigma_2 = \sigma_3 = 0$ (i.e. the state of stress that exists in the uniaxial tensile test), $\sigma_{eff} = \sigma_1$ and the effective stress is equal to the stress that exists in the tensile test. The combinations of principal stress that have the same effect as any uniaxial stress σ_{eff} can be plotted as a three-dimensional surface in σ_1, σ_2, σ_3 axes. An important special case is the *yield locus* given by $\sigma_{eff} = \sigma_y$, which shows all the combinations of stress at which the material will yield. As Fig. 9.20 shows, the yield locus for a von

Fig. 9.20

Yield locus, inside which the material remains elastic, for a von Mises material under combined stress. (a) If $\sigma_3 = 0$, the locus is an ellipse centred on the origin; (b) otherwise it is a cylinder centred on the line $\sigma_1 = \sigma_2 = \sigma_3$.

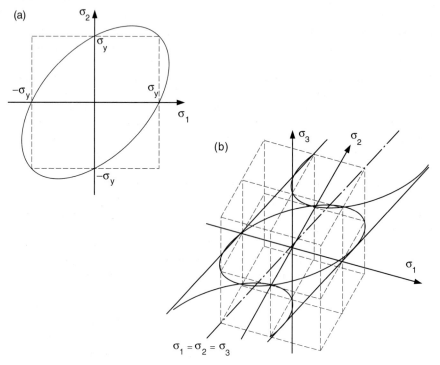

Mises material takes the form of a cylinder whose axis is the line $\sigma_1 = \sigma_2 = \sigma_3$, representing pure hydrostatic stress. One can never cross the yield locus by moving along this line, because a von Mises material will never yield under positive or negative hydrostatic pressure alone.

It must be emphasized that a von Mises model material is both isotropic and homogeneous, i.e. has the same properties in all directions and in all places. This implies that the stress–strain curves in compression and tension are equal and opposite, whereas for plastics in particular they are actually very different. Furthermore, a preferred orientation may be induced in a material during manufacture, resulting in mechanical anisotropy.

Plasticity theory goes on to define an equivalent plastic strain corresponding to σ_{eff} so that the stress–strain curve can be used to predict deformation. We shall not pursue it here. To solve a real problem it is necessary to treat each cubical element separately and to ensure that the forces between adjacent cubes are in equilibrium. Hence an additional *equation of equilibrium* is required. Similarly, the strains of neighbouring cubes must be matched to one another, for their faces must meet or there would be voids in the solid. This condition is met by an *equation of compatibility*. Another set of equations called boundary conditions ensure that the boundary surfaces of the specimen are of the shape defined in the problem and that the stresses on them conform to the actual applied stresses. Only rarely can such a complicated set of equations be solved algebraically to give an analytical expression describing the stress and strain fields. All practical cases resort to numerical analysis, using computational methods that are explained in more advanced texts.

9.6 Fracture

Fracture, the separation of a material by creation of a new surface, is the other principal way in which a material can respond mechanically to applied stress. At one extreme of this broad definition lie the *ductile* fractures often inflicted on aluminium drink cans and polyethylene bags. Here separation is achieved only after the surrounding material has been deformed plastically, a process that is also sometimes described as *rupture*. At the other extreme are the classical brittle fractures typical of inorganic glass. Here the separation process has been so finely localized that the fragments can be reassembled so as to make the new surface plane almost invisible. In general, the term 'fracture' is reserved for separation under stresses that are not high enough to cause yield in an unflawed component of the same material.

Fracture is intensively studied for its potential to cause sudden catastrophic failures of life-threatening severity, for example of a nuclear reactor pressure vessel, an aircraft fuselage or a fuel gas pipeline. Other fracture events are equally significant but much more insidious, e.g. the fatigue failures that can cause electrical discontinuities in microelectronic devices. Sometimes, however, fracture is studied for those situations in which it is a desired outcome, perhaps under carefully defined conditions, e.g. the programmed rupture of plastic containments for car safety airbags.

Whether ductile or brittle, fracture may occur under impact, fatigue, creep or environment-assisted conditions. Sometimes the mechanism may involve a combination of causes, e.g. a corrosion-fatigue fracture. In this section we will concentrate on distinguishing ductile and brittle fractures and introduce the concept of fracture toughness. But first let us estimate the strength of an ideal, defect-free material.

9.6.1 Ideal fracture stress

Consider applying an increasing tensile stress to an ideal atomic material containing no flaws or defects. When the stress reaches some critical value – the ideal fracture stress – the atomic bonds will fail across a plane perpendicular to the tensile stress. We can calculate the ideal fracture stress using similar reasoning to that employed to estimate Young's modulus in Section 9.2.2. Referring again to Fig. 9.3, if the tensile force at point d is F_{max} and the stress σ_{max}, then

$$F_{max} \approx \sigma_{max} r_0^2$$

In order to make the mathematics easier, we approximate the force-displacement curve by a half sine wave; therefore, in terms of stress, we have

$$\sigma \approx \sigma_{max} \sin \frac{2\pi r_0 e}{\lambda}$$

where λ is the wavelength of the sine wave. Now σ_{max} is a constant and the simplest way to evaluate it is to consider the situation at small strains, but this does not mean the answer is only applicable at small strains. At small strains $\sin \theta \approx \theta$ and Hooke's law is obeyed, hence

$$\sigma \approx \sigma_{max} \frac{2\pi r_0 e}{\lambda}$$

then substituting $\sigma = Ee$ (Hooke's law) and rearranging we get

$$\sigma_{max} \approx \frac{E\lambda}{2\pi r_0} \tag{9.28}$$

For most materials this result predicts that σ_{max} lies approximately within the range $E/5$ to $E/30$. In practice, materials fracture at stresses much lower than σ_{max}; for example, ordinary glass has a fracture stress of about $50\,\mathrm{MN\,m^{-2}}$, which is $E/1000$, while alloys of iron and nitrogen can even be heat treated to give a fracture stress as low as $5\,\mathrm{MN\,m^{-2}}$ ($E/40,000$).

9.6.2 Brittle fracture: Griffith's theory

Brittle fracture occurs suddenly, at a stress well below the ideal fracture stress, and is preceded by very little, if any, plastic deformation. Griffith was the first to offer an explanation for the low fracture strength of brittle materials such as glass. He postulated that such materials contain many small cracks, each of which concentrates stress at its tip. We will now outline the theory that he developed from this idea and that bears his name.

Figure 9.21 represents a crack-like, elliptical hole in a large sheet of material. The plate is under a remote tensile stress σ applied across the major axis of this ellipse, i.e. tending to open it. Figure 9.7 illustrated that for a circular hole the stress is concentrated at the equator. It can be shown that the stress σ_{tip} at the tip of this elliptical hole is given by

$$\sigma_{tip} = 2\sigma \left(\frac{a}{\rho} \right)^{1/2} \tag{9.29}$$

Fig. 9.21

Model for the Griffith crack theory.

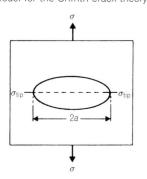

where $2a$ is the length of the major axis and ρ is the radius of curvature at each tip. If ρ is very small, the stress is very highly concentrated in a region within approximately ρ of the tip. If σ_{tip} exceeds the internal cohesive strength of bonds we can assume that crack extension is possible: in other words, a *mechanism* for crack extension exists.

However we must also consider a *thermodynamic criterion* for crack growth. This was the approach taken by Griffith, who argued that there were two energies to be taken into account when a crack propagated: a release of elastic strain energy U_E and an increase in surface energy U_S. The reader will recall from the previous chapter that solid–gas interfaces have an energy associated with them. As the material separates along a crack, new surfaces are being created and a certain amount of energy must be provided to create them (i.e. some work must be done). Now before the crack propagates, elastic strain energy is stored in the stressed material. Some of this is released when the material relaxes as the crack extends. Griffith supposed that the crack propagates when the released strain energy is just sufficient to provide the surface energy necessary for the creation of the new surfaces.

The elastic strain energy for unit volume has been given in Eq. (9.12) as $\frac{1}{2}\sigma^2/E$. To simplify the analysis we take the plate thickness to be unity. Now although the stress is very highly concentrated near each crack tip and the strain energy density there is high, the volume is very small. Near to most of the crack surface, on the other hand, the stress falls to zero while very far from the crack it remains σ. We can therefore assume that the presence of the crack relieves stress and therefore releases elastic strain energy from an approximately circular region of the plate, of radius a. This gives a total elastic strain energy release due to the crack of

$$\frac{\sigma^2}{2E} \times \text{area} \times \text{width} = \frac{\sigma^2}{2E}\pi a^2$$

per unit thickness. More exactly, the strain field can be integrated from infinity to the surface of the crack, which makes the elastic energy U_E released per unit width twice the above value, i.e.

$$U_E = \frac{\sigma^2 \pi a^2}{E} \tag{9.30}$$

If the surface energy of the material per unit area is γ joules per square metre then the surface energy of a crack of length $2a$ and unit width will be

$$U_S = 4\gamma a \tag{9.31}$$

where we multiply by 2 because there are two faces. The total energy U associated with the presence of the crack is therefore given by

$$U = 4\gamma a - \frac{\sigma^2 \pi a^2}{E}$$

and these terms are plotted as a function of crack length a in Fig. 9.22. The total energy is a maximum at length a_{critical} and consequently a crack of that length is unstable and reduces the total energy by growing. The maximum in U occurs when

$$\frac{dU}{da} = \frac{dU_S}{da} - \frac{dU_E}{da} = 0 \tag{9.32}$$

Fig. 9.22

Variation in the surface energy, strain energy and total energy with crack length.

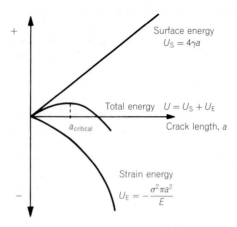

From this equation we can see that the crack will grow when the strain energy release rate dU_E/da is equal to, or greater than, the rate of change of the surface energy with crack length dU_S/da. Substituting for U_S and U_E we obtain

$$\frac{dU}{da} = \frac{d}{da}(4\gamma a) - \frac{d}{da}\left(\frac{\sigma^2 \pi a^2}{E}\right) = 0$$

therefore

$$4\gamma - \frac{2\sigma^2 \pi a}{E} = 0 \tag{9.33}$$

and so

$$\sigma = \left(\frac{2\gamma E}{\pi a}\right)^{1/2} \tag{9.34a}$$

This is the Griffith equation. Although the factor $1/\pi^{1/2}$ will not be the same for a different geometry (e.g. for a surface flaw rather than a through crack), the stress necessary to make the crack propagate will always vary inversely as the square root of the crack length. Hence the fracture stress of a brittle material is determined by the length of the largest crack existing before loading. The relationship between fracture stress and crack length was first confirmed by measuring the strengths of glass containing sharp cracks of known lengths. It explains why very thin whiskers or fibres of glass and other brittle materials are so strong – it is because they are only large enough to contain very small cracks. This fact, as we will see in Chapter 13, is the foundation on which modern composite materials science is based.

The above analysis applies to a crack in a *thin* plate under stress – these are known as *plane stress* conditions. Because of the very high stresses and strains near the crack tip, the effect of Poisson's ratio, ν, pulls the plate surfaces together, forming a 'dimple' on each surface that is often visible. In a thin plate this dimpling can take place freely because the plate surface is nearby and unrestrained. Within a thicker plate, however, the material surrounding a crack tip is restrained from contracting in the through-thickness direction by the less highly stressed material around it. Under these *plane strain* conditions the crack front is held in a state of tension that adds to the strain energy of the plate, and Eq. (9.34) becomes

$$\sigma = \left(\frac{2\gamma E}{\pi a(1 - \nu^2)}\right)^{1/2} \tag{9.34b}$$

Although Eq. (9.34b) suggests that plane strain conditions slightly increase the fracture stress, they often have a deleterious effect on the material itself so that γ is effectively reduced. To understand this we will need to look more closely at the way materials resist crack extension.

9.6.3 Fracture mechanics

Griffith's theory treats fracture as a thermodynamically *reversible* process – the separated surfaces are so clean that if the stress were reduced to zero they would meet and heal again. Spontaneous crack healing is possible in some systems, particularly if a little extra energy is supplied by heating. However, most real materials behave as if they had a much higher surface energy γ because work is absorbed by plastic deformation of material just *beneath* the fracture surface. This is often revealed by roughening and irregularity of the surface, so that the area itself increases.

This additional pseudo-surface work does not make the Griffith theory any less useful, because crack healing is seldom of any practical interest. In place of a recoverable surface energy 2γ per unit area of material separated, we simply characterize the material by the *total* energy needed to separate the two surfaces, including that to deform nearby material. To drive crack propagation the strain energy release rate dU_E/da must reach this critical total value, which is given the symbol G_c and replaces 2γ in Eq. (9.34):

$$\sigma = \left(\frac{EG_c}{\pi a}\right)^{1/2} \tag{9.35a}$$

for plane stress and

$$\sigma = \left(\frac{EG_c}{\pi a(1 - \nu^2)}\right)^{1/2} \tag{9.35b}$$

for plane strain conditions. G_c is a measure of the fracture resistance of a material and as such it is a material property just like Young's modulus, yield stress, etc. G_c is usually given in units of $kJ\,m^{-2}$ and the smaller it is the less tough, or more brittle, is the material, as illustrated by the values given in Table 9.1.

The subject of *fracture mechanics* constructed on these foundations is concerned largely with assessing the stability and remaining strength of loaded structures containing cracks. In most cases these structures cannot be represented as large, centrally cracked plates under uniaxial stress. A difference in either the crack shape or the structure geometry will lead to a different expression for the strain energy release rate term $-dU_E/da$

Material	G_e (kJ m^{-2})	K_e (MN m$^{-3/2}$)
High-strength steel	40	92
Co–Cr–Mo–Si alloy	2.5	25
Polymethyl methacrylate	0.6	1.5
Epoxy resin	0.2	0.8
Alumina	0.06	4.5
Glass	0.003	0.4

Table 9.1
Fracture toughness of a variety of materials

for a given stress σ. This term, the *strain energy release rate*, is now denoted as G. For the particular geometry of Fig. 9.21

$$G = \sigma^2 \pi a \qquad (9.36a)$$

under plane stress conditions or

$$G = \sigma^2 \pi a (1 - \nu^2) \qquad (9.36b)$$

under plane strain conditions. Many other similar expressions exist for geometries simple enough to be analysed exactly, while computation using finite element or similar methods can determine G for the more complicated geometries typical of real components.

Once G has been defined in terms of applied load or stress for a given geometry in this way, a single test using that geometry is sufficient to determine the material property G_c, because at the point of fracture

$$G = G_c \qquad (9.37)$$

For the same material the point of fracture can then be calculated – as a critical crack length or a critical stress – in any other geometry for which G is a known function of stress and geometry, like Eq. (9.36).

Other parameters are used for quantifying the fracture toughness of a material and using the result to predict fracture in another geometry. The most commonly used in engineering is the *stress intensity factor K*, and its critical value K_c. There is a simple relationship between G and K:

$$K^2 = EG \qquad (9.38a)$$

for plane stress conditions and

$$K^2 = \frac{EG}{(1 - \nu^2)} \qquad (9.38b)$$

for plane strain. Because plane strain conditions are the most severe the value of K_c usually quoted is the *plane strain fracture toughness*, given the symbol K_{1c} or K_{Ic}.

Combining Eq. (9.38) with Eq. (9.36) we find that the stress intensity factor for the geometry of Fig. (9.21) is

$$K = Y\sigma\sqrt{a} \qquad (9.39)$$

where $Y = \sqrt{\pi}$. Stress intensity factors for many other simple geometries have the same form as Eq. (9.39), being proportional to applied stress or load and differing only in the factor Y. As for the strain energy release rate parameter G, the critical point of crack length or stress for a loaded structure is defined by the expression

$$K = K_c \qquad (9.40)$$

which is analogous to Eq. (9.37). The material property K_c is known as the fracture toughness.

9.6.4 Cohesive models

The stress intensity factor K is more directly related to load and more convenient for strength calculations than G. However, G_c is usually easier to interpret than K_c in terms of the physical origin of a material's fracture

(a)

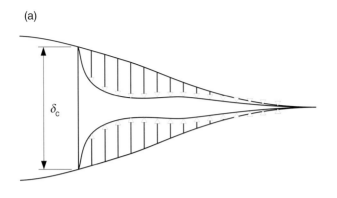

Fig. 9.23

A cohesive model explains fracture behaviour in terms of the variation in stress as internal surfaces are separated.

(b)

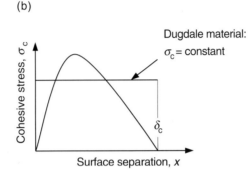

resistance, i.e. the ability of the surface creation process to dissipate energy. This is well illustrated by the *cohesive zone* fracture model, which applies directly to materials in which the damage is confined to a relatively shallow layer beneath each fracture surface. This is often the case for polymers and, at a larger scale, for concrete.

The model represents this damaged zone by an imaginary extension to the crack, and replaces the damaged material within it by tensile 'cohesive' forces that act to prevent its surfaces from drawing apart (Fig. 9.23a). Now each material can be characterized by a *cohesive law* describing the way in which cohesive stress σ_c varies with surface displacement x (Fig. 9.23b). We met an idealized example of a cohesive law in the Condon–Morse force–displacement curve for separation of two atoms. The total energy needed to separate unit area of the surfaces is simply the area under this curve, and this total energy per unit area is equal to G_c so that

$$G_c = \int \sigma_c \, dx \qquad (9.41)$$

A special case arises where σ_c is constant. This is a good representation of a polymer *craze* in which, as we will see in Section 12.10.3, a layer of polymer is converted to a spongy, cavitated form and drawn out under almost constant stress to a critical distance δ_c. For this case Eq. (9.41) reduces to $G_c = \sigma_c \delta_c$. Furthermore, detailed stress analysis leads to an expression for the *length* of this constant-stress cohesive zone:

$$c = \frac{\pi}{8} \frac{EG}{\sigma_c^2} \qquad (9.42)$$

The cohesive zone length before fracture can often be measured directly or from features left on the fracture surface.

9.6.5 Ductile fracture

Fig. 9.24

Cavities linking up perpendicular to the applied tensile stress in the necked region of a copper specimen [K. E. Puttick, *Phil. Mag.* 4, 964 (1959)].

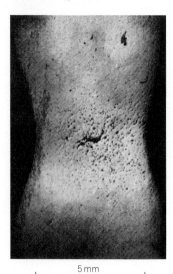

5 mm

Ductile failure occurs after extensive plastic deformation. In fact, much of the energy absorbed by a ductile failure is not essential to the fracture process at all, but is absorbed in deformation of the material surrounding the propagating crack tip. The damage inflicted by concentrated stresses around the crack tip is often so severe that it is difficult to identify such a 'point' at all. First small cavities or voids are formed, and in the case of a tensile test this takes place in the neck where the deformation is concentrated. This is followed by the linking of the cavities to form a crack that spreads in a plane approximately perpendicular to the applied tensile stress (Fig. 9.24). Finally the crack propagates to the surface of the sample by shearing in a direction approximately 45° to the tensile axis to give a *'cup and cone'* type fracture (Fig. 9.25).

The cup and cone fracture illustrates that ductile materials have less resistance G_c to fracture under plane strain conditions. Because the stress state near an internal crack tip involves additional tension in a third direction (along the crack front), shear stresses – which, remember, arise from differences between stresses – are actually reduced for the same value of externally applied stress. The level of shear stress that could drive plastic flow and blunt the crack near the specimen surface cannot develop internally before the material separates under direct stress. Thus ductile materials are usually much tougher as thin films. This fact is very apparent for the often spectacularly tough polymer films used for packaging – a 5 mm thick stack of these films may show a toughness many times greater than that shown by a 5 mm plate of the same polymer. We saw earlier that the smaller the cross-section of a brittle material, the stronger it becomes; we see now that the thinner a ductile material is, the tougher it becomes. These facts, as we shall see in Chapter 13, underlie the design of strong and tough man-made *composite* materials.

A ductile fracture surface appears dull when viewed with the naked eye, whereas the fracture surface of a metal or plastic that has failed in a brittle manner is silky or shiny. Closer examination of the surfaces with a scanning electron microscope reveals marked topographical differences. The ductile failure surface has a characteristic 'dimpled' appearance which is a manifestation of the cavities that merged to form it. It is now accepted that the nucleation of the cavities in metals during plastic deformation is associated with second phase particles and inclusions; this relationship between

Fig. 9.25

A typical cup and cone ductile fracture resulting from a tensile test on steel.

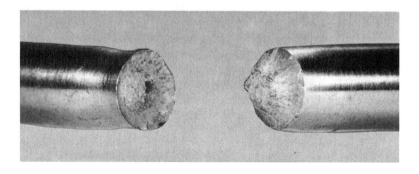

(a)

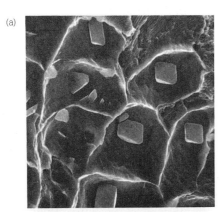

(b)

15 μm

Fig. 9.26

Scanning electron micrographs of fracture surfaces in a mild steel: (a) ductile fracture, (b) brittle cleavage failure. (Courtesy of T. J. Baker.)

cavity and particle is clearly shown in Fig. 9.26(a). This explains why the ductility of a metal increases as the purity is increased, i.e. the purer the material the fewer second phase particles and inclusions. A simple model for cavity nucleation at particles due to interfacial decohesion has been proposed by Gurland and Plateau. Their approach was similar to that of Griffith for brittle fracture, in that they assumed that the release of elastic strain energy must be sufficient to create the new surfaces. Indeed the following expression, which they derived for the tensile stress required for particle–matrix decohesion, has similarities to the Griffith equation [see Eq. (9.34)]:

$$\sigma = \frac{1}{q}\left(\frac{E\gamma}{d_{\mathrm{p}}}\right)^{1/2} \qquad (9.43)$$

where q is the stress concentration factor at the particle of diameter d_{p}, γ is the surface energy and E is the weighted average of the elastic moduli of the particle and the matrix. If this simple model is extended to take account of the energy of plastic deformation local to the particle, Eq. (9.43) is modified to

$$\sigma = \frac{1}{q}\left(\frac{E\gamma}{d_{\mathrm{p}}}\right)^{1/2} + \frac{\sigma_{\mathrm{y}}}{q}\left(\frac{\Delta V}{V}\right)^{1/2} \qquad (9.44)$$

where σ_Y is the yield stress of the matrix with particles present, V is the volume of a particle and ΔV is the volume plastically deformed around the particle. Both of these equations indicate the importance of particle size on the initiation stress.

In contrast to the 'dimpled' topography of the ductile fracture, a metal that has failed in a brittle manner has a surface consisting of flat, shiny facets [Fig. 9.27(b)]. This is because the fracture has taken place by the separation of grains along specific crystallographic planes, called the *cleavage planes*, e.g. in b.c.c. iron the cleavage planes are the {100}.

9.6.6 Environment-assisted cracking

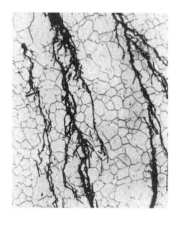

Fig. 9.27

Transgranular stress-corrosion cracking of stainless steel due to chlorides (*ASM Metals Handbook*, 8th edn, Vol. 7, p. 135).

Even though an environment may not cause general corrosion, it can markedly reduce the resistance to crack propagation under stress – whether the stress be externally applied or a residual stress due to cold working. In metals this is termed *stress corrosion* and most structural metals are susceptible to it in some environments, e.g. chloride solutions for stainless steel and solutions containing ammonia for brass. A stress corrosion failure may be transgranular, as in the case of chloride solutions and stainless steel (Fig. 9.27), or intergranular, as exemplified by the fracture of brass in an ammonia solution. In plastics a similar phenomenon is known as *environmental stress cracking* and will be discussed in Chapter 12.

Fracture mechanics has been applied to the analysis of stress corrosion failures. In a corrosive environment, crack growth commences at *subcritical* values of $K < K_c$ or $G < G_c$. In these circumstances either K or G still serves to characterize how severely the crack is loaded, so it is not surprising that there is a relationship between the crack velocity and K, as depicted in Fig. 9.28. This has three distinct regions of crack growth. K_0 is the stress corrosion limit and between K_0 and K_T, i.e. region I, crack propagation is controlled by the rate of chemical reaction at the crack tip. At intermediate values of K, in region II, it is the rate of diffusion of the corrosive species to the crack tip that controls its rate of propagation. Finally, in region III, the crack propagates in a mixed mode with both stress corrosion and normal fracture contributions.

Hydrogen cracking or *hydrogen embrittlement* is another form of environment-assisted fracture in metals. As the name suggests, this type

Fig. 9.28

The variation in crack velocity with stress intensity factor for corrosive and inert environments.

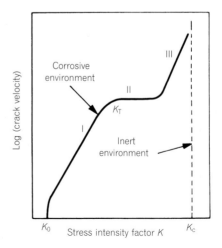

of fracture is the result of hydrogen from a corrosion reaction being absorbed by the metal and reducing its ductility. Many steels are prone to hydrogen embrittlement, e.g. oil or gas contaminated with H_2S has been responsible for innumerable failures in high strength steels used for well casings and risers. However hydrogen cracking is not confined to ferrous alloys and can occur in refractory metals, titanium alloys and aluminium alloys. In many ferrous and aluminium alloys the absorbed hydrogen segregates to grain boundaries and reduces G_c, leading to intergranular failure.

Other forms of environment-assisted cracking are also encountered, such as liquid metal embrittlement and corrosion fatigue, but these will not be discussed here because sufficient information has been given in this section to demonstrate the significant effect that the environment can have on the mechanical behaviour of materials.

9.7 Mechanical testing

We have now surveyed some of the general ways in which materials deform and fracture under stress, and introduced some continuum 'models' by which this behaviour can be represented. Although we looked at a few examples of the *mechanisms* within materials that give rise to particular deformation and fracture phenomena, more detailed consideration will be deferred to later chapters, which review each main class of materials in turn. For the time being, we will look at the ways in which materials are tested to reveal this behaviour under different forms of loading.

9.7.1 The tensile test

As described earlier in this chapter, in the tensile test the specimen is gradually elongated under tension, the load and the extension being recorded continuously. The load–extension curve obtained in this way is identical to a plot of engineering stress against engineering strain.

Some typical engineering stress–strain curves for different materials are shown in Fig. 9.29. For all materials the stress–strain relationship depends on the chemical composition, the heat treatment and the method of manufacture. In brittle materials such as cast iron, high strength steel, tungsten carbide and high strength aluminium alloys, necking does not occur and failure occurs after very limited extension. Glassy polymers such as polymethyl methacrylate and polystyrene also behave in this brittle manner, but fail at much lower stresses. The properties of concrete, which is brittle, depend on the mix (i.e. the proportions of sand, cement, aggregate and water) and the curing time, as described in more detail in Chapter 13.

In ductile materials like aluminium, copper and annealed mild steel, which exhibit large strains before failure, the engineering stress at fracture is lower than the tensile strength although the actual (true) stress in the neck will be higher. Rubber shows exceptional behaviour, exhibiting an extended region in which the stress increases only slightly with strain, followed by a region of rapidly increasing hardening; even more remarkable is the fact that rubber will recover elastically from strains up to its breaking point. The fundamental nature of rubber elasticity will be considered in Chapter 11. Finally, many polymers show plastic behaviour in which necking is followed by a drop in stress to a lower, constant value.

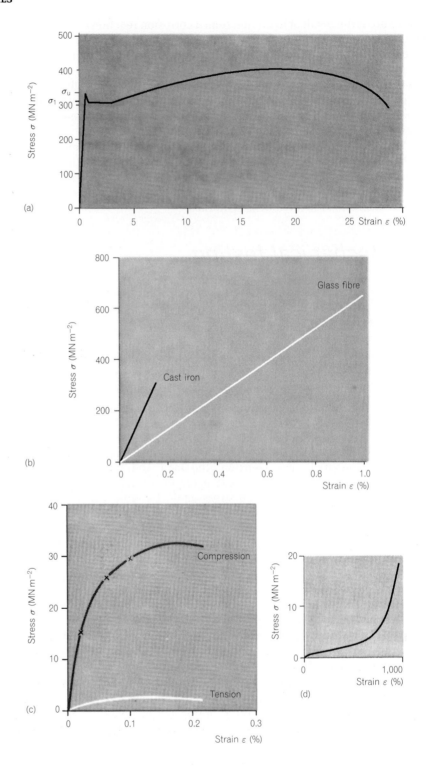

Fig. 9.29

Engineering stress–strain curves for (a) mild steel (σ_u = upper yield stress and σ_l = lower yield stress); (b) brittle solids; (c) concrete in tension and compression; (d) vulcanized rubber.

As extension proceeds, these plastics undergo *cold drawing*, during which the neck does not reduce further in cross-section but instead extends steadily by drawing in fresh material until the entire gauge length of the specimen has been consumed (Fig. 9.30).

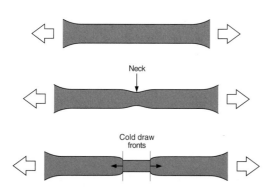

Fig. 9.30

Necking and cold drawing. After extending to a certain strain, the neck stabilizes and the 'shoulders' move outwards to draw in new material.

As cold drawing illustrates quite clearly, the engineering stress–strain curve from a tensile test does not always provide a clear picture of fundamental material properties. The object of the test is to subject a defined volume of material to a steadily increasing uniaxial strain at constant rate. In this case it succeeds in continuously feeding material through a rather uncontrolled process, which begins and ends with a strain rate of nearly zero.

To develop a more general understanding of the plastic flow of materials, which is fundamental to their mechanical properties, we need to develop the concepts of the true stress and true strain. If the current (instantaneous) load is F and the instantaneous cross-sectional area is A then the true stress σ_T is [from Eq. (9.16)]

$$\sigma_T = \frac{F}{A}$$

Similarly, the engineering strain e does not give a meaningful measure of the overall strain when the specimen has lengthened considerably. Two alternatives are in common use: the extension ratio introduced through Eq. (9.15), and the *true strain*, ε, defined as the integral of the elemental incremental strain $d\ell/\ell$, where ℓ is the instantaneous length, between the initial length ℓ_0 and the current (instantaneous) length. Thus

$$\varepsilon = \int_{\ell_0}^{\ell} \frac{d\ell}{\ell} = \ln \frac{\ell}{\ell_0} \tag{9.45}$$

$$= \ln \lambda$$

For example an extension ratio of 2 (100% engineering strain) is equivalent to a true strain of $\ln 2 = 0.69$, while an extension ratio of $\frac{1}{2}$ (−50% engineering strain, or a compression ratio of 2) is equivalent to a true strain of $\ln \frac{1}{2} = -0.69$.

The true stress–true strain curve for a metal always has a continually rising characteristic, as would be expected when consideration is given to the mechanism of work-hardening, which is the interaction of mobile dislocations with each other and with dislocation tangles (Fig. 9.31). However, some polymers *strain soften*.

The occurrence of necking can be predicted from the true stress–true strain curve by an analysis originally due to Considère. The instantaneous load is

$$F = \sigma A \tag{9.46}$$

Fig. 9.31

True stress–true strain curve for copper.

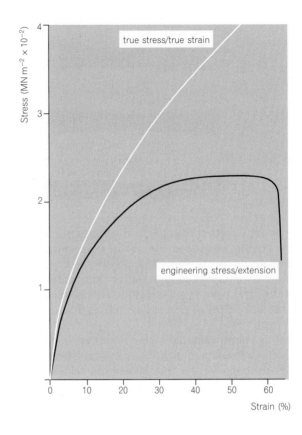

At necking the load reaches a maximum value, so that $dF/F = 0$. Differentiating Eq. (9.46) gives

$$\frac{dF}{F} = \frac{d\sigma}{\sigma} + \frac{dA}{A} \tag{9.47}$$

Now, because most materials undergo plastic flow at substantially constant volume, $A\ell = $ constant; differentiating this condition gives

$$\frac{dF}{A} = -\frac{d\ell}{\ell} = -d\varepsilon_T \qquad \text{[from Eq. (9.45)]}$$

Substituting this into Eq. (9.47) and putting $dF/F = 0$ we find that, at maximum load,

$$\frac{d\sigma_T}{d\varepsilon_T} = \sigma_T \tag{9.48}$$

Hence necking will occur in a tensile test at a strain at which the work-hardening rate equals the effective (true) stress.

An empirical relationship that gives an approximate fit to the behaviour of some metals is

$$\sigma_T = C\varepsilon^n \tag{9.49}$$

where C and n are constants. The exponent n is often called the strain-hardening exponent and lies between 0.1 and 0.5 for most metals. Applying Eq. (9.48) to Eq. (9.49), we find that necking occurs when

$$nC\varepsilon^{n-1} = C\varepsilon^n \tag{9.50}$$

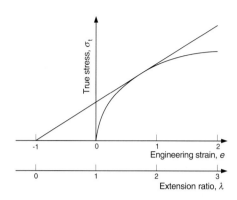

Fig. 9.32

Considère's construction for the point of neck formation.

i.e. when $\varepsilon = n$, so the physical significance of n is that it is a measure of the effective (true) strain at necking. The value of the engineering strain at necking, which can be derived from Eq. (9.50), is referred to as the *uniform elongation*.

The point at which necking will occur can be predicted directly from the true stress versus engineering strain (or extension ratio) curve using a simple construction. The relationships between engineering strain e, true strain ε and extension ratio λ are given using Eq. (9.15) as:

$$\lambda = 1 + e$$

and

$$\varepsilon = \ln \frac{\ell}{\ell_0} = \ln(1 + e) \tag{9.51}$$

Combining Eqs (9.48) and (9.51) we can easily show that, when necking begins

$$\frac{d\sigma_T}{de} = \frac{\sigma_T}{1 + e} = \frac{\sigma_T}{\lambda} \tag{9.52}$$

which is satisfied where a tangent drawn from $e = -1$ meets the true stress–strain curve or one drawn from $\lambda = 0$ meets the true stress–extension curve, as illustrated in Fig. 9.32.

Cold drawing and superplasticity

Materials that yield abruptly at a sufficiently high stress and then strain harden strongly can show not just one but *two* tangents of this kind, as

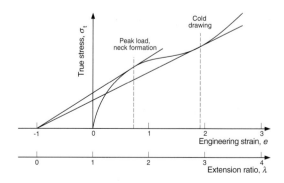

Fig. 9.33

The second Considère condition provides a condition for cold drawing.

shown in Fig. 9.33. Because the Considère construction was devised from the condition that the engineering stress–strain curve passed over a maximum, it is clear that the second tangent point must correspond to a *minimum* on this curve. A neck forms between the first and second tangent points, and it is within this neck that all deformation is concentrated while the engineering stress (the load) falls. The load can pull out of this dive if strain hardening is able to compensate for the continuing reduction in area.

If strain hardening is sufficient we may observe the remarkable phenomenon of cold drawing. In order to extend further, the material at the centre of the neck must be subjected to a higher engineering stress, i.e. a higher load. The material adjacent to it, however, has yet to reach this strain. The result is that as the specimen is further extended one or even two drawing 'fronts' propagate away from the original neck site as successive planes of material are brought to the second Considère tangent condition. In Section 12.10.1 we will see that the conditions for a second tangent are satisfied by many polymers whose internal viscosity gives them a high yield stress while their rubber-like elasticity gives them strong strain hardening. This phenomenon is also seen in a few metals such as titanium and aluminium, where it is known as *superplasticity* and originates from a rate of work hardening that increases strongly with strain rate.

Ductility

Other terms referred to in the tensile test are the ductility, defined as the strain at fracture and usually quoted as a percentage; the reduction in area at the neck (another measure of the ductility); and the proof stress. The 0.1% proof stress, for example, is the point on the engineering stress–strain curve from which, after removal of the stress and elastic recovery (as illustrated for example by the lines SS' and RR' in Fig. 9.1), there would remain 0.1% permanent plastic strain.

Referring to Fig. 9.20 it can be seen that brittle materials such as cast iron show little or no plastic deformation before fracture, i.e. they are not ductile. Materials like this can be dangerous because high local tensile stresses occur in structures at points of *stress concentration*, such as the hole in the plate of Fig. 9.7, and can lead to brittle failure. Ductile materials such as copper exhibit considerable plastic flow before fracture in tension, so that stress concentrations are 'relieved' by plastic flow. Note that the diagram for mild steel is different from other materials in that it exhibits a sharply defined *upper yield point* followed by a period of deformation at a *lower yield stress*. During this phase the plastic deformation is non-uniform along the specimen. It begins at points of stress concentration and propagates along the specimen in the form of bands of deformation (bands of moving dislocations) known as *Lüder's bands*. When the Lüder's bands have covered the whole specimen, the deformation becomes uniform and the material work hardens in the normal way.

This phenomenon can be explained in terms of an interaction between the interstitial carbon and nitrogen atoms, which are always present in steel, and dislocations. The carbon and nitrogen atoms preferentially occupy interstitial sites near dislocations, and in so doing hinder the movement of the dislocations. The dislocations are said to be *pinned* by interstitial *atmospheres*. When the upper yield point is reached, at points of stress concentration the dislocations either break away from their interstitial atmospheres or multiplication occurs producing mobile (unpinned) dislocations.

Fig. 9.34

Photograph of barrel-shaped distortion produced in a compression test.

The stress required to move the dislocations (the lower yield stress) is less than the unpinning or multiplication stress (the upper yield point).

9.7.2 The compression test

Brittle materials such as concrete are generally tested in compression because this is the kind of stress that they are most frequently used to support. Necking does not occur but problems may arise because of friction between the ends of the specimen and the loading platens. Friction leads to non-uniform deformation and the specimen tends to form a barrel shape, as shown in the photograph of Fig. 9.34, especially when ductile materials (which undergo large deformation) are tested.

A big advantage of the compression test is that the true stress–true strain curve can be estimated from the results, up to much larger strains than can be obtained in the tension test. This is often valuable when we need to predict the forces needed to form metals by processes such as forging, rolling and extrusion. Various methods have been devised to reduce or mitigate the effects of friction and non-uniform deformation at the loading platens, and hence reduce barrelling of the specimen.

9.7.3 The hardness test

The hardness test is used as a quick, inexpensive method of assessing the mechanical properties of a material. The *hardness* of a material is determined by pressing an indenter into its surface and measuring the size of the impression. The bigger the impression, the softer the material. The early Brinell hardness test used a spherical ball of 10 mm diameter, made from hard steel or tungsten carbide. More recent methods employ pyramidal or conical indenters, which have the virtue that the impression remains geometrically similar irrespective of its size. The Vickers test used a diamond

Fig. 9.35

Indentation from a Vickers hardness test.

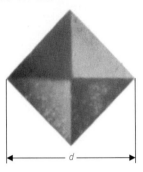

cut in the form of a square pyramid having an apex angle of 136°. The Vickers hardness number (VHN) is defined as the load/surface area of the impression, which can be shown to be

$$\text{VHN} = 1.854 \, \frac{P}{d^2} \, \text{kgf mm}^{-2} \qquad (9.53)$$

where P is the applied load (kgf) and d is the distance across the diagonal of the impression (mm), as shown in Fig. 9.35. The traditional units given in Eq. (9.53) are still commonly quoted for metals, but the hardness test is increasingly used for ceramics and generally these hardnesses are given in SI units of MN m^{-2}.

In the Brinell test, the Brinell hardness number (BHN) may be found from the expression

$$\text{BHN} = \frac{2P}{\pi D[D - \sqrt{(D^2 - d^2)}]} \, \text{kgf mm}^{-2} \qquad (9.54)$$

where D is the diameter of the ball (mm) and d the diameter of the impression (mm). For consistent results the diameter of the indentation d should be between $0.3D$ and $0.6D$. Under these conditions fairly soft materials have the same Brinell and Vickers hardness numbers, because the geometries of the impressions are not too dissimilar.

The hardness test involves indenting a material by a process of plastic deformation that has been shown to correspond to approximately 8% strain of the material. Thus the hardness is a function of the yield stress σ_y and the work-hardening rate of the material. As a rough guide the VHN is between 0.2 and 0.3 times σ_y, where σ_y is in MN m^{-2} for hard materials and $0.3\sigma_y$ for metals (see Table 9.2).

In the Rockwell test, a diamond conical indenter of 120° included angle is used and the depth of the indentation is measured.

9.7.4 Fracture toughness tests

In general terms the 'toughness' of a material is its ability to maintain strength after damage. Because the most severe form of damage is a sharp crack, this is measured by the applied stress σ needed to extend a crack. The force tending to extend a crack of length a is expressed by the stress intensity factor $K = Y\sigma\sqrt{a}$, where Y is a factor that depends on the geometry of the cracked body. The crack will extend when K reaches a critical

Table 9.2

Typical values of VHN and yield stress

Material	VHN (kgf mm⁻²)	Yield stress (MN m⁻²)
Diamond	8400	54 100
Alumina	2000	11 300
Boron	2500	13 400
Tungsten carbide	2100	7 000
Beryllia	1300	7 000
Steel	210	700
Annealed copper	47	150
Annealed aluminium	22	60
Lead	6	16

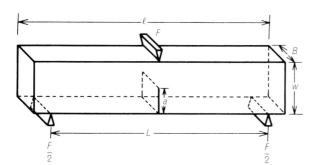

Fig. 9.36

An SENB specimen for the determination of K_c.

value K_c, the fracture toughness, so that K can be calculated by substitution of the fracture stress into this equation.

The foundations of fracture mechanics were developed in terms of an infinitely large plate containing a crack of length $2a$, for which Y simply has the constant value $\sqrt{\pi}$, but which is obviously not a very convenient geometry to test in a laboratory! However, the power of fracture mechanics lies in the fact that elastic stress analysis can be used to obtain the Y factor for any other specimen or any component geometry under a given external load, F. An example is the widely used single edge notch bend (SENB) specimen illustrated in Fig. 9.36, for which K is given by

$$K = \frac{3FL}{2BW}Y\sqrt{a} \tag{9.55}$$

where L, B, W and a are defined in Fig. 9.36 and Y is a function of a/W. As the load on an SENB specimen is increased during a fracture test, K will increase in accordance with Eq. (9.55) and the crack will propagate when K reaches the critical value K_c, the fracture toughness. Thus K_c can be calculated by substituting the fracture load into Eq. (9.55); it has units of $MN\,m^{-3/2}$. If the Young's modulus E is known, G_c, with units of $kJ\,m^{-2}$, can be calculated from the same result using Eq. (9.38). However, materials will rank differently on K_c and G_c scales because those whose fracture mechanisms give them a high fracture energy may show lower toughness (i.e. a lower failure load for a given crack size) owing to a relatively low modulus.

9.7.5 The impact test

We know that materials can separate in a brittle or a ductile manner, the latter occurring only after appreciable plastic flow. Plastic flow in metals, as has already been explained, occurs by the movement of dislocations under the action of a shear stress. Brittle fracture involves tensile separation with little or no plastic flow, so that stress systems that involve high ratios of tensile to shear stress are likely to favour brittle failure. Similarly any material composition or heat treatment giving a low ratio of tensile strength to shear strength will tend to promote brittle fracture.

Materials such as cast iron or marble are normally brittle but can be made to exhibit ductile behaviour by superimposing very high all-round hydrostatic pressure. Apart from the stress system, other important variables affecting the tendency for a material to fail in a brittle manner are the rate of straining and the temperature. Pitch flows slowly even at room temperature but will break into pieces if hit with a hammer (showing the effect of strain rate). Glasses will flow and can be moulded into almost any

Fig. 9.37

Schematic diagram of an impact
testing machine.

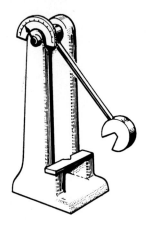

shape at elevated temperatures but are usually brittle at room temperature (showing the effect of temperature). A normally ductile material may become brittle if there is a crack-like stress concentration at a point where the material section is thick. As we saw in Section 9.6.5, this situation sets up plane strain conditions that prevent plastic flow from relieving a high level of direct stress.

A commonly used test for brittle behaviour is the Charpy test, illustrated in Fig. 9.37. In this test a standard notch is cut in a standard test specimen that is struck under impact conditions by a heavy weight forming the end of a pendulum. The notch helps to set up plane strain conditions that encourage brittle fracture. The bar of the material is 55 mm long, with cross-section 10×10 mm, and has a V-notch 2 mm deep of 45° included angle and a root radius of 0.25 mm. The bar is clamped in the machine as shown in Fig. 9.37 and the weight released from a known height so as to strike the specimen on the side opposite the notch and induce tensile stresses in it. After breaking the specimen the pendulum swings on and the height to which it rises on the other side is measured. Thus the energy absorbed in breaking the bar may be determined and if this is low the specimen is brittle. The notch brittleness of the specimen can also be assessed from the appearance of the fracture – a flat, featureless surface usually indicates that less energy has been expended in creating it.

Typically, a tough steel absorbs in the region of 130 joules of energy in a Charpy test at room temperature. A typical result from notch tests on mild steel over a range of temperatures is shown in Fig. 9.38. It will be seen that as the test temperature is raised there is a rapid rise in energy absorbed at 280 K. This is the *ductile–brittle transition*: for V-notched specimens the *transition temperature* is taken to be that at which 13–15 J of energy are absorbed. Above this temperature much more plastic flow occurs in the notch so that more energy is absorbed. Any metallurgical treatment that will move the test curve shown in Fig. 9.38 over to the left of the diagram will improve the usefulness of the steel. The unfortunate characteristic of having a ductile–brittle transition is not typical of all metals, as shown by the curve for aluminium in Fig. 9.38. Aluminium is ductile at all temperatures, although at ambient and elevated temperatures less energy is required to fracture aluminium than mild steel – we say that the mild steel is tougher at these temperatures.

Fig. 9.38

Notched Charpy test curves for
mild steel and aluminium.

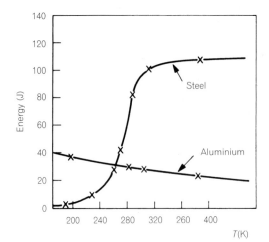

9.7.6 Fatigue testing

It is estimated that 90% of all mechanical failures are caused by fatigue, i.e. the failure of a structure under the repeated application of a load far smaller than that required to cause failure in one application. Fatigue failures often occur in a catastrophic manner with no gross distortion preceding collapse. The size and location of the cracks formed in a structure by the fatigue process often make their detection during routine inspection almost impossible.

Traditional laboratory fatigue test machines reproduce typical situations in which these failures occur: (1) push–pull (axial loading); (2) rotating bending (rotating a cantilever beam with a stationary end load); (3) reversed bending (applying an alternating bending moment to a flat plate specimen); (4) torsion (applying alternating torque). The stress cycle is often of sinusoidal form owing to the nature of these situations, and the *stress amplitude* σ_a is usually kept constant [Figs 9.39(a) and (b)]. Most tests are performed with the *mean stress* $\sigma_m = 0$, as shown in Fig. 9.39(a), but a non-zero value of σ_m may also be used [Fig. 9.39(b)].

In practice the fluctuating loads imposed on a structure during service may be very different from the regular load cycling so far described. This can be overcome to a certain extent in the laboratory by using servo-hydraulic testing machines that are programmed to give a good approximation to random loading such as that given in Fig. 9.39(c). In 'safety-critical' cases, e.g. in the aircraft industry where a fatigue failure could lead to a major disaster, large-scale models or even full-size structures are tested under

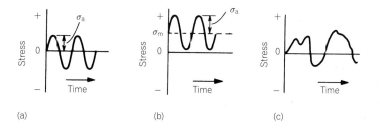

(a)　　　　　　　　(b)　　　　　　　　(c)

Fig. 9.39

Stress cycle for fatigue testing: (a) $\sigma_m = 0$; (b) σ_m is positive; (c) random σ_m.

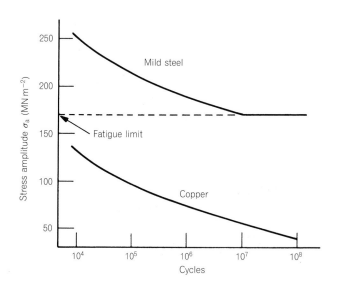

Fig. 9.40

Results of fatigue tests on steel and copper presented as S–N curves.

simulated service conditions. A complete Concorde airframe was tested under fluctuations of load and temperature designed to simulate service conditions.

To investigate the fatigue behaviour of a material a series of specimens of the material is tested to failure at different values of the stress amplitude σ_a and each test gives one point on a graph of σ_a versus the logarithm of the number of cycles to failure (this is known as an S–N curve). Most materials have S–N curves of the type shown by copper in Fig. 9.40, and the fatigue strength or *endurance limit* must be quoted relative to a specified number of cycles, the most common being the stress amplitude that will cause failure in 10^7 cycles. On the other hand, some materials, e.g. most steels and titanium alloys, have a fatigue curve of the type shown by mild steel in Fig. 9.40. For such materials there is a definite value of stress amplitude below which fatigue failure will never occur and this value, the *fatigue limit*, is quoted in some material specifications.

Fatigue mechanisms

Let us now turn our attention to what is happening on the microscopic scale during cyclic loading. We can consider a fatigue failure to consist of three stages: stage 1 is crack initiation, stage 2 is fatigue crack growth, and stage 3 is the final catastrophic fracture. For large structures such as bridges and oil rigs there are many nucleation sites for a fatigue crack, e.g. surface imperfections and defects associated with welds. In such circumstances, the growth stage of a fatigue crack is the most important. In other cases, where the surface finish is good, the initiation of the fatigue crack may occupy much of the fatigue life. It has been estimated that in well-finished laboratory specimens, which are used for S–N curve determination, the initiation stage may be as much as 90% of the lifetime at low stress amplitudes, reducing to 10% at higher stress levels. Stage 3 of a fatigue failure is, of course, always rapid compared with the preceding stages.

In specimens with smooth surfaces the initiation of the fatigue crack involves the localized movement of dislocations on slip planes at approximately 45° to the tensile axis. This localized deformation gives rise to slip bands that are similar in appearance to the slip lines described in Section 8.4.1. There is, however, a major difference between the fatigue slip bands and slip lines; on re-polishing the surface of a specimen, slip lines are removed as they are surface steps, whereas the slip bands remain. For this reason the fatigue slip bands are known as *persistent slip bands*. The persistent slip bands are associated with intrusions and extrusions at the surface from which a fatigue crack propagates for a few grain diameters (Fig. 9.41).

As shown in Fig. 9.41, in this stage the crack changes direction and becomes roughly normal to the stress axis. The fracture has a series of fine, non-crystallographic ridges known as *fatigue striations* (Fig. 9.42). Each striation is produced by a single stress cycle, and there are several models to account for their formation. One such model is illustrated in Fig. 9.43 and shows that a striation is a consequence of an increment of crack growth and concomitant blunting and resharpening of the crack tip by plastic deformation. Figure 9.43(a) corresponds to the zero load condition after a number of cycles; striations 1, 2 and 3 were formed during previous cycles. When a small tensile stress is applied the crack opens up and shear stresses are set up at the crack tip [Fig. 9.43(b)]. As the applied tensile

(a) Cyclic stress (b)

Crack

3 2 1

Intrusion

Extrusion

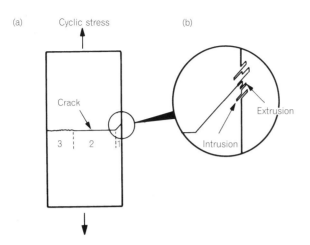

Fig. 9.41

Fatigue failure. (a) The three stages. (b) Stage 1.

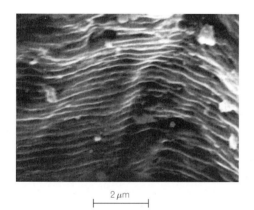

2 μm

Fig. 9.42

Fatigue striations in stainless steel.

stress increases, plastic deformation occurs due to the shear stress and this leads to crack extension and blunting [Fig. 9.43(c)]. Once the compression period of the cycle is entered the crack begins to close and the shear stresses are reversed [Fig. 9.43(d)]. Finally, at the point of maximum compressive stress the crack is almost closed and the reverse plastic flow that has taken place during the compression half-cycle has resulted in the formation of a new striation and resharpening of the crack tip [Fig. 9.43(e)].

Clearly the propagation of a fatigue crack is closely related to the stress situation at the crack tip and we know from the previous section in this chapter that all crack tip stresses are directly proportional to the stress intensity factor, K. It is not surprising, therefore, that a relationship between K and stage 2 fatigue crack growth has been found. In cyclic loading K will vary throughout the cycle and the stress intensity factor range ΔK is defined as the difference between K at the maximum load (K_{max}) and K at the minimum load (K_{min}), i.e.

$$\Delta K = K_{max} - K_{min} \qquad (9.56)$$

If the minimum stress is compressive, K_{min} is normally taken to be zero. A schematic log–log plot of the crack growth rate da/dN against ΔK is given in Fig. 9.44. The lower limit to the curve ΔK_{T} is called the threshold stress intensity factor range for crack growth. Whether there is a true threshold or

Fig. 9.43

Formation of a fatigue striation during one cycle: (a) zero load, (b) small tensile stress, (c) large tensile stress, (d) small compressive stress, (e) maximum compressive stress, (f) zero load.

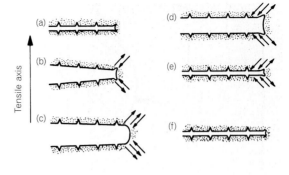

Fig. 9.44

Graph showing the relationship between fatigue crack growth rate and ΔK.

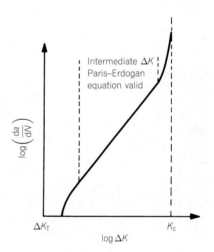

not is under debate, but certainly at low ΔK values the rate of growth becomes vanishingly small. At high ΔK values, where K_{max} is approaching the critical stress intensity factor K_c, crack growth is very rapid and the failure modes are characteristic of normal static failures, e.g. brittle cleavage failure. The normal fatigue stage 2 range is for intermediate values of ΔK where the plot is linear. We therefore have for stage 2:

$$\frac{da}{dN} = C(\Delta K)^m \tag{9.57}$$

which is known as the *Paris–Erdogan equation*. C and m are constants for a particular material; m is usually between 2 and 4 for most ductile metals.

Stage 3 occurs when the fatigue crack has grown to such an extent that the component finally fails in a catastrophic manner. The final fracture mode varies with material and service conditions; it may be a simple ductile overload failure due to the cross-sectional area of the component having been reduced so much by the fatigue crack, or it may be brittle cleavage failure as a result of K_{max} reaching K_c.

A consequence of the different stages is that a fatigue failure can often be recognized from the macroscopic appearance of the fracture. Two distinct zones can often be distinguished, as shown by the examples in Fig. 9.45. There is a relatively smooth zone that corresponds to stage 2 fatigue crack growth. This zone has been smoothed by the continual rubbing together of the cracked surfaces during the cyclic loading but it also

Fig. 9.45
Fatigue failures.

(a)

2 μm

(b)
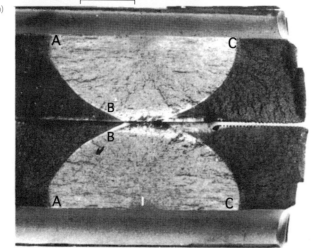

sometimes contains concentric rings, called *clam shell markings*, which indicate successive positions of the crack. The clam shell markings are usually made visible by corrosion and the significance of the markings is that they focus back to the point of initiation of the fatigue crack. The second zone is produced by the final failure and has a rougher surface, which may be shiny or dull depending on the fracture mode.

9.7.7 Creep testing

Under certain combinations of stress and temperature, all materials when subjected to a constant stress will exhibit an increase of strain with time. This phenomenon is called *creep* and most materials creep to a certain extent at all temperatures, although the engineering metals such as steel, aluminium and copper creep very little at room temperature. High temperatures lead to rapid creep, which is often accompanied by microstructural changes. The application for which creep assumes its greatest technological importance, involving intensive research and routine testing, is that of gas turbine blades, particularly those used for aircraft engines. The blades of a high-pressure turbine must survive very high stresses, for extended periods, and at red heat – and must do so not only without failing, but also

Fig. 9.46

Schematic diagram of a creep machine.

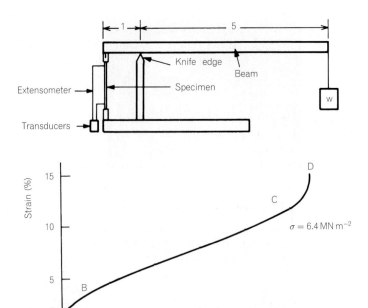

Fig. 9.47

Creep curve of strain against time for lead at room temperature.

without compromising the tight clearances that guarantee efficiency and low pollution.

Creep tests are normally carried out in tension for ductile materials such as metals and in compression, or three point bending, for brittle materials like ceramics. The force on the specimen is usually applied via a simple beam system as illustrated in Fig. 9.46. For this machine the force on the specimen is five times the load W because the ratio of beam lengths on either side of the knife edge is $5:1$. Such a simple machine gives a constant load throughout the test, but more sophisticated beams and attachments are available that keep the stress constant on the specimen as its cross-sectional area changes as a result of creep. The instrumentation must be sensitive to small displacements as often the strains in a creep test are not large. The results from a creep test are generally presented as a graph of creep strain against time, as shown for lead in Fig. 9.47. The strains experienced during creep are often much less than those shown by lead but the shape of the creep curve of Fig. 9.47 is typical for most materials.

Creep mechanisms

A typical creep curve is shown in Fig. 9.47. It can be conveniently divided into sections:

(a) OA, the initial 'instantaneous' strain, which is usually elastic.
(b) AB, known as *primary creep*, a period of continuously decreasing creep rate.
(c) BC, *secondary* or *steady state creep*, a period of dynamic steady state where the creep rate is constant and at a *minimum*.
(d) CD, known as *tertiary creep*, a period of accelerating strain culminating in fracture.

All materials, under specific test conditions, can exhibit a creep curve of the form shown in Fig. 9.47, and we will now discuss some of the equations used to describe the curve. At the same time, a physical explanation of the three stages of creep will be given for crystalline materials.

Primary creep

Primary creep in crystalline materials is solely a result of dislocation movement. Work-hardening occurs more rapidly than any concomitant recovery processes so the creep rate decelerates with time.

Two alternative equations are used to describe primary creep. The first is

$$e = \alpha \log t \qquad (9.58)$$

where α is a constant and t is time. This equation usually only applies at very low temperatures, but it does hold for a variety of materials, e.g. aluminium, rubber and glass. A more widely obeyed equation, which is applicable at higher temperature and stress, is

$$e = \beta t^m \qquad (9.59)$$

where β is a constant and m ranges from 0.03 to 1.0 depending on the material, stress and temperature.

During this phase the rate of recovery is sufficient to balance the rate of work-hardening, so that the material creeps at a steady rate. Structural observations show that polygonization is an important recovery process during secondary creep. Grain boundary sliding is also a feature of polycrystalline materials deforming by steady-state creep. However, the contribution to the total strain from grain boundary sliding is generally small, less than 10%.

The equation for secondary creep is, of course, simply the equation for a straight line:

$$e = Kt \qquad (9.60)$$

where K is a constant, which is temperature- and stress-dependent. This is because recovery processes are temperature- and stress-dependent; the temperature dependence, for example, arises from the change in the rate of diffusion with temperature (discussed in Chapter 7). In fact the creep rate, which is equal to the constant K, can often be represented by an expression of the form

$$\frac{de}{dt} = A\sigma^n \exp\left(-\frac{Q}{RT}\right) \qquad (9.61)$$

where A and n are constants; n is usually in the range 3 to 7 for metals and 1 to 2 for polymers. Q is called the *activation energy for creep* and is approximately equal to the activation energy for diffusion.

At very high temperatures in metals and alloys creep can take place by vacancy migration and no dislocation motion is involved. Stress-directed diffusion of vacancies from grain boundaries that are under tension to compression-stressed boundaries occurs as shown in Fig. 9.48(a) and leads to a creep strain. This mechanism is known as *Herring–Nabarro creep* and the creep rate is given by

$$\frac{de}{dt} = \frac{\sigma a^3 D}{2d^2 kT} \qquad (9.62)$$

Fig. 9.48

Schematic diagram of diffusion creep in a grain of size *d*.
(a) Herring–Nabarro creep; vacancy flow through the grain.
(b) Coble creep; vacancy flow along the grain boundaries.

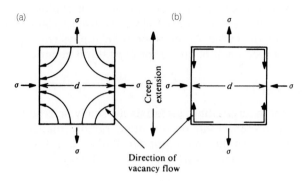

where σ, k and T have their usual meanings, a^3 is the atomic volume, D is the diffusion coefficient, and d is the grain diameter. The significant feature of this equation is the d^{-2} creep rate dependence.

Instead of the stress-directed vacancy migration taking place through the grain it can proceed along the grain boundaries [Fig. 9.48(b)]; this is called *Coble creep* and leads to a d^{-3} creep rate dependence. It follows that if diffusion creep – either Herring–Nabarro or Coble – is likely during high-temperature service then a small grain size is undesirable.

The accelerating creep rate is associated with the formation of voids or microcracks at the grain boundaries. The voids are formed either by vacancy coalescence at the boundaries, which produces rounded voids, or by grain boundary sliding. Two models for the initiation of microcracks by sliding are illustrated in Fig. 9.49 and as can be seen from this figure, and also from the micrograph of Fig. 9.50, grain boundary sliding results in angular or wedge-shaped voids. These tend to be formed at lower creep temperatures or at higher creep stresses than the vacancy voids. The voids grow and link up and the material eventually fails in an *intercrystalline* manner, i.e. fails at the grain boundaries. The elongation to fracture in creep is often much less than in the conventional tensile test, e.g. nickel

Fig. 9.49

Formation of microcracks by grain boundary sliding.

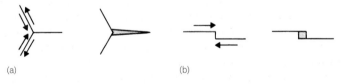

Fig. 9.50

Micrograph of creep voids in stainless steel.

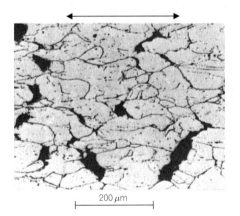

200 μm

Fig. 9.51

Effect of temperature and stress on the creep curves.

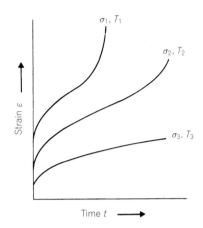

and aluminium alloys may fail after 1 to 3% strain under creep conditions, whereas they exhibit ductilities of 30% in tensile tests at room temperature.

The creep curve of Fig. 9.47 is a generalized creep curve and whether or not a material exhibits the three stages of creep depends on the stress and temperature of testing. The effect of temperature and stress is illustrated in Fig. 9.51. If this figure is taken to represent a series of tests at constant stress, i.e. $\sigma_1 = \sigma_2 = \sigma_3$, then the temperatures are in the order $T_1 > T_2 > T_3$; alternatively, for a series of constant temperature tests, i.e. $T_1 = T_2 = T_3$, the relative magnitude of the stresses is $\sigma_1 > \sigma_2 > \sigma_3$.

High-temperature creep is usually of most concern to engineers, particularly in the design of aircraft, gas turbine components, pressure vessels for high-temperature chemical processes, etc. For most practical purposes, creep is usually unimportant in steels below 300°C, or high-temperature nickel-base alloys below 500°C, but must be considered in aluminium alloys at 100°C and in polymers at room temperature. Design criteria usually specify that a particular strain (say 0.1%), or fracture, must not occur in less than the anticipated lifetime of the part.

If the lifetime of a component is specified in terms of strain, then equations such as (9.59) and (9.60) would be used to obtain the approximate combination of material, stress and temperature to satisfy the specification. However if fracture (which is often referred to as *creep rupture*) is specified then the selection of the material would involve the use of creep rupture time (t_r)–temperature–stress extrapolation procedures of which the *Larsen–Miller parameter* is an example. The Larsen–Miller parameter (LM) is

$$\text{LM} = T(C + t_r)$$

and it has a constant value at a given stress for a particular material (C is a constant). Hence from a master curve for a particular material of stress versus LM it is possible to predict the creep rupture time for any combination of stress and temperature. For example, Fig. 9.52 gives the stress dependence of the Larsen–Miller parameter for two high-temperature nickel alloys.

Let us estimate the creep rupture time of Nimonic 80 A at a stress of $300\,\text{MN}\,\text{m}^{-2}$ and a temperature of 800°C. From the graph $\text{LM} = 22.8 \times 10^3$ at a stress of $300\,\text{MN}\,\text{m}^{-2}$; but also $\text{LM} = T(C + \log t_r) = 1073(20 + \log t_r)$ for Nimonic 80 A at 800°C, therefore $22.8 \times 10^3 = 1073(20 + \log t_r)$, which gives $t_r = 14.3\,\text{h}$.

Fig. 9.52

The stress dependence of the Larsen–Miller parameter for two high-temperature nickel alloys. The units of T and t_r are K and h^{-1} respectively.

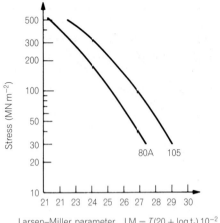

Larsen–Miller parameter $LM = T(20 + \log t_r) \, 10^{-2}$

We have seen that many different deformation mechanisms may occur when a material is stressed, e.g. dislocation glide, dislocation creep, diffusion creep; and that equations which relate strain rate, temperature and stress are available for these mechanisms. From these equations, together with experimental data, it is possible to determine which is the dominant mechanism, i.e. which mechanism permits the fastest strain rate, under specific conditions of stress and temperature for a given material. It is convenient to be able to summarize the information from this type of calculation in a diagrammatic form and this is achieved by diagrams called *deformation maps*. A deformation map is a diagram, with axes of normalized stress (stress/shear modulus) and homologous temperature (temperature/melting point), divided into areas within each of which a particular mechanism is dominant. An example of a deformation map is given in Fig. 9.53, in which the elastic regime is where the strain rate predicted for the various flow mechanisms would be too small to be measured. Often, superimposed on the maps are contours of constant strain rate so that the maps can also be used to give an approximate value for the strain rate under given conditions. Therefore the diagrams present the relationships between stress, temperature and strain rate – if any two of these are specified the map may be used to determine the third and to identify the dominant deformation mechanism.

Fig. 9.53

Deformation map for pure silver of grain size 32 μm [M. F. Ashby, *Acta Met.* **20**, 887(1972)].

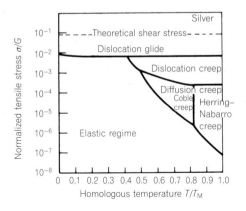

Problems

9.1 A 20,000 N tensile load is applied axially to a steel bar of cross-sectional area 8 cm². When the same load is applied to an aluminium bar it is found to give the same elastic strain as the steel. Calculate the cross-sectional area of the aluminium bar. (Young's modulus E for steel $= 2.0 \times 10^5$ MN m^{-2} and for aluminium it is 0.7×10^5 MN m^{-2}.)

9.2 The bulk modulus of water is 2300 MN m^{-2}. At the deepest point in the ocean (the Mariana trench, about 11 km below sea level) the pressure has increased to 110 MPa. What is the resulting volumetric strain in water at this point?

9.3 The steel bar of Problem 9.1 is 1 m long. Calculate its change in volume under the 20,000 N applied axial load. Take Poisson's ratio ν to be 0.3.

9.4 Using the table of physical properties in Appendix 3, together with the periodic table of the elements, plot (a) boiling point against Young's modulus; (b) Young's modulus against valency; (c) density against valency for all the elements that you would regard as metals. What general conclusions can you draw concerning the relationship between these various factors and the strength of the metallic binding forces?

9.5 For copper Young's modulus $= 1.26 \times 10^5$ MN m^{-2} and the shear modulus $= 0.35 \times 10^5$ MN m^{-2}. Calculate the ratio of the tensile strain to the shear strain if the same energy density is produced by a tensile stress alone and a shear stress alone.

9.6 A particular grade of rubber has a Young's modulus of 1 MN m^{-2} and a bulk modulus of 1 GN m^{-2}. Calculate Poisson's ratio. If a block of rubber is prevented from contracting laterally, so that Poisson's ratio is effectively zero, what then will be its effective Young's modulus?

9.7 A round bar of metal is 9 mm in diameter and it is observed that a length of 250 mm extends by an amount of 0.225 mm under a load of 11.8 kN. At the same time its diameter contracts by 0.00227 mm. Determine Young's modulus and the shear modulus for the metal.

9.8 The maximum torque that can be applied by hand to the handle of a tool (e.g. a screwdriver) at a rotational speed of 1 radian s^{-1} is of the order of 10 Nm. If a sample of viscous material in a cone-and-plate viscometer of cone angle 9° and diameter 10 cm is sheared by hand at this torque and speed, what is its viscosity?

9.9 Describe the phenomena that indicate viscoelastic behaviour in a creep and a relaxation test. When the torque applied to the viscometer cone of Problem 9.8 is suddenly removed, the cone springs back as the viscous fluid recovers in the manner characteristic of a Maxwell material. If this recovery has a relaxation time of 1.5 s, what is the shear modulus of this material?

9.10 An empirical equation representing the strain-hardening behaviour of many engineering metals is

$$\bar{\sigma} = A(B + \bar{e})^n$$

Show that the 'uniform elongation' for such a material would be

$$\bar{e} = n - B$$

9.11 In a state of pure shear it can be shown that, if the maximum shear stress is τ_s, the principal stresses are $\sigma_1 = -\sigma_2 = \tau_s$, $\sigma_3 = 0$. In pure tension the principal stresses are $\sigma_1 = 2\tau_t$, $\sigma_2 = \sigma_3 = 0$, where τ_t is the maximum shear stress in the tensile specimen. Using Eq. (9.27) show that yielding occurs in a tension specimen when the maximum shear stress in it is $(\sqrt{3})/2$ times the maximum shear stress for yielding in pure shear.

9.12 How would you distinguish between a brittle and a ductile fracture? Why is the brittle fracture strength of real materials lower than the ideal breaking strength?

9.13 Explain why the strength of a glass plate may be increased by etching off its surface with hydrofluoric acid. A glass plate has a sharp crack of length 1 μm in its surface. At what stress will it fracture when a tensile force is applied perpendicular to the plane of the crack? (Young's modulus $= 70\,\text{GN}\,\text{m}^{-2}$ and surface energy $= 0.3\,\text{J}\,\text{m}^{-2}$. Assume plane stress conditions.)

9.14 Explain why the level of the true stress–strain curve from a tensile test is higher than that of an engineering stress–strain curve. What

would be the relative positions of the two curves in a compression test in which barrel-shaped distortion occurs, and why?

9.15 For what sorts of material is a hardness test most useful? How would a knowledge of the hardness of a material be of help to the machinist, the design engineer, the testing engineer and the mineralogist?

Self-assessment questions

1 If a force of 6 N is applied to a flat tensile specimen of length 60 mm, width 10 mm and thickness 3 mm, the stress on the specimen is

(a) $1.8\,\text{N}\,\text{cm}^2$

(b) $1.8 \times 10^{-12}\,\text{MN}\,\text{m}^2$

(c) $20\,\text{N}\,\text{m}^{-2}$

(d) $0.2\,\text{MN}\,\text{m}^{-2}$

(e) $1\,\text{MN}\,\text{m}^{-2}$

2 The elastic strain obtained on applying a stress to a material is

(a) time-dependent

(b) instantaneous

(c) partially permanent

(d) reversible

(e) directly proportional to the stress

(f) inversely proportional to the stress

3 If Poisson's ratio during elastic deformation under uniaxial tensile stress is less than 0.5

(a) there is a contraction in each of the two lateral directions

(b) there is an expansion in each of the two lateral directions

(c) there is a decrease in specimen volume

(d) there is no change in specimen volume

(e) there is an increase in specimen volume

4 The stresses acting on a cubical element of a material are shown the diagram on the right. (i) The stresses σ_1, σ_1 and σ_1 are (ii) The stresses τ_1, τ_2 and τ_3 are ...

(a) tensile stresses

(b) shear stresses

(c) often denoted as σ_{11}, σ_{22} and σ_{33}

(d) often denoted as σ_{12}, σ_{13} and σ_{23}

(e) often denoted as τ_{11}, τ_{22} and τ_{33}

(f) related such that $\sigma_1 = \sigma_2$, $\sigma_2 = \sigma_3$, etc.

(g) related such that $\tau_{12} = \tau_{13}$, $\tau_{21} = \tau_{23}$, etc.

(h) related such that $\tau_{21} = \tau_{12}$, $\tau_{13} = \tau_{31}$, etc.

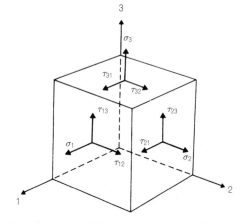

5 The elastic resilience of a material is

(a) the stored energy per unit volume during elastic deformation

(b) the stored energy per unit volume associated with dislocations

(c) given by $\frac{1}{2}e^2E$

(d) given by $\frac{1}{2}\sigma e$

(e) given by $E\dfrac{\partial e}{\partial t}$

where σ, e and E are stress, strain and Young's modulus respectively.

6 Anelastic deformation, like elastic deformation, is instantaneous

(a) true   (b) false

7 The critical resolved shear stress is a material property independent of orientation

(a) true   (b) false

8 The graph shows a generalized stress–strain curve for a single crystal. During Stage II:

(a) only the primary slip system operates

(b) the primary slip system operates

(c) slip occurs on more than one slip system

(d) the work-hardening rate is less than in stage III

(e) the work-hardening rate is greater than in stage I

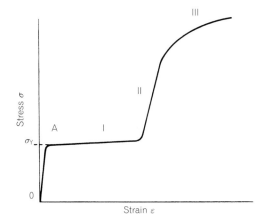

9 The stress–strain curves of polycrystalline materials do not exhibit stages I and II shown by single crystals

(a) true   (b) false

10 Work-hardening is a useful strengthening mechanism but it has the following disadvantages

(a) only useful for two-phase materials

(b) decreases the ductility of the material

(c) not suitable if the material is to be used at an elevated temperature

(d) only applicable to single crystals

11 The presence of a dispersion of small particles increases the strength of a material by an amount that depends on the

(a) volume fraction of the particles

(b) valency of the solute atoms

(c) size of the particles

(d) viscoelastic deformation rate of the particles

12 In the plasticity theory for an isotropic material the effective value of Poisson's ratio is

(a) $\frac{1}{2}$ (b) $\frac{1}{3}$ (c) 0   (d) 1   (e) 2

13 The effective stress $\bar{\sigma}$ is a function of the principal stresses σ_1, σ_2 and σ_3

(a) true   (b) false

14 According to Griffith's theory, a crack propagates when the released elastic energy is just sufficient to provide the surface energy necessary for the creation of the new surfaces

(a) true   (b) false

15 According to Griffith's theory for brittle fracture, the stress required to propagate a brittle crack is

(a) less than the ideal fracture stress

(b) more than the ideal fracture stress

(c) proportional to $a^{1/2}$, where $2a$ is the crack length

(d) proportional to $a^{-1/2}$

(e) proportional to γ, the surface energy

16 This fracture resulted from a tensile test. The fracture

(a) is a brittle fracture

(b) is a ductile fracture

(c) was accompanied by considerable plastic deformation

(d) probably started at a surface crack

(e) is called a 'cup and cone' fracture

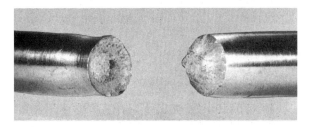

17 Necking occurs in a tensile test at a strain at which the work-hardening rate equals the effective (true) stress

(a) true   (b) false

18 The true stress at necking of a material with a strain-hardening exponent of 0.2 is

(a) 0.1   (b) 0.2   (c) 0.4   (d) 0.8

19 The true strain is given by

(a) $\ln\dfrac{\ell_1}{\ell_0}$ (b) $\dfrac{\Delta\ell}{\ell_0}$ (c) $\dfrac{e}{\ell_0}$ (d) $\ln(1+e)$

20 The stress–strain curve below

(a) is for pure copper

(b) is for mild steel

(c) shows an upper yield point

(d) shows a lower yield point

(e) is for a brittle material

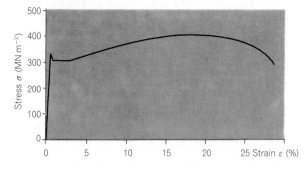

21 The fatigue resistance of a material is reduced by

(a) permanent residual compressive stresses

(b) a mean positive (tensile) stress

(c) chemically or mechanically hardening the surface

(d) poor surface finish

22 In primary creep the work-hardening occurs more rapidly than any concomitant recovery processes

(a) true   (b) false

23 In secondary creep the

(a) recovery rate is greater than the work-hardening rate

(b) recovery rate is equal to the work-hardening rate

(c) creep strain e is given by $e = Kt$ where K is a constant and t is the time

(d) creep strain is given by $e = Kt^{1/3}$

24 The Paris–Erdogan equation

(a) relates creep rate and temperature

(b) relates fatigue crack growth and ΔK

(c) is applicable to tertiary creep

(d) is applicable to stage 2 of a fatigue failure

(e) gives the stress for void formation during a ductile failure

25 For a material exhibiting brittle behaviour with some local plastic deformation the critical strain energy release rate is given by $G_c = 2\gamma + \gamma_p$

(a) true   (b) false

26 In fatigue the mean stress σ_m

(a) is equal to twice the stress amplitude

(b) is equal to one-quarter of the stress amplitude

(c) does not affect the fatigue life

(d) has an effect on fatigue life that can be analysed by means of the Goodman diagram

(e) is determined by the ease of dislocation climb

27 The diagram below shows three strain–time plots for the same material at different stresses (σ_1, σ_2 and σ_3) and different temperatures (T_1, T_2 and T_3). These plots

(a) are creep curves

(b) are fatigue curves

(c) are conventional tensile test curves for a single crystal

(d) indicate that $\sigma_1 > \sigma_2 > \sigma_3$ (assuming $T_1 = T_2 = T_3$)

(e) indicate that $T_3 > T_2 > T_1$ (assuming $\sigma_1 = \sigma_2 = \sigma_3$)

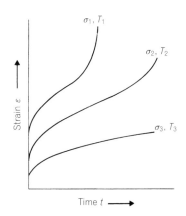

Each of the sentences in Questions 28–38 consists of an assertion followed by a reason. Answer as follows:

(a) If both assertion and reason are true statements and the reason is a correct explanation of the assertion.

(b) If both assertion and reason are true statements but the reason is *not* a correct explanation of the assertion.

(c) If the assertion is true but the reason contains a false statement.

(d) If the assertion is false but the reason contains a true statement.

(e) If both the assertion and the reason are false statements.

28 Elastic strain is directly proportional to the applied stress *because* the force–displacement curve is parabolic near to the equilibrium spacing.

29 Single crystals are elastically anisotropic *because* the interatomic spacing, and in some cases the bonding, varies with direction.

30 Over a limited temperature range the viscosity of a solid increases exponentially with increasing temperature *because* viscous flow normally involves the thermally activated movement of atoms or molecules within the material.

31 In a tensile test on a single crystal, slip commences in the slip system with the highest Schmid factor *because* this is the system with the highest resolved shear stress for a given tensile stress.

32 Interstitial atoms cause a rapid increase in yield stress with concentration *because* their strain fields are symmetrical.

33 In a tensile test the true stress is always greater than the engineering stress *because* the cross-sectional area decreases with strain.

34 Brittle materials are generally tested in compression *because* barrelling of the specimen in compression enhances necking.

35 The hardness test is a slow, expensive method of assessing the mechanical properties of a material *because* the hardness is a function of the yield stress and the work-hardening rate of the material.

36 Specimens for impact testing are never notched *because* a notch increases the level of hydrostatic tensile stress, which encourages brittle fracture.

37 All materials have a fatigue limit *because* fatigue cracks are never initiated at the surface.

38 Creep failure in polycrystalline metals is normally intercrystalline *because* the secondary creep rate is stress dependent.

Answers

1 (d)	**2** (b), (d), (e)	**3** (a), (e)	**4** (i) (a), (c), (ii) (b), (d), (h)
5 (a), (c), (d)	**6** (b)	**7** (a)	**8** (b), (c), (e)
9 (a)	**10** (b), (c)	**11** (a), (c)	**12** (b)
13 (a)	**14** (a)	**15** (a), (d), (f)	**16** (c), (c), (e)
17 (a)	**18** (b)	**19** (a), (d), (e)	**20** (b), (c), (d)
21 (b), (d)	**22** (a)	**23** (b), (c)	**24** (b), (d)
25 (a)	**26** (d)	**27** (a), (d)	**28** (c)
29 (a)	**30** (d)	**31** (a)	**32** (c)
33 (a)	**34** (c)	**35** (d)	**36** (d)
37 (e)	**38** (b)		

10 | Phase diagrams and microstructure of alloys

10.1 Introduction

Applications of metals are so widespread that a list would fill pages. It might be thought that metals, having been in use for centuries, would have developed little in the last 50 years. Among the myriad of metallic alloys that have been improved by the methods of materials science, we highlight here just two late 20th century examples. First, High Strength Low Alloy (HSLA) steels, which, by making clever use of cold-rolling and heat treatments, have enabled industry to reduce vehicle weights without sacrificing strength or low cost, and second, the so-called superalloys developed for use at temperatures of 1000–1100°C in jet engines. The latter are expensive nickel-based and cobalt-based alloys, used in the combustion chamber and just downstream of it, for the turbine blades, which are exposed to the hottest exhaust gases and where the rotational stresses are extreme. At these high temperatures, the blades glow red and creep in service, their length extending under the force of tonnes exerted on each blade by rotation. A small clearance at the blade tip is vital to the turbine action, so that their working life is limited primarily by creep. These expensive alloys allow the fuel to be burned at the highest possible temperature, raising the thermodynamic efficiency and reducing the CO and CO_2 expelled at high altitude. Further improvements are expected as a result of ongoing research into these materials.

Most materials of industrial significance consist of more than one atomic or molecular species, i.e. they are not single-component materials. For example, the basic components of steel are iron and carbon, although there are other components (elements) present. The components may be distributed throughout the material in a variety of ways and, under equilibrium conditions, a material may consist of only one crystalline structure, and so termed *single-phase*, or it may be *multi-phase*, depending on the temperature and composition. In practice, the situation is further complicated by the fact that materials may often exist in metastable conditions for long periods of time without changing to their equilibrium structure. It is useful to categorize metallic alloys under the broad headings listed in Table 10.1.

Because many properties, e.g. plastic and magnetic properties, are structure-sensitive, it is essential to know the microstructure of a material, i.e. the number of phases present, their distribution through the solid, their volume fractions, shapes and sizes. Much of this chapter is concerned with understanding the formation of phases (Section 10.2), how the stability of

Table 10.1
A broad classification of alloys

Ferrous metals	Plain carbon steels	Includes cast iron
	Low-alloy steels (non-carbon additions < 5%)	Steels named by the major added element, e.g. manganese steels, chromium steels, nickel steels
	High-alloy steels (non-carbon additions > 5%)	Stainless (corrosion resistant) types: austenitic stainless steels, ferritic stainless steels, martensitic stainless steels
		Tool steels (cutting applicatins), types: plain carbon steel, alloy steels
		Superalloys (high-temperature strength)
Non-ferrous metals	Alloys named by the principal element(s)	E.g. aluminium alloys, copper alloys, nickel alloys, aluminium–copper alloys, etc.
	Superalloys (high-temperature strength)	Nickel-based, cobalt-based

phases may be represented in diagrams called *equilibrium phase diagrams* (Sections 10.3 and 10.4), and how we can use these diagrams to interpret microstructures. These ideas are equally useful in discussing metals and ceramics, but we leave the discussion of ceramics to Chapter 11, and concentrate on metallic alloys here.

We find that the atomic radii of the elements present in a metallic alloy usually control the possible phases that appear. The *miscibility* of those phases with one another, and with the liquid phase, determines major features of the equilibrium phase diagram, which in turn determines the way that the microstructure changes when the liquid is allowed to solidify. We explain in Sections 10.4–10.6 various reactions and transformations of structure or composition that can occur in a solid as it cools, and the effect that these have on the microstructure.

We then turn our attention in Section 10.7 to mechanisms that help to strengthen metals by hindering dislocation motion. In Section 10.8 we make use of all the above ideas in looking at the structure and mechanical properties of some commercial alloys of iron, aluminium, nickel and copper, including the superalloys mentioned above. The chapter concludes with an introduction to metallic corrosion mechanisms, and a discussion of some protection methods.

We begin by considering what happens when two components are mixed together.

10.2 Solid solutions and intermediate phases

When two liquids A and B (which may also be molten metals or oxides) are mixed together in varying proportions, they do so in one of the following two ways:

1. The liquids may be *completely miscible* in one another over the whole composition range from pure A to pure B, i.e. at any composition a

homogeneous single-phase solution is formed with the atoms or molecules of one of the components randomly dispersed in the other. The component (liquid) in excess is called the solvent and the other the solute. An example of two completely miscible liquids is water and ethyl alcohol (whisky can be diluted as much as one likes!).

2. The liquids are only partially soluble, or miscible, in each other. Thus, if there is only a small amount of solute and a large amount of solvent, a single-phase, homogeneous solution is formed. If more solute is added, the limit of solubility is reached and two solutions form, which, on standing, separate into two layers. We now have a two-phase system. In each phase (layer) one component is the solvent, with a limited quantity of the other dissolved in it. Phenol and water behave in this way at room temperature. However, if we heat up the two-phase mixture, it will change into a single-phase, homogeneous solution at a specific temperature that depends on the overall composition. We can represent the range of temperature and composition over which the two-phase mixture is stable on a diagram of temperature against composition. Such diagrams are called *equilibrium phase diagrams* and the phenol–water diagram is given in Fig. 10.1. We can see from this diagram that if we had, say, a 20% phenol mixture, it would be two-phase at room temperature, but would change to a single-phase solution above about 43°C.

Generally, the more dissimilar are the components, both chemically and in atomic or molecular size, the more restricted is the partial solubility. Consequently, very dissimilar liquids are nearly, but not quite, completely insoluble in each other, e.g. oil and water.

The discussion so far has been concerned with liquids but the concepts introduced, and indeed the terminology, are equally applicable to solids. Thus, two components may be completely or partially soluble in each other in the solid state.

In a *solid solution* the solute atoms are distributed throughout the solvent crystal, the crystal structure of the solvent being maintained. As described in Chapter 8, the solute atoms can be accommodated in two different ways. If they occupy interstitial positions, as shown in Fig. 10.2(a), we have an *interstitial solid solution*. Alternatively, if they replace the solvent atoms as shown in Fig. 10.2(b), the resulting arrangement is called a *substitutional solid solution*. Which of these is formed depends mainly upon the relative

Fig. 10.1

Equilibrium phase diagram for water–phenol, showing that a two-phase mixture can be transformed into a single-phase homogeneous solution by raising the temperature.

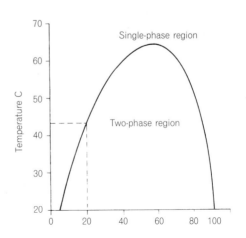

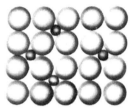 

Fig. 10.2

(a) Interstitial and (b) substitutional solid solutions.

sizes of the ionic solids discussed in Chapter 6. In general, interstitial solid solutions can only form when the solute atom diameter is 0.6 or less of the atomic diameter of the solvent.

The distances between atoms in metals, as between ions in ionic crystals, approximately obey an additive law, each atom or ion being packed in a structure as if it were a sphere of definite size, as discussed in Chapter 6. Actually, the radius of the sphere for any atom or ion is not a constant size but varies according to the number of neighbours it has, that is, it depends on the coordination number, defined in Chapter 6. There is, for example, a 3% contraction in the radius when passing from 12-fold (close-packed) to 8-fold coordination. However, for purposes of comparison, atomic radii given in books of tables are based on the size for 12-fold coordination. These radii are given in Table 10.2; the figures marked with an asterisk are elements that form crystals of very low coordination. In these cases the quoted figure is half the smallest interatomic distance.

Table 10.2

Atomic radii

Element	Radius (Å)	Element	Radius (Å)	Element	Radius (Å)
H	0.46	Ir	1.35	Mg	1.60
O	0.60	V	1.36	Ne	1.60
N	0.71	I	1.36[†]	Sc	1.60
C	0.77	Zn	1.37	Zr	1.60
B	0.97	Pd	1.37	Sb	1.61
S	1.04[†]	Re	1.38	Tl	1.71
Cl	1.07[†]	Pt	1.38	Pb	1.75
P	1.09[†]	Mo	140	He	1.79
Mn	1.12[†]	W	1.41	Y	1.81
Be	1.13	Al	1.43	Bi	1.82
Se	1.16[†]	Te	1.43[†]	Na	1.92
Si	1.17[†]	Ag	1.44	A	1.92
Br	1.19[†]	Au	1.44	Ca	1.97
Co	1.25	Ti	1.47	Kr	1.97
Ni	1.25	Nb	1.47	Sr	2.15
As	1.25[†]	Ta	1.47	Xe	2.18
Cr	1.28	Cd	1.52	Ba	2.24
Fe	1.28	Hg	1.55	K	2.38
Cu	1.28	Li	1.57	Rb	2.51
Ru	1.34	In	1.57	Cs	2.70
Rh	1.35	Sn	1.58	Rare }	{ 1.73 to
Os	1.35	Hf	1.59	Earths }	{ 2.04

[†] Estimated from half the interatomic distance in the pure material.

Note that Table 10.2 gives the radii of complete (neutral) atoms; where an atom has lost one or more electrons it becomes an ion and, naturally, has a smaller radius. It is ionic radii that are given in Table 6.1.

Because the commercially important metals range from cobalt (1.25 Å) to lead (1.75 Å), it will be seen that the atoms which can go into interstitial solutions in these metals must have radii less than 0.75 to 1.05 Å. This effectively limits the possibilities to the first six elements, hydrogen to sulphur. It should be noted that this includes carbon and the interstitial solid solution of carbon in iron is the basis of steel.

When atoms of the two components are more nearly the same size a substitutional solid solution is formed. For complete solid miscibility the two components must have the same crystal structure, but even when this is so, if the atoms differ in size by more than 14%, the solubility will be restricted. A simple view of the thermodynamics of the situation will explain this behaviour. As solute is added to the solvent the degree of disorder increases, therefore the entropy of the material increases (see Section 7.16). This increase in entropy is independent of the sizes of the atoms and is determined solely by the concentration. At the same time, the solute atoms introduce elastic strains into the solvent lattice, i.e. some of the solvent atoms are slightly displaced from their equilibrium positions by an amount that depends on the relative sizes of the solute and solvent atoms. This gives an increase in the enthalpy which is dependent on atomic size – the greater the size difference the larger the enthalpy increase. Now, increases in enthalpy and entropy, act in opposite sense on the free energy as $\Delta G = \Delta H - T\Delta S$ [Eq. (7.24)]. For low solute concentrations the entropy term outweighs the enthalpy, leading to a decrease in free energy and the formation of a stable, homogeneous solid solution. However, at higher concentrations the entropy does not increase so rapidly with solute content and the enthalpy predominates. This results in the single-phase solid solution becoming energetically unfavourable compared with a two-phase mixture as the concentration is increased.

We not only have to consider crystal structure and atomic size when examining the extent of solid solubility, but also oxidation number. When the oxidation numbers of the components differ markedly there will be a tendency, as mentioned in Chapter 6, to form *intermediate* or *intermetallic compounds*. These compounds are normally formed at, or near, compositions corresponding to a simple ratio of components, such as AB, AB$_2$, and may exist over a composition range around the simple ratio. They often have a crystal structure which differs from that of either of the components.

The effects of atomic size and oxidation state are well illustrated by the equilibrium phase diagrams for silver (f.c.c.) alloyed with the f.c.c. elements gold, copper and aluminium. Silver and gold are completely miscible in each other as they have the same atomic sizes and are both Group I elements. Copper is also in Group I, but the Cu and Ag atoms differ in size by about 11%, with the result that they are only partially soluble in each other. Aluminium has atoms of similar size to silver but is from Group III; thus intermediate compounds are formed in the Ag–Al system.

10.3 Equilibrium phase diagrams

In the previous section we have seen that the number of phases present in a material may be affected by the temperature and the composition, and that the regions of phase stability can be represented on an equilibrium phase diagram. In this section the more important features found in the equilibrium phase diagrams for solids will be described. Furthermore, we will show how these diagrams may be used (1) to determine the relative amounts and compositions of the phases and (2) sometimes to deduce the morphology of the phases present.

In referring to compositions in this chapter, we shall normally quote the percentage by weight of each component, sometimes abbreviated to wt%. Occasionally, however, it will be useful to refer to the percentage by number of atoms, or atomic percent (at%).

In the following it is assumed that there is complete miscibility in the liquid phase, although there are some notable exceptions to this, e.g. Al–Pb, Zn–Pb.

10.3.1 Complete solid miscibility

It has already been mentioned that it is possible to form a continuous range of solid solutions between two components, e.g. Ag–Au, Cu–Ni, NiO–MgO, and now we want to study the equilibrium phase diagram for this situation, termed complete solid miscibility. Generally, information from a number of experimental techniques is brought together in order to construct a phase diagram. However, *thermal analysis* is the most widely used technique in the determination of phase diagrams. Thermal analysis is the monitoring of the heat that is evolved during an *exothermic* transformation or absorbed during an *endothermic* transformation. The simplest form of thermal analysis is the recording of heating and cooling curves; these curves exhibit a change in slope or a plateau when a transformation is occurring. More sensitive to small thermal changes are the techniques of *differential thermal analysis* (DTA) and *differential scanning calorimetry* (DSC). In DTA the temperature of the specimen under investigation is compared with that of an inert standard as the temperature is raised or lowered at a controlled rate. DSC also uses an inert standard but instead of recording temperature differences, the heat evolved or absorbed during a transformation is quantified. We will employ the simple cooling curve technique in our study of phase diagrams, although DTA plots will be presented for comparison in the discussion on the eutectic system. First, let us look at the cooling curves for some alloys in the Cu–Ni system and from the data obtained construct the equilibrium phase diagram.

Referring to Fig. 10.3(a), the cooling curve for pure copper contains a flat portion, AB, during which complete solidification takes place at constant temperature with liberation of latent heat. When 20 wt% nickel is added the flat portion is no longer present but there is a transition region of temperature, A_1B_1, with completely liquid mixture at A_1 going to complete solid at B_1. At temperatures in between there is a mixture of solid and liquid which will remain in equilibrium so long as the temperature is held constant. With 60 wt% nickel similar behaviour occurs over the transition region A_2B_2, but when 100% nickel is reached the transition from liquid to solid, A_3B_3, is again flat, occurring at a fixed temperature. From these results the points A, A_1, A_2 and A_3, which correspond to the lowest temperatures at which the solution is entirely liquid, can be plotted as in Fig. 10.3(b). This line is called the *liquidus* curve. Similarly, a curve through the points B defines the highest temperature at which the solution is a solid and is called the *solidus* curve. The points B are not well defined on cooling

Fig. 10.3

(a) Cooling curves and
(b) equilibrium phase diagram
for the copper–nickel system.

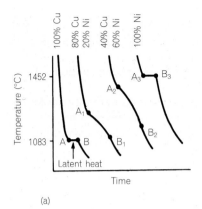

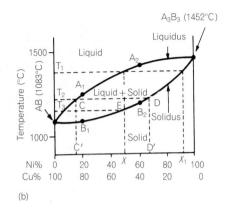

(a)   (b)

curves and the solidus is usually determined from heating curves. The resulting graph is the equilibrium phase diagram for Cu–Ni and is typical for a system exhibiting complete solid miscibility.

Not only does the equilibrium phase diagram tell us how many phases are present in a given material at a particular temperature, but also the compositions and relative proportions of the phases. For example, at temperature T_2 an alloy of X% Ni consists of liquid and solid in equilibrium. To obtain the compositions of these phases a horizontal, isothermal line is drawn, called a *tie-line*, and the intercepts of the tie-line with the phase boundaries, i.e. the solidus and liquidus, give the relevant compositions, namely liquid C'% Ni and solid D'% Ni. The proportions by weight of the two phases are given by the *lever rule* and are in the ratio of the lengths CE and ED. In fact:

$$\text{weight of solid} \times \text{ED} = \text{weight of liquid} \times \text{CE}$$

The rule can be stated in the following way. Let the point representing the composition and temperature be the fulcrum of a horizontal lever. The lengths of the lever arms from the fulcrum to the boundaries of the two-phase field multiplied by the weights of the phases present must balance. We can take our alloy or composition X% Ni and derive the lever rule from first principles. Let the weight of the alloy be W_A and the weights of the liquid of composition C'% Ni and the solid of D'% Ni at temperature T_2 be W_L and W_S respectively. Clearly the weight of the alloy W_A is just the sum of the weights of the liquid and solid phases, i.e.

$$W_A = W_L + W_S$$

Similarly, the weight of nickel in the alloy is the sum of the weight of nickel in the two phases, therefore,

$$\frac{X}{100} W_A = \frac{C'}{100} W_L + \frac{D'}{100} W_S$$

Combining these two equations gives

$$W_S (D' - X) = W_L (X - C')$$

which is

$$W_S \times \text{ED} = W_L \times \text{CE}$$

Fig. 10.4

Cored grains in a chill cast Cu–5% Sn solid solution.

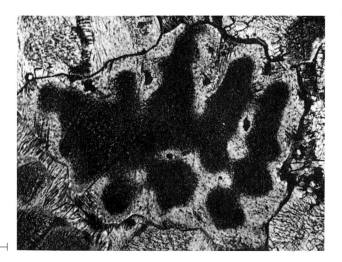

75 μm

Now let us look in detail at the solidification of the alloy containing $X\%$ Ni, which will be typical of the solidification of any alloy in the system. Solidification begins at temperature T_1 and the appropriate tie-line shows that the first solid is of composition $X_1\%$ Ni. If a series of tie-lines is drawn at lower temperatures in the two-phase region, it can be seen that the proportion of solid increases with decreasing temperature, and that the composition of the solid forming at any given temperature changes with temperature along the solidus. Correspondingly, the composition of the remaining liquid follows the liquidus. When the cooling rate is slow enough to maintain equilibrium, the solid formed earlier changes composition by diffusion so that at any temperature all the solid is of the same composition. Thus, when the temperature has fallen to T_2 all the solid is of composition $D'\%$ Ni and the remaining liquid is of composition $C'\%$ Ni. Solidification continues in this manner and is completed at temperature T_3. The microstructure of the alloy so formed will consist of grains, or crystallites, of homogeneous solid solution of composition $X\%$ Ni.

It is common for the cooling rate to be too rapid for the composition of the solid to change significantly by diffusion during solidification. This results in a non-equilibrium structure in which the composition varies throughout the grains, i.e. the grains are heterogeneous. The grains are said to be *cored*. Figure 10.4 is a micrograph of a cored solid solution; the variations in composition are shown by the different tones in the micrograph within each grain.

Cored structures are common in practice as the rate of cooling in chill casting, and even with sand casting, is generally much too rapid for equilibrium to be maintained. A cored structure will generally persist indefinitely at room temperature; it is therefore a *metastable* structure. However, it will revert to the equilibrium condition of homogeneous grains when annealed at an elevated temperature.

10.3.2 Partial solid miscibility

Although in many systems the components are completely soluble in the liquid state, often they are only partially miscible in each other in the solid state, e.g. Cu–Al and MgO–CaO. Partial solid miscibility results in two distinct types of equilibrium phase diagram, namely the eutectic and the peritectic.

(a) *Eutectic*. To study the eutectic phase diagram, we will use the same approach as was used for complete solid miscibility, namely, we will examine the cooling curves [Fig. 10.5(a)] for a number of alloys in a particular system, lead–tin, and from these plot the liquidus and solidus.

Fig. 10.5

Use of thermal analysis for the determination of the phase diagram for the lead–tin system: (a) cooling curves; (b) phase diagram; (c) DTA (heating) plots.

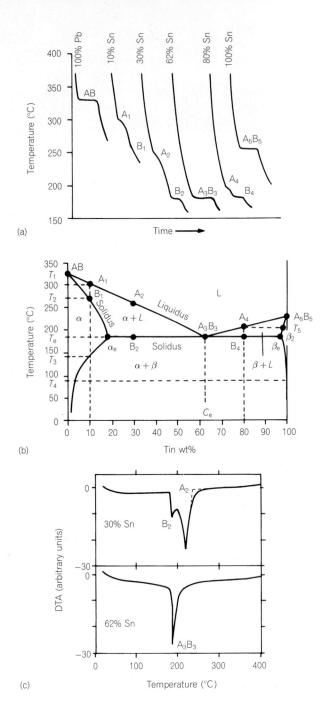

For comparison DTA plots, that is, graphs of the temperature difference between the sample under investigation and an inert standard as a function of temperature, are also given for two of the alloys [Fig. 10.5(c)]. The DTA plots show distinct minima corresponding to the endothermic melting processes that occur on heating.

The distinctive feature of the cooling curves of Fig. 10.5(a) is the *arrest* shown by a number of the alloys at about 180°C. The equilibrium phase diagram constructed from the thermal analysis data is given in Fig. 10.5(b). In this diagram α and β are solid solutions of tin in lead and lead in tin respectively. It can be seen that over a considerable composition range the solidus is horizontal, corresponding to the arrests at 180°C. The temperature of the horizontal portion of the solidus is called the *eutectic temperature* and the composition at which the liquidus meets this section of the solidus is the *eutectic composition*, C_e. It is important to note that the eutectic composition is very unlikely to be the equiatomic (50%–50%) composition, but varies from system to system, being about 62 wt% Sn ($\sim$73 at% Sn) in this case.

Let us follow the solidification of some alloys in the Pb–Sn system. The first alloy we will consider is Pb–10 wt% Sn, which is typical of any alloy containing less tin than the composition shown as α_e in Fig. 10.5(b). Under equilibrium conditions this alloy will solidify in the manner already described for complete solid miscibility, i.e. solidification will start at temperature T_1 and finish at T_2, giving homogeneous grains of α solid solution. The interesting point about this alloy is that as it cools in the solid state, the *solubility limit* of tin in lead is reached at temperature T_3. Therefore, at temperatures below T_3 the β phase begins to be *precipitated* out of the solid solution. At T_4 the compositions of the α and β phases that will exist in equilibrium are those at the *boundaries* of the α phase region and the β phase region, namely Pb–4 wt% Sn and virtually 100% Sn respectively. The proportions will be given by the lever rule, i.e.

$$(\text{weight of } \alpha)(10 - 4) = (\text{weight of } \beta)(100 - 10)$$

i.e. $(\text{weight of } \alpha)(6) \quad = (\text{weight of } \beta)(90)$

Thus, the amount of β precipitate is $[6/(90 + 6)]100\% \approx 6\%$.

Any alloy with more tin than the composition β_e will behave in a similar manner to the Pb–10 wt% Sn alloy except that β solid solution forms first and α precipitates out if the solid solubility limit is exceeded.

The significance of precipitation is that in some alloy systems it can lead to strengthening. This will be discussed further in Section 10.7 on strengthening mechanisms.

The phase diagram and the cooling curve show that an alloy of the eutectic composition completely solidifies at the eutectic temperature T_e. At the eutectic temperature all the liquid, which is of course of the eutectic composition C_e, solidifies into an intimate mixture of the α and β phases. The compositions of the α and β phases in the eutectic mixture are α_e and β_e respectively, so the eutectic reaction may be written as

$$L_{C_e} \rightleftharpoons (\alpha_e + \beta_e) \tag{10.1}$$

where L_{C_e} represents the liquid.

As the temperature falls below T_e the compositions of the phases will, under equilibrium conditions, change by solid state diffusion. When temperature T_4 is reached the compositions of the phases in equilibrium will be the same as for the Pb–10 wt% Sn alloy but the proportions will be different. There will be far more β in the eutectic alloy. According to the lever rule we have, at T_4,

$$\text{(weight of } \alpha)(62 - 4) = \text{(weight of } \beta)(100 - 62)$$

i.e. $\text{(weight of } \alpha)(58) = \text{(weight of } \beta)(38)$

so the amount of β present is $[58/(38 + 58)]100\% \approx 60\%$.

It will be noted that the eutectic alloy has the lowest melting point, much lower than the melting points of the pure components. For this reason the eutectic alloy was the basis of soft solder, which was used in the electrical industry for joining wires in circuits.

Now consider an alloy lying between the compositions C_e and β_e, say one containing 80 wt% Sn. Solidification will start at temperature T_5 with the formation of the solid, having composition β_5 [Fig. 10.5(b)]. Solidification down to the eutectic temperature is similar to that described for complete solid miscibility, i.e. as the temperature falls the amount of solid β increases and the compositions of the solid and liquid follow the solidus and liquidus respectively. Therefore at T_e we have *primary* β solid of composition β_e and the remaining liquid is of composition C_e. This liquid then undergoes the eutectic reaction, hence

$$\beta_e + L_{C_e} \rightleftharpoons \beta_e + (\alpha_e + \beta_e) \tag{10.2}$$

The microstructure of the alloy consists of crystallites of primary β within a background, called the *matrix*, of the eutectic mixture. This characteristic microstructure of a tin-rich alloy is shown in Fig. 10.6(a).

Alloys in the composition range α_e to C_e solidify in an analogous manner but with α as the primary phase. Figure 10.6(b) is a micrograph of an alloy containing 40 wt% Sn, showing the primary α phase (black) in a eutectic mixture matrix. The exact form of the primary phase, and of the eutectic mixture, varies from system to system. For example the primary phase in antimony-rich alloys of the lead–antimony system has cubic shape (Fig. 10.7) while the eutectic in the copper–phosphorus system is more lamellar (plate-like) than the eutectics shown in Figs 10.6 and 10.7 (see the micrograph of Fig. 10.8).

(b) *Peritectic.* As we are now more familiar with the construction and use of equilibrium phase diagrams, we need not spend so much time on the peritectic diagram. A generalized peritectic diagram is shown in Fig. 10.9 and, as for the eutectic, it can be seen that a section of the solidus is

Fig. 10.6

Microstructures of lead–tin alloys: (a) tin-rich (Pb–70% Sn); (b) lead-rich (Pb–40% Sn). (From J. Nutting, and R.G. Baker, *The microstructure of metals*, Institute of Metals, London, 1965).

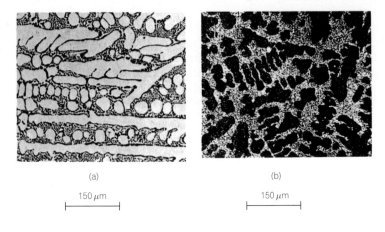

(a)

150 μm

(b)

150 μm

Fig. 10.7

Antimony-rich lead–antimony alloy, showing a cubic primary phase in a eutectic mixture matrix.

50 μm

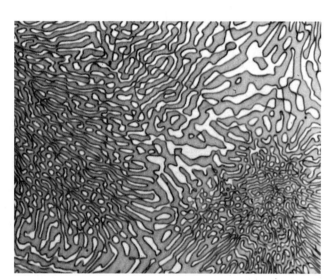

Fig. 10.8

The eutectic mixture in the copper–phosphorus system.

40 μm

horizontal. The temperature corresponding to this section is the *peritectic temperature*, T_p. An example of a simple peritectic phase diagram occurs in the Co–Cu system. The solidification and precipitation behaviour of an alloy lying between pure A and α_p is the same as that for alloys between O and α_e in the eutectic diagram of Fig. 10.5(b). Furthermore, alloys in the range L_p to pure B solidify in an identical manner to an alloy in a system that exhibits complete solid miscibility. Therefore, only between the compositions α_p and L_p do we encounter new solidification behaviour.

First let us consider an alloy of the *peritectic composition* C_p. Solidification commences at T_1 with the formation of primary α of composition α_1. Solidification of α continues until T_p is reached, when the solid will be of composition α_p and the remaining liquid L_p. At the peritectic temperature, all the liquid reacts with all the α_p to form solid β of composition $\beta_p = C_p$, that is

$$L_p + \alpha_p \rightleftharpoons \beta_p \qquad (10.3)$$

This is the *peritectic reaction*.

Fig. 10.9

Generalized peritectic equilibrium
phase diagram.

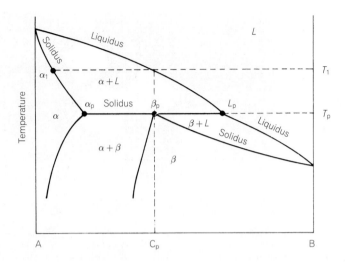

The solidification of an alloy of composition between α_p and β_p is similar; at T_p we again have L_p and α_p but in different proportions (more α_p present). This time all the liquid reacts with some of the α_p to form solid β_p

$$L_p + \alpha_p \rightleftharpoons \alpha_p + \beta_p \tag{10.4}$$

Thus the microstructure consists of primary α surrounded by a partial, or complete, network of the β phase.

Finally, for an alloy lying in the composition range β_p to L_p, the reaction that occurs at the peritectic temperature is

$$L_p + \alpha_p \rightleftharpoons L_p + \beta_p \tag{10.5}$$

In other words, some of the liquid reacts with all of the α_p to give solid β_p. After this reaction the remaining liquid solidifies as β phase and, under equilibrium conditions, the final structure will consist of homogeneous grains of β solid solution.

It must be emphasized that in this section on partial solid miscibility we have been dealing with solidification under ideal equilibrium conditions. Fast rates of cooling, which are often found in practice, can result in (1) coring, as described in Section 10.3.1, (2) modifications of the form of the phases, and (3) changes in the relative proportions of the phases.

10.3.3 Eutectoid reaction

So far we have concentrated on transformations involving a liquid phase, the only solid state transformation we have discussed being the precipitation that takes place when the solid solubility limit is exceeded. There is, however, a number of structural changes that occur in the solid state, e.g. those called eutectoid, peritectoid, and miscibility gap. We will study the eutectoid reaction as a knowledge of the reaction is essential if the heat treatment and microstructures of steels are to be understood.

The copper–aluminium system exhibits a eutectoid reaction and the appropriate section of the phase diagram is given in Fig. 10.10, where T_E and C_E are the *eutectoid temperature* and the *eutectoid composition* respectively. The similarity between the eutectoid and eutectic transformations is obvious when Fig. 10.10 is compared with Fig. 10.5(b). The

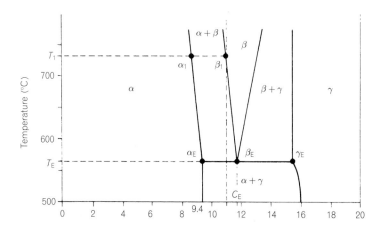

Fig. 10.10

Part of the equilibrium phase diagram of the Al–Cu system, which displays a eutectoid reaction.

important difference is that all the phases involved in the eutectoid reaction are solid, either solid solutions or intermediate phases (in the case of Cu–Al, β and γ are intermediate phases and α a solid solution).

Consider cooling the eutectoid alloy, Cu–11.8 wt% Al, from the β phase field. The alloy will exist as single phase β until the eutectoid temperature T_E is reached. At T_E the following reaction occurs by solid state diffusion

$$\beta_E \rightleftharpoons \alpha_E + \gamma_E \tag{10.6}$$

That is, the β phase decomposes into an intimate mixture, called the *eutectoid mixture*, of the α and γ phases. The eutectoid mixture has a characteristic lamellar morphology as shown in Fig. 10.11(a). The proportions of α_E and γ_E are

$$\text{(weight of } \alpha)(11.8 - 9.4) = \text{(weight of } \gamma)(15.6 - 11.8)$$

i.e. $\text{(weight of } \alpha)(2.4)$ $= \text{(weight of } \gamma)(3.8)$

therefore the percentage of α present is $3.8/(3.8 + 2.4)100 \approx 61\%$.

Let us now turn our attention to the structural changes that occur during the cooling of an alloy removed from the eutectoid composition, e.g. a Cu–11 wt% Al alloy. The boundary of the β phase field is cut at temperature T_1 and α, of composition α_1, begins to be precipitated. As the temperature falls, more and more α is precipitated, the α and β phases changing composition by diffusion down the curves $\alpha_1 - \alpha_E$ and $\beta_1 - \beta_E$ respectively. Hence, when the eutectoid temperature is reached we have α_E and β_E. The β_E then decomposes by the eutectoid reaction

$$\alpha_E + \beta_E \rightleftharpoons \alpha_E + (\alpha_E + \gamma_E) \tag{10.7}$$

and the final structure consists of pro-eutectoid α (that which existed before the eutectoid reaction) in a matrix of the eutectoid mixture. The amount of α present, counting both pro-eutectoid and eutectoid α, is clearly greater than for the eutectoid alloy. In fact

$$\text{(weight of } \alpha)(11.0 - 9.4) = \text{(weight of } \gamma)(15.6 - 11.0)$$

i.e. $\text{(weight of } \alpha)(1.6)$ $= \text{(weight of } \gamma)(4.6)$

giving approximately 74%, by weight, of α.

Alloys, such as the 11 wt% Al alloy, which contain less solute than the eutectoid composition, are termed *hypo-eutectoid*. If an alloy contains more solute than the eutectoid composition, it is said to be a *hyper-eutectoid*.

Fig. 10.11

The microstructure of
(a) a Cu–11.8% Al eutectoid alloy;
(b) a hypereutectoid alloy.

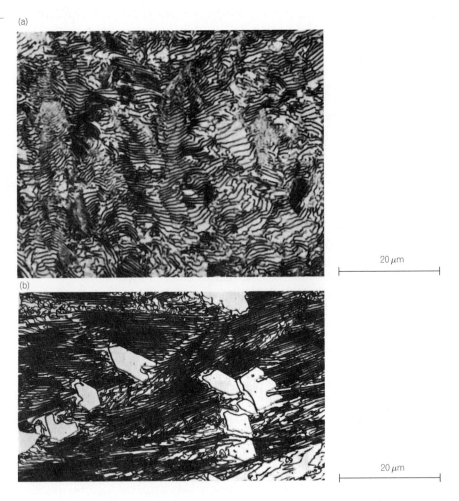

(a)

(b)

20 μm

20 μm

(The prefixes 'hypo-' and 'hyper-' may also be used in connection with the eutectic transformation.) The structural changes in a hyper-eutectoid alloy, e.g. Cu–13 wt% Al, are similar to those in a hypo-eutectoid alloy except that the pro-eutectoid phase is different. Thus the structure of the 13 wt% Al will be pro-eutectoid γ in a eutectoid matrix [Fig. 10.11(b)].

We have now considered the more important features of phase diagrams in some detail. Some systems, for example copper and zinc (i.e. brass), show several different phases with variations of temperature and composition and the equilibrium phase diagram appears very complex indeed (Fig. 10.37). However, the principles discussed above can be applied to all diagrams, and the compositions and proportions of the various phases present under equilibrium conditions for any alloy composition can be determined at any temperature.

10.4 Free energy and equilibrium phase diagrams

In Chapter 7 we learned that the equilibrium condition of a material was that it should have the lowest possible free energy. We have also seen that equilibrium phase diagrams are a means of representing the equilibrium condition of a material as a function of temperature and composition. It follows that there must be a close connection between the temperature and composition dependences of the free energy and equilibrium phase diagrams. This relationship will be demonstrated for the complete solid miscibility and eutectic systems. Before we do this, however, we must define a new thermodynamic parameter called the *chemical potential* μ.

Consider the case of two components A and B which form a solid solution α. When A and B are placed in contact, they spontaneously dissolve in one another, because the free energy of the solution is lower than the sum of the free energies of the components. It is possible to define a quantity called chemical change when in the phase α formed by A and B. A similar chemical potential can be defined for component B. In fact the chemical potential is the free energy of a component (e.g. A or B) in a phase (e.g. α or β) and is also known as the partial molar free energy. Now consider a case in which A and B form two phases, α and β, and let these be placed in contact with one another. If α and β are in equilibrium then the free energy of the mixture must be a minimum with respect to changes in the compositions of α and β, and the chemical potentials of A and B in the phase α are equal to the respective chemical potentials in the phase β. If, on the other hand, the chemical potentials differ, the mixture is not in equilibrium, and its free energy can be reduced by a rearrangement of the components A and B between the phases. Thus component A is redistributed until its chemical potential is the same in all the phases present, i.e. equilibrium is reached. Similarly for component B. This may result in a reduction in the number of phases (e.g. $\alpha + \beta \rightarrow \alpha$) or further phases may become involved (e.g. $\alpha + \beta \rightarrow \alpha + \gamma$).

Figure 10.12 is a curve of free energy against composition for two components, A and B, which are completely soluble in each other. It can be shown that the intercepts on the free energy axis of the tangent at any composition C give the chemical potentials, μ_A and μ_B, of the components A and B in the solution of composition C. This concept is true for all free energy curves and is basic to the following discussion.

The equilibrium phase diagram for a system exhibiting complete solid miscibility is given in Fig. 10.13(a). A number of temperatures are marked on the phase diagram and the corresponding free energy curves for the

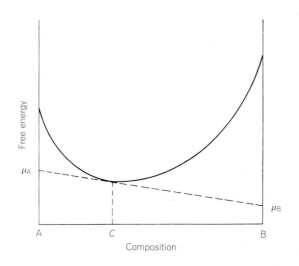

Fig, 10.12

Free energy versus composition for two completely miscible components. The tangent at C gives the chemical potentials of the components A (μ_A) and B (μ_B) in a solution of composition C.

Fig. 10.13

The phase diagram and corresponding free energy curves for a system exhibiting complete solid miscibility.

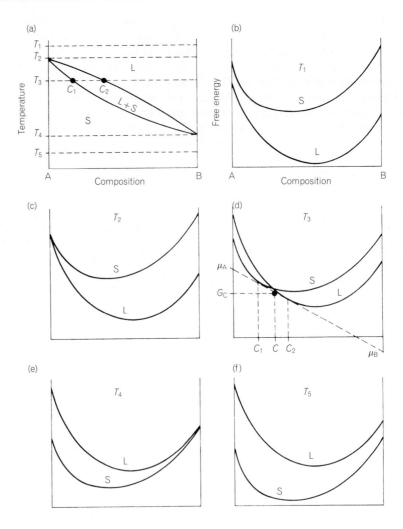

liquid and solid phases at these temperatures are given in Figs 10.13(b), (c), (d), etc. At the temperature T_1, the free energy curve for the liquid is lower than for the solid, therefore the equilibrium state is liquid over the complete composition range A to B. Temperature T_2 is the melting point of component A, and we see that at this temperature solid A and liquid A have the same free energy. The free energy curves cross at temperature T_3 and a common tangent has been drawn to the two curves at compositions C_1 and C_2. This means that the chemical potentials of components A and B in solid of composition C_1 are equal to those in liquid of composition C_2, i.e. these two phases are in equilibrium. Thus any alloy of composition between C_1 and C_2, such as alloy C, is two-phase and its free energy, G_c, lies on the tangent and is lower than the free energy of liquid, or solid, of the same composition. At T_4, which is the melting point of component B, the free energies of solid B and liquid B are the same. For all other compositions the solid has the lower free energy and so is the equilibrium phase. Finally, at T_5 the free energy curve for the solid is below that for the liquid at all compositions, hence only solid can exist under equilibrium conditions.

If two components have the same crystal structure and yet are only partially miscible in the solid state, it is a consequence of the free energy

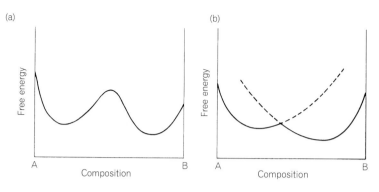

Fig. 10.14

The free energy curves that lead to partial solid miscibility;
(a) components A and B have the same crystal structure;
(b) components A and B have different crystal structures.

curve for the solid having two minima, as shown in Fig. 10.14(a). Alternatively, when two components have different crystal structures, and so are only partially miscible, we have two superimposed free energy curves, which effectively give a curve with two minima [Fig. 10.14(b)]. The analysis that follows, which is based on a free energy curve for the solid with two minima, is therefore applicable to either type of system.

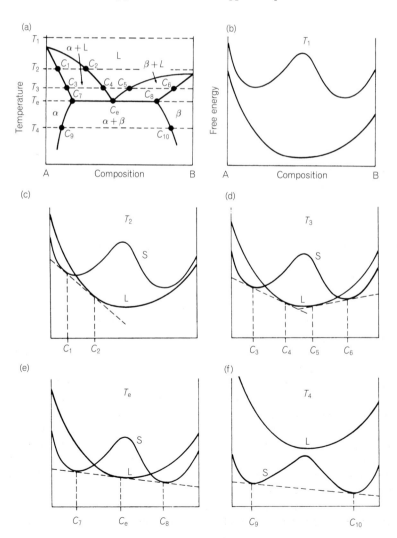

Fig. 10.15

The phase diagram and corresponding free energy curves for a eutectic system.

The phase diagram and free energy curves at various temperatures for a eutectic are given in Fig. 10.15. Liquid has the lower free energy for all compositions at temperature T_1 and is therefore, as shown by the phase diagram, the equilibrium phase over the complete composition range A to B. At temperature T_2 there is a common tangent to the solid and liquid curves at C_1 and C_2 respectively. Consequently, the equilibrium state of an alloy between compositions A and C_1 is solid α, between C_1 and C_2 is two-phase solid plus liquid, and between C_2 and B is liquid. At lower temperatures, e.g. T_3, the free energy curve for the solid cuts the curve for the liquid twice, giving two common tangents corresponding to the solid α + liquid and solid β + liquid phase fields. There is only one tangent at T_e showing that, at the eutectic temperature, liquid of composition C_e is in equilibrium with solid of composition C_7 and C_8. Finally, at T_4, no liquid is present under equilibrium conditions and the common tangent touches the curve for the solid at C_9 and C_{10}. Thus at all compositions between C_9 and C_{10} a two-phase structure exists consisting of solid of compositions C_9 and C_{10}.

10.5 Nucleation and growth

We have seen that when the temperature of a material is altered a phase transformation may occur. The transformation may involve only a change in structure, examples being solidification and allotropic changes in a pure element such as iron. In other cases, however, the transformation may result in a change in composition as well as structure, e.g. the precipitation that takes place when the solid solubility limit is exceeded. Both types of transformation commonly proceed by the *nucleation and thermally activated growth* mechanism.

As the name suggests, the nucleation and thermally activated growth mechanism can be conveniently divided into two stages, namely a nucleation stage and a growth stage. The first stage is *nucleation*, which is the formation of small grains, or *nuclei*, of the new phase, each just a few atoms in size, in the old phase. *Growth* of these nuclei then occurs by material being transferred, generally by diffusion, from the old phase and into the new phase.

In practice a phase transformation does not always take place exactly at the 'transition temperature', for example, liquids may undercool before solidification commences, and precipitation may not occur immediately the solid solubility limit is exceeded. Let us look at the latter in more detail; the relevant part of the phase diagram showing the solid solubility limit is given in Fig. 10.16. When an alloy of composition C is rapidly quenched to room temperature from the α phase field, e.g. from temperature T_S, the precipitation of β is suppressed and a metastable state is attained, called a *supersaturated* solid solution. This procedure is called solution treatment. If the metastable solid solution is then heat treated at an elevated temperature T_A in the two-phase region, precipitation of the β phase occurs and the alloy reverts to the equilibrium two-phase condition. This latter treatment, termed *aging*, can be carried out at various temperatures below the 'transition temperature' T_T and this affects the final microstructure. A quantitative thermodynamical analysis for the microstructural dependence on the aging temperature is possible, but complex, and a simple qualitative approach will suffice for our purposes.

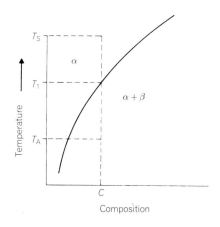

Fig. 10.16

Part of a phase diagram showing the solid solubility limit and the solution treatment (T_S) and ageing (T_A) temperatures.

Over a normal range of aging temperatures, the rate of nucleation increases with decreasing temperature. In contrast, the growth rate, because it is generally diffusion controlled, is slower the lower the ageing temperature. As a result, the temperature dependence of the *overall transformation rate*, which is a function of the nucleation and growth rates, is as shown in Fig. 10.17. Near to T_T, say at T_1, the overall transformation rate is slow, even though the growth rate is fast, because the nucleation rate is low. Aging at this temperature will produce a coarse dispersion of β particles as only a few nuclei are formed and these grow rapidly. At much lower temperatures, such as T_2, the nucleation rate is high but the growth rate very low, again giving a slow overall transformation rate. This time, however, because of the many nuclei present and the low growth rate, ageing produces a fine dispersion of the β phase.

The overall transformation rate curve of Fig. 10.17, and our deductions from it, are applicable to all nucleation and thermally activated growth processes, thus for solidification the temperature T_T in Fig. 10.17 is the melting point of the material. We can now understand why rapidly cooled liquids generally produce a finer microstructure than slowly cooled liquids. Rapid cooling causes considerable undercooling, resulting in a large number of solid nuclei, small in size and densely packed.

If, instead of plotting rate against temperature, we plot the start and finish times for a transformation at various temperatures on a temperature–logarithm (time) graph we obtain a curve that is a mirror image of the

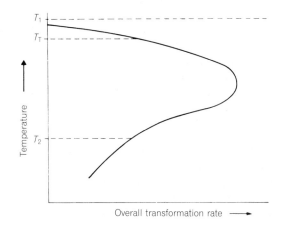

Fig. 10.17

Overall transformation rate as a function of temperature for a transformation occurring by the nucleation and growth mechanism.

Fig. 10.18

Time–temperature–transformation
diagram.

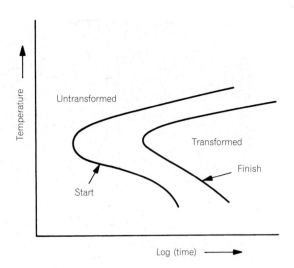

overall transformation curve (Fig. 10.18). Such a curve is known as a Time–Temperature–Transformation (TTT) diagram and is commonly employed to present the temperature dependence of a transformation; see, for example, the discussion later in this chapter on steels.

Up to now we have not discussed where the nuclei appear in a material undergoing a transformation. When the nucleation is completely at random throughout the material, it is said to be *homogeneous*. Homogeneous nucleation rarely occurs in practice as it is energetically easier for nucleation to commence at structural imperfections. For example, during solidification in a mould nucleation takes place at the walls and on solid impurities in the melt, whilst dislocations, non-metallic inclusions and grain boundaries are preferred sites for nucleation of precipitate particles in a solid. Nucleation on structural imperfections is called *heterogeneous* nucleation and is the rule rather than the exception.

10.6 Martensitic transformation

Phase transformations that occur by the nucleation and thermally activated growth mechanism proceed relatively slowly, as shown by the fact that we can prevent the decomposition of a solid solution by rapidly quenching through the solid solubility limit. Furthermore, these transformations usually, but not always as we will see in the following section, result in the formation of the equilibrium phase or phases. There is, however, another type of transformation, called the *martensitic transformation*, which is very rapid and often produces a metastable phase.

In contrast to the nucleation and growth transformation, no diffusion is involved in the martensitic transformation. Instead, it occurs by the spontaneous and systematic coordinated shearing of the lattice of the old phase in such a way that the distance moved by any atom is less than one atomic spacing. This means that an atom retains the same neighbours; hence a martensitic transformation can only lead to changes in crystal structure, and not in composition, of the phases.

The electron micrograph of Fig. 10.19 shows the typical plate-like microstructure formed by a martensitic transformation. Note also the large

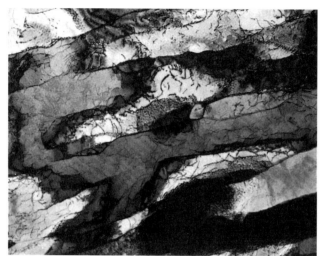

Fig. 10.19
Electron micrograph showing martensite plates with high dislocation density in a Ti–1 % Si alloy.

2 μm

number of dislocations (the thin black lines) within the plates. The time taken for an individual martensite plate to grow to its final size may be less than 10^{-14} s, even at low temperatures, which confirms that diffusion cannot be involved. The martensitic transformation is therefore extremely rapid and it is impossible to prevent it by rapid quenching. We will come back to the martensitic transformation when we discuss the heat treatment of steels later in this chapter.

10.7 Strengthening mechanisms

We now turn to the ways in which the strength of a metal is enhanced by the introduction of imperfections, in which we include precipitates of new phases, as described above.

Plastic deformation occurs in metals by the movement of dislocations, so that anything that hinders their movement increases the stress required for the onset of plastic flow. It has already been mentioned in Chapter 8 that dislocations interfere with each other's motion, and that grain boundaries act as barriers to dislocations, so both of these may be considered as strengthening mechanisms and will be discussed in more detail in this section. We also describe the strengthening achieved by alloying (solution and dispersion hardening).

Work-hardening

As plastic deformation proceeds, dislocations interact with one another making further deformation more and more difficult. This phenomenon is called work- or strain-hardening and is best explained by reference to the stress–strain diagram in Fig. 9.1(b), repeated here as Fig. 10.20. Consider plastically deforming a material by $S\%$, i.e. deform to the point S on the stress–strain curve, and then unload. The material will have been work-hardened and on retesting will not yield until the stress σ_S is reached.

Work-hardening is a commonly used strengthening mechanism but it has certain disadvantages. For example, by deforming by $S\%$ we achieved a considerable increase in the yield stress from σ_y to σ_S. However, if we

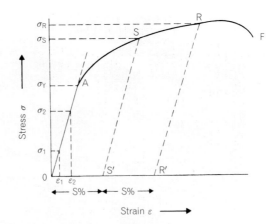

deform a further $S\%$ (to point R on the curve of Fig. 10.20), because the work-hardening rate decreases with strain, the increment in the yield stress is smaller, i.e. $(\sigma_S - \sigma_Y) > (\sigma_R - \sigma_S)$. Moreover, at point R on the stress–strain curve, the material is approaching the point of fracture at F, giving little safety margin if the material were used for structural purposes in this condition.

Materials often have to withstand raised temperatures during service, and work-hardening is not a good strengthening mechanism in these circumstances. The reason for this is that dislocations are non-equilibrium defects and therefore, when solid state diffusion is sufficiently rapid, dislocation rearrangement and annihilation will take place in order to reduce the free energy of the material. At temperatures in the range 0.3–$0.4T_m$, where T_m is the melting point of the material in degrees Kelvin, *recovery* takes place. Recovery results in a slight decrease in the yield stress due to a reduction in the number of dislocations and the rearrangement of the dislocations, but the grain structure remains unaffected. Two recovery processes, both of which involve the thermally activated climb of edge dislocations, are illustrated in Fig. 10.21. These are the mutual annihilation of two edge dislocations, thereby reducing the dislocation density [Fig. 10.21(a)], and the rearrangement of dislocations (known as *polygonization*) into configurations with lower energy [Fig. 10.21(b)].

At higher temperatures recrystallization takes place, and the yield stress falls to its original value of σ_Y. *Recrystallization* is the nucleation and growth of *new grains of low dislocation density* (Fig. 10.22). The free energy of the material is lowered by the significant reduction in the dislocation density that occurs during recrystallization; in other words, the energy of the dislocations is the 'driving force' for recrystallization. It follows that a heavily deformed material with a large dislocation density will recrystallize more rapidly than a material that has been only slightly deformed (Fig. 10.23). Recrystallization is a thermally activated process, and, like the recovery process, it follows the rise in the rate of self-diffusion.

It is interesting to note that some materials recrystallize at room temperature, e.g. lead for which room temperature is approximately $0.5T_m$. Consequently, it is impossible to work-harden lead at room temperature and this is why lead is so soft and malleable. If a material is deformed, or worked, at a temperature that is sufficient for concurrent recrystallization it is said to be hot-worked. Thus lead is hot-worked at room temperature,

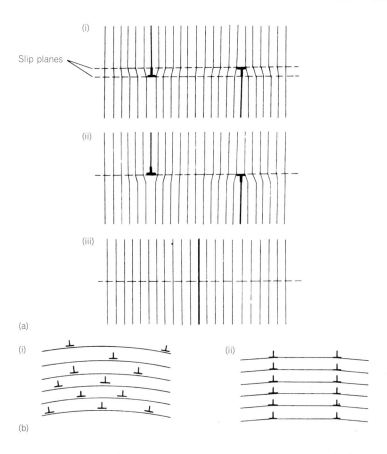

(i)

Slip planes

(ii)

(iii)

(a)

(i)   (ii)

(b)

Fig. 10.21

Some recovery processes.
(a) Annihilation of two
edge dislocations (i) two edge
dislocations on parallel slip planes,
(ii) climb of dislocations onto the
same slip plane, (iii) dislocations
come together and annihilate.
(b) Polygonization (i) 'random'
distribution of edge dislocations,
(ii) the same dislocations in a
low-energy configuration of low
angle grain boundaries.

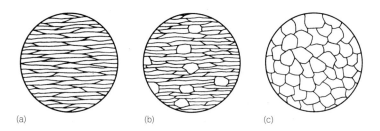

(a)   (b)   (c)

Fig. 10.22

Recrystallization: (a) deformed
grains of high dislocation density;
(b) new grains of low dislocation
density are nucleated and begin to
grow; (c) specimen consists solely
of the new grains, i.e. it is fully
recrystallized.

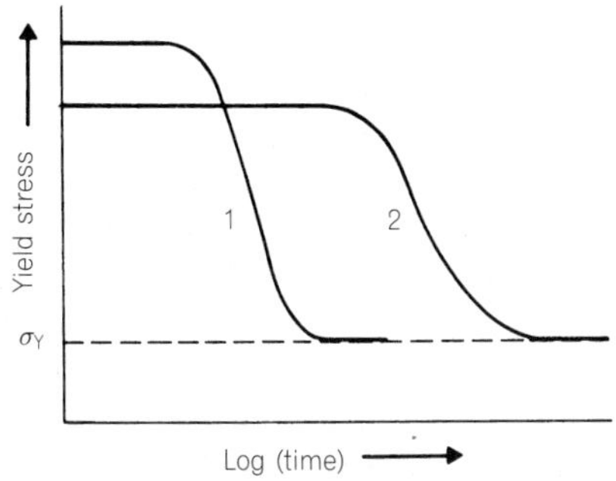

Fig. 10.23

Effect of the amount of prior
deformation on the rate of
recrystallization. Specimen 1 was
deformed more than specimen 2.

while a temperature in excess of 900°C is required to hot-work mild steel. *Cold-working* is the process of deforming at temperatures at which recrystallization does not occur.

Grain boundary hardening

Grain boundaries act as physical barriers to dislocations, causing them to 'pile up' and hindering their motion. Obviously, the greater the grain boundary area per unit volume the more difficult is dislocation movement and the higher the yield stress. The effect of grain boundaries on the yield stress is given by the *Petch* equation:

$$\sigma_y = \sigma_i + k_y d^{-1/2} \tag{10.8}$$

where d is the grain diameter, and σ_i and k_y are constants at a given temperature: σ_i is a measure of the intrinsic resistance of the material to dislocation motion, while k_y is associated with the ease of operation of dislocation sources. Figure 10.24 shows that a linear relationship between yield stress and $d^{-1/2}$ is obtained for a variety of metals in accordance with this expression.

Solution hardening

A common way of increasing the hardness and yield stress of a material is to introduce solute atoms into the solvent lattice, so forming a *solid solution* or alloy. The solute atoms, whether they be substitutional or interstitial, will cause *local elastic strains* in the material. These local strains hinder the motion of dislocations and hence increase the strength of the material.

Substitutional atoms produce *symmetrical* strain fields and only give a gradual increase in yield stress with concentration, typically $Gc/20$–$Gc/10$, where c is the fraction of substitutional atoms present (the *atom fraction*), and G is the shear modulus. Theoretical models for the hardening due to

Fig. 10.24

Effect of grain size on yield stress.

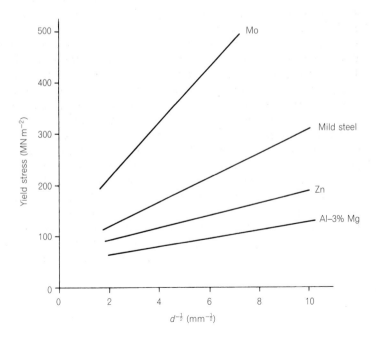

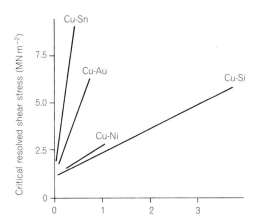

Fig. 10.25

Effect of substitutional alloying additions on the critical resolved shear stress of Cu single crystals (J.O. Linde and S. Edwards, *Arkiv Fysik*, 1954, **8**, 511).

substitutional atoms predict that the yield stress (or the critical resolved shear stress in the case of single crystals) is directly proportional to the concentration for concentrations less than about 10%; this is confirmed by experiment, as illustrated by Fig. 10.25. The reason that the different solutes in Fig. 10.25 vary in their effectiveness as strengtheners is because of their differing atomic sizes relative to that of the solvent. If the size difference is large, as for Sn in Cu (see Table 10.2), the strengthening is more marked than when the solute and solvent atoms are of similar size, for example Si in Cu.

Some interstitial atoms, e.g. carbon in b.c.c. iron, are associated with strain fields that are high in one direction compared to the transverse directions. These cause rapid strengthening of the order of 2G–3G per atom fraction. The greater the difference between the longitudinal and transverse strains caused by the interstitial, the more considerable is the strengthening. It has been suggested that the dependence on concentration of the yield stress σ_y for interstitial additions differs from that for substitutional atoms and is $\sigma_y \propto c^{1/2}$.

In the solid solutions discussed so far there has been no preference for like, or unlike, atoms to be nearest neighbours, i.e. the solutions have been random. This is not always the case and the situation can arise where solute atoms prefer solvent atoms to be their neighbours. This leads to an ordered arrangement of the atoms. If the ordering is slight and on a local scale, it is called *short-range order*. However, in the extreme case, the solute and solvent atoms arrange themselves on two distinct lattices throughout the crystal and the material is said to be exhibiting *long-range order* [Fig. 10.26(a)]. The plastic deformation of ordered materials, especially those with long-range order, is complex, but a simplified view of dislocation motion in these materials will be adequate for our purpose. When a unit dislocation moves through an ordered material it breaks up the special distribution of solute atoms and produces a fault in the ordered stacking sequence, known as an *anti-phase domain boundary*, in its wake [Fig. 10.26(b)]. This requires a great deal of energy and if the deformation of ordered crystals took place by the motion of single unit dislocations then these materials would have much higher flow stresses than random solid solutions. In reality the unit dislocations in ordered materials move in pairs: the first dislocation creates the anti-phase domain boundary and the

Fig. 10.26

Plastic deformation of a long-range ordered material (a) before deformation, (b) after the passage of a unit dislocation that has created an anti-phase domain boundary, and (c) a superlattice dislocation.

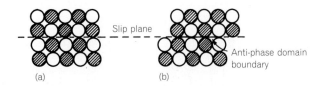

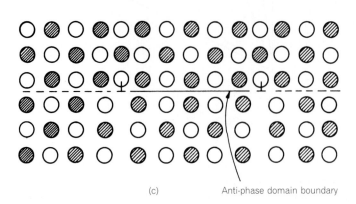

(c)   Anti-phase domain boundary

second destroys it [Fig. 10.26(c)]. Consequently the stress required for the plastic deformation of an ordered material does not differ greatly from that for the same material in the disordered condition. This arrangement of two unit dislocations is called a *superlattice* dislocation and the similarity between a superlattice dislocation and an extended dislocation in an f.c.c. lattice (see Section 8.4.5) is apparent. A superlattice dislocation is two unit dislocations separated by a ribbon of anti-phase domain boundary, and an extended dislocation is two partial dislocations separated by a ribbon of stacking fault.

Dispersion hardening

Many materials consist of two or more phases and often one of the phases is in the form of small particles distributed throughout the material. The presence of a dispersion of small particles increases the strength of the material by an amount that depends on the volume fraction and size of the particles, and hence on the interparticle spacing S_p. Figure 10.27(a) shows a dislocation approaching an array of particles. If the dislocation cannot cut the particles, it has to bow between them and the maximum stress required for this is reached when the radius of curvature $R = S_p/2$ [Fig. 10.27(b)]. Thus, using Eq. (8.15), the stress required is

$$\tau = \frac{Gb}{R} = \frac{2Gb}{S_p} \tag{10.9}$$

The sequence of events shown in Figs 10.27(c) and (d) occurs at stresses less than $2Gb/S_p$, therefore the dislocation passes the particles leaving loops around them (note the similarity to the operation of the Frank–Read source described in Chapter 8). This process of bowing past particles and leaving a loop is called the *Orowan* mechanism. Figure 10.28 is an electron micrograph of a deformed copper specimen containing a dispersion of silica particles, and the dislocation loop around the silica particle in the centre of the micrographs is clearly visible.

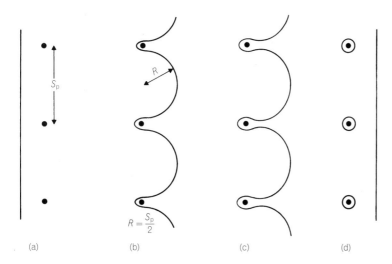

Fig. 10.27

The sequence of events when a dislocation bows past particles in the Orowan mechanism.

A dispersion of particles may be produced by a number of methods. For example, the silica dispersion in the copper of Fig. 10.28 was produced by *internal oxidation*. A solid solution of Cu–Si was made, then oxygen diffused into the material and allowed to react with the silicon to form silica. Silver with a dispersion of Al_2O_3 is produced this way and is often used for contacts in electrical equipment. Another common method of forming a dispersion is by heat treating a metastable solid solution so that a second phase precipitates. This is called *precipitation hardening* and certain Al–Cu and nickel alloys, which will be discussed later in this chapter, are hardened this way.

In the case of precipitation-hardened systems the particles may be metastable and have a *coherent* interface with the matrix. An interface is said to be coherent when there is matching at the interface of the crystal planes in the matrix with those in the precipitate (Fig. 10.29). The metastable precipitates may be small clusters of only about 100 solute atoms, or

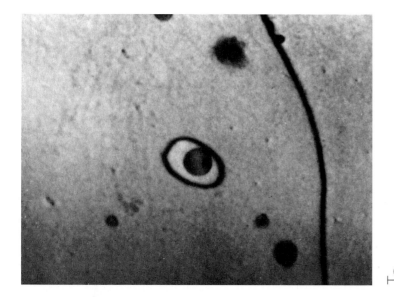

0.5 μm

Fig. 10.28

Electron micrograph of deformed copper containing a dispersion of silica particles. A dislocation loop has been left around the particle in the centre of the micrograph. (Courtesy of F.J. Humphreys.)

Fig. 10.29

The interface between a precipitate particle and the matrix.
(a) Coherent interface with coherency strains (the difference between a_p and a_m is exaggerated to emphasize the coherency strains). (b) Incoherent interface with no matching of planes at the interface.

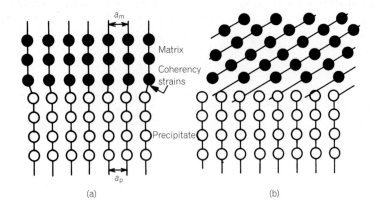

they may have a well-developed crystal structure different from that of the matrix. Both types of particle are cut by the dislocations rather than the dislocations bowing past the particles by the Orowan mechanism. Particle cutting can give considerable strengthening and it has been shown that there are several different mechanisms contributing to the hardening. A detailed discussion of all the contributions is not necessary at this stage and we will just mention two mechanisms in order to demonstrate the general principles. First, as seen in Fig. 10.29, a coherent interface can lead to elastic strains in the matrix because the interplanar spacings a_p and a_m are not exactly the same. Dislocations interact with these elastic strains, known as *coherency strains*, in a similar way (but more strongly) to the elastic strains around a solute atom. The second contribution is illustrated in Fig. 10.30, which shows that new precipitate–matrix surfaces are produced

Fig. 10.30

Diagram illustrating the formation of new precipitate–matrix interfaces when particles are cut by a dislocation (a) plan view of slip plane; (b) side view of slip plane.

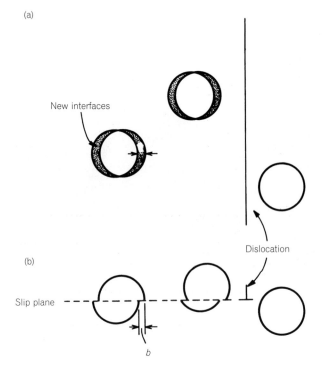

when particles are cut. There is an energy per unit area associated with these surfaces and hence additional work, i.e. a higher stress, is required to move the dislocations. Some of the superalloys mentioned in the introduction and discussed later make good use of these mechanisms.

10.8 Some commercial alloy systems

10.8.1 Iron–carbon system (cast iron and plain carbon steels)

Because of its considerable importance in engineering, the iron–carbon system has received much detailed study. The iron–carbon, or more properly the iron–iron carbide phase diagram, is given in Fig. 10.31.

The carbon atom is smaller than the iron atom (see Table 10.2) and dissolves interstitially in all three phases (α, γ, and δ) of iron. The solubility in f.c.c. γ-iron is at a maximum near 2.0 wt% at 1130°C, the solid solution being known as *austenite*. The solubilities in the b.c.c. phases are much smaller, the maxima being 0.1 wt% at 1492°C in δ-iron and 0.03 wt% at 723°C in α-iron, the latter being called *ferrite*.

Iron and carbon form an intermediate compound, iron carbide, which contains 6.67 wt% carbon. This has the formula Fe_3C, is called *cementite*, and is extremely hard and brittle. Cementite is involved in both of the important transformations that occur in the iron–iron carbide system. First, as shown by the phase diagram (Fig. 10.31), there is a eutectic reaction at 1130°C where liquid of composition 4.3 wt% C solidifies into a eutectic mixture of austenite and cementite. The eutectic mixture is known as *ledeburite*. The other important transformation is the eutectoid transformation that

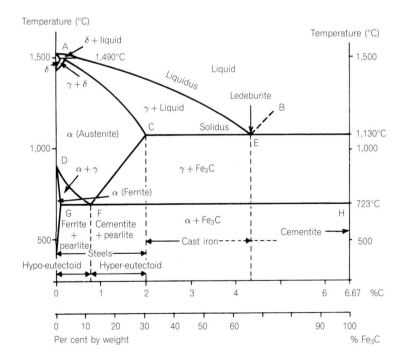

Fig. 10.31

Phase diagram for the iron–iron carbide system.

occurs at 723°C; austenite of composition 0.8 wt% C decomposes into a eutectoid mixture of ferrite and cementite, called *pearlite*.

Let us examine the microstructure and properties of some ferrous alloys. In this discussion we shall first consider *plain carbon steels and irons* then later steels containing alloying additions such as nickel, chromium, tungsten and molybdenum.

Pig iron is product of the blast furnace and has a carbon content in the range 2–4% but generally about 3.5 wt%. *Wrought iron* is made by refining pig iron and contains as little as 0.05 wt% C and a small amount of refining slag, principally iron silicate. When pig iron is just remelted, but not refined, and cast in its final form it is known as *cast iron*. The microstructure, and hence properties, of cast iron vary with the rate of cooling and the presence of elements such as sulphur, manganese and silicon.

When the rate of cooling is fast, nearly all the carbon in a cast iron exists as cementite. This makes the material hard and brittle and it is only used in applications where hardness and wear resistance are important, e.g. grinding and crushing machinery. The fracture surface of such iron has a white, or silvery, appearance and consequently it is called *white cast iron*.

Most white cast irons are hypo-eutectic and a typical microstructure is given in Fig. 10.32. This microstructure is best understood by following the phase transformations that occur as the iron is cooled. Primary austenite is formed in increasing amounts until the eutectic temperature is reached, at which point the remaining liquid solidifies as the eutectic mixture, ledeburite. Using the lever rule in conjunction with the phase diagram, it can be seen that the proportion of cementite increases as the temperature falls from the eutectic temperature to the eutectoid temperature, i.e. precipitation of cementite occurs between 1130° and 723°C. On cooling to below the eutectoid temperature all the austenite transforms to the eutectoid mixture, pearlite. The lamellar structure of the pearlite is not resolved in the micrograph and it appears grey; the larger grey regions correspond to the pearlite formed from the primary austenite, and the finer regions to that formed from the austenite in the ledeburite. The cementite remaining from the ledeburite is white. The transformations that take place in a hyper-eutectic white cast iron are very similar, except that primary cementite is formed instead of primary austenite. The microstructure, which is shown in

Fig. 10.32

Microstructure of a hypo-eutectic white cast iron.

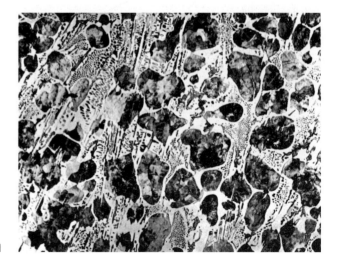

75 μm

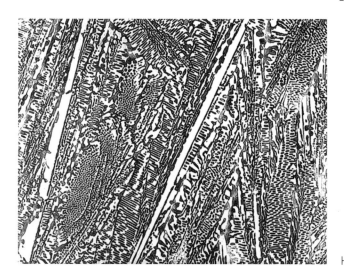

Fig. 10.33
Microstructure of a hyper-eutectic white cast iron.

75 μm

Fig. 10.33, consists of primary cementite platelets in a matrix of cementite and pearlite from the ledeburite, which is not resolved by the microscope.

When pig iron is cooled slowly, or if it contains for example more than 3 wt% Si, much of the carbon occurs in the free form as flakes of graphite. The fracture surface is dark, as fracture takes place along the soft graphite flakes, and so the iron is referred to as *grey cast iron*. Grey cast iron is relatively soft and weak, but it is easily machined. The microstructure of a grey cast iron is shown in Fig. 10.34. The large, black graphite flakes are the result of the eutectic reaction $L \rightleftharpoons \gamma + \text{graphite}$, which is the true equilibrium reaction and is therefore favoured by slow cooling. The matrix in Fig. 10.34 is mainly pearlite.

Now let us turn our attention to lower carbon contents, i.e. the steels. The solidification of steel is generally less important than the changes that occur on cooling from the austenitic region and so we will concentrate on the latter. If a steel of the eutectoid composition is air-cooled from the austenite phase field, it is said to be *normalized*, and it will have a 100% pearlitic structure similar to the eutectoid structure shown in Fig. 10.11(a).

Fig. 10.34
Micrograph of a grey cast iron showing the black graphite flakes.

75 μm

Table 10.3

Effect of heat treatment on the hardness and ductility of eutectoid (0.8% C) steel

Heat treatment	Microstructure	Hardness VHN (kgf/mm^2)	Ductility (%)
Air-cooled from γ-region (normalized)	Pearlite	280	15
Furnace-cooled from γ-region	Coarse pearlite and some spheroidization	210	20
Heated just below eutectoid temperature	Completely spheroidized	175	
Quenched from γ-region	Martensite	850	Negligible
Quenched then tempered for 1 h at 500°C	Precipitates of iron carbide in ferrite	400	13

Slower cooling rates, such as experienced in furnace cooling, give coarser structures and there may even be some spheroidization. Subsequent prolonged heating at temperatures near to, but below, the eutectoid temperature produce complete spheroidization and coarsening of the cementite. In this condition, for a given carbon content, the steel has the minimum hardness and maximum ductility (see Table 10.3).

The pearlitic transformation, as it involves solid state diffusion, can occur at temperatures well below the eutectoid temperature of 723°C. This is true for all eutectoid reactions and represents a major difference between eutectoid and eutectic transformations – eutectic transformations always take place at temperatures very close to the eutectic temperature. Thus it is possible to quench a eutectoid steel from the austenite phase field to, say, 650°C and then isothermally transform to pearlite at that temperature. In fact pearlite in plain carbon steels may be obtained by quenching to and then transforming at temperatures in the range 450°C to 723°C. As we will see later in this section, a TTT curve can be constructed for the isothermal transformation of austenite to pearlite.

The preceding discussion has concentrated on steels of the eutectoid composition. Steels of other compositions behave in a similar manner except that on cooling from the austenite condition a pro-eutectoid phase is formed before the pearlitic transformation starts. For hypo-eutectoid steels, i.e. less than 0.8 wt% C, the pro-eutectoid phase is ferrite, which is soft and ductile (Fig. 10.35), whereas for high carbon, hyper-eutectoid steels the pro-eutectoid phase is hard, brittle cementite. Because of this change in the type and proportion of phases with carbon content, the mechanical properties of pearlitic steels with a given thermal history vary considerably with composition. As shown by the stress–strain curves for steels in the normalized condition (Fig. 10.36), as the carbon content increases the steels have higher yield stresses and tensile strengths but lower ductilities.

We have seen that austenite in plain carbon steels, that is steels containing no major alloying additions, isothermally transforms to pearlite in the temperature range 450°C to 723°C. If austenite is quenched to and then transformed at lower temperatures, in the range 250°C to 550°C, a different reaction occurs, namely the *bainitic transformation*. Note that there is an overlap of the temperature ranges for the pearlitic and bainitic transformations and both reactions occur when isothermally transforming in the range 450°C to 550°C.

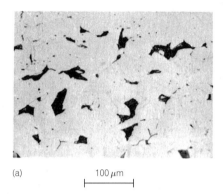

(a)   100 μm

(b)   10 μm

Fig. 10.35

Microstructure of hypo-eutectoid steel in the normalized condition. (a) Light micrograph. The pro-eutectoid phase is ferrite and is the lighter colour. The dark regions are pearlite, which is unresolved at this magnification. (b) Scanning electron micrograph with the pearlite resolved into lamellar ferrite and cementite. (Courtesy H.M. Flower.)

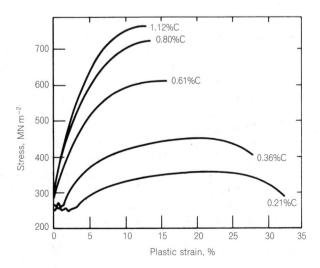

Fig. 10.36

Effect of carbon content on the stress–strain curves of pearlitic steels.

The product of the bainitic transformation is termed *bainite* and essentially it consists of lath-shaped ferrite and cementite, although the exact details of the microstructure vary with transformation temperature. The bainitic reaction is unusual in that it has some characteristics of a martensitic transformation and other characteristics typical of a nucleation and thermally activated growth process. For example, as observed in martensite, bainite has a relatively high dislocation density, although generally not as high as found in martensite. On the other hand, in common with nucleation and growth transformations, a TTT curve can be constructed for the bainitic reaction.

When a steel is rapidly quenched from the austenitic condition, the pearlitic and bainitic transformations are suppressed and the *martensitic transformation* occurs, resulting in a typical martensitic structure as shown in Fig. 10.37. As no long-range diffusion is involved in the martensitic transformation all the carbon that was in solution in the austenite remains in solution in the martensite. Martensite is, therefore, supersaturated with respect to carbon and consequently the normal b.c.c iron structure is distorted to a body-centred tetragonal (b.c.t.) structure [Fig. 10.38(a)]. The greater the carbon content of the martensite the greater the distortion, as can be seen by the increase in the c/a ratio for the b.c.t. structure with increasing carbon [Fig. 10.38(b)].

On rapidly cooling a steel the martensitic transformation commences at the martensite start temperature, M_s. The transformation continues as the temperature falls and is completed at the martensite finish temperature, M_f. The extent of the transformation to martensite depends on the temperature, but is independent of time. In other words, if we quench a steel to a temperature of between M_s and M_f a certain amount of martensite will rapidly be formed, but if we continue to hold the steel at that temperature the amount of martensite will not increase. To increase the proportion of martensite we have to lower the temperature. The M_s and M_f temperatures are composition dependent, and in the case of carbon they decrease with increasing carbon content (Fig. 10.39). An important feature to note about the M_f curve is that it is below room temperature for plain carbon steels containing more than 0.7 wt% C. This means that if we

Fig. 10.37

Martensitic structure in a 1.5% carbon steel quenched from the austenite phase field. There is some retained austenite (white in this structure) which would not be present for a hypo-eutectoid steel.

50 μm

Fig. 10.38
Crystallography of martensite in steels. (a) The unit cells of the b.c.c. and b.c.t. structures (not all the interstitial sites shown would be occupied). (b) The increase in the c/a ratio of the b.c.t. martensite with increasing carbon content.

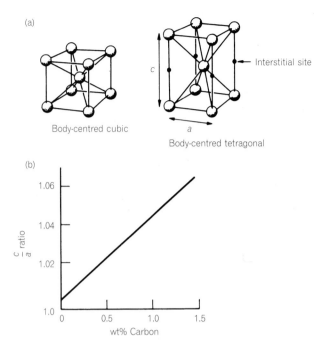

(a)

Interstitial site

Body-centred cubic

a

Body-centred tetragonal

(b)

c/a ratio

wt% Carbon

quench a eutectoid or hyper-eutectoid steel to room temperature the martensitic transformation will not go to completion and some austenite will be retained as shown in the micrograph of Fig. 10.37. The retention of austenite affects the hardness of quenched steels. The hardness of quenched steels containing up to 0.7 wt% C increases markedly with carbon content due to the strong solid solution hardening effect of the interstitial carbon atoms in the martensite. However, above 0.7 wt% C the increasing proportion of retained austenite, which is relatively soft, balances the increasing hardness of the martensite and as a result the hardness of the steel remains essentially constant (Fig. 10.39).

Steel in the martensitic condition is too hard and brittle for normal service (Table 10.3). However, the properties are improved by reheating, that is *tempering*, to a temperature below the eutectoid temperature. During tempering carbon is removed from solution by the precipitation of iron carbides, mainly cementite, and the matrix reverts to the b.c.c. structure. At the higher tempering temperatures some spheroidization and coarsening of the carbides occurs and also the dislocation substructure associated with the martensitic transformation is modified by processes analogous to recovery and recrystallization. The temperature ranges of these structural changes are marked on Fig. 10.40, which gives the mechanical properties as a function of tempering temperature. From this figure we can see that tempering is accompanied by an increase in ductility at the expense of strength, as shown by the decrease in yield stress and hardness.

Mild steel, which contains less than 0.25 wt% C (Fig. 10.35), is used in the ferrite–pearlite condition and is easily cold-formed and machined. It is a general purpose steel for applications where strength is not particularly important, such as nails and car bodies. Higher carbon steels may be used in the pearlitic state or in the quenched and tempered condition, depending on the properties required. *Medium carbon steel* (0.25–0.5 wt% C) is

Fig. 10.39

The effect of carbon content on the martensite start M_s and finish M_f temperatures and on the hardness of plain carbon steels.

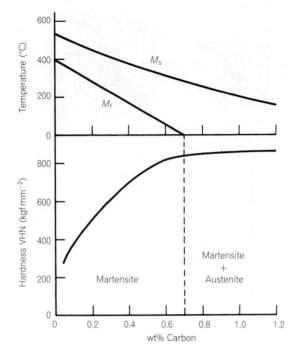

Fig. 10.40

Graph showing the decrease in hardness and yield stress and the increase in ductility on tempering a 0.55% C steel.

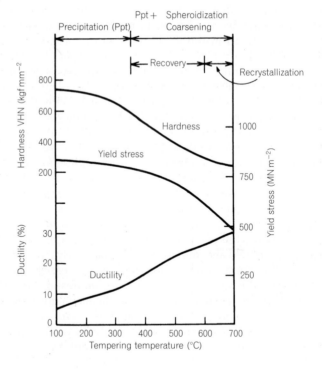

employed for components needing a higher strength than that of mild steel coupled with good toughness, e.g. hand tools, crankshafts. *High carbon steel* (>0.50 wt% C) is hard but lacks the ductility and toughness of the lower carbon content steels. It is, therefore, often found in applications where wear is the prime consideration, as in cutting tools.

10.8.2 Alloy steels

So far our discussion of steels has concentrated on plain carbon steels, that is iron–carbon alloys containing only small amounts of impurities, e.g. less than 0.5% Mn and 0.5% Si. However the microstructure and properties of steels can be altered significantly by the introduction of alloying elements. Alloying elements have various different effects but the most important are associated with:

(a) stabilization of ferrite or austenite,
(b) changes in the rates of transformations, and
(c) carbide formation.

Let us look at each of these in turn.

Elements that are more soluble in ferrite than in austenite, such as silicon, chromium, vanadium and molybdenum, tend to favour the formation of ferrite and hence are known as α-*stabilizers*. In contrast elements that are more soluble in austenite tend to stabilize that phase and are called γ-*stabilizers*, e.g. manganese, nickel, cobalt. The well known *18/8 austenitic stainless steel*, which contains 18% Cr and 8% Ni, is a good example of the effect of a γ-stabilizer. Chromium is present because it reduces corrosion and oxidation by forming a thin, chromium-rich film on the surface of the steel. A binary Fe–18% Cr alloy would be ferritic at room temperature, but on adding 8% Ni the γ-phase field is expanded to such an extent that an 18/8 stainless steel is austenitic at room temperature. Austenitic stainless steels are non-magnetic and exhibit good ductility.

The effect of alloying elements on the rates of the pro-eutectoid, pearlitic and bainitic transformations is best studied by means of TTT diagrams (see Fig. 10.18). First let us examine the diagram for a plain carbon steel of the eutectoid composition. In this case there is no pro-eutectoid phase but the diagram must take into account the pearlitic and bainitic reactions, and the individual TTT curves for these reactions are shown in Fig. 10.41(a). The two curves overlap so much that the resulting combined TTT diagram for a eutectoid steel consists, by chance, of one single continuous curve [Fig. 10.41(b)]. Figure 10.41(b) shows that the decomposition of austenite is most rapid at about 550°C, where the transformation to a mixture of

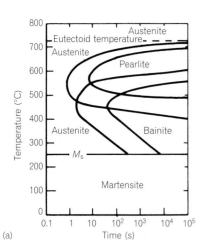

(a)

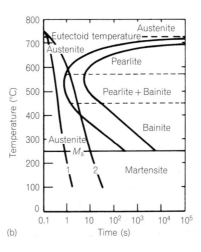

(b)

Fig. 10.41
(a) Schematic diagram of the overlapping TTT curves for the pearlitic and bainitic transformations in plain carbon steel of the eutectoid composition (0.8 wt% C). (b) TTT diagrams for the same steel showing a single continuous curve with two superimposed cooling rates designated 1 and 2.

pearlite and bainite starts in just under 1 s. Thus it is possible by rapid quenching to suppress the pearlitic and bainitic transformations and produce a martensitic structure, with a little retained austenite, throughout a small (say less than 5 mm diameter) sample of a eutectoid steel; see cooling rate 1 on Fig. 10.41(b). The cooling rate varies with the size of the sample and the quenching medium, e.g. water gives a more rapid quench than oil. It follows that the resultant microstructure will depend on these two factors. When we quench a larger sample (>5 mm diameter) in the same medium as before, the cooling rate will decrease from the surface to the centre of the sample and for this larger sample the rate in the centre is represented by cooling rate 2 on the TTT diagram. This will be slow enough for some pearlite to be produced as well as the martensite.

A typical TTT diagram for a plain carbon hypo-eutectoid steel is shown in Fig. 10.42(a). The essential differences between this diagram and that for the eutectoid steel are (a) the rates of the pearlitic and bainitic transformations are faster, i.e. the curve is displaced to shorter times, and (b) the addition of a curve for pro-eutectoid ferrite. The TTT diagram for a hyper-eutectoid steel is very similar, except, of course, the pro-eutectoid phase is cementite. As the reactions are faster for hypo- and hyper-eutectoid steels it is even more difficult to obtain predominantly martensitic structures in samples of reasonable size. Furthermore, problems can arise if very fast quenches, such as a brine quench, are employed due to thermal stress induced cracking, which is termed *quench cracking*.

The ability of a steel to form martensite on quenching is termed the *hardenability* – the greater the hardenability the easier it is to form martensite. In order to increase the hardenability the rates of the pro-eutectoid, pearlitic and bainitic reactions must be slowed down and this may be achieved by alloying. All the common alloying elements in steel, with the exception of cobalt, decrease the reaction rates and displace the TTT curve to longer times, as illustrated by the diagram for a Ni–Cr–Mo steel [Fig. 10.42(b)]. It follows that martensite can be produced in this alloy steel with less severe quenching, and hence throughout components of larger sections, than in the case of the plain carbon steels. Note also that the alloying additions have separated the curves for the pearlitic and bainitic reactions.

Fig. 10.42

TTT diagrams for (a) a plain carbon steel containing 0.35 wt% C and (b) a 1.5% Ni–1.1% Cr–0.3% Mo alloy steel of the same carbon content.

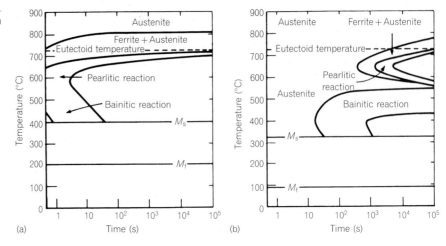

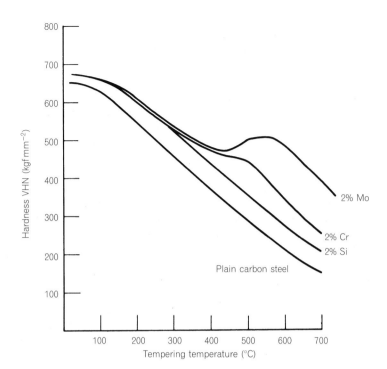

Alloying additions enable us to obtain steels in the martensitic condition more easily but they also affect the tempering characteristics. Firstly, the precipitation and coarsening of the iron carbides may be altered by the presence of an alloying addition. For example, silicon slows down the precipitation and coarsening of cementite and this can slightly reduce the fall-off in hardness observed for plain carbon steels. More important is that some alloying elements called the strong *carbide formers*, e.g. Cr, Mo, Ti, V and W, form alloy carbides such as V_4C_3 and Mo_2C, which are more stable than the iron carbides. Due to the slow rate of diffusion of these substitutional alloying elements, the alloy carbides do not form until tempering temperatures in the range 450–600°C are reached. The precipitation of the alloy carbides results in an increase in hardness, termed *secondary hardening*, at high tempering temperatures (Fig. 10.43). Secondary hardening enables us to produce steels with a high hardness and strength but with a better ductility than that exhibited by a plain carbon steel tempered to the same hardness by lower temperature tempering. Also, because of the slow diffusion of the carbide forming elements their carbides do not readily coarsen, consequently alloy steels so hardened may be used for high-temperature applications.

10.8.3 Aluminium–copper system

If resistance to dislocation motion were all that was required of a strong material we could use intrinsically hard solids like silicon carbide for engineering applications. Such materials are, however, brittle and it is more often a combination of ductility and strength that is required. The ductility not only makes it possible to work the material to a required shape but gives a safety factor during service, i.e. if it is accidentally stressed the material will deform plastically rather than fracture in a catastrophic manner. We have already seen (Section 9.5) how a dispersion of small particles can strengthen an intrinsically soft material, but what is also important is that, with careful control of microstructure, this strengthening can be achieved without a marked loss in ductility. Aluminium alloys containing small amounts of copper (2–4.5 wt% Cu) are capable of being strengthened this way, the dispersion being produced by *precipitation* from a supersaturated solid solution. These include the well-known alloy 'duralumin', which is widely used where strength coupled with light weight is required.

We discussed precipitation earlier in this chapter and saw that it was possible to control the precipitate dispersion by solution treatment and subsequent aging in the two-phase region (refer to Fig. 10.16). The appropriate part of the Al–Cu phase diagram is similar to that shown in Fig. 10.16 and is given in Fig. 10.44. Thus, we can produce a super-saturated solid solution of, say, an Al–4% Cu alloy, by rapidly quenching from the temperature T_s. The behaviour of this alloy differs from the general precipitation behaviour described earlier in the structural changes that occur during aging. On aging, the Al–Cu alloy does not always transform directly to the equilibrium two-phase condition of $\alpha + \theta$, where θ is $CuAl_2$. Instead, the precipitation process may take place via inter-mediate, metastable phases; the complete sequence of events is

$$\alpha_{ss} \rightarrow \alpha + GP \rightarrow \alpha + \theta'' \rightarrow \alpha + \theta' \rightarrow \alpha + \theta$$

where α_{ss} is the supersaturated solid solution, while GP, θ' and θ'' are metastable phases. For interest, GP stands for Guinier–Preston zones, after the investigators who discovered them by using X-ray diffraction tech-niques. Further details, such as the compositions and structures of these intermediate phases, are not essential to this discussion.

An explanation for the precipitation behaviour of Al–Cu can be given in terms of an activation energy that has to be provided for a phase change to proceed. The concept of activation energy as an energy barrier was fully

Fig. 10.44

The aluminium-rich section of the aluminium–copper equilibrium phase diagram.

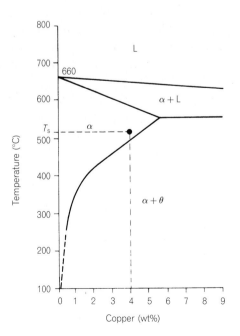

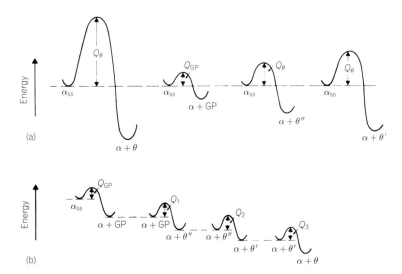

Fig. 10.45

The activation barriers for transformations in aluminium-rich Al–Cu alloys: (a) the barriers when transforming from the supersaturated solid solution directly to another phase; (b) the barriers corresponding to the sequence of events $\alpha_{ss} \to \alpha +$ GP $\to \alpha + \theta'' \to \alpha + \theta' \to \alpha + \theta$.

developed in Chapter 7 (Section 7.18) and the various barriers in the Al–Cu system are represented in Fig. 10.45. It can be seen that a large activation energy, Q_θ, is associated with the change from the supersaturated solid solution to the equilibrium two-phase condition of $\alpha + \theta$. Much smaller activation barriers have to be surmounted for changes from the supersaturated solid solution to intermediate conditions (namely, Q_{GP}, $Q_{\theta''}$ and $Q_{\theta'}$), and from one intermediate phase to another (Q_1, Q_2 and Q_3). Thus, when the alloy transforms via the metastable, intermediate phases it is following the path of lowest activation energy at each stage.

Whether or not the complete sequence of events is followed in a given alloy depends on the ageing temperature. At high ageing temperatures, considerable thermal energy is available to aid transformations and hence the changes with high activation energy, namely $\alpha_{ss} \to \alpha + \theta$ and $\alpha_{ss} \to \alpha + \theta'$, occur readily. At low ageing temperatures only changes with low activation energy can take place. Figure 10.46 illustrates this point as well as showing the marked effect that microstructure has on the hardness of an Al–4 wt% Cu alloy. An $\alpha + \theta'$ structure is shown in Fig. 10.47.

Maximum hardness in this system is associated with the precipitation of particles of the metastable θ'' which are cut by dislocations during

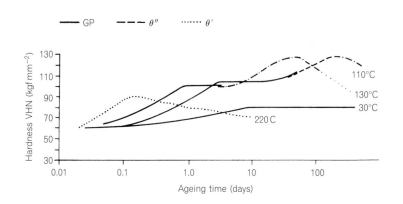

Fig. 10.46

Ageing curves for an Al–4% Cu alloy.

Fig. 10.47

Electron micrograph of the $\alpha + \theta'$ structure in an Al–4% Cu alloy. (Courtesy H.M. Flower).

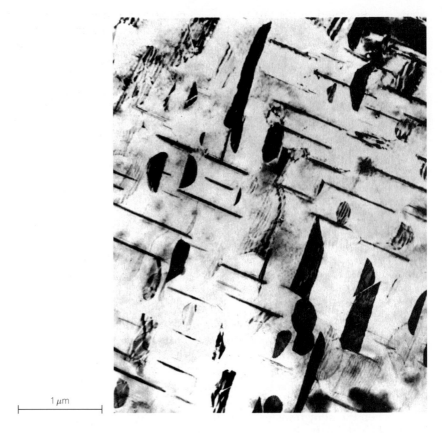

1 μm

deformation. The Orowan mechanism (see Section 10.7) is operative when θ' or the equilibrium phase θ are present and a lower hardness results.

10.8.4 Ni-based and Co-based superalloys

The strengthening mechanisms introduced in the two previous sections are combined in the *superalloys* based on iron, nickel and cobalt to give high tensile and yield strengths at temperatures up to 1100°C. They include some of the stainless steels already mentioned, so here we only discuss the other types.

The three mechanisms involved are (i) precipitation hardening as used in Al–Cu, (ii) refractory carbide precipitation as used in steels and described in connection with Fig. 10.43, and (iii) solution hardening, as described in Section 10.7.

The simplest such alloy is Ni 82%–Mo 28% by weight, known as Hastelloy B-2. It has a tensile strength of 896 MPa (130 kpsi) and a yield strength of 414 MPa (60 kpsi). A study of the phase diagram shows that this is a single-phase alloy, and solution hardening is the principal hardening mechanism involved.

Nickel alloys containing Al or Ti display precipitation hardening on ageing, arising from precipitates of a highly ordered phase of either Ni_3Al or Ni_3Ti (the γ' phase), embedded in a random γ phase of similar composition. While the random γ phase alloy is relatively ductile, dislocations are strongly pinned as long as the ordered precipitates, which have a higher density than the matrix, are small enough to generate coherency strain fields as illustrated in Fig. 10.29. The precipitates are grown to optimum size during a carefully controlled ageing process. Simultaneously, carbides precipitate within the grain boundaries of the γ phase matrix, stabilizing their locations, and so toughening the matrix. The addition of Cr to the alloy inhibits corrosion.

Such Ni-based superalloys are widely used in jet engines at temperatures up to 1000°C, while for temperatures up to 1100°C the more expensive Co-based alloys containing Mo, W and C are needed. Co-based superalloys are dominantly carbide-strengthened, and have measured yield strengths up to more than 690 MPa (100 kpsi). For example Stellite 6B (60%Co–30%Cr–4.5%W) has a remarkable tensile strength of 1220 MPa (177 kpsi) and a yield strength of 710 MPa (103 kpsi).

Careful processing further increases the strength of a jet engine turbine blade made from a suitable superalloy. A technique called *directional cooling* encourages the individual grains to grow in a direction parallel to the blade length (called *columnar growth*), eventually extending over its whole length. Even greater strength is conferred by a subsequent heat treatment designed to cause one grain to grow at the expense of others, until the blade becomes a single crystal.

10.8.5 Copper–zinc system

The phase diagram of the copper–zinc system is complex (Fig. 10.48), but we need only concern ourselves with the composiiton range 0 to about 50 wt% zinc, which represents the series of alloys known as *brasses*, widely used where ductility or ease of machining are important.

Copper is capable of dissolving up to 38 wt% zinc in substitutional solid solution, so forming the α brasses. The hardness of α brass as a function of zinc content is shown in Fig. 10.49. Up to about 10 wt% zinc the relationship between hardness and concentration is approximately linear, in

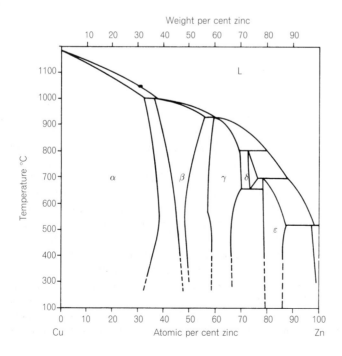

Fig. 10.48

Equilibrium phase diagram for the copper–zinc system.

Fig. 10.49

The hardness of brasses as a function of zinc content.

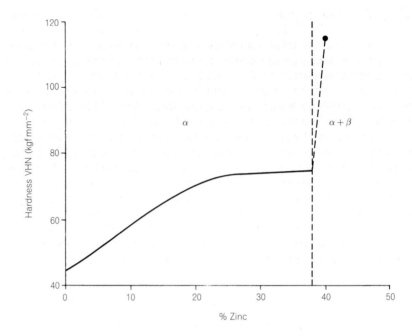

accordance with the discussion on solution hardening earlier in the chapter. At greater zinc concentrations individual zinc atoms are less effective as hardeners due to the overlapping of their individual elastic strain fields. All α brasses are ductile and can be severely cold-worked, consequently they are widely used for pressings and drawn tubing.

Zinc in excess of 38 wt% leads to the formation of the β phase and most of the cast brasses have a two-phase $(\alpha + \beta)$ structure. A cast and annealed $(\alpha + \beta)$ structure is shown in Fig. 10.50. The β phase is hard at room temperature but soft at elevated temperatures, therefore $(\alpha + \beta)$ brasses are commonly hot-worked, at a temperature at which they are predominantly β phase. Because of the presence of the β phase, $(\alpha + \beta)$ brasses are harder than α brasses (see Fig. 10.49).

Fig. 10.50

Cast and annealed $(\alpha + \beta)$ brass. α is dark (etched in ammonium persulphate).

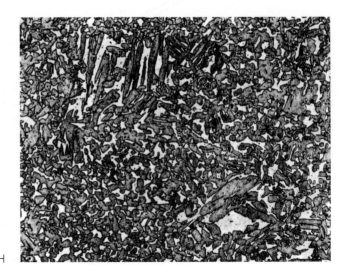

75 μm

10.9 Corrosion

Many metals react readily in the environment, with oxygen, water, common salt (in, for example, sweat or seawater), and with a variety of industrial pollutants such as sulphur in the form of SO_2 and H_2S. Indeed, the reactivity of most metals is the reason why they are most commonly found as compounds in the Earth's crust such as alumina (Al_2O_3), haematite (Fe_2O_3), brown iron ore ($Fe_2O_3.H_2O$), copper pyrites ($CuFeS_2$) or copper glance (CuS). Corrosion is the term we use for the reaction of metallic components with atmospheric oxygen and water. It determines the lifetime of many components, leading to costs that have been estimated at about 4% of gross domestic product, hence understanding and preventing it is a major concern for materials scientists and engineers. Thus the oldest and cheapest engineering metals need protective coatings, whereas most ceramics and many polymers degrade much less rapidly. Nevertheless, coatings for metals must be chosen with care because silicates also react with atmospheric moisture, and polymers are attacked by various solvents.

10.9.1 Metallic corrosion and the electrochemical series

Several forms of corrosion can be identified, all involving water. For example, steel-hulled ships corrode by galvanic action unless protected, and many steels rust in a humid atmosphere. Atmospheric water is retained in crevices, giving rise to *crevice corrosion* and to *pitting corrosion*. The latter is usually caused by pinpoint failures in a protective coating (whether natural or artificial), resulting in tiny pits scattered across the metal surface. The role of water in corrosion is to dissolve metal ions.

Metals act as reducing agents in their reactions – they donate electrons to a non-metal, e.g. water or H^+ ions, i.e. a reaction involving the donation of n electrons begins with the process:

$$M \rightarrow M^{n+} + ne^- \tag{10.10}$$

where M denotes an atom at the surface of a metal that goes into solution as an M^{n+} ion, where $n = 1, 2$, etc. This is called an *anodic* reaction. As we shall explain, the corresponding *cathodic* reaction:

$$M^{n+} + ne^- \rightarrow M \tag{10.11}$$

invariably occurs simultaneously, but at a different rate, as explained for chemical reactions in Section 7.21.

The ease with which a metal loses its electrons and dissolves as an ion is directly linked to the value of its *standard electrode potential* and hence to its position in the *electrochemical series*, which is listed in Table 10.4. The table also shows the heat evolved in a reaction with chlorine, an indication of the metal's reactivity, and which mostly, but not invariably, follows in the same sequence as the standard potentials.

This standard potential is measured in an electrochemical cell, and is just the electric potential measured on the metal relative to a 'standard half-cell' or electrode, under particular conditions of ionic concentration, pressure and temperature, as we shall explain in the next section. The table shows that amongst those metals potassium is the most reactive (more properly called the most *active*), while gold, the least reactive, is termed the most *noble*. When comparing two metals, the standard potential

Table 10.4

The electrochemical activity series for pure metals, listed in order of their standard (or reduction) potentials, compared with their heats of reaction with chlorine

Metal	Standard potential (V)	Heat evolved in reaction with 1 mol Cl_2 (kJ)	
K	−2.93	873	
Ca	−2.87	795	More anodic
Na	−2.71	826	or active
Mg	−2.37	642	↑
Al	−1.66	464	
Zn	−0.76	416	
Fe	−0.45	270	
Pb	−0.13	359	
Cu	0.34	206	↓
Hg	0.76	230	More cathodic
Ag	0.80	255	or noble
Pt	1.19	142	
Au	1.50	79	

of the more active of the two is termed more *anodic*, while the more noble metal's potential is called more *cathodic*.

While the electrochemical activity series indicates the potential for reactions, it says nothing about the rate at which a reaction occurs. Factors such as naturally formed protective surface layers are also important (see Section 10.10 on passivation). However, the electrochemical series is useful for predicting *relative* corrosion rates in water containing dissolved impurities such as atmospheric gases, or seawater. The same series is thus useful for considering the need for protection. It is also relevant to assessing the de-alloying of materials like brass and cast iron. In de-alloying, an element that is more active (i.e. electrochemically negative) than the metal into which it is alloyed has a tendency to leach out to the surface. Brass dezincifies, while grey cast iron loses its iron, leaving a weak graphitized network. The composition of an alloy can determine whether such processes occur – brass only dezincifies if it contains more than 15 wt% zinc, and even then the process may be significantly slowed by the addition of Sn, As, Sb or P.

Alloys behave very differently from pure metals, and in engineering applications it is better to refer to empirical lists of their relative tendency to corrode in specific environments. An example is the 'galvanic series in seawater', which is used by marine engineers to choose compatible metals for use in close proximity to one another.

10.9.2 Galvanic corrosion and galvanic protection

Galvanic corrosion occurs when two distinct metals are both immersed in an *electrolyte*, an aqueous solution containing ionic salts, and are electrically connected together. Such a combination, idealized in Fig 10.51, results in corrosion of one of the two metals when the switch is closed, and is known as a *galvanic cell*.

Accelerated corrosion is also found where differing concentrations of ions in an electrolyte occur in contact with different parts of the surface of a single metal, creating a *concentration cell*. In both types of corrosion, the

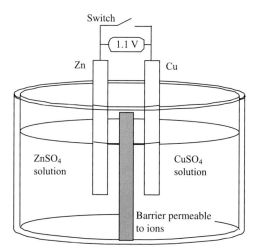

Fig. 10.51

A galvanic cell in which Zn and Cu electrodes may be electrically connected by closing the switch. When the switch is open, the high-impedance voltmeter registers a 1.1 V potential difference.

metallic connection completes an electrical circuit around which flows a current – a current of positive and/or negative ions flows within the electrolyte, while electrons flow in the metal.

Consider first the case when the switch in Fig. 10.51 is closed, enabling electrons to flow away from the zinc anode, the more active of the two metals. Zinc atoms at the metal surface then undergo the anodic reaction:

$$Zn \rightarrow Zn^{2+} + 2e^- \qquad (10.12)$$

Simultaneously, the Zn^{2+} ions go into solution in the electrolyte, i.e. the metal is continuously corroded. Note that the anode loses an equal number of positive and negative charges – the net electrical charge on the anode remains unchanged. Indeed the reaction only proceeds because the electrons are free to leave the metal. The electrons flow via the switch to the copper cathode, where they discharge positive Cu^{2+} ions arriving at the cathode from the electrolyte, leading to the cathodic reaction

$$Cu^{2+} + 2e^- \rightarrow Cu \qquad (10.13)$$

The copper atoms, no longer soluble, are deposited on the surface of the cathode, i.e. copper is *electroplated* there.

It must be emphasized that during these processes the electrolyte remains everywhere electrically neutral, for the negative ions in it are mobile, and move around so that their concentration is equal to that of the Zn^{2+} or Cu^{2+} ions.

Thus when two metals with different standard potentials such as Zn and Fe are both in contact with seawater, one corrodes significantly, while the other is protected from corrosion. This fact is used to protect the steel hulls of ships, by electrically bonding a large zinc sacrificial anode to the hull, usually near the propeller, which is particularly vulnerable to corrosion because of the turbulence it creates. The zinc anode alone corrodes by the reaction

$$Zn \rightarrow Zn^{2+} + 2e^-$$

and the two electrons then flow to any point where the steel hull has contact with the seawater (for example at a crack or pinhole in paintwork), where they take part in one or more cathodic reactions, the most important of which is:

$$2H_2O + 2e^- \rightarrow H_2 + 2OH^-$$

Hence, at the cathodic steel hull, gaseous hydrogen is produced, but no corrosion occurs thanks to the supply of negative electrons from the zinc anode.

Even the use of graphite as a lubricant can sometimes result in galvanic corrosion, for graphite, being a metallic conductor, forms a potent galvanic cell in combination with some metals, especially magnesium alloys.

Steel in contact with water is often protected by a coating of anodic zinc (*galvanized steel*), which protects even if scratches or pinholes expose the underlying steel. The zinc coating itself corrodes to a white carbonate, which is visually less objectionable than rust.

If galvanic corrosion cannot be avoided by an appropriate choice of metal, protection may be achieved by coating the cathode with an insulating paint. Painting an anode to protect it should be avoided, because it usually results in accelerated attack at pinholes or other defects in the coating, where the galvanic current density is high because of the small area of anode exposed.

Protection of buried pipe work is sometimes ensured by installing an auxiliary anode in the ground nearby, which is held at a positive potential relative to the pipes by connecting each to the appropriate terminals of an electrical voltage source. The supply of electrons from the voltage source ensures protection.

10.9.3 Electrochemical basis of corrosion

It is important to note that the continuous loss of electrons from a metal is always accompanied by corrosion. An example of a different way in which this can happen is when an Fe anode is isolated electrically, and is immersed in an aqueous solution containing many H^+ ions. Two electrons may then combine at the metal surface with two H^+ ions, to form a molecule in the reaction:

$$2H^+ + 2e^- \rightarrow H_2 \text{ (gas)} \qquad (10.14)$$

In this case, putting together Eq. (10.14) with the anodic reaction Fe $\rightarrow$ $Fe^{n+} + ne^-$, we have simply a description of an (electro-)chemical reaction of the metal with a surrounding acidic solution containing the hydrogen ions in question, which is accompanied by the evolution of hydrogen gas.

To understand how both corrosion and deposition of the same metal are possible, consider what happens in the simple arrangement shown in Fig. 10.52(a), sometimes referred to as a *half-cell*. A single metal, iron in this case, in contact with an ionic solution containing only Fe^{2+} and negative ions, undergoes *both* anodic and cathodic reaction processes at its surface, which just balance one another – they are in dynamic equilibrium. Thus metal atoms, having lost electrons, go into solution as ions. In the case of iron this anodic reaction is

$$Fe \rightarrow Fe^{2+} + 2e^- \qquad (10.15)$$

Because the electrons are left in the metal, it therefore carries an electrical charge. These electrons spread across the metal surface, attracting

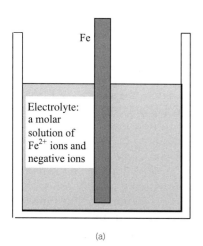

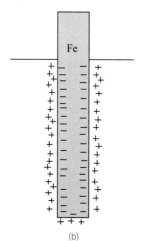

Fig. 10.52

(a) A half-cell consists of a metal in an electrolyte. (b) The equilibrium distribution of electric charges that results – electrons in the metal, and ions in solution. The metal is negatively charged, and its electrical potential is negative, relative to the uncharged state.

Fe

Electrolyte: a molar solution of Fe^{2+} ions and negative ions

Fe

(a)

(b)

towards the surface the positive ions in the solution, which form a layer of high ionic concentration close against the metal surface, as in Fig. 10.52(b). The raised ionic concentration results in deposition of ions onto the metal, becoming neutral atoms, by the cathodic reaction

$$Fe^{2+} + 2e^- \rightarrow Fe \tag{10.16}$$

No further loss of metal from the surface occurs once the concentration of ions in the surface layer rises to the point where the rate at which they redeposit just balances the rate at which reaction (10.15) occurs. In this state of dynamic equilibrium the negative charge on the metal lowers the electrical potential there by an amount V, so that an additional amount of energy $2eV$ is needed to put another Fe^{2+} ion into solution. The potential can be measured electrically, and, as might be expected, varies with the chemical composition of the electrolyte according to the *Nernst equation*, as follows

$$V = V_0 + \frac{kT}{ne} \ln[M^{n+}]$$

where V_0 is the value of V under standard conditions, T is temperature, k as usual is Boltzmann's constant, e is the electronic charge, and $[M^{n+}]$ is the *concentration* in moles per litre of the ion M^{n+}.

By standardizing the electrolyte for each metal to a molar solution (containing 1 mole of metal ions per litre), the measured values shown in Table 10.4 are determined. It is important to note that the potentials listed in Table 10.4 apply only in the special case of molar ionic solutions, and not to any other electrolyte composition.

We may sum up the above by noting that corrosion occurs at any location where the rate of the anodic reaction (10.10) exceeds the reverse, cathodic reaction, while deposition of metal occurs if the cathodic reaction, like reaction (10.11), is the faster of the two. For example in Fig. 10.51, where Zn and Cu are both immersed in a solution, the zinc, being the more active (having the more negative standard potential) undergoes the anodic reaction faster than the reverse cathodic reaction, once the switch is closed to complete the electrical circuit.

10.9.4 Concentration-cell corrosion

Now consider Fig. 10.53, where each of two identical metal electrodes is in contact with a different neutral electrolyte solution. The solutions differ in having very different concentrations of, for example, chloride ions, Cl^-. The two electrodes are at different potentials, the one in the higher Cl^- concentration being more anodic (i.e. negative). Such cases occur in real-life corrosion where a crevice occurs beneath a washer or bolt, or in a crack in a metal surface. The crevice allows the build-up of a higher than average concentration of soluble reaction products containing the Cl^- ions, while at the air–water interface their concentration is lower. The resulting difference in ionic concentration creates an electrical potential difference, indicated in Fig. 10.53 by the voltmeter reading, when the switch isolates the two metal parts. Closing the switch, as in Fig. 10.51, allows electrons to flow from the anodic region, where there is an excess of Cl^- ions, through the metal to the more cathodic region, where they discharge positive ions in solution, and as before, corrosion results at the anode. This is the mechanism behind crevice corrosion and also *pitting corrosion*, mentioned later.

A similar situation arises when rusting is initiated by differences in oxygen concentration near the waterline in steel or zinc water tanks, as illustrated in Fig. 10.54. The oxygen concentration is highest at the water surface, where it is readily replenished from the air. Slightly below this, oxygen is consumed by combination with Fe^{2+} or Zn^{2+} ions, which are released into solution by the reaction $Fe \rightarrow Fe^{2+} + 2e^-$ at an anodic region yet further down. Nearest the surface of the water, the metal is protected (we say *passivated* – see Section 10.10) by the supply of electrons conducted via the metal from the anodic region below, resulting in the chemical reduction of the excess dissolved oxygen near the surface by the cathodic reaction

$$O + H_2O + 2e^- \rightarrow 2OH^-$$

In this galvanic cell, the dissolved Fe^{2+} (or Zn^{2+}) ions move upwards from the anodic region, while OH^- ions move downwards from the cathodic surface region. The positive and negative ions react together where they meet between those two regions, as follows:

$$Fe^{2+} + 3OH^- \rightarrow Fe(OH)_3 \text{ (rust)}$$

Fig. 10.53

Concentration-cell corrosion occurs where identical electrodes are immersed in electrolytes containing very different concentrations of negative ions.

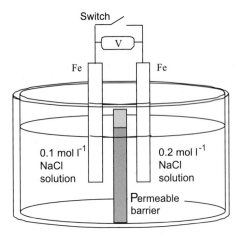

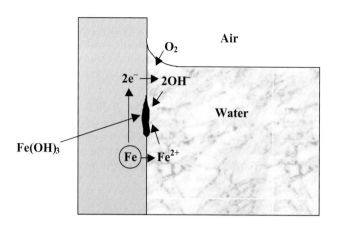

Fig. 10.54
A steel vessel rusts near its waterline, caused by a decrease in the dissolved oxygen concentration with increasing depth below the surface. Fe^{2+} and OH^- ions are driven together both by diffusion and by the resulting electric potential gradient.

The deposit of red–brown rust forms just below the waterline, above the anodic region, which remains unprotected from further corrosion. The ions are driven towards one another by a combination of diffusion and a super-imposed electrical force, which arises from the electric voltage gradient.

A similar concentration cell builds when salt-laden drops of water remain for long enough on a horizontal steel surface – at the centre of the drop the oxygen concentration falls, encouraging corrosion there, while the concentration at the periphery remains high. A ring of rust eventually surrounds a central, disk-shaped, corroded pit in the metal. Once this mechanism has created a pit or crevice where Cl^- ions congregate without being washed away, the Cl^- concentration difference between pit and surface maintains the potential difference, which drives both the flow of current, and the corrosion process itself. Such concentration cells are common, because of the plentiful supply of NaCl in natural environments, in many chemical process plants, and in mammalian body fluids.

Corrosion of this type may be slowed or prevented by reducing the dissolved oxygen concentration below harmful levels by one of two methods. Either the water may be heated above 80°C, or the oxygen content may be lowered chemically by dissolving sodium sulphite (with $CoCl_2$ added as a catalyst), which oxidizes to the sulphate:

$$2Na_2SO_3 + O_2 \rightarrow 2Na_2SO_4$$

10.9.5 Atmospheric corrosion and its prevention

At normal environmental temperatures, the combination of high humidity, high average temperature, and the presence of sea salt or an industrial pollutant such as SO_2 increases corrosion rates. Steel forms a layer of Fe_3O_4 covered by an outer layer of FeOOH (rust), which develops fissures that allow O_2 and water vapour to penetrate and to oxidize the Fe_3O_4 to Fe_2O_3 or FeOOH. Traces of SO_2 create acidic conditions in the oxide, which opens pores in the Fe_3O_4 layer.

Atmospheric corrosion involving gaseous oxygen is invariably promoted by high humidity or splashing with water, and so is an electrochemically assisted process. It is often accelerated by pollutants such as SO_2 or NO_2. There is usually a critical humidity level below which atmospheric pollutants cause little corrosion, so that rates of corrosion vary widely with differing ventilation, even in similar climatic conditions. High temperatures may either accelerate the corrosion reaction, or they may reduce it if they are high enough to dry up rapidly all local sources of moisture.

Protection results from the presence of a surface coating, which can either be a natural oxide, or may be applied to the surface in the form of a polymer (e.g. paint), or of a ceramic enamel (a glassy oxide) 'fired' onto the surface by melting the glass in an oven. Natural surface oxides occur on high-strength low-alloy steels containing a few tenths % of Cu, Cr, Si, Ni, P, and these show improved resistance to atmospheric corrosion except in the wettest situations. The oxidation rate of an alloy is reduced if the natural oxide coating possesses a combination of several of the following factors:

- good adhesion
- high melting point, low vapour pressure
- similar thermal expansion to the metal
- low electrical conductivity
- low diffusion coefficients for oxygen and metal ions
- ability to accommodate stresses

Because oxides are stronger in compression than in tension, a small compressive stress might be expected to assist in maintaining a protective oxide coating. A compressive stress would arise during growth of the oxide, if the volume of oxide that is created is greater than the volume of metal consumed. Thus in 1923 Pilling and Bedworth proposed that the ratio of these volumes would be indicative of protective behaviour. This *Pilling–Bedworth ratio* can be expressed as

$$\text{Pilling–Bedworth ratio} = \frac{Wd}{nDw}$$

where W and D are the molecular weight and density of the oxide respectively, w is the atomic weight of the metal, d its density, and n is the number of metal atoms per oxide molecule. Values of this ratio just above unity should be ideal, but high values might be expected to lead to excessive stress, causing the coating to buckle. Some values are quoted in Table 10.5, together with the protective, or otherwise, nature of the oxide. You can see that, by itself, the ratio is not predictive, so that the other factors listed above must be as important.

Table 10.5

Pilling–Bedworth ratios for some metals and their oxides

Metal	Oxide	P–B ratio	Protective?
Cr	Cr_2O_3	2.02	Yes
Fe	FeO	1.78	Yes
Ti	TiO_2	1.76	No
Zn	ZnO	1.58	No
Zr	ZrO_2	1.57	Yes
Cd	CdO	1.42	No
Pb	PbO	1.28	No
Al	Al_2O_3	1.28	Yes
Mg	MgO	0.81	Yes
Ca	CaO	0.64	No

Oxides that form a stable, uniform layer are protective if the rate at which they grow in thickness decreases rapidly with time. Some examples are discussed in Section 10.10. Growth is a result of diffusion of one of the reactants, metal or oxygen, through the coating, driven by the concentration gradient across the layer. As the gradient is inversely related to thickness, the rate of growth decreases with time as required for protection. Only an oxide that is porous is not protective, for then oxygen can continue to penetrate the pores.

Metals that corrode rapidly in air can be protected temporarily by a *vapour inhibitor*. Some organic liquids, which have a vapour pressure lying between about 1 Pa and 10^{-5} Pa, will condense onto all the internal surfaces inside a closed volume, and so protect a metal. The liquid can be impregnated into a wrapping (paper, cardboard or plastic) that is then used to preserve raw materials or small machine parts while they are in storage. The protection lasts for periods from a few months, up to several years if an impervious outer layer is added to the wrapping.

10.10 Passivation in stainless steels

The surface of many metals (Fe, Ti, Ni, Cr, Co) is said to be *passivated* when the corrosion rate falls dramatically – by a factor between about 10^3 and 10^6 – despite a high driving force for corrosion. Passivity results from the creation by corrosion of a protective oxide coating, typically less than 10 nm thick, that acts as a barrier to the anodic reaction. Passivation depends on the oxidizing power of the corrosive solution that is in contact with the metal. Even pure iron is passivated in concentrated nitric acid, HNO_3, as was noted far back in the 1840s, by Michael Faraday.

Chromium, Cr, readily forms an excellent passive layer but its brittleness prevents its use for engineering components, except as an electroplated coating. However, Cr is a key element in achieving passivation in many alloys, helping to form a resistant, passive oxide surface film. Chromium is particularly useful in conjunction with nickel and iron to form the many stainless steels. The ranges of compositions used in three main types of stainless steel are listed in Table 10.6.

The ultra-thin surface oxide on stainless steels can, however, be punctured by microscopic pockets of impurity. Once the impurity has corroded, *pitting corrosion* can occur in the surrounding stainless steel, driven, like crevice corrosion, by a concentration cell maintained by Cl^- ions harboured within the pit. Prevention of pitting corrosion in stainless

Table 10.6

Composition ranges typical of three types of stainless steel

Elements alloyed with Fe		C (wt%)	Mn (wt%)	P (wt%)	S (wt%)	Si (wt%)	Cr (wt%)	Ni (wt%)	Mo (wt%)	Cu (wt%)
Austenitic	min	0.03	2.0	0.045	0.03	1.0	17	6	–	0.0
types	max	0.08	2.0	0.2	0.03	3.0	30	12		4.0
Ferritic	min	0.08	1.0	0.04	0.03	1.0	10	–	0.0	–
types	max	0.2	1.5	0.06	0.15	1.0	23		1.25	
Martensitic	min	0.15	1.0	0.025	0.025	0.5	11.5	0.0	0.0	–
types	max	1.2	1.25	0.06	0.15	2.0	18	2.5	0.6	

steels depends upon maintaining very low levels of impurities during the manufacturing process.

A related problem in stainless steels is *intergranular corrosion*, which occurs when a grain boundary is corroded as a result of precipitation within the grain boundary of either non-metallic or intermetallic compounds. This causes a composition change in or adjacent to a grain boundary, making it either anodic or cathodic relative to the surrounding metal. For example, *weld decay* in austenitic stainless steels can result from the precipitation of a mixed carbide, probably $(FeCr)_{23}C_6$, within grain boundaries during the welding together of two stainless parts in a hydrocarbon flame. The consequent loss of Cr in a thin layer adjacent to each grain boundary in the welded region renders the layer anodic relative to the grain, so destroying its stainless quality. Severe attack results when the weld is subsequently exposed to a humid atmosphere. The steel can nevertheless be stabilized against weld decay by the addition of titanium, Ti, or niobium, Nb, either of which metal forms carbides more readily than Cr and prevents the occurrence of Cr depletion.

Problems

10.1 Using the Sn–Pb equilibrium phase diagram of Fig. 10.5(b) and applying the lever rule, answer the following questions:

(a) What is the eutectic composition?

(b) What are the compositions of the solid phases in equilibrium at the eutectic temperature?

(c) For an alloy containing 90 wt% Sn, what fraction exists as the α phase at 220°C?

(d) What fraction of this alloy is liquid just above the eutectic temperature?

(e) Determine the fractions of α and β phases just below the eutectic temperature in the 90 wt% Sn alloy and the eutectic alloy

(f) What fraction of the total weight of the 90 wt% Sn alloy will have the eutectic structure, just below the eutectic temperature?

10.2 Draw free energy curves versus composition for the solid and liquid phases in a peritectic system at:

(a) a temperature above the peritectic temperature T_p, where a solid phase is present

(b) T_p

(c) a temperature below T_p where a liquid phase is present

10.3 Using the iron–carbon diagram of Fig. 10.31 answer the following questions:

(a) A 0.4 wt% C steel is cooled from 1600°C to room temperature under equilibrium conditions. State the sequence of phases and the temperatures at which they occur.

(b) Determine the relative weights and compositions of the phases present at 800°C.

(c) What are the weights of ferrite and cementite present in pearlite?

(d) What is ledeburite?

10.4 Sketch and label TTT diagrams for a hypo-eutectoid plain carbon steel and a eutectoid plain carbon steel. By reference to the diagrams explain how you would produce samples with the following microstructures:

(a) martensite with some retained austenite,

(b) ferrite and martensite (a small amount of pearlite may be present),

(c) ferrite and pearlite, and

(d) pearlite and martensite.

Self-assessment questions

1 For complete solid miscibility the two components must have

(a) similar melting points

(b) the same crystal structure

(c) similar size atoms

(d) identical free energies at room temperature

(e) similar hardness

(f) the same valency

2 An intermediate or intermetallic compound, such as AB_2, always has the crystal structure of one of the components A or B

(a) true   (b) false

3 The proportions by weight of two phases in equilibrium can be determined from the phase diagram using a simple law of mixtures

(a) true   (b) false

4 An alloy of the eutectic composition does not solidify over a temperature range but completely solidifies at the eutectic temperature

(a) true   (b) false

5 The eutectic reaction is

(a) $S_1 \rightleftharpoons S_2 + S_3$

(b) $L \rightleftharpoons S_1 + S_2$

(c) $L_1 + S_1 \rightleftharpoons L_2 + S_2$

(d) $L_1 + S_1 \rightleftharpoons S_2 + S_3$

(e) a martensitic transformation

(f) found in systems that exhibit complete solid miscibility

(S = solid; L = liquid)

6 With reference to the lead–tin phase diagram below, which of the following statements apply to a Pb–80 wt% Sn alloy?

(a) the first solid to solidify is of composition α_e

(b) the first solid to solidify is of composition β_e

(c) the first solid to solidify is of composition A_4

(d) the first solid to solidify is of composition β_5

(e) when solidification is complete the microstructure of the alloy consists of an $(\alpha + \beta)$ eutectic mixture

(f) when solidification is complete the microstructure of the alloy consists of primary β crystals in an $(\alpha + \beta)$ eutectic mixture matrix

(g) when solidification is complete the microstructure of the alloy consists of primary α crystals in an $(\alpha + \beta)$ eutectic mixture matrix

(h) the cooling curve shows one discontinuity

(i) the cooling curve shows two discontinuities

(j) the cooling curve shows three discontinuities

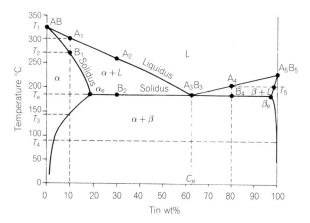

7 An alloy of the peritectic composition does not solidify over a temperature range but completely solidifies at the peritectic temperature

(a) true   (b) false

8 A solid is produced from two components A and B. The equilibrium condition of this solid is two-phase, $\alpha + \beta$, when

(a) the α phase and the β phase have the same free energy

(b) the α phase and the β phase have the same crystal structures

(c) the chemical potentials of A and B in the α phase are equal to the chemical potentials in the β phase

(d) A and B have similar melting points

(e) the two-phase $\alpha + \beta$ structure corresponds to the lowest free energy state

9 The free energy curves at temperature T_3 for the solid and liquid phases in a system that exhibits complete solid miscibility are given in the diagram. The diagram shows that

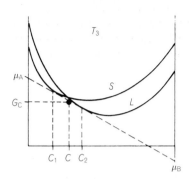

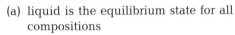

(a) liquid is the equilibrium state for all compositions

(b) solid is the equilibrium state for all compositions

(c) only alloys of composition between C_1 and C_2 are completely liquid

(d) only alloys of composition between C_1 and C_2 are completely solid

(e) alloys of composition between C_1 and C_2 are two-phase $(S + L)$

(f) μ_A is the chemical potential of component A in solid of any composition

(g) μ_A is the chemical potential of component A in solid of composition C_1

(h) μ_A is the chemical potential of component A in liquid of composition C_2

10 The overall transformation rate as a function of temperature for a nucleation and growth process is shown below. At high temperatures, such as temperature T_1,

(a) the nucleation rate is high

(b) the nucleation rate is low

(c) the growth rate is high

(d) the growth rate is low

(e) the resulting microstructure will be coarse

(f) the resulting microstructure will be fine

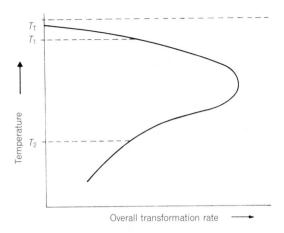

11 A martensitic transformation

(a) involves the precipitation of an equilibrium precipitate

(b) is extremely rapid

(c) can only lead to a change in the crystal structure of the phases

(d) can lead to a change in composition of the phases

(e) does not involve diffusion

12 The micrograph below

(a) is a light (optical) micrograph

(b) is an electron micrograph

(c) shows a two-phase structure

(d) shows a martensitic structure

(e) is of an annealed α-brass

2 μm

13 The eutectoid mixture in steel is

(a) a mixture of ferrite and cementite

(b) a mixture of ferrite and austenite

(c) a mixture of austenite and cementite

(d) called pearlite

(e) called ledeburite

14 A simple cast iron with no major alloying additions

(a) is remelted and cast pig iron

(b) is product of the blast furnace

(c) has a carbon content in the range 0.2 to 2.0 wt% C

(d) has a carbon content in the range 2.0 to 5.0 wt% C

(e) always contains pearlite

(f) always contains martensite

15 The micrograph below

(a) is a light (optical) micrograph

(b) is an electron micrograph

(c) is of a white cast iron

(d) is of a grey cast iron

(e) shows large black flakes of cementite

(f) shows large black flakes of graphite

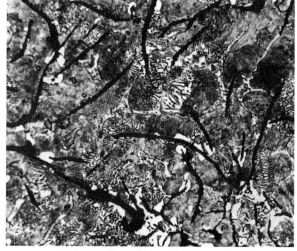

75 μm

16 An Al–4 wt% Cu alloy can be hardened by solution treatment and subsequent ageing

(a) true   (b) false

17 This is the aluminium-rich end of the Al–Cu phase diagram. Rapidly quenching an Al–4 wt% Cu alloy from temperature T_s to room temperature produces

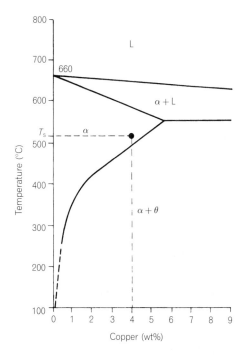

(a) a single-phase material

(b) a two-phase material

(c) a supersaturated solid solution

(d) a metastable material

(e) the room temperature equilibrium state of the alloy

(f) a martensitic structure

18 Martensite in steel is

(a) hard   (b) ductile   (c) b. c. c.   (d) b. c. t.

(e) an equilibrium phase.

19 Tempering of martensite in steels is accompanied by an increase in ductility and a decrease in hardness

(a) true   (b) false

20 Precipitation hardening in the Al–Cu system

(a) is due to the presence of metastable phases

(b) is due to the presence of the equilibrium phase, θ

(c) is accompanied by a marked loss in ductility

(d) involves a solution treatment and an ageing treatment

21 The ductility of a normalized plain carbon steel increases with carbon content

(a) true   (b) false

22 Bainite

(a) is commonly found in brasses

(b) consists of lath-shaped ferrite and cementite

(c) is a product of the decomposition of austenite

(d) is only stable at temperatures above 800°C

23 Brasses are copper–zinc alloys

(a) true   (b) false

24 The ability of a steel to form martensite on quenching is termed the hardenability

(a) true   (b) false

25 Alloying additions to steel

(a) generally decrease the pearlite and bainite reaction rates

(b) are important because of their effect on the peritectic reaction

(c) always decrease ductility and hardness

(d) may produce secondary hardening

(e) promote quench cracking

(f) may be α or γ stabilizers

Each of the sentences in Questions 26–35 consists of an assertion followed by a reason. Answer:

(a) If both assertion and reason are true statements and the reason is a correct explanation of the assertion.

(b) If both assertion and reason are true statements but the reason is *not* a correct explanation of the assertion.

(c) If the assertion is true but the reason contains a false statement.

(d) If the assertion is false but the reason contains a true statement.

(e) If both the assertion and reason are false statements.

26 Cored solid solutions are common in practice *because* the rate of cooling in chill casting and even with sand casting is generally too rapid for equilibrium to be maintained.

27 Only solid phases are involved in a eutectoid reaction because it occurs in the FeC system.

28 Homogeneous nucleation rarely occurs *because* it is energetically easier for nucleation to take place at structural imperfections.

29 The overall transformation rate in a nucleation and thermally activated growth process is temperature dependent *because* the nuclei of the new phase are just a few atoms in size.

30 The fracture surface of a grey cast iron is dark *because* failure takes place along the weak cementite plates.

31 A supersaturated Al–Cu solid solution does not always transform directly to the equilibrium two-phase condition of $\alpha + \theta$ *because* the transformation follows the path of maximum enthalpy.

32 Quenched plain carbon hyper-eutectoid steels have some retained austenite *because* their M_f temperatures are below room temperature.

33 A brass containing 40% zinc cannot be hot-worked *because* it has a two-phase $\alpha + \beta$ structure.

34 Secondary hardening is observed on tempering certain alloy steels *because* of the precipitation of alloy carbides which are more stable than iron carbides.

35 Martensite in steel is hard and brittle *because* of the large number of substitutional atoms present.

Answers

1	(b), (c), (f)	**2**	(b)	**3**	(b)	**4**	(a)
5	(b)	**6**	(d), (f), (i)	**7**	(b)	**8**	(c), (e)
9	(e), (g), (h)	**10**	(b), (c), (e)	**11**	(b), (c), (e)	**12**	(b), (d)
13	(a), (d)	**14**	(a), (d), (e)	**15**	(a), (d), (f)	**16**	(a)
17	(a), (c), (d)	**18**	(a), (d)	**19**	(a)	**20**	(a), (d)
21	(b)	**22**	(b), (c)	**23**	(a)	**24**	(a)
25	(a), (d), (f)	**26**	(a)	**27**	(b)	**28**	(a)
29	(b)	**30**	(c)	**31**	(c)	**32**	(a)
33	(d)	**34**	(a)	**35**	(c)		

11 | Ceramics

11.1 Introduction

The word 'ceramic' derives from the Greek 'keramos', which means 'burnt-stuff' or pottery. In fact ceramics date from before the ancient Greeks, making them the oldest man-made materials. House bricks, earthenware pots and porcelain cups are everyday examples of the use of these traditional ceramic materials. However, today the word ceramic is applied to a much wider range of materials than those used to make these common items. We will classify as ceramics all man-made non-metallic and inorganic solids. As well as the traditional clay ceramics, this wide classification embraces the new 'purer' ceramics, e.g. alumina and silicon nitride, glasses, glass-ceramics and cement.

These advanced ceramics also originate from raw materials mined or quarried from the earth (clays, sands, etc.) and processed at high temperatures, but they are being used at the leading edge of technology. New ceramics have dramatically extended the operating limits for use in applications such as heat-resistant bricks for furnaces, but they have also found new applications, e.g. tiles for the space shuttle, fibres and laser sources for optical communications and components for the high-temperature end of next-generation aircraft engines. In the rapidly expanding fields of medical and biomedical engineering, ceramics have found uses as strong, biocompatible, wear-resistant implants, and even as radiotherapy media – irradiated glass spheres, a few tens of microns in size, can be targeted precisely on the tumour under attack.

All of these applications exploit the incomparable physical and chemical stability that ceramics have as a result of their structure. We will begin our detailed investigation with the nature of this structure.

11.2 Structure of ceramics

Ceramics are usually of a hard, brittle nature and the bonding is ionic or covalent, or has mixed ionic–covalent characteristics (Table 11.1).

Most ceramic materials are crystalline, although their crystal structures are generally more complex than the simple metallic structures (see for example Figs 6.9 and 11.1). Familiar naturally occurring ceramic materials that are obviously crystalline to the naked eye are the large calcite ($CaCO_3$) crystals found in limestone rock, and the gemstone ruby, which is no more than Al_2O_3 with some chromium impurity. However, ceramics can also exist in an amorphous or glassy state and often the microstructure

Table 11.1

Bonding in ceramics (where a single bond type is given it comprises over 70% of the bonding)

Material	Bonding
Si	Covalent
SiC	Covalent
Si_3N_4	Covalent
NaCl	Ionic
MgO	Ionic
Mica	Ionic
Al_2O_3	Covalent–ionic
SiO_2 (quartz)	Covalent–ionic
Soda-lime glass	Covalent–ionic

of a ceramic is complex, with both crystalline and glassy phases present (Fig. 11.2). Silica (SiO_2) is an example of a ceramic that may be produced in either the crystalline or the glassy state, and the following discussion of this material will show the relationships between these states.

If silica is melted and cooled very slowly it will crystallize at a particular temperature T_m, called the freezing or melting point, in an identical manner to that of a metal. The specific volume as a function of temperature exhibits a discontinuity at the melting point, as shown by the full curve in Fig. 11.3(a). Silica can crystallize in a number of forms, all of which can be regarded as a network of silicon ions, following a cubic or hexagonal type of lattice, with oxygen ions in the tetrahedral spaces between them. This is shown schematically in two dimensions in Fig. 11.4(a). If the silica is cooled more rapidly from the molten state, it is unable to attain the long-range order of the crystalline state and the temperature dependence of the specific volume is given by the dashed curve of Fig. 11.3(a). The temperature T_g on this curve is called the *glass transition temperature*, which is not a well-defined temperature and depends on the cooling rate [Fig. 11.3(b)]. The slope of the curve between T_g and T_m is the same as that above T_m, indicating that there is no change in structure at T_m. This means that between T_g and T_m the material is a *supercooled liquid*. There is a change in slope at the glass transition temperature but no marked discontinuity as shown by the slowly cooled material at the melting point. Furthermore, again unlike the slowly cooled material at T_m, there is

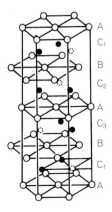

O oxygen
● aluminium
☉ vacant Al site

Fig. 11.1

Crystal structure of α-Al_2O_3 (hexagonal close packed structure). The aluminium sites (layers C_1, C_2 and C_3), which are only two-thirds full, are sited between the hexagonal layers (A, B) of oxygen atoms.

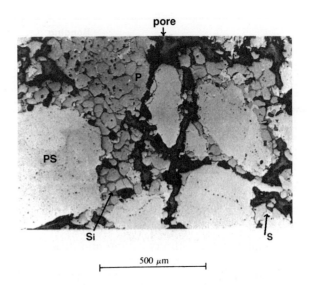

no evolution of latent heat at T_g. These observations suggest that the state
below T_g, termed the glassy state, is very similar to the liquid state. This
is indeed the case – the glassy state consists of a short-range ordered
network as shown in Fig. 11.4(b). This is a metastable structure and will
very slowly tend to change to the lower free energy crystalline form.
A glass is called a *vitreous* solid and if the material transforms to the
crystalline state it is said to have been *devitrified*. At room temperature
the rate at which devitrification to the crystalline state occurs is infinitely
slow. Consequently glass articles manufactured thousands of years ago are
not significantly, if at all, more crystalline than they were when they were
first produced. However, if molecular mobility is increased by raising the
temperature, devitrification can proceed at a faster rate. Devitrification is
also enhanced by the presence of foreign particles in the glass, which act
as nucleation sites for crystallization. Unplanned-for partial devitrification,
due to heterogeneous nucleation during glass production, is undesirable
as the glass may become opaque and lose strength.

One of the important features of the structure of glass is that it is a very
open network and can easily accommodate atoms of different species,

Fig. 11.3

(a) Specific volume versus
temperature curves showing the
relationship between the liquid,
crystalline and glassy states.
(b) Effect of cooling rate on T_g
(cooling rate 3 > 2 > 1).

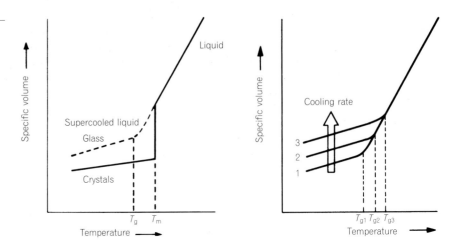

Fig. 11.4

(a) Crystalline structure of silica, (b) network structure of glassy silica, (c) soda-silica glass. Open circles indicate oxygen atoms, black dots silicon atoms, and large circles sodium atoms.

such as sodium, potassium and calcium. We will illustrate the effect of such atoms by concentrating on two specific examples, namely soda-silica glass and Pyrex glass.

First let us look at soda-silica glass, which contains monovalent sodium ions. The addition of sodium to silica decreases the silicon/oxygen ratio as, in order to maintain electrical neutrality, one Si^{4+} ion must be removed for the addition of every four Na^+ ions. It follows that, whereas in pure silica every oxygen atom is bonded to two silicon atoms [Fig. 11.4(b)], when sodium is present some of the oxygen atoms are only bonded to one silicon atom and are said to be *non-bridging*. Thus, the continuity of the network is disrupted, and as a consequence sodium is called a *network modifier*. The non-bridging oxygen atoms and the disrupted network are illustrated in Fig. 11.4(c). As sodium breaks up the network structure it produces significant changes in the properties of the glass. For example, at high temperatures the viscosity of a soda-silica glass is much less than that of pure silica and so it is easier to fabricate into shape. The change in viscosity is very marked – at 1400°C the viscosities of silica and silica with a 20% Na_2O addition are 10^{11} Ns m^{-2} and 10 Ns m^{-2} respectively. Ordinary window, plate and container glass all have about 15% Na_2O in them, as well as other additions. It is not normally possible to make a conventional soda-silica glass with more than 50% Na_2O because above this amount there are so many non-bridging oxygen ions that a network structure cannot form. The material crystallizes instead of forming a glass.

In contrast to soda-silica glass, Pyrex glass contains only small amounts of network modifiers. Instead it has about 14–15% of *glass formers*, mostly B_2O_3, in addition to silica. Glass formers, as their name suggests, contribute to the network formation. The characteristics of Pyrex glass, e.g. high viscosity, resistance to chemical attack and low coefficient of expansion, are due to the network being undisrupted.

We have seen that defective glass can result from unplanned devitrification but if crystallization is controlled, new types of material, called *glass-ceramics*, can be produced. Most 'oven-to-table' ware and ceramic hobs found in the modern kitchen are manufactured from a particular range of glass-ceramics that have low coefficients of thermal expansion. Glass-ceramics are polycrystalline materials formed by the controlled devitrification of glasses of carefully chosen compositions that provide a large number of nuclei for crystallization. After shaping the component while the material is still in the glassy state, a two-stage heat treatment is followed to transform the glass to a glass-ceramic. First the glass is held at

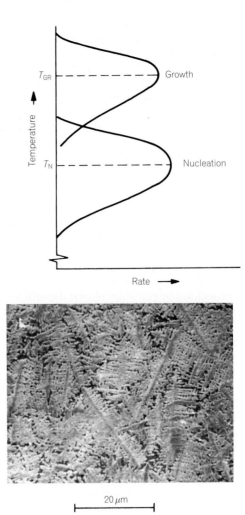

a low temperature, T_N, to produce a large number of well dispersed nuclei (Fig. 11.5). The glass is then heated to a higher temperature, T_{GR}, at which the crystal growth rate is a maximum. On holding at this temperature the crystalline phase, which may be of different structure and composition to the nuclei, grows upon the nuclei until crystallization is almost complete, with only a small amount of residual glass remaining (Fig. 11.6). A glass-ceramic, like a glass, has negligible porosity and its mechanical properties are intermediate between those of a glass and of the new 'purer' ceramics such as Al_2O_3 and Si_3N_4 (Table 11.2, p. 325).

11.3 Production of ceramics other than glass and cement

11.3.1 Raw materials

The most widely used raw materials are *clays*, and they are generally used without any major chemical purification. The term clay covers a huge range of natural substances differing greatly in appearance, texture, chemical composition and properties. Clays contain many phases, but the most important phases are the clay minerals such as kaolinite ($A_2O_3.2SiO_2.2H_2O$). The amount and type of clay minerals in a clay varies considerably – English china clay may contain up to 90% of kaolinite, whereas some brick clays have only 30% of this mineral with considerable quantities of other minerals such as montmorillonite and illite. The common characteristics of all clay minerals are a sheet-like structure and the ability to absorb water on the surface and between these sheets. The water molecules act in a similar way to a plasticizer in a polymer (see Chapter 12), and so wet clays are malleable and said to be *hydroplastic*.

Many of the newer ceramics are produced from raw materials that have to be processed beforehand. We will look in detail at the processing of *alumina* (Al_2O_3) as being typical of this type of material because of the large quantities produced throughout the world. Alumina is produced chiefly from bauxite, which is a mixture of oxides of aluminium, silicon and iron, by the Bayer method. First the ore is ground, then it is treated with hydroxide solution under pressure at about 165°C to give sodium aluminate:

$$Al_2O_3 + 2NaOH = 2NaAlO_2 + H_2O$$

The ferric oxide impurity is removed at this stage by filtering, but the silica forms sodium silicate and remains in solution with the sodium aluminate.

The sodium aluminate is unstable and precipitation of aluminium hydroxide takes place on passing carbon dioxide gas through the solution:

$$2NaAlO_2 + CO_2 + 3H_2O = Na_2CO_3 + 2Al(OH)_3$$

The $Al(OH)_3$ is separated from the sodium silicate, which remains in solution, by filtration. Finally the $Al(OH)_3$ is calcined at around 1100°C to form alumina:

$$2Al(OH)_3 = Al_2O_3 + 3H_2O$$

11.3.2 Forming processes

Once prepared, the raw material must be fabricated into the required shape. Several shaping techniques are available, and the one chosen depends largely on the characteristics of the raw material and on the size, shape and desired properties of the finished component. We will discuss the three most widely used processes – hydroplastic forming, slip casting and powder pressing.

We have seen how water makes clays malleable and it follows that they are most easy to form in this state. Forming of wet clays by applying an appropriate stress is called *hydroplastic forming*. The most common method of hydroplastic forming is to *extrude* the malleable clay through a die orifice of the required cross-sectional shape. This process is also used extensively for metals and for thermoplastics. For thermoplastics and clays the extrusion force is generated by an Archimedean screw (Fig. 11.7), which also provides a mixing and homogenizing action. After extrusion into shape, the clay has enough strength to maintain that shape and to be carefully handled. Pipes and bricks are produced by extrusion.

In *slip casting* the ceramic material must be available as a suspension, or *slip*, of particles in a liquid. Slip casting depends on the ability of a mould, made of porous material such as plaster of Paris, to absorb the

Fig. 11.7

Hydroplastic extrusion of clay. The
feed is clay that has been
shredded and the air removed.

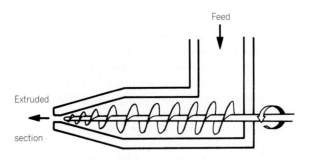

liquid (which is usually water) from the slip, leaving an even layer of
ceramic particles on the mould walls. When this layer has reached the
required thickness the remaining slip is poured off [Fig. 11.8(a)]. The
article is then allowed to dry until it is strong enough to be removed from
the mould. This may take only a few minutes for a thin-walled article but
can extend into several hours for components with thick sections.

Slip casting is particularly suitable for complex and irregular shapes,
e.g. household articles such as pots and jugs (the famous Delft pottery is
made this way). Variations of the basic slip casting process enable the
production of solid components, or of components with a high surface
quality on internal as well as external surfaces. Solid components are built
up in layers, as shown in Fig. 11.8(b). To achieve the smooth and well-
defined finish required on both internal and external surfaces of sanitary-
ware, a more complex mould is required, which forms all of the component
surfaces except that at a relatively small slip drain point [Fig. 11.8(c)].

Now let us turn our attention to the third of the forming techniques,
powder pressing. This process consists of compacting dry or slightly damp

Fig. 11.8

Slip casting: (a) the basic process,
(b) production of a solid article,
(c) double wall casting.

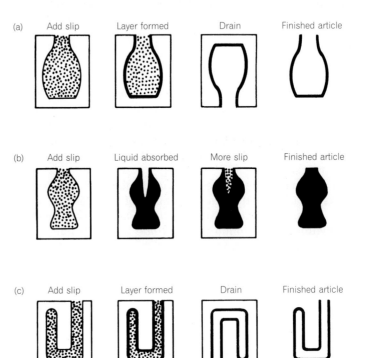

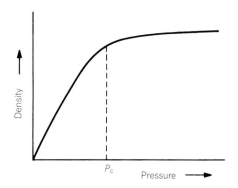

Fig. 11.9

Effect of pressure on the density of
a powder compact.

powder at a sufficiently high pressure so that a relatively dense and strong
article, which can be handled, is formed. The density of the pressed compact increases with applied pressure up to a certain pressure P_c, but raising
the pressure further has little effect (Fig. 11.9). In general the density, and
hence other properties, will vary throughout the article because of pressure differences arising from friction at the die walls. This is illustrated for
unidirectional pressing in Fig. 11.10(a). The density variation can be
reduced by pressing simultaneously from both ends and/or by reducing the
ratio of length, L, to diameter, D [Fig. 11.10(b), (c)]. However, the greatest
uniformity of density is achieved by the application of pressure from all
directions, which is known as *isostatic pressing*. The powder to be compacted is encased in a rubber mould and the pressure is applied via a
surrounding fluid [Fig. 11.10(d)]. One advantage of this technique for clay
is that it tends to align the plate-like particles to form ordered layers
parallel to the surface, and this has a beneficial strengthening effect.

There are limitations on the size and shape of articles that can be formed
by powder pressing. The process tends to be used mainly for the production of high-density components from the newer ceramics, e.g. alumina
spark plugs.

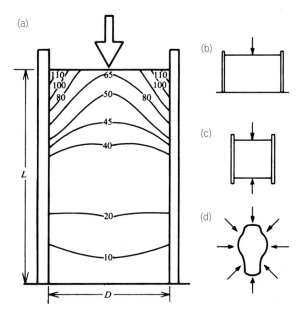

Fig. 11.10

Powder pressing. (a) Pressure
distribution, and hence density
variation, in a unidirectional
pressing [after P. Duwez and
L. Zwell, *Met. Trans.*, **185**,
137(1949)]. Methods of reducing
the density variation are (b) reduce
L/D ratio, (c) press from both ends
and (d) isostatic pressing.

Fig. 11.11

The effect of sintering: (a) powder with high-energy solid–gas interfaces; (b) solid with lower energy solid–solid interfaces. Note also the porosity.

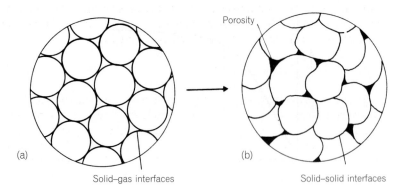

11.3.3 Post-forming processes

Ceramics formed by slip casting, hydroplastic forming, and in some cases powder pressing, have to be dried before final firing at an elevated temperature. After shaping, the particles of the ceramic are not in intimate contact, being separated by a thin continuous layer of water. During the *drying process* the water evaporates. Initially the rate of drying is rapid and a considerable amount of shrinkage takes place. This rapid drying stage has to be carefully controlled by using environments of constant humidity and temperature if warping and cracking of the formed component are to be avoided. Once enough water has evaporated so that the ceramic particles are in contact the movement of the residual water is restricted, and therefore the rate of drying decreases and further shrinkage is minimal.

We now come to *firing*, which is the final stage in the production of a ceramic. Firing is the densification of a dried component by heating to an elevated temperature. Densification during firing occurs by the processes of sintering and vitrification.

Sintering is the consolidation of a powder by means of prolonged use of elevated temperatures that are, however, below the melting point of any major phase of the ceramic. Figure 11.11 shows that sintering involves the replacement of high-energy solid–gas interfaces by lower energy solid–solid interfaces (grain boundaries). It is this reduction in total interface energy that is the driving force for the sintering process. Clearly sintering requires the movement of atoms or molecules through the component and it has been found that the mechanisms of mass transport, e.g. lattice diffusion, surface diffusion and evaporation–condensation (Fig. 11.12), vary from ceramic to ceramic.

Equations that tell us how the contact area between the particles changes as a function of important variables such as time t and temperature T of firing, have been derived for each of these transport mechanisms. One such equation, which is for the densification via lattice diffusion, is

Fig. 11.12

Sintering mechanisms: (a) transport by lattice diffusion and (b) transport by evaporation–condensation.

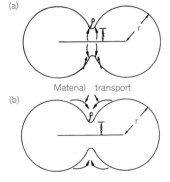

$$\frac{x}{r} = \left(\frac{40\gamma a^3 D}{kT} \right)^{1/5} r^{-3/5} t^{1/5} \tag{11.1}$$

where x is the interparticle neck radius and r the particle radius [see Fig. 11.12(a)], D is the self-diffusion coefficient, γ is the surface energy, a^3 is the atomic volume and k is Boltzmann's constant. This equation shows that the rate of sintering decreases with time but increases on raising the temperature through the exponential temperature dependence of D. Indeed, this is true no matter which mechanism is operative and Eq. (11.1) may be written in a general form applicable to all mechanisms, i.e.

$$x = Ct^m \qquad (11.2)$$

The coefficient of proportionality C depends on various factors such as diffusion coefficient, vapour pressure, etc., which are specific to a particular mechanism, but it always increases on raising the temperature. The exponent m is less than unity and its value is characteristic of the transport mechanism, e.g. 1/3 for evaporation–condensation, 1/5 for lattice diffusion and 1/7 for surface diffusion. It follows from Eq. (11.2) that in most practical situations the amount of sintering that has taken place, and hence the final microstructure (percent porosity, etc.), is more sensitive to the temperature than the time of sintering.

The sintering mechanisms discussed so far have not involved a liquid. There is an industrially important mechanism, known as *reactive liquid-phase sintering*, which occurs in the presence of a small amount of liquid. For this mechanism to be successful, it is essential for the liquid to wet the solid, i.e. low contact angle between the liquid and solid, and for the solid to be reasonably soluble in the liquid. Material transfer takes place by solution, transport through the liquid and precipitation. The transport through the liquid by diffusion is responsible for the temperature dependence of this sintering process and therefore, as the diffusion coefficient in a liquid usually changes less markedly with temperature than in a solid, reactive liquid-phase sintering is less sensitive to temperature variations than solid-state sintering. Another advantage of this mechanism is that it is generally more rapid than its solid-state counterparts.

As sintering proceeds the density of the ceramic increases, but the final product will always contain some porosity – although with a good quality modern ceramic such as alumina it may only be a few percent. Unfortunately, grain growth, which has a detrimental effect on the mechanical properties, may also occur during sintering. Therefore, a compromise has to be reached between maximizing the density and minimizing grain growth. A modified procedure sometimes employed is *hot pressing*, which is simultaneous forming and firing. This has the advantage over normal sintering in that higher densities and finer grain sizes may be achieved at lower temperatures.

Vitrification is densification in the presence of a viscous liquid and is the process that takes place in the majority of ceramics produced on a large scale, e.g. all clay-based ceramics. It differs from reactive phase sintering in many ways but particularly in the amount of liquid present during firing. More liquid is present in vitrification, although the proportion has to be kept to less than about 40% to prevent distortion (*slumping* and *warping*) of the component under the forces of gravity. Initially there is a rapid increase in density as the individual solid particles are drawn together by the liquid phase, but the rate of densification falls off once a solid skeleton has been established. Vitrification is increased by reducing the particle size, lowering the viscosity and increasing the surface tension. In most cases surface tension is not changed significantly by composition or temperature and therefore the important variables are the particle size and the viscosity of the liquid. Clearly we have some control over the particle size but the viscosity of the liquid phase is the variable that can be altered the most. Not

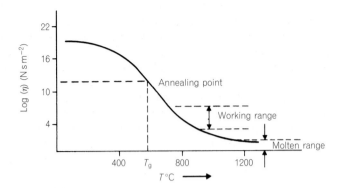

only is viscosity changed by composition (Section 11.2) but, as you will
recall from Chapter 9, it is very temperature dependent (Fig. 11.13). This
means that the firing temperature has to be carefully controlled. On cooling
from the firing temperature the liquid phase solidifies as a glass, thereby
further increasing the bonding between particles and so forming a solid
material, albeit with some porosity.

A solid compact of a few specialized ceramics is produced not by
sintering or vitrification but by a chemical reaction, the process being
termed *reaction bonding*. This technique is used particularly for silicon-
based ceramics such as silicon nitride and silicon carbide, and we will
briefly discuss the first of these. The raw material is silicon, which is mixed
with a polymer binder and formed by pressing or extrusion. After forming,
the binder is burnt out at a relatively low temperature to give porous
silicon. The component is then fired in a nitrogen atmosphere at just below
the melting point of silicon. As the silicon is porous the nitrogen penetrates
to the interior of the component and the following reaction occurs:

$$3\mathrm{Si} + 2\mathrm{N_2} = \mathrm{Si_3N_4}$$

Once a solid skeleton of silicon nitride has been formed the temperature
is raised to increase the reaction rate, but even so the process may take a
few days. The porosity content of the finished product is high (10–30%).

11.4 Production of glass

11.4.1 Melting of glass

To produce molten glass we heat the raw materials in either a number of
ceramic pots, each holding up to 750 kg, or in a large, refractory-lined tank
furnace with a capacity of up to 1000 tonnes. The major constituent of glass
is silica, which is obtained from sand. Other raw materials are added to
silica in appropriate proportions. For example, in the production of soda-
silica glass the additions are the carbonates of calcium and sodium; these
decompose on heating, evolving carbon dioxide and leaving the oxides.
It is also usual to add to the raw materials up to 30% of recycled scrap glass
(*cullet*) of the same composition as the glass being produced. The addition
of cullet not only gives an economic benefit but also improves the rate of
fusion as it conducts heat better, and has a lower melting temperature,
than the raw materials.

The melting and reaction of these constituents to produce a glass melt
involves a succession of many complicated reactions. We can view the

process as consisting of three distinct stages. The first involves the evaporation of water and the formation of a low-viscosity eutectic melt, consisting mainly of sodium and calcium carbonates, which wets and reacts with the silica grains. Various chemical reactions, such as the decomposition of carbonates, sulphates and nitrates, evolve gas and the end product of this first stage is a more viscous liquid containing many small bubbles. The second stage, known as *refining*, is concerned with the removal of these bubbles by raising the temperature and adding chemical refining agents, which cause small bubbles to grow. At this stage the melt is relatively fluid (with a viscosity of about $10\,\mathrm{N\,s\,m^{-2}}$) and the large bubbles can rise to the surface. Finally the temperature is reduced until the melt is sufficiently viscous to be formed (10^3 to $10^7\,\mathrm{N\,s\,m^{-2}}$, see Fig. 11.13). During this stage the refining agents may still be active in helping small bubbles to dissolve.

11.4.2 Glass forming and annealing

Everyone is familiar with the use of glass for bottles and windows and so we will take these as examples of forming processes.

Bottles are produced by *blow moulding*, a technique that is also used in a modified form for plastics (see Section 12.4.4). In its oldest and simplest form, blow moulding uses compressed air to inflate a blob of molten glass within a mould so that the outside of the glass 'bubble' takes on the internal shape of the mould. The problem with this one-stage process is that the walls of the article, in this case a bottle, vary in thickness [Fig. 11.14(a)]. Fortunately we can overcome this problem by using a two-stage operation. The first stage produces a partially blown bottle, known as a *parison*, which has a large but controlled variation in wall thickness. The parison is so designed that when blown in a second (finishing) mould a bottle is produced with a uniform wall thickness [Fig. 11.14(b)].

As in all glass-forming processes the viscosity of the melt is important at every stage of the operation. The initial melt temperature and the speed of the operation must be controlled so that the viscosity is low enough to

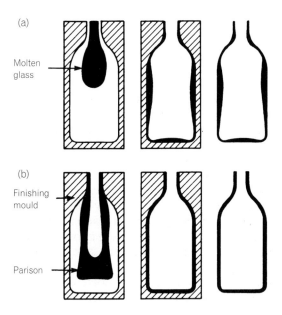

(a)

Molten glass

(b)

Finishing mould

Parison

Fig. 11.14

Blow moulding of a glass bottle. (a) Single-stage operation giving a variation in wall thickness. (b) Two-stage operation giving an even wall thickness.

permit easy blowing, i.e. the viscosity has to remain within the working range shown in Fig. 11.13. However, once blowing is completed and the bottle is ready to be removed from the finishing mould the viscosity must be high enough for the shape to be maintained – in fact the outer surface must be below T_g.

Glass for windows has to be of constant thickness and free from surface blemishes to avoid distorting the view. For a long time the only way to achieve these properties was by an expensive procedure of grinding and polishing of sheet glass. In 1959, high-quality flat glass was produced for the first time by the then revolutionary *float process*, which does not involve mechanical finishing. In the float process the molten glass leaves the furnace in a continuous ribbon up to 4 m wide and then floats on the surface of a bath of molten tin. The bath is surrounded by an atmosphere of nitrogen to prevent oxidation of the tin, because tin oxide imparts a 'bloom' (a slightly reflective surface discolouration) to glass. With good temperature control it is possible to produce glass whose surfaces are accurately parallel and which both have an excellent finish, as the top surface is protected by the nitrogen atmosphere and the bottom surface by the liquid tin.

Glass is a poor conductor of heat. As a result, it is inevitable that the cooling rate during forming will not be uniform throughout the component and that marked temperature gradients will appear. As a result of thermal expansion, these temperature gradients develop significant stresses, which are more easily understood if we imagine the cooling of a glass sphere. If the sphere is very slowly heated to a uniform temperature below T_g, it expands uniformly throughout and no stresses arise. If we now cool the surface relatively quickly, it will try to contract more than the interior which remains hotter. While there is still a temperature gradient the interior remains 'too big' for the outside surface, and it holds the surface in tension while itself being squeezed into compression. Too high a cooling rate may actually fracture the outside surface during cooling but, if it does not, then the stresses will disappear once the sphere again reaches thermal equilibrium at room temperature.

However, if we cool the sphere from *above* T_g, the situation is more complicated. As the outer layer of liquid cools through T_g and becomes a rigid glass, it forms a rigid shell around the core which is at a higher average temperature. As each successive layer of the core solidifies and contracts, it does so under increasing negative (i.e. tensile) pressure, and it pulls the outer layers which solidified earlier into increasing compression. This pattern of surface compression and internal tension remains as a *residual* thermal stress after thermal equilibrium is restored. In glasses, residual thermal stresses are exaggerated because, as seen in Fig. 11.3(b), slow cooling is associated with larger volume changes. Thus the slowly-cooling liquid core will actually 'want' to contract even more in total than the surface layers did.

Residual stresses may augment any stress applied during the final stages of production or even in service, and lead to premature failure. It is therefore often necessary to eliminate them by *annealing*: heating to a temperature at which there is sufficient molecular motion to allow stresses to relax – namely, near to T_g – and then cooling very slowly. In the manufacture both of bottles and of plate glass the annealing is carried out immediately after forming by passing the glass component through a long furnace called a *lehr*, which has an entry temperature near to T_g and an exit temperature near to room temperature.

11.5 Mechanical properties of ceramics

The reader will recall from the section on fracture in Chapter 9 that when a stress is applied to a material with a crack or flaw the stress is concentrated at the crack tip. The stress concentration may be relieved by flow (permanent deformation) of the material, by propagation of the crack or by a combination of these. If the material flows easily at the stress required, crack propagation requires more energy, and is therefore difficult, and the material behaves in a ductile manner. On the other hand, if flow is difficult crack propagation is favoured and the material is brittle. The most noticeable characteristic of ceramics is that they are hard and brittle at room temperature (see Table 11.2). Therefore the question that we must answer is 'why is flow in ceramics so restricted?'

Let us first consider crystalline ceramics, in which any flow would take place by dislocation motion. For a dislocation to move it must overcome the intrinsic lattice friction stress τ_{PN} (the Peierls–Nabarro stress), which is given by Eq. (8.10):

$$\tau_{PN} \sim G \exp(-2\pi w/b)$$

where G and b have their usual meaning and w is the width of a dislocation. As w appears in an exponential term it markedly affects the value of τ_{PN}; the greater w the lower is τ_{PN} and the easier it is for dislocation motion. The width of a dislocation is determined by the shape of the energy versus atomic displacement curve, which in turn depends on the type of bonding. Metals are bonded primarily by a free-electron gas and the energy of an atom is relatively insensitive to displacement. This leads to wide (5 to 10 atomic spacings) dislocations, a low Peierls–Nabarro stress and ductile materials. In contrast, covalent bonds are directional and the bond energy is sensitive to the angle between the atoms and therefore displacement. This gives narrow (1 to 2 atomic spacings) dislocations, a high Peierls–Nabarro stress, limited dislocation mobility and brittle behaviour. Some crystalline ceramics are ionically bonded, which is nondirectional, and results in relatively wide dislocations and a moderate Peierls–Nabarro stress. This is why ceramics that exhibit primarily ionic bonding, such as MgO, show some ductility in single crystal form.

Table 11.2

Room temperature mechanical properties of some ceramics

Material	Crystal	Slip system		No. of independent systems	Hardness[†] (GN m^{-2})	K_{1C}[†] (MN m$^{-3/2}$)
Glass-ceramic[‡]	–	–		–	7.0	2.0
Soda-lime glass	–	–		–	5.5	0.7
Al$_2$O$_3$	Hexagonal	{0001}	$\langle 11\bar{2}0 \rangle$	2	14.0	4.9
Si$_3$N$_4$	Hexagonal	{10$\bar{1}$0}	$\langle 0001 \rangle$	2	18.5	4.0
MgO	NaCl	{110}	$\langle 1\bar{1}0 \rangle$	2	9.3	1.2
SiC	Cubic (ZnS)	{111}	$\langle 1\bar{1}0 \rangle$	5	20.0	4.0
Porcelain	Multiphase	–		–	–	1.0

[†] The hardness and K_{1C} are structure-sensitive and will vary around the values quoted according to the composition, grain size and porosity content of the ceramic.
[‡] Silceram.

All ceramics are brittle at room temperature in the polycrystalline condition. This is a consequence of the magnitude of τ_{PN} and the lack of slip systems. For a polycrystalline material to plastically deform each grain must be able to accommodate the deformation of the neighbouring grains. This means that each grain must be capable of undergoing an arbitrary change in shape – otherwise voids and cracks will form at the grain boundaries leading to failure. *Von Mises* has shown that five independent slip systems are needed to satisfy this requirement and so permit a polycrystalline material to deform plastically. A slip system was defined in Section 8.3.1 and an independent slip system is one whose operation produces a change in shape that cannot be produced by a combination of slip on the other independent slip systems.

In general for metals, the slip direction is the closest packed direction and the slip plane the closest packed plane. The same is also applicable to ceramics, with the following provisos:

(a) slip must replace the ions on the appropriate sub-lattice
(b) the slip plane should be such that like ions are not brought into juxtaposition during gliding.

These provisos can reduce the number of slip systems in a ceramic. For example, for an f.c.c. metal there are twelve slip systems of the $\{111\} \langle110\rangle$ type, of which five are independent, giving ductile behaviour in the polycrystalline state. However the NaCl structure, which is taken up by NaCl and MgO, may be viewed as two interpenetrating f.c.c. structures (see Section 6.5), but the slip systems are not the $\{111\} \langle110\rangle$ but $\{110\} \langle110\rangle$, and only two of these are independent. It follows that MgO, although exhibiting some ductility when a single crystal, is brittle in the polycrystalline form due to insufficient independent slip systems (Table 11.2).

If we now turn our attention to the non-crystalline ceramics, i.e. glasses, we find that the possibility of flow by dislocation motion does not exist as there is no periodic lattice. Flow in glass has to take place by the thermally assisted motion of molecules past one another. However by definition a glass is below T_g at ambient temperature and hence thermally activated molecular motion cannot occur to a significant extent. Thus on the application of a stress molecules are unable to move any distance and consequently glass is brittle.

As the temperature is raised, thermal activation causes the molecular vibrations to become more and more violent. At temperatures above T_g the thermal energy plus the applied stress are sufficient to move molecules past one another and dramatic changes occur in the mechanical properties of a glass, as illustrated by the viscosity versus temperature curve of Fig. 11.13. Such large changes in mechanical behaviour with increasing temperature do not occur for most crystalline ceramics. For some crystalline ceramics additional slip systems operate at elevated temperatures and these, combined with the primary slip systems, can give five independent systems. For example in MgO the $\{001\} \langle1\bar{1}0\rangle$ slip systems become active, which gives an additional three independent slip systems, making five independent systems in all. Thus polycrystalline MgO, which is ionically bonded, is reasonably ductile at temperatures in excess of 1700°C; however, this is unusual and most polycrystalline ceramics exhibit very little ductility even at elevated temperatures.

Let us return to the brittle behaviour of ceramics. The literature provides ample evidence that if identical mechanical tests are carried out on a number of specimens of a given ceramic a large variation will be obtained in the measured fracture stresses. In contrast similar experiments performed on a ductile metal would give only a small scatter in the tensile strength. This difference is due to fracture in the ceramic occurring from flaws, whereas the ductile failure of a metal is a more homogeneous process involving dislocation motion (see Section 9.6.5 on ductile failure). In most cases the flaws that lead to the catastrophic failure of ceramics are surface defects that propagate under the action of a tensile stress. In a ceramic specimen or component there will be a distribution of flaw sizes and flaw orientation and location with respect to the applied stress. The variation in strength is a direct consequence of the flaw distribution. In order to characterize fully the fracture strength of a ceramic it is clearly necessary to define in some way the variation in strength. This is done by describing the strength variation by an appropriate statistical distribution function. The most widely applicable, and hence the most commonly used, distribution function is the empirical *cumulative distribution function* due to Weibull. According to this function the survival probability, *P*, of a specimen of volume *V* is given by

$$P = \exp\left[\left(\frac{-V}{V_0}\right)\left(\frac{\sigma}{\sigma_0}\right)^m\right]$$

where σ is the fracture stress, V_0 is unit volume, and σ_0 and m are constants. The *Weibull modulus m* is the important constant as it characterizes the width of the fracture stress distribution; the higher m, the narrower the distribution, i.e. the smaller the scatter in the measured fracture stresses. For modern high-quality engineering ceramics m is usually in the range 5 to 10 and is never significantly more than 20. Metals on the other hand have much higher values, typically 50 to 100.

In principle the strength may be improved by increasing the fracture toughness K_{1c} (which is usually difficult), by reducing the number and size of the flaws, or by having compressive residual stress acting against the applied tensile stress. The strength of glass, for example, can be increased by improving the surface finish or by putting the surface into compression. The latter is the basis for *toughened glass*: any applied tensile stress has first to overcome the residual compressive stress before surface flaws are put into tension. We discussed earlier how thermal residual stresses set up during the cooling of glass have a detrimental effect on strength and that glass components are generally annealed to relieve these stresses. Toughened glass is produced by cooling the surface at a high but carefully controlled rate. Too high a rate will cause fracture under the temporary stresses set up during cooling, while too low a rate will eliminate the differences in cooling rate that make interior material want to contract more during the glass transition. The correct rate allows the interior to draw the surface into compression while itself remaining under a tensile stress that, being protected by the surface layer from scratching, it can survive indefinitely. Not only can toughened glass withstand high stresses, but also when it eventually does fail it disintegrates into many small cubes as the uniform internal tensile stresses are relieved, and these cubes are much less harmful than the large jagged fragments produced by normal glass. This desirable failure behaviour will be familiar to readers who have experienced breakage of a toughened glass car windscreen.

Another novel way of improving the surface properties of glass is by altering the composition of the surface layer. There are now many elaborate methods for doing this, but the most widely used are those applied to multiple-use, recyclable drinks bottles, which often suffer their roughest treatment within the bottling plant itself. Tin oxide, SnO_2, deposited as a surface coating 5–10 nm thick, is sufficient to increase the abrasion resistance and the toughness of the glass itself. This is then protected by an even thinner polymer layer that reduces the coefficient of friction, and thereby reduces the abrasive stresses that generate surface cracks.

11.6 Wear and erosion resistance

Wear and erosion are commonly used terms and everyone has encountered a solid component that shows obvious signs of wear or erosion. It is somewhat surprising therefore to find that the mechanisms of wear and erosion are not well understood and that there is not even any consensus as far as terminology is concerned. Wear is sometimes categorized into four main mechanisms, namely abrasive, adhesive, fatigue and corrosive. The first of these, *abrasive wear*, which is the loss of material from the surface due to rubbing against the asperities on the surface of another solid or due to the grinding action of small particles, is the most relevant to ceramics. Particles are also responsible for *erosion*, which is defined as the loss of material due to the impingement of particles onto a surface.

In general ceramics have good wear and erosion resistance and consequently are used in applications such as valve faces and for lining chutes and pipework. The good resistance of ceramics is often attributed to their high hardness, and certainly there is a general correlation between wear resistance and hardness when materials exhibiting a wide range of properties are compared. However the wear and erosion resistance also depends on other material properties, as well as a number of external parameters such as surface roughness, presence of a liquid, size of abrading particles, angle of impact of impinging particles, etc. In order to illustrate some of the factors involved let us examine erosion resistance in more detail.

Erosion, which is usually quantified by the mass or volume removed from a surface by unit mass of impinging particles, varies markedly with

Fig. 11.15

Comparison of the effect of impact angle on the erosion of a ductile metal (aluminium) and a brittle ceramic (alumina).

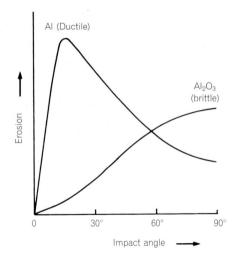

the angle of impingement or impact. This is shown for a ductile metal and a ceramic in Fig. 11.15. We can see from this figure that the erosion of the two classes of material have very different impact angle dependencies. Brittle materials, such as ceramics, characteristically exhibit maximum erosion under conditions of normal incidence, whereas for a metal the erosion reaches a maximum at an oblique angle of around 20° to 40°. A simple explanation is that the maximum erosion occurs at oblique angles for metals because material is removed by the gouging action of the particles. In contrast, material is removed by microfracture in the case of ceramics; the extent of the microcracking associated with a single impact increases with the depth of penetration of the erodant particle, which in turn is a maximum at normal incidence. The depth of penetration, and hence the erosion, also increases as the velocity of the impacting particles is raised. A number of equations relating the erosion, V_E, at a given impact angle to the characteristics of the impinging particles (e.g. velocity ν, density ρ and radius R) and the mechanical properties of the ceramic have been proposed. Although these equations differ in detail they are essentially of the following form:

$$V_E \propto \nu^a R^b \rho^c K_{1C}^d H^e$$

where the exponents a, b, c, d and e are constants. The exponent d is negative, hence the higher the fracture toughness K_{1C} the better the erosion resistance. There is some uncertainty about the effect of hardness on the erosion of ceramics and both negative and positive values of e may be found in the literature. Finally, turning our attention to the eroding particles, as the exponents a, b and c are all positive, the greater the velocity and density, and the larger the size of the particles, the more significant is the erosion.

11.7 Thermal shock

Earlier in the chapter we discussed how changes in temperature result in thermal stresses in glasses as a consequence of the glass transition. Rapid fluctuations in temperature can induce stresses not only in glasses but in all ceramics. In these circumstances ceramics, with a few notable exceptions such as silicon nitride and certain glass-ceramics, are prone to cracking and a loss in strength, that is, ceramics have poor *thermal shock resistance*. When we bear in mind that most ceramic production involves a high-temperature stage and that ceramics are often used in high-temperature applications, it is clear that during production and service there is the possibility of a component being subjected to thermal shock.

Thermal shock resistance is usually assessed by measuring the retained strength after quenching into water from elevated temperatures. A typical set of results, in this case for two aluminas, is shown in Fig. 11.16. It can be seen that there is no degradation in strength until the temperature difference between the elevated and quench-medium temperatures exceeds a critical value ΔT_c. Above ΔT_c the thermal stresses are sufficient for the formation of microcracks that reduce the strength. The reduction in strength can be drastic and the reader will appreciate that it is therefore important to have an understanding of the factors controlling the thermal shock characteristics.

Fig. 11.16

Retained strength of two
aluminas (A 99.5% and D 87%
purity) as a function of ΔT
(courtesy I. Thompson).

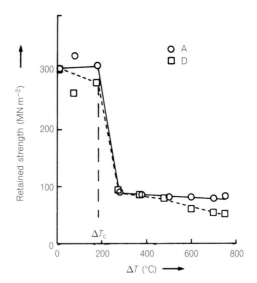

The maximum surface stress, σ_m, developed when a material experiences an infinitely fast quench over a temperature range ΔT is given by

$$\sigma_m = \frac{E\alpha\Delta T}{1 - \nu} \qquad (11.3)$$

where E is Young's modulus, α is the coefficient of thermal expansion and ν is Poisson's ratio. If σ_m is less than the fracture stress, σ_f, of the ceramic then no microcracking will occur; this is the situation for temperature changes of less than ΔT_c. Conversely, if $\sigma_m \geq \sigma_f$, which applies for temperature changes equal to or greater than ΔT_c, then microcracking will be initiated. Thus we can write

$$\sigma_f = \sigma_m = \frac{E\sigma\Delta T_c}{1 - \nu}$$

which on rearranging gives for the critical temperature change

$$\Delta T_c = \frac{\sigma_f(1 - \nu)}{E\alpha} \qquad (11.4)$$

From this equation we can deduce that the requirements for good thermal shock resistance, i.e. a high ΔT_c, are a high fracture stress and low Young's modulus and coefficient of thermal expansion. Practical experience has established that silicon nitrides are able to tolerate more drastic temperature changes without damage than aluminas and this is consistent with the values of ΔT_c calculated from Eq. (11.4) for these materials (Table 11.3).

The reader may have noticed that the calculated ΔT_c for alumina is less than that obtained from the experimental data presented in Fig. 11.16. This is because Eqs (11.3) and (11.4) are only applicable for infinitely fast quenches that are not achieved in practice. In order to take this into account two further parameters have to be considered, namely, the heat transfer coefficient, h, between the ceramic and the quench medium and the thermal conductivity, K, of the ceramic. These are introduced into our calculation for thermal stress and ΔT_c via a dimensionless parameter, β, known as *Biot's modulus*

$$\beta = ah/K$$

Table 11.3

Comparison of the material properties relevant to shock resistance for a silicon nitride and an alumina

	Reaction-bonded silicon nitride	99.5% purity alumina
Young's modulus E (GN m^{-2})	166	382
Coefficient of thermal expansion α (K^{-1})	2×10^{-6}	8.5×10^{-6}
Fracture stress σ_f (MN m^{-2})	210	332
Poisson's ratio ν	0.27	0.27
Thermal conductivity at 50°C K (Wm^{-1} K^{-1})	15	9
Critical temperature change[†] ΔT_c (K)	462	75

[†] Calculated from Eq. (11.4).

Table 11.3

Comparison of the material properties relevant to shock resistance for a silicon nitride and an alumina

where a is the heat transfer length and is usually taken as the ratio of specimen volume to surface area. The 'infinitely fast quench' model is modified by the addition of some function of β into Eqs (11.3) and (11.4), for example,

$$\Delta T_c = \sigma_f \frac{(1 - \nu)}{E\alpha} \left(A + \frac{B}{\beta} \right)$$

which is applicable at small values of β; A and B are constants. The function of β is such that a high thermal conductivity is beneficial as far as shock resistance is concerned. Reference to the values for K in Table 11.3 shows that this is another factor that contributes to the better performance of silicon nitride compared to alumina.

11.8 A commercial ceramic system: the silica–alumina system

Equilibrium phase diagrams are also useful for studying non-metallic materials and the diagram for the silica–alumina system is given in Fig. 11.17. Hydrated silica–alumina compounds occur widely in natural form in the Earth's crust, and the various phases have therefore been identified by the names given to the natural minerals.

Materials based on the SiO_2–Al_2O_3 system are refractory, i.e. they are capable of withstanding high temperatures, and are used in high-temperature environments such as kilns and furnaces. The performance of a refractory material under load at high temperatures depends mainly on the amount of liquid phase present at the working temperature, and on its viscosity. Bearing this in mind let us examine some materials in the silica–alumina system.

Silica refractories contain 1 wt% Al_2O_3 and can be used under load at temperatures up to 1650°C, decreasing with increasing Al_2O_3 content.

Fig. 11.17

Equilibrium phase diagram for the
SiO$_2$–Al$_2$O$_3$ system.

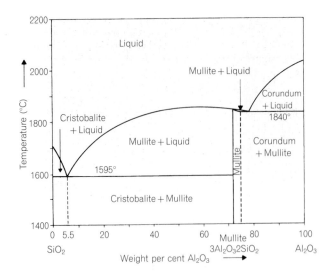

Fig. 11.17

Equilibrium phase diagram for the SiO$_2$–Al$_2$O$_3$ system.

Silica refractories are resistant to attack by iron oxide and acid slags and therefore have been of special importance to the steel industry.

Refractories containing up to about 50 wt% Al$_2$O$_3$ are produced from blends of naturally occurring clays called *fireclays*. From the phase diagram it can be seen that alumina contents near to the eutectic composition of 5.5 wt% Al$_2$O$_3$ should be avoided, as the material would be useless under load as soon as the eutectic temperature was reached. As the Al$_2$O$_3$ content is increased so the proportion of liquid decreases and the performance of the material in service improves. For example, a superduty fireclay brick contains typically 42 wt% Al$_2$O$_3$ and will deform much less: 5% in 10 minutes at 1600°C under a compressive stress of 0.2 MN m^{-2}.

To improve the refractory properties still further we have to increase the Al$_2$O$_3$ content. This is achieved by adding a non-clay ingredient, rich in alumina, to the fireclay. Such an ingredient is calcined bauxite and the final product is termed a *bauxite* or *high alumina refractory*. A bauxite refractory may contain from 50 to 80 wt% Al$_2$O$_3$, and the best quality bricks can operate at 1800°C, although if they are under load the service temperature may have to be reduced to 1650°C.

Refractories containing more than 80 wt% Al$_2$O$_3$ are made from fused pure alumina bonded by a little fireclay. They have better refractory properties than bauxite refractories and good resistance to both oxidation and reduction. However, their resistance to *spalling* (breaking of corners and flaking) is poor.

Finally we come to *alumina*, a term applied to materials containing more than about 90% Al$_2$O$_3$ – some of the best quality, high purity commercial aluminas consist of greater than 99.5% Al$_2$O$_3$. Alumina is produced in a fine polycrystalline form by sintering. The fine polycrystalline nature of alumina is clearly revealed on examining a fracture surface as the fracture is intercrystalline (Fig. 11.18). Alumina is very hard and exhibits good wear and erosion resistance. In addition it maintains its strength to high temperatures and can stand for short periods working temperatures in the region of 1900°C. Not only is it a useful material on account of its mechanical properties but also because it is a good electrical and thermal insulator. Al$_2$O$_3$ is not exceptional in having good insulating properties

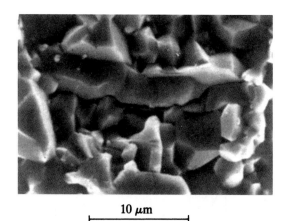

Fig. 11.18
Fracture surface of alumina showing the fine grain structure.

10 μm

and one of the chief uses of ceramics is as insulators. Their good insulating properties arise because all the valency electrons are occupied in the covalent–ionic bonds and none are left free to conduct electricity. Alumina is the white ceramic used in spark plugs for internal combustion engines.

11.9 Two technical ceramics – zirconias and Sialons

Although ceramics have been used by man for thousands of years there have been considerable advances made in ceramic technology recently. A new class of ceramic materials has been developed that has come to be known as *engineering, special* or *technical ceramics*. Zirconias and Sialons are examples of this new class of ceramics.

Zirconia can exist in three crystalline forms, cubic (c), tetragonal (t) and monoclinic (m). At normal cooling rates the high-temperature c-phase transforms at 2680°C by a nucleation and thermally activated growth process (see Section 10.5) to the t-phase. In contrast the lower temperature tetragonal-to-monoclinic transformation is martensitic and therefore has many features in common with the well-known martensitic transformation in metals, and particularly in steel – the reader is referred to Sections 10.6 and 10.7. The transformation to the m-phase is accompanied by a 3% increase in volume and in pure zirconia the transformation commences at about 1150°C, i.e. the M_s temperature is 1150°C. In pure zirconia the transformation proceeds unabated and the volume expansion causes severe cracking; consequently it is impossible to produce components from pure zirconia. In order to produce crack-free components from zirconia ceramics it is necessary to control the martensitic transformation by means of alloying. The alloying additions used are other oxides, such as magnesia (MgO), calcia (CaO) and yttria (Y_2O_3), which are termed *stabilizing oxides*. We now discuss the effect of one of these stabilizing oxides, Y_2O_3, in more detail.

The yttria-rich section of the ZrO_2–Y_2O_3 phase diagram of Fig. 11.19 shows that yttria additions in excess of about 9 mol% completely stabilize the c-phase to room temperature. A zirconia with the cubic structure stabilized by a large yttria addition is called a *fully stabilized zirconia*

Fig. 11.19

Zirconia-rich section of the
ZrO_2–Y_2O_3 phase diagram.

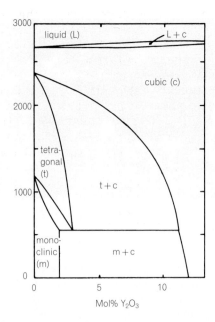

(FSZ). The main uses of FSZs are in electrical applications, such as oxygen sensors, fuel cells, etc.

An intermediate amount of yttria, e.g. 6 mol%, gives a *partially stabilized zirconia* (PSZ), which has a microstructure containing the three main crystal forms c, t and m. A PSZ is normally fired at an elevated temperature, which depends on the composition but is of the order of 1750°C and in the cubic phase field; it is then heat treated at a lower temperature (1300–1400°C) in the cubic plus tetragonal phase field to produce a fine dispersion of particles of the t-phase. On cooling to room temperature some, but not all, of the t-particles transform martensitically and become monoclinic. Those particles that remain in the tetragonal state do so because the elastic constraint of the cubic matrix opposes the volume change associated with the martensitic transformation. The t-particles are therefore in a metastable state. It has been found that the inhibition of the martensitic transformation to the monoclinic depends on the size of the particles; the smaller the particle the more likely it will remain in the metastable tetragonal state. PSZs have a low thermal conductivity, which is effectively invariant with temperature, and a high coefficient of thermal expansion. These characteristics make them suitable for thermal barrier coatings on metal substrates. The low conductivity results in a large temperature difference across the PSZ coating, thus protecting the metal substrate in high-temperature environments, and the high thermal expansion coefficient reduces the thermal expansion mismatch between the ceramic coating and the substrate.

Finally we come to low yttria additions, as typified by ZrO–3 mol% Y_2O_3. These zirconias have a fine-grained microstructure (less than 1 μm) and are fully tetragonal; the metastable tetragonal state being a consequence of the fine grain size. These zirconias are known as *tetragonal zirconia polycrystals*, TZPs.

Both TZPs and PSZs have good mechanical properties, especially the TZPs, which have K_{1C} values in the range 10–15 MN m$^{-3/2}$. The good toughness of these two types of zirconia is associated with the martensitic t → m transformation, which increases the toughness by two distinct mechanisms. Firstly, if a restricted number of particles undergo the transformation during cooling from the fabrication temperature, a fine distribution of microcracks is produced. The microcracks increase the toughness by interacting with a propagating crack, causing deflection and blunting of the crack [Fig. 11.20(a)]. However, this toughening mechanism, known as *microcrack toughening*, occurs at the expense of strength, which is reduced by the microcracks acting as flaws. The stress field at a crack tip can induce a metastable t-particle to transform to the monoclinic. This is the basis of the second toughening mechanism, *transformation toughening*, where the propagation of a crack is hindered by both the transforming particles at the crack tip and by the compressive back-stress due to the transformed particles in the crack wake [Fig. 11.20(b)]. Transformation toughening, unlike microcrack toughening, does not have a detrimental effect on strength.

The second example of a technical ceramic is the Sialons, which are closely related to silicon nitride, Si_3N_4. The reader will recall that Si_3N_4 produced by reaction bonding has a high porosity content, which is obviously undesirable. Denser material can be made by hot-pressing but the process is limited to simple shapes. Thus, although Si_3N_4 is a successful technical ceramic in its own right (its particularly good thermal shock resistance was discussed earlier in the chapter), there are considerable problems with production. The attraction of the Sialons is that they have similar properties to Si_3N_4 but are easier to manufacture.

The crystal structure of β-Si_3N_4 is built up of SiN_4 tetrahedra, as shown in Fig. 11.21. This structure is capable of accommodating the simultaneous substitution of aluminium atoms for silicon and oxygen for nitrogen to give

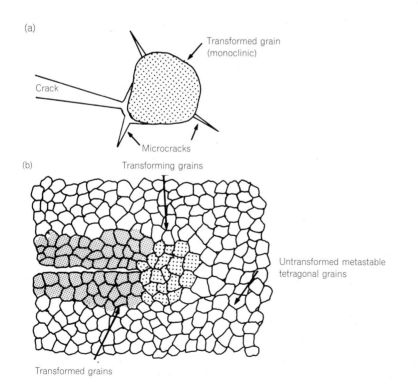

(a)

Transformed grain (monoclinic)

Crack

Microcracks

(b) Transforming grains

Untransformed metastable tetragonal grains

Transformed grains

Fig. 11.20

Toughening mechanisms associated with the tetragonal-to-monoclinic transformation in zirconias: (a) microcrack toughening; (b) transformation toughening.

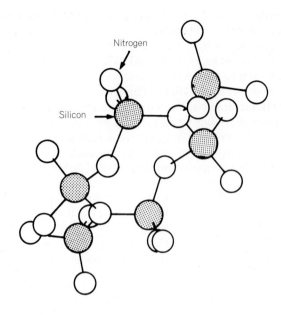

a range of solid solutions termed β-Sialons. The chemical formula of the β-Sialons is

$$Si_{6-z}Al_zO_zN_{8-z}$$

where z denotes the level of substitution of aluminium for silicon and oxygen for nitrogen and has a maximum value of 4.5. By appropriate control of composition, including the addition of oxides such as magnesia, a reasonable amount of liquid is formed at relatively low temperatures, which facilitates densification by vitrification so that hot-pressing is not necessary.

A Sialon produced in this manner would have a good combination of properties at room and intermediate temperatures (Table 11.4) but could not be used in applications where the operating temperature continuously exceeds about 1000°C. This operational limit is due to the presence of a continuous intergranular glassy phase that softens at high temperatures. We may obtain Sialons with improved high-temperature performance through the addition of yttria, Y$_2$O$_3$. As before, densification takes place via a liquid phase, but in this case there is Y$_2$O$_3$ in the liquid, and under normal cooling conditions an intergranular glassy phase is again formed. However by a post-firing heat treatment at about 1400°C, or by controlled cooling from the firing temperature, the glass reacts to give crystalline

Table 11.4

Properties of Sialons and reaction-bonded (RB) and hot-pressed (HP) silicon nitride

	RB Si$_3$N$_4$	HP Si$_3$N$_4$	Sialon + glass	Sialon + YAG
Strength, 20°C (MN m^{-2})	210	800	945	
Strength, 1200°C (MN m^{-2})	175	490		600
Weibull modulus, 20°C	15	20	11	
Toughness (MN m$^{-3/2}$)	4.0		7.7	
Thermal conductivity (W m^{-1}K^{-1})	10–15	15–20	22	
Thermal expansion coeff. (K^{-1})	2–3 × 10^{-6}	2–3 × 10^{-6}	3 × 10^{-6}	

yttrium–aluminium–garnet (YAG, $Y_3Al_5O_{12}$) at the grain boundaries and a slightly changed Sialon composition:

$$Si_5AlON_7 + Y{-}Si{-}Al{-}O{-}N = Si_{6-z}Al_zO_zN_{8-z} + Y_3Al_5O_{12}$$

Sialon with   glass   Sialon   YAG

$z = 1$

The YAG results in the Sialons retaining strength up to higher temperatures of the order of 1400°C.

11.10 Cement and concrete

In terms of quantity the most extensively used man-made structural material is concrete, which is produced by mixing mineral lumps (stones), called *aggregate*, with water and cement. The aggregate, which consists of coarse and fine lumps, may comprise about three-quarters of the volume of the concrete. The rest of the volume is occupied by the *cement paste*, which is formed by the reaction of water with the cement, and air voids. In many ways concrete may be viewed as a composite material with the cement paste as the matrix and the aggregate as the filler material. We will consider more conventional engineering composites in Chapter 13.

To manufacture cement, calcareous deposits such as limestone, and clays, are ground together and fed into a rotary kiln maintained at temperatures of 1400–1600°C. During the heating a broad sequence of reactions occurs between the raw materials, the most important being the formation of a liquid in the temperature range 900–1250°C and compound formation at temperatures above 1280°C. This heating process is known as *clinkering* and the product is mainly calcium silicates as clinker. Finally the clinker is ground with a few percent of gypsum to give the fine powder that is so familiar to us.

The composition of cement varies but the most commonly used formulation is that of *Portland cement*, the constitution of which is given, together with a shorthand notation for the constituents, in Table 11.5. In this

Constituent	Symbol	Weight %
Dicalcium silicate ($2CaO.SiO_2$)	C_2S	28
Tricalcium silicate ($3CaO.SiO_2$)	C_3S	46
Tricalcium aluminate ($3CaO.Al_2O_3$)	C_3A	11
Tetracalcium alumino ferrite ($4CaO.Al_2O_3.Fe_2O_3$)	C_4AF	8
Gypsum ($CaSO_4$)	–	3
Magnesia (MgO)	M	3
Calcium oxide (CaO)	C	0.5
Sodium oxide (Na_2O)	N	0.5
Potassium oxide (K_2O)	K	

Table 11.5

The constitution of Portland cement

notation, which is universally used by cement scientists, each oxide is designated by a single letter corresponding to the first letter of the chemical symbol for the cation, e.g. $CaO = C$, $SiO_2 = S$, $Al_2O_3 = A$, $H_2O = H$, etc.

Mixing cement with water produces a plastic workable paste. For some time the characteristics remain unchanged, and this period is known as the *dormant* or *induction* period. At a certain stage the paste begins to stiffen to such a degree that, though still soft, it becomes unworkable. This is known as the *initial* set. The *setting* period follows, in which the paste continues to stiffen until it can be regarded as a rigid solid, i.e. *final set*.

Setting and hardening are brought about by the hydration of the cement constituents. On adding water there is an initial fast reaction with the tricalcium aluminate

$$C_3A + 6H \rightarrow C_3AH_6$$

which evolves considerable heat of hydration. If allowed to proceed unhindered this reaction would cause rapid setting and a rise in temperature without any significant contribution to the strength of the concrete (Fig. 11.22). In Portland cement we control this reaction by the addition of gypsum. In the presence of gypsum C_3A hydrates to form a high-sulphate calcium sulphoaluminate known as *ettringite*. This coats the C_3A grains, retarding further hydration and so regulating the rapid set that would otherwise occur.

The principal subsequent hydration reactions involve the calcium silicates C_2S and C_3S, which make up about 75% of the cement. The hydration of these compounds contributes significantly to the strength of the concrete. The calcium silicates hydrolyse to form calcium hydroxide and less basic calcium silicate hydrate, which is assumed to have the composition $C_3S_2H_3$ on complete hydration. The reactions are approximately represented by

$$2C_3S + 6H \rightarrow C_3S_2H_3 + 3CH$$

and

$$2C_2S + 4H \rightarrow C_3S_2H_3 + CH$$

where CH is calcium hydroxide. The C_3S takes about 30 days to reach 70% of its ultimate strength, while the C_2S only reaches about two-thirds of its

Fig. 11.22

Compressive strength and rate of hydration of the pure constituents of cement.

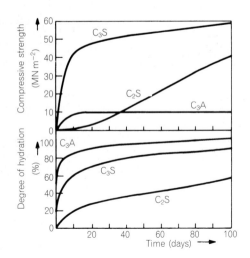

final strength in six months at normal temperatures (Fig. 11.22). Thus concrete goes on hardening for years.

The resulting hydrate is poorly crystallized and produces a porous solid defined as a rigid gel. This gel is sometimes called *tobermorite* gel, but its similarity to this naturally occurring structure is weak and so it is preferable to refer to the hydrates simply as calcium silicate hydrate (CSH).

On the microstructural level the hydration products consist of a heterogeneous mixture of hexagonal platelets of $Ca(OH)_2$, needles of ettringite and thin foils with colloidal dimensions of CSH [Fig. 11.23(a), (b)]. The volume of the hydration products is more than twice that of the anhydrous cement, and so as hydration proceeds the products gradually grow and fill up the spaces between the cement grains. The formation of points of contact causes stiffening of the paste and eventually their number so increases that the mobility of the cement grains is restricted and the paste becomes rigid, i.e. sets. With this setting mechanism in mind we can see that the choice of the term concrete, which is derived from the Latin word 'concretus', meaning to grow together, is particularly apt.

The composition of Portland cement can be varied to give the appropriate properties for a specific application. There are four main types of Portland cement commonly available: ordinary (the composition of which is given in Table 11.5), rapid hardening, low heat and sulphate resisting. The *rapid hardening cement* reaches about twice the strength of the ordinary cement over the first 24 hours of setting and this is useful when manufacturing pre-cast components or when working in a low-temperature environment. Rapid hardening is achieved by increasing the proportion of C_3S and grinding the cement more finely to increase the surface area and so enhance the hydration reactions. If, however, we are building a large structure from concrete, such as a dam, care must be taken to minimize the

(a)

(b)

Fig. 11.23

Microstructure of cement (a) Transmission electron micrograph of 12-h-old cement showing fine CSH gel surrounding dark grains of unhydrated cement; (b) scanning electron micrograph of the fracture surface of a two-day-old cement, showing CSH together with crystals of $Ca(OH)_2$ and needles of ettringite (examples labelled). (Courtesy K. Scrivener)

Approximate chemical composition	$3CaO \cdot SiO_2$ (C_3S)	$2CaO \cdot SiO_2$ (C_2S)	$3CaO \cdot Al_2O_3$ (C_3A)	$4CaO \cdot Al_2O_3 \cdot Fe_2O_3$ (C_4AF)
Rate of hydration	Rapid	Slow	Very rapid	Rapid
Rate of strength development	Rapid	Slow	Rapid	Medium
Ultimate strength	High	High	Low	Low
Heat of hydration	Medium	Low	Very high	Medium
General comments	Proportion increased in rapid hardening cement, reduced in low-heat cement	Responsible for the continued hardening of cement over long periods of time	Proportion reduced in low-heat and sulphate-resisting cements	Gives grey colour to cement. Proportion increased in sulphate-resisting cement

temperature rises associated with the evolution of the heat of hydration as these may set up thermal stresses and damage the structure. It follows that a rapid rate of hydration is undesirable and that hydration reactions that evolve little heat are preferable. A *low-heat cement* therefore has reduced amounts of C_3S and C_3A as these hydrate rapidly (see Fig. 11.22) and have high heats of hydration compared to C_2S. Finally, it has been found that concrete structures deteriorate when in contact with water or soil containing sulphates. The hydration products of C_3A are responsible for the deterioration and hence the problem is overcome in *sulphate-resisting cements* by reducing the C_3A content to below 5% and increasing the proportion of C_4AF. A summary of the main constituents of cement and their characteristics and roles in concrete technology is given in Table 11.6.

In order to obtain high strength it is important to use the correct ratio of water to cement. If there is too little water, incomplete hydration and entrapped air gives a porous, and therefore weak, structure. Full hydration requires a water/cement (w/c) ratio by weight of about 0.3, but the strength falls if too much water is added as evaporation of the free water leaves pores. The dependence of the compressive strength, σ_c, on w/c is illustrated in the

Fig. 11.24

Compressive strength of concrete at 28 days as a function of the water/cement ratio.

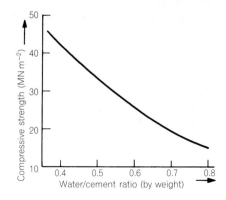

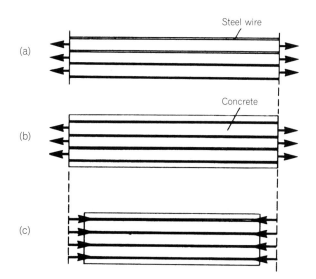

(a)

Steel wire

(b)

Concrete

(c)

Fig. 11.25

Production of prestressed
concrete: (a) tensile stress applied
to steel wires; (b) concrete added
and allowed to set whilst tensile
stress is maintained; (c) tensile
stress removed – the concrete is
now in compression.

graph of Fig. 11.24 and may be mathematically expressed by an equation
of the form

$$\sigma_c = A/B^{1.5(w/c)} \tag{11.5}$$

where A is an empirical constant usually taken to be about $100\,\mathrm{MN\,m^{-2}}$
and B is a constant that depends on the type of cement. If you have ever
made concrete you will be well aware that hand mixing and pouring is
easier at higher w/c ratios but unfortunately, as just described, this leads to
reduced strengths. Nowadays mechanical mixing and consolidation by
applying vibrations are common practice; this allows stiffer mixes to be
used and produces concrete with fewer voids and less entrapped air, which
results in about a 15% improvement in compressive strength over concrete
made in the traditional manner.

Concrete has many desirable features, e.g. castable into any shape, economical, fire-resistant, but has the disadvantage of a low tensile strength
and low ductility. Although concrete is poor in tension, its performance is
improved if it is allowed to set around steel wire in the form of mesh or rods
welded together. This is called *reinforced concrete* and the steel wire
hinders the propagation of cracks through the concrete. The reader may
recall the stress–strain curves given in Chapter 9, which showed that
concrete was about eight times as strong in compression as in tension
[Fig. 9.29(c)]. Thus, even greater strengthening may be obtained if the
concrete is put permanently into a state of compression. This is achieved
by elastically deforming the steel reinforcing wires in tension while
allowing the concrete to set around them. When the concrete has set the
tensile stress on the wires is removed and the wires contract elastically.
This elastic contraction produces compressive stresses in the concrete,
which is known as *prestressed concrete*. The production of prestressed
concrete is illustrated in Fig. 11.25. Prestressed concrete is a very attractive
structural material as concrete is inexpensive and, because it also protects
the steel from rust, maintenance is eliminated.

Problems

11.1 What basic mechanical properties distinguish ceramics from metals? Discuss the factors that determine these properties, from the atomic and microstructural viewpoint.

11.2 Use the SiO_2–Al_2O_3 phase diagram of Fig. 11.17 to answer the following:

(a) What fraction of the total ceramic is liquid for a 30 wt% Al_2O_3 composition at 1600°C?

(b) What are the compositions of the phases involved in the eutectic reaction at 1595°C?

(c) Describe the changes that take place as a 60 wt% Al_2O_3 composition cools from 2000°C to room temperature under equilibrium conditions.

11.3 What is the origin of the residual stresses in a glass component cooled from the melt, and how may these stresses be reduced? Describe the principal features of a process in which thermal stresses are used to advantage in glass technology.

11.4 Define the following terms: clinker, cement and concrete. Discuss the main reactions occurring during the hydration of cement, making reference to their heats of hydration and their contribution to strength.

11.5 During service a ceramic valve that is operating at 250°C is likely to be suddenly quenched to room temperature by a flowing aqueous solution. The valve could be manufactured from either of two ceramics, designated A and B, whose properties are listed below. Determine, giving detailed reasons, which ceramic should be selected for this application.

	A	B
Young's modulus, E (GN m^{-2})	120	350
Coefficient of thermal expansion, α (K^{-1})	2×10^{-6}	9×10^{-6}
Fracture stress, σ_f (MN m^{-2})	230	270
Poisson's ratio, ν	0.25	0.27
Thermal conductivity at 50°C K (W m^{-1} K^{-1})	14	8

Self-assessment questions

1 The glass transition temperature is

(a) a characteristic of all metals

(b) associated with the change from a supercooled liquid to the glassy state

(c) a constant temperature for a given material

(d) cooling rate dependent

(e) identical to the melting point

2 A glass-ceramic is predominantly crystalline with a small amount of residual glass

(a) true   (b) false

3 The addition of a network modifier to silica

(a) disrupts the network structure

(b) enhances the network structure

(c) produces vacancies

(d) produces non-bridging oxygen atoms

(e) increases the viscosity

4 Slip casting

(a) is the casting of a molten material

(b) is the casting of a suspension of particles

(c) requires a constant temperature for a given material

(d) involves slip by dislocation motion

(e) is commonly used for the production of glass articles

(f) involves the use of a porous mould

5 Unfortunately the density variation obtained by isostatic pressing is greater than that from unidirectional pressing

(a) true   (b) false

6 Vitrification is

(a) the same as devitrification

(b) the densification in the presence of a viscous liquid phase

(c) the transformation from a glassy to a crystalline state

(d) important during the firing of clays

(e) associated with the hydration to tobermorite gel

7 The viscosity of glass decreases linearly with increasing temperature

(a) true   (b) false

8 Sintering is the consolidation of a powder at an elevated temperature that is below the melting point of any major phase

(a) true   (b) false

9 SiC is

(a) covalently bonded

(b) ionically bonded

(c) produced by the controlled devitrification of glass

(d) produced by reaction bonding

(e) soft for a ceramic

10 The microstructure shown in this micrograph is that of a

(a) complex, clay-based, multiphase ceramic

(b) glass

(c) glass-ceramic

(d) alumina

(e) sintered material

(f) devitrified material

20 μm

11 Refractories in the SiO_2–Al_2O_3 system containing 5 to 50 wt% Al_2O_3 are

(a) produced from blends of naturally occurring clays called fireclays

(b) produced from fused alumina bonded by a little fireclay

(c) produced by melting SiO_2 and casting

(d) more refractory the greater the Al_2O_3 content

(e) only of commercial importance if they are of the eutectic composition

12 Concrete continues to harden for years after setting

(a) true   (b) false

13 The production of prestressed concrete is shown in the sequence of events (a), (b) and (c)

(a) the steel wires are deformed elastically in (a) and (b)

(b) the steel wires are deformed plastically in (a) and (b)

(c) in (c) the length of the steel wires is reduced by an externally applied compressive force

(d) in (c) the length of the steel wires is reduced due to elastic contraction as the applied tensile force is removed

(e) in (c) the concrete is put into tension

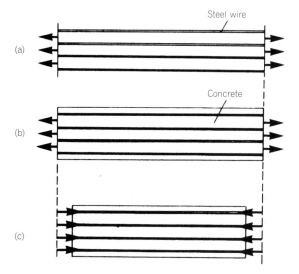

14 The strength of concrete in compression is about eight times that in tension

(a) true   (b) false

15 C$_3$A is a constituent of concrete and has the following characteristics:

(a) very rapid rate of hydration

(b) slow rate of strength development

(c) high ultimate strength

(d) high heat of hydration

(e) hydrates, in the presence of gypsum, to form ettringite

16 The proportion of C$_3$S is increased in rapid hardening cement

(a) true   (b) false

17 A high thermal coefficient of expansion is beneficial to thermal shock resistance

(a) true   (b) false

18 The tetragonal-to-monoclinic transformation in zirconia is:

(a) a process of nucleation and thermally activated growth

(b) a martensitic transformation

(c) accompanied by an increase in volume

(d) unaffected by the addition of oxides such as zirconia

(e) a high-temperature transformation, i.e. takes place at temperatures in excess of 2000°C

Each of the sentences in Questions 19–26 consists of an assertion followed by a reason. Answer as follows:

(a) If both assertion and reason are true statements and the reason is a correct explanation of the assertion.

(b) If both assertion and reason are true statements but the reason is *not* a correct explanation of the assertion.

(c) If the assertion is true but the reason contains a false statement.

(d) If the assertion is false but the reason contains a true statement.

(e) If both the assertion and the reason are false statements.

19 The dislocations in a covalently bonded ceramic are wide *because* the covalent bond is directional.

20 During sintering the total energy associated with interfaces is reduced *because* high-energy solid–gas interfaces are replaced by lower energy solid–solid interfaces.

21 When making concrete it is important to have the correct ratio of water to cement *because* too much water dissolves the aggregate.

22 Ceramics are generally good electrical insulators *because* all the valency electrons are occupied in the covalent-ionic bonds.

23 After manufacture most glass articles are annealed at a temperature near to T_g *because* at T_g there is a change of slope of the specific volume versus temperature curve.

24 During the production of cement care is taken to remove all traces of gypsum from the clinker *because* gypsum increases the rate of hydration of C$_3$A.

25 The float process produces plate glass with a good surface finish *because* tin oxide from the tin bath diffuses into the glass and hardens the surface.

26 The Weibull modulus of a ceramic is lower than that of a metal *because* a ceramic is less tough and more sensitive to flaws.

Answers

1 (b), (d)	2 (a)	3 (a), (d)	4 (b), (e), (f)
5 (b)	6 (b), (d)	7 (b)	8 (a)
9 (a), (d)	10 (c), (f)	11 (a), (d)	12 (a)
13 (a), (d), (f)	14 (a)	15 (a), (d), (e)	16 (a)
17 (b)	18 (b), (c)	19 (d)	20 (a)
21 (c)	22 (a)	23 (b)	24 (e)
25 (c)	26 (a)		

Polymers and plastics | 12

12.1 Introduction

The use of metals and ceramics is almost as old as civilization itself. Iron, bronze and pottery artefacts mark milestones in the evolution of early technology. The true ancestors of plastics, however, date back even further – they include horn, rubber, wood, cotton and hardened oils. These natural materials have always been valued for qualities that we now know to be conferred by their *macromolecular* structure – they consist of exceptionally large molecules whose internal bonding is strong but whose bonding to neighbours is generally weak. The first synthetic materials to share this structure, organic *polymers*, appeared only in 1910. In less than a century polymers have come to dominate so many applications, not only as plastics and rubbers but also as fibres, paints and adhesives, that their consumption is an important barometer for an industrial economy. In this chapter we concentrate mainly on the use of synthetic polymers, compounded with various additives (e.g. pigments and fillers), as solid *plastics* from which components and structures can be formed.

The earliest plastics were *thermosets* such as phenol formaldehyde ('Bakelite') and melamine formaldehyde. Thermosets cure permanently, during forming into their final shape, to become rigid, durable and heat-resistant. Rubbers or *elastomers* (elastic polymers) are also cured to prevent flow but these materials retain a spectacular ability to recover elastically from extremely large extensions. Most structural polymers used nowadays are *thermoplastics*, principally the 'commodity' thermoplastics polyvinyl chloride (PVC), polyethylene (PE), polypropylene (PP) and polystyrene (PS). These can be repeatedly remoulded under heat and pressure, making their conversion into finished products particularly rapid and economical. The price for this reformability is a relatively strong sensitivity to temperature, and in particular a tendency to creep under load at temperatures that hardly affect metals. Outside the electrical industry, in which their intrinsic high resistivity and dielectric properties were exploited relatively early on, thermoplastics originally gained popularity less for their excellence in performance than for their convenience as substitutes for other materials. They offered lower cost per unit volume, easier processability to their final shape and better decorative qualities or transparency.

The oil price crises of the 1970s, however, brought the use of plastics to maturity. As the commodity thermoplastics became more expensive, they were used more carefully. Meanwhile, greater understanding of the relationships between molecular structure and bulk properties guided the steady introduction of better *grades* (variants in formulation) of existing polymers. No new commodity polymers emerged, and it is unlikely that any more will, but structural modification using new synthesis methods has brought steady improvement in the properties even of such familiar plastics as polyethylene. The elite of *engineering polymers* has gained many new members, some of which offer mechanical performance or physical properties that would have seemed unattainable before 1970.

The properties of existing plastics can be enhanced, without changing the structure of the base polymer, by compounding with different additives or by blending with other polymers. Formulating a variant of an engineering material merely by mixing in additives is a possibility almost unique to plastics. Often the conferred property is needed only on the surface, for example the colour given by a *dye* or *pigment*. The fact that a component moulded from a plastic has uniform properties throughout and that its surface needs little or no finishing greatly simplifies its production but greatly complicates its recycling. There are tens of thousands of grades of plastics, and identifying and separating them is almost impossible.

Underlying this enormous diversity are common structural features, which distinguish polymers as a class of material very different from those we have considered so far. It is the relationships between these structural features and properties that this chapter will introduce. Because the ultimate properties of these materials are heavily influenced by the manner in which they are formed, we will also look at the techniques used to do so.

12.2 Molecular structure

The atoms and molecules in materials we have considered up to now are bonded to their neighbours in essentially three-dimensional structures. The basic structural units of polymers are *one-dimensional* – long, thread-like, covalently bonded chain molecules of repeated subunits. These chains are sometimes branched, and sometimes interconnected at points along their length (by long branch chains or by short crosslinks) to form a network.

12.2.1 Monomers and polymers

The repeat units that make up a synthetic polymer chain are (or are each assembled from) molecules of *monomer*. For polyethylene (Fig. 12.1), whose structure was introduced in Section 6.14, the monomer is ethylene gas, C_2H_4. The polyethylene repeat unit is an ethylene molecule whose C=C double bond has opened up, so that each end of the unit can bond to the next. The chain ends could consist most simply of methyl groups, $-CH_3$, the 'extra' carbon valency having been terminated with a hydrogen atom. In poly(ethylene terephthalate) – well known as PET, the material of many high-transparency plastic bottles – *two* monomer molecules react to form each identical repeat unit. In any case, the chemical reaction that assembles the chain from its monomers is known as *polymerization*. Because polymer chains are so long, the nature of the chain ends usually has little influence on properties, so long as they are stable enough to prevent the chain from depolymerizing spontaneously.

Polyethylene is a *homopolymer*, constructed from a single kind of repeat unit. As outlined in Chapter 6, linear polyethylene can be regarded as a high molecular weight member of the linear alkane series, $C_{2i}H_{4i+2}$ (Fig. 12.1). The alkanes show steadily changing physical properties as i

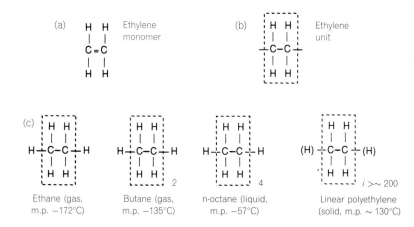

Fig. 12.1

Structure of polyethylene. Opening of the double bond in ethylene gas (a) forms an active ethylene unit (b) that can form chains. Chains with hydrogen end-caps are alkanes (c). Above a critical chain length neither the end-groups nor the chain length affect properties very much; these are polyethylenes.

increases and the chain becomes longer. In particular, the crystalline melting points and boiling points increase with the size of the molecule, and for $i > 50$ we meet substances that are semi-solid greases and waxes at room temperature. However, at a critical i (for this series, about 100) there is a qualitative transition and many physical properties, such as melting point and density, settle to values substantially independent of i. The nature of the chain ends now has little influence on properties – and this was certainly not the case for lower alkanes. The substance has become a true polymer, in this case high-density polyethylene (HDPE); i is the *degree of polymerization*. Members of the polymer series with $i < 100$ are termed *oligomers*. Most commercial polyethylenes, on the other hand, are based on much longer polymer chains because some mechanical properties do continue to improve with increasing molecular weight.

The essential nature of polymers arises from the combination of strong bonding along the chain *backbone* (which in polyethylene is simply a covalently bonded chain of carbon atoms), and weak bonding between *neighbouring* chains (often, as in this case, van der Waals bonding). Bonding ethylene gas molecules into a chain greatly restricts their mobility – but the mobility that remains, and the lack of constraint by its neighbours, gives each chain great flexibility. Because it has only weak, secondary inter-chain bonding, HDPE is also *thermoplastic* – heat or solvent action allow the chain molecules to slide past each other and the polymer to flow.

Table 12.1 shows the chemical structures of some other thermoplastics. Most are based on quite simple repeat units and are synthesized from commonly available monomers; those that are not are usually relatively expensive. This table also shows the abbreviated (non-systematic) chemical name by which each polymer is known, and the abbreviation allocated to it by the International Standards Organization, ISO (and which often appears in the familiar recycling symbol on moulded products). A few well-known trade names used by manufacturers are also shown.

12.2.2 Synthesis

We can see that in order to form a chain, each monomer molecule must have at least two points at which it can bond chemically to another. Essentially, there are two kinds of sequence by which this reaction event can result in the construction of a long chain polymer.

Table 12.1
Structure and properties of common thermoplastics

ISO abbreviation and name	Repeat unit structure	T_g(°C)	T_m(°C)	Notes and typical applications
PE Polyethylene	$+CH_2-CH_2+$	−90*	130	LDPE (low density, branched): packaging (film, soft containers, flexible container lids). HDPE (high density, unbranched): semi-rigid bottles and mouldings. Extruded pipe.
PP Polypropylene	$+CH_2-CH(CH_3)+$	−5	180	Usually copolymerized with PE. Low-cost, semi-rigid moulded products: housewares, car interior components. Bottle caps (with hinges, springs, pump action, etc.). Extruded pipe. Carpet fibre.
PS Polystyrene	$+CH_2-CH+$ (with phenyl group)	95	240 (isotactic) 270 (syndiotactic)	Low-cost transparent mouldings, e.g. CD cases, ballpoint pens. Rubber modified as ABS: moulded housings for TVs and computers. Rigid packaging foam (Styrofoam).
PVC Poly(vinyl chloride)	$+CH_2-CH+$ with Cl	80	200	Unplasticized: extruded pipe, guttering and window frames; drinking water bottles. Plasticized: cable covering, floor tiles.
PMMA Poly(methyl methacrylate) ('acrylic')	$+CH_2-CH+$ with CH_3 and $COOCH_3$	105	–	Exterior glazing (Perspex, Lucite). Lenses, auto rear-light covers. Baths and other rigid formed components.
PTFE Poly(tetrafluoroethylene)	$+CF_2-CF_2+$	20	340	Low-friction engineering mouldings (seals, bellows, gaskets), coatings and additives for other polymers (Teflon).
PA6 Polyamide ('nylon')	$+NH(CH_2)_5CO+$	50	215	Engineering components (gears, bearing cages). Fibres. Reinforced with glass fibre for engineering housings (e.g. power drills).
PA6,6 Nylon 6,6	$+NH(CH_2)_6NH-CO(CH_2)_4CO+$	65	255	
PA11 Nylon 11	As for PA6 with $(CH_2)_{10}$	57	195	
PA12 Nylon 12	As for PA6 with $(CH_2)_{11}$	40	180	
POM Poly(oxymethylene) ('polyacetal')	$+CH_2-O+$	−75	180	Very high crystallinity gives white colour and high modulus. Engineering mouldings.

		Structure			Applications
PET	Poly(ethylene terephthalate)	$+CH_2-CH_2-O-CO-$⬡$-CO-O+$	65	255	Soft drinks bottles. Fibres (Terylene).
PC	Polycarbonate	structure	150	(260)	Amorphous form: CDs, car headlamp mouldings, interior safety glazing. Crystalline form: electric kettle mouldings, crash helmets, auto bumper bars.
PPO	Polyphenylene oxide		205		High-temperature amorphous polymer. Computer components, high-temperature consumer electrical components, automotive parts.
PPS	Poly(phenylene sulphide)	structure	85	290	Similar to PPO but usable to higher temperatures.
PES	Poly(ether sulphone)	structure	220	–	Microwave cooker components, auto engine components, sterilizable medical components.
PUR	polyurethane				Foam.
PEEK	poly(ether ether ketone)	structure	143	334	High-temperature engineering components.

* Approximate value for amorphous phase: increases to −20°C in semicrystalline form.

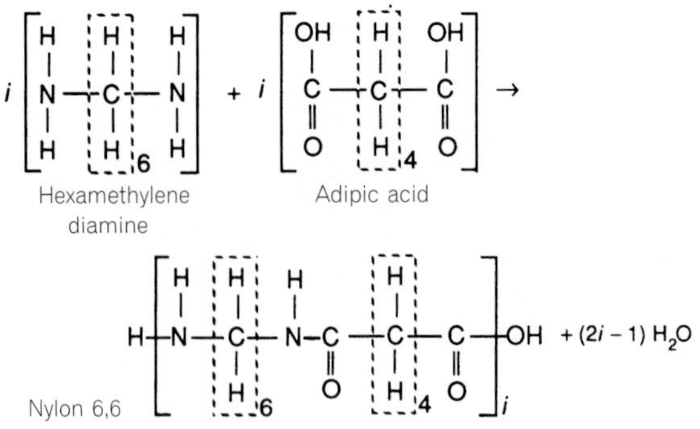

Fig. 12.2

Nylon 6,6, a polyamide (PA), is formed by a step-growth reaction in which a molecule of water is condensed out from between reacting monomers.

$$i \begin{bmatrix} H & H & H \\ | & | & | \\ N & C & N \\ | & | & | \\ H & H \end{bmatrix}_6 + i \begin{bmatrix} OH & H & OH \\ | & | & | \\ C & C & C \\ \| & | & \| \\ O & H & O \end{bmatrix}_4 \rightarrow$$

Hexamethylene diamine   Adipic acid

$$H \begin{bmatrix} H & H & H & H \\ | & | & | & | \\ N & C & N-C & C & C \\ | & | & \| & | & \| \\ H & & O & H & O \end{bmatrix}_i OH + (2i-1) H_2O$$

Nylon 6,6

The first is *addition polymerization*, by which polyethylene and many other similar polymers are synthesized. An initiator starts the reaction by attacking a few monomer units, opening the double bond to provide active seeds from which a reaction proceeds. As each monomer unit joins the growing chain, it becomes the new reaction site, while the point – the other end of the growing chain – moves away. Often the reaction takes place at a single site on a catalyst particle. This gives the catalyst extremely precise control over the way in which the new monomer unit is 'fitted' onto the chain, and we will see the importance of this in Section 12.2.4. The reaction may terminate in any of several ways: for example, exchange of a hydrogen atom from one active chain end to another will terminate one by a methyl group (CH_3) and the other by a double bond ($=CH$). The active lifetime of a growing chain, from initiation to termination, might be a few minutes.

The other major group of polymerization processes is more commonly used for higher-performance engineering polymers. In *step growth polymerization* every monomer unit is activated at the two or more sites where it will bond to its neighbours, and these units polymerize at a rate that falls as they are consumed. Step-growth polymers include the polyester PET and the polyamides (nylons), which polymerize by a reaction between acid and base end-groups on the monomers (Fig. 12.2). Because such reactions often liberate water or other gases, the polymers they produce are sometimes referred to as condensation polymers.

Industrial control of these complex, barely stable synthesis reactions, involving phase changes from liquid or gaseous monomers, is a major challenge. Polymers have low heat diffusivity and the reactions by which they are synthesized are exothermic. Hundreds of millions of tonnes of polymers are synthesized each year, and the motivation to perfect efficient processes is intense. The details of such processes are given in more specialized texts, but some are of direct interest to the polymer materials science, e.g. the way in which different processes produce polymer chains whose chemistry is similar but whose geometrical configurations differ. One feature of polymerization that is already easy to appreciate is that its rate is strongly controlled by inherently random events. This leads to an almost universal feature of polymers – molecular weight dispersion – which we will consider next.

12.2.3 Molecular weight

When we compared the properties of alkanes, ethylene oligomers and polyethylenes we considered *monodisperse* HDPE in which every molecule

Fig. 12.3
Distribution of molecular weight in a typical thermoplastic.

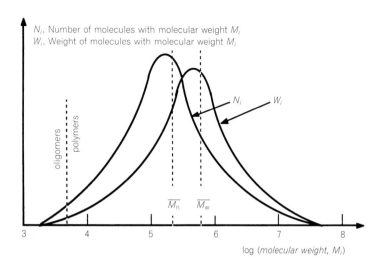

N_i, Number of molecules with molecular weight M_i
W_i, Weight of molecules with molecular weight M_i

log (*molecular weight, M_i*)

has the same molecular weight (or relative molecular mass[†]) $M = iM_r$, M_r being the molecular weight of the repeat unit. For polyethylene, $M_r = 0.028\,\text{kg mol}^{-1}$, and the critical molecular weight at which the monodisperse oligomer acquires the characteristics of a polymer is $M_c \approx 3\,\text{kg mol}^{-1}$. In practice, polymerization creates a *polydisperse* material – a mixture containing different quantities of I monodisperse polymers with molecular weight M_i, I usually being a large number.

Because the exact distribution has an important influence on some properties, it is important to be able to measure, describe and control it. There are several ways of doing so. The most complete methods involve *fractionating* a sample of the polydisperse polymer, i.e. separating the mixture back into its monodisperse constituents. In principle, the number of chains N_i having each degree of polymerization i, and thus molecular weight $M_i = M_r i$ can then be measured. Figure 12.3 shows a typical distribution, plotted as N_i versus i. This bell-shaped curve is very similar to the *Gaussian distribution* that appears in many situations governed by essentially random processes.

If the bell-shaped distribution of chain lengths (Fig. 12.3) is approximately symmetrical it will peak near the *number-average molecular weight*:

$$M_n \equiv \frac{\sum_i N_i M_i}{\sum_i N_i} \qquad (12.1)$$

M_n is the molecular weight of a chain whose length is the simple arithmetic mean of all the individual chain lengths. Alternatively, we could plot the weight of each fraction, $W_i = N_i M_r i$. This distribution would look rather different, emphasizing the fractions of longer and heavier chains, and would peak near the *weight-average molecular weight*:

$$M_w = \frac{\sum_i W_i M_i}{\sum_i W_i} \qquad (12.2)$$

[†] Note that the term *relative molecular mass* is rarely used – it is more correct, but much less widely used than the almost equivalent *molecular weight*.

The distributions shown in Fig. 12.3 represent a typical commercial HDPE. The very shortest chains are oligomers having only about 10 repeat units. Few as they are, these contribute to bulk properties, which is why some commercial polyethylenes feel and smell waxy. On the other hand there are still many chains 10 000 times longer than this, so that it makes sense to plot the distribution on a logarithmic basis, $\log(M_i)$. For a monodisperse material $M_n = M_w$, but for polydisperse materials the ratio $M_n = M_w > 1$. The value of this ratio, the *polydispersity*, typically 5–10 for HDPE, provides a useful measure of the width of dispersion.

The molecular weight distribution of a particular polymer is its most basic structural property, but even this is far from straightforward to measure. The most direct methods begin with the preparation of a dilute solution. Although many amorphous plastics dissolve readily in common solvents (e.g. both PS and PMMA dissolve in acetone), polyethylene will dissolve in only a few solvents (xylene, biphenyl) at high temperatures, and handling these solutions is not easy. Some polymers, notably fluoropolymers such as PTFE, are virtually insoluble.

Measurement of molecular weight

The most widely used method of molecular weight characterization is *gel permeation chromatography* (GPC). Chromatographic techniques fractionate polymer chains of different molecular weights rather as a marathon separates runners, by spreading out their time of passage through a process of diffusion (see Section 7.2.2) through a gel. As we shall see in Section 12.9.2, a *gel* is itself usually a polymer consisting of chain sections, of which some have been connected into a network while the rest have been extracted and replaced by a suitable solvent. This leaves the gel with pores of a controllable size, here in the range 0.01–1 μm, containing solvent through which the polymer can diffuse.

This gel (in fact, a sequence of gels ordered in pore size) is packed into a column and a very dilute solution of polymer is pumped in at one end, as shown in Fig. 12.4. Isolated from each other in this solution, polymer

Fig. 12.4

Principle of gel permeation chromatography. Chains in a homogeneous solution (shown as rigid rods) are sorted by length as they pass through a gel-packed column.

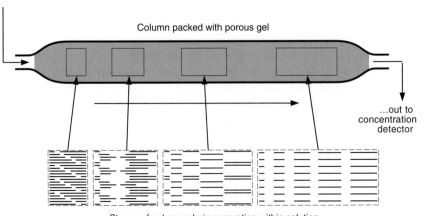

Polymer solution in...

Column packed with porous gel

...out to concentration detector

Stages of polymer chain separation within solution

chains wriggle busily by a form of Brownian motion. For each chain, the gel is an obstacle course – the smaller it is, the longer it will be detained by the pores that it has wriggled into. Thus the gel column, unlike a marathon course, is a racetrack on which the heaviest species win. The concentration of polymer in the solution that emerges is usually measured by the degree to which the solution scatters laser light. The concentration is no longer constant but varies with time as progressively lighter fractions emerge, tracing out the graph of Fig. 12.3 from right to left. The use of polymer gels to separate chain species of varying weight, combined with some means of detecting and measuring the fractions, has become well known for its application to the analysis of DNA, which is itself a long-chain polymer.

The molecular weight distribution of a polydisperse polymer is the most basic parameter that can be adjusted to control its properties – for example, increasing M_n makes a thermoplastic tougher but more difficult to mould. Even at this basic level, however, we meet the problem that, because polymer structure is inherently complex, it is often only imprecisely known. GPC is not an absolute method. The fractionation it achieves depends on the length and mobility of the chain rather than simply on its weight, so that the results must be calibrated for each polymer structure. This is a major problem. Other methods that do yield absolute results tend either to yield only one measure of the distribution (e.g. M_n) or to be usable only for some polymers.

12.2.4 Branching and tacticity

If a *side group* on a linear polymer (in the case of HDPE just —H—) is replaced by another complete chain to form a branched polymer as in Fig. 12.5, a polymer of quite different properties is formed. Low-density polyethylene (LDPE) would typically have a long branch every 100 or so C atoms and a short branch (of one or two repeat units) about every 100 C atoms. Repeated branching in some polymers can lead to a tree-like structure. We will see later that branching restricts the ability of molecules to pack efficiently, and that this affects not only the density but many other properties as well. Different chain architectures can be produced during polymerization of the same monomer by controlling the reaction pressure and the action of catalysts, and this provides another way of tailoring polymers to have specific bulk properties.

The most important of the other *polyolefins* – polymers constructed from simple hydrocarbon monomers – is polypropylene (PP), which is similar

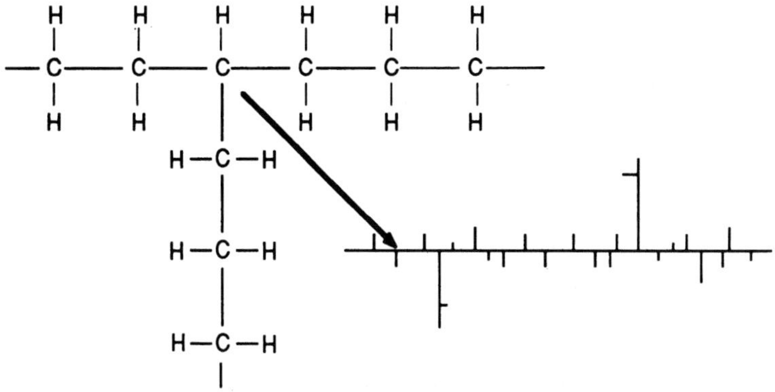

Fig. 12.5

Branching in LDPE.

Fig. 12.6

Tacticity of PP: (a) isotactic,
(b) syndiotactic and (c) atactic
forms.

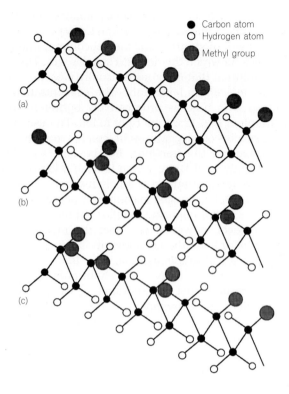

● Carbon atom
○ Hydrogen atom
● Methyl group

in many of its properties to HDPE. It is formed by polymerization of propylene gas into linear chains of the repeat unit $-$ [C(CH$_3$)H$-$CH$_2$]$-$, i.e. an ethylene unit in which one of the hydrogen atoms has been replaced by a methyl group. The substitution of a hydrogen atom by chlorine, on the other hand, produces polyvinyl chloride (PVC), while its replacement by a benzene ring yields polystyrene (PS). PE, PP, PVC and PS account for most of the market in thermoplastics.

Polymers whose repeat units, unlike that of PE, are asymmetrical, share the important structural property of *tacticity*, possessing chemically distinct configurations or *stereoisomers*. In PP, for example, each propylene repeat unit could join the growing polymer chain either head (methyl group end) to head, or head to tail; in practice the first form is rare because the side groups tend to interfere with each other. In the head-to-tail form (Fig. 12.6) the methyl groups can be aligned regularly on the same side of the backbone plane (the *isotactic* form) or can alternate regularly from one side to the other (the *syndiotactic* form). The *atactic* form has no significant regularity at all. These three structures give rise to very different properties.

Recent advances in catalyst technology have enabled unprecedented control over the tacticity of addition polymers. The polymer chains grow from isolated sites on the surface of the catalyst, and these sites can act as assembly 'templates' that correctly align an arriving monomer molecule before it is activated and added.

12.2.5 Copolymers and blends

Sometimes two monomers, say A and B, which can be polymerized individually to form polyA and polyB, can be polymerized together to form a *copolymer*. Copolymerization extends the range of available molecular structures even further by producing a material that blends the properties of its constituents. Many commercial polymer grades described as polypropylene, for example, are in reality ethylene–propylene copolymers whose exact recipes have been adjusted to optimize properties such as impact resistance.

After the basic choice of *comonomers* A and B, the most basic structural variable in a copolymer is the order of A and B units along the chain. In the *random* copolymer designated 'poly(A-*ran*-B)', chains have no detectable regularity (say ABBAABABAABAAA). Each repeat unit finds itself in a perfectly mixed environment of A and B units and the copolymer properties will usually reflect a weighted average of those for 'polyA' and 'polyB'. In a statistical copolymer ['poly(A-*stat*-B)'] there is some degree of short-range order, with a periodic pattern of A and B repeats. A limiting case of such local order is the *alternating* copolymer ABABAB ['poly(A-*alt*-B)']. *Block copolymers* ('polyA-*block*-polyB') contain substantial A and B sequences. These are sometimes referred to as 'short block' copolymers if these sequences constitute oligomers, or 'long block' copolymers if each A or B section is a substantial polymer chain. At the same level of structure is the *graft copolymer* polyA-*graft*-polyB, in which entire polyB side branches are covalently bonded onto a polyA backbone.

Long block and graft copolymers introduce the possibility of *multi-phase* polymer structures. In Chapter 10 we saw the benefits that a secondary phase, in the form of a hardening precipitate, can bring to metallic materials. A very important feature of polymers is that, of the many homopolymers that can economically be synthesized, very few mixtures are *miscible*. If mechanically mixed they will tend to separate spontaneously at the micron scale, usually forming two distinct phases with a weak interface. However, if immiscible polymers polyA and polyB are bonded into a copolymer, e.g. a *triblock* copolymer in which polyB sequences are bonded onto each end of polyA chains, the resulting structure will probably have a two-phase structure of polyB regions securely and covalently bonded into a polyA matrix. This polyA matrix will be relatively free of the chain ends whose existence often detracts from polymer properties. We will meet multiphase polymers again in Chapter 13.

12.2.6 Crosslinking

Many solid thermoplastic polymers cohere as solids largely because their chain ends are few and far between. The mere fact that chains cannot pass through each other tends to lock them against flow: they are *entangled*. However, the ends of each chain still allow it to wriggle past or slip through its neighbours under stress. This is an advantage while a thermoplastic is being moulded into shape but it becomes a disadvantage later, making its properties sensitive not only to temperature but also to *time*. An early example of this was seen in the properties of natural rubber (NR) whose raw material, *latex*, is a viscous fluid containing a linear polymer of polyisoprene. Processing of latex to form natural rubber produced a material whose tendency to creep under load made its extraordinary elasticity difficult to exploit.

The first great step forward in polymer technology (although it was not understood at the time) was Goodyear's discovery of a solution to this problem: *vulcanization*. Mixing a small amount of sulphur into rubber during processing was found to prevent latex flowing under sustained load by what is now known to be a *crosslinking* process. The two ends of short chains (2–3 atoms) of sulphur atoms open double bonds on neighbouring polyisoprene chains, forming a covalently bonded bridge that ties the polymer chains together. A very few crosslinks, typically one per hundred or so repeat units, can stabilize each chain against slip – as we will see in Section 12.3.2, it is still entanglements that are mainly responsible for holding a mass of isolated, extensible chains together into a strong, elastic solid.

Early experiments on vulcanization revealed that a wide range of properties could be achieved by controlling the degree of reaction, i.e. the crosslink density. Light crosslinking preserved the soft elasticity of the rubber, while dense crosslinking converted it to a hard, brittle glass. Using different chemical processes, polyethylene and many other thermoplastics can be crosslinked during or after melt processing. Two neighbouring polyethylene chain backbones can be crosslinked directly, using reactive peroxides or by using γ irradiation to strip hydrogen atoms from them. Again, the objective is to stabilize their mechanical properties and it is important not to crosslink too much. Excessive crosslinking at the exposed surfaces of rubbers or ductile polymers tends to make them brittle and this is one mechanism by which they weather and degrade.

These observations provide valuable insights into how polymer chains serve as structural elements within a strong, elastic solid. We will now look at this subject in more depth.

12.3 *Mechanics of flexible polymer chains*

When we considered the structure of a polyethylene chain in Chapter 6 we noted that the minimum energy position of each carbon–carbon bond gave the backbone a zigzag conformation in its most extended state, but that there were two other minimum-energy *gauche* positions. If a chain formed at absolute zero temperature with one bond rotated into a *gauche* state, it would remain kinked. A random succession of *trans* and *gauche* bonds along the chain would give its backbone a *conformation* (i.e. a path through space) that could be traced along the bonds of a diamond lattice (Fig. 12.7).

At any non-zero temperature, random thermal vibrations will allow bonds to switch between states. As the temperature increases the rotations,

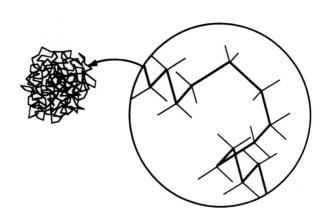

which are no more than Brownian motions of the ethylene units bonded into the chain, become frequent. As a result, an isolated chain allowed to conform freely will not remain in the stretched out *trans* conformation of Fig. 6.24, but will always crumple into a wriggling ball: a *random coil*.

12.3.1 Freely jointed chains

Although every polymer chain tends to collapse into a random coil, the relative radius of the coil differs and is a fundamentally important property. To characterize it, we replace the apparently smooth curve of the coil by an equivalent sequence of straight, infinitely thin 'link' sections, each of length ℓ, connected by joints that are free to pivot in any direction (Fig. 12.8). If ℓ has the correct value (the *Kuhn length*), a 2ℓ long section of the real chain it represents can loop back on itself, just as a pair of jointed links can be folded in half. The Kuhn length therefore represents a basic property of the chain – if it is short, the chain can easily conform to or wrap around its neighbours. This ability depends not only on the nature of bonds along the backbone (which cannot always rotate as freely as the C—C bond) but also on inter-ference between side groups. Each link represents several real repeat units. For example, the two C—C bonds enclosing a 110° angle give the —C_2H_4 repeat unit an end-to-end length of 0.254 nm. The Kuhn length for linear PE is more than 1 nm. The PS repeat unit also has two C—C bonds and a similar repeat unit length, but the bulky side group makes the chain much less flexible and gives it about twice the Kuhn length of PE.

Having thus disposed of the chemical details, imagine now construct-ing a single chain of ideal links from an end tethered at the origin of coordinates. The chain floats freely in space without interacting with any neighbours. A close approximation to this ideal state is given by forming a very dilute solution of the real polymer in a suitable solvent – and this is the state in which molecular weight and other properties of polymers are usually characterized. As chain construction proceeds, the free choice of direction at each new link takes us on a *random walk* (also known

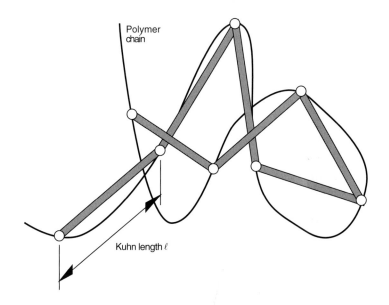

Polymer chain

Kuhn length ℓ

Fig. 12.8

A linear polymer molecule represented as a freely jointed chain. The length of each link characterizes chain flexibility.

picturesquely as a *drunkard's walk*) through three dimensions. Determining the end position of an *n*-link chain is equivalent to determining where a drunkard will arrive after *n* random steps of length ℓ. This is a well-known statistical problem. Its solution is best described using a probability density function equivalent to those introduced in Chapter 2 to describe electronic charge clouds. Like the low-energy electron states in a hydrogen atom, this distribution is spherically symmetrical, and is represented by a function $p(r)$, the probability of finding the chain end in a particular volume of space at radius r from the origin.

An effective though tedious method of determining the probability distribution is to choose a radial line from the origin and count the number $\Omega(r)$ of possible chain conformations (i.e. drunkard's paths) that could lead to any point at radius r on that line. To make the number countable we might allow each 'free' chain joint only six positions: straight on, straight back alongside the previous link, and left, right, up or down 90°. Now we stretch the chain end as far as it will go – to $r = n\ell$ (the *contour length*). Only one straight line path leads to this point [i.e. $\Omega(n\ell) = 1$] and none lead to anywhere beyond $\Omega(r > n\ell) = 0$. The next possible chain length r is two links shorter, either because a link has doubled back by 180° or because it has taken a 90° turn and another link has done the opposite – even for a few links there is a large but countable number of possibilities. The next possible radius is four links shorter and could be formed by an even larger number of allowable joint rotations. The greatest (and very large) number of possible paths leads back to the origin, $r = 0$; then the number falls away symmetrically until $r = -n\ell$.

If we allow each chain joint a less restricted number of rotations we will arrive at even larger numbers Ω for each r, but when plotted as in Fig. 12.9 $\Omega(r)$ will always resemble a Gaussian distribution:

$$P(r) = \frac{1}{\pi^{1/2} R_0} \exp\left[-\left(\frac{r}{R_0}\right)^2\right] \tag{12.3}$$

The probability of the chain end being at *some* point r is unity, so to obtain the probability density function $p(r)$ we simply divide every $\Omega(r)$ by a number large enough to make the area under this curve equal to one. The curve represented by Eq. (12.3) reaches a peak at $r = R_0$, a constant that remains to be determined. The probability of finding the chain end is $p(r)\,dr$, where dr is the width of the radial interval in which we search; $p(r)$ therefore has dimensions of m^{-1}. Note that we are simply *assuming* that

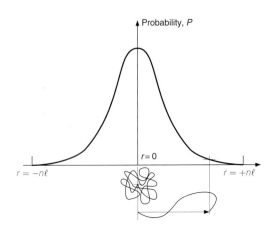

Eq. (12.3) represents the probability distribution. The assumption cannot be quite correct because, unlike $\Omega(r)$, it does not vanish at any r; the correct function must be *zero* for $|r| > n\ell$.

Allowing the chain end to lie on *any* radial line, so that the point $r=$const. becomes a spherical shell, reduces the probability density to

$$P(r) = \frac{1}{(\pi^{1/2}R_0)^3} \exp\left[-\left(\frac{r}{R_0}\right)^2\right] \tag{12.4}$$

which has dimensions of m^{-3}. The shape of this distribution along a radius remains the same, but for each radial interval dr chain ends can now be found throughout the shell volume $4\pi r^2\,dr$ in which the probability of finding them becomes

$$P(r) = 4\pi r^2 p(r)\,dr = \frac{4\pi^{1/2}}{(R_0)^3} r^2 \exp\left[-\left(\frac{r}{R_0}\right)^2\right] dr \tag{12.5}$$

which is *zero* at $r = 0$ and has a maximum at $r = R_0$. By keeping track of the number we divided by in order to obtain the probability distribution of Eq. (12.3) we can verify the important result that the peak appears at

$$R_0 = n^{1/2}\ell \tag{12.6}$$

For a typical polymer chain characterized by 1000 freely jointed links, the most probable distance R_0 between the chain ends is just a few percent of the stretched out length.

A chain whose ends are R_0 apart has an 'average' conformation in two equivalent senses. If the temperature is low, each chain lies frozen in a *glassy* state with its conformation almost undisturbed by Brownian motion. However, any representative volume of solution contains so many molecules that R_0 is the most probable end-to-end distance when averaged over their *number*. If, on the other hand, the temperature is high, Brownian motion will change the chain conformation so rapidly that observation would reveal a fuzzy 'probability density' cloud comparable to those associated with electrons in Chapter 2. The polymer is then in a *melt* state or a *rubbery* state and the R_0 is the most probable end-to-end distance averaged over *time*. We will soon explore the second situation in more detail to explain rubber elasticity.

12.3.2 Entanglements

Although a stretched out polymer chain would occupy negligible volume, an isolated polymer chain in a dilute solution takes up a randomly coiled conformation winding through a relatively large volume. If we begin to concentrate the polymer solution, chains start to interpenetrate each other's space and to hinder each other's movement. An *entanglement* is best visualized as a conflict caused by the inability of two chains to pass through each other as their centres of mass are drawn apart; the *disentanglement* needed to resolve this conflict must involve changes in chain conformation. A surprising result of Eq. (12.6) is that the longer the Kuhn length is, i.e. the less flexible the chain, the more entanglements there are per unit volume, because the chain sprawls a larger volume shared by more neighbours. This shows how different entanglements are from 'knots'. A polymer with infinitesimal Kuhn length would condense into loosely bonded one-molecule spheres.

The appearance of entanglements makes even a polymer solution much more viscous and cohesive; if a glass rod is dipped in the solution and withdrawn, it will usually draw a thread containing many molecules. It is because entanglement demands a certain minimum chain length that there is a transition in physical properties at a molecular weight M_c, marking the transition from oligomer to true polymer. As the most basic cause of cohesion, entanglements are of fundamental importance to the properties of polymer solutions, melts and solids.

12.3.3 Rubber elasticity

Let us return to the freely jointed chain model and to the 'thought experiment' of counting chain conformations. We began by fully extending the chain and allowing a 'kink' to shorten it [Fig. 12.10(a)]. For this initially extended conformation any such motion pulls the chain end inwards. If we now kink the shorter conformation, we find that some kinks restore the original chain length while many more will decrease it further; the one illustrated in Fig. 12.10(b) leaves it unchanged. The average effect will be an end-to-end contraction, but a weaker one. In this way, kinks generated by Brownian motion along a randomly conformed, freely jointed polymer chain generate a contraction force between its ends. The contraction force decreases linearly as the chain contracts, as if the chain were a perfect Hookeian spring. It also increases with the absolute temperature which generates the Brownian motion.

This tendency of a randomly conformed chain to contract is the origin of rubber elasticity. The mechanism of the contraction force has nothing to do with bond extension, whose contribution to Young's modulus and bulk modulus was analysed in Section 9.2.2, but it is just as physical. It is more closely related to the mechanism that generates pressure in a compressed gas. However, whereas the equation of state for a gas can be obtained using either a 'mechanical' approach (the kinetic model of Section 7.4) or a 'thermodynamic' approach, rubber elasticity can only be explained using the latter. The analysis is not difficult but here we will outline only its principal results.

We begin by taking a representative unit cube of rubber. A single force F, acting through some kind of friction-free linkage as shown in Fig. 12.11,

Fig. 12.10

The origin of rubber elasticity in a freely jointed chain. For an extended chain, random rotation of a single link is certain to reduce the end-to-end distance, but for a less extended chain the probability is weaker.

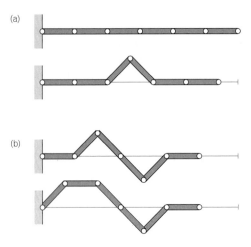

(a)

(b)

Fig. 12.11

A rubber cube stretched by direct forces – the stiffness depends on strain in the other two dimensions.

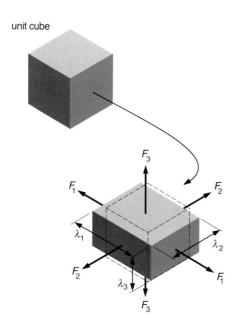

unit cube

stretches the rubber in all three directions so that its dimensions change to λ_1, λ_2 and λ_3 under imposed forces F_1, F_2 and F_3. At each value of F a small displacement $\mathrm{d}L$ at its point of application does work $F\,\mathrm{d}L$; this occurs isothermally (i.e. slowly) so that the temperature T of the rubber does not change. Now although the flexibility of chain sections in a rubber allow it to change shape very easily – i.e. the shear modulus is low – the bulk modulus K remains high, so that the volume of the unit cube will remain essentially unchanged:

$$\lambda_1 \lambda_2 \lambda_3 = 1 \qquad (12.7)$$

Because we observe that the shape change is elastic – completely reversible – we can identify the work $F\,\mathrm{d}L$ with an increase in free energy ΔG where

$$\Delta G = \Delta U + \Delta w - T\Delta S \qquad \text{[from Eq. (7.23)]}$$

This kind of free energy is known as *strain energy*. In metals and crystalline solids, as we saw in Chapter 9, it consists almost entirely of an increase ΔU in internal energy, stored in stretched interatomic bonds. The resulting elasticity is characterized by a high shear modulus and a small elastic limit, usually just a few tenths of one percent. The small forces that stretch rubber, however, barely change its internal energy and are insufficient to change the volume, so that $\Delta w = p\,\mathrm{d}v = 0$ and Eq. (7.23) reduces to

$$F\,\mathrm{d}L = -T\Delta S \qquad (12.8)$$

Thus rubber elastic behaviour arises when the increase in free energy is associated with a reduction ΔS in *entropy*. The same kind of 'entropic elastic' behaviour is shown by a gas constrained to occupy a smaller volume. In both cases, increased constraint has reduced the number of available internal states. Only materials that resemble a gas in being complicated enough to have many such states can show large changes in entropy. Entropic elasticity in solids is characterized by a relatively low shear modulus but also by an impressive ability to recover elastically from extension by up to ten times the original length.

Rubber elasticity is of enormous importance not only for the materials that we know as rubbers or elastomers, but also for nearly all other polymers and for other macromolecular materials, many of biological origin. These materials have several structural features in common. Firstly, they consist at least partly of chain-like molecules. Secondly, the chain backbones are flexible, in the sense that that they can be 'kinked' at many points. Thirdly, bonding between chain sections is weak enough for the internal energy U to be neglected. This weak secondary bonding gives rubbery materials some rather liquid-like properties – they have a bulk modulus typical of organic liquids, and very low shear moduli (up to five orders of magnitude lower than those of metallic or ionic solids). The feature that distinguishes rubbers from liquids is that the flexible chain sections are somehow anchored together to form a network. In polymeric rubbers, these anchor points are entanglements stabilized by occasional chemical crosslinks.

Now we turn to the task of calculating the change in entropy of the block as it changes shape. The block consists of N randomly conformed chains. For each chain we know that the probability $P(r)$ of finding its ends separated by a distance r is given by Eq. (12.5). Now we also know that entropy and probability are related through Boltzmann's relationship:

$$S = k \ln W \qquad \text{[from Eq. (7.24)]}$$

where k is Boltzmann's constant. Because Eq. (7.24) is as true for every chain within the rubber as it is for the entire block, we can use it to derive the force f needed to extend the end-to-end distance r of a single chain:

$$f = -T \frac{dS}{dr} \qquad (12.9)$$

which using Eqs (12.5) and (12.11) becomes

$$f = \left[\frac{2kT}{R_0^2} \right] r \qquad (12.10)$$

Because this force arises from random thermal vibrations that modify chain conformation, rubber elasticity gives the surprising property that stiffness *increases* with temperature, T.

We will now go on to derive an expression for the strain energy U_e of the block in terms of the imposed extension ratios and, from this expression, to derive stress–strain relationships. The chains that make up the block are anchored at randomly distributed nodes to form a network. Their end-to-end vectors are randomly distributed. Whatever the vector of any chain, it deforms in an *affine* manner, i.e. in the same way as the whole block – its length in each principal direction increases by the extension ratio in that direction (Fig. 12.11). This eventually leads to an expression for the strain energy per unit volume

$$U_e = \tfrac{1}{2} NkT [\lambda_1^2 + \lambda_2^2 + \lambda_3^2 - 3] \qquad (12.11)$$

where N is the number of chains per unit volume.

In a conventional uniaxial tension test of the kind introduced in Section 9.1.1, we apply a force $f = f_1$ to extend the block by $\lambda = \lambda_1$ in one direction and allow it to contract equally in the other two so that from Eq. (12.7), $\lambda_2 = \lambda_3 = \lambda^{-1/2}$. Hence Eq. (12.11) becomes

$$f = \frac{dU_e}{d\lambda} = \tfrac{1}{2}NkT\left[2\lambda - \frac{2}{\lambda^2}\right] = NkT\left[\lambda - \frac{1}{\lambda^2}\right] \qquad (12.12)$$

Fig. 12.12

Pure shear deformation of a unit cube at constant volume. The dimension into the paper remains unity while any square element ABCD in the plane of the paper undergoes a shear strain γ.

Before discussing this relationship further we will look at a second mode of deformation: pure shear. This is important because the ability of rubber to deform easily in shear is widely exploited in engineering. Pure shear is defined by fixing the length of face 3, so that $\lambda_2 = 1/\lambda_1 = 1/\lambda$ (Fig. 12.12). Substituting these values into Eq. (12.11) gives

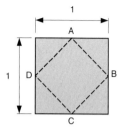

$$U_e = \tfrac{1}{2}NkT\left[\lambda^2 + \frac{1}{\lambda^2} + 1 - 3\right] \qquad (12.13)$$

$$= \tfrac{1}{2}NkT\left[\frac{(\lambda^2 - 1)}{\lambda}\right]^2$$

It can be shown using simple geometry that the shear strain γ is given in terms of the principal extension ratio by

$$\gamma = \lambda - \frac{1}{\lambda} \qquad (12.14)$$

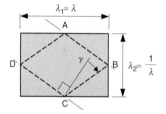

while in Chapter 9 we showed that the strain energy per unit volume under shear stress is

$$U_E = \tfrac{1}{2}G\gamma^2 \qquad \text{[from Eq. (9.14)]}$$

where G is the shear modulus. Combining Eqs (9.14), (12.12) and (12.14) we find that the shear modulus of an ideal rubber is constant (i.e. independent of strain) and is given by

$$G = NkT \qquad (12.15)$$

Rewriting the uniaxial force/extension relationship in this form simplifies it to

$$F = G\left[\lambda^2 - \frac{1}{\lambda}\right] \qquad (12.16)$$

whose slope at zero strain, $\lambda = 1$, which is the nearest equivalent to Young's modulus for a Hookeian material, is

$$E = \left.\frac{d\sigma}{d\lambda}\right|_{\lambda=1} = \left.\frac{d\sigma}{de}\right|_{\lambda=1} = 3G \qquad (12.17)$$

As we saw in Section 9.24, this is the expected result for an incompressible material. Partly because the stress versus strain curve for polymers is seldom linear or 'Hookeian', Young's modulus for a polymer is usually simply referred to as the *tensile modulus*.

Now recall that in Chapter 9 engineering stress was defined as force per unit area of unstrained block, and here is simply equal to F, while the engineering strain is equal to $(\lambda - 1)$. Figure 12.13 shows Eq. (12.17) as an engineering stress–strain curve; the stress σ has been divided by E so that the shape is seen clearly for an initial slope of unity. This ideal rubber seems to become softer as it extends: the ratio of stress to strain has fallen

Fig. 12.13

Fig. 12.13

Stress–strain curve for an ideal rubber.

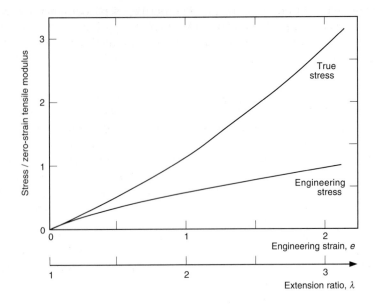

from unity to about 0.58 at unit strain (i.e. at an extension ratio $\lambda = 2$). However, to maintain constant volume at $\lambda = 2$ the cross-sectional area has fallen by a factor of 2. A more fundamental picture of material behaviour is given by plotting *true stress*, σ_t, equal to the force divided by the current cross-sectional area. Because for constant volume deformation $\sigma_t = \lambda\sigma$ this clearly shows the true hardening effect of chain extension (Fig 12.12): the more extended the rubber is, the more stress is needed to strain it further.

Figure 12.14 shows how the volumetric rigidity and the shear flexibility of rubber affect its use for springs. The two configurations shown use the same volume of rubber and give the same shear stiffness. However, under vertical load the block of Fig. 12.14(a) can expand laterally between a single pair of rigid plates, whereas in Fig. 12.14(b) this is prevented by the many interleaved rigid sheets (usually of steel) and the block is vertically rigid. Rubber springs like that shown in Fig. 12.14(b) are used to support loads as great as the weight of buildings and bridges while allowing lateral

Fig. 12.14

The same volume of rubber used in different configurations has the same shear stiffness but very different vertical stiffness.

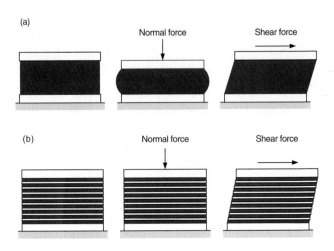

movement, e.g. during earthquakes. The rigidity of rubber constrained in this way also helps to explain why rubbery regions between rigid crystalline sheets in polymer microstructures are much less flexible than we might expect.

12.4 Thermoplastic melts

The term 'plastic' means 'mouldable', i.e. formable under pressure. Above all other properties it has been the mouldability of plastics, and in particular of thermoplastics, that has sustained the extraordinary increase in their use. This remains true even if the leading edge of development now involves polymers that offer extremely high performance (e.g. the high strength of Kevlar fibres) or unique properties (the low coefficient of friction of polytetrafluoroethylene, PTFE) at the cost of much greater difficulty in moulding. The way in which thermoplastic melts deform and flow under stress (their *rheology*) is an important field of polymer engineering. Polymer melts bear little resemblance to the simple fluids of molten metals, which can be poured and cast under gravity; they are usually more like rubbery dough.

12.4.1 Viscosity

Viscous deformation has been discussed in Section 9.4. The shear viscosity, η, is the ratio of shear stress to shear strain rate in an ideally viscous (Newtonian) material. Just as a shear viscosity η can be defined by analogy with the shear modulus G, viscous fluids also have an extensional viscosity, η_e, analogous to the Young's modulus E: it is the tensile stress required to extend a sample of the 'fluid' at unit strain rate. For an ideal 'Troutonian' fluid, these constants are related, rather as E and G are for an elastic material, by

$$\eta_e = 3\eta \tag{12.18}$$

The extensional viscosity of most liquids is insignificant, but polymer melts differ in this respect mainly because their viscosity is very high.

This high viscosity does not arise purely from a high molecular weight. To see this, we can again examine the way in which the viscosity η increases with molecular weight M. For simple liquids with similar intermolecular forces, e.g. lower-M alkanes, the origin of viscosity lies in the frictional force that a short chain experiences as it slides through its surrounding chains. As we would expect, η increases linearly with chain length, i.e. with molecular weight. Above the critical molecular weight, however, the dependence is much stronger. On plotting viscosity versus molecular weight on logarithmic axes this change is seen as an increase in gradient from 1 to about 3.5 (Fig. 12.15), i.e. there has been a change from $\eta \propto M$ to $\eta \propto M^{3.5}$.

This transition coincides with the onset of rubber elasticity, and arises from the same cause: the appearance of entanglements. A given chain can now barely move laterally because of the sheer length of its neighbours. Short lengths (*segments*) of it can still thrash around from side to side, because many of the surrounding chain segments lie parallel or nearly parallel to it and are doing the same thing. The problems arise from sections of adjacent chain that *cross* the path – our chain cannot pass through these.

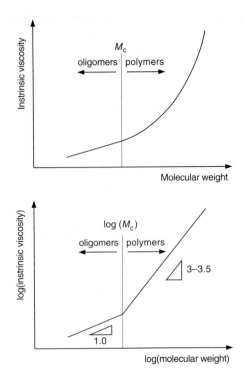

These chains, which loop round the chain path, are the same features that caused entanglement in the solid polymer. Here the 'path crossing' (i.e. entangling) chains confine the motion of each and every chain to a tube. Each molecule's domain of freedom has shrunk from three dimensions to one, because in a global sense it can only move along the tube axis. It does so by a snake-like wriggling motion known as *reptation*. The entangling chains – whose nearby path crossings form the tube – are themselves unable to move any faster, so the tube disappears only after the chain has moved out of it. The reduction in exercise space from three dimensions to one corresponds to an increase from one to three in the power law exponent relating viscosity to chain length. The result is that doubling the molecular weight of a real polymer can increase its melt viscosity by a factor of ten.

We will see in Section 12.4.4 that the limited freedom of polymer chains to diffuse even in the melt state causes problems in polymer processing. An example of the trade-off between melt processability and mechanical properties is the linear polymer poly(methyl methacrylate), PMMA, known as Perspex in the UK and Lucite in the USA. PMMA plate can be made by casting the liquid monomer before it is polymerized. This gives the polymer a high molecular weight and good mechanical properties but leaves it with a melt viscosity too high for moulding. For moulding components such as automobile rear light filters, the molecular weight must be significantly lower to reduce the melt viscosity – but this leaves the material more vulnerable to fracture.

Polymer melt viscosity depends on other chain structure features such as the degree of branching, inherent flexibility and mutual interference between its side groups. Other factors that influence viscosity are temperature (increasing temperature increases mobility, promotes diffusion and thus decreases viscosity) and pressure (which forces chains closer together and thus increases viscosity).

Measurement of viscosity

A crude but widely used industrial method for comparing polymer melt viscosities is the *melt index* or *melt flow index* test (Fig. 12.16). A sample of the polymer is heated in a cylindrical chamber 10 mm in diameter. When the melt has reached a controlled temperature a weighted piston moves downward to extrude it like toothpaste through a smaller (2 mm diameter) capillary tube at the bottom of the chamber. The rate of extrusion is normalized to units of 'grams per 10 minutes'. Clearly, the higher this result (the *melt index*) is, the lower the viscosity. Although it does not 'measure' viscosity, this method is universally used for plastics and is useful for describing differences between grades, or for detecting changes in composition – particularly of changes in molecular weight. A *capillary rheometer* is a much more sophisticated variant: the apparatus can be fitted into a compression testing machine to control the extrusion rate and the result can be interpreted accurately as a viscosity. The simple cone-and-plate viscometer introduced in Section 9.3 has a significant disadvantage here – the same sample of melt is tested continuously, whereas in capillary methods it is constantly replaced. One result of this is that the work done in shearing the melt causes a rapid temperature increase, and there is a pronounced decrease in melt viscosity with increasing temperature – for most polymers, 1–2% per °C. Shear heating of melts is an important factor in several polymer processing methods.

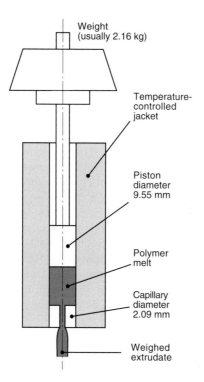

Fig. 12.16

Apparatus for measuring the melt flow index of a polymer melt.

12.4.2 Shear thinning

If we increase the speed of the viscometer while testing a *Newtonian* material [see Eq. (9.15)], the torque should increase in proportion. This behaviour is observed in tests at low speed on polymer melts, but at higher rates the gradient of the torque versus speed graph begins to decline. The viscosity calculated from the ratio of shear stress to shear strain (the *apparent* viscosity) is falling with increasing shear rate, a property mentioned in Section 9.4 and known as *pseudoplasticity* or *shear thinning*. The origin of polymer melt pseudoplasticity lies in chain extension under the shear stress. Chains are dragged into the direction of flow and are extended, removing entanglements so that they slip easily alongside each other. However, this extension is neither complete nor permanent because of the continuing tendency of polymer chains – more than ever in the hot, melt state – to contract back into random coils. To keep the melt thin it must be sheared continuously, and the faster it is sheared the more complete is the disentangling and straightening process.

Most polymer melts also *tension thin*: like the shear viscosity, the extensional viscosity η_e decreases with increasing extension rate. However, the extensional viscosity usually decreases much more slowly so that the ratio of η_e to η increases from its Newtonian value of 3 at low shear rate – sometimes to several orders of magnitude more. Some polymers even *tension stiffen*. An interesting comparison is that between linear polyethylene, HDPE, and branched polyethylene, LDPE. The melts of these two polymers are similar in shear viscosity and elasticity but they differ dramatically in extensional viscosity: HDPE tension thins while LDPE tension stiffens. Again, the reason is entanglement. This time, it is the side branches in LDPE that interfere with chain slippage, and this effect does not vanish as the chain backbone is extended and straightened by shear flow at increasing rates. We will see, however, that tension stiffening can be useful: it accounts partly for the extensive use of LDPE for low-cost packaging film.

12.4.3 Melt elasticity and strength

Because we have seen that entanglements can act like crosslinks and introduce rubbery behaviour, and we know that entanglements exist in thermoplastic melts as well as solids, it is not surprising that polymers show pronounced *melt elasticity*. This appears in a variety of ways during processing to final shape; virtually all are unwanted because, of course, the whole point of forming is to produce a change of shape that is permanent.

The cone-and-plate viscometer provides a good illustration of melt elasticity. If the rotating cone is suddenly stopped and locked, the torque does not disappear instantaneously but decays in time as the melt relaxes like a Maxwell material (see Section 9.4.1). Conversely, if the torque driving the cone is suddenly removed, the cone will not simply stop but will spring back as the material recovers. A melt flow indexer, meanwhile, can illustrate *die swell* – one of the most inconvenient effects of melt elasticity. This 'springback' after deformation causes the material extruded from a melt indexer (referred to in Section 12.4.1) to swell towards the barrel diameter from which it was compressed. Although even Newtonian materials show slight die swell, that shown by polymers is much greater. It increases further with extrusion speed, because the polymer has less time to relax within the capillary.

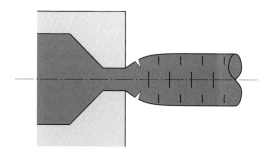

Fig. 12.17
Die swell and melt fracture.

   A melt flow indexer can illustrate another polymer property that can cause problems in the processing of thermoplastics: *melt fracture*. The tensile strengths of polymer melts are in the region of $1\,MN\,m^{-2}$. This is not dissimilar to those of other organic liquids, but the high viscosity of polymer melts means that high stresses must be applied to process them at economically high rates. A common manifestation of melt fracture is the break-up of extrudate emerging too rapidly from an extruder die, under the tensile stresses that accompany die swell (Fig. 12.17).

12.4.4 Processing of thermoplastics

Having investigated the structural origin of polymer melt properties it is interesting to survey the main processes used to form thermoplastics, to which these melt properties are so important. Before doing so, two more properties must be mentioned. Firstly, polymers are often chemically unstable at temperatures significantly greater than their glass transition and melt temperatures. This decomposition can be controlled using additives, but the second property tends to exacerbate it: low thermal diffusivity, which makes it extremely difficult to heat a body of melt quickly without overheating some regions. Moreover, the mixing needed to homogenize viscous melts can itself generate heat within the material. If so, *removing* heat becomes a problem.

   In the first few processes we will look at, a *plasticating screw* plays an essential role. This is essentially an Archimedean screw, turning in a heated barrel, which draws in polymer at one end, melts and mixes it under increasing pressure along its length, and expels it at high pressure from the other. The key advantage of a plasticating screw is its ability to mix the melt thoroughly, avoiding 'hot spots' and ensuring that individual granules melt thoroughly rather than merely welding to each other. The critical effect that the screw shape, channel profile and surface texture have on screw efficiency, and the punishment that the surface of the screw must take in service, make this a complex and expensive component that must be carefully matched to the polymer being processed.

Extrusion

Extrusion (Fig. 12.18) is used to produce long shapes of constant cross-section. These range from simple forms like pipe and medium-thickness sheet to the complex, internally webbed cross-sections of window frames. A plasticating screw pumps the melt at constant rate through a die of the required shape: the principle is similar to that of the hydroplastic extrusion

Fig. 12.18

Extrusion.

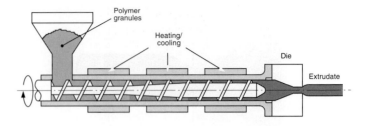

of clay, illustrated in Fig. 11.7. The section emerging from the die (the extrudate) is drawn off, often through a sizing die to confer more accurate dimensions, into a cooling water bath. Because the extrudate must usually hold its shape for a short period while unsupported, extrusion of a controlled profile is usually best suited for high viscosity, i.e. high molecular weight, grades. This immediately introduces the need for a typical engineering trade-off, because the higher molecular weight leads to increased die swell unless the extrusion speed is reduced.

Virtually every molecule of every thermoplastic produced passes through an extruder at some time. The initial *compounding* stage for many plastics is melt extrusion, which uses the powerful mixing action of a plasticating screw to homogenize base polymer fresh from the reaction process, and to blend it with essential additives. Most processes by which thermoplastics are formed to their final shape use polymer in the preprocessed form of granules 2–3 mm in size. These, too, are produced by high-rate extrusion of many parallel strands that are immediately chopped to length by rotating blades. The processes used to cover electrical cable or optical fibre and to coat cardboard are also modified extrusion processes.

Extrusion is the first step in a number of other processes. *Extrusion blow moulding* is a discontinuous process, similar to that previously described for a glass (Section 11.4.2, Fig. 12.19). For plastics the process begins with the downwards extrusion of a short tubular section of melt (a *parison*). The viscosity of thermoplastic melts is sufficient for the parison to hold its shape until a cold mould has closed around it and air is then injected to inflate the polymer against the mould surface. Pressure is maintained until the melt has solidified sufficiently (usually less than a second or two). Extrusion blow moulded components – mainly bottles – are easily recognized by the thick region at the base where the parison was pinched and sealed.

Another variant of extrusion is used to produce most of the vast quantity of general-purpose, low accuracy plastic film, e.g. for plastic bags. *Film*

Fig. 12.19

Blow moulding.

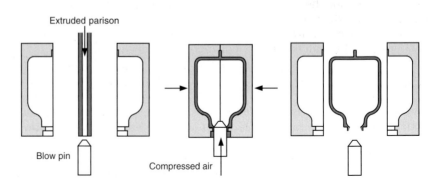

Fig. 12.20

Film blowing.

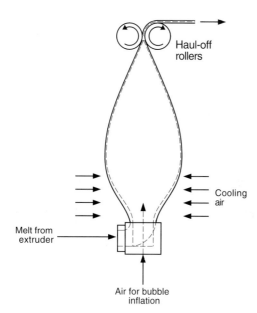

blowing (Fig. 12.20) is a continuous process in which the polymer is extruded upwards as a very thin-walled tube that is inflated by a relatively low internal air pressure. Having cooled sufficiently, the tube meets a pair of rollers that pinch the bubble to seal it. These rotating rollers maintain a speed that usually adds longitudinal stretching to the circumferential stretching induced by blowing. The majority of polymer used by both these processes is polyethylene, particularly LDPE. Both blow moulding and film blowing require the melt to extend in a stable way without rupturing; this is where tension stiffening, described above, comes into play, and where LDPE gains its advantage for bulk use as blown film. Films of more precisely controlled thickness or finish are produced by calendering between rollers.

The shear strain rates involved in extrusion are not high enough to gain much benefit from shear thinning. Large-sectioned extrusions must partly support their own weight while cooling, so that high melt viscosity here becomes a real advantage. Extrusion grades therefore tend to have a higher average molecular weight with a wider distribution.

Injection moulding

The plasticating screw unit that forms the heart of an extruder is also used almost universally for injection moulding (Fig. 12.21). In this immensely important process – a masterpiece of production engineering, involving every complexity of engineering science – the screw offers a matchless ability to very rapidly prepare a controlled quantity of homogeneous polymer melt. This *shot* is then injected through the *gate*, a small orifice, into a closed, cold mould. Pressure is maintained for a short time while the moulding cools and solidifies. While cooling continues, the next charge of melt is prepared. Finally the mould is opened to allow ejection of a finished component.

With careful control, injection moulding can be used to produce parts of astonishing complexity and precision, at relatively high speed. In this

Fig. 12.21

Injection moulding.

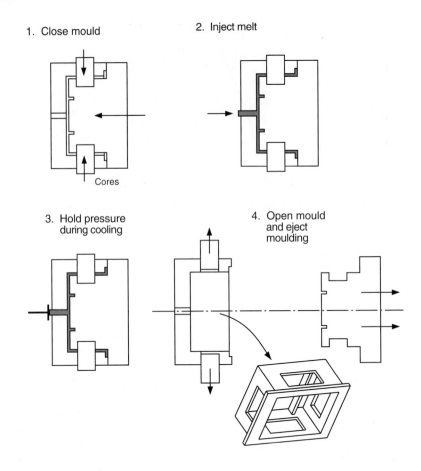

1. Close mould

2. Inject melt

Cores

3. Hold pressure
   during cooling

4. Open mould
   and eject
   moulding

process more than any, care must be taken to avoid excessive shear heating of the melt as it flows through extremely fine channels. Grades of polymer designed for injection moulding will normally have a low average molecular weight with a narrow distribution. The very high shear rates inflicted by rapid mould filling reduce the viscosity further: the melt temperature might increase by 20°C in a few thousandths of a second as it passes through the gate. Other danger points arise if the advancing polymer can trap air in corners of the mould cavity; it may then compress the air to extremely high temperatures and suffer scorching. Several inherent material weaknesses are exposed where a hole must be moulded into a component. The flow of polymer splits at the mould 'core' that forms the hole, and the separate streams meet on the other side. The slowness with which polymer chains reptate to fuse the melt fronts together creates permanent and visible lines of weakness known as *weld* or *knit lines*.

12.5 Amorphous polymers

The transition from melt to solid as a thermoplastic cools is never as well defined as for a pure metal. Many polymers, as we saw in Chapter 6, are able to crystallize, and these do show a melting 'point', although it is often quite broad. In an amorphous polymer (a better but more rarely used term is 'non-crystallizing'), the ability to flow is lost much more gradually and it

Fig. 12.22

The variation of shear modulus with temperature for an amorphous polymer.

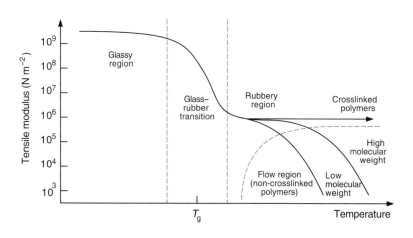

leaves the material in a rubbery state. We will deal with this class first. Our task is simplified by the fact that, in many respects, all amorphous polymers behave in the same way.

An effective way of observing the changes that occur as a polymer is cooled is to shear a sample with oscillatory motion (e.g. in a cone-and-plate viscometer), and to measure the corresponding load or torque. This idea was introduced in Section 9.3, where we saw that as a viscous melt becomes a solid the phase angle between displacement and load will diminish from 90° to zero. Of course, the amplitude of the load will also change from a value that characterizes the viscosity of the melt to a value that characterizes the shear modulus of the solid. If the temperature continues to decrease, we expect the load to increase further with the increasing stiffness of the solid polymer. If – for the time being – we ignore the change in phase angle and simply infer a shear modulus from the ratio of load to displacement, we will arrive at a curve of the form shown in Fig. 12.22.

12.5.1 Solidification

We have seen that the rubbery state is characterized by a stiffness that is proportional to the density of effective crosslinks and is relatively insensitive to temperature. The crosslinks that give a thermoplastic melt its rubbery character are physical entanglements, whose density does not change during solidification. The transition from a melt to a rubbery solid, then, is the loss of a polymer chain's ability to wriggle through entanglements.

This transition is rather well represented by a Maxwell model (Section 9.4.1), which has two significant constants: a tensile modulus E and a time constant $\tau = \eta/E$. The modulus can be taken to represent the 'rubbery' stiffness arising from entanglements, and the viscosity η to represent the resistance to chain slip by reptation through these entanglements. Solidification can be seen purely as a continuous increase in η as temperature decreases, and thus as an increase in relaxation time τ. Because for a given polymer the chain length between entanglements is almost constant (corresponding to the critical molecular weight M_c), the rubber modulus does not depend much on molecular weight, but the viscosity certainly does – varying with M to the very high power of about 3.4 (see Section 12.4.1). As the molecular weight increases, the solidification seen within a particular time occurs at a higher and higher temperature.

12.5.2 The glass transition

As a rubbery polymer is cooled further, its stiffness begins to increase again (Fig. 12.22). Eventually a temperature range is reached at which a very large increase in tensile or shear modulus, from the order of $MN\,m^{-2}$ to that of $GN\,m^{-2}$, takes place over a temperature range of little more than 20–30°C. This increase is centred on the glass transition temperature, T_g, at which rotational movements of bonds in each chain backbone virtually cease, the structure freezes and the rubber becomes an amorphous, rigid glass. Why?

The abrupt nature of the glass transition does *not* arise from any sudden change in the mechanics of a single chain, but from the fact that for chains to extend in a rubbery way all of the surrounding chains must cooperate. So rapid are the Brownian movements in a chain segment between entanglements in the rubbery state that, recorded over any observable time, the segment would be a blurred probability 'cloud' in space. Neighbouring chains do constrain movement but, because they are in the same state of motion, they do so only in the sense of one cloud passing through another. As the temperature falls, each chain experiences its neighbours more strongly as constraints on extension – it must wait longer and longer for the right combination of surrounding movements before a segment can move. We can already see that if we define the glass transition as a full-scale change from rubber to glassy behaviour, T_g will depend on the timescale on which the determination is made. We will return to this time dependence in Section 12.5.5. As the temperature continues to fall the molecular movement visible within a particular timescale diminishes from free motion of entire chain sections between entanglements, through large-scale rotation of sequences of carbon–carbon bonds ('crankshaft' movements) and finally to small, local rotations of individual bonds.

Polymer chain structures differ widely in their ability to pivot at backbone bonds and T_g shows a corresponding range. One basic classification of a polymer is whether T_g is below or above room temperature. Polymethylmethacrylate ($T_g = 105°C$) and polystyrene ($T_g = 95°C$) are amorphous glassy polymers at room temperature, because backbone rotations in both are severely hindered by bulky side groups. An amorphous thermoplastic with T_g below room temperature is seldom useful because of its tendency to flow by slippage at entanglements (although, as we will see in Section 12.11.1, T_g may be deliberately reduced using chemical additives). To make use of the softness and extensibility of polymers above T_g they must be crosslinked to prevent flow, but these crosslinks have little effect below T_g and all elastomers become glassy at very low temperatures – a fact often exploited to machine them or to clean superfluous material from mouldings.

12.5.3 Viscoelastic models for amorphous polymers

The concept of using spring and dashpot models to represent viscoelastic behaviour was introduced in Chapter 9. It was shown that a Maxwell model (Fig. 9.15) provides quite a good model for relaxation in a viscoelastic material (i.e. where the stress is diminishing at constant extension), but a poor one for creep (i.e. extension under constant stress). On the other hand a Voigt model (Fig. 9.16) provides an adequate model for creep but a poor one for relaxation. The representation of polymers using spring and dashpot models can be particularly useful; slightly more complicated models not only give much more realistic behaviour – they also provide a real and clear insight into the underlying viscous and elastic processes and their interdependence.

Because the strengths and weaknesses of the Maxwell and Voigt models complement each other, it seems natural to combine them in three-element models. For example, if we replace the dashpot of a Maxwell model by a Voigt model, i.e. we connect a spring in parallel with the dashpot, we prevent it from responding like a liquid to long times under load, and provide a much better model for a transition between glassy and rubbery moduli. Likewise, if we replace the dashpot in a Voigt model with a Maxwell model, i.e. we insert a spring in series with it, we soften its unrealistically rigid short-time behaviour. These two three-element models are actually represented by differential equations of the same form and therefore show similar behaviour. Both models are referred to as the *standard linear solid* [Fig. 12.23(a)].

The standard linear solid shows rich and realistic behaviour, which can be determined by solving the governing differential equation under particular conditions, e.g. constant stress (creep) or constant strain (relaxation conditions), but which can often be imagined just by looking at the mechanical model. If a load is applied quickly, the material shows a finite and instantaneous elastic response that can be interpreted as the extension of secondary bonds in the glassy state (Fig. 12.24). If the same load is applied for a period very much longer than τ_2, the second spring has time to extend completely, motion of the dashpot ceases and the extension remains constant. This is the behaviour of a crosslinked rubber.

Because the Maxwell series combination of a spring and a dashpot provided a qualitative model for melting, it is clear that we can extend the standard linear solid to model melting simply by adding a sufficiently viscous dashpot in series [Fig. 12.23(b)]. Effectively, the material will flow after a time long enough for the other two time constants τ_1 and τ_2 to have become insignificant.

(a)

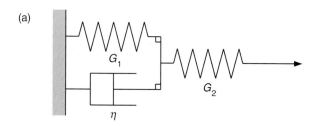

(b)

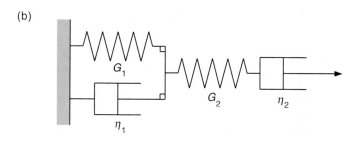

Fig. 12.23

Viscoelastic models for polymers: (a) the standard linear solid exhibits a glass–rubber transition and (b) the connection of another dashpot in series introduces melting.

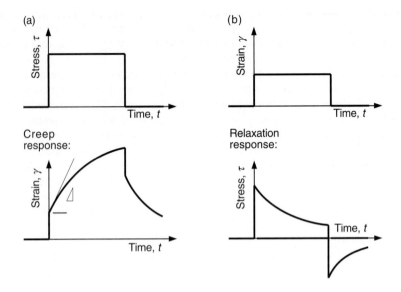

Models can be constructed that represent creep or relaxation response to any desired degree of accuracy, by connecting large numbers of Maxwell elements in parallel or Voigt models in series, so that as each reaches its relaxation time others remain partially elastic. It is possible to identify elements (or groups of elements) in such models with specific relaxation processes in the polymer, such as coordinated crankshaft rotations of chain segments to allow rubbery extension, or reptation (slip at entanglements) to allow flow. Each process has its own relaxation time and modulus. However, such realism must be paid for in mathematical complexity.

12.5.4 Time–temperature correspondence

We have already mentioned that for a chain to be able to change conformation, neighbouring chains must move cooperatively. As a result, any deformation that requires a segment to move relative to another must wait for a coincidence of thermal vibration, which will allow them to side-step each other. Only the uniform stretching of primary and secondary interatomic and intermolecular bonds, which determine the glassy modulus, can occur instantaneously. It is therefore not surprising that when the stress versus strain curves from a long series of creep tests are reviewed, a relationship is found between the effects of time under stress and of temperature, which increases the average vibration amplitude.

Figure 12.25(a) illustrates a set of results from creep tests on the amorphous polymer polycarbonate, expressed as log[creep modulus, $E_c(t)$] versus log(time, t), the creep modulus being the constant stress applied divided by the strain accumulated at time t. Each test was carried out at a fixed temperature and the temperatures ranged from below the glass transition to above the rubber plateau. For each, extension was recorded at times from a few seconds up to one hour. Experimenters have found that by choosing some reference temperature (e.g. the glass transition temperature T_g) and shifting these curves along the log(time) axis, they can be made to overlap and form a *master curve* as shown. In principle the same curve could have been plotted from creep tests at a single temperature $T = T_g$

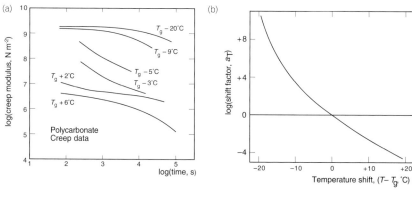

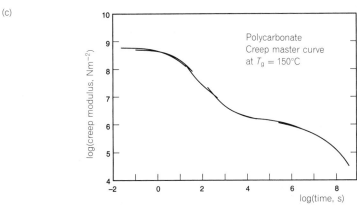

Fig. 12.25

Time–temperature correspondence. (a) Tensile modulus versus time curves from creep tests at different temperatures *T*. (b) Each curve is shifted along the time axis by a factor a_T that depends on $T - T_g$; (c) the result is a single master curve.

(150°C). Thus we find that polymers at temperatures near T_g at ordinary testing rates can behave more like glassy polymers under rapidly applied stresses (such as impact loads or high-frequency oscillations), and more like elastomers under static loads applied for very long times. However, even to replace the limited range of temperatures actually used, these imaginary tests would have had to cover the time range from 10 ms to more than 3 years! Note that this concept of an equivalence between the effects of temperature and time is very similar to that of the Larson–Miller parameter introduced in Section 9.7.7.

Figure 12.25(b) records the shift a_T along the log(time) axis that is needed to compensate for a change in temperature from T_g to T, i.e. to reverse the shift made to form the master curve. Williams, Landel and Ferry discovered that the relationship is remarkably similar for all amorphous polymers, and can be expressed by the 'WLF' equation:

$$\log a_T = \frac{-C_1(T - T_g)}{C_2 + T - T_g} \qquad (12.19)$$

where $C_1 = 17.4$ and $C_2 = 51.6\,\text{K}$. For example, to obtain creep results for $T = T_g - 20°C$ we must shift the master curve by $(-20) \times (-17.4) \div (51.6 - 20) = +11.0$ along the time axis. Whereas an extension under stress typical of a rubber is seen in polycarbonate at T_g after about 10^4 s (about 3 hours), we would have to wait $10^{(4+11)}$ s $= 3170$ years at a temperature just 20°C lower!

Chain segments in a polymer at any temperature, like the atoms in all substances from monoatomic gases to crystalline solids, have a wide spectrum of different vibrational energies (see Section 7.10). Thus even in a cold, glassy polymer a few chain segments will occasionally find the energy needed to shift under stress. For a polymer melt, at the other extreme of temperature, we can sketch the timescales involved. Relative movement of individual repeat units by rotation of side groups or short side chains may occur within 10^{-10} s. Rubbery deformations requiring large-scale motion (crankshaft or skipping-rope like rotations of short chain sections) will take at least 10^{-6} s. Viscous flow involves translation of the entire chain (reptation) by one segment and takes at least 10^{-4} s, so that a typical chain in a polymer melt will need several seconds to diffuse through its own length. This is why the self-diffusion needed for welding to take place is so inconveniently slow.

12.5.5 Dynamic mechanical thermal analysis

The way in which the stiffness of a polymer falls with increasing time can provide a great deal of information about its internal structure. However, we have seen that the range of timescales needed to observe these changes at constant temperature are enormous, whereas the equivalent range of temperature is quite modest. This is the basis for another widely used method for characterizing polymers – dynamic mechanical thermal analysis, known as DMTA or DMA. A small sample of the polymer is loaded and unloaded repeatedly as its temperature is scanned over a pre-set range. A common method is to apply a sinusoidally varying deflection, of constant (and very small) amplitude, to the end of a small beam specimen. The amplitude of the sinusoidally varying reaction load at the support falls with increasing temperature.

Section 9.3 introduced the idea of measuring the *phase angle*, δ, between load and deflection. We applied this idea to shear deformation, which is more convenient for liquids; however, it is equally applicable to direct stress applied to solids. The phase angle is $0°$ for linear elastic (Hookeian) behaviour and $90°$ for viscous (Newtonian) behaviour. Not surprisingly for viscoelastic solids it lies somewhere between these extremes, and is usually small.

Now if we apply sinusoidal vibration at a given frequency to the load point of a standard linear solid material, we will observe the dashpot to be effectively frozen solid at temperatures well below T_g, while well above T_g its viscosity is too low to affect the load. Near T_g the dashpot moves far enough for its viscous resistance to be apparent, but it contributes force when the speed is greatest, rather than when the displacement is greatest as does a spring, so that in each cycle the load maximum occurs slightly

Fig. 12.26

As the viscosity of a standard linear solid material falls with increasing temperature, there is a transition from one limiting modulus to the other as the damping factor tan δ passes through a peak.

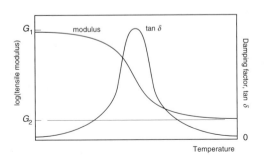

ahead of the displacement maximum. The magnitude of the *phase angle*, δ, represents the significance of viscous forces relative to elastic forces, and in this simple model material δ reaches a peak in the middle of the only transition – at T_g itself. DMTA results for real polymers, plotted as the tangent of the phase angle – $\tan\delta$ – versus temperature, show a series of peaks as each new molecular motion becomes possible within the period of the load cycle (Fig. 12.26).

12.6 Crystalline polymers

Polyethylene has a very low T_g, and if it were an amorphous polymer it would be a rubbery material with poor creep resistance. On the other hand ethylene (paraffin) waxes, which are no more than low molecular weight polyethylenes, are hard, rigid, highly crystalline solids, but are very brittle. In practice, PE has many more useful properties than either of these materials because it is neither completely amorphous nor highly crystalline, but *semi-crystalline* ('crystallizable' would be a more accurate term but the most commonly used is just *crystalline*).

The ability of many polymers to crystallize has a profound influence on their properties. Within crystallites the closer packing and stronger bonding increase strength and modulus. At temperatures between the glass transition temperature and the melting point of the crystalline phase, the amorphous phase is reinforced by *tie molecules* – chains that pass through it but are anchored in adjacent crystallites. This gives many crystalline polymers great toughness and ductility.

12.6.1 Crystallization from a melt

As described in Chapter 6, almost perfect crystals of many polymers can be crystallized from dilute solutions, where there are few conflicts between the individual polymer chains that form them. Thin, plate-like lamellae of the order of 10 nm thick (i.e. just 100–200 repeat units) form by repeated folding of each chain that joins them. These lamellae can grow as wide as several micrometers as the isolated and relatively mobile chains participate in an orderly construction process. The driving force for this ordering process (which involves a substantial decrease in entropy) is the increase in intermolecular attraction when chains are densely packed.

Crystallization from a cooled melt is more complicated but, given the densely entangled state of polymer chains, it is surprising that it can take place at all. It begins by a *nucleation* process at many sites where, without changing conformation very much, several short lengths of chain can line up side by side. Attractive forces between these closely packed chain lengths are high but in a melt, unlike a dilute solution, they are not high enough to reel in the rest of each chain at a reasonable rate. In a melt, each chain winds through a volume shared by many entangled neighbours. The overall conformation of each chain, and the volume of space it occupies, can hardly change during crystallization; instead, crystalline domains nucleate and grow independently at several points along it. Lamellae then grow from these nuclei, as they do from a solution, but their surfaces are not flat planes of neat folds. Entanglements are not reduced in number but are driven out into amorphous regions that remain trapped between lamellae. The mass fraction χ of crystalline material in a given volume – the *degree of*

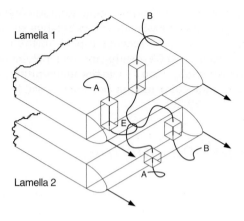

crystallinity – can therefore never reach 100%. In PE χ can be up to 90% and in polyacetal even higher, but a more typical level would be 50%.

Lamella formation from a melt is not easy to represent in a two-dimensional picture, and is not fully understood. Figure 12.27 visualizes it showing just two chains, but many others fill the space. Each chain can only join the lamella when it is sufficiently extended and oriented in the direction shown, but chains in the melt are very mobile and their conformations change very quickly. The lamella fronts advance, organizing chain sections and pushing entanglements (and any other species present, such as additive molecules) into the amorphous regions between them.

For several reasons, lamellae tend to diverge and to multiply as they extend. Lamellae are typically 15% more dense than amorphous polymer, so that amorphous polymer must drift towards them (i.e. from the right of Fig. 12.27) and this tends to wedge them apart. As the lamellae diverge, new ones may be nucleated (or previously nucleated ones may be trapped) within the thickening amorphous region between them. Each lamella may also thicken until straightening of chain regions on the growth front can no longer be organized, a region of disorder develops within it, and the lamella splits apart. The result is that a spreading stack of lamellae thickens until, as shown in Fig. 12.28, it becomes an expanding sphere: a *spherulite*.

Spherulites are formed by many – but not all – crystalline polymers. Typically 100 μm in size – large enough to be observed under an optical microscope – they can grow to millimetres before impinging upon one another. The direction of chain organization on the spherical growth surface is circumferential; when the spherulite has grown to such a size there will be thousands of sites on its surface at which lamella growth is taking place.

A typical lamellar thickness is of the order of 10 nm while the end-to-end radius of a chain in the melt is several times larger. A typical chain contour length is 1 μm (1000 nm); 50–90% of this length is destined to arrive packed into lamellae in 10 nm sections. The many tie molecules that end up woven through several crystallites have a supremely important role in maintaining the mechanical strength of a crystalline polymer; it is tie molecules that make linear polyethylene so much stronger and tougher than candle wax.

12.6.2 Factors affecting crystallinity

Many factors affect the degree of crystallinity in a sample of polymer. Some of these involve the inherent ability of a polymer to crystallize while others depend on external conditions.

The most important factor is simply the time available for crystallization – which nearly always, in practice, takes place during forced cooling. Lamellae, and therefore any spherulites composed from them, grow into the melt at a rate that is always much lower than in metals. This rate reaches a maximum of the order of 1 μm s^{-1} at some temperature below the crystalline melting point T_m but high enough above T_g for chains still to be quite mobile (Fig. 12.29). Rapid cooling – *quenching* – from a melt tends to freeze the polymer in its amorphous state. If cooling is slow enough to allow significant crystallization, the final degree achieved increases with the number of nucleation sites (which can be increased by seeding the polymer with nucleating agents) and with the maximum crystallization rate. The more mobile the chain, the easier become the conformational changes allowing a chain section to join a crystallite. This mobility is much greater within a long segment of chain near each end. As a result, the reduction in crystallinity seen as molecular weight increases from paraffin wax to linear polyethylene continues as we move from linear polyethylene to *ultra high molecular weight polyethylene* (UHMWPE). These materials, whose best known application is in artificial hip joints, have average molecular weights ten times greater than those of conventional PE, but their crystallinity is not much lower.

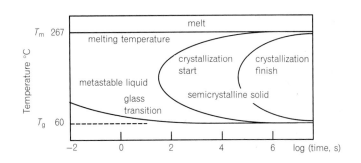

Fig. 12.29

Schematic time–temperature–transformation diagram for a crystalline polymer.

The driving force for crystallization increases with two main factors. Firstly, chain regularity. The backbone of linear HDPE is very flexible (allowing the chain to reconform easily onto a crystallite surface) and the chain itself is very regular, allowing it to pack well. Branched LDPE is just as flexible but can only crystallize to around 60% because packing is disrupted by side branches. Isotactic and, to a lesser extent, syndiotactic polymers are regular in structure and usually crystallize, while polymers with irregularly spaced bulky side groups such as PMMA are necessarily amorphous. Thus atactic polypropylene (Fig. 12.5) is amorphous, while the isotactic and syndiotactic forms are crystalline; most commercial grades are based on the isotactic form, with a crystallinity of about 70%. In copolymers, alternating structures (ABABABAB) and copolymers with large block lengths can generally crystallize while random, graft and block copolymers lacking long-range regularity cannot. The second important factor is the strength of inter-chain bonding. Polyvinyl chloride (PVC), although atactic, can achieve sparse crystallinity (about 10%) because polar bonding provides strong intermolecular packing forces. PVC chemically decomposes at such a low temperature that it is commonly processed below the melting point as a gel, i.e. a network of chains 'crosslinked' by the small crystallites.

The degree of crystallinity is of such importance to polymer properties that it is often necessary to measure it accurately. Two common methods are of general interest: *densitometry*, a simple method that illustrates a general law relevant to many material properties, and *differential scanning calorimetry*, an advanced analytical technique of much more general application.

Densitometry

A 'law of mixtures' determines the value of some property for a mixture of phases, where the value of that property is known for each phase in isolation. The law can only apply to systems in which the value of this property on one side of the interface between phases does not affect the value on the other, and density is usually such a property. Because the total specific volume v (the volume per unit mass, i.e. the inverse of density ρ) is the sum of the crystalline and amorphous volumes

$$v = \chi v_c + (1 - \chi)v_a \tag{12.20}$$

where v_c and v_a are the specific volumes of the crystalline and amorphous phases and χ is the mass fraction of crystallinity. Crystallinity can therefore be determined by measuring density ρ – *densitometry* – using a rearranged form of Eq. (12.20):

$$\chi = \frac{v_a - v}{v_a - v_c} = \frac{1 - \rho_a/\rho}{1 - \rho_a/\rho_c}$$

The density of the amorphous phase of most crystallizable polymers is much less than that of perfect crystals, making densitometry a relatively straightforward method for determining χ. The simplest technique uses a density-gradient column containing mixtures of liquids of differing densities spanning a suitable range. The density of a polyethylene, for example, usually lies within the range 930–970 kg m^{-3}. The height within the column at which a sample of the polymer (free from attached air bubbles or internal voids) floats indicates its density.

The measurement of v_a and v_c is not straightforward, because it is seldom possible to prepare large enough pure samples of either phase. For the crystalline phase, accurate density estimates can be made from a knowledge of the crystalline structure and lattice constants. For the amorphous phase, the simplest method is to extrapolate the density of the melt back to the temperature of interest (the importance of measuring melt density is discussed in the next section).

Differential scanning calorimetry

Another method for measuring the degree of crystallinity provides a much broader picture of processes that take place as the polymer is heated or cooled. Differential scanning calorimetry (DSC) measures the changing rate at which heat must be supplied in order to increase the temperature of a fixed mass at a constant rate, e.g. 2°C per minute. For a simple material having a constant specific heat, this rate of heat supply would be constant. An exothermic reaction like those involved in curing of thermosets registers as a sudden decrease in rate of heat supply because the reaction alone can maintain most of the temperature increase demanded. During a phase change such as melting, on the other hand, the required heat will increase sharply in order to provide the latent heat, ΔH_f kJ kg^{-1}, and a 'peak' will be registered. Because the latent heat of fusion for the pure crystalline phase (ΔH_{fc} kJ kg^{-1}) can often be calculated with high precision, the magnitude of this peak can be used to determine the mass crystallinity: $\chi = \Delta H_f / \Delta H_{fc}$.

A typical DSC apparatus consists of two identical chambers, each containing an aluminium sample cup (Fig. 12.30). Each cup is fitted with an independent electrical heater and is instrumented by a precision thermocouple to measure its temperature. One cup contains a small sample of the polymer under investigation – just a few milligrams, so that heat is transferred to or from it very quickly. Using precisely controlled heating systems, both cups are now made to follow the same linear increase in temperature with time. The same power must be delivered to both chambers in order to heat the pan and to compensate for losses. However, only one chamber contains the polymer sample, so only the *difference* in power delivered to the two cups is recorded.

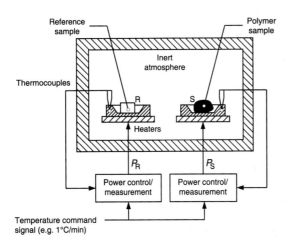

Fig. 12.30

DSC apparatus for determining enthalpy versus temperature. The additional power, $P_S - P_R$, required to keep a weighed polymer sample at the same temperature as that of a reference pan, is measured as the temperature is increased at a programmed rate.

Fig. 12.31

Schematic DSC traces for heating of (a) polyethylene, showing a broad peak corresponding to the endothermic melting process, and (b) an uncured epoxy resin, showing the exothermic curing process.

(a)

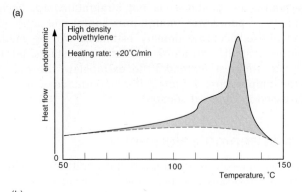

(b)

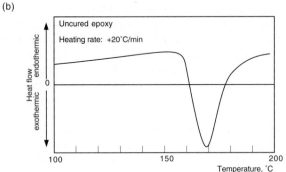

Figure 12.31(a) shows a DSC trace for a crystalline polymer: a high density polyethylene. The broad peak represents the latent heat of fusion absorbed by the melting process, and the crystallinity can be calculated from the area enclosed by it. Melting is very far from being the clearly defined event expected from a simple organic liquid such as one of the lower molecular weight alkanes. Because the crystalline structure is so complex and so imperfect at each level of size, melting takes place over a temperature range. The melting peak appears at a typical T_m of 128°C, but the melting process is well under way at 80°C and there are signs of it even at 60°C. The lower end corresponds to a few small local regions of relative order, the upper end to a few large lamellae of near perfect structure, while the large majority of crystalline structure lies between these extremes. The thickest lamellae are those that began to form early at the highest temperatures; thinner lamellae then form at lower temperatures within the remaining amorphous regions. Because the rate of temperature increase can be chosen at will, DSC is particularly useful for studying the often quite low rates at which crystallization and other processes proceed.

Another feature often visible in DSC traces (although not shown in Fig. 12.31) is the glass transition of the amorphous phase at T_g. This is seen as an abrupt change in specific heat corresponding to the greater number of vibrational degrees of freedom available to chain segments. The effect is the same as that of an increased number of degrees of freedom on the specific heats of a gas (Chapter 7). There is a strong empirical correlation between T_g and T_m:

$$T_g = \tfrac{2}{3} T_m$$

where both are expressed as absolute temperatures in degrees K.

Other thermal properties

The physics underlying thermal properties of solids was introduced in Chapter 7. We are now familiar with the fact that the overall structure of polymers, even crystalline ones, is quite loose compared to that of a truly crystalline solid such as diamond. Each segment along a chain is strongly bonded only to its two immediate neighbours. We can therefore visualize both that a segment will only efficiently transmit vibration to these neighbours (i.e. there is poor 'thermal contact' between non-crosslinked chains), so that overall thermal conductivities in polymers are low; and that its freedom for independent vibration is high, so that the specific heats C_p are high. The ratio of thermal conductivity to ρC_p, the *thermal diffusivity*, characterizes the rate at which a volume of the material will reach thermal equilibrium with its environment. For all polymers, despite their low density ρ, this ratio is very low. This has very important economic consequences, because long times are needed for mouldings and extrusions to cool to a temperature at which they will support their own weight. Pipes can be extruded with walls as thick as 100 mm but they take several hours to cool.

PVT diagrams for thermoplastics

Injection moulding of thermoplastics is an expensive and complicated process. Achieving the astonishing precision of the components illustrated in Fig. 1.2, in a crystalline polymer whose volume contracts by up to 15% during cooling, is a delicate task demanding advanced technology. The material data needed for computer simulations of injection moulding has made the task of accurately determining the specific volume of a polymer over a range of temperatures and pressures increasingly important. The materials for which the greatest moulding precision is required are the engineering polymers, which derive their good thermal and strength properties from high melting points and levels of crystallinity – which conspire to produce high shrinkage. Some of the shrinkage that occurs as the moulding freezes from the surface inwards can be accommodated by the slow, high pressure injection of fresh polymer from the gate. When the gate freezes, however, the mould cavity must still remain 'over-packed' with melt under the correct *dwell pressure*. As cooling continues the pressure will fall until, ideally, it just falls to zero as the mould opens to eject the moulding.

Essential information for analysing this problem is the 'PVT' (pressure-specific volume–temperature) diagram shown schematically in Fig. 12.32. A PVT diagram is presented as a series of specific volume versus temperature lines at constant pressure. Remember that most processes in polymers – and especially crystallization – are very rate-sensitive, so that the heating or cooling rate should also be specified.

The PVT diagram illustrates a number of interesting features. For a solid polymer whose coefficient of linear thermal expansion is α in all three directions the volume will expand by a factor

$$(1 + \alpha \Delta T)^3 = 1 + 3\alpha \Delta T + 3\alpha^2 (\Delta T)^2 + \alpha^3 (\Delta T)^3 \qquad (12.21)$$

Because α is always relatively small, the right-hand side of Eq. (12.22) is approximately $1 + 3\alpha$, so that the gradient of the specific volume versus temperature line is approximately 3α. For all polymers, α is much higher

Fig. 12.32

PVT diagrams for crystalline and
amorphous polymers.

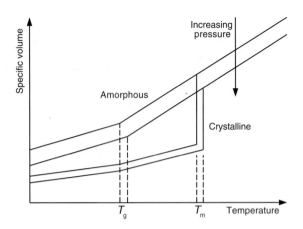

than for metals and, as seen for inorganic glass in Fig. 11.3, there is an
abrupt increase at the glass transition temperature T_g. Though not really as
sharp as shown in Fig. 12.32, this discontinuity can be found by extra-
polating the linear regions at lower and higher temperatures, and this is a
good method for defining T_g. The difference is seen in the fact that the
lowest values of α are usually shown by glassy polymers (e.g. PMMA:
$\alpha = 60$–$80 \times 10^{-6}\,\mathrm{K}^{-1}$) and the highest by elastomers ($300 \times 10^{-6}\,\mathrm{K}^{-1}$).

Next, notice that the specific volume contraction on crystallization is
very large indeed – for example 15%, much greater than for metals – and
that both T_g and T_m increase with pressure. The inability of polymer chains
to pass through each other leads to much poorer packing in polymers than
in inorganic glasses. Polymer melts are particularly compressible: a typical
bulk modulus would be $0.8\,\mathrm{GN\,m}^{-2}$, which is little more than one-half that
of light oils, despite the broad similarity in molecular and intermolecular
bonding. Looking at other liquids we find that water has a bulk modulus of
$2.3\,\mathrm{GN\,m}^{-2}$, and metals of about $25\,\mathrm{GN\,m}^{-2}$.

Finally, the thermal expansion coefficient of the crystalline phase is much
lower than that of the amorphous phase. For HDPE $\alpha = 100$–$130 \times 10^{-6}\,\mathrm{K}^{-1}$
whereas for LDPE $\alpha = 160$–$220 \times 10^{-6}\,\mathrm{K}^{-1}$, partly because it is less crystalline and partly because the free volume around branch points makes the
amorphous phase even more compressible. Another interesting example of
the influence of structure on thermal expansion is that of polyacetal (POM)
which, despite its high crystallinity, has an α exceeding $200 \times 10^{-6}\,\mathrm{K}^{-1}$.
This value places POM firmly and correctly in the class of low-T_g crystal-
line thermoplastics – like HDPE – a fact that its high yield stress, high
tensile modulus and brittleness tend to hide.

12.7 *Crosslinked polymers*

In reviewing the structure and properties of thermoplastics formed by
polymerization of monomers in which a double bond was opened up to
form a linear chain, we have seen the importance of physical crosslinking
between chains in holding the structure together into an elastic solid. Two
important classes of polymers with very different properties arise if these
crosslinks are fixed chemically – the rubbery elastomers, and the thermosets, which are generally glassy.

Table 12.2
Structure and properties of common elastomers

Abbreviation and name	Repeat unit structure	T_g (°C)	Notes and typical applications
NR	Poly(cis-isoprene) Natural rubber	$\{CH_2-C=CH-CH_2\}$ $\quad\quad\;\; CH_3$	−70	Low cost; used where resistance to ozone, oil or high temperatures is not critical. Tyres (especially aircraft tyres). Engine mountings and suspension system components.
BR	Polybutadiene	$\{CH_2-CH=CH-CH_2\}$	−85	Tyres and general-purpose moulded components.
SBR	Poly(styrene-co-butadiene) ('styrene butadiene rubber')	Copolymer of butadiene (see above) with styrene (see Table 12.1)	−85 upwards	Copolymer. The most widely used rubber; mainly for tyre sidewalls and treads.
NBR	Poly(acrylonitrile-co-butadiene) ('nitrile butadiene rubber')	$\{CH_2-CH\}$ $\quad\quad\;\; C$ $\quad\quad\;\; \equiv$ $\quad\quad\;\; N$ co-butadiene (see above)	−85 upwards	Copolymer with high resistance to swelling: used for oil and fuel system seals.
CR	Polychloroprene (Neoprene)	$\{CH_2-C=CH-CH_2\}$ $\quad\quad\;\; Cl$	−50	Belts, hoses, moulded goods, seals, cable covering. Heavy-duty bearings for buildings and bridges.
EPDM	Ethylene propylene diene monomer	Copolymer of ethylene, propylene (see Table 12.1) and complex diene(s)	−60	High resistance to water and ozone. Automobile applications, e.g. radiator hoses.

12.7.1 Elastomers

Natural rubber, whose crosslinking by vulcanization was described in Section 12.2, remains one of the most widely used elastomers. The natural polymer cis-polyisoprene is a soft, elastic fluid, typical of elastomers in having weak intermolecular bonding so that the transition temperature is very low ($-70°$C). Synthetic rubbers are polymerized from *dienes*, monomers that have two double bonds; polymerization opens one of them while the other remains available for the crosslinking process. As shown in Table 12.2, most common synthetic rubbers are based on polybutadiene (BR).

An interesting exception is EPDM – ethylene propylene diene monomer – which is in fact a group of rubbers based on ethylene–propylene copolymers. Ethylene and propylene form copolymers across the entire range of proportions, those at each extreme, i.e. polyethylene and (isotactic) polypropylene, being crystalline, as we have already seen. However, the inherent asymmetry of the propylene units and the lack of regularity in their position prevents crystallization when the proportions of ethylene and propylene are roughly equal, very much as the same factors did for atactic PP, as shown in Fig. 12.6. Introducing a third diene comonomer gives this rubbery polymer its oddly unsystematic name and allows it to be crosslinked into a low-cost rubber with outstanding resistance to water.

12.7.2 Thermosets

Thermosets are no longer widely used for the moulded components that originally brought 'plastics' into use. Excellent electrical and thermal properties, and a remarkable ability to retain strength and dimensional stability over long periods and at high temperatures, have meant that the traditional thermosets phenol formaldehyde and melamine formaldehyde still have a place in construction and engineering applications. These traditional thermosets are still used in very large quantities as bonding materials for chipboard and similar products and as wear-resistant and fire-resistant surfacing materials for laminates. For moulded components in electrical equipment, however, they have largely given way to injection moulded thermoplastics.

Other thermosets have a much brighter future, and we will review these in turn below: they include the common thermosetting polyesters and polyurethanes, and the 'engineering thermosets', which include epoxies and polyimides. Although thermosets are mouldable, they are now most widely used as matrix materials for composites, as we will see in Chapter 13.

Epoxies

Each of the groups listed above is characterized by great diversity. The epoxies, a class of polymers united only in the presence of reactive epoxide groups:

$$\begin{array}{c} \text{O} \\ \diagup \diagdown \\ +\text{C}-\text{C}+ \end{array}$$

along the chain, are no exception. Some epoxies are widely used as consumer adhesives, and these provide a familiar illustration of the way thermosets are formed.

Epoxy adhesives are usually supplied in packs of two separate tubes containing the viscous resin itself, and a thinner liquid 'hardener', which must be mixed with the resin before use. The resin is an example of a *precursor* polymer, in which repeat units have already polymerized to produce a polymer of low molecular weight. The higher molecular weight epoxy precursors used in engineering materials are usually solid and must be melted before mixing.

The hardener initiates a process by which the epoxy chains are cross-linked into a network. Observing the mixture in bulk shows interesting and typical behaviour. At first, there is little observable change; the liquid mixture remains formable and can easily be moulded or squeezed into an adhesive film between the surfaces to be bonded. Then an exothermic reaction process begins, as seen clearly at 160–180°C in the DSC trace of Fig. 12.31(b), where the temperature of a small sample of uncured epoxy was increased at a constant rate. The mixture becomes more viscous as the constituents crosslink into larger molecules, but it remains mouldable until the distinct transition (*gelation*) at which the first giant molecule spans the mixture. The polymer has become a *gel*, a solid that can only be fractured by breaking chemical bonds (though as yet very few of them, so that it remains very weak). Many precursor polymer and crosslinking agent molecules remain unreacted within this network molecule and, as more of them join it, the remainder have ever-increasing difficulty in diffusing to reaction sites. Usually crosslinking can only be brought near completion either by waiting for many times longer than the gelation time, or by heating the mixture (now a rubbery solid) to *postcure* it. The end result is a glassy solid.

Because crosslinking is irreversible, the resulting solid cannot be melted or dissolved without chemically destroying it. This confers valuable properties of heat and chemical resistance and dimensional stability, but by locking chains against extension it greatly reduces ductility. Nevertheless, carefully formulated epoxies can retain enough chain extensibility not only to show a glass transition (usually at about 150°C) but also even to be quite ductile below T_g.

Polyesters

Polyesters are characterized by the presence of the ester group:

$$\begin{array}{c} O \\ \| \\ {+}C{-}C{+} \end{array}$$

in the backbone. *Saturated polyesters*, which have no double bonds in their backbones, form a group of linear thermoplastics of which PET (Table 12.1) is the best known example. The *unsaturated polyesters* can be crosslinked and are used more commonly (and less expensively) than epoxies as matrix materials for composites. They begin as unsaturated oligomers, with molecules less than ten repeat units long. These viscous liquids are usually cured by copolymerization with styrene, a thinner liquid that also makes the dissolved polyester considerably easier to form. Heat or the addition of a catalyst initiates a reaction in which the styrene crosslinks at several points along each polyester chain.

Polyurethanes

Although scrap formed during forming of thermosetting plastics cannot be recycled, as it can for thermoplastics, some thermosets have the great advantage of low processing temperatures and pressures. This accounts for the growing use of *reaction injection moulding* (RIM) to make large panels for road vehicle bodies using relatively light moulds. This process is suitable for thermosets whose precursors are low in viscosity and high in reactivity, because the precursor and crosslinking agent are mixed continuously just before being pumped into a closed mould where they rapidly cure to shape.

The polyurethanes fill these requirements perfectly. This very diverse group is formed by condensation-like reactions between isocyanates (containing the group —NCO) to which crosslink-forming reactive groups are attached, and hydroxyl groups. The large majority of polyurethane is polymerized as flexible or rigid foam, as we will see in the next chapter.

Polyimides

All thermosets are characterized by good stability at high temperatures, but the polyimides are exceptional. The branched nature of the functional group that characterizes them:

$$\text{+N} \begin{array}{c} \overset{O}{\underset{\|}{}} \\ C\text{+} \\ \\ C\text{+} \\ \overset{\|}{O} \end{array}$$

allows a rigid backbone to be constructed from benzene ring structures (which bond directly to the right-hand side) and then lightly crosslinked. The resulting polymers remain strong and ductile up to temperatures of over 300°C. They are, however, very difficult indeed to process, and are most often used in the form of film (well known as Kapton) or machined to shape from solid block.

12.7.3 Thermoplastic elastomers and ionomers

Chemically crosslinked rubbers, like thermosets, cannot be reprocessed because their covalent crosslinks cannot reform once broken. Rubbery thermoplastics, on the other hand, tend to flow during large extensions. Thermoplastic elastomers combine some of the mechanical properties of vulcanized rubbers with the processability of thermoplastics, and have become very widely used for 'soft to the touch' areas on rigid mouldings used in consumer goods.

Most thermoplastic elastomers currently available are *triblock* copolymers of styrene and a rubber, in which a centre block of the rubber is terminated by smaller polystyrene end blocks. Because these two components are immiscible, on cooling from the melt a phase separation takes place, the polystyrene agglomerating into distinct glassy amorphous domains within the rubber matrix. Each blob of polystyrene may entangle the ends of several hundred polydiene chains into a sort of crosslink. On reheating, the polystyrene crosslinks melt and the rubber flows. The most common material of this type is styrene–butadiene–styrene (SBS) rubber, although newer variants use polyisoprene or other rubbers to form the matrix. In the linear form a rubber chain is terminated by polystyrene segments, while in the radial form strong primary crosslinks bond up to twenty SBS chains together near the centre.

In another system, a polyether and a polyester are copolymerized. Short sections of the relatively hard polyester crystallize into lamellae, while the intervening polyether sections remain soft and amorphous. Here it is the crystalline lamellae that adopt the crosslinking role, and the structure becomes thermoplastic at the crystalline melting point. Similar mechanisms account for the rubber-like properties of thermoplastic polyurethanes, which are true block copolymers of chemically distinct sequences.

Another way to provide 'temporary' crosslinks between chains, strengthening and stiffening the polymer without making reprocessing more difficult, is to use ionic bonding (see Section 5.4). This is achieved by making a copolymer, most commonly based on ethylene or styrene, containing up to 10% of repeat units with an anionic group (e.g. the acid carboxyl group —COOH) and with a metal cation (usually sodium, zinc or lithium). As the polymer crystallizes the cation and anionic groups associate to form clusters that, like the glassy polystyrene 'blobs' in thermoplastic elastomers, tie together many chains.

One effect of these physical crosslinks is to disrupt the assembly of crystallites into spherulites large enough to scatter light. As we will see in Section 12.2.1, this can yield a polymer that is strong, tough, resistant to diffusion and yet highly transparent – a valuable and unusual combination. Ethylene-based ionomers have exactly these properties and have found a special application as moulded bottles for perfumes. However, we must remember that physical crosslinks in any of these polymers can never achieve the stability of chemical crosslinks so that thermoplastic elastomers cannot be expected to offer the long-term mechanical properties of rubber.

12.8 Liquid crystal polymers

We have seen that many properties of solid polymers result from the tendency of flexible chain molecules to wriggle into an entangled, amorphous mass while in the melt state. Even 'crystalline' polymers contain amorphous intercrystalline regions that strongly influence their properties. We will now look at a class of polymers based on linear chain segments whose backbones are extremely rigid. Even in the melt state (in which thermal agitation has overcome intermolecular forces enough for them to slip freely against each other) or in solutions concentrated enough for molecules to be forced to approach one another, these rod-like chains form small regions of almost uniform orientation separated by sharp boundaries. Polymers containing such regions are *liquid crystal polymers*. In the solid state, this structure of small, highly anisotropic crystalline domains freezes by the interlocking of adjacent chains so that they cannot slip along each other. As we would expect, each crystalline domain is extremely strong in the chain direction.

The most well known of all LCPs is *Kevlar* (polyparaphenylene terephthalamide, PPTA), which is well known for its outstanding strength: $5\,\text{GN}\,\text{m}^{-2}$ or more. The repeat unit structure of PPTA is

$$\left[C\text{-}\langle\!\!\bigcirc\!\!\rangle\text{-}\overset{\overset{\displaystyle O}{\|}}{C}\text{-}\underset{\underset{\displaystyle H}{|}}{N}\text{-}\langle\!\!\bigcirc\!\!\rangle\text{-}\underset{\underset{\displaystyle H}{|}}{N}\right]$$

The origins of its remarkable properties (and those of other lyotropic liquid crystalline polymers) lie mainly in the density of *aromatic* units, i.e. benzene rings, along the backbone. These units give the backbone great rigidity [Fig. 12.33(a)] and, as Table 12.1 shows clearly, increase both T_g and T_m. Another important factor in PPTA is the strong hydrogen bonding between side groups on adjacent chains, which locks them against slip. As a result this is a *lyotropic* LCP whose melt temperature exceeds its decomposition temperature. It can only be processed after solution in concentrated sulphuric acid, and then only into fibres or thin sheet.

In solid form, however, these materials still suffer from the relative weakness of lateral bonding between chains within crystalline regions and of bonding between these regions. *Thermotropic* LCPs, which can be processed in melt form, overcome this difficulty with the aid of ingenious molecular architecture and of copolymerization [Fig. 12.33(b)]. Increasing the proportion of flexible units along the backbone can reduce T_m because they act as 'impurities' in the crystalline phase. Another strategy, shown in

Fig. 12.33

Chain structures of some liquid crystalline polymers: (a) rigid segments in the main chain of lyotropic LCPs give too high a melting point for melt processing; (b) a higher proportion of flexible units reduces the melting point; (c) alternative structures include side-chain LCPs.

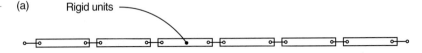

(a)   Rigid units

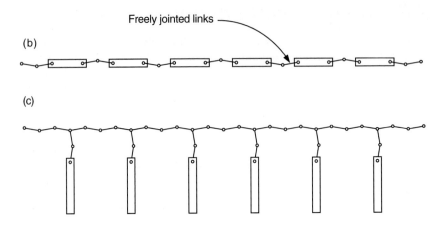

(b)

(c)

Fig. 12.33(c), is to connect groups of rigid chain sections onto a flexible backbone via flexible side branches – 'spacers' – that allow the rods freedom to gather into crystalline regions while tying them securely to each other and to other such regions. The flexible backbones and spacer sections thus act like tie chains in crystalline polymers.

The properties of thermotropic LCPs do not match those of lyotropics, but they remain outstanding; and these materials are actually easier to process than other engineering thermoplastics. Because entanglements are rare, the melts are much less viscous than those of amorphous polymers. Moreover, because the crystalline structure persists in the melt state, there is little contraction during or after solidification, so that moulded dimensions remain very stable. This combination of properties makes a thermotropic LCP the natural choice, for example, optical system support components in a lightweight virtual reality headset.

12.9 Mechanical properties

The mechanical behaviour of crystalline solids was introduced in Chapter 9. Some of the concepts introduced there also apply to crystalline polymers, but in general plastics must be dealt with as a distinct group. We can also now distinguish polymers by class – elastomers, amorphous glasses, crystalline polymers with glassy or rubbery amorphous phases, and thermosets. Most plastics in each class have comparable mechanical properties.

12.9.1 Stress–strain behaviour

The conventional tensile test described in Chapter 9 was developed for metals. Tensile tests on polymers are done in a similar way, although polymer specimens are usually made from more readily available flat sheet rather than rod. This gives them a shape for which they are known as 'dog-bone' specimens.

Elastic behaviour

The polymers that show the most straightforward tensile behaviour – almost linearly elastic – are glassy thermoplastics (e.g. PS and PMMA), in which rubbery behaviour is frozen, and thermosets in which it is prevented by dense crosslinking. These have Poisson ratios of around 0.35–0.4, not dissimilar to those of metals, and shear moduli of the same order as the bulk modulus: thousands of $MPa\,m^{-2}$. At the other end of the scale are the crosslinked elastomers, with shear moduli G of just a few $MN\,m^{-2}$. The elastic behaviour of rubber was explained in Section 12.3.3. Real elastomers broadly follow the expected behaviour; however, if the extension ratio λ exceeds 2, secondary bonds may start to induce crystallization. The microstructure, as revealed by X-ray diffraction, then resembles that of a crystalline polymer. This considerably increases the stiffness, and also contributes significantly to the great toughness and tearing resistance of rubbers.

In any polymer, crystallinity increases tensile modulus. Lamellae are stiff, strong structures. An extended carbon–carbon backbone is very stiff: the modulus in the chain direction depends largely on the size of the side groups that occupy cross-sectional area but for polyethylene it is about $320\,GN\,m^{-2}$, which exceeds that of steel. However, the stiffness in the other two directions remains typical of secondary bond stretching, and is therefore similar to the bulk modulus.

The stiffness of a crystalline polymer also depends partly on the nature of the amorphous phase. If its T_g is low, as in PE, the modulus usually increases sharply with the degree of crystallinity. This is partly because the crystallites effectively 'crosslink' the rubbery phase, but also because the rubber itself is too constrained by crystallites to deform freely, as discussed in Section 12.3.3. Because the strain applied to a crystalline polymer is concentrated within these rubbery regions, their elastic non-linearity shows itself at much lower strains than in a pure elastomer. This is one reason why the elastic behaviour of PE, PP, etc. is much less linear than that of glassy and thermosetting polymers. Another reason is that as the load increases in a tensile test, the amorphous phase begins to creep viscoelastically at an ever-increasing rate. If T_g is relatively high, as it is in the engineering polymers such as nylon (PA) and PEEK, the amorphous phase is glassy, the stiffness is higher and the stress–strain curve is more linear.

Yield

Loaded plastic components are not usually designed to criteria based on their yield stress, because brittle fracture and surface degradation mechanisms pose a much greater threat. The plastic flow that follows yield is also of less interest in polymers than in metals, because plastics are seldom formed to shape in the solid state. It is, anyway, much more difficult to define 'yield' and the onset of plastic flow in polymers, for two main reasons. Firstly, because all polymers show viscoelastic response at service temperatures, all are to some extent non-linearly elastic when tested in a conventional stress–strain test. This makes the definition of a yield point more difficult than for a metal, in which a small departure from linearity is a good indication that permanent flow has taken place. Secondly, even when ductile polymers such as PE are extended to several times their original length, this extension is almost completely recovered if the material is heated, i.e. deformation has not been irreversible. All of these factors lead us to see plastic flow in polymers as fundamentally different from that in metals.

The point chosen to define yield of a polymer is simply the maximum on an engineering stress–strain (i.e. load–displacement) curve. As we saw in Section 9.7.1, this corresponds to the first Considère tangent to the true stress versus strain curve which better represents underlying properties. Common glassy polymers such as PMMA and PS fracture in tension before reaching this yield point, but careful compression tests can deform them well beyond it. Under combined stresses, most plastics conform reasonably well to a von Mises yield criterion (see Section 9.5.2) if account is taken of the fact that their yield stress is usually higher in compression than in tension.

Plastic flow

Remember that amorphous polymer glasses are essentially 'frozen rubbers'. In terms of the standard linear solid model (Fig. 12.22), chain extension represented by spring E_1 has been almost completely locked by the high 'viscosity' η represented by the parallel dashpot, although even under low stresses the chain will still extend given sufficient time. The most straightforward way of understanding plastic flow in glassy polymers is to regard it

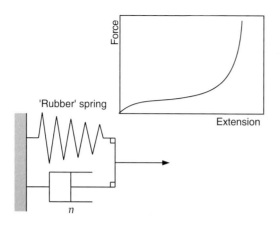

'Rubber' spring

n

Fig. 12.34

The Haward–Thackray model represents polymers extended to large strains. The spring strain-hardens like a rubber while the dashpot adds a constant viscous stress.

as rubbery behaviour in which the stress needed to extend chains is far exceeded by the stress needed to overcome internal viscosity. The *affine* nature of rubber deformation, involving little slip between neighbouring chain segments, is very different from plastic flow in metals. If this model is correct, when plastic flow is continued up to large extensions we will see the hardening behaviour typical of rubber, caused by the straightening of extended chains, and we should replace the Hookeian spring of the standard linear solid by a 'rubber' spring with the nominal stress–strain relationship derived in Section 12.12 (Fig. 12.34). As usual for plastic behaviour we can assume constant volume, so that the cross-sectional area varies inversely with λ. From Eq. (12.16) the true stress versus strain relationship for an ideal rubber becomes

$$\sigma_t = G\left[\lambda - \frac{1}{\lambda^2}\right] \tag{12.22}$$

Now we add the stress σ_y needed to extend the dashpot of the model, which is simply the stress needed to make the solid flow like a viscous Newtonian liquid and is independent of λ. The dependence of true stress on strain expected from such a material is therefore

$$\sigma_t = \sigma_y + G\left[\lambda - \frac{1}{\lambda^2}\right] \tag{12.23}$$

When plotted as a curve of true stress versus $\lambda - 1/\lambda^2$, this should give a straight line through the origin from which the constant parameters σ_y and G can be determined (Fig. 12.35).

Glassy polymers do indeed generally show this stress versus strain relationship, suggesting that plastic flow involves the extension of chain segments between physical crosslinks formed by entanglements, rather than reptation or slippage of chains. We can also understand why plastic flow reverses spontaneously on heating. The viscous forces that locked chains in their extended conformations diminish, allowing weak entropic forces from extended chains to pull the solid back into its former shape. Plastic flow in metals (Section 9.5) is very different – hardening is caused by interactions between dislocations as slip proceeds, and there is no structure to 'remember' where atomic planes slipped from.

Some crystalline polymers too are well represented by Eq. (12.23). This suggests that crystallites in polymers act as crosslinks that stabilize and

Fig. 12.35

Stress–strain relationship for a
Haward–Thackray material.

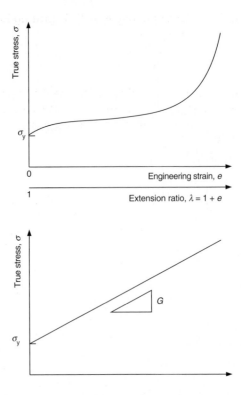

supplement those caused by entanglements in the amorphous phase.
However, during plastic deformation lamellae do begin to deform more
like those in metals, shearing in the two directions that the strong chains
allow them to. They may separate into blocks that shear past each other
and rotate into the extension direction. These movements are very small
and local but they are enough to erase any spherulitic structure. The fact
that heating the polymer to just below its melting point allows almost
complete reconstruction and recovery shows, yet again, how persistent
is the network of entangled chains, and how much it dominates mechanical properties.

Cold drawing

In a tensile test, elastomers extend homogeneously – there is no yield point,
and the entire central region of a dog-bone specimen extends (and its cross-
section contracts) uniformly. Rubbery crystalline polymers such as LDPE
show a yield point at which they form a neck, and this neck then extends
to fracture. HDPE and many other crystalline polymers – and even some
glassy polymers, notably polycarbonate – show the remarkable phenomenon of cold drawing, which was introduced in Section 9.7.1. We saw that
this can occur if a second tangent can be drawn to the true stress versus
extension ratio curve, i.e.

$$\frac{d\sigma_T}{de} = \frac{\sigma_T}{1+e} = \frac{\sigma_T}{\lambda} \qquad \text{[(from Eq.(9.52)]}$$

This curve must have a particular shape in order to accommodate the second tangent. Using the Haward–Thackray representation we can explore

the conditions for this important phenomenon. Substituting Eq. (12.23) into Eq. (9.52) we arrive at the expression

$$\sigma_Y = \frac{3}{\lambda^3} G \tag{12.24}$$

Figure (12.31) shows that a first Considère condition for a Haward–Thackray material occurs at the yield point $\lambda = 1$, so that the material will always neck. However, a second tangent can only exist for $\lambda > 1$ if the yield stress is greater than $3G$.

12.9.2 Orientation

A sample of, for example, HDPE that has been cold drawn in a tension test will have been stretched by a ratio of 4–8. It is now *orientated*: its network structure and the crystalline regions within it have been rotated towards one preferred direction. This gives the material several striking properties. If coloured, it will probably have acquired a bleached (*stress whitened*) appearance with a visibly fibrous texture in the axial direction. Having strain hardened, it may be reloaded to several times the original yield stress before breaking in a brittle 'fibrous' manner, like string. The polymer will also be noticeably stiffer. Orientation is of great importance to polymer properties, and it may result from stretching in the melt and/or the solid phase and may arise accidentally or by design.

In a polymer melt, each chain continuously reptates through its neighbours while remaining randomly coiled. Shearing or stretching the melt temporarily extends chains in the tensile direction. They will normally recover their random conformations within a short relaxation time, but if the melt is meanwhile cooled the viscosity increases, the relaxation time lengthens and the melt becomes a stretched rubber. The external forces holding this rubber in its stretched state are exerted by, for example, an injection mould, or even – if cooling was not uniform – just by the surrounding colder material. If cooling continues to below T_g, this orientation will be 'frozen in' semi-permanently. Because injection moulding is carried out rapidly, every moulded component is orientated to some degree. This is why, when reheated to above T_g, most moulded components distort – they retain a 'memory' of the unstrained, unformed melt.

Extension of a rubber increases the number of fully extended chain sequences lying in that direction and its stiffness increases. A high degree of orientation increases the tensile modulus of an amorphous polymer in the same way. Because covalent bonds are also much stronger than secondary bonds between chains, strength is also increased. Naturally, these gains are paid for by reduced stiffness and strength in the perpendicular direction. The strengthening effect is therefore more efficiently exploited by stretching film and even plate in *two* directions; only the through-thickness direction suffers, and this is seldom important. The high strength and toughness of PMMA aircraft windows is achieved by casting plate 50 mm or more thick, reheating it, stretching it successively in two directions until it is just a few millimetres thick, and quickly cooling it to freeze the biaxially orientated structure.

In crystalline polymers, orientation has other aspects. If the orientation arises from stretching in the melt state, the crystallites that form will usually be different from those that form from an isotropic melt. Because these polymer chains, unlike those of amorphous polymers, naturally pack efficiently, stretching accelerates the crystallization process (an effect known as *strain crystallization*). However, the chains within these early crystallites are themselves oriented in the stretch direction, locking in an additional level of orientation and further increasing strength. This reduces the tendency of crystallites to grow into spherulites that are inherently *isotropic* structures, identical in every direction.

Polymer fibres

The crystallites developed by stretching a crystallizing polymer in its melt state are large, efficiently packed, stiff and strong. The entire artificial fibres industry is constructed on this fact. *Melt spinning* (Fig. 12.36) begins with the extrusion of polymer melt through a die perforated by many tiny holes. These fibres are usually drawn off at a speed much higher than that at which the melt emerges, producing extremely high stretching of the melt, which is frozen in by rapid cooling through the melt temperature. Stretching may then be continued after freezing, aligning chains still further by cold drawing. This process normally produces fibres with a tensile modulus exceeding $100\,\text{GN}\,\text{m}^{-2}$, two orders of magnitude greater than the same polymer might achieve in its spherulitic, moulded form. To gain most from this treatment the polymer needs long, straight molecules having few chain end defects and a strong tendency to pack. However, entanglements still exist, so that oriented regions in drawn fibres remain separated by densely entangled, oriented amorphous regions (Fig. 12.37). As in solid polymers, crystalline and amorphous regions are bonded together by tie molecules.

The polymers most often used to form melt-spun fibres are crystalline PA (nylon 6 and 6,6), PP, polyester (PET) and polyacrylonitrile (PAN). Liquid crystal polymers such as Kevlar, which are more rarely used as textiles and more often as reinforcement, are produced by *gel spinning*. The gel is formed by evaporation of solvent from a solution until each chain has entangled with its neighbours just enough for minimal cohesion. It can be spun into fibres of matchless strength and stiffness. This method is now also used for other polymers, e.g. linear PE and PP.

Orientation of bulk crystalline polymers

Orientation of crystalline thermoplastics produces such outstanding properties that methods have been sought to achieve these in bulk components. Tube and other profiles that need to be strong and rigid in the axial direction can be oriented by cold drawing below the melting temperature, after extrusion and cooling. This is especially suitable for PE, whose amorphous phase remains quite rubbery at room temperature. Another new material, hot compacted polypropylene, is a composite of highly oriented polypropylene fibres within a conventional, crystalline, isotropic matrix. The composite is formed by heat and pressure for a time sufficient to bond the fibres into the matrix without destroying them. This material retains many features of a fibre-reinforced composite but it is much easier to recycle.

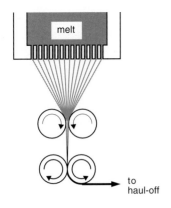

Fig. 12.36

Melt spinning of polymer fibres.

melt

to haul-off

Fig. 12.37

Fibres drawn from crystalline polymers contain highly oriented regions separated by disordered amorphous zones.

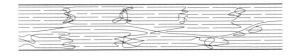

12.9.3 Craze formation and crack growth

The ability of polymers to extend to extremely high strains, and to harden so much while doing so, accounts for another deformation mode that is unique to them – the formation of *crazes*. The term 'crazing' originally referred to the fine network of cracks sometimes seen on the glazed surface of pottery. Glassy polymers sometimes acquire a similar surface appearance under stress, particularly in certain environments (e.g. alcohol), which are not in themselves aggressive. On closer inspection the surface haze reveals many crack-like defects oriented perpendicular to the maximum tensile stress, but very high magnification reveals the 'crack' faces to be bonded together by fibre-like *fibrils*. This structure develops when a crack front, moving inwards from the surface, develops local corrugations that cut off islands behind it (Fig. 12.38). These islands then extend into fibrils by cold drawing further material through the 'crack' faces. Each fibril is extremely fine and, having been highly drawn, extremely strong. As a result, the craze as a whole may be as strong as the parent polymer. However the fibril drawing process, by which a craze thickens as it extends, cannot continue forever. At some point each fibril breaks at its mid-point (where it has been stressed for longest), or is uprooted, and the craze becomes an extending crack carrying a craze at its tip.

Surface crazing is not seen in crystalline polymers, but crack tip crazes are so commonly seen that fracture in thermoplastics can be understood purely in terms of their survival or failure, i.e. it is the crack tip craze that resists crack propagation. Although crosslinked thermosets are often brittle, and brittle fracture in thermosets must involve the scission of backbone bonds, brittle fracture in thermoplastics involves very little chain scission. If chains remain intact, how then do the craze fibrils break?

In fact crystalline polymers are generally very tough under the extension rates applied in a tensile test. A sharp crack simply becomes blunt, and the material around it yields and ultimately tears. There are essentially two situations under which sharp cracks are seen to propagate in a truly brittle manner. One is when loading is applied over very long periods – this leads to *slow crack growth* under static stress or *fatigue crack growth* under cyclic stress. The second is when the load is applied extremely rapidly, leading to *impact fracture*. A typical feature of thermoplastics is that a crack may jump from a sharp notch under impact and then continue to run at high speed (over $100 \, \mathrm{m \, s^{-1}}$).

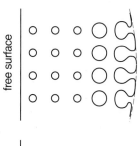

free surface

Fig. 12.38

Craze formation.

One mechanism for craze fibril failure unites these two situations at opposite ends of the timescale: *disentanglement*. Given sufficient time under stress at operating temperatures, the constant thermal motions of segments of chain will pull it even from the relatively secure anchorage of a crystallite (and will easily pull it through the amorphous phase). If, on the other hand, the stress is applied very quickly, drawing of craze fibrils generates heat that cannot diffuse away. The resulting increase in temperature increases thermal motion, melts crystallites and may even cause thermal decomposition.

12.10 Chemical properties

12.10.1 Solution, diffusion, swelling and plasticization

Broadly speaking, polymers resist agents that attack metals (aqueous acids and bases) but may succumb to organic solvents that metals resist. If solvent molecules are attracted to sites on the chain at least as strongly as they are attracted to each other, the polymer may dissolve completely, so that each chain floats freely amongst them. If the attraction is weaker, more subtle and sometimes more dangerous possibilities arise.

Many applications of polymers – from polyethylene bottles to paints and silicon rubber gaskets – exploit their ability to seal fluids in or out. It is therefore perhaps surprising that polymers are inherently less effective barriers to gas and liquids than are metals or ceramics. The poorly packed structure of the amorphous phase, and the mobility of chain segments within it, allow small, compatible molecules to diffuse through the polymer by dodging from one free volume site to another. In many applications low resistance to diffusion is unimportant; in others (e.g. bottles for carbonated drinks) it is a severe disadvantage, while for some (such as food packaging films) it is, when controlled, a feature to be exploited.

As we would expect, crystalline polymers are generally more effective barriers to diffusion than amorphous ones, the crystallites themselves being effectively impenetrable structures. The HDPE grades used for car fuel tanks are chosen for their high crystallinity but will still absorb, during their lifetime, several percent of their weight in fuel – and some fuel will diffuse right through the wall. PE pipe for potable (drinkable) water must not be buried in soil contaminated by any potentially harmful chemical. Even where PE water pipe is laid alongside PE gas pipe, the chemicals added to make fuel gas leaks detectable by smell may diffuse through both pipe walls to be tasted in the water!

Small molecules that have diffused into free volume sites usually have very little effect on mechanical properties but larger molecules can, without actually dissolving the polymer, have very significant effects indeed. The first is simply that the specific volume increases: the polymer *swells*, a well-known effect of oil on certain elastomers. We will see in the next section that this can cause problems when adjacent regions of a rigid polymer absorb different concentrations. A second effect arises when molecules of swelling agent create more free volume than they occupy. Chain segments between entanglements can then manoeuvre more easily at a given temperature, reducing the glass transition temperature – an effect known as *plasticization*.

The most familiar example of deliberate plasticization is that of PVC. In its unplasticized form PVC is a glassy polymer with a T_g of about 80°C. The addition of a *plasticizer* such as dioctyl phthalate (chosen for the ease with which it penetrates the polymer and the low rate at which it diffuses out again) reduces T_g by tens of degrees Celsius and effectively transforms the glass into a rubber whose most familiar application is for electrical cable covering. PVC is perfectly suitable for this modification because the crystallinity, about 10%, is high enough to provide the 'crosslinks' needed to partially stabilize entanglement, but low enough to reduce the modulus by a factor of nearly 1000. A less convenient example of this effect is PA, nylon, which (like other hydrogen bonded chains) absorbs and is plasticized by water; the effect on its stiffness is too large to be ignored and too small to be useful.

12.10.2 Degradation, oxidation and chemical ageing

Polymers are synthesized chemicals, in a sense that metals and ceramics are not. This fact underlies many of the special ways in which polymer surfaces can deteriorate in service under the action of radiation, water and atmospheric pollutants, i.e. relatively mild agents that the surfaces of our own bodies can usually tolerate. Where the polymer surface is decorative or functional, this may mean the end of service life. Worse, as we shall see in the next section, deterioration of the surface may trigger fracture of a plastic component under load, or even under internal stresses 'frozen in' during processing.

The seconds or minutes during which a polymer is processed into its final form are the most destructive it will ever experience. Temperatures high enough to melt a thermoplastic, and the high shear stresses inflicted by extrusion and injection moulding, will always break some covalent bonds. Scission of main chain bonds is particularly destructive because relatively few such events can have a disproportionate effect on molecular weight. To appreciate this, remember the importance of long chains as tie molecules in crystalline polymers, and consider the destruction that could be wrought by cutting chains at random points in a mass of polymer: the longest chains will suffer most heavily. In some polymers (e.g. nylon, PA) points on the main chain are susceptible at high temperatures to hydrolysis, so that thermal degradation is greatly accelerated by traces of water absorbed by the polymer before processing. These polymers are often also hygroscopic and must be carefully stored and dried before processing.

Chemical decomposition at processing temperatures is a particularly severe problem for unplasticized PVC. The reaction can evolve corrosive HCl and must be blocked by stabilizer additives. The first sign of this, as Eq. (12.5) suggests, may be a reduction in melt viscosity, but mechanical properties, particularly toughness and impact strength, are also affected. This is a matter of concern to processors, who naturally want to exploit the reformability of thermoplastics by reprocessing scrap. In practice the proportion of scrap must be kept low to avoid cumulative degradation, which can be detected by measuring the melt viscosity.

Some polymers, for example polyacetal, tend to depolymerize at high temperatures by breakage of monomer groups from chain ends. A much more stable material is provided by copolymerizing with a little ethylene oxide, which forms much more stable chain ends. Even if a chain end is lost, decomposition only continues until the next ethylene oxide group is reached. This 'acetal copolymer' is the basis for an important engineering plastic widely used for gears and other small machine parts.

Oxidation reactions, some of them extremely complicated, can occur in polymers either during processing or in service. Many of these have the effect of forming crosslinks which, as we have seen, tend to have an embrittling effect. Remember that the ductility of polymers originates from the combination of soft amorphous region, hard crystalline regions and tie chains to hold the structure together. Remember too that the end result of cross-linking carbon backbone chains will be diamond which, although certainly strong, is extremely brittle! This crosslinking is a particular problem in rubbers, which rely on their tendency to form crosslinks for their mechanical properties. At normal service temperatures oxidation can be catalysed by light – particularly by sunlight – and can be reduced by the dual strategy of incorporating both antioxidants and light-blocking pigments such as carbon black. Processes such as these in which degradation progresses over long periods of time in normal environments are known as chemical ageing.

12.10.3 Weathering and environmental stress cracking

The action of stress on a chemically damaged polymer surface may very easily lead to failure. A particular problem for polymers is that the stress need not even be applied by loads that it would be expected to meet in service. Just as we saw for inorganic glass in Section 11.4.2, stresses can be 'locked in' during processing.

Under the high hydrostatic tensile stresses near a crack tip, the already open structure of a polymer opens up much further, allowing greater access to diffusion of fluids. This results in the particularly insidious problem of slow *environmental stress cracking* (ESC). ESC is the equivalent of stress corrosion cracking in metals, although the mechanisms are quite different. Stress cracking agents, which are usually organic liquids, inflict little or no damage on the bulk material but may, even in extremely small quantities, cause a dramatic reduction in ductility and crack resistance under stress. They probably do so by plasticizing the crack tip, allowing faster disentanglement. An example is the effect of aqueous detergent solution on LDPE.

12.11 *Other polymer properties and applications*

Up to this point, we have concentrated mainly on plastics as materials for solid components, occasionally touching on their use as films and fibres. In this short section we will look at some of the diverse applications in which polymers serve a *non*-structural function – either as a result of some special physical activity or by modifying the behaviour of other materials.

12.11.1 Optical properties

For a material to be optically transparent, it must be uniform on a scale greater than that of a wavelength of light ($\sim 0.5\,\mu$m). This is about a thousand times larger than a single repeat unit, but smaller than a typical crystallite. Because crystallites usually have a different refractive index to that of the amorphous phase, refraction at interfaces scatters light and crystalline polymers are generally white or translucent in their pure state. Melting of the crystalline phase in polymers such as LDPE reduces them to a single amorphous phase and is clearly visible as a sudden increase in transparency, giving rise to a visible 'freeze line' in film blowing (Fig. 12.20) for example.

The amorphous glasses retain this transparency in the solid state and find many applications on their transparency alone. Polymethyl methacrylate (PMMA), with excellent transparency and moderate strength, is exploited for its optical properties in both cast sheet form and in injection mouldings such as auto rear-light filters. Although tougher than glass, PMMA is more vulnerable to surface scratching and crazing. Impact-resistant grades of both PMMA and PS have been developed by incorporating a compatible rubber with the same refractive index, but this further compromises hardness (the principles of rubber toughening will be discussed in Chapter 13). Polycarbonate (PC) has displaced PMMA in many applications for which toughness and impact resistance are important, e.g. aircraft windows and canopies. Unfortunately it is slightly less transparent and tends to become yellow and brittle in ultraviolet (UV) light unless a UV block is incorporated in the plastic. PVC has some crystallinity and is much less transparent but is widely used, on the basis of its lower cost, for applications such as packaging and bottles. Compact discs themselves are injection moulded from specially formulated PC grades with high purity and low melt viscosity. A secondary finishing process then covers the surface with a protective layer of PVC. Very few thermosetting polyesters and epoxies are significantly transparent. A thermoset with an outstanding combination of optical and mechanical properties is polydiallyldiglycolcarbonate (CR39), used for moulded lenses.

We noted in Section 12.10.1 that crystallites resist diffusion of gases. This leads to an interesting conflict in the packaging of carbonated drinks, which must keep the carbon dioxide gas in while maintaining high transparency. The solution has been found using PET, a thermoplastic that crystallizes at an unusually low rate. Bottles are blow moulded not from melt-extruded tube, as described in Section 12.4.4, but from an injection moulded preform that has been cooled quickly enough to freeze the amorphous state. This preform is reheated to a temperature between T_g and the crystalline melting point T_m, and blow moulded to its final form. The orientation that accompanies stretching develops enough crystallinity to resist gas diffusion, but processing below T_m prevents the organization of spherulitic crystals large enough to scatter light.

12.11.2 Electrical properties

Because the electrons that are shared covalently within chains are highly localized, most polymers are very good insulators; indeed, many early applications of both thermoplastics and thermosets exploited this property. Offset against this is their generally low resistance to diffusion, which can allow the absorption of water and other polar fluids and a corresponding increase in conductivity. Plasticized PVC, polyethylene and polytetrafluoroethylene are all flexible insulators with low enough permeability to win them high-volume applications in shielding electrical wire and cables.

Where insulation is *not* wanted, e.g. in antistatic foams and packaging for electronic components, useful levels of conductivity can be produced by filling with graphite particles. Beyond this typical use of additives to manipulate properties lies the much more exciting frontier beyond which the electrical and physical properties are built into the repeat unit and molecular architecture. Methods of achieving intrinsically high electrical conductivity and controllable semiconductivity will be outlined in Chapter 14. As well as offering some potential for lightweight accumulators for vehicle and aerospace applications, these materials have been cited as potential elements in 'molecular computers', with electronic components a few nanometres in size. Also useful in the electronics field, as well as for transducers, are piezoelectric polymers, which develop an electric field when stressed, and pyroelectric polymers which do so when heated. Polymer electrolytes have also been developed.

Electroactive polymers can be macroscopically aligned, to give a polarizing effect or a colour change, by an external electrical or magnetic field. The chain architecture of Fig. 12.33(c) can be used to attach active polymer segments to a polymer backbone through flexible spacers so that they are only free to align when these spacers are in the melt state. They can therefore be written on when melted by a finely focussed laser beam while in a magnetic field, for use as optical information storage media. *Photoactive* polymers can be constructed in a similar way by fixing light-activated charge transfer groups to the main chain; these materials are used for xerography, in copiers and in laser printers.

12.11.3 Adhesives and coatings

Adhesives are always applied in liquid form, because part of their function is to fill the surfaces to be adhered (the substrates). This demands good wetting qualities and low viscosity. For some applications, such as pressure-sensitive adhesive tapes and films, amorphous rubbery polymers are used because they remain permanently in a highly cohesive liquid state. Alternatively, *hot melt* adhesives are essentially thermoplastics that are applied at high temperatures and pressures in a sort of moulding operation and resolidify on cooling; the EVA (ethylene–vinyl acetate co-polymer) adhesives used for bookbinding and packaging are an example. Other adhesives are solutions of polymers that set by evaporation or diffusion of a volatile solvent, and then perhaps crosslink.

Most uncrosslinked systems tend, as we have seen, to be vulnerable to temperature and solvent action, and the development of strong and stable adhesives for use at high temperatures and in adverse environments has focused on thermosets, in particular the epoxies. With good surface preparation these materials can attain remarkable bond strengths, and have won use in demanding structural applications in the automotive and aerospace industries. For high-strength structural joints, the toughness of the cured adhesive layer is as important as the strength of the adhesive-to-substrate interface; strong epoxy adhesives therefore make increasing use of rubber-toughening (see Chapter 13).

Another well publicized class of polymer adhesives is the cyanoacrylates, which are based on a liquid monomer stabilized by small amounts of a weak acid. On encountering a slightly basic surface the inhibitor is neutralized, and polymerization and cure occur with prodigious speed to give a strong bond. However, there is little crosslinking and the temperature and solvent resistance of the joint are not exceptional.

12.11.4 Gels

A 'gel' is a distinctive state of matter, like a solid, a liquid or a plasma. Several examples and applications of gels have already been mentioned in this chapter. The clearest example was provided by the crosslinking process in which a thermoset forms from relatively short polymer chain units; the thermoset temporarily passes through a gel state as a single, giant, crosslinked molecular network spans the 'melt' of many small uncrosslinked chains. If, at this stage, we were to wash the unreacted polymer out and replace it with a compatible, non-reactive and non-volatile liquid, we would obtain a stable gel. These materials have many fascinating properties and many potential applications. There is much to explore in this field.

Gels are usually 'soft' materials of low mechanical strength. These properties are exploited in their use for implants in plastic surgery and for components (e.g. ear-piece seals for headsets) that must conform to variable or changing shapes. In these applications gels must usually be contained in soft shells of conventional polymers or elastomers. Other applications of gels exploit the nature of their structure at a molecular level – as described in Section 12.2.3, they are widely used in chromatographic methods of molecular 'sorting' by size or mobility.

Problems

12.1 Which of the following characteristics are shared by *all* polymers?

(a) a backbone consisting of carbon atoms

(b) large molecules

(c) regions of crystalline order

(d) molecules containing repeated sequences of the same structure

(e) a glass transition above room temperature

Provide one example of a polymer that does *not* have each of the characteristics which you exclude.

12.2 An experimental grade C of a linear polyethylene is produced by mixing a monodisperse polymer A, having molecular weight $100\,kg\,mol^{-1}$, with an equal mass of monodisperse polymer B having molecular weight $400\,kg\,mol^{-1}$. What are the number-average and weight-average molecular weights of grade C?

12.3 Describe how the molecular weight distribution of a sample of polystyrene might be measured using GPC.

12.4 A freely jointed chain representing polyethylene at its melting point has a Kuhn length of $1.05\,nm$. For freely jointed chains of (i) 10 and (ii) 100 links calculate

(a) the fully extended length

(b) the most probable end-to-end distance

(c) the probability of finding the chain end between the fully extended length and a radius that is 10% greater. Use the Gaussian approximation of Eq. (12.5) to calculate this probability: what should be the exact result?

12.5 A sample of crosslinked butadiene rubber at $27°C$ has a shear modulus of $1\,MN\,m^{-2}$ and a density of $930\,kg\,m^{-3}$. Each butadiene chain section between crosslinks can be represented as a freely jointed chain, each link of which has a molecular mass of $0.35\,kg\,mol^{-1}$. Calculate:

(a) the number of crosslinked chain sections per cubic metre

(b) the number of freely jointed links in each of these chain sections

(c) the number of repeat units in each chain section (see Table 12.2 for the repeat unit structure of polybutadiene)

12.6   What would be, at low shear rates, the ratio of the melt viscosity of polymer grade B to that of grade A in Problem 12.2? Calculate the ratio of viscosity C to A assuming that viscosity depends on (a) number-average and (b) weight-average molecular weight. Which of the three grades would be more suitable for injection moulding and which grade more suitable for injection moulding? Explain your answers.

12.7   A sample of polymer melt at 150°C is tested using a cone-and-plate viscometer and shows a relaxation time of 1 s. If its viscosity decreases by 1% per K, estimate:

   (a) the relaxation time of the same polymer sample at 160°C

   (b) the relaxation time at 150°C of a sample of the same polymer having a molecular weight 10% lower

   The entanglement density of the polymer, and thus its shear modulus, can be assumed to remain constant.

12.8   Samples P and Q of polyethylene are scanned using DSC and these show melting peaks of 138 and 199 kJ kg^{-1} respectively. Estimate, using the data below, the mass fraction of crystallinity and the density of each polyethylene grade. If the two samples are known to have been moulded in a similar way, in what way would you expect their molecular structures to differ?

   Density of crystalline phase 1000 kg m^{-3}, density of amorphous phase 856 kg m^{-3}, latent heat of fusion of crystalline phase 285 kJ kg^{-1}.

12.9   Polystyrene can be synthesized in both atactic and isotactic forms. Explain which form of the two you would expect to crystallize more readily, and estimate its melting temperature.

12.10 A thermoplastic polymer can be represented as a Haward–Thackray solid. The yield stress is 35 MN m^{-2} and when the extension ratio has reached $\lambda = 2$ the engineering stress has fallen to 24.5 MN m^{-2}.

   (a) What is the true stress when $\lambda = 2$?

   (b) What is the hardening modulus G?

   (c) What will be the true stress at $\lambda = 4$?

12.11 Explain in detail why crystalline polymers never fracture with 'cleavage planes' like those seen in crystalline materials.

Self-assessment questions

1   In a thermoplastic polymer, adjacent molecules are bonded by

   (a) primary bonds

   (b) secondary bonds

   (c) covalent bonds

   (d) van der Waals or hydrogen bonds

2   In a thermoplastic polymer, side groups are bonded to the backbone by

   (a) primary bonds

   (b) secondary bonds

   (c) covalent bonds

   (d) van der Waals or hydrogen bonds

3   A typical molecule of low density polyethylene has:

   (a) one branch per ethylene repeat unit

   (b) 10–100 branches in total

   (c) one branch   (d) no branches

4   The glass transition corresponds to the onset of

   (a) scission of backbone bonds

   (b) free rotation at backbone bonds

   (c) melting of crystallites

   (d) breakage of crosslinks

   (e) slippage between adjacent chains

SELF-ASSESSMENT QUESTIONS **407**

5 The glass transition involves a significant change in

(a) density of entanglements

(b) specific heat

(c) elastic modulus

(d) density of crosslinking

(e) tacticity

(f) latent heat

6 Which of the following polymers could show variation in tacticity:

(a) PP   (b) PE   (c) PTFE   (d) PVC

7 The large extensions possible in elastomers are primarily associated with

(a) stretching of primary bonds

(b) stretching of secondary bonds

(c) bending of primary bonds

(d) rotation at primary bonds

8 A melt-spun polymer fibre has a higher tensile modulus than that of the bulk polymer because

(a) it can contain only very small flaws

(b) polymer chains within it are extended along the fibre direction

(c) its glass transition temperature is higher

(d) it contains reinforcing additives

9 Thermosetting polymers

(a) consist of a network of polymer chains

(b) are crosslinked by crystalline regions

(c) cannot be remoulded

(d) are generally more rigid than thermoplastic polymers

(e) have very temperature-sensitive mechanical properties

10 The glass transition is almost undetectable in thermoplastics with very low crystallinity

(a) true   (b) false

11 The melt flow index of a polymer decreases with increasing viscosity

(a) true   (b) false

12 Polymers show rubber elasticity above T_g only if they are crosslinked

(a) true   (b) false

13 The effect of adding a plasticizer to a thermoplastic is to

(a) increase the melting point

(b) reduce the glass transition temperature

(c) reduce the melt viscosity

(d) reduce the rate of chemical crosslinking

(e) increase the molecular weight

14 Polymer chains tend to form random coils because

(a) their backbone bonds can rotate relatively easily

(b) their backbones are very long

(c) addition polymerization produces a wide range of molecular weights

(d) repeat units are subject to disturbance by Brownian motion

15 Most glassy polymers are hard and brittle

(a) true   (b) false

16 In a polymer that can crystallize the degree of crystallinity is decreased by

(a) the presence of side branches

(b) absorption of water

(c) fast cooling from the melt

(d) slow cooling from the melt

17 The diagram below represents part of the chain structure of a polymer that consists of two repeat units, A and B. This polymer is:

(a) a branched polymer

(b) a block copolymer

(c) a graft copolymer

(d) a linear copolymer

(e) a random copolymer

```
            B
            B
            B
            B
  – A A A A A A A A A A A A A –
            B
            B
            B
            B
```

18 Isotactic polypropylene is

(a) a polyolefin

(b) a semicrystalline thermoplastic

(c) a glassy thermoplastic

(d) a thermoplastic elastomer

19 There is a linear relationship between the density of a polyethylene and

(a) the degree of polymerization

(b) the molecular weight

(c) the melt viscosity

(d) the degree of crystallization

(e) the density of crosslinking

20 The effect that increasing the temperature of a thermoplastic polymer has on its tensile modulus is similar to that of

(a) increasing its crystallinity

(b) decreasing the time under load

(c) increasing the time under load

(d) decreasing the density of crosslinks

21 Crosslinks between chains in an elastomer have the effect of

(a) resisting permanent creep under load

(b) reducing the elastic modulus

(c) preventing softening on heating

(d) increasing crystallinity

(e) preventing oxidation

22 The viscosity of a polymer melt falls as

(a) molecular weight increases

(b) flexibility of the chain backbone increases

(c) pressure increases

(d) shear strain rate increases

(e) temperature increases

23 Liquid crystal polymers are characterized by

(a) rigid chain backbones

(b) high electrical conductivity

(c) high tensile modulus

(d) high melt viscosity

24 A 'tan delta' peak measured during DMTA of a thermoplastic at increasing temperature indicates a temperature at which

(a) there is a sharp maximum in the tensile modulus

(b) the specific heat shows a sharp maximum

(c) the tensile modulus is decreasing most rapidly

(d) the crystalline phase melts

25 On cooling, thermoplastics weld more slowly than metals because

(a) crystallization occurs more slowly

(b) diffusion occurs more slowly

(c) melting occurs more slowly

(d) oxidation occurs more quickly

26 The PVT diagram for a thermoplastic can be used to estimate

(a) the melting point

(b) the latent heat of fusion

(c) the bulk modulus

(d) the linear coefficient of thermal expansion

(e) the degree of crystallinity

27 The presence of aromatic units in a polymer chain results in

(a) a long Kuhn length

(b) high chemical reactivity

(c) an ability to form crosslinks

(d) high glass transition and melting temperatures

28 All amorphous polymers above T_g show rubbery behaviour because

(a) traces of peroxide cause crosslinks to form

(b) entanglements between chains create 'effective' crosslinks

(c) long sections of polymer chain can extend freely

(d) crystallites survive here and there to crosslink adjacent chains

29 Which of the following properties depend only on the chemical structure of a polymer and are independent of the manner in which the sample has been prepared:

(a) chain configuration

(b) chain conformation

(c) crystallinity

(d) crosslink density

30 *Pseudoplasticity* is shown by almost all thermoplastics as

(a) decreasing stress with time under constant strain

(b) reduction of melt shear stress with increasing shear strain rate

(c) reduction of melt viscosity with increasing shear strain rate

(d) apparent yield of the solid caused by viscoelasticity

31 The ratio M_w/M_n of weight-average to number-average molecular weight

(a) is always <1

(b) is always >1

(c) is known as the polydispersity

(d) is known as the degree of polymerization

(e) can be measured by DSC

32 The following material properties can be estimated from a DSC scan:

(a) molecular weight distribution

(b) latent heat of fusion

(c) melting temperature (d) tan delta
(e) glass transition temperature

33 The abrupt transition from viscous liquid to weak solid seen in an epoxy adhesive while it sets is

(a) due to crystallization

(b) due to the onset of crosslinking

(c) due to the appearance of a single molecule spanning the sample

(d) referred to as the glass transition

(e) referred to as gelation

34 Plastic deformation in ductile polymers is fundamentally different from that in metals and ceramics because:

(a) it is usually reversible on heating

(b) it necessarily involves the breakage of chemical bonds

(c) it takes place by reptation and slippage of chains

(d) it takes place within a stretched network with nearly fixed links

Each of the sentences in Questions 35–38 consists of an assertion followed by a reason. Answer as follows:

(a) If both assertion and reason are true statements and the reason is a correct explanation of the assertion.

(b) If both assertion and reason are true statements but the reason is *not* a correct explanation of the assertion.

(c) If the assertion is true but the reason contains a false statement.

(d) If the assertion is false but the reason contains a true statement.

(e) If both the assertion and the reason are false statements.

35 Commercially available HDPE is monodisperse *because* it consists only of a single kind of repeat unit.

36 Weld lines can be a problem in injection moulded components *because* polymer chains diffuse very slowly.

37 Low molecular weight polymers crystallize more rapidly *because* chain ends are more mobile.

38 PMMA used for aircraft windows is biaxially stretched *because* this encourages crystallization while preventing the formation of spherulites large enough to scatter light.

Answers

1 (b), (d)	**2** (a), (c)	**3** (b)	**4** (b)
5 (b), (c)	**6** (a), (d)	**7** (d)	**8** (b)
9 (a), (c), (d)	**10** (b)	**11** (a)	**12** (b)
13 (b), (c)	**14** (a), (b), (d)	**15** (a)	**16** (a), (c)
17 (c)	**18** (a), (b)	**19** (d)	**20** (c)
21 (a), (c)	**22** (d), (e)	**23** (a), (c)	**24** (c)
25 (b)	**26** (a), (c), (d)	**27** (a), (d)	**28** (b), (c)
29 (a), (d)	**30** (c)	**31** (b), (c)	**32** (b), (c), (e)
33 (c), (e)	**34** (a), (d)	**35** (d)	**36** (a)
37 (a)	**38** (c)		

Multiphase materials and composites | 13

13.1 Introduction

In surveying each of the three main classes of structural materials – metals, ceramics and polymers – we met some consisting of more than one phase. Usually the secondary phase emerged spontaneously, under thermodynamic forces, from a 'pure' material or a mixture of them. Precipitation hardening of aluminium alloys, for example, shows that some mechanical properties may then be much better than those of either phase in isolation. Such *synergistic* effects are not accounted for by a simple law of mixtures, which applies only when a property such as density in one phase does not affect that in the other phase. Strength and toughness, unlike density, gain so much in many cases from cooperation between phases that where these properties are critical *multiphase* materials are almost always used. Their inherent superiority is well illustrated by natural structural materials, e.g. cartilage and wood, which have adapted to their function by evolution. These natural materials develop their structure by self-organization but we can usually only achieve comparable results using *composite* materials, which are 'put together' – if only by mixing – from separate constituents.

We will begin by looking at some of the simplest material mixtures: filled and extended polymers. Here a particulate second phase is just mixed in mechanically, and the aim is sometimes no more ambitious than a reduction in cost. However, we quickly find that the distinction between such mixtures and true composites that have gained from some form of 'cooperation' between their constituents, is not a simple one. Filled polymers can show real gains in toughness, which lead us to consider the spectacular and industrially important toughening effect of *rubber* particles in plastics. We then turn to a detailed review of conventional composites – plastics, metals and ceramics reinforced by fine fibres. It is here that we find materials such as glass and carbon fibre-reinforced plastics, whose strength exceeds the bulk strength of either constituent by a factor of twenty or more.

Finally we will explore another group of multiphase materials for which there is a natural prototype. In the performance of our own bodies, as of many mechanical systems such as aircraft, the ratio of strength to weight is of paramount importance. Natural structural members maximize strength to weight ratio by developing as a solid shell and a foam or *cellular* core. Foam plastics are very familiar from their use in furnishing and packaging but important cellular solids can also be made from ceramics and metals. The latter in particular are relatively new materials with varied and exciting applications.

13.2 Particulate fillers in plastics

Low-cost particulate fillers have been used extensively since the earliest days of plastics. Originally the primary objective was to reduce cost – as suggested by the alternative term *extender* (now used mainly for paints). However, the very first commercial plastic, Bakelite, showed that fillers could provide benefits in other properties. Phenol formaldehyde is a glassy solid with very low toughness until reinforced by a suitable fibrous material such as wood flour or textile fibre. Indeed, when reinforced by woven cotton cloth, phenol formaldehyde becomes a tough, easily machined, high-performance engineering material that is still very widely used. The plastics most commonly filled nowadays are the low-cost, flexible commodity plastics, such as polyethylene and especially polypropylene (PP). These materials have low values of tensile modulus (Young's modulus) E – about $1 \, \text{GN m}^{-2}$ – and, as we saw in Chapter 12, they also tend to suffer from slow viscoelastic creep under load. The objectives of adding filler are to increase and stabilize modulus at minimum cost, so that rigid components can be made without having to mould thick sections. If the strength of the material can at least be preserved, this is an additional benefit.

A wide range of particulate fillers is used, the most common being finely ground minerals, i.e. ceramic powders, such as calcium carbonate, talc (hydrated magnesium silicate) and kaolin. The particles may be roughly spherical, or needle- or plate-like, with sizes ranging from less than $1 \, \mu\text{m}$ to more than $50 \, \mu\text{m}$. These fillers are commonly added in proportions of up to 30 vol% (i.e. a filler *volume fraction*, V_f, of 0.3) or even higher. Now the total mass M of a volume V of mixture is the sum of the mass of filler and the mass of *matrix* (the primary 'mother material' that contains the secondary phase), so that

$$M = \rho_f V_f V + \rho_m V_m V$$

Dividing throughout by V, we find that the mixture density is

$$\rho = \rho_f V_f + \rho_m V_m = \rho_f V_f + \rho_m (1 - V_f) \tag{13.1}$$

where subscripts m and f represent matrix and filler parameters respectively. Equation (13.1) is a typical example of the *law of mixtures*. It rests on the assumption that the density of each phase is not significantly affected by the presence or absence of the other phase – an assumption that is often not valid for strength or toughness.

Mineral fillers have bulk densities of about $3–4 \, \text{kg m}^{-3}$, about twice those of polymers, so that filled polymers are significantly denser. However, fillers have Young's modulus values of around $100 \, \text{GN m}^{-2}$, two orders of magnitude higher than those of the base polymer, and this increase in density is far outweighed by the gain in stiffness. If the simple law of mixtures also applied to modulus, we might expect the filled polymer to have a modulus of about $30 \, \text{GN m}^{-2}$. In fact the gain in modulus, although real and worthwhile, is a factor of 2–3 rather than 30. This is because Young's modulus represents the strain per unit stress, and the contributions from strains in different phases cannot simply be added as the law of mixtures assumes. The rigid filler particles in fact concentrate stress at the interface with their soft embedding matrix, and the increased strain in the surrounding matrix outweighs the reduced strain in the region occupied by the particles themselves.

We might expect that if the particles are to stiffen the mixture, they must remain very strongly adhered to the matrix. However, the main reason why the polymer remains locked against the surface of the filler particles is not good adhesion, but the large difference in thermal expansion coefficient between polymer and filler. The filler is mixed into the polymer melt and cools with it, but it contracts much less, so that each particle is left permanently under pressure. Now a uniaxial stress σ, applied externally to the mixture in a tension test for example, generates an average hydrostatic stress, i.e. a negative pressure, of $\sigma/3$ (see Section 9.2.4). Thus before σ can debond the polymer from the rigid filler particles, $\sigma/3$ must cancel out this *thermal mismatch* pressure, and once it has done so the stiffening effect of the rigid filler is lost (Fig. 13.1). Unlike most adhesive bonds, this one is

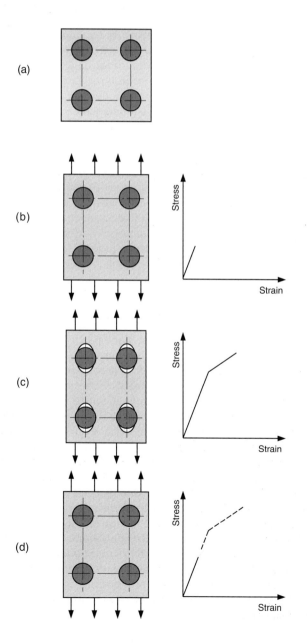

Fig. 13.1

Hard filler particles (represented here as cylinders in a two-dimensional array) debond reversibly from the polymer matrix as the stress is increased, reducing the stiffness of the mixture.

reversible, because as soon as the external stress is reduced again the polymer contracts to clamp back around the filler particles. Nevertheless, the effective transformation under stress of filler particles into voids means that most filled polymers have less strength than the unfilled polymer even if, as we shall see in the next section, they can have greater toughness.

In Section 13.3 we will consider fibre-reinforced composites whose second phase has a reinforcing effect, increasing both modulus and strength. An outstanding example of a particulate filler with a reinforcing action is carbon black, as used in very large proportions in rubber. Generated by incomplete combustion of hydrocarbons, carbon black consists of near-spherical particles 10 nm to $1\,\mu$m in size, each of which has an onion-like structure of graphite sheets. These particles tend to cluster into strongly bonded and irregularly shaped clumps (*aggregates*) into which many polymers bond very strongly. At processing temperatures, the chain molecules of a polymer are able to burrow in by their typical wriggling motion, *reptation*. There they anchor to the carbon surfaces by a strong combination of physical and chemical bonding. The benefits are many. Used very heavily (50% or more by weight) in tyres, carbon black improves strength by a factor of up to 10, with more modest increases in both stiffness and wear resistance. Ultraviolet resistance, which is generally low for rubbers, is increased spectacularly; this is also exploited for many other less sensitive polymers that must survive long periods in direct sunlight.

Finally, we must mention fillers that are used *functionally*, contributing their own physical properties to those of the mixture. Aluminium is widely used for its appearance, but may also induce useful levels of electrical conductivity if the volume fraction is sufficient for conductive paths to form. Aluminium nitride, which is as effective an electrical insulator as it is a thermal conductor, finds use in the plastics moulded around integrated circuits, where it must efficiently dissipate heat both from the circuit itself and from soldering operations.

13.3 Rubber-toughened polymers

The affinity that polymers show for carbon black particles is unusual – they show little or no such compatibility with inorganic fillers. We might expect to find far greater compatibility between different polymers, so that we could expand the range of available properties by mixing them to form *alloys*. This is not the case, and true alloys are rare between amorphous polymers and virtually unknown between crystalline ones. Because every chain segment in a true alloy finds itself in a well mixed environment, there is a single glass transition temperature T_g that lies between the T_g values of the constituents. Mutual solubility between different chain structures, however, is usually low, and even when they are thoroughly blended the mixture may spontaneously separate into distinct phases again. Multiphase blends have the physical properties of a simple mixture (e.g. two T_g values) but the mechanical properties may well be better than either constituent. In polymer blends, unlike filled polymers, the properties of the two phases are generally rather similar, but synergistic effects still appear and blending has become the main route for developing new polymeric materials without developing new base polymers.

Every possible combination of glassy, rubbery, thermoplastic and thermosetting polymers – and many other more exotic combinations, as well as many *ternary* (three component) blends – has now been explored. Here, to introduce the principles, we will focus on the most common and economically important polymer blends – rubber-toughened (or 'elastomer-toughened') thermoplastics.

Polystyrene, the glassy polymer used for low-cost transparent components such as CD cases, is the most brittle of the commonly used plastics. It was therefore the first for which a toughened form was produced in bulk. To produce 'high impact' polystyrene (HIPS) an uncrosslinked diene rubber, usually butadiene (see Table 12.2), is dissolved in styrene monomer and this solution is polymerized. As polymerization proceeds the rubber begins to separate out as small spherical particles, 1 μm or less in diameter. Vigorous stirring is used to keep these particles isolated within a thermoplastic polystyrene matrix; otherwise the rubber surrounds regions of polystyrene and crosslinks permanently. Usually some of the butadiene forms side branches chemically bonded to the main polystyrene chain (i.e. *grafts* – see Section 12.2.3). These grafts strengthen adhesion at the interface between the two polymers by providing primary chemical bonds across it. The rubber inclusions are large enough to scatter light and the copolymer loses its transparency but it gains much toughness. The stress–strain curves of PS and HIPS are dramatically different, the rubber-toughened plastic yielding at about one-third of the fracture stress for PS, but showing a ductility of perhaps 40% instead of fracture at a strain of less than 1%. HIPS is usually produced in sheet form and is used to make disposable containers for foodstuffs.

Acrylonitrile–butadiene–styrene (ABS), which takes the same concept to a higher level, has become the most common of this class of materials. It is very widely used for components such as desktop computer housings, where toughness is as important as rigidity and low cost. In ABS the matrix is a single-phase random copolymer, styrene acrylonitrile or SAN, more systematically known as poly(styrene-*co*-acrylonitrile). The elastomer component is polybutadiene, and as in HIPS this separates out as a second phase of micron-sized particles. SAN readily grafts onto polybutadiene, providing the compatibility needed to bond these particles strongly into the matrix. Amongst other polymers that benefit from rubber toughening is unplasticized PVC, which illustrates a higher level of complexity in that the toughening phase may itself be ABS or another rubber-toughened thermoplastic compatible with the PVC matrix. Virtually every other engineering plastic, notably nylon, is produced in rubber-toughened grades.

13.3.1 Toughening mechanisms

To understand the effect of the rubber in toughened plastics, it is useful to remind ourselves that the thermoplastic elastomers described in Section 12.7.3 share a rather similar molecular structure. A number of thermoplastic elastomers, too, are phase-separated copolymers of styrene with elastomers, but in these materials the glassy polystyrene is a minority phase that 'crosslinks' a soft rubber matrix. Toughened plastics seldom contain more than 25–30 vol% of rubber, whose presence reduces the stiffness of the matrix by about the same proportion. Even this small stiffness reduction is an unwanted side-effect for what are used mainly as structural materials. It is *not* the reduction in stiffness that toughens the polymer. How, then, does rubber toughening work?

Fig. 13.2

Rubber toughening: cavitation of rubber particles encourages energy-absorbing craze formation and shear band formation near a crack tip.

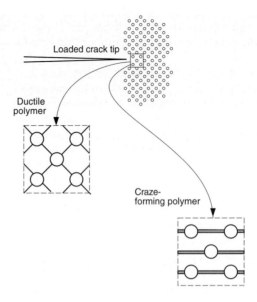

A number of different mechanisms contribute to toughness, and their relative contributions differ between polymers. We noted in Chapter 9 that materials which are ductile in thin sheets often become brittle in thick sections, because high hydrostatic tensile stress can prevent yielding near a critically loaded crack tip. The basis of rubber toughening is that the rubber particles *cavitate* under high hydrostatic stress, becoming rubber-lined pores. Between these pores the hydrostatic stress is relieved and the polymer recovers its ability to yield and blunt a crack. In brittle polymers, e.g. polystyrene, crazes (Section 12.9.3) grow between particles. In either case (Fig. 13.2) the material around the crack tip becomes softer, and stresses in the rigid bulk polymer cannot be concentrated sufficiently to extend the crack before it becomes too blunt. This toughening mechanism, by which small particles are 'sacrificed' to distribute damage over a larger area, also accounts for the toughening effect of some hard particulate fillers, for example calcium carbonate in polypropylene and polyethylene. This is why their toughening effect can be increased by treating the surface of the filler with a 'non-stick' *release agent*, e.g. calcium stearate.

Because both rubber and mineral particles can toughen a polymer by becoming 'effective voids', the reader may wonder why the same effect cannot be achieved more simply by putting in voids in the first place. The toughening effect of voids does indeed prove to be comparable, but well bonded rubber particles, although soft in shear, have a high enough bulk modulus to preserve most of the stiffness of the matrix polymer. Voids certainly do not; in addition, they are difficult to introduce initially and cannot survive processing in the way that rubber particles can.

A secondary mechanism by which rubber and some particulate fillers toughen polymers is *bridging*, which is illustrated in Fig. 13.3. Although rubber particles cavitate internally ahead of a crack, the rubber 'lining' of the cavity remains well bonded to the crack faces and this adds a small *cohesive* element to the fracture toughness (see Section 9.6.4). For fillers that are strong in themselves and whose particles take the form of plates (e.g. mica, aluminium) this toughening mechanism, which is quite distinct

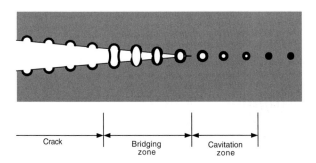

Fig. 13.3
The bridging mechanism in rubber-toughened polymers.

Crack — Bridging zone — Cavitation zone

from that discussed above, can be very significant. We now turn to a class of materials that gain virtually *all* of their spectacular toughness from the same effect.

13.4 Fibre-reinforced composites

The breaking strength of strong solids is, as was seen in Chapter 9, sensitive to the presence of cracks. If the material takes the form of a bundle of fibres then the only cracks that matter must necessarily be short – across the width of one fibre. A solid made up from such fibres may therefore be expected to have greater fracture strength than would a solid

	Specific gravity	Young's modulus ($GM\,m^{-2}$)	Tensile strength ($GM\,m^{-2}$)
Glasses			
E-Glass	2.5	73	3.5
S-Glass	2.5	86	4.6
D-Glass	2.2	52	2.4
SiO_2	2.2	74	5.9
Polycrystalline ceramics and multiphase			
Alumina	3.2	173	2.1
Carbon	1.8	544	2.6
Boron	2.6	414	2.8
Silicon carbide	4.1	511	2.1
Whiskers			
Alumina	3.9	1550	20.8
Boron carbide	2.5	448	6.9
Graphite	2.2	704	20.7
Silicon nitride	3.2	379	7.0
Metals (cold worked)			
Tungsten	19.3	345	2.9
Molybdenum	10.2	335	2.2
Austenitic stainless steel	7.9	200	2.4
Eutectoid steel	7.8	240	4.0
Organic			
Kevlar (aromatic nylon)	1.45	130	2.7

Table 13.1
Properties of various fibres used in composites

Fig. 13.4

Examples of composites:
(a) continuous unidirectional
fibres, (b) aligned discontinuous
fibres, (c) randomly orientated
discontinuous fibres.

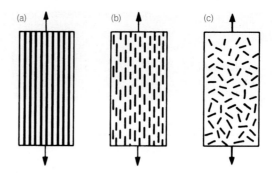

lump of the strong material of the same total volume. The fibres must
somehow, of course, be joined together; hemp fibres, for example, can be
twisted together to form long strands that are then coiled together to form a
rope. Strong brittle fibres can be produced (Table 13.1) but their strength is
very dependent upon surface damage and they cannot be twisted into a
rope form.

However, if a suitable embedding material is used it can serve to hold the
fibres together and at the same time protect the fibre surfaces from damage
caused by abrasion. The embedding matrix has two other functions –
it separates the fibres so that cracks cannot run from one fibre to another,
and it binds to the fibre surface so that the load can be transferred to the
fibres. Both metals and polymers have been found to be satisfactory matrix
materials, and currently there is much interest in the development of
ceramic matrix composites.

Although we have seen that hard particles can have a reinforcing effect,
the greatest improvement in mechanical performance is obtained by the
use of fibres and hence most composites are fibre-reinforced. The fibres
may be *continuous* or *discontinuous*, i.e. in short lengths, and may be
aligned or randomly orientated (Fig. 13.4). Analysis of the mechanical
properties is simplest for a unidirectionally reinforced composite with
continuous fibres [Fig. 13.4(a)] and we will therefore study this first and
then modify the results for an aligned discontinuous fibre composite.

13.4.1 Mechanical properties of continuous fibre composites

Strength

Let us consider the situation where the continuous fibres are aligned in the
direction of the applied force. Some of the force F_c applied to the composite
is taken by the fibres (F_f) and the remainder by the matrix (F_m)

$$F_c = F_f + F_m$$

Writing this equation in terms of stress σ and cross-sectional area A
gives

$$\sigma_c A = \sigma_f A_f + \sigma_m A_m \qquad (13.2a)$$

where the subscripts c, f and m refer to composite, fibres and matrix
respectively. It is usually more convenient to use the volume fraction of
fibres V_f (V_f – volume of fibres/volume of composite $= A_f/A$ and the volume

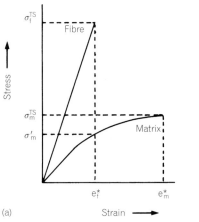

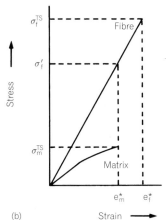

Fig. 13.5

Stress–strain curves for two different fibre–matrix combinations: (a) ductile matrix, i.e. $e_f^* < e_m^*$; (b) brittle matrix, i.e. $e_f^* > e_m^*$.

fraction of matrix V_m ($V_m = 1 - V_f$ = volume of matrix/volume of composite $= A_m/A$, hence we rearrange Eq. (13.2a) into the following:

$$\sigma_c = \sigma_f V_f + \sigma_m V_m \qquad (13.2b)$$

It should be emphasized that σ_c is the tensile stress on the composite at any specified strain and is not the stress at failure of the composite, i.e. it is not the fracture stress or the tensile strength, σ_c^{TS}, of the composite. To determine expressions for the tensile strength we have to consider the relative mechanical properties of the fibres and the matrix. In particular it is the relative values for the strains to failure that are paramount. Figure 13.5 shows two sets of stress–strain curves for fibres and matrix; it can be

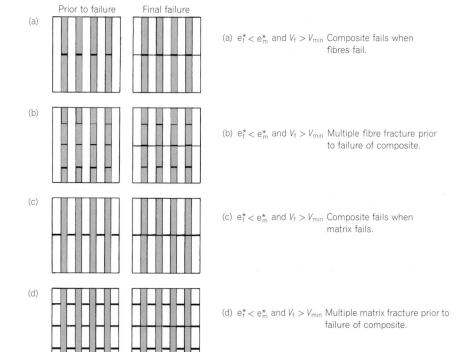

Fig. 13.6

Variation in failure mechanism of composite with relative values of strains to failure of the fibres (e_f^*) and the matrix (e_m^*) and with volume fraction of the fibres (V_f).

seen that in (a) the tensile failure strain of the fibre, e_f^*, is less than that of the matrix e_m^*, whereas in (b) the fibre has the higher tensile failure strain.

Let us first study a fibre–matrix combination where $e_f^* < e_m^*$, as shown in Fig. 13.4(a). The failure sequence will depend on the volume fraction of the fibres. Provided that the volume fraction of fibres is above a certain minimum value V_{min} the composite will fail when the fibres fail, thus the tensile strength σ_c^{TS} of the composite is given by

$$\sigma_c^{TS} = \sigma_f^{TS}V_f + \sigma_m'V_m \tag{13.3}$$

where σ_c^{TS} is the tensile strength of the fibres and σ_m' is the tensile stress borne by the matrix at the failure strain of the composite. This type of failure is illustrated schematically in Fig. 13.6(a) and a specimen that has failed in this manner is shown in Fig. 13.7(a).

For small volume fractions of fibres below V_{min} there is sufficient matrix material present to carry the load as the fibres break. The composite, therefore, only fails when the stress on the matrix reaches the tensile strength of the matrix σ_c^{TS} and multiple fracture of the fibres can occur

Fig. 13.7

Fracture of two polymer matrix composites: (a) high volume fraction of fibres with a brittle matrix; (b) as (a) but with a ductile matrix.

(a)

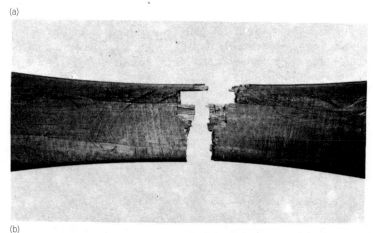

(b)

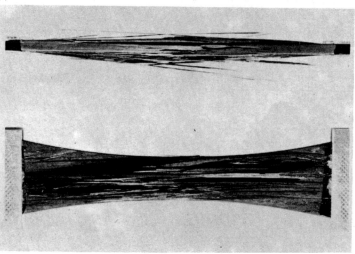

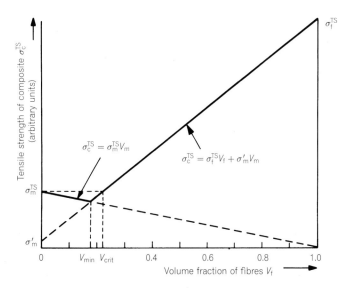

Fig. 13.8

Variation in the tensile strength of a composite for which $e_f^* < e_m^*$, as a function of the volume fraction of continuous unidirectional fibres, according to Eqs (13.3) and (13.4).

[Fig. 13.6(b)]. In these circumstances the fibres do not contribute to the strength and the tensile strength of the composite is

$$\sigma_c^{TS} = \sigma_m^{TS} V_m \tag{13.4}$$

Equations (13.8) and (13.9) are plotted in Fig. 13.8, and from the graph we can see that V_{min} may be evaluated by equating these two equations, namely

$$\sigma_{c(min)}^{TS} = \sigma_f^{TS} V_{min} + \sigma_m' V_m = \sigma_m^{TS} V_m$$

Now $V_m = 1 - V_f$, so that for conditions in which the fibre breaking strain is the critical factor

$$V_{min} = \frac{\sigma_m^{TS} - \sigma_m'}{\sigma_f^{TS} + \sigma_m^{TS} - \sigma_m'} \tag{13.5}$$

In fact the fibres do not lead to strengthening until a *critical volume* V_{crit}, which is slightly larger than V_{min}, has been exceeded (Fig. 13.8). To determine V_{crit} we put $\sigma_c^{TS} = \sigma_m^{TS}$ and $V_m = 1 - V_{crit}$ into Eq. (13.3):

$$\sigma_m^{TS} = \sigma_c^{TS} = \sigma_f^{TS} V_{crit} + \sigma_m' (1 - V_{crit})$$

$$\therefore \quad V_{crit} = \frac{\sigma_m^{TS} - \sigma_m'}{\sigma_f^{TS} - \sigma_m'} \tag{13.6}$$

As σ_m' is generally much smaller than both σ_m^{TS} and σ_f^{TS} this equation approximates to

$$V_{crit} \approx \sigma_m^{TS} / \sigma_f^{TS} \tag{13.7}$$

Now we will consider the situation where $e_f^* > e_m^*$. Again two different fracture sequences may be envisaged depending on V_f. The matrix, being the most brittle component, fractures before the fibres and the load is therefore transferred to the fibres. At low volume fractions of fibres, they are unable to support the load and hence the composite fails [Fig. 13.6(c)]; the tensile strength of the composite is given by

$$\sigma_c^{TS} = \sigma_f' V_f + \sigma_m^{TS} V_m \tag{13.8}$$

Fig. 13.9

Variation in the tensile strength of a composite for which $e_f^* > e_m^*$, as a function of the volume fraction of continuous unidirectional fibres, according to Eqs (13.8) and 13.9).

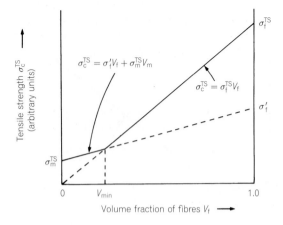

where σ_f' is the stress on the fibres at the failure strain of the matrix [see Fig. 13.5(b)]. In contrast, if there is a high volume fraction of fibres the large number of fibres is capable of sustaining the transferred load when the matrix first fractures. Multiple fracture of the matrix occurs [Figs 13.6(d) and 13.7(b)] and failure of the composite does not take place until the stress on the fibres reaches the fibre fracture stress. Thus at large V_f the tensile strength of the composite is given by

$$\sigma_c^{TS} = \sigma_f^{TS} V_f \tag{13.9}$$

The fracture strength as a function of V_f for a brittle matrix is illustrated in Fig. 13.9. The change in fracture mechanism is at V_{min}, which is obtained by equating the two previous equations

$$\sigma_c^{TS} = \sigma_f' V_{min} + \sigma_m^{TS}(1 - V_{min}) = \sigma_f^{TS} V_{min}$$

Therefore where the matrix breaking strain is the critical factor

$$V_{min} = \frac{\sigma_m^{TS}}{\sigma_f^{TS} - \sigma_f' + \sigma_m^{TS}}$$

So far in the analysis we have assumed that the fracture strength of all the fibres is the same, namely σ_f^{TS}. However, brittle fibres exhibit a considerable scatter in strength due to a variation in flaw size in the same manner as previously discussed in Section 11.5 for monolithic ceramics. The simplest way to attempt to account for the variation in strength is to employ the mean fibre strength σ_{mean} in the equations instead of σ_f^{TS}. A further improvement would be to take into account the distribution in fibre strength by incorporating the Weibull modulus m (see Section 11.5) into the analysis. There are a number of models available that include m but we will only consider the *cumulative weakening model*. In this model it is assumed that the weak fibres will fail and lead to lengths, l_i, of fibres that will not support the full load; l_i is known as the *ineffective length* because fibres of that length are ineffective as far as load bearing is concerned. According to this model the appropriate fibre strength σ_{cum}^* to insert in the equations for the tensile strength of the composite is

$$\sigma_{cum}^* = \sigma_{mean} \left(\frac{L}{l_i me}\right)^{1/m} \frac{1}{\Gamma(1 + 1/m)} \tag{13.10}$$

where Γ is a tabulated gamma function, e is the base of natural logarithms, and L is the gauge length of the composite sample. It is difficult to quantify l_i accurately, but inserting typical values for l_i and the other parameters into Eq. (13.10) shows that σ^*_{cum} exceeds σ_{mean}.

Young's modulus

Initially, on applying a force the composite deforms elastically and the strains experienced by the composite (e_c) and fibres (e_f) and the matrix (e_m) are the same:

$$e_c = e_f = e_m \tag{13.11}$$

Remembering that Young's modulus is equal to stress divided by strain and that the stress on the composite is given by Eq. (13.2b) we have for the Young's modulus of the composite

$$E_c = \frac{\sigma_c}{e_c} = \frac{\sigma_f V_f + \sigma_m V_m}{e_c}$$

But all the strains are equal [Eq. (13.11)], therefore:

$$E_c = \frac{\sigma_f}{e_f} V_f + \frac{\sigma_m}{e_m} V_m = E_f V_f + E_m V_m \tag{13.12}$$

where E_f and E_m are the Young's modulus of the fibres and matrix respectively. This equation is another example of the law of mixtures in that the contribution of the fibres and matrix is additive and in proportion to the volume occupied by each.

13.4.2 Mechanical properties of discontinuous fibre composites

Many composites contain discontinuous fibres that have either been chopped to length or were initially produced in short lengths, e.g. whiskers. It has been found experimentally that the strength of an aligned discontinuous fibre composite is always less than that of an aligned continuous fibre composite with the same volume fraction of fibres. However, the longer the discontinuous fibres the stronger the composite, and for very long fibres the strength of a continuous unidirectional composite is approached

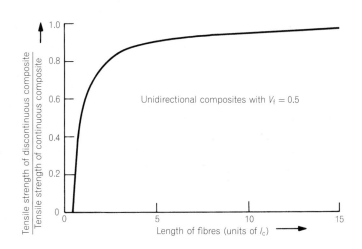

Unidirectional composites with $V_f = 0.5$

Tensile strength of discontinuous composite / Tensile strength of continuous composite

Length of fibres (units of l_c)

Fig. 13.10

The ratio of the tensile strengths of composites containing discontinuous and continuous fibres, as a function of fibre length.

Fig. 13.11

Shear stress in the matrix (a) and the tensile stress in the fibre as a function of fibre length l (b, c and d).

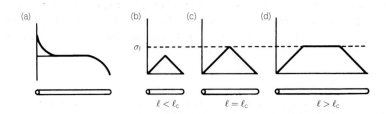

(Fig. 13.10). To understand this behaviour of discontinuous fibre composites we must study the stress situation around and in the fibres.

Equation (13.11) is not obeyed for a discontinuous fibre composite, and a consequence of the difference in strain in the fibres and in the matrix is that a shear stress is induced in the matrix at the matrix–fibre interface parallel to the fibres. This shear stress has a maximum at the fibre ends and a minimum at the middle of the fibre [Fig. 13.11(a)]. In turn, the shear stress sets up a tensile stress in the fibres that is always zero at the ends and increases towards the mid-length of the fibres [Fig. 13.11(b)]. The maximum tensile stress attainable in the fibres is, of course, the tensile strength σ_f^{TS} and, as shown in Figs 13.11(c) and (d), this is only achieved when the fibres are equal or greater in length than l_c, which is known as the *critical length*. It can be shown that l_c is related to the shear strength of the matrix τ_m, the tensile strength of the fibres and the diameter d of the fibres by

$$\frac{l_c}{d} = \frac{\sigma_f^{TS}}{2\tau_m} \tag{13.13}$$

The ratio l_c/d is called the *critical aspect ratio* and this ratio, and l_c, vary considerably from composite to composite (Table 13.2).

We can see from Fig. 13.11 that the average tensile stress $\bar{\sigma}_f$ on a fibre of length l, even when l is greater than l_c, must be less than σ_f^{TS}. It follows that the equations derived in the previous section for the strength of aligned continuous fibre composites have to be modified to take into account the reduced load-bearing contribution of the fibres in discontinuous fibre composites. As for continuous fibre composites, the behaviour of discontinuous fibre composites depends on V_f and on the relative values of e_f^* and e_m^*. We will only consider the case of an aligned discontinuous fibre composite with a high V_f and $e_f^* < e_m^*$.

For such a composite the strength may be determined by the initiation of failure by the fracture of the fibres or by the fracture of the matrix. First, let us concentrate on the former failure mode, i.e. the composite fails when the

Table 13.2

Typical values of the critical length (l_c) and the critical aspect ratio (l_c/d)

Matrix	τ_m (MN m^{-2})	Fibre	σ_f^{TS} (MN m^{-2})	d (μm)	l_c/d	l_c (mm)
Ag	55	Al$_2$O$_3$ (Whisker)	20 800	2	189	0.38
Cu	76	Tungsten	2 900	2000	19	38
Al	80	Boron	2 800	100	18	1.75
Epoxy	40	Boron	2 800	100	35	3.5
Polyester	30	Glass	2 400	13	40	0.52
Epoxy	40	Carbon	2 600	7	33	0.23

tensile stress on the fibres reaches σ_f^{TS}. Clearly for this to occur ℓ must be greater than ℓ_c; the relationship between l and the average tensile stress on a fibre when the tensile strength of the fibre is reached can be shown to be

$$\bar{\sigma}_f = \sigma_f^{TS}\left(1 - \frac{\ell_c}{2\ell}\right) \tag{13.14}$$

The equation for the strength of a continuous fibre composite [Eq. (13.3)] is modified by the substitution of $\bar{\sigma}_f$ for σ_f^{TS} to give

$$\sigma_c^{TS} = \sigma_f^{TS}\left(1 - \frac{\ell_c}{2\ell}\right)V_f + \sigma_m'V_m \tag{13.15}$$

It is this equation that is plotted in Fig. 13.10; note that when $\ell = \ell_c$ the tensile strength of a discontinuous fibre composite is approximately half that of the corresponding continuous fibre composite as the last term in Eq. (13.15), $\sigma_m'V_m$, is generally small.

The other failure mode occurs when the tensile stress on the fibres is insufficient to cause fracture of the fibres but when the stress in the matrix reaches σ_m^{TS}. The average stress $\bar{\sigma}$ when the matrix stress is σ_m^{TS} is

$$\bar{\sigma}_f = \left(\frac{\ell\tau_m}{d}\right)$$

when $\ell < \ell_c$. In this situation the strength of the aligned discontinuous composite is given by the following modification of Eq. (13.3):

$$\sigma_c^{TS} = \left(\frac{\ell\tau_m}{d}\right)V_f + \sigma_m^{TS}V_m$$

It is apparent from the preceding discussion on strength that discontinuous fibres are less efficient than continuous fibres. It comes as no surprise therefore that this reduced reinforcing efficiency also applies to the elastic properties and the law of mixtures equation [Eq. (13.12)] becomes

$$E_c = \eta E_f V_f + E_m V_m$$

where η is a constant that depends on the dimensions of the fibres and on the properties of the fibres and the matrix. It has a value of less than unity, although it approaches unity for long fibres, say greater than about 1 mm, provided there is a strong fibre–matrix bond.

Anisotropy

So far we have only considered the behaviour of composites with aligned fibres, and when the tensile stress is applied along the fibre axis. Of course, in practice the situation can be more complicated as the applied stress system may be complex and the fibres need not be aligned. Still, concentrating for the moment on aligned fibre composites, let us see what happens when the tensile stress is applied normal to the fibre axis. In this case the composite is extremely weak, i.e. the *transverse strength* is very low. In fact the transverse strength is usually less than the strength of the matrix – in other words the fibres, rather than reinforcing, have a detrimental effect on the transverse strength. When properties vary with orientation the material is said to be *anisotropic*.

Shear failure of the composite may occur with other combinations of stress. In these circumstances the strength is normally dominated by the matrix properties because failure can take place by shear of the matrix without fracturing the fibres.

If we took an aligned fibre composite and measured the tensile strength as a function of the angle ϕ between the stress and fibre axes, we would obtain results similar to those presented in Fig. 13.12. This figure shows that a ϕ value of only a few degrees is sufficient to reduce the tensile strength dramatically for both polymer and metal matrix composites. Analysis of the failure modes indicates that tensile fracture occurs only at low angles of less than about 5°. Over an intermediate ϕ range, which depends on the composite under consideration but is typically 5–25°, shear failure dominates. Finally, at high ϕ values, the failure mode changes yet again to transverse fracture.

We have discussed previously the law of mixtures equation [Eq. (13.12)] for the Young's modulus of an aligned composite when the tensile stress is applied parallel to the fibre axis. In order to see whether the elastic properties of a composite are also orientation dependent, we will derive an analogous equation for the *transverse modulus* E_T. When a stress σ_T is applied normal to the fibre axis it is assumed that it acts equally on the fibres and on the matrix, so that

$$\sigma_m = \sigma_f = \sigma_T$$

This contrasts with the assumption made when the stress is parallel to the fibre axis, namely $e_c = e_f = e_m$ [Eq. (13.11)]. The transverse strains of the fibre $e_{f(T)}$ and the matrix $e_{m(T)}$ are

$$e_{f(T)} = \frac{\sigma_T}{E_f}$$

and

$$e_{m(T)} = \frac{\sigma_T}{E_m}$$

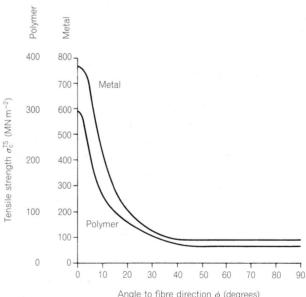

Fig. 13.12

Anisotropy of the tensile strength (σ_c^{TS}) of composites with aligned fibres. The metal matrix composite is aluminium with silica fibres, and the polymer matrix composite is epoxy resin with glass fibres.

respectively. The transverse strain of the composite, $e_{c(T)}$, is

$$e_{c(T)} = e_{f(T)}V_f + e_{m(T)}V_m = \frac{\sigma_T V_f}{E_f} + \frac{\sigma_T V_m}{E_m} \tag{13.16}$$

But

$$E_{c(T)} = \frac{\sigma_T}{e_{c(T)}}$$

and substituting for $e_{c(T)}$ from this equation into Eq. (13.16) gives for the transverse modulus

$$E_T = \frac{E_f E_m}{E_f V_m + E_m V_f} \tag{13.17}$$

To illustrate the difference in the Young's modulus with orientation let us substitute typical values for a polyester resin–glass fibre composite into Eq. (13.12) for the longitudinal modulus and Eq. (13.17) for the transverse modulus. The values used are $V_f = 0.5$, $E_m = 3\,\text{GN}\,\text{m}^{-2}$ and $E_f = 73\,\text{GN}\,\text{m}^{-2}$, and these give calculated moduli $38\,\text{GN}\,\text{m}^{-2}$ for E but only $6\,\text{GN}\,\text{m}^{-2}$ for E_T – a factor of more than 6 in difference.

It is clear from the preceding discussion that aligned fibre composites are very anisotropic, and this is an undesirable characteristic. It might be thought that the solution to this problem is to randomize the orientation of the fibres. This is indeed often done to varying extents and the properties do become more isotropic. Unfortunately, the isotropic properties of a randomly orientated fibre composite are inferior to the properties parallel to the fibre axis in an aligned composite with the same volume fraction of fibres.

Toughness

The various toughening mechanisms that may be operative in a composite are illustrated in Figs 11.20 and 13.13. We discussed *transformation toughening* and *microcrack toughening* in Chapter 11 (Fig. 11.20) and will not consider them any further except to point out that microcracks may be produced by differences in thermal expansion coefficients of the components as well as by phase transformations.

Figure 13.13(a) shows the *crack bowing* mechanism. Resistance of the reinforcing phase to fracture hinders the progress of the crack and causes it to bow into a non-linear crack front. Bowing reduces the stress intensity, K, on the matrix while producing an increase in K on the reinforcing phase. As the extent of the bowing increases so K on the toughening phase rises until fracture of the toughening phase occurs and the crack advances. *Crack deflection* [Fig. 13.13(b)] is similar to crack bowing in that the reinforcing phase perturbs the crack front. Deflection results in a non-planar crack that requires an increase in the applied stress to maintain sufficient stress intensity at the crack tip for crack propagation. The effectiveness of the crack bowing and crack deflection mechanisms depends on the morphology of the toughening phase; the greatest improvement in toughness is obtained with constituents, such as fibres, which have high aspect ratios.

When a crack causes *debonding* [Fig. 13.13(c)], extra energy is required because of the creation of new interfaces. We can measure this energy by pulling a fibre that is partially embedded in the matrix material and recording the stress–strain curve. Debonding takes place at point A on the stress–strain curve and the energy of debonding is given by the area OAB

Fig. 13.13

Toughening mechanisms in composites: (a) crack bowing; (b) crack deflection; (c) debonding; (d) pull-out; (e) wake toughening (fibre bridging).

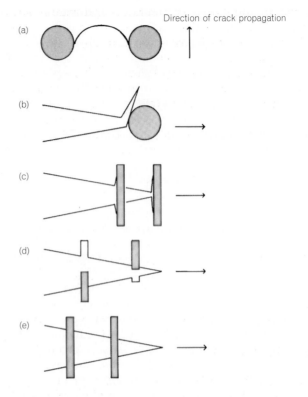

(Fig. 13.14). The energy varies with the length x embedded in the matrix and it can be shown that the maximum energy of debonding, W_D, for a single fibre of diameter d occurs when $x - l_c/2$, or in other words when the tensile stress in the fibre reaches the fracture stress. W_D has been calculated to be

$$W_D = \frac{\pi d^2 (\sigma_f^{TS})^2 l_c}{48 E_f} \qquad (13.18)$$

Fig. 13.14

Pull-out test and the resulting stress–strain curve showing the difference in magnitude of the energies of debonding (area OAB) and pull-out (area OBCD).

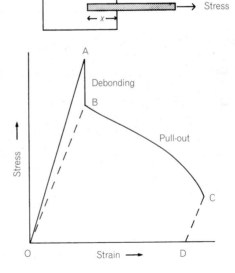

where E_f and σ_f^{TS} are the Young's modulus and fracture stress of the fibre, respectively. If the embedded length is less than $l_c/2$ we can pull the debonded fibre out of the matrix [Fig. 13.13(d)], but a stress is required to overcome frictional forces. These frictional forces arise from residual stresses and from the Poisson expansion of the fibre diameter as the tensile stress is reduced in accordance with Eq. (9.6). The energy associated with *pull-out* is given by the area OBCD under the stress–strain curve. Comparison of the areas OAB and OBCD indicates that the energy of pull-out is greater than the energy of debonding. This difference may be quantified as the maximum energy of pull-out, W_P, given by

$$W_P = \frac{\pi d^2 \sigma_f^{TS} l_c}{16} \qquad (13.19)$$

(a)

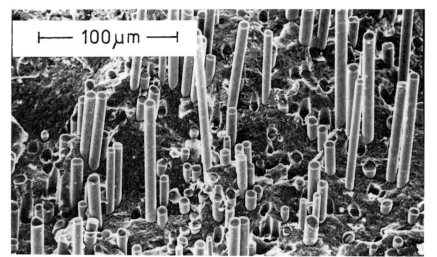

(b)

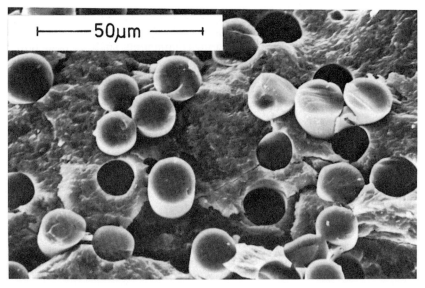

Fig. 13.15

Scanning electron micrographs of the fracture surfaces of two glass–ceramic matrix composites: (a) tough composite exhibiting pull-out; (b) less tough composite with planar fracture and negligible pull-out (courtesy H. S. Kim, P. S. Rogers and R. D. Rawlings).

Therefore

$$\frac{W_P}{W_D} = \frac{3E_f}{\sigma_f^{TS}}$$

Now E_f/σ_f^{TS} is always greater than unity – sometimes considerably so, e.g. for SiC fibres this ratio can be over 200 (see Table 13.1) – and hence pull-out contributes more significantly to toughness than does debonding. Reference to Eqs (13.18) and (13.19), remembering that l_c is inversely proportional to τ_m [Eq. (13.13)], demonstrates that strong, thick fibres and a low matrix shear strength (or interfacial strength) are needed to obtain high energies of debonding and pull-out. Pull-out can clearly be seen in the fracture surface of the tough composite shown in Fig. 13.15(a). However, if the correct balance between σ_f^{TS} and τ_m is not obtained, fibre fracture occurs with negligible pull-out and consequently there is little toughening [Fig. 13.15(b)].

If some debonding has taken place, but the fibre is strong and has not fractured, then the fibre will bridge the faces in the wake of the propagating crack [Fig. 13.35(e)]. For the crack to continue to grow its surfaces must open but, in simple terms, crack opening is hindered by the bridging fibre that has to deform elastically [Fig. 13.13(e)]. This bridging mechanism is most effective when strong fibres are used in as high a proportion as possible, but even then the contribution to toughness is thought to be rather limited.

13.4.3 Comparison of polymer, metal and ceramic matrix composites

Polymer matrix composites, especially glass-reinforced plastics (GRPs), are now commonplace. The matrix may be a thermosetting or a thermoplastic polymer: the former, usually a polyester or epoxy, is more common, but the use of fibre-reinforced thermoplastics is increasing much more rapidly. Carbon fibre-reinforced plastics (CFRPs) are considerably more expensive than GRP but can offer much greater strength and stiffness. CFRPs are usually based on an epoxy matrix.

Because the properties of composite materials are so heavily influenced by the method of production, let us consider techniques used for forming GRP. The most familiar GRP is glass-reinforced thermosetting polyester, formed by the *hand lay-up* method into components such as boat hulls. The viscous unsaturated polyester base resins are usually thinned by dilution in styrene; mixing in a small quantity of catalyst starts the slow polymerization reaction during which the constituents crosslink and set. A former is coated with a release agent and a thin initial layer of catalysed polyester is applied. This will form a protective *gel coat* to act as a barrier to water, which can otherwise diffuse along and weaken the polymer/fibre interface. The core of the composite is then overlaid in successive layers of glass fibre mat and/or tape into which the resin is rolled or brushed. Curing usually takes place slowly at room temperature, but may be accelerated by heating. Although an advantage of this method is that the thickness and direction of reinforcement can be varied at will, many hand lay-up composites use randomly oriented chopped strand mat. A faster alternative for the production of randomly oriented fibre composites is the *spray-up* method. Here the mould/former and its preparation are similar to that in the hand lay-up method but, instead of using mats of fibres, bundled and chopped fibres are blown on through a spray gun. Several more refined techniques, which are capable of achieving better properties and higher production rates, involve compression moulding of pre-impregnated mats (*prepregs*) or of premixed resin/fibre *dough moulding compounds.*

Tubular components such as pressure pipes, in which control of the reinforcement direction is more critical, are usually produced by *filament winding* of resin-impregnated strands or rovings onto a rotating mandrel or liner. This simple process can be given great flexibility by numerically controlling the rate of traverse of the winding head along the axis (pressure pipes, for example, may be laminated by winding alternate layers at opposite 45° directions to the axis) and has been extended to produce a variety of complex hollow shapes. Beams of uniform cross-section can be produced continuously by *pultrusion*. The required lamination stack of fibre yarn, mat, veils, etc., converges from separate rolls, passes through a resin bath for impregnation, and is pulled through a heated die that compacts and shapes it in the precise external cross-section required. Curing is completed in a continuous oven.

Whereas these production processes for reinforced thermosets evolved to exploit the directional properties of the material, reinforced thermoplastics were originally developed to exploit the speed and flexibility that the processes used for thermoplastics (see Section 12.4.4) already offered. As a result, glass-filled nylon grades developed for injection moulding, for example, were restricted to very short fibre lengths – little greater than the critical length defined in the last section – by the need for the melt to flow freely. More recently, however, long-fibre injection moulding grades and techniques have been developed. Another current trend is towards the compression moulding of glass-mat thermoplastics, consisting of longer fibres in high-performance thermoplastic matrix materials such as polyetheretherketone (PEEK).

Because most fibres have a greater tensile strength and Young's modulus than polymers, reinforcing a polymer can markedly increase the stiffness and improve the strength. Fibre reinforcement may, or may not, be beneficial as far as toughness is concerned.

Metal matrix composites are probably not as familiar to the reader, but development is continuing in this area. Several methods are used to produce metal matrix composites, but we will only briefly mention three. Two closely related techniques are *powder metallurgy* and diffusion bonding. In powder metallurgy a metal powder is mixed with whiskers or chopped fibres, pressed into the desired shape and fired. During firing, the metal powder will sinter into a solid compact, as previously described for ceramics (see Section 11.3.3). *Diffusion bonding* employs metal foil instead of powder. Alternate layers of metal foil and fibres are pressed and heated so that sufficient diffusion occurs to bond the foils together in a similar manner to the sintering of a powder. Low melting point metals can be poured into a mould and allowed to infiltrate around the fibres previously arranged in the mould. This process is known as *liquid metal infiltration* and is particularly suitable for producing small components. Liquid infiltration is usually carried out under pressure applied mechanically or via a gas; when the pressure is applied mechanically the process is known as *squeeze casting*. Because most metals have much higher Young's moduli than polymers, this constant is not generally increased too much by fibre reinforcement as for polymer matrices. There is usually an improvement in strength, but this is often achieved at the expense of toughness.

Let us consider the mechanical behaviour of polymer and metal matrix composites in more detail by studying the stress–strain curves for aligned fibre composites with $e_f^* < e_m^*$ and $V_f > V_{min}$. These curves are qualitatively the same for both continuous and discontinuous fibre composites, provided the discontinuous fibres exceed the critical length. In general the deformation proceeds in four stages, although not every stage is present for all composites.

Stage 1: Initially, both the matrix and fibres deform elastically and the Young's modulus of the composite, E_c, is given by Eq. (13.12), i.e. a simple law of mixtures. The Young's modulus of a metal matrix is high and so the matrix makes a significant contribution to the modulus of the composite. In contrast, the Young's modulus of a polymer is much lower than that of a metal, and consequently the modulus of the fibres is normally one or two orders of magnitude greater than that of a polymer matrix. It follows that the polymer matrix contributes little to the modulus of the composite, and Eq. (13.12) reduces to

$$E_c \approx E_f V_f \qquad (13.20)$$

In practice the Young's modulus of a polymer matrix composite is even less than $E_f V_f$, even when the fibres are continuous. The reasons put forward for this are (a) the breaking of weak fibres at low strains, and (b) the buckling of fibres during curing. The fact that the polymer matrix does not contribute to the modulus of the composite does not matter as long as fibres with a high Young's modulus are used, e.g. carbon fibres. Glass fibres, however, have a low E_f so GRP has a low Young's modulus. The data of Table 13.3 show that an epoxy–carbon fibre composite has a Young's modulus intermediate between that of metal matrix composites and GRP. In many applications the weight of material has to be kept to a minimum; therefore a parameter that is sometimes used to assess elastic characteristics is the *specific modulus* E/ρ, where ρ is the density. The density of polymer matrix composites is low and this means that GRP has a reasonable specific modulus, while a carbon fibre composite has a high specific modulus that is much greater than $0.03\,(\mathrm{GN\,m^{-2}})/(\mathrm{kg\,m^{-3}})$, which is typical of conventional metals such as steel and aluminium alloys (Table 13.3).

Table 13.3

Comparison of the mechanical properties of polymer-matrix composites, metal-matrix composites and metals

Composite	Density (kg m^{-3})	Young's modulus (GN m^{-2})	Specific modulus [(GN m^{-2})/ (kg m^{-3})]	Tensile strength (MN m^{-2})	Specific strength [(GN m^{-2})/ (kg m^{-3})]
Aluminium–50% boron	2 700	207	0.077	1137	0.42
Copper–50% tungsten	14 130	262	0.018	1207	0.09
Epoxy–58% carbon	1 660	165	0.099	1517	0.91
Epoxy–72% E glass	2 170	56	0.026	1640	0.76
Epoxy–72% S glass	2 120	66	0.031	1896	0.89
Epoxy–63% Kevlar	1 380	83	0.060	1310	0.95
Aluminium alloy	2 800	71	0.025	600	0.21
Steel (mild)	7 860	210	0.027	460	0.06
Steel (eutectoid)	7 800	210	0.027	3000	0.38
Brass (Cu–30% Zn)	8 500	100	0.012	550	0.06

Stage 2: Stage 2 commences at low strains and is associated with the matrix deforming plastically, while the fibres are still elongating elastically. On removal of the load the matrix and the fibres first contract elastically, then the matrix deforms plastically in compression due to the continuing elastic contraction of the fibres. This results in a quasi-elastic behaviour and the composite almost regains its initial shape and size. The stress–strain curve is not linear in Stage 2 as the matrix is deforming in a non-linear manner. The slope of the stress–strain curve at a given strain is called the *secondary modulus*, E_s, and depends on the elastic properties of the fibres and the work-hardening rate, $d\sigma_m/de_m$, of the matrix according to

$$E_s = E_f V_f + \frac{d\sigma_m}{de_m} V_m \qquad (13.21)$$

This stage is particularly significant for metal matrices but is often difficult to detect in the stress–strain curves of polymer matrix composites. This is because the non-linearity of the polymer matrix is not marked and $d\sigma_m/de_m$ is small compared with E_f hence, from Eq. (13.21), E_s is approximately equal to $E_f V_f$.

Stage 3: This stage, in which both the matrix and fibres deform plastically, only occurs in certain composites. It is absent in composites containing brittle fibres and is only prominent in composites consisting of strong metal wires in a metal matrix.

Stage 4: Finally, the composite fails at a stress σ_c^{TS}, given by Eq. (13.3) if the fibres are continuous and Eq. (13.15) if they are discontinuous. Just as the specific modulus is sometimes used to characterize elastic behaviour, so the specific strength σ_c^{TS}/ρ is employed for assessing strength. Because of their low densities, and indeed good strengths, polymer matrix composites have high specific strengths that are well in excess of those for conventional metals (Table 13.3).

Finally, let us turn our attention to ceramic matrix composites. Generally ceramics have sufficient stiffness and strength and hence the prime objective in producing ceramic matrix composites is not to enhance these properties. We have seen in Section 11.5 that the major problem with ceramics is their brittleness and sensitivity to flaws. Thus the main aim of reinforcing a ceramic, whether with fibres or particles, is to improve the fracture toughness. This improvement is illustrated by the stress–strain curves of Fig. 13.16. The failure of a ceramic is catastrophic and the small area under the stress–strain curve is indicative of the low toughness. The

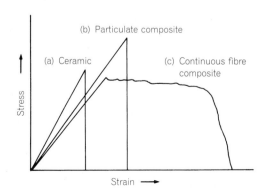

Fig. 13.16

Comparison of the stress–strain behaviour of (a) ceramic, (b) particulate, or short random fibre, composites and (c) a continuous fibre composite.

failure of a particulate-reinforced, or short random fibre-reinforced, ceramic composite is similar to that of the ceramic matrix in that it is catastrophic; note however that the toughness has increased, as demonstrated by the larger area under the stress–strain curve. In contrast the failure of a continuous fibre composite is not catastrophic and the composite maintains a substantial load-carrying capacity well after failure has commenced. Therefore not only has the fracture toughness increased but the failure mode is more benign. Thus the attraction of both particulate- and fibre-reinforced ceramic composites is the improved toughness, which makes them less sensitive to flaws and more reliable during assembly and service.

Ceramic matrix composite production is the least established composite production technology. Because of the high melting point of most crystalline ceramics, techniques involving a melt are unsuitable. Melt infiltration is even difficult with precursor melts for glasses and glass-ceramics as these are more viscous and more reactive than molten metals. The routes that at this stage of development look most promising are:

- Conventional mixing of the components followed by cold pressing and firing or hot pressing.
- Mixing using a slurry followed by sintering.
- Mixing of a ceramic-producing polymer and the toughening phase followed by pyrolysis of the polymer.
- Chemical vapour deposition of the matrix, i.e. impregnation of a preform with a gas that deposits the ceramic.

13.4.4 Some commercial composites
Carbon–carbon and zirconia-toughened alumina

Two examples of ceramic matrix composites will be discussed in this section. These have been chosen because of the marked differences in method of production, structure and properties.

Carbon–carbon composites consist of fibrous carbon in a carbonaceous matrix. Although both constituents consist of the same element their morphology and structure differ greatly from composite to composite. The carbon in these composites may vary from highly crystalline graphite through a wide range of quasicrystalline forms such as turbostratic carbon, to amorphous, glassy carbon. By controlling the proportions, morphology and crystallography of the constituents, together with the porosity content, it is possible to produce a series of composites ranging from a soft, porous material used for furnace insulation to a hard, high temperature, structural material.

The composite is produced by impregnating the fibrous reinforcement, which is available in many forms from unidirectional lay-ups to multi-directional weaves. Impregnation is achieved by two basic methods – *carbonization* of an organic solid or liquid, or *chemical vapour deposition* (CVD) of carbon from a hydrocarbon.

Examples of the organic precursor materials for the carbonization method are pitch and phenolic resin. A typical manufacturing procedure using a phenolic resin would be to cure after impregnation of the fibrous reinforcement and then to pyrolyse at a higher temperature. Shrinkage occurs as the resin decomposes during pyrolysis, and therefore, in order to obtain high densities, a number of impregnation, cure and pyrolysis cycles are required. In the CVD process carbon is deposited by passing a hydrocarbon gas through the fibrous reinforcement, which is held at a temperature of about 1100°C.

The main feature of structural carbon–carbon composites is that they maintain their strength and toughness up to high temperatures – of the order of 2000°C. The superior high-temperature mechanical properties of these composites (especially when assessed in terms of specific strength) compared to alumina, to silicon nitride and to a high-temperature nickel alloy (*superalloy*) is illustrated in Fig. 13.17.

The problem with carbon–carbon composites is that they oxidize and sublime in the presence of oxygen at temperatures in excess of 600°C. Therefore for high-temperature applications we must protect the composite by using either an inert atmosphere or a coating. For example carbon–carbon composites were selected for the leading edges and nose cap of the Space Shuttle, in order to survive the stresses of re-entry to the atmosphere at temperatures that can far exceed 600°C. For this application the surfaces are sealed by a coat of silicon carbide. However, SiC does not give complete protection – microcracks develop in the coating because of a mismatch in thermal expansion coefficients between it and the underlying composite. Additional protection is imparted by a process that impregnates the composite surface with tetraethylorthosilicate. This enhances oxidation resistance by depositing silica in microcracks as they develop.

This section has been mainly concerned with fibre-reinforced composites for the simple reason that, in general, fibres produce a greater improvement in mechanical properties than particles. Nevertheless there

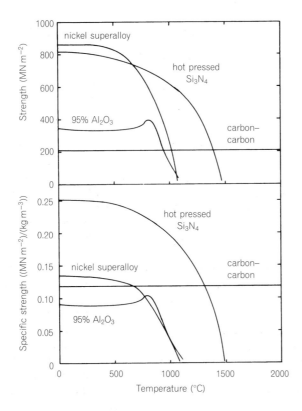

Fig. 13.17

High-temperature mechanical properties of a carbon–carbon composite and some other high-temperature materials.

Fig. 13.18

Scanning electron micrograph of ZTA showing a fine dispersion of small metastable tetragonal zirconia particles (white) in a coarser polycrystalline alumina matrix (dark).

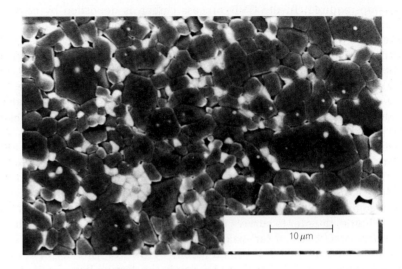

are some successful particulate-reinforced composites in commercial production, of which *zirconia-toughened alumina* (ZTA) is an example.

ZTA is a composite consisting of a fine-grained polycrystalline alumina matrix reinforced with about 15 vol% ZrO_2 (Fig. 13.18). The ZrO_2 particles, which typically contain 3 mol% of the stabilizing oxide Y_2O_3, are less than 1 μm in size. In contrast to the carbon–carbon composites, ZTA is manufactured by the conventional ceramic technology of pressing and sintering. The Al_2O_3–ZrO_2 mixture for sintering may be obtained by mechanically mixing Al_2O_3 and ZrO_2, but this tends to give inhomogeneous products of variable particle size. Better quality mixtures are prepared by a *sol–gel* route. A ZrO_2–Y_2O_3 *sol* (which is a dispersion of small particles of less than 100 nm, termed a colloid, in a liquid) is added to an Al_2O_3 slip and a gel produced. In this context a *gel* is simply a sol that has lost some liquid to give a significant increase in viscosity. The gel is then dried, and after some further treatment, a fine homogeneous powder suitable for compaction is obtained.

Because of their size, and the stabilizing effect of the Y_2O_3, the zirconia particles in ZTA are in the metastable tetragonal condition. As previously explained in our study of zirconia (Section 11.9), the metastable tetragonal particles undergo a martensitic transformation to the monoclinic structure under the action of the stress at a crack tip. Thus the same transformation toughening mechanism (Fig. 11.20) found in some zirconias is also operative in ZTA. The improvement in the toughness and strength of ZTA over the corresponding values for the alumina matrix is shown by the mechanical property data of Table 13.4.

Table 13.4

Room temperature mechanical properties of ZTA and the alumina matrix

	ZTA	Alumina
Strength (MN m^{-2})	461	287
Weibull modulus m	9.0	10.4
Toughness K_{1C} (MN m$^{-3/2}$)	6.5	3.8
Hardness (GN m^{-2})	16.1	12.5

Alumina-reinforced aluminium alloys

Fibre-reinforced aluminium alloys are produced by powder metallurgy or by squeeze casting. We will only consider those produced by the latter method and, as this is a casting technique, most of the aluminium alloys that have been reinforced are conventional casting alloys, such as Al–Si and Al–Mg base alloys. In squeeze casting of these materials the preform, die and ram are preheated before the introduction of the molten alloy at a controlled degree of superheat. Impregnation is then achieved at pressures normally in the range 50–100 MN m^{-2}. A variety of fibres have been used for the preforms but we will discuss the effects of reinforcing with short, α-alumina fibres that have a tensile strength and Young's modulus of 2000 MN m^{-2} and 300 GN m^{-2} respectively.

The Young's modulus of aluminium and its alloys is low compared to most structural metals and one of the major benefits of reinforcement with high modulus fibres is the significant increase in stiffness (Fig. 13.19). In contrast there is only a minor improvement in the room temperature tensile strength although, because of the superior properties of the ceramic

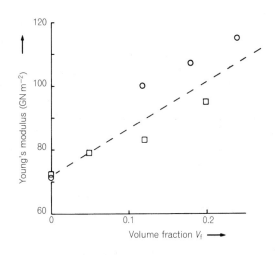

Fig. 13.19

Graph illustrating the increase in Young's modulus of two aluminium alloys obtained by reinforcing with α-alumina fibres (circles: Al–9% Si–3% Cu, J. Dinwoodie *et al.*, 5th Int. Conf. Composite Materials, 1985; squares: Al–12.5% Si–1% Cu–1% Ni, L. Ackermann *et al.*, 5th Int. Conf. Composite Materials, 1985).

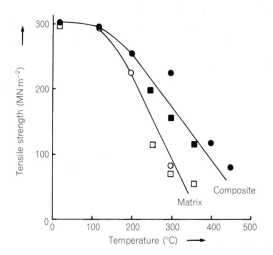

Fig. 13.20

Graph demonstrating the superior strength of aluminium alloy matrix composites at elevated temperatures (circles: Al–9% Si–3% Cu, J. Dinwoodie *et al.*, 5th Int. Conf. Composite Materials, 1985; squares: Al–12.5% Si–1% Cu–1% Ni, L. Ackermann *et al.*, 5th Int. Conf. Composite Materials, 1985).

Matrix	Ductility (%)
Al	4.0
Al–2.5% Mg	3.3
Al–10% Mg	1.3
Al–12% Si–1% Cu–1% Ni	<1.0

fibres at elevated temperature, the composites maintain their strength to higher temperatures than the matrix alloys (Fig. 13.20). In common with most fibre-reinforced metals the toughness and ductility of the matrix is degraded; Table 13.5 gives some examples of the low ductilities shown by alumina-reinforced aluminium and aluminium alloys. As far as other mechanical properties are concerned, these composites can show an improvement in both wear and fatigue resistance, e.g. the 10^7 cycles endurance limit of Al–12% Si–1% Cu–1% Ni casting alloy may be increased by 30% by reinforcing with 0.2 volume fraction of alumina.

One of the first commercial applications of these composites was for the pistons of heavy-duty diesel engines. This component is subjected to wear and to thermal and mechanical cyclic stresses at elevated temperatures which, in certain areas, may approach 500°C. Other potential applications for these lightweight cast composites are brake drums, gears and rotary compressors.

Polymer matrix composites

A wide variety of polymer matrix composites are commercially available. Rather than discuss one or two of these in detail we will therefore expand on some of the points made in Section 13.3.3. The first point to emphasize is that the polymer matrix may be a thermoset (e.g. epoxy, polyester), an amorphous thermoplastic (e.g. polycarbonate, acrylic), or even a crystalline thermoplastic (e.g. polyetheretherketone, nylon). In fact virtually all types of polymers have been used as the matrix for some form of composite.

The reinforcement most widely used with polymers is fibrous. The fibres are normally carbon, glass or 'Kevlar', although boron fibres find some applications – mainly in the aerospace industry – and gel-spun polyethylene and polypropylene fibres are making steady progress. In addition, there are many types of particulate reinforcement, which differ greatly in size, shape and composition, and consequently in their effect. An important feature of many new systems is the combination of different fibres, sometimes with other reinforcing or toughening agents, to form hybrid composites. The aim is to capitalize on the best properties of each component, and particularly to benefit from any synergistic effect.

There has been a very rapid growth of fibre-reinforced polymer matrix composites over the last few years, which shows every sign of continuing. The main reason for this growth is the high specific modulus and specific strength of these materials (Table 13.3), which mean that in many cases the weight of components can be reduced considerably. Limiting the weight is an important factor in moving components, particularly in transport and aerospace, where weight reductions save energy and give greater efficiency. The wide range of applications for polymer matrix composites is well illustrated by the examples given in Table 13.6.

Industrial sector	Examples
Aerospace	Wings, fuselages, radomes, antennae, tail-planes, helicopter blades, landing gear, seats, floor, interior panels, fuel tanks, rocket motor cases, nose cones, launch tubes
Automobile	Body panels, cabs, spoilers, consoles, instrument panels, lamp housings, bumpers, leaf springs, drive shafts, gears, bearings
Boats	Hulls, decks, masts, engine shrouds, interior panels
Chemical	Pipes, tanks, pressure vessels, hoppers, valves, pumps, impellers
Domestic	Interior and exterior panels, chairs, tables, baths, shower units, ladders
Electrical	Panels, housings, switchgear, insulators, connectors
Leisure	Motor homes, caravans, trailers, golf clubs, racquets, protective helmets, skis, archery bows, surfboards, fishing rods, canoes, pools, diving boards, playground equipment

Table 13.6
Examples of applications of polymer–matrix composites

There are many manufacturing possibilities for fibre-reinforced plastics, but it is important to recognize that the particular means chosen, and the processes involved, can have a profound effect on the final properties of the composite owing to their effect on microstructure and internal stresses. Those processes that allow the placement of long fibres in known preferred directions have a clear advantage in terms of the composite mechanical properties over other processes that sacrifice the high potential offered by the fibres for reasonable isotropy of the composite. Processes that allow good control over fibre alignment include vacuum bag, pressure bag and auto-clave, filament winding, pultrusion, compression moulding, resin injection, and, to a lesser extent, hand lay-up. Reaction injection moulding (described in Section 12.7.2) can be used to produce composite components with controlled fibre alignment by lay-up of glass fibre mats in the mould before injection. The use of conventional injection moulding for thermoplastics filled with short fibres less than 1 mm long is now very common. Fibres tend to become oriented in the direction of flow, which can itself be controlled to some extent by the placement of the injection point (the *gate*); however, problems often arise where two melt streams diverge around a flow obstacle and meet again to form a weak *weld line*.

Processes that give little control over fibre alignment include centrifugal casting and spray-up. These latter techniques are, however, ideally suited for short fibres, and, in general, confer good isotropy to the product. Of the former processes only compression moulding and hand lay-up are routinely used for both long and short fibre composites. Many of the techniques referred to above can cope with both thermoset and thermoplastic resin composites, but care is required in identifying the best method for economical production.

13.5 Cellular solids

We now turn to a very different class of multiphase materials in which the second phase has neither the incompressibility of rubber nor the strength

Fig. 13.21

Honeycomb, usually made from aluminium and clad with rigid skin, is widely used for lightweight structures such as aircraft flooring.

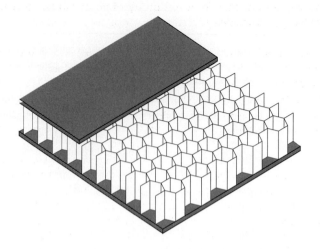

of reinforcing fibres: it is a gas, usually air. This class of *foams* or *cellular solids* covers a wide spectrum. On the one hand we find materials whose void content emerges from within with a self-organized structure, like that of a reinforcing precipitate in an alloy. As well as structural materials such as polymer foams we find many foodstuffs, e.g. bread, in this class. Other foams are produced by mechanical whisking: here we find the earliest synthetic foam, natural rubber, in which air was enfolded by a whisking process before crosslinking. On the other hand we find cellular materials that are 'constructed' either by mixing in hollow particles or by assembling the components in an ordered structure. The former, *syntactic foams*, are essentially isotropic in the sense that short-fibre reinforced plastics are. The latter are, like long-fibre reinforced plastics, deliberately manufactured to be anisotropic. The most common examples are *honeycombs*, which are usually clad with another material to form a 'sandwich' (Fig. 13.21) that has significant strength and rigidity only in the through-thickness direction.

The properties of a foam depend not only on those of the matrix solid from which it is made, but also on the size, proportion, shape, distribution and interconnection (if any) of the voids within it. This range of variables allows a vast range of properties to be achieved. The only feature common to all cellular materials is relatively low mass density. The reduction in density from the parent matrix material is usually achieved without an equivalent loss in strength properties and with gains in other properties such as thermal insulation.

13.5.1 Geometrical properties

The most basic property of a foam is its relative density, i.e. the ratio of the mass density ρ^* of the foam to the mass density ρ_m of the bulk solid material from which it is formed. Because the density of the void content is negligible, we can again invoke the law of mixtures [Eq. (13.1)] to show that

$$\frac{\rho^*}{\rho_m} = 1 - V_p$$

where V_p is the volume fraction of voids (the *porosity*). Virtually no other properties of a foam are so simply related to those of the bulk solid because, as we have come to expect from our study of composites, most properties depend strongly on the *form* rather than simply the *quantity* of the remaining solid.

Let us recall that rubber-toughened polymers usually contain less than about 30 vol% of rubber, and that cavitation of these small rubber particles still leaves the material recognizable as a solid. Cellular solids, on the other hand, seldom have a porosity of less than 70%, i.e. a relative density of 0.3. At this end of the relative density range we find, for example, the ceramic heat insulation foam used for some insulating firebricks. Near the lightest end, the familiar polystyrene foam used for packaging may have a relative density less than 0.02. Even the densest foams do not have the character of a solid containing voids or pores, but that of a framework of struts as in Fig. 13.22(a), like the 'space frames' used extensively in architecture, or of a fine structure of box-like cells as in Fig. 13.22(b). Classifying the many possible geometries in which cell edges and walls can make up a homogeneous solid is an advanced subject that has many similarities to crystallography, and it is central to understanding how solid properties contribute to those of the foam.

Closed-cell foams are usually generated by expansion of voids within a viscous liquid. Competition between adjacent voids determines the position and connection of edges, while surface tension and viscous forces act to keep stretched cell faces between these edges flat. Most often, four cell edges meet at each vertex, like the four bond directions radiating from a carbon atom (Fig. 6.4). The cell edges are often thicker than the cell walls, and the vertices at which they meet are usually even more substantial 'blobs'. The most common closed-cell shape is that of a dodecahedron (Fig. 13.23) with flat pentagonal faces; in this geometry each pair of faces intersects with an internal angle of 120° and three faces meet at each corner.

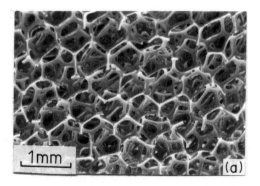

Fig. 13.22

(a) An open-cell ceramic foam and (b) a closed cell polyethylene foam. Reproduced with permission, from Gibson, L.J. and Ashby, M.F. (1997) *Cellular Solids: Structure and Properties*, 2nd edn, Cambridge University Press.

Fig. 13.23

The cell shape most commonly seen in both open- and closed-cell foams is a regular dodecahedron.

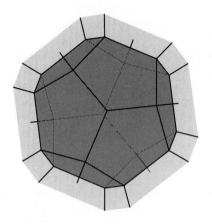

Open-cell foams are skeleton-like structures of beams and struts. Usually they develop from closed-cell foams as the still liquid wall ruptures and contracts onto the cell edges – like a soap film on the wire frame of a bubble-blowing toy. This process is known is *reticulation* (from the Latin *reticulum*, net), and the structures that it produces as *reticulated foams*, although the term is used rather loosely to describe skeletal networks even if they are generated by other means.

13.5.2 Polymer foams

Polymer foams are usually produced by addition to the base polymer of a *blowing agent*, which generates gas either on heating or, in the case of thermosets, during curing. In an open-cell structure this gas usually diffuses out quite rapidly and is replaced by air. The polymers most widely used in cellular form are polystyrene, familiar as rigid and brittle Styrofoam[®], and the thermosetting polyurethanes. The latter yield a vast range of properties from rigid, closed-cell blocks for thermal insulation to flexible open-cell types used in furnishing. Increasingly, polyethylene and polypropylene are used for packaging and for demanding shock-absorbing applications. These usually have a characteristic waxy feel to the surface, and often have a solid skin.

The high stiffness-to-weight ratio of some rigid structural polymer foams is widely exploited in sandwich structures, where they form a lightweight core with more isotropic properties than the honeycomb of Fig. 13.21. Look around at a range of load-bearing structures, from skeletons to buildings, and you will find that they contain two basic kinds of load-bearing elements – those that are extended or compressed (putting their material under uniform axial stress) and those that are subjected to bending: *beams*. The deck of a bridge, a typical beam, is supported at points and loaded across rather than along the axis. The load is still communicated to the support by stresses acting along the axis, but this stress varies from compressive on the upper surface through zero at the midplane to tensile on the lower surface. It is principally the upper and lower surfaces of the beam that carry the stress and which give the beam its resistance to deflection. This is why beams are often constructed as 'box' sections or 'I' sections [Figs 13.24(b) and (c)], which have much less effective cross-sectional area near the midplane. *Structural foam* is the term used to describe a cellular

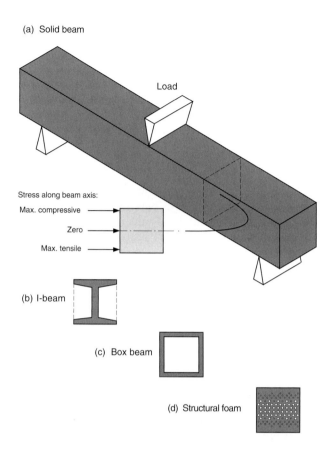

(a) Solid beam

Load

Stress along beam axis:

Max. compressive

Zero

Max. tensile

(b) I-beam

(c) Box beam

(d) Structural foam

Fig. 13.24

The material of a loaded beam (a) experiences maximum stress at the surface and minimum at the centre. Box sections (b) and I-beams (c) as well as structural foam (d) remove material near the midplane where stress is low.

material that reflects this load-bearing distribution – its relative density is nearly unity at the loaded surface skin but falls to a minimum at the mid-plane. Many human and animal bones have evolved, in order to resist bending, towards this structure.

In this respect thermoplastic polymers offer unique possibilities: they can be moulded in a single operation into shapes of great complexity throughout which the skin is solid and the core is foamed. This kind of heterogeneous material, known as *structural foam*, is initially homogeneous – as usual, the blowing agent that will generate gas to form the foam becomes active only at moulding temperatures. When this polymer is injection moulded (see Section 12.4.4), the pressure is high enough to prevent gas evolution until it enters the mould and the skin of the moulding freezes rapidly onto the cold mould surfaces. Beneath the surface of the moulding, however, the material remains warm enough, for sufficient time and under a sufficiently reduced pressure, for gas evolved by the blowing agent to transform the polymer into foam. The resulting component is thick and rigid, but it is lighter and has cooled more quickly than it would have done if solid.

13.5.3 Ceramic foams

Open-cell ceramic foams, with cell sizes ranging from 100 μm to over 1 mm and densities of 0.1–0.3, are well established materials. The production of ceramics such as alumina by the high-temperature firing of slips – suspensions of powder in water or some other fluid – was described in Section 12.3.2. To make a ceramic foam the powder is supported in a liquid thermosetting polymer, usually polyurethane, which is then foamed; the polymer itself is sacrificed during firing. Applications for ceramic foams include furniture for high-temperature kilns, support structures for catalysts in catalytic converters, and filter elements for hot gases and liquids, e.g. molten aluminium. Advanced ceramic foams can replace asbestos fibre, which is a health hazard, in many applications from which it is now excluded.

Glass foams are mainly used in construction. They are produced by adding additional gases or blowing agents to the glass melt. The resulting fine-pore, closed-cell foam has a very low density but a high compressive strength and dimensional stability, making it particularly suitable for thermally and acoustically insulating construction materials. One attractive feature of the glass foam is that high proportions of scrap can be recycled.

13.5.4 Metal foams

Metal foams are amongst the newest of all materials. Whereas polymers are often exploited as foams by virtue of their low thermal conductivity, open-cell metal foams are excellent heat transfer media, and found early applications in heat exchangers and in high-power electronic systems.

Both closed-cell and open-cell metal foams can be produced. A variety of methods have been used, although the number of metals they have been applied to is relatively small. Bubbling air or gas through the molten metal as it solidifies is effective but produces very large cell sizes – up to tens of millimetres across. A better method is to seed molten metal with carbon under an inert atmosphere or, conversely, to infiltrate the metal into a porous carbon matrix; in either case the carbon is then burnt off. Probably the most promising conventional technique for making aluminium foam is to sinter, at relatively low temperature, a mixture of metal particles and a blowing agent: titanium or zirconium hydride. The temperature is then increased towards the melting point of the aluminium, which combines with the titanium or zirconium to form intermetallic compounds while the evolved hydrogen gas generates voids. Whereas few other methods can reduce the relative densities of metal foams to much less than 0.5, this method can reduce it to 0.1.

13.5.5 Syntactic foams

Syntactic foams are 'constructed': each void is inserted in prefabricated form as a rigid bubble with its own shell – a *microsphere*. In this sense they are special cases of filled materials. One of their advantages is the ability to control much more precisely the size of voids and the distribution of void sizes. The voids may be so small that the material resembles a solid. Materials created by mixing a solid with minute (30–200 μm) spheres of glass, ceramic, polymer or even metal itself are finding an increasing range of uses in industrial and technological applications. Most of the several thousand tonnes of syntactic foams produced each year are used for flotation in offshore drilling rigs, buoys, small boats and submarines. These applications make use of the high compressive strength of the higher-density foams and of their generally excellent water resistance.

13.5.6 Mechanical properties

Even when they are used as functional materials (e.g. as supporting structures for catalysts), the structure of foams is usually determined by the need for enough strength to survive. In several applications, however, mechanical properties of foams are of primary importance:

- Padding in furniture, car headrests, etc., where they must show elastic response to very high compressive strains.
- Shock-protection foams – mainly in packaging but also in some exotic applications such as flak-jackets – where the maximum energy absorption is needed and is usually achieved by *plastic* deformation.
- Structural foam, in which the highest possible stiffness and strength must be achieved at minimum volume.

A typical stress–strain curve from the uniaxial compression of a foam is shown in Fig. 13.25. At low strains (typically 10–15%) all deformation in the cellular structure is linear elastic, showing a modulus E^*. This is followed by a plateau region during which the cells collapse by a combination of buckling and brittle crushing of cell walls, with plastic hinge formation at the edges. The stress at this stage remains more or less constant, at the *collapse stress* σ^*_{col}, up to large strains (typically 0.7–1.0). Finally, increasing numbers of cells have completely collapsed so that their opposing cell walls touch and further compression is effectively compression of the solid cell wall material. This phase is signified by the steep, abrupt rise in stress.

Thus E^* and σ^*_{col} effectively correspond to the stiffness and yield strength of the foam and are its two principal mechanical properties. They depend on the properties of the bulk solid (Young's modulus E_m, Poisson ratio ν_m, etc.) and on the architecture and shape of the cells. If the cells are even slightly elongated in one direction, the foam becomes considerably anisotropic with the properties dependent on direction. Most foams, whether man-made or natural, are anisotropic and become more so when deformed.

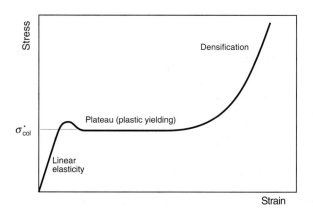

Fig. 13.25

Stress–strain curve for a typical polymer foam.

Analysis leads to the following expressions for Young's modulus and collapse stress of open-cell foams:

$$\frac{E^*}{E_m} \approx \left(\frac{\rho^*}{\rho_m}\right)^2 \tag{13.22}$$

and

$$\frac{\sigma^*_{col}}{E_m} \approx 0.05 \left(\frac{\rho^*}{\rho_m}\right)^2 \tag{13.23}$$

For closed-cell foams:

$$\frac{E^*}{E_m} \approx \phi^2 \left(\frac{\rho^*}{\rho_m}\right)^2 + (1-\phi)\frac{\rho^*}{\rho_m} + \frac{p_0(1-2\nu_m)}{E_m(1-\rho^*)} \tag{13.24}$$

and

$$\frac{\sigma^*_{col}}{E_m} \approx 0.05 \left(\frac{\rho^*}{\rho_m}\right)^2 + \frac{p_0 - p_{at}}{E_m} \tag{13.25}$$

where p_0 is the initial pressure of the cell gas (usually equal to atmospheric pressure, p_{at}, for man-made foams). ϕ is the fraction of the solid contained in the cell edge 'struts', so that the fraction of the solid contained in faces of closed-cell foams is $1 - \phi$. The relative modulus predicted for open-cell foams differs from the result for closed-cell foams because in the latter, apart from bending of the cell edges, two other deformation modes are taking place: compression of the cell gas and stretching of the cell face membranes. The contribution to the modulus of closed-cell foams from the membrane stretching decreases as ϕ increases.

13.5.7 Thermal conductivity

One of the most important applications of cellular materials is for thermal insulation, the most familiar example being the rigid polyurethane foam used between double-skinned steel walls of domestic refrigerators and hot-water tanks. However, the thermal conductivity of air is about one order of magnitude less than that of solid polymers and about two orders of magnitude less than that of ceramics. Because air is rather a good insulator, why do we bother with insulation foam – or, for that matter, blankets and clothes?

The answer lies in the other two modes of heat transfer – convection and radiation. The 'active ingredient' in closed-cell thermal insulation foams is indeed air, which the cell walls totally immobilize. The use of foam with a low relative density reduces the heat conducted through the cell walls and edges. On the other hand, the foam itself must be strong enough to survive intact and dense enough to be opaque to radiation.

Problems

13.1 Explain the principal mechanism by which well-bonded rubber particles toughen a polymer, commenting on the role of crack bridging.

13.2 Discuss the significance of the 'critical length' with reference to composites with discontinuous fibres. From the following data, calculate the critical length of the fibres and hence the tensile strength of the composite:

Fibre			Matrix		
Diameter	Tensile strength	Length	Shear strength	Tensile stress at failure strain	Volume fraction
(μm)	(MN m^{-2})	(mm)	(MN m^{-2})	(MN m^{-2})	(%)
100	3000	5	100	250	50

13.3 Discuss the elastic and quasielastic behaviour of composites.

Self-assessment questions

1 The effect of rubber inclusions in a thermoplastic such as ABS is:

(a) to decrease the modulus

(b) to increase the strength

(c) to increase the toughness

(d) to increase the yield stress

(e) to crosslink the thermoplastic and prevent flow

2 Polypropylene grades used for large automotive components are often heavily filled with talc

(a) to increase the toughness

(b) to increase the rigidity (tensile modulus)

(c) to reduce the cost

(d) to improve the appearance

(e) to reduce component weight

3 The reinforcing effect of a filler or fibres in any matrix always increases with the strength of the fibre/matrix interface

(a) true   (b) false

4 Rubber is usually filled with carbon black

(a) to increase the toughness

(b) to increase the shear modulus

(c) to increase strength

(d) to reduce the cost

(e) to increase weather resistance

5 The Young's modulus, E_c, of a composite with a polymer matrix is:

(a) generally higher than E_c for a metal matrix composite

(b) independent of the properties of fibres used for a given polymer matrix

(c) approximately proportional to the Young's modulus of the fibres

(d) often called the secondary modulus

6 The strength of an aligned discontinuous fibre composite is always less than that of an aligned continuous fibre composite of the same volume of fibres

(a) true   (b) false

7 In order to strengthen a matrix, fibres for which $e_f^* < e_m^*$ must form more than the critical volume of the composite

(a) true   (b) false

8 The figure shown refers to composites and it shows:

(a) the shear stress in a fibre

(b) the shear stress in the matrix

(c) the tensile stress in a fibre

(d) the tensile stress in the matrix

(e) the shear stress along the length of the composite

(f) the tensile stress along the length of a composite

$\ell > \ell_c$

9 In a tensile test on any composite the deformation proceeds in four distinct stages

(a) true   (b) false

10 During the initial stage of the deformation of a composite with a metal matrix:

(a) there is a hyperbolic relationship between stress and strain

(b) both the matrix and the fibres deform elastically

(c) the Young's modulus of the composite is given by a simple law of mixtures

(d) the Young's modulus of the composite is determined primarily by the yield stress of the metal matrix

(e) the majority of the fibres break and rotate so that their longitudinal axes are parallel to the stress axis

11 A composite with aligned fibres is mechanically anisotropic

(a) true   (b) false

12 The transverse strength of an aligned fibre composite is very low – usually less than the strength of the matrix

(a) true   (b) false

13 Carbon–carbon composites:

(a) are produced by mixing in a slurry and sintering

(b) are produced by carbonization

(c) are produced by chemical vapour deposition

(d) have good oxidation resistance

(e) maintain their strength and toughness up to temperatures of the order of 2000°C

Each of the sentences in Questions 14–16 consists of an assertion followed by a reason. Answer as follows:

(a) If both assertion and reason are true statements and the reason is a correct explanation of the assertion.

(b) If both assertion and reason are true statements but the reason is *not* a correct explanation of the assertion.

(c) If the assertion is true but the reason contains a false statement.

(d) If the assertion is false but the reason contains a true statement.

(e) If both the assertion and the reason are false statements.

14 A composite with a polymer matrix has a very low specific modulus *because* the density is low.

15 The critical aspect ratio varies from composite to composite *because* polymers are less dense than metals.

16 Wake toughening is the most effective and most widely observed toughening mechanism in composites *because* it always occurs before any debonding.

Answers

1	(a), (c)	**2**	(b), (c)	**3**	(b)	**4**	(a), (b), (c), (e)
5	(c)	**6**	(a)	**7**	(a)	**8**	(c)
9	(b)	**10**	(b), (c)	**11**	(a)	**12**	(a)
13	(b), (c), (e)	**14**	(d)	**15**	(b)	**16**	(e)

part III

Electromagnetic properties and applications

Electrical conduction in metals | 14

14.1 Introduction: role of the valence electrons

Many readers may have learned quite a lot about electricity and electric currents without knowing just why it is that some materials will conduct readily while others are insulators, although these can acquire a static electric charge. In the foregoing chapters we have seen how all matter is built up of charged constituents, both positive and negative, and obviously conduction of electricity must be associated with motion of those charges. Because the protons in the nucleus of an atom are firmly fixed they can only move when the whole atom moves. Now when electrical conduction occurs in metals, we know that no matter is transported, so that the motion of protons cannot be involved and the loosely bound electrons must be responsible for the passage of current. In contrast, we saw in Chapter 10 that electrolytic conduction involves the movement of both positive and negative ions.

For the moment we confine ourselves to metals, and remember that, when discussing metallic bonding in Section 5.6, we pointed out that only the *valence* electrons could be readily removed to take part in bonding. Similarly, we would expect only the valence electrons to be able to take part in conduction. Thus the number of conducting electrons per atom is determined by the atomic structure. Looking at Table 14.1 we can see that copper, silver and gold have only one such electron per atom, zinc and cadmium have two, while aluminium has three. The Group I metals sodium and potassium are far too reactive to be of much engineering importance. If there are more than three valence electrons both the character of the bonding and the electrical properties change completely, except in the case of the heavy metals such as Sn and Pb. These two are included in the table, because when they are alloyed together they are used as solder. Other practical materials, and some transition elements (which of course have incomplete inner shells) are also included. Table 14.1 also gives the *resistivities* (equal to the resistance of a cube of unit dimensions) of those elements measured at 20°C. A comparison with the resistivities of alloys of the same elements in Table 14.2 shows that alloying generally makes for a higher resistance than in the pure metals – the reason for this is explained in Section 14.8.

Because alloying is often used to increase mechanical strength, we see that the lowest resistance, and hence the least power dissipation (current × voltage across the wire) in power lines and all electrical connections is only achievable at the expense of tensile strength. Suspended power cables are

Table 14.1

Resistivities of some metallic elements at 20°C (some at 300 K)

Chemical symbol	Resistivity ($\mu\Omega$ m)	No. of valence electrons/atom	Applications
Cu	0.0174	1	Most electrical wiring
Ag	0.0163	1	Coatings for VHF inductor wire
Au	0.0239	1	Corrosion-resistant wiring
Cd	0.0684	2	
Zn	0.060	2	
Al	0.027	3	Cheaper alternative to Cu
Sn	0.115	4	
Pb	0.207	4	
Bi	1.069	5	
Cr	0.130	1 (4s shell) + 5 (3d shell)	
Mn	1.85	2 (4s shell) + 5 (3d shell)	
Ni	0.0691	2 (4s shell) + 8 (3d shell)	
Pt	0.106	1 (6s shell) + 9 (5d shell)	Accurate thermometry, corrosion-resistant wiring
W	0.056	6	Very high temperature heating wire, e.g. lamp filaments, cathode ray tubes, etc.

Table 14.2

Resistance of some alloys at 20°C

Chemical composition	Resistivity ($\mu\Omega$ m)	Applications
Ni80%–Cr20% (Nichrome)	1.1	Heating wire
Cu65%–Zn35% (brass)	0.07	Switch contacts, plugs, sockets
Pb60%–Sn40%	0.15	Lead–tin solder*
Sn95.5%–Ag3.8%–Cu 0.7%	0.11	Lead–free solder

*Banned from use as solder in Europe from 2004, because lead is highly toxic.

thus made around steel cores for high tensile strength. However, high ductility makes both for a flexible wire, and for ease of manufacture by drawing (i.e. pulling) the metal through a hole in a circular die. It is fortunate that copper has these qualities, for superconductors generally do not.

The special requirements of corrosive environments mean that gold, a gold alloy or platinum are preferred in such cases, sometimes as a surface coating.

When a high-frequency alternating current flows in a wire, Faraday's law of induction predicts that the rapidly changing magnetic flux associated with the current produces an induced electromotive force that pushes the flowing electrons out towards the surface of the wire. The current flows mostly in a thin 'skin' at the surface, within a so-called *skin depth* δ, which is related to the frequency f of the alternating current, the resistivity ρ and the magnetic permeability $\mu_r \mu_0$ of the wire, by the equation

$$\delta = \sqrt{\frac{\rho}{\pi f \mu_r \mu_0}} \qquad (14.1)$$

The effective resistance of the wire to the alternating current equals the DC resistance of a skin of thickness δ. In copper, Eq. (14.1) predicts that δ

is about 7 mm at a frequency of 100 Hz (somewhat above the power distribution frequency), while it is $2\,\mu$m at the frequency 1 GHz (10^9 Hz, a mobile phone frequency). The power lost as heat in a wire increases as δ decreases. Because of its low resistance, a thin coating of silver on a wire reduces the electrical power lost[†]. In some very high frequency (VHF) radio applications this is useful, and even necessary, because it not only reduces power wastage, but also improves the *selectivity* of a resonant electrical circuit to a narrower frequency range than is otherwise possible.

The interconnecting wires inside electronic 'chips' (silicon integrated circuits), which are discussed in detail in Chapter 17, have until recently been made predominantly from aluminium in preference to copper, because even the smallest trace of copper must be prevented within silicon, to avoid destroying its electrical behaviour. But since 2001, new materials technology has enabled the use of copper wiring, which is rapidly taking over in power-hungry chips such as fast central processors for computers.

The next section begins with a discussion of how valence electrons move freely around in a regular crystal lattice, and shows how their kinetic energies cover a surprisingly wide range of values, even when the thermal energy kT is a small fraction of an electron volt. The 'electron gas' model for a metal is introduced in Section 14.3, and is further developed in Sections 14.4 and 14.5, where the concept of a *mean time between collisions* of electrons is explained. In Sections 14.6 and 14.7 the effect of temperature on resistivity is explained through the concept of phonons, first introduced in Chapter 7. The effects of alloying and magnetic fields are covered in Section 14.8, and in Section 14.9 we give a simplified explanation of the topic of superconductivity and its applications.

Looking again at the data for the elements in Table 14.1, we can see that the number of valence electrons alone does not determine resistivity, so it does not follow that more electrons lead to higher conductivity. To understand this, we shall need to look in more detail at the mechanism of conduction.

14.2 Electrons in a field-free crystal

We begin by considering the behaviour of the valence electrons when there is no potential gradient in the metal and no current is flowing. Having understood this, we then consider how the behaviour is modified when a potential drop is applied across the solid.

The valence electrons in a metal are not only able to leave their parent atoms but they can also wander freely through the lattice of ions. One might expect them to collide with each immobile ionic core, and it was once thought that this would cause a large resistance to their flow. However, this conclusion was drawn without taking account of the wave nature of electrons, as we shall now show.

An electron moving in a given direction at a constant velocity behaves as a plane wave. The wave interacts with each ionic core and is thereby 'scattered'. Each ion then becomes the 'source' of spherical secondary wavelets in phase with the incident wave (Fig. 14.1), exactly as was assumed in the discussion of Bragg diffraction at the end of Chapter 6.

[†] For frequencies between about 300 kHz and 3 MHz, the surface area of the wire is commonly increased instead, by the use of ultra-fine stranded wire, called Litz wire.

Fig. 14.1

A plane wave cannot be diffracted
by a row of atoms when $\lambda > 2d$.

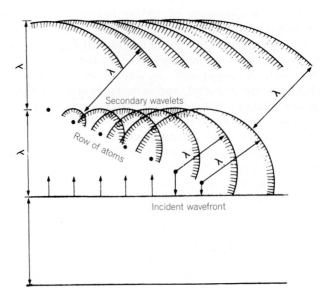

It was concluded there that the secondary wavelets reinforced one another when the condition

$$n\lambda = 2d \sin \theta$$

was satisfied. In this equation, the expression $2d \sin \theta$ represents the difference in total path length for two adjacent wavelets. But if $\lambda > 2d$, as in the case for electrons of sufficiently low speed, there is no value of θ for which the path difference can be as much as λ, so that reinforcement can only occur where the path difference is zero. This can only occur in the direction of travel of the original wave which therefore proceeds, undiminished in intensity by its encounter with the ions.

A low-energy electron, therefore, is able to move undeflected by a perfectly regular lattice, and behaves as though the ions were not there! We can thus calculate the wave functions for the conduction electrons and determine the expected energy levels by ignoring the presence of the lattice. (A more exact treatment shows that this is not strictly valid. However, the elementary approach outlined here gives the main features of metallic conduction, while the more advanced method merely enables the finer details to be understood.)

Naturally, in a real metal there are many electrons, all moving in different directions and having different wave functions, but we begin by asking what are the possible wave functions for a single electron in the metal. Then we can add the other electrons, taking care not to violate Pauli's principle.

If, as explained in earlier chapters, the electrons are spread uniformly throughout the lattice they cannot be standing waves since in standing waves there are local maxima and minima in the probability distribution. They must therefore be travelling waves because these have uniform intensity everywhere.

Now each electron must be confined within the solid but it is not localized near any atom or group of atoms. It can be regarded as 'orbiting' all the atoms in the solid, moving in a curved path of very large radius so

Fig. 14.2

The valence electrons in a metal move along a path that encompasses the whole crystal.

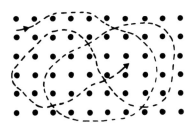

that it is (almost) a plane wave. The path will link every atom in the solid so that the overall path length is very great.

An illustration of this in two dimensions is shown in Fig. 14.2. The path must eventually close upon itself for the same reason as in the hydrogen atom (Chapter 3), namely that if it were not closed it would eventually lead off to infinity. A mathematical description of this picture, in one dimension for simplicity, is to introduce boundary conditions (see Section 2.6), which describe the conceptual model. The main point is that the long, meandering path of the electron must eventually close up on itself, so at the junction the wave functions must be equal and we can write

$$\psi(x) = \psi(x + L) \tag{14.2}$$

where L is the length of the meandering path and x is measured along the path. This is called a 'periodic' boundary condition.

Because the electron is free to wander anywhere in the solid unaffected by the ions, its potential energy with respect to the lattice must be zero, i.e. $V = 0$ in Schrödinger's equation. Thus, in one dimension, we want to solve

$$\frac{d^2\psi}{dx^2} + \frac{8\pi^2 m}{h^2} E\psi = 0 \tag{14.3}$$

and the solution has to obey the boundary condition of Eq. (14.1). We assume a solution (see Appendix 5)

$$\psi = A\exp(jkx) \tag{14.4}$$

in which A is a constant. Differentiating twice and substituting in Eq. (14.2) we obtain

$$E = h^2 k^2 / 8\pi^2 m \tag{14.5}$$

which is the expression for the kinetic energy of the electron [see Eq. (2.20)]. But we require, for the boundary condition,

$$A\exp(jkx) = A\exp(jk(x + L))$$

$$= A\exp(jkx)\exp(jkL)$$

so that

$$\exp(jkL) = 1 \tag{14.6}$$

or

$$\cos(kL) + j\sin(kL) = 1$$

using de Moivre's theorem. This is fulfilled if $kL = 2n\pi$, i.e.

$$k = 2n\pi/L \tag{14.7}$$

where n is an integer.

Substituting this in Eq. (14.5)

$$E = n^2 h^2 / 2mL^2 \qquad (14.8)$$

and we see that the energy is quantized, n being the quantum number, and that if the path length, L, is large the quantum steps in energy will be small. Normalizing the solution [see Eq. (2.21)] we have

$$\int_0^L A \exp(jkx) A \exp(-jkx) dx = 1$$

i.e.

$$\int_0^L A^2 \, dx = 1$$

and

$$A = 1/\sqrt{L} \qquad (14.9)$$

Thus, the full solution is

$$\psi = \frac{1}{\sqrt{L}} \exp(jkx) \qquad (14.10)$$

The probability distribution is given by $|\psi|^2 = A^2 = 1/L$, which is independent of position x, so there is an equal probability of finding an electron anywhere, which fits our picture of 'free' electrons in a metal.

A full treatment would show that, just as in the hydrogen atom, there will be one quantum number for each dimension, because the solid is three-dimensional, so we require three quantum numbers to describe the energy levels. These are, in fact, so closely spaced that the electron can have almost any energy and the range of permitted energy forms an essentially continuous *band*. The energy level diagram is shown in Fig. 14.3, where the spacing between the levels is greatly exaggerated for clarity.

Now, if we add all the other valence electrons one by one to the solid we shall gradually fill these energy levels from the lowest up. Only two electrons, with opposite spins, may have the same set of quantum numbers so that even where several sets of quantum numbers give the same energy there is only a finite number of electrons in each energy level.

Eventually each of the available electrons will have been allocated to an energy level, and all those levels up to a given value, known as the *Fermi level*, will be filled (Fig. 14.3).

This is the Fermi energy that appears in the Fermi–Dirac probability distribution given in Eq. (7.12). At absolute zero, $kT = 0$ and so any value of energy above the Fermi level, which makes $(E - E_F)$ finite and positive will give $\exp(\infty)$ in the denominator of Eq. (7.12) and the probability of occupancy of levels above E_F is, at absolute zero, itself equal to zero.

We now summarize the important results obtained so far and which will be used in subsequent sections:

(a) The valence electrons in a metal behave as if the ion cores were absent and the solid were composed of free space.
(b) The spacing between the energy levels available to the valence electrons is so close that their energy may be regarded for practical purposes as continuously variable.

Fig. 14.3

The energy levels available to the valence electrons.

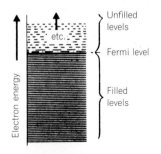

(c) In spite of this there is a finite number of available levels, and these are filled by the valence electrons, up to a highest level called the Fermi level. The energy of this level is typically about 4 eV above the lowest level in the band.

14.3 'Electron gas' approximation

Because the kinetic energy of the valence electrons in a metal is almost continuously variable the electrons may be treated as if they were particles in empty space. This important deduction enables us to simplify greatly the calculation of electrical resistivity. Before proceeding with that calculation it is useful to think a little more about the electron in this light. The electrons have a wide range of kinetic energies, as we have seen, owing to Pauli's principle. Thinking of them as particles, we see that this implies a correspondingly wide range of velocities. However, if there is no overall motion of electrons in any direction (that is, no net current) there must be as many electrons moving in one direction as in the opposite direction. The velocity may thus have both positive and negative values but the *average* velocity is zero.

If we plot the fraction, f, of the total number of electrons with a velocity, v, against the value of v, we obtain a curve like that in Fig. 14.4. All velocities between $-v_{max}$ and $+v_{max}$ are represented. The value v_{max} is that corresponding to the energy, E_F, of the Fermi level, mentioned in the previous section. By putting $\frac{1}{2}mv_{max}^2 = E_F$ it is found that, when E_F is 4 eV, v_{max} is about 1.2×10^6 m s^{-1} – a very high velocity.

We conclude that the valence electrons in a metal may be pictured as particles moving randomly around at a very high average speed, although also with a wide range of speeds. This suggests that the electrons are very much like a gas of atoms, in which the atoms move about randomly with high thermal energy. The analogy must not be allowed to cloud the fact that the reason why the electrons have high energy (velocity) is quite different from that for the gas atoms. The latter are in rapid motion because of thermal agitation. In the case of the electrons, the effects of thermal agitation are secondary: the velocity of an electron with kinetic energy $\frac{3}{2}kT$ is very tiny indeed compared to the average velocity in Fig. 14.4.

There is one further important effect that must be considered and this is that the electrons fail to 'collide' with the lattice atoms only when the lattice is perfectly regular. If any irregularity is present the moving electron may be scattered and its direction altered. In a real metal there are many such irregularities (see Chapter 8). The most important, however, is due to the vibration of atoms about their mean positions as a result of thermal

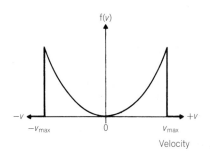

Fig. 14.4

The idealized distribution of velocities among the valence electrons.

Fig. 14.5

Each valence electron takes a zig-zag path, due to frequent collision with vibrating atoms.

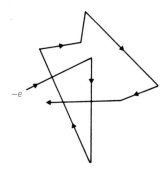

$-e$

agitation. Because these vibrations are random they upset the regularity of the lattice structure and the electrons can thus 'collide' with them. We shall see later how it is we know that this is normally the most commonly occurring kind of collision and that the presence of impurity atoms, grain boundaries, and so on, is often of secondary importance.

In between these collisions the electrons move undisturbed in a straight line (Fig. 14.5). Their paths are thus random zig-zag patterns threading the lattice.

It should be noted that the speed of the electron may also change when it makes a collision – the loss or gain in its kinetic energy is transferred to or from the vibrating atom with which it collides. However, the overall distribution of electron velocities is still as shown in Fig. 14.4 because for every electron that gains in velocity by collision there is another which loses an equal velocity.

We see from the above that the analogy between the electron and a gas is quite close. Just as in a gas, it is possible to define an average path length between collisions, which is usually called the *mean free path*, and a corresponding *mean free time*, namely, the average time spent in free flight between collisions. It is found that the mean free path is about 400 Å in copper at room temperature so that, on average, the electron passes about 150 atoms before colliding with one.

14.4 Electron motion in applied electric fields

Having established the behaviour of electrons in the absence of an electric field, we now turn to the effects produced by applying a field. Because the electrons act between collisions as if they were in free space they are readily accelerated by a potential difference across the crystal, and a large current flows.

The field creates a drifting motion of the whole cloud of electrons in the direction opposite to the field – we often say that the electrons acquire a drift velocity, v_d. This gives rise to an electric current whose magnitude we now calculate in terms of v_d.

Let the current be flowing in the x direction and imagine a cylinder of unit cross-section whose axis is also parallel to the x direction (Fig. 14.6). The current flowing down it is equal to the amount of charge crossing any plane, say plane A, in unit time. This must be just equal to the amount of charge contained in the cylinder in a distance v_d upstream of the plane A. This volume is shown shaded in the figure.

If there are n electrons per unit volume, each of charge $-e$, the total charge in the shaded volume is just $-nev_d$. They are moving in the +x

Fig. 14.6

Calculation of the current carried by electrons moving with a drift velocity v_d.

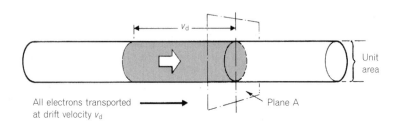

v_d

Unit area

All electrons transported at drift velocity v_d →   Plane A

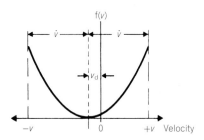

Fig. 14.7

Distribution of velocities among the valence electrons in the presence of an electric field.

direction but by convention the current is in the −x direction, and therefore has a negative sign. Thus we obtain

$$-J = -nev_d \quad \text{or} \quad J = nev_d \qquad (14.11)$$

Because J is the current per unit cross-sectional area it is called the current density. When the current density in a piece of copper is say $1\,\text{A}\,\text{cm}^{-2}$ the drift velocity v_d is about $7.4 \times 10^{-7}\,\text{m}\,\text{s}^{-1}$ (see Problem 14.2) so that v_d is only a very small fraction of the average random velocity of the electron gas. It is therefore permissible to assume that the distribution of velocities among the electrons is scarcely affected by the flow of current. All that happens is that each electron acquires an additional velocity equal to v_d; if the original velocity was v the new value is $(v + v_d)$. Note that if v is negative there is a reduction in the magnitude of the velocity. We show in Fig. 14.7 the new graph of $f(v)$, which is the same shape as before but shifted horizontally by an amount v_d in the direction *opposite* to the field ξ. The size of v_d has been exaggerated to clarify the illustration.

14.5 Calculation of drift velocity v_d

So far we have assumed that each electron acquires a drift velocity, v_d, without enquiring why it is that the field does not accelerate the electron without limit. To understand this point we must look again at the details of the electron motion. In Fig. 14.5 the path of the electron was depicted as a zig-zag course due to collisions with vibrating atoms. In the interval between collisions the electron experiences a force due to the electric field. Figure 14.7 shows how an electron already travelling in the −x direction is thereby accelerated, while an electron travelling in the +x direction is decelerated. We show in Fig. 14.8 how the path becomes curved when the electron is travelling in some other direction – the curvature is greatly exaggerated for clarity. In each case, however, the force, whose magnitude is $-e\xi$, is in the same direction. During free flight the electron is accelerating; the rate of change of v_d is then equal to the force divided by the electron mass so that in free flight

Fig. 14.8

The path of each electron becomes curved in an electric field.

$$\left(\frac{dv_d}{dt}\right) = -\frac{e\xi}{m} \qquad (14.12a)$$

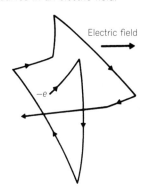

On the other hand, because the system is in equilibrium we know that, averaged over a significant period of time, $dv_d/dt = 0$. This means that the rate of change of v_d given by Eq. (14.3a) is counterbalanced by the rate at which the drift velocity is lost by collision with the lattice vibrations. The latter can be calculated as follows.

The essentially random nature of a collision is not affected by the fact that the electron has acquired a little extra velocity while in free flight. The direction of the electron velocity immediately after a collision is thus entirely random, as it would have been if the electric field were not present. Immediately after collision, then, the electron's velocity is completely random, and all memory of the drift velocity appears to have been lost. (Note that the *energy* associated with the drift velocity is not lost. It is in part transferred to the lattice.) At the beginning of the next free flight we may assume that $v_d = 0$.

At each collision the change in v is thus equal to v_d. If we let τ be the average time spent in free flight between collisions, then each electron collides $1/\tau$ times per second, each time losing a velocity v_d.

The total amount of velocity lost in one second is therefore $v_d \times 1/\tau$ and this is the rate of change of v_d.

Thus the average collision loss is

$$\left(\frac{dv_d}{dt}\right)_{loss} = \frac{v_d}{\tau} \qquad (14.12b)$$

Note that, unlike the gain in velocity, the loss occurs only at the time of the collision. The rate of loss is therefore zero for the time, τ, between collisions and very large for the negligible time it takes to complete the collision. The average rate is given by Eq. (14.12b).

We have already noted that the total rate change of v_d is zero so we must add Eqs (14.11a) and (14.11b) and equate to zero, giving

$$\frac{dv_d}{dt} = \left(\frac{dv_d}{dt}\right)_{gain} + \left(\frac{dv_d}{dt}\right)_{loss} = 0$$

Thus

$$\frac{v_d}{\tau} = \frac{\xi e}{m} \quad \text{and hence} \quad v_d = \frac{e\xi\tau}{m} \qquad (14.13)$$

Thus v_d is proportional to ξ, and it is customary to call the drift velocity in unit electric field the *mobility* of the electron, μ.

Using Eqs (14.10) and (14.12) we obtain the equation for current density

$$J = \frac{ne^2\xi\tau}{m} \qquad (14.14)$$

Because the resistivity, ρ, of the metal is just ξ/J (see Problem 14.1) we have

$$1/\rho = \frac{ne^2\tau}{m} = ne\mu \qquad (14.15a)$$

where

$$\mu = \frac{e\tau}{m} \qquad (14.15b)$$

The two quantities that determine the resistivity are therefore the electron density, n, and the mean free time, τ. Because the electron density cannot vary with temperature the whole of the temperature dependence of ρ must be due to changes in τ. In the next section we shall consider how τ depends on temperature and also how it can be influenced by the defect structure of a metal.

For the moment let us return to consider the implications of the increase in the random velocity that occurs between each collision. Naturally this cannot be a cumulative process and the electrons actually lose some of this extra energy during the collision, transferring it to the vibrating atoms whose vibration therefore increases. This corresponds to an increase in temperature of the solid. Indeed this is just what is observed; the passage of current through a metal causes heating. The amount of energy transferred to the lattice increases as v_d^2, and so the heating effect is proportional to the square of the current.

14.6 Phonon scattering

So far, in discussing collisions between electrons and atoms we have treated each atom as vibrating independently. This was the basis of the theory of resistivity developed by Drude in 1900 and subsequently refined by Lorentz. However, as was pointed out in Chapter 7, a single atom in a solid, when it vibrates, exerts forces on its neighbours and the atomic vibrations must be treated in terms of lattice waves (see Section 7.10).

As with electron waves, we can allocate a wave vector, $q = 2\pi/\lambda$, to the lattice wave. Using the wave-particle duality concepts that we applied to electromagnetic and to electron waves, we represent the lattice waves as a stream of 'particles' each carrying a momentum $hq/2\pi$ and with an energy quantized in units of $h\nu$ where ν is the frequency of the lattice wave. These 'particles' are the *phonons* described in Chapter 7 and the interaction of an electron wave with a lattice wave can be regarded as a 'collision' between an electron and a phonon. In the collision process either the phonon energy will be transferred to the electron and the phonon disappears, or else the electron will lose energy to the lattice thereby creating a phonon. In thermal equilibrium at a fixed temperature the rate of creation equals the rate of destruction and the phonon population is constant.

Energy must be conserved in such an interaction, and so we can write

$$E(k) - E(k') = \pm h\nu \tag{14.16}$$

where $E(k)$ is the electron energy of wave vector k before the interaction and $E(k')$ afterwards. Also, by conservation of momentum

$$hk'/2\pi - hk/2\pi = \pm hq/2\pi \tag{14.17}$$

where the positive sign will apply if the electron has taken up a momentum $hq/2\pi$ from the lattice and the minus sign if momentum is transferred to the lattice.

We saw in Section 14.5 that the effect of an electric field in (say) the $-x$ direction is to increase the velocity of the electron in the $+x$ direction and to decrease it in the $-x$ direction. Through electron–phonon 'collisions' the $+x$ electron interacting with a lattice wave can lose its extra velocity by creating one or more phonons, whilst the $-x$ electron can regain the velocity it lost by destroying one or more phonons. These processes result in a change in velocity and direction of travel for an electron at each 'collision' and are referred to as phonon scattering of the electrons, each 'collision' being a scattering event.

14.7 Dependence of resistivity on temperature

The reader is probably familiar with the fact that the resistivity of a metal increases linearly with temperature. It is possible to demonstrate this using the theory outlined in this chapter and we now do so. To simplify the derivation we merely deduce the form of the relationship and the constant of proportionality (that is, the temperature coefficient of resistance) will not be calculated.

In Eq. (14.14) the only temperature-dependent quantity is the mean free time. This is connected with the probability of a phonon–electron interaction (i.e. a 'collision' or 'scattering event'). If the probability is small a long time will elapse before a collision occurs and the mean free time will be long. It is, in fact, a standard result of kinetic theory (see the end of this chapter) that if the mean free time is τ, then the probability of a collision occurring in time δt is $\delta t/\tau$ or, putting it another way, $1/\tau$ is the probability per unit time of a collision occurring.

The probability of a phonon–electron interaction occurring will be proportional to the product of the numbers of electrons and of phonons present. Now the number of electrons is constant and independent of temperature but the number of phonons is given by Eq. (7.19). At ordinary temperatures, $h\nu$ is very small ($< 4 \times 10^{-4}$ eV) whilst the value of $k_B T$ at room temperature is 0.025 eV (k_B is Boltzmann's constant). Thus the fraction inside the exponential bracket in Eq. (7.19) is very small and we can replace the exponential by the first two terms in the exponential series, i.e. by $(1 + h\nu/kT)$. Thus Eq. (7.18) can be approximated to

$$n_q \approx k_B T/h\nu$$

We see, therefore, that the probability p of an interaction will be given by $p \propto n_e n_q \propto n_e k_B T/h\nu$, i.e.

$$p = 1/\tau = CT$$

where C is a constant. But from Eq. (14.15a) resistivity ρ is proportional to $1/\tau$ and so we deduce that the resistivity of a metal will be proportional to temperature over the range of temperature for which $kT \gg h\nu$. This precludes low temperatures (below, say, 20 K) where, experimentally, ρ is no longer proportional to T. At extremely low temperatures, a metal may exhibit superconductivity (see Section 14.9).

14.8 Dependence of resistivity on structure and magnetic fields

When measurements are made at low temperatures the resistivity tends to a finite value [Fig. 14.9(a)] as the temperature is lowered. This is because the number of phonons becomes so small at these temperatures that electron and phonon collisions become less important than those with crystal defects such as impurities, grain boundaries, vacancies, and other irregularities in the ordered pattern of a perfect crystal. Because the number of these is fixed by the structure, the mean free time no longer depends so strongly on temperature, as we shall see below.

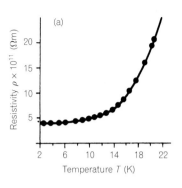

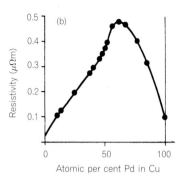

Fig. 14.9

Resistivity of (a) sodium at low temperature and (b) palladium–copper alloy at room temperature.

The effect of foreign atoms on resistance is most clearly seen by studying the *increase* in resistivity with increasing concentration of *small* amounts of a solute. This can be small or large: for example 1 atomic percent of tin in otherwise pure copper raises the resistivity to $0.26\,\text{m}\Omega\,\text{m}$ from $0.18\,\text{m}\Omega\,\text{m}$. Matthiessen (1864) showed that the increase does not depend on temperature, proving that it is caused by structural changes alone. This result also implies that the temperature coefficient of resistivity, $d\rho/dT$, is independent of concentration, which is found by experiment to be true except in Co and Cr host metals.

A foreign atom is said to 'scatter' the electron waves, reducing the mean free path. If the mean time between collisions with phonons is τ_{ph}, while that between collisions with impurities is τ_i, then the addition of the impurity raises the overall frequency of collisions from $1/\tau_{ph}$ to $(1/\tau_{ph} + 1/\tau_i)$. The new mean time τ between collisions is thus given by

$$\frac{1}{\tau} = \frac{1}{\tau_{ph}} + \frac{1}{\tau_i} \qquad (14.18)$$

provided that the addition of the solute atoms does not affect the number of free electrons available, and that the amplitude of thermal vibration of solute and solvent atoms is similar.

In such cases, Eq. (14.15a) leads us to expect the solute to make a temperature-independent addition to the resistivity, proportional to $1/\tau_i$, i.e.

$$\rho = \rho_{ph} + \rho_i \qquad (14.19)$$

where ρ_{ph} is the temperature-dependent resistivity. Equation (14.18) shows that if τ_i is very small compared with τ_{ph} then ρ_i becomes large, so that the fractional change in resistivity with temperature becomes much smaller.

From a practical viewpoint, the temperature coefficient of resistivity, $(1/\rho)\,d\rho/dT$, is more important than just $d\rho/dT$, because it determines the fractional change in resistance. The discussion above suggests that by simply introducing many defects it is possible to decrease the temperature coefficient of resistivity, because this raises the resistivity ρ without at the same time increasing $d\rho/dT$. This is the case in many disordered solid solutions, whose resistance may be much higher than that of either constituent. An example of this is shown in Fig. 14.9(b) for the copper–palladium alloys. The copper–nickel alloys (for example, 'Constantan'), which are used for making electrical resistances with low temperature coefficients, show similar behaviour, partly for the same reason.

On the other hand, quite different behaviour is found in two-phase alloys that have a eutectic structure at intermediate compositions. In this case the material is essentially a mixture of two phases with different resistivities, and from an electrical viewpoint it may be regarded as a network of resistances with randomly varying values in a complicated series-parallel combination with one another. It can be shown that the resistance of such a structure varies linearly with the volume fraction of either constituent. Thus, if the resistivities of the two separate phases, α and β, are ρ_α and ρ_β then the resistivity ρ of the two-phase material is

$$\rho = \rho_\alpha V_\alpha + \rho_\beta V_\beta$$

where V_α and V_β are the volume fractions of the α and β phases.

14.8.1 Magnetoresistance

Because electric currents experience forces in magnetic fields, it is no surprise to find that magnetic fields can change the voltage across a piece of conducting material. Two important effects can be distinguished: the Hall effect, which will be discussed in connection with semiconductors in Chapter 16, and magnetoresistance. As its name implies, magnetoresistance describes a change in the resistivity of a material when a magnetic field is applied. The effect arises from an increase in the frequency of collisions, requiring the addition of an extra term to the right-hand sides of Eqs (14.18) and (14.19). In most metals, and in many other materials, the effect is small, typically a few percent at most when the magnetic flux density B reaches 1 Tesla. Because the explanation of this behaviour depends strongly on the details of their magnetic properties, we shall defer this topic to Chapter 17.

14.9 Superconductivity

The electrical resistance of some metals and compounds drops suddenly to zero at some, very low, temperature called the *critical temperature*, T_c. This was first observed by Kammerlingh Onnes in 1911 in mercury at a temperature of 4.2 K, and since then it has been found that, if magnetic elements are excluded, about half of all elemental metals and thousands of alloys and intermetallic compounds, including metallic oxides, borides and nitrides as well as some semiconductors, polymers, and even, under suitable conditions, C_{60} and DNA molecules, exhibit the effect at low enough temperatures. In the zero-resistance state they are called *superconductors*. Of the pure metals, niobium has the highest critical temperature, 9.2 K; others have values all the way down to 0.001 K.

The search for materials with high critical temperatures has been intense. Until 1986 the highest known critical temperature was 23 K for a complex alloy of niobium, tin and other constituents.[†]

But in 1986 a very significant revolution occurred when a totally new class of metallic oxides with very much higher critical temperatures was announced by Mueller and Bednorz of IBM, who were awarded the 1987 Nobel Prize for Physics. One example is $YBa_2Cu_3O_7$, with a critical temperature of 92 K. The great significance of this development is that it is no longer necessary to use expensive liquid helium to cool below T_c, because liquid nitrogen boils at 77 K and costs about the same as a cheap

---

[†] As recently as 2001 the rather simple compound MgB_2 was unexpectedly found to have $T_c = 40$ K.

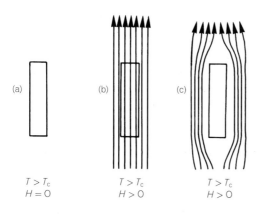

Fig. 14.10

The expulsion of magnetic flux from a superconductor – the Meissner effect. Flux lines are drawn for three cases: (a) no magnetic field; (b) with applied magnetic field and $T > T_c$; (c) as in (b) but with $T < T_c$.

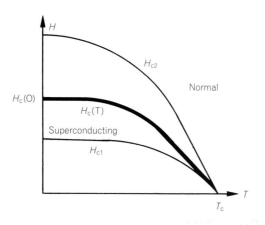

Fig. 14.11

The critical magnetic field H_c of a superconductor varies dramatically with temperature.

beer! At the time of writing the highest reported critical temperature is about 135 K. These materials are not easy to manufacture and are mostly brittle, but they allow many practical applications of superconductivity at economic prices.

In the superconducting state the electrical resistivity is zero so that, if a current is induced in the superconductor, it will continue to flow, without diminution, so long as the material remains at a temperature below T_c. As the value of the induced current is increased, a *critical current density*, J_c, is reached, at which the material loses its superconductivity and reverts to normal behaviour. In Section 14.9.3 we shall explain that superconducting electrons move around in pairs, called *Cooper pairs*.

Associated with superconductivity there is an equally dramatic magnetic effect, discovered in 1933, called the *Meissner effect*. If the superconductor material, in its normal state, is placed in a magnetic field and then cooled through the critical temperature it is found that, at T_c, the whole of the magnetic flux is suddenly ejected from the material – it behaves while superconducting as a perfect *diamagnet* (see Chapter 18), as illustrated in Fig. 14.10. However, if the strength of the applied magnetic field strength is steadily increased, we reach a *critical field*, H_c, at which the superconductivity is destroyed and the material reverts to its normal, resistive, state. This effect was an important clue in developing the theory of superconductivity, discussed later. The critical field H_c is related to the critical current density mentioned above, because any electric current generates a magnetic field. The magnitude of the critical field at which superconductivity disappears varies with the temperature, being lower the nearer the temperature approaches the critical temperature T_c from below, as illustrated in Fig. 14.11. The value of H_c quoted for a superconducting element is usually the value of field at which the curve cuts the H axis at $T = 0$ K.

As either the temperature approaches T_c or the applied magnetic field approaches H_c, the magnitude of the critical current density approaches zero. The critical current J_c is not strictly an intrinsic property of the superconductor and depends on the metallurgical state of the material, i.e. crystallite size, defect density and so on.

14.9.1 Type I and Type II superconductors

The flux density B in a material experiencing a magnetic field of strength H is given by Eq. (17.1):

$$B = \mu_0 (H + M)$$

where μ_0 is the magnetic permeability of free space and M is the *intensity of magnetization* of the material. Because of the Meissner effect, the flux density B is exactly zero inside a superconductor and hence $H = -M$. In other words, the superconductor can be thought of as having a negative intensity of magnetization equal to the applied field strength H. In Fig. 14.12 this negative magnetization is plotted against field for typical elemental and typical alloy superconductors. For the former, the magnetization rises in magnitude up to the critical field and then abruptly drops to zero; this is called *Type I* superconducting behaviour.

In a few elemental and most alloy superconductors it is found that, above a certain value of flux density, B_{c1}, the magnetic flux can penetrate the material without destroying the superconductivity. As the applied field is increased further, the magnitude of the negative magnetization decreases until, at an *upper critical field*,[†] B_{c2}, the material reverts completely to the normal state. The material is called a *Type II* superconductor and the value of B_{c2} may be up to 2000 times the value of B_{c1}. Between these values of field

Fig. 14.12

Intensity of magnetization, M, of Types I and II superconductors, plotted against magnetic field strength H.

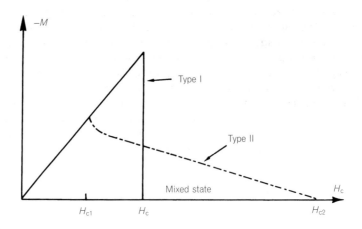

[†] The term *critical field* is normally used, even though B_{c2} is a flux density, measured in Teslas.

Elements	T_c (K)	B_c $(T=0\,K)$ (Tesla)	Alloys	T_c (K)	B_{c2} $(T=4.2\,K)$ (Tesla)
Aluminium	1.19	0.01	$Pb_{0.9}Mo_{5.1}S_6$	14.4	60.3
Indium	3.41	0.029	V_3Ga	14.8	23.9
Mercury	4.15	0.041	NbN	15.7	10.0
Vanadium	5.3	0.132	Nb_3Sn	18.0	26.4
Lead	7.18	0.090	$Nb(Al_{0.7}Ge_{0.3})$	20.7	41.5
Niobium	9.46	0.196	Nb_3Ge	22.5	36.4
			MgB_2	40	>17

High-temperature alloys	T_c (K)	B_{c2} (Tesla) $B//a,b$*	B_{c2} (Tesla) $B//c$*
$YBa_2Cu_3O_7$	92	240	34
$Bi_2Sr_2Ca_2Cu_3O_{10}$	109	high	high
$Pb_2Sr_2(Y,Ca)Cu_3O_8$	76	590	96
$La_{0.95}Ca_{0.05}CuO_4$	14	>20	>13

* Note: $B//a,b$ means a field applied in the *ab* basal plane of the crystal and $B//c$ means a field along the *c*-axis, where the crystal lattice constants are *a*, *b* and *c*. The *ab* plane contains the CuO_2 groups.

the material is in a mixed state in which the specimen is penetrated by long filaments of magnetic flux, each of which contains the same, quantized, amount of magnetic flux equal to $hc/2e$, where h and e have their usual meanings, and c is the velocity of light. Each filament contains a cylindrical core of 'normal' (resistive) material surrounded by a cylinder around which a vortex of supercurrent flows. Each cylinder is referred to as a *vortex* and adjacent vortices repel each other. As the magnetic field increases, the number of filaments per unit cross-sectional area increases, and the filaments pack together in a two-dimensional lattice. Between the vortices the material remains superconducting until the field B_{c2} is reached, when the core regions of the vortices are forced into contact with one another, and the whole specimen reverts to the normal state. The high-temperature superconductors are Type II and have usefully high values of B_{c2}.

Thus B_{c2} represents an upper limit to the flux density in practical situations where the flux is unchanging – in other words, for direct current use. In an alternating magnetic flux resulting from an alternating current, the vortices move around, resulting in energy dissipation, unless the material contains just enough crystallographic defects that the vortices are 'pinned' in place and are unable to move. The highest magnetic flux density in which the vortices remain pinned is called the *irreversibility field*, and is usually somewhat less than the value of B_{c2}. Materials scientists thus face the challenge of finding how to introduce defects that pin the vortices most effectively, at a low enough cost to make large quantities of useful materials.

A short list of superconducting materials with their critical fields, where known, is given in Table 14.3. We have left C_{60} (buckminsterfullerene) out of the table, because its superconducting properties, only recently discovered, appear to vary remarkably strongly with the separation of the 'buckyball' molecules.

14.9.2 Practical high-temperature superconductors

Since their discovery in 1986, much effort has gone into solving the technical challenges of turning high-temperature superconductors into practical materials. Their large-scale commercial use in power distribution (cables and transformers), in electrical motors and in large magnets for industrial processes, and magnetic resonance imaging units for medical applications, is only just beginning at the time of writing. On the other hand, they are already in use in hundreds of microwave transmitters in cellphone base stations, where they are built into frequency-selective filters – small components of high value.

The technical challenge is to make long flexible wires at sufficiently low cost. The long-term aim is for about $US50 per metre for 1000 amps carried, but in 2002 the price is about six times higher. Materials technologists face demanding microstructural requirements for making wires, outlined below. Two materials that have been turned into practical, flexible, 'wire' are $Bi_2Sr_2Ca_2Cu_3O_{10}$ (abbreviated Bi-2223) and $Bi_2Sr_2CaCu_2O_8$ (Bi-2212). Their crystal structures are similar – a cross-section through the structure of Bi-2212 is shown in Fig. 14.13, where you can see that layers of copper oxide, flanked on either side by calcium and strontium oxide layers, are separated by a double layer of bismuth oxide. As in all the cuprate superconductors, transfer of electrons between the atoms results in 'missing' electrons, referred to as 'holes' in the copper oxide layer, and these are responsible for enabling conduction in this oxide structure, which would otherwise be an insulator (conduction by holes will be discussed in detail in connection with semiconductors in Chapter 15). The CuO_2 layers superconduct readily along their planes, but the planes are so far apart that the flow of superconducting electron pairs from layer to layer is easily disturbed by structural defects that are larger than about 0.1 nm. This is because the motion of the two electrons in such a Cooper pair must cohere with one another over a distance known as the *coherence length*, which is about 0.1 nm when measured perpendicular to the CuO_2 layers. The coherence length measured in any direction *parallel* to the CuO_2 layers is however very much larger, about 2 nm, which is fortunate, for the following reasons.

The coherence length represents essentially the extent of the wave function of a Cooper pair. Thus, within the CuO_2 layers, the pair is shaped either like a well-worn coin, or the discus used in athletics. If a pair encounters a grain boundary edge-on, as in grain 1 in Fig. 14.14, the long dimension of the wave function, about 2 nm, allows the pair to 'feel' the presence of the neighbouring grain through the disordered structure of the

Fig. 14.13

A cross-section through the crystal structure of $Bi_2Sr_2CaCu_2O_8$. The seven atoms shown in dashed outline are oxygen atoms that occupy a similar plane of atoms to the one shown, but which lies nearer the viewer. They are included to indicate how the copper and oxygen atoms are coordinated within the horizontal conducting layers near the top and bottom of the diagram.

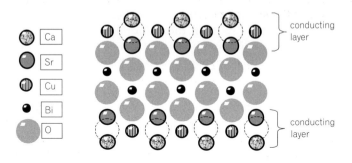

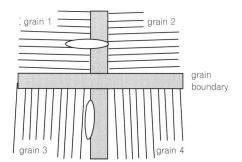

Fig. 14.14

One disc-shaped Cooper pair meeting a grain boundary (top) edge-on, and another (bottom), face-on. The lines represent the CuO₂ layers.

grain boundary, and this allows it to flow unhindered into grain 2. On the other hand, if the CuO_2 layers in the grain are oriented so that the Cooper pair presents its flat face to a grain boundary (grain 3 in Fig. 14.14), the wave function amplitude falls away so rapidly with distance that, at the other side of the grain boundary, it is too weak to enable electrons to flow across as pairs. They must revert to 'normal' electron behaviour to do so, and a finite electrical resistance results. As a result, when grains have their CuO_2 layers lying at right angles to the current flow, the grain boundaries form resistive regions that cause heating when a current flows. The consequent rise in temperature turns neighbouring grains non-superconducting when the current is large.

To avoid this, most grains must be aligned with the CuO_2 layers parallel to the current. The most successful process to date for doing this aligns the grains using cold flow. A powder containing Bi-2212, which will eventually be converted to Bi-2223, is inserted into silver tubes. These are then drawn through a die into wires, which are bundled together and drawn down again and again, before finally heating the wire to convert the powder grains to Bi-2223. The cold drawing aligns the grains by making use of the anisotropic strength: the material fractures most easily along the weak Bi-O layers rather than perpendicular to the superconducting planes. This mechanical feature is what makes the process practical.

The coherence length is inversely related to the upper critical field, and hence to the critical current density. Improvements in grain alignment, which increases the coherence length along the current direction, therefore is linked to a rise in the critical current density at which superconductivity disappears. This has risen in Bi-2223 wires from about $5000\,A\,cm^{-2}$ at the temperature of 77 K in 1991 to over $70\,000\,A\,cm^{-2}$ by 2000.

Much higher critical currents of $10\,MA\,cm^{-2}$ or more are obtainable in wires based on $YBa_2Cu_3O_7$, but manufacture was still expensive in 2001. They are made at the time of writing by coating a supporting wire with a thin layer of the superconductor.

14.9.3 Advanced topic: The theory of superconductivity – 'dancing in pairs'

For a long time after its first discovery, superconductivity was considered the most mysterious and difficult phenomenon to account for theoretically and it was not until the work of Bardeen, Cooper and Schrieffer in 1957 that a satisfactory theory (the BCS theory) was produced. The discovery of high-temperature superconductors has again challenged theoreticians, for they are not fully explained by the BCS theory.

The essential basis of BCS theory is that there is an attractive force between two electrons having nearly the same kinetic energy, a force that, in the right circumstances, causes them to move in pairs. Superconductivity occurs only if this attraction exceeds the electrostatic repulsion that always occurs between two electrons. It should be noted that the repulsive force between electrons in the solid is not as great as the force between two isolated electrons, because the lattice of positive ions acts to 'screen' the electrons from each other to some extent, reducing the repulsion. The attractive forces arise because, in a crystal lattice of positive ions, an electron will produce a small distortion of the lattice by attracting the positive ions towards itself, making the lattice slightly more dense in its vicinity. To a passing electron, this distortion will behave like a local increase in positive charge density and will attract the second electron towards it. Thus the electrons tend to form pairs through their electrostatic interaction with the lattice. If the atoms are vibrating sufficiently with thermal energy, this pairwise interaction will be obscured, but at very low temperatures where thermal agitation is small, the attractive force can assert itself. It can be shown that the attractive force will be a maximum for electrons having opposite spins, and moving with equal velocities in opposite directions.

The effect of this force is to produce an ordering of the electron gas. Just as atoms of a gas are ordered on condensing into a crystalline solid, so the electrons are ordered into pairs. The attractive force makes the two move in a coordinated, but not identical way, rather like a dancing couple. The distance over which the two electron wave functions are coordinated is known as the coherence length, referred to earlier, typically a few nm. This ordered motion is associated with a lowering of the total energy of the electron gas by an amount ΔE, and in order to return to the 'normal' state an electron must be supplied with this amount of energy. It can come from an increase in temperature above T_c, at which point the thermal energy kT becomes greater than ΔE, or from magnetic energy due to a field greater than B_c.

In terms of the energy level concept, introduced in Section 14.2, ΔE represents a gap in the allowed energy states for electrons, a so-called 'energy gap', lying at the Fermi energy level. Unless the electrons gain sufficient energy to cross the gap (and this is not possible at temperatures below T_c) there is no possibility of electrons changing their state of motion, because there are no vacant energy states that they can move to. Thus the normal mechanism of resistance to current flow cannot operate and the resistance of the material will be zero. This will remain true until the drift velocity of the electrons that carry the current exceeds a critical value, i.e. until the critical current is exceeded. The critical velocity corresponding to this critical current density is given approximately by

$$v_c = \frac{\lambda_F \Delta E}{2h}$$

where λ_F is the electron wavelength for an electron at the Fermi level.

The width of the energy gap is a function of temperature, being a maximum at $T = 0$ and falling to zero at $T = T_c$, as shown in Fig. 14.15. Just below T_c it rises steeply from zero, varying as

$$\Delta E(T) = 3.2kT_c(1 - T/T_c)^{1/2}$$

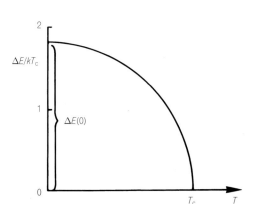

Fig. 14.15

Variation of energy gap with temperature.

and at lower temperatures varies more slowly, flattening off to a constant value $\Delta E(0) = 1.76kT_c$ at about $T = T_c/2$.

The coherence length, c_0, can be related to the energy gap through the approximate relationship

$$c_0 \simeq \frac{h^2}{m\pi\Delta E\lambda_F}$$

and is of the order of $1\,\mu\mathrm{m}$ in most elemental superconductors. This means that the electron pairing is extended over thousands of atomic spacings so that atomic scale defects and impurities are too small to have much effect on superconducting behaviour. However, in the 123 high-temperature superconductors the coherence length is of the order of a few nanometres, which means that the pairing behaviour is almost on an atomic scale and the superconducting behaviour will be much more dependent on atomic scale defects.

At the time of writing there is considerable doubt about the mechanism by which pairing of electrons comes about in the 123 superconductors. There is evidence that the mechanism may not involve phonons but is, rather, a combination of charge and magnetic interactions. It is generally agreed that the crucial role is played by the interaction between copper and oxygen ions, but the exact form of the interaction is, as yet, unknown. The only certain thing is that the next decade will see a tremendous level of activity in superconductor research and an upsurge of interest in its applications.

14.10 Collision probability and mean free time of electrons

The probability of an electron undergoing collision within time δt is proportional to δt, so we may let it equal $p\delta t$. Take a group of N_0 atoms at time $t = 0$. Some of these will collide only a short time later, others will be in free flight for longer. To calculate the *mean* free time we must know how many are left (i.e. have not collided) after an arbitrary time t; for the moment we assume that there are N of these.

Between times t and $(t + \delta t)$ a further small number, δN, will have collided. This number δN will be equal to the number N left in the group, multiplied by the probability of collision, $p\delta t$, for each electron, that is

$$\delta N = -Np\delta t \qquad (14.20)$$

The minus sign enters because δN represents a decrease in N.

We now find the relationship between N and t by integrating Eq. (14.20) between times $t = 0$ (when $N = N_0$) and t (when $N = N$), thus,

$$\int_{N_0}^{N} \frac{dN}{N} = -\int_{0}^{t} p\, dt$$

Performing the integration we find that

$$N = N_0 \exp(-pt) \qquad (14.21)$$

Now Eq. (14.20) tells us the number of electrons, δN, which collide between t and $(t + \delta t)$, and which therefore have a time of free flight equal to t. The average flight time, τ, is given by

$$\tau = \frac{1}{N_0} \int_{0}^{t=\infty} t\, dN$$

and with the help of Eq. (14.20) this becomes

$$\tau = -\int_{0}^{\infty} \frac{N}{N_0} pt\, dt$$

Using Eq. (14.21) this may be integrated by parts as follows

$$\tau = -\int_{0}^{\infty} pt \exp(-pt)\, dt$$

$$= \left[t \exp(-pt) \right]_{0}^{\infty} - \frac{1}{p} \int_{0}^{\infty} p \exp(-pt)\, dt$$

$$= \left[t \exp(-pt) \right]_{0}^{\infty} + \left[\frac{1}{p} \exp(-pt) \right]_{0}^{\infty}$$

Because the first term is zero at both $t = 0$ and $t = \infty$ we finally discover that

$$\tau = \frac{1}{p}$$

so that the collision probability is just $1/\tau$ per unit time.

Problems

14.1 Show that the following two definitions of the resistivity ρ are equivalent

(a) Resistance of a bar $=$

$$\rho \times \frac{\text{length of bar}}{\text{cross-sectional area of bar}}$$

(b) $\rho = \dfrac{\text{electric field}}{\text{current density}}$

14.2 The density of copper is $8.93 \times 10^3\,\text{kg m}^{-3}$. Calculate the number of free electrons per cubic metre and hence deduce their drift velocity when a current is flowing whose density is $1\,\text{A cm}^{-2}$.

14.3 The Fermi energies and interatomic distances in several metals are given below. Check in each case that the electrons of highest energy have a wavelength greater than twice the distance between atomic planes, and hence are not diffracted by the lattice.

Metal	Na	K	Cu	Ag
Fermi energy (eV)	3.12	2.14	7.04	5.51
Interplanar spacing (Å)	3.02	3.88	2.09	2.35

14.4 What is the kinetic energy of an electron that has a wavelength just short enough to be diffracted in each of the metals listed in Problem 14.3?

14.5 From the data given in Table 14.1 and Appendix 3, calculate the mobility of an electron in each metal at room temperature.

14.6 What is the mean free time between collisions in the metals Cu, Ag and Au? Their resistivities are given in Table 14.1 and their densities are respectively $8.93\,g\,cm^{-3}$, $10.5\,g\,cm^{-3}$ and $19.35\,g\,cm^{-3}$.

14.7 Deduce the maximum random velocity of electrons in Cu, Ag and Au if their Fermi energies are respectively 7.04, 5.51 and 5.51 eV. Using these velocities and the mean free times obtained in Problem 14.6, find approximate values for the electron mean free path in the three metals.

14.8 When about 1 atomic percent of a monovalent impurity is added to copper, the mean free time between collisions with

the impurities is about $5 \times 10^{-14}\,s$. Calculate the resistivity of the impure metal at room temperature.

Using the electron velocity deduced in Problem 14.7, find the mean distance between collisions with impurities. Compare this with the average distance between the impurity atoms and explain why the two distances are different.

14.9 If the electron mean free path were independent of temperature (as a result, say, of a large impurity content) and the electrons behaved as if they constituted a perfect classical gas, show that the resistivity would be proportional to $T^{1/2}$.

14.10 In Chapter 9 the changes in a metal resulting from plastic deformation were described. Suggest what changes you would expect in the conductivity of copper as a result of cold working. How could you restore the resistivity to its original value?

Self-assessment questions

1 The conductivity of a metal is determined only by the number of valence electrons per atom

   (a) true   (b) false

2 In a perfectly regular metal lattice the valence electrons behave as though the ions were not there

   (a) true   (b) false

3 The energies of the valence electrons in a metal are quantized because

   (a) the electrons are each localized near an atom

   (b) the electron wave is a plane wave

   (c) the electron's path must eventually close on itself

4 The spacing between the energy levels of the valence electrons is small because

   (a) there is a large number of valence electrons

   (b) the average length of closed path traversed by the electron wave is very large

   (c) Pauli's principle applies to the electrons

5 The Fermi level is

   (a) the highest occupied energy level at absolute zero of temperature

   (b) the average of the available energy levels

   (c) the highest available energy level

6 The valence electrons in a metal may be treated as an 'electron gas' because

   (a) they exert a pressure at the boundaries of the solid

   (b) there is a very large number of them

   (c) the kinetic energy of each electron is almost continuously variable

7 The velocity of an electron at the Fermi level, if the Fermi energy is 5.0 eV, is

   (a) $3 \times 10^{10}\,m\,s^{-1}$ (b) $1.33 \times 10^{6}\,m\,s^{-1}$

   (c) $1.33 \times 10^{3}\,m\,s^{-1}$

8 Electrons will 'collide' with the lattice atoms if the atoms are

   (a) very large   (b) close together

   (c) displaced from their regular positions

9 An applied electric field causes an electron gas cloud to

(a) drift in the direction of the field

(b) drift in the direction opposite to the field

(c) become smaller

10 If a metal contains a density of 10^{28} valence electrons per metre^3 and carries a current of $10^6\,A\,m^{-1}$, the drift velocity is

(a) $6.25 \times 10^{-4}\,m\,s^{-1}$ (b) $6.25 \times 10^4\,m\,s^{-1}$

(c) $6.25\,m\,s^{-1}$

11 The distribution of velocities amongst the electrons is hardly affected by the flow of current

(a) true   (b) false

12 The current density in a metal having 10^{28} valence electrons/m^3 with a mean free time of $10^{-14}\,s$, in a field of $10^4\,V\,m^{-1}$, is

(a) $2.8 \times 10^{10}\,A\,m^{-2}$ (b) $2.8 \times 10^6\,A\,m^{-2}$

(c) $2.8\,A\,m^{-2}$

13 The mobility of an electron in a metal is the drift velocity per unit applied field

(a) true   (b) false

14 The mobility of an electron in a solid in which it has a mean free time of $10^{-14}\,s$ is

(a) $17.6\,m^2\,V^{-1}\,s^{-1}$ (b) $5.69 \times 10^2\,m^2\,V^{-1}\,s^{-1}$

(c) $1.76 \times 10^{-3}\,m^2\,V^{-1}\,s^{-1}$

15 The mean free time of an electron in a metal is

(a) proportional to

(b) inversely proportional to

(c) proportional to the square of

the temperature

16 The resistivity of a metal is a function of temperature because

(a) the electron velocity varies with temperature

(b) the electron gas density varies with temperature

(c) the number of phonons varies with temperature

17 The resistivity of all normal metals as temperature is lowered

(a) tends to zero

(b) tends to a constant value

(c) at first decreases and then increases

18 A disordered solid solution between two metals has a resistivity that is

(a) higher than that of either pure component

(b) lower than that of either pure component

(c) equal to that of the pure component which has the highest resistivity

19 A two-phase polycrystalline alloy contains 10% of the α phase and 90% of the β phase by volume. If the resistivities are $1.7 \times 10^{-8}\,\Omega\,m$ and $7.0 \times 10^{-8}\,\Omega\,m$ respectively, the resistivity of the alloy is

(a) $8.7 \times 10^{-8}\,\Omega\,m$ (b) $6.47 \times 10^{-8}\,\Omega\,m$

(c) $2.23 \times 10^{-8}\,\Omega\,m$

20 The collision probability per unit time for an electron in a metal is equal to the reciprocal of the mean free time

(a) true   (b) false

Answers

1	(b)	2	(a)	3	(c)	4	(b)
5	(a)	6	(c)	7	(b)	8	(c)
9	(b)	10	(a)	11	(a)	12	(a)
13	(a)	14	(c)	15	(b)	16	(c)
17	(b)	18	(a)	19	(b)	20	(a)

Semiconductors | 15

15.1 Introduction

By the end of the 20th century the semiconductor silicon had become the commonest material available in single crystal form. This was a result of our ability to make electronic switches and amplifiers by modifying crystalline silicon's conductivity with the introduction of minute quantities of impurities. Other semiconductors can be made into light-emitting diodes (LEDs), and lasers. LEDs are both more efficient and more reliable than incandescent lamps, and are becoming cheap enough to replace them. Yet another use of silicon is as a mechanical component – the accelerometer used in cars to sense deceleration and to trigger the airbags is made of silicon using some of the methods described in the next chapter, where we shall discuss the processes used to turn semiconductors into electronic products.

This chapter begins by examining the connection between interatomic bonding and conductivity, then explains how conductivity in p-type and n-type semiconductors is controlled, and how these materials function in electronic diodes and transistors. Although we concentrate here largely on the properties of silicon, many other useful semiconducting materials are introduced along the way. Table 15.1 shows the section of the periodic table we will need to consider in this chapter.

15.2 Bonding and conductivity

If we examine the electrical resistivities of the elements, particularly of the elements in Groups 11, 12, 13 and 14, it will be seen that they tend to increase with an increase in the number of valence electrons. At first sight, this is surprising, because an increase in the number of loosely bound electrons in the outer shell of an atom might be expected to lead to an

Group 12		Group 13		Group 14		Group 15		Group 16	
		Boron	B	Carbon	C	Nitrogen	N	Sulphur	S
		Aluminium	Al	Silicon	Si	Phosphorus	P	Selenium	Se
Zinc	Zn	Gallium	Ga	Germanium	Ge	Arsenic	As	Tellurium	Te
Cadmium	Cd	Indium	In	Tin	Sn	Antimony	Sb		
Mercury	Hg			Lead	Pb	Bismuth	Bi		

Table 15.1

Elements that are important in semiconductors and their technology

Fig. 15.1

A two-dimensional representation of the covalent bonds in a crystal of C, Si or Ge. The electron clouds are concentrated between the atoms.

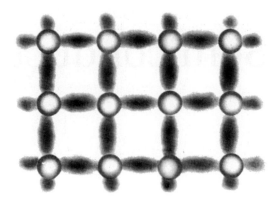

increase in electrical conductivity, especially in view of Eq. (14.15a). The resistivity of copper (Group 11) is about $1.7 \times 10^{-8}\,\Omega\,\mathrm{m}$ at room temperature, while that of germanium (Group 14 in the same period of the table) can be anywhere in the range 10^3 to $10^4\,\Omega\,\mathrm{m}$, depending on the impurities present. This phenomenon is directly related to the interatomic bonding in these materials.

It will be recalled that the bonding in a metal is visualized in terms of a fixed array of positive ions held in position by a 'sea' of negative charge. This sea consists of the valence electrons, which are distributed throughout the solid, i.e. they are shared by all the atoms. In the Group 14 materials covalent bonds are formed – their distinguishing feature is that the electrons are *localized* in the bonds. This means that the wave function of a bonding electron is confined to a region of space between adjacent atoms, as depicted in Figs 5.9 and 5.12. In fact, the maximum probability density for the bonding electrons is just midway between the atoms. We show this in the diamond lattice in Fig. 15.1, where for clarity the three-dimensional structure has been 'unfolded' and drawn in two dimensions. Each bond contains two electrons with opposite spin directions – more than this is not possible owing to Pauli's exclusion principle. Do not think, however, that these electrons are immobile, condemned forever to stay in the same bond. The probability distributions are smeared out so that those of neighbouring bonds overlap slightly, and this makes it possible for electrons in adjacent bonds to change places with one another. Indeed, all the electrons are continually doing so, moving through the network of bonds at great velocity. In spite of this rapid motion, Pauli's principle still ensures that at any one time there are no more than two electrons in any particular bond.

Now although the electrons are moving through the network of bonds, they are not strictly to be regarded as travelling waves. Travelling waves have a constant intensity at every point, while we know that in covalent bonds the electron clouds are concentrated midway between the atoms. The electron density (Fig. 15.1) is more like the distribution in a set of standing waves, with nodes at the ion centres and antinodes between them, as suggested in Fig. 15.2.

All the evidence confirms that the bonding electron waves are standing waves, not running waves. However, this does not mean that the electrons cannot change places with one another, because a standing wave can be mathematically decomposed into two running waves travelling in opposite directions. Because the amplitudes of the two oppositely travelling waves

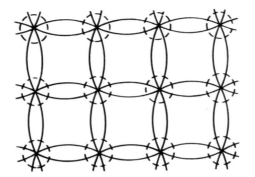

Fig. 15.2

The valence electron distribution in the diamond structure is like a system of standing waves.

are equal, exactly the same amount of charge is continuously transported in each direction. In this way, the charge density at each point remains unchanged, and Pauli's principle is satisfied. From this we deduce that no net flow of charge in any direction is possible, i.e. a perfect covalent crystal cannot carry an electric current.

15.3 Semiconductors

The above argument suggests that a crystal with covalent bonding should be a good insulator, and in many cases this is substantially correct. Our prototype covalent solid, diamond, has a resistivity of $10^{12}\,\Omega\,m$ at room temperature, while polymers such as polyethylene and polyvinyl chloride have resistivities of about $10^{15}\,\Omega\,m$ in their pure state. However, two elements, silicon and germanium, both of which are of great technological importance, have much lower resistivities. Why is this? If we compare the bond strengths of diamond, silicon and germanium, we find that they decrease as the atomic weight increases. (This has already been discussed in Section 5.11.) It is thus easier to break the bonds in germanium than in silicon, while it is hardest of all to break them in diamond. If the bond is sufficiently weak, the thermal vibration of the atoms may occasionally be sufficient to sever a bond, even when the temperature is not very high. Naturally, the higher the temperature, the more likely it is that any one bond will be broken, and hence the more bonds will be broken at a given time.

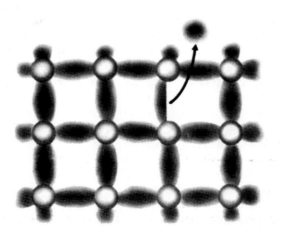

Fig. 15.3

Thermal vibration causes an electron to be freed from one of the bonds.

Now, breaking one bond does not greatly affect the atoms that it joins together – each still has three bonds to keep it in place. But breaking a bond does entail releasing one of the electrons that form it, as shown in Fig. 15.3. Thus at any temperature a small proportion of the large number of bonding electrons will be set free. Being no longer confined in the bonds, these electrons may wander freely through the lattice and conduct an electric current. At room temperature the freed electrons in a piece of pure silicon constitute only about one in 10^{13} of the total number of valence

Fig. 15.4

As the bonding electrons move about, the hole is filled and appears to move.

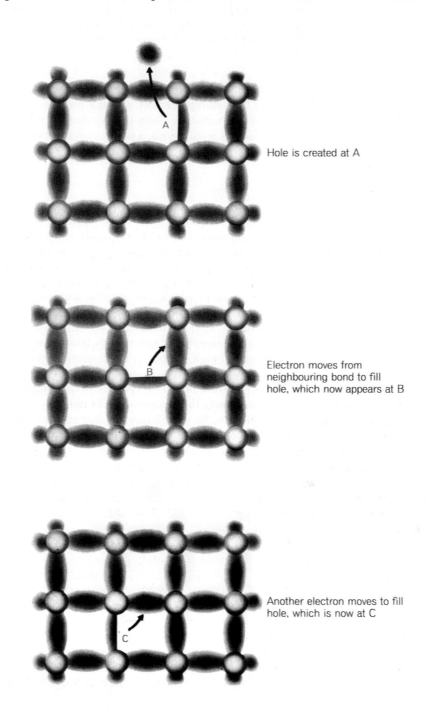

Hole is created at A

Electron moves from neighbouring bond to fill hole, which now appears at B

Another electron moves to fill hole, which is now at C

electrons. This is in complete contrast to the situation in a metal, in which all the valence electrons are free to conduct current.

The number of free electrons is essentially fixed at a given temperature, although the process of bond rupture by thermal agitation is continuous. The rate at which electrons are freed is exactly balanced by the rate at which those already free are trapped by an empty bond and lose their freedom.

We shall postpone the calculation of the conductivity generated by the presence of free electrons to the next section, for we must now take note of a curious phenomenon that occurs in the bonds. The freed electrons leave behind them gaps in the bonds, called *holes*. Because the bonded electrons are in constant motion, these holes are soon filled by nearby electrons. This necessarily causes the holes to appear in the bonds vacated by the itinerant electrons, and the holes appear to move (Fig. 15.4). In fact the holes move about in much the same random way and at much the same velocity as the electrons in the filled bonds. We shall now show that the presence of these holes enables the bonded electrons to carry a current in such a way that it appears as if it is conducted by positive charges carried by the holes!

15.4 Conduction by holes

We have seen that a hole moves by virtue of an electron jumping into it from an adjacent bond. We have also stressed that, when all the bonds are full, the electron distribution is not changed by an electric field. However, when a field is applied and a hole is present, the probability of an electron from an adjacent bond entering the hole will be increased if its movement is assisted by the field; conversely, the probability of motion in the opposite direction is lowered. Thus, if the field favours the motion of electrons from left to right (say) in Fig. 15.4, the hole will tend to drift from right to left and thus its motion is influenced by the field in the same way as a mobile *positive* charge. Moreover, there is one more positive charge e on the Si nucleus in its new locality than the number of negative charges on electrons located in the bonds. In other words, it effectively carries a positive charge from right to left. Its speed of movement between bonds may be divided into two components – a random velocity arising from the random electron motion, and a drift velocity that comes from the fact that the field makes the average electron velocity slightly greater in one direction than in the other. Thus, a hole experiences an acceleration in an electric field equal but opposite in direction to that of a free electron, that is, it behaves like a free electron with *positive* charge. Its drift velocity is limited by collisions in just the same way as is the drift velocity of an electron, because its movement is determined by the drift of all the real electrons in the opposite direction.

Because the drift velocity is a direct result of the application of an electric field, the hole makes a direct contribution to the conductivity of the crystal. If there are p holes present the current they carry is p times the current carried by one hole – it is as if there were p positively-charged 'electrons'. The validity of this point of view has been demonstrated in experiments in which both magnetic and electric fields are applied, as follows. Consider a charged particle that is accelerated to a drift velocity v_d in an electric field of strength ξ. If a magnetic flux B is also present, lying perpendicular to the direction of ξ, then the magnetic force F on the particle, as given in Eq. (1.13), is $F = Bev_d$ in a direction perpendicular to both

Fig. 15.5

The forces on electrons and holes
moving through a magnetic field.

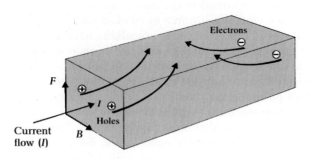

v_d and B (Fig. 15.5). Note that if the sign of e is changed, the direction of
motion caused by the electric field is reversed, i.e. v_d becomes negative.
The magnetic force F, however, remains in the same direction because
both e and v_d have changed signs. The particle, whether hole or electron,
tends to move in the direction of the force F, and by detecting this transverse motion electrically it is possible to determine the sign of its charge.
Such experiments, when carried out on semiconductors, confirm the existence of positive holes and the phenomenon illustrated in Fig. 15.5 is
known as the *Hall effect*.

15.5 Energy bands in a semiconductor

We have seen that there are two kinds of electrons in a semiconductor –
those in the bonding system and those that are 'free'. The free electrons
move about in the crystal and contribute to electrical conductivity in just
the same way as the free electrons in a metal. They can be treated as plane
waves and their energies must be quantized in a large number of closely
spaced energy levels, as were the electrons in a metal. The band of
energies formed by these levels is called the *conduction band*. Because the
electrons in the covalent bonds also move about at varying speeds, they
also have a range of permitted energies. Detailed consideration of the
quantization conditions shows that these levels are also closely spaced and
form an energy band called the *valence band*.

There is, in fact, a minimum value for the amount of additional energy a
bonding electron must acquire in order to leave the bond. The absolute
minimum for this energy is that which the most energetic bonding electron
needs in order just to become free, with no kinetic energy left over to carry it
away. This will be represented in the energy level diagram by a gap between
the highest level in the band of energies occupied by the bonding electrons,
i.e. the valence band, and the lowest level in the conduction band, as shown
in Fig. 15.6. This range of energies that is prohibited to the electrons is
called the *forbidden energy gap*, usually just the *energy gap* (or sometimes
band gap), and is a direct measure of the amount of energy an electron
needs to leave the bonds. It is different for different materials. In germanium
it is 0.65 eV, in silicon 1.15 eV, while in diamond it is 5.3 eV. The arrow in

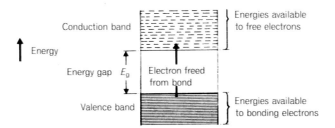

Fig. 15.6
The energy band diagram for a semiconductor, showing the energy gap E_g.

Fig. 15.6 represents the transition of an electron from the valence band to a state in the conduction band, i.e. from a bond to freedom. This leaves vacant a state in the valence band, i.e. a 'hole'. A downward arrow would represent an electron falling back into a lower level, and hence recombining with a hole. This diagram gives us another way of looking at conduction by electrons. Clearly, electrons in the conduction band can accelerate readily because empty, higher energy levels are available to accommodate them. If a hole (i.e. a vacant state) exists at the top of the valence band, an electron in a lower level can be given energy to fill it and the accelerating electron carries a current. On the other hand, if there were no vacant levels in the valence band, no electrons could be accelerated.

15.6 Excitation of electrons

When an electron near the top of the valence band (i.e. in a bond) collides with a vibrating atom, it can absorb some of the atom's energy and jump into the conduction band. This is just another (more accurate) way of describing how thermal vibrations can break a bond and create an electron-hole pair. The probability of this occurring is just proportional to the probability of an atom having thermal energy greater than E_g, the amount of energy needed to excite the electron. This probability was stated in Chapter 7 to be almost exactly proportional to $\exp(-E_g/kT)$ when $E_g/kT \gg 1$. The rate of generation of electron-hole pairs is therefore proportional to the same factor. In steady state, the rate of generation of electron-hole pairs must be exactly equal to the rate at which holes and electrons recombine to form fully occupied bonds again. This latter rate must be proportional both to the number of electrons per unit volume, n, and to the number of holes per unit volume, p. It is therefore proportional to the product np. By equating the rates of generation and recombination, we therefore find that

$$np = N_C^2 \exp(-E_g/kT) \qquad (15.1)$$

where N_C^2 is a constant of proportionality.

Because $n = p$, we then have

$$n = p = N_C \exp(-E_g/2KT) \qquad (15.2)$$

This equation can be interpreted as follows. In the discussion of thermal energy distributions in Chapter 7 it was remarked that electrons, because they obey Pauli's principle, cannot be distributed exponentially over the different energy states. A thorough analysis of the problem shows that in an intrinsic semiconductor the probability of an electron being in a quantum state at the bottom of the conduction band is just $\exp(-E_g/2kT)$. This factor also appears in Eq. (15.2). Of course, not all the electrons are in the lowest quantum state in the conduction band, but the vast majority have only a little higher energy. Hence, if we regard the quantity N_C as an effective number of states (per m^3) at the bottom of the band, then the total number of electrons in the conduction band, n per m^3, is the number of states N_C multiplied by the probability $\exp(-E_g/2kT)$ that a state is occupied. This gives, as before,

$$n = N_C \exp(-E_g/2kT)$$

Because N_C is not a number of actual states, it is not a constant, independent of temperature. However, its temperature dependence is very small compared to that of the exponential factor in Eq. (15.2) and may safely be ignored for our purposes. N_C also varies little among the common semiconductors, from the typical value of $2.5 \times 10^{25}\,\mathrm{m}^{-3}$ when $T = 300\,\mathrm{K}$. The electrons in the conduction band and the holes in the valence band may carry current in a similar way to that described for metals in Chapter 14. Thus we can use Eq. (14.11) to write down the current density (current per unit area) J_n carried by n free electrons per cubic metre whose drift velocity is v_n:

$$J_n = nev_n$$

Note that J_n has a positive sign, because both v_n and e are negative. Similarly the current density J_p due to the motion of holes in the valence band may be written in terms of their drift velocity $+v_p$

$$J_p = pev_p$$

where p is the number density of positive holes. The total current density J is then the sum of these:

$$J = J_n + J_p = nev_n + pev_p \tag{15.3}$$

The number of free electrons n and the number of holes p per unit volume in a *pure* semiconductor are usually equal (they are not always so; see Section 15.7) and given by Eq. (15.2), so that we may write the current density

$$J = N_C(ev_n + ev_p)\exp\left(-\frac{E_g}{2kT}\right) \tag{15.4}$$

Assuming that both the free electrons and the holes behave like the free electrons in a metal, we may use the results of Chapter 14 to express their drift velocities. Thus both v_n and v_p will be proportional to the electric field ξ [see Eq. (14.13)] and we may divide Eq. (15.3) by ξ to obtain the conductivity σ, expressed in *siemens/metre* (S m^{-1}) in SI units (the siemens is the unit of conductance: 1 siemens $= 1\,\Omega^{-1}$):

$$\frac{J}{\xi} = \sigma = N_C(e\mu_n + e\mu_p)\exp\left(-\frac{E_g}{2kT}\right) \tag{15.5}$$

where μ_n and μ_p are the drift velocities of electrons and of holes, respectively, in a unit electric field. As before we call μ_n and μ_p the *mobilities* of electrons and of holes respectively. In general they will not be equal to one another, but we may expect them both to vary inversely with temperature as we showed for electrons in metals in Chapter 14. However, the exponential in Eq. (15.5) varies more rapidly with temperature and completely dominates the temperature dependence of the whole expression. Neglecting the temperature dependence of μ_n and μ_p, we see that σ rises approximately exponentially with temperature, in marked contrast to metals, for which it falls with rising temperature. The reason is that for semiconductors the *number* of available charge carriers grows exponentially as the temperature rises, while for metals it is roughly constant. In a semiconductor, a rising temperature gives more and more valence electrons enough energy to jump into the conduction band, freeing them and the corresponding holes to conduct an electric current.

The experimental values of the resistivity of silicon measured at different temperatures T are plotted on a log scale against $1/T$ in Fig. 15.7. Exactly as predicted by Eq. (15.5), the points lie very close to a straight line, the slope of which can be seen by differentiation of that equation to be equal to $E_g/2k$. Using the value for Boltzmann's constant k we can find from the graph that for silicon $E_g = 1.15$ eV, as stated earlier.

There are more accurate ways than this of determining energy gaps, and the values for several materials are given in Table 15.2. Among these, silicon has been the material most used for transistors and diodes, although germanium was used in the 1950s, and several semiconducting compounds are now used for special purposes. Compound semiconductors are discussed in Section 15.8.

Among these materials a wide range of properties is available – different energy gaps, different electron and hole mobilities – which makes each suitable for a different application. In this chapter we shall just consider diodes and transistors, whose operation depends upon the controlled introduction of very small amounts of impurities into an initially very pure crystal.

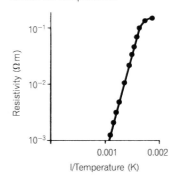

Fig. 15.7
Dependence of the resistivity of silicon on temperature.

15.7 Doped semiconductors (n-type semiconductors)

Each silicon or germanium atom in a semiconducting crystal has four valence electrons that it shares with its neighbours to form four covalent bonds. From another viewpoint, these electrons fill energy levels in an energy band (the valence band) so that this band is just filled at the zero of absolute temperature.

Suppose that a pentavalent atom (Group 15) were to be substituted for one of the silicon atoms. An arsenic or antimony atom would be suitable, although a phosphorus atom would fit better than either of these into the

Table 15.2

Energy gaps, electron densities and conductivities of pure materials

Material	E_g (eV)	Calculated value of n at 300 K (m^{-3})	Measured value of n at 300 K (S m^{-1})
Copper	None	8.5×10^{28}	5×10^{7}
Aluminium	None	1.8×10^{29}	3.3×10^{7}
Grey tin	0.08	5.4×10^{24}	10^{6}
Germanium	0.66	9.3×10^{19}	2.2
Silicon	1.11	6.0×10^{15}	5×10^{-4}
Diamond	5.3	1.1×10^{-20}	$10^{-12}\ast$
Lithium fluoride	11.0	2.0×10^{-20}	$10^{-11}\ast$

* Conductivity in these materials does not occur by the motion of electrons because other processes occur more readily.

Fig. 15.8

Energy level diagram of an n-type semiconductor.

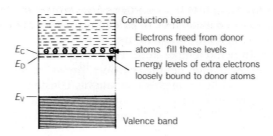

silicon lattice, because its atomic radius is very similar to that of silicon. Four of the five valence electrons belonging to this atom are required to form the four covalent bonds with the neighbouring silicon atoms. The fifth electron is bound to the atom by the extra positive charge on the nucleus, but, as we will shortly show, it requires very little energy to break this bond, and at room temperature the thermal energy in the lattice is more than sufficient to do so. The extra electron is then free to move through the silicon lattice. The silicon is said to be *doped* by the introduction of the Group 15 impurity.

Let us now build up the same picture using the energy band model. The freed electron has been excited to an empty state in the conduction band. Because the energy required to remove it from its parent nucleus was very small, it must have come from an energy level *just below* the bottom of the conduction band – in the energy gap! (Fig. 15.8). Thus the introduction of the Group 15 atom has created a new energy level, near to the conduction band, which is normally vacant at room temperature because the electron is donated to the conduction band – the impurity is called a *donor* for this reason.

In order to indicate that the valence electron is localized while attached to the donor atom, we make the horizontal axis of the band diagrams represent distance through the crystal – the donor levels are then shown as short lines.

We may show that the binding energy of an electron in this level is small by regarding the extra electron and the excess positive nuclear charge as if they were like a hydrogen atom. In both situations a single, negatively-charged electron moves in the electric field of a single, fixed positive charge (Fig. 15.9). But in the present case the electron moves through a silicon lattice that we may treat approximately as a medium with a large

Fig. 15.9

Path of the fifth electron around a donor atom in the lattice of a semiconductor.

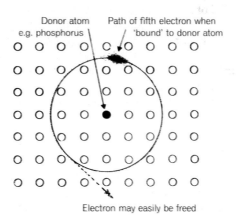

relative permittivity, $\varepsilon_r = 11.7$. The expression derived in Chapter 3 for the hydrogen atom can be used now to derive the binding energy, but we must replace ε_0 by $\varepsilon_r \varepsilon_0$, giving the result

$$\text{binding energy} = \frac{me^4}{8\varepsilon_0^2 \varepsilon_r^2 h^2} \qquad (15.6)$$

This is the energy difference between the donor level and the bottom of the conduction band, i.e. it is equal to $E_C - E_D$ (see Fig. 15.8).

Because the relative permittivity of silicon is 11.7, the binding energy is *smaller* than the ionization energy of hydrogen by a factor of 11.7^2, so that

$$E_C - E_D = \frac{13.6}{11.7^2} = 0.1\,\text{eV}$$

Naturally this result is only approximate (the measured value is nearer 0.05 eV), but it serves to show that the binding energy is indeed small compared with the width E_g of the energy gap, which in silicon is 1.15 eV. The corresponding values for germanium are $E_C - E_D = 0.01\,\text{eV}$ and $E_g = 0.65\,\text{eV}$.

The density of free electrons in pure silicon, n_i (called the *intrinsic* value) can be calculated from Eq. (15.2). Using the value for N_C given earlier, the density of electron-hole pairs at $T = 300\,\text{K}$ is found to be

$$n_i = N_C \exp(-E_g/2kT) = 2.5 \times 10^{25} \times \exp(-1.15/2 \times 0.026)$$

Thus

$$n_i = 6.3 \times 10^{15}\,\text{m}^{-3} \qquad (15.7)$$

On the other hand there are 5×10^{28} atoms per m^3 and about 2×10^{29} valence electrons per m^3 in solid silicon. This confirms our earlier statement that only a few electrons are thermally excited across the band gap.

If now only one silicon atom in 10^{10} were replaced by a donor impurity, the density of donors would be about $5 \times 10^{18}\,\text{m}^{-3}$ and the same number of extra electrons would be available to conduct an electric current. This is far in excess of the number of carriers in the intrinsic material, so the conduction process is dominated by the presence of the electrons, which are called the *majority carriers*. Because the current carriers are almost exclusively negative, the material is termed an *n-type* impurity semiconductor.

We have glossed over one point in the argument. Not all the donor electrons appear in the conduction band. Just as in the case of intrinsic electrons, the number of electrons in the conduction band must be calculated from the probability of a donor electron being excited across the gap between the donor level E_D and the conduction band edge E_C. This probability is calculated in more advanced textbooks (see list of Further reading) with the result that, at room temperature, nearly all the donor atoms are found to be ionized, unless the density of donors becomes too large (greater than about $10^{25}\,\text{m}^{-2}$ in germanium).

For present purposes we may therefore treat a piece of n-type semiconductor containing N_D donors per unit volume as a conductor in which the free electron density is approximately equal to N_D. As the density of electrons is increased by the presence of donors, so the density of holes is decreased. This is because a hole has a greater probability of meeting an electron with which it can recombine. The number of holes therefore reduces so that the net recombination rate, which is proportional to the product np, is unchanged by the increase in n. We can be sure of this, because the generation rate is not affected by the presence of donors, as it depends solely on the transfer of thermal energy from atoms to bound electrons. The number of donor atoms is so small that the effect they have on this process is negligible. Because the generation rate must equal the recombination rate, we must still have the relation:

$$np = N_C^2 \exp(-E_g/kT)$$

exactly as in Eq. (15.1). But in the present case n and p are not equal, for n is approximately equal to N_D, so that we can now find an approximate value for p:

$$p = \frac{N_C^2}{N_D} \exp(-E_g/KT)$$

Note that this result is quite different from that for an *intrinsic* semiconductor: there are far fewer holes here.

15.8 Doped semiconductors (p-type semiconductors)

By contrast with the action of Group 15 elements, small traces of a Group 13 element such as boron or aluminium create a deficiency of electrons. Each impurity atom has only three valence electrons, while four are required for bonding. The incomplete covalent bond contains a hole that can, when it escapes from the impurity atom, take part in conduction.

The escape of the hole occurs by the capture of a valence electron from a neighbouring silicon atom, i.e. from the valence band on the energy diagram. For this reason the Group 13 impurity is called an *acceptor* atom. The captured electron now has a slightly higher energy than the other valence electrons, because it has upset the neutrality of the impurity atom, which has now become a negative ion. We therefore allocate to the captured electron an energy level called an acceptor level, just above the top of the valence band (Fig. 15.10). Because the energy gap between the top of the valence band and the acceptor level is very small, electrons from the valence band are readily excited across it and most of the acceptor levels will be filled at normal temperatures. A corresponding number of holes is created in the valence band and these may conduct an electric current.

As with electrons in n-type silicon, the number of holes created by only a small concentration of acceptors may greatly exceed the number of holes in intrinsic silicon [given in Eq. (15.7)] and conduction thus occurs primarily by the motion of these positive holes. The material is called a p-type impurity semiconductor, and in contrast to n-type material the majority of carriers are now holes.

Fig. 15.10

Energy level diagram for a p-type semiconductor.

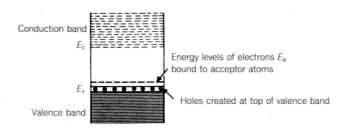

The energy gap between the valence band and the acceptor levels may be calculated in a similar way to that for donor levels. Because the captured electron gives an acceptor atom a charge $-e$, it has an 'attraction' for holes. So a hole can become bound to an acceptor as if it were an 'inverted' hydrogen atom: a positive charge e orbiting a fixed negative charge $-e$. Equation (15.6) can again be used to obtain an approximate value for the binding energy, which is thus about 0.05 eV in silicon.

Many other impurities may be put into silicon or germanium to create new energy levels at almost any point in the energy gap, but the Group 13 and Group 15 elements are by far the most important for practical purposes. Lattice defects of many kinds can also act as donor or acceptor levels, or as traps for current carriers. It is therefore important to prepare the materials for diodes and transistors with the greatest care, to reduce both unwanted impurities and lattice defects below very low levels, as described in Chapter 16. As a result of the invention of ingenious technologies for doing this, silicon and other semiconductors have become among the purest obtainable materials, and the largest single crystals available (about $1.5\,\text{m} \times 30\,\text{cm}$ in 2002).

Finally, what happens if we put *equal* quantities of Group 13 and Group 15 impurities into a semiconductor? In that case, all the electrons donated by the donors end up by filling *all* the vacant states in the acceptor, leaving the material with no extra electrons or holes! The material is said to be *compensated* rather than intrinsic, but it behaves in many respects like the intrinsic material. This fact is put to good use in making integrated electronic circuits in silicon, as the next chapter explains.

15.9 Compound semiconductors and their uses

It was noted in Section 6.7 that covalent bonding, with the four-fold coordination exhibited in the diamond structure, is found in compounds of Group 13 and Group 15 atoms like GaAs and InP, and also in some 12–16 compounds like ZnS. As in silicon, there are just enough valence electrons to fill the covalent bonds, leaving none (in the perfect crystal) in the conduction band to carry a current. GaAs, InP and ZnS are therefore semiconductors, with many properties like silicon. Table 15.3 lists these and a number of other compound semiconductors that have unique and useful properties. The absence of a Group 15 (or 16) atom from the crystal lattice has the effect of removing a valence electron, creating a hole, while the absence of a Group 13 (or 12) atom creates a free electron. The imperfect crystal behaves as if doped with donor or acceptor impurities, making it somewhat more difficult to control their electrical properties.

Further differences show up in the optical properties. As explained in Section 19.6, light is absorbed in a semiconductor only if the photon energy hf of the light is equal to or greater than the energy gap E_g. This is because only then can the absorption of a photon create an electron-hole pair. Conversely, the recombination of an electron with a hole can, in the right circumstances, result in the emission of a photon of light with energy E_g or greater. This is called direct recombination. However, in silicon or germanium, recombination almost invariably occurs indirectly, by first trapping the hole or electron at an impurity. This stems from a peculiar

Table 15.3

Some important compound
semiconductors and their
properties at 300 K

Chemical formula	Energy gap E_g (eV) and type*		Electron mobility† ($m^2 V^{-1} s^{-1}$)	Applications
GaAs	1.42	D	0.85	LEDs, high-frequency FETs for mobile communications
$Ga_{1-x}Al_xAs$ for $x = 0$ to 0.37 for $x = 0.37$ to 1.0	1.42 to 1.92 1.92 to 2.16	D I	0.85 to 0.1 0.03	Red and infrared LEDs, lasers
$Ga_{1-x}In_xAs_{1-y}P_y$	0.36 to 2.2	D	0.011 to 3.3	LEDs, lasers for fibre optic communications
$Si_{1-x}Ge_x$ for $x = 0$ to 1.0	1.11 to 0.66	I	0.145 to 0.39	High-frequency (>10 GHz) transistor circuits (replacement for GaAs)
InSb	0.18	D	8.0	Infrared light detectors/meters
$Ga_{1-x}In_xN$	2.0 to 3.4	D		Blue and green-emitting LEDs
CdS	2.42	D	0.034	Visible light meters

* 'D' indicates a direct band gap, 'I' an indirect band gap.
† Hole mobilities are always smaller.

property possessed by the energy bands in these materials, which causes free electrons with energy near E_C to have either much greater momentum (in Si) or much less momentum (in Ge) than holes with energy near E_V. Thus so-called radiative recombination, involving the emission of a photon, almost never occurs, because momentum cannot be conserved during the process. If a free electron is to 'fill' a hole in silicon, it must first lose its momentum, so that recombination occurs in two stages. First, the electron becomes trapped at an impurity atom, losing its momentum to the atom as it does so. Subsequently, a hole happens to pass by, whereupon recombination occurs, the energy being transferred to the lattice, in the form of phonons.

Indirect recombination dominates in materials with energy bands similar to those in silicon or germanium, e.g. AlAs, GaP. They are said to have an *indirect (energy) band gap*. Materials such as GaAs or InP, in which radiative recombination occurs easily, are said to have a *direct* band gap. Both of these last materials can be used as the basis for sources of infrared light such as light-emitting diodes (LEDs) and lasers, described in Section 19.11. In lasers for fibre optic communications, the four-component semiconductor GaInAsP is frequently used, because its energy gap can be adjusted so that the wavelength of the light emitted on recombination, at 1.55 µm, matches the wavelength at which fibres are most transparent (see Section 19.9).

Indium antimonide has a very small energy gap, and is used for the detection of far-infrared radiation. Cadmium sulphide, on the other hand, has an energy gap optimized for absorption of visible light, and is used in photographic light meters and other kinds of light detector.

Materials scientists have devised ways of growing compound semi-conductor crystals with compositions varying from atomic layer to layer, making the energy band gap vary in such a way as to satisfy a particular engineering requirement. This 'band gap engineering' has enabled many novel applications of semiconductors in electronics.

15.10 Organic semiconductors

The Nobel Prize for Chemistry in 2000 was awarded to a physicist, Alan Heeger, and chemists Alan MacDiarmid and Hideki Shirakawa, for their discovery that polymers can be doped to make them into useful semi-conductors. A single polymer molecule having alternate double and single covalent bonds ('conjugated' bonds) allows electrons to move freely along its length. As described in Section 5.8, conjugated bonding results in electrons becoming *delocalized* in molecular, rather than atomic, orbi-tals, extending along the molecular backbone. Thus free movement of the valence electrons is possible along the chain of bonds. In such molecules, the occupied valence electron states (the 'molecular orbitals' in chemical language) are the equivalent of the valence band states in a covalently bonded crystal of silicon, germanium or gallium arsenide. Likewise, the lowest *un*occupied molecular orbitals form an equivalent to the conduction band. Exactly as in macromolecular inorganic solids, these may be sepa-rated by an energy gap, or they may have overlapping energies, resulting in either metallic-like or semiconductor-like molecules. They may even become superconducting at temperatures near 0 K.

Examples of metallic-like molecular behaviour are discussed in Section 19.4, where we explain how the colour of light absorbed by some organic dye molecules can be simply described in terms of the length that the

Table 15.4

Energy gaps and dopants (where known) and uses of some organic semiconductors

Material and abbreviated name	Effective energy gap (eV)	Dopants and type		Actual or potential uses
Anthracene Pentacene	4	None effective		Field-effect transistor
cis-Polyacetylene	2	ClO_4^-, I_3^- Na$^+$	Acceptors Donor	Field-effect transistor
Poly(p-phenylene vinylidene) (PPV)	2.5			LED emitting layer
Cyano-PPV (CN–PPV)	2.2			
Poly(dioxyethylene thienylene) (PEDOT)		Poly(styrene sulphonic acid) (PPS)	Acceptor	Hole-injecting electrode for LEDs
σ-sexithiophene		Naturally p-type as prepared		Field-effect transistor
Buckminsterfullerene (C_{60})		n-Type as prepared		

electrons can travel along the zigzag chain of bonds. Where an energy gap exists, promotion of an electron to the 'conduction band' leaves a hole (or an equivalent charge carrier) free to move along the π bonds. Energy gaps ranging from about 1.5 to 3 eV are common in large conjugated molecules, including many conjugated polymers. Table 15.4 shows just a few of them and their uses, both existing and potential.

If semiconducting behaviour is to result, it is necessary that electrons and holes must cross readily from one molecule to the next. This is possible, given suitable molecules. As long as the wave functions of the 'conduction' electrons in one molecule overlap in space with those of its neighbour, electrons may transfer between those states in the two molecules, and holes likewise (but of course via overlapping wave functions of *valence* electrons).

Two classes of such materials can be distinguished, with different degrees of overlap. In one, the overlap is sufficient to allow extended but narrow energy bands to develop, and conduction can be described using the same model as described above for inorganic semiconductors like silicon. In the other, the separation between molecules is just large enough (typically 0.5 nm or greater) that electron wave functions do not extend over more than one or two molecules at most. A different description is then needed for conduction. We begin with the first category, in which the intermolecular separation is generally less than 0.4 nm.

15.10.1 Organic materials with electronic energy bands

Because the overlap of electronic wave functions in neighbouring molecules is weak, the splitting of the electronic energy levels is smaller than in a macromolecule such as a silicon crystal. The spread in energy of the resulting energy band is therefore smaller – typically around 100 meV, contrasting with the 1 eV or so in inorganic crystals. This is what happens with fairly large planar molecules such as anthracene and pentacene (composed of three and five benzene rings respectively – see Fig 15.11). A large area of physical overlap between neighbouring molecules gives a sufficiently strong interaction between them to create energy bands. Smaller molecules like benzene bind their electrons too tightly to allow bands to develop – benzene's low melting point is evidence of weak intermolecular interactions. Anthracene has a conduction and a valence band separated by a large energy gap, and is an insulator, but pentacene, with a smaller energy gap, is among the simplest of organic semiconductors. The mobilities of electrons and holes in these materials are typically around $2 \times 10^{-4} \, \text{m}^2 \, \text{Vs}$, compared with values near $0.1 \, \text{m}^2 \, \text{Vs}$ in silicon.

For the semiconducting behaviour to be observable when making electrical contacts, it is necessary that (i) all impurities in the material should be at low enough concentrations not to affect conduction and (ii) that injection of electrons and holes be possible via suitable contact electrodes. A conducting electrode material suitable for injecting electrons must, when in contact with the organic material, have its Fermi level close in energy to the

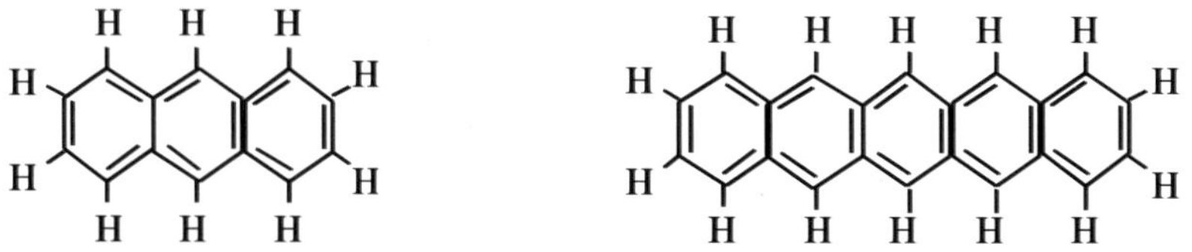

Fig. 15.11

The chemical structures of anthracene (left) and pentacene molecules. Each hexagonal ring consists of six carbon atoms (not drawn) and is structured like the benzene molecule in Fig. 5.13.

electronic conducting states of the organic molecules, in order that little energy is needed to inject either an electron or a hole. The search for suitable contact materials, stimulated by the possibility of replacing inorganic semiconductors in transistors and LEDs, has been intense.

As a practical example, consider the case of semiconducting polyphenylene vinylene (PPV), whose structure is shown in Fig. 15.12(a). In the same diagram, (b) illustrates in cross-section a thin sandwich of PPV between an electron-injecting calcium electrode, and a hole-injecting electrode composed of a transparent glassy material, indium-tin oxide, coated with a thin layer of poly(ethylenedioxy)thiophene. This sandwich has the current–voltage relationship shown by the solid line in Fig. 15.12(c). At low currents this resembles the current–voltage graph of a silicon p–n junction diode (see Section 15.11), but with a voltage at the onset of current that is just over 2 V, rather than 0.6 V in silicon p–n diodes. The difference can be understood in terms of the energy barriers that electrons and holes must surmount at the interfaces between either electrode and the semiconductor. However, the current does not continue to rise exponentially with the voltage as it does in a silicon diode, because there is a build-up of charge inside the semiconducting PPV that does not occur in silicon, and which limits current flow.

This sandwich arrangement is particularly interesting because of what happens when a hole injected from one electrode meets an electron injected from the other. Because they carry opposite charges, such a pair are attracted to one another, and go into orbit around a common centre, forming an *exciton*. After a few nanoseconds, they join together, and the energy released (nearly equal to the energy gap) goes to create a photon of light, whose frequency f is related to the energy so released by Planck's formula, $E = hf$. This process is of great interest at the time of writing because polymer diodes like this are fairly efficient, can be made in regular arrays by

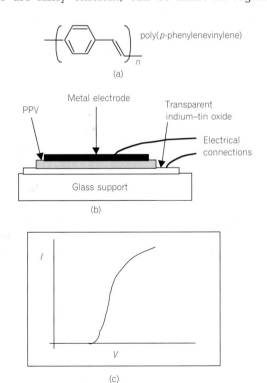

(a) poly(p-phenylenevinylene)

PPV
Metal electrode
Transparent indium–tin oxide
Electrical connections
Glass support
(b)

I
V
(c)

Fig. 15.12

(a) Structure of polyphenylene vinylene (PPV). (b) A thin slice of PPV sandwiched between electron-injecting and hole-injecting electrodes. (c) Current–voltage behaviour of the sandwich in (b). (Adapted from *Physics World*, June 1999, Institute of Physics.)

inexpensive printing techniques, and could eventually replace both cathode ray tubes and liquid crystal display screens. The colour of the emitted light is readily 'pretuned' by adjusting the composition of the polymer molecule to select the required energy gap and consequently the wavelength of the photons. The resulting flat display panel can be energy-saving when compared to a liquid crystal display, and is potentially much cheaper to make because the polymer can be deposited with an inkjet printer.

15.10.2 Conduction by 'hopping' in organic semiconductors

If the separation between molecules is slightly larger than in the above materials – typically 0.5 nm or greater – then electron wave functions do not extend over more than one or two molecules at most. A different description is then needed to explain electrical conduction, for an energy barrier exists to prevent electrons from transferring freely between molecules. To surmount that energy barrier, an electron (or hole) must acquire a thermal activation energy E_a, leading to low mobility of holes or electrons, and high resistivity. The mobility of electrons, and often, too, the resistivity, then has a rather strong temperature dependence of the form $\exp(-E_a/kT)$, and the conduction mechanism is referred to as 'hopping' of electrons or holes. Often the molecules are not stacked in regular arrays, making the activation energy E_a differ slightly for each intermolecular jump in a random way, and analysis of the resistivity must take this variation into account. The mobility of charge carriers may then not vary with temperature simply as $\exp(-E_a/kT)$.

Such materials are unlike silicon, GaAs and germanium, because they have no extended conduction or valence energy bands. So the name 'semiconductor' is here used only in its original sense, to describe a material having a conductivity between that of a good conductor and that of a good insulator. The preparation methods for organic semiconductors also frequently result in many more impurities than in their inorganic cousins, making the tasks of both explaining and controlling their behaviour more difficult.

These materials can, however, conduct electric currents either by holes hopping between the highest occupied states[†] in neighbouring molecules, or by free electrons hopping between the 'conduction' energy levels. Organic 'hopping' semiconductors of this type have been used for many years in laser printers and photocopiers that use the xerographic process. In these applications, photons of light irradiate the semiconductor and release electron-hole pairs called *excitons*, in which the hole and electron remain in close proximity because of their mutual attraction. The pair can nevertheless be separated by an applied electric field and then flow as an electric current to an external circuit via suitable electrodes. Triphenyl amine, TPA, is an example of a material used to make photosensitive surfaces for the applications mentioned. It is held in solid solution in an electrically inert matrix of bisphenol-A polycarbonate. Conduction in TPA occurs primarily by holes that hop between molecules with an activation energy lying between 0.32 and 0.45 eV. Practical materials like this usually contain more than one chemical species, making their conduction mechanisms too complex to explain here in detail.

[†] This state is usually referred to as the *highest occupied molecular orbital* (HOMO), while electrons are excited into the *lowest unoccupied molecular orbital* (LUMO).

15.11 Semiconductor devices: the p–n junction diode as an electric current rectifier

The importance of semiconductor electronics in modern society leads to the employment of many materials scientists in the manufacture of silicon 'chips'. We therefore include a description here of the working of the three principal devices used in silicon chips, which are made using the manufacturing processes described in Chapter 16.

A piece of semiconductor in which a section of p-type material is joined to one of n-type material is called a p–n junction. The properties of this junction are used in both the semiconductor diode, which can convert alternating current to direct current, and in transistors. The p–n junction is not usually made by joining two separate pieces of material, but by taking a single crystal of silicon and introducing donor and acceptor impurities (see Chapter 16) into the appropriate parts of the solid. We shall now consider the electrical properties of such a junction.

Let us imagine what happens to the electrons and holes in two pieces of semiconductor, one p-type and one n-type, as they are placed in contact. The initial distribution of holes and electrons at the moment of contact is shown in Fig. 15.13(a).

On one side of the junction is a material containing electrons moving freely and randomly at high velocities through the lattice, while on the other side is a material with free holes moving at similar velocities. It is natural that the excess carriers on either side will be able to cross the junction into the 'foreign' material, just as two gases will diffuse into one another. But, unlike gases, the free electrons and holes can combine self-destructively with one another. Far from the junction on either side the normal, undisturbed carrier densities must persist. This means that the concentration gradients of free electrons and holes across the junction persist in spite of the diffusion, as shown in Fig. 15.13(c). Moreover, the diffusion rates of holes and electrons, which are proportional to the concentration gradient (just like the diffusion of atoms, discussed in Chapter 7), are high near the junction, so that close to the junction the number of carriers becomes depleted. This depletion is so marked that there is a thin layer on either side of the junction that is virtually empty of mobile carriers – the densities are several orders of magnitude lower than in the rest of the crystal. This region is illustrated again in Fig. 15.13(b), and is termed the *depletion* layer. In practice it is about 10^{-4} cm thick.

Although there are very few mobile charge carriers in the depletion layer, the region is actually highly charged [Fig. 15.13(d)], owing to the presence of the ionized impurity atoms. Thus the donor atoms in the n-type material are positively charged, while the acceptors on the other side carry a negative charge. Outside of the depletion layer, these charges are neutralized by the presence of the mobile majority carriers. Because the whole crystal has to be neutral, the total numbers of positive and negative charges must be equal. The distribution of the charge on these impurities across the junction is roughly as shown in Fig. 15.13(e). This charge distribution is rather like that on a parallel-plate capacitor each of whose plates is situated midway between the edge of the depletion layer and the junction. The charge per unit area on each 'plate' is equal to the volume concentration of donors (or acceptors) multiplied by the width W of the

Fig. 15.13

The functioning of a depletion layer in a p–n junction.

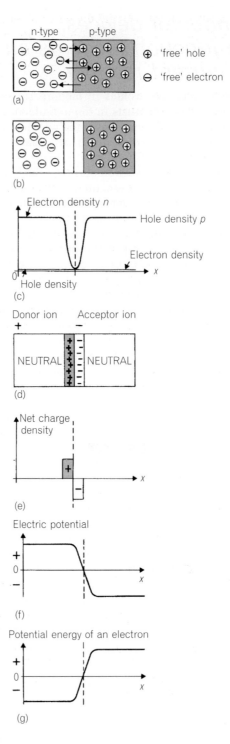

depletion layer. The built-in voltage V_0 across this 'capacitor' is a function of the carrier densities on either side of the junction, and an expression for V_0 will be calculated shortly.

This voltage is equal to the work required to be done in moving unit positive charge across the junction, and the voltage distribution across the junction, shown in Fig. 15.13(f), therefore gives the potential energy distribution for a unit positive charge. Because electrons are negatively charged, their potential energy distribution is inverted, as shown in Fig. 15.13(g). In the light of this diagram, the energy band diagram for the p–n junction can be redrawn as in Fig. 15.14(b), where the energy bands are shown bent to take account of the electrical potential variation across the junction.

It is clear that this potential distribution prevents the free passage of electrons and holes across the junction without work being done on them. Only a few, those with exceptional amounts of thermal energy, can cross this barrier in the 'uphill' direction. The holes in the p-type region are thereby hindered in moving across to the n-type region, and similarly, motion of free electrons towards the p-type region is impeded. Equilibrium is achieved when the flows of both electrons and holes in the two opposite directions just balance one another.

To calculate the built-in voltage difference across the junction we shall make use of the fact that the two halves of the crystal are in thermal equilibrium. The condition for thermal equilibrium between two bodies, which was derived in Section 7.7, is that the probability that quantum states of a given energy are occupied is the same in both bodies. This condition was stated in Chapter 7 for the quantum states of atoms, but it must also be true for electrons (and holes) if they are in thermal equilibrium with the atomic lattice. As the free electrons are present in the conduction band because they have acquired thermal energy from the lattice, they must be in thermal equilibrium, and therefore we may assume that the probability of occupation of electron states in the conduction band in Fig. 15.14(b) is independent of position.

Fig. 15.14

The energy band diagrams for a p–n junction with and without voltage bias.

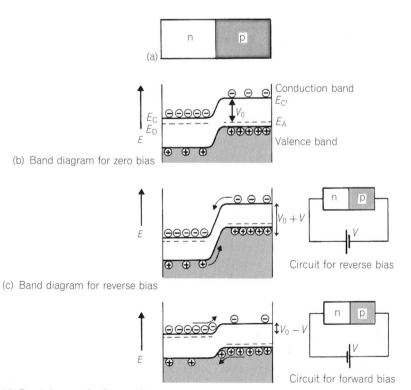

(a)

(b) Band diagram for zero bias

(c) Band diagram for reverse bias

Circuit for reverse bias

(d) Band diagram for forward bias

Circuit for forward bias

The probability distribution appropriate for electrons is the Fermi–Dirac expression of Eq. (7.12), but it was pointed out in Section 7.6 that, for electrons with energy E for which $(E - E_F) \geq 2kT$, the relative probabilities for different energies are given almost exactly by the Boltzmann factor. We have not discussed the Fermi energy E_F for a semiconductor, but we have a clue available from impurity semiconductors. We know that the impurity energy level, E_D, is just below the conduction band edge for electrons in an n-type semiconductor and that release of electrons from E_D to the conduction band is a result of thermal agitation. So, near absolute zero of temperature, the probability that the donor level E_D is occupied should be very close to unity, i.e. we find $E \leq E_F$ in Eq. (7.12). On the other hand, because the conduction band is practically empty near 0 K, E_F must be less than E_C. Thus we can deduce that E_F must be somewhere in the vicinity of the donor energy E_D for an n-type semiconductor and for normal doping densities N_D this is so. Similarly it is in the vicinity of the acceptor energy, E_A, for the p-type material.

In order that the occupation probability of any quantum level can be independent of position in thermal equilibrium, the Fermi energy, E_F, must be the same everywhere in the silicon. Thus, if E_F lies close to E_A in the p-type silicon but close to E_D in n-type material, we have a situation as shown in Fig. 15.14(b), where the edge of the conduction band has energy E_C on one side of the junction, but energy E'_C on the other. Now the probability $f(E_C)$ of occupation for an energy level, E_C, lying at the conduction band edge in the n-type material will be proportional to $\exp[-(E_C - E_F)/kT]$ while in the p-type material it will be proportional to $\exp[-(E'_C - E_F)/kT]$.

But, looking at Fig. 15.14(b), we can see that in the p-type silicon

$$E'_C - E_F = [eV_0 + (E_C - E_F)]$$

so that, in the p-type material, the probability $f(E_C)$ is proportional to the product of exponentials:

$$\exp[-(E_C - E_F)/kT] \times \exp(-eV_0/kT)$$

The actual numbers of free electrons per unit volume close to E_C in the two materials are therefore in the ratio

$$n_p/n_n = \exp(-eV_0/kT) \tag{15.8}$$

where n_n is the concentration of electrons on the n-type side and n_p the concentration on the p side. Inverting Eq. (15.8) and taking the natural logarithm, we obtain the result

$$V_0 = (kT/e)\ln(n_n/n_p) \tag{15.9a}$$

This is the equation that determines the height of the potential barrier that is set up in the p–n junction. To express V_0 in terms of the doping concentrations N_A and N_D, we make use of Eq. (15.1), which implies that the product np is always a constant at a given temperature and is equal to n_i^2. On the p-type side, therefore,

$$n_p p_p = n_i^2 \tag{15.10}$$

where the hole concentration p_p as usual equals the acceptor impurity concentration N_A. Similarly, on the n-type side, $n_n p_n = n_i^2$ and n_n is determined by the donor concentration N_D.

Thus, with a little manipulation, Eq. (15.9a) can be rewritten as

$$V_0 = (kT/e)\ln(N_A N_D/n_i^2) \tag{15.9b}$$

In Eq. (15.7) the value of n_i^2 is given as $6.3 \times 10^{15}\,\mathrm{m}^{-3}$ when $T = 300\,\mathrm{K}$. If N_A and N_D each have the typical value of $10^{22}\,\mathrm{m}^{-3}$ we find that $V_0 = 0.7\,\mathrm{V}$. It is important to recognize that V_0 cannot be measured directly by connecting a voltmeter across the junction, as there is no source of energy to supply current.

15.12 Current flow through a p–n junction with voltage bias

The dynamic equilibrium set up by the interdiffusion of holes and electrons is upset when a voltage is applied across the ends of the crystal. To understand the new situation we must first decide what form the energy band diagram takes when the voltage is applied. Let us take the case where the n-type region is made electrically positive with respect to the p-type region. The first thing to note is that, compared to the bulk of the crystal, the depletion layer has a higher resistance because of the absence of free carriers. Thus, remembering that the depletion layer is electrically connected in series with the rest of the crystal, we deduce that the applied voltage drop appears almost entirely across the depletion layer. This gives a new energy band diagram, as shown in Fig. 15.14(c). The effect of the applied voltage is to increase the potential difference across the junction. This *reduces* the flows of holes into the n-type region and of electrons into the p-type region, while making little difference to the reverse flows, because the latter are limited mainly by the scarcity of holes in the n region and electrons in the p region. The result is a very small net current flow from p to n, nearly independent of applied voltage.

On the other hand, reversing the sign of the bias voltage V reduces the height of the potential drop across the junction [Fig. 15.14(d)]. Once again the flow of minority holes from the n side and electrons from the p side are scarcely affected – remember that these flows are limited by the total numbers of minority carriers in the crystal and not by the rate at which they cross the depletion layers. But the flow in the other direction – up the potential hill – is increased, because the hill now causes less hindrance to the motion of the large number of majority carriers. This current can increase continuously as the applied voltage increases, so that the junction responds quite differently to voltages applied in opposite senses.

Just how the current depends upon voltage can be determined by reconsidering Eq. (15.8). That equation expresses the fact that the two halves of the crystal are in thermal equilibrium and the passage of a current through it does not upset this equilibrium. However, the potential energy difference between electrons on either side of the junction is no longer eV_0, but $e(V_0 \pm V)$. Reference to Fig. 15.14 shows this can be written just as $e(V_0 - V)$, if we give the voltage V a *positive* sign when the p region is made electrically positive and a *negative* sign when it is negative with respect to the n region.

This change in the potential energy difference means that Eq. (15.8) must be modified when a voltage is applied, giving

$$n_p' = n_n' \exp[-e(V_0 - V)/kT] \qquad (15.11a)$$

where the concentrations n_p' and n_n' have been primed to indicate that they are different from the values given by Eq. (15.8), which applies when $V = 0$.

An exactly similar equation holds for holes:

$$p'_n = p'_p \exp[-e(V_0 - V)/kT] \qquad (15.11b)$$

Because the crystal is electrically neutral outside the depletion layers, any change in n_p must be associated with a numerically equal change p_p. But because p_p is perhaps 10^5 times greater than n_p, the *percentage* change in p_p is negligible compared to the *percentage* change in n_p. Hence the interpretation of Eq. (15.11a) is that the dominant change occurring is that n_p, the concentration of electrons, the minority carriers in the p-type region, is dramatically altered, while the majority carrier concentration, n'_n, remains almost unchanged from the value n_n. In the same way, Eq. (15.11b) tells us that in the n-type material, p_n is changed by the same factor as n_p. In forward bias (when V is positive) the value of n_p is increased. The extra electrons have come from the n-type region, because the potential barrier has been lowered by the applied voltage. But these extra electrons can travel only a short distance, about 0.1 mm on average, before recombining with holes so that, far from the junction, the electron concentration is back to its normal value. The concentration gradient is illustrated by the circled charges in Fig. 15.14(d), and it is this gradient that gives rise to the flow of current, by diffusion of the electrons, where the voltage gradient is very small *outside* the depletion layer. According to Fick's Law (Chapter 7), the magnitude of the current is proportional to the concentration gradient, which in turn is proportional to the *excess* concentration of carriers, $(n'_p - n_p)$.

Using Eqs (15.8) and (15.11a) and approximating $n'_n \approx n_n$ we find that

$$n'_p - n_n = n_p[\exp(eV/kT) - 1] \qquad (15.11c)$$

so that the current can be written

$$I = I_s[\exp(eV/kT) - 1] \qquad (15.12)$$

where I_s is a constant, independent of the applied voltage.

Figure 15.15 shows that the *I–V* characteristic for a real junction does indeed have this form – at least for low values of applied voltage. The reverse current is so small that for practical purposes the junction behaves very nearly as a one-way conductor, i.e. as a rectifier.

The reason why semiconductor rectifiers and transistors rapidly superseded vacuum valves after 1950 lies largely in their reliability. There is little that can go wrong with the simple p–n junction and it deteriorates only if so much current is passed that the crystal becomes excessively hot. When this happens, kT increases, and with it the current, according to Eq. (15.12), causing further heating and eventually destruction of the device. The current at which this occurs depends on the product of voltage and current (i.e. the power dissipated) and the ease with which heat can be removed.

Fig. 15.15

The voltage–current relationship for a p–n junction.

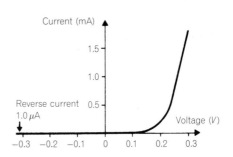

15.13 Field effect transistors (FETs)

A transistor is a device that can be used to amplify small alternating voltages or currents. It amplifies by converting power injected from a d.c. source (e.g. a battery) into a.c. power at the frequency of the input signal. In the following simplified description we show how power amplification is achieved.

The p–n junction described above depends for its operation on the presence of both holes and electrons and is sometimes referred to as a *bipolar* device. The field effect transistor (FET) on the other hand could be described as a *monopolar* transistor because it employs only one carrier type. The principle of operation is to vary the number of electrons (or holes) in a semiconductor by application of an electric field – this is the 'field effect'.

To understand how it works let us consider a parallel-plate capacitor made up of two rectangular metal electrodes placed a short distance apart, as in Fig. 15.16(a). When we apply a voltage, V, between the plates a charge, Q, is stored in the capacitor, of value $Q = CV$, where C is its capacitance (see Section 18.3). The charge is in the form of a net positive charge, $+Q$, on one plate and a net negative charge, $-Q$, on the other, but what are these actual charges? They must, in fact, be electrons and ions in the metal electrodes themselves. Recalling our picture of metallic bonding (Chapter 5), a metal is made up of an array of positive ions held in fixed positions with a mobile 'sea' of electrons acting as a glue. The electrons will move in an applied electric field, but the ions will not. Now the applied voltage produces an electric field between the capacitor plates. Referring to Fig. 15.16(b) this field will try to move the electrons away from the surface (deeper into the plate) at the top and to attract extra electrons to the surface at the bottom (remember that electrons move in the *opposite* direction to the electric field). Thus the positive charge on the top plate is due to pushing back the electrons to reveal the fixed positive ions. The negative charge on the lower plate is due to pulling extra electrons to the surface. As this happens, immediately after closing the switch [Fig. 15.16(a)], electrons are displaced round the circuit, from the top plate to the bottom plate and a transient, charging current flows.

We will now determine how many electrons are likely to be moved. Let our capacitor have plates of area $1\,\text{cm}^2$, the space between them being filled with a dielectric whose thickness is $1\,\mu\text{m}$ ($10^{-4}\,\text{cm}$) and whose permittivity equals ϵ_0. The charge Q can now be found from the voltage and the capacitance. Because capacitance equals $\epsilon \times$ area $\div$ spacing, it has the value $8.84 \times 10^{-10}\,\text{F}$, or $884\,\text{pF}$. If the battery has an e.m.f. of $10\,\text{V}$ we find $Q = 8.84 \times 10^{-9}\,\text{C}$, which is equivalent to 5.5×10^{10} electrons. This is the number of ions 'revealed' per cm^2 on the positive plate and the number

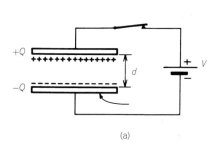

(a)

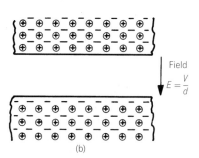

(b)

Fig. 15.16

The field effect in a semiconductor.

of extra electrons per unit area on the negative plate. Now a typical metal has $\sim 5 \times 10^{15}$ atoms cm^{-2} in its surface so that, on average, an electron is displaced from the layer of atoms at the surface in only one in every 10^5 atoms. We see, therefore, that the electric field barely penetrates the surface of the metal, because of the very high concentration of free electrons it contains.

Now suppose that one of the plates of the capacitor is made instead of an n-type semiconductor containing 10^{15} donor impurity atoms per cm^3. The ionized donors provide the fixed positive charge and there will be very nearly the same concentration of free electrons as there are ionized donor atoms. Each plane of the semiconductor will, on average, contain $(10^{15})^{2/3} = 10^{10}$ impurity atoms per cm^2. Using the same figures as before, to obtain the necessary 5.5×10^{10} electrons for the negative charges and 5.5×10^{10} positive charges, and assuming all the donor atoms to be ionized, we would have to penetrate about 5 atomic layers below the surface. We conclude that the electric field must penetrate the semiconductor to a depth determined by the impurity concentration, for a given applied voltage and capacitance. This is the basis of the field effect.

An FET is illustrated in schematic form in Fig. 15.17(a), while Fig. 15.17(b) shows its actual structure in cross-section through a silicon slice. The gate electrode is the top plate of the capacitor and the semiconductor slab, shown as p-type in Fig. 15.17(a), is the bottom plate. The insulator between them is a layer of silicon dioxide, SiO$_2$. The gate electrode may be made of a metal and, for this reason, the device is called a MOSFET, meaning metal–oxide–semiconductor FET. Nowadays the gate electrode is most often made of polycrystalline silicon instead of metal, but the terminology MOS is still used. The transistor operates by passing a current through a shallow *channel* in the semiconductor substrate, just below the oxide–semiconductor interface, between two contacts called the *source* and *drain* electrodes. These are, in fact, heavily doped n-type regions. When a voltage is applied to the gate it sets up an electric field that penetrates the semiconductor and causes the number of electrons carrying the current in this channel to be increased or decreased, according to the polarity of the voltage. Thus the source–drain current is modulated by the gate voltage. A positive gate potential will induce negative charge in the semiconductor and will repel holes, which are majority carriers in the p-type substrate.

Thus a depletion layer, in which the free charge concentration is reduced below that of the bulk semiconductor, is produced in the surface channel region. If an increasing, positive potential is applied to the gate then, at the surface, the concentration of minority carrier electrons will grow. Electrons from the n-type source and drain contacts will move into the channel under the attraction of the positive gate potential. Eventually, for a sufficiently positive gate potential, the electron concentration locally at the surface will exceed the hole concentration and the surface region is

Fig. 15.17

Cross-sectional views of the FET structure: (a) schematic, (b) as constructed.

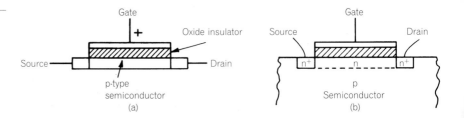

said to be *inverted*. This inverted region, referred to as an *n-channel*, bridges the gap between the n-type source and drain contacts so that current may flow between them. A transistor operating in this way is called an *n-MOS transistor*.

Using instead an n-type substrate and a negative gate voltage, a *p-channel* would be formed in just the same way and such a device is called a *p-MOS transistor*. In each case the source and drain contacts are of the opposite semiconductor type to the substrate, as in the example quoted with n-type contacts in a p-type semiconductor. Thus the source/substrate and drain/substrate areas are, in fact, n–p junctions and, unless forward biased, they prevent current flowing from source to substrate or drain to substrate.

We choose the value of the voltage at the drain and source so that these junctions are never forward-biased, in order that virtually no current flows between source and drain until the gate 'turns on' the inverted channel. In this way the current is confined to the channel region, immediately below the gate electrode.

If the gate voltage is varied in accordance with the magnitude of an a.c. electrical signal, the source–drain current also varies. The input power is equal to the gate–source voltage multiplied by the a.c. current charging and discharging the gate, which is exceedingly small; the output power is greater, being equal to the (much larger) a.c. source–drain current multiplied by the alternating drain–source voltage. The transistor is thus a power amplifier.

The power gain of a transistor is related to the change in the source–drain current for a given gate voltage change, called the *transconductance*, in units of $A V^{-1}$. Typically, MOS transistors have transconductances of about $0.25\,\text{mA V}^{-1}$, rather smaller than the value for a junction transistor. However, this disadvantage is offset by the fact that MOS transistors are physically very small and that, unlike the junction transistor, no current need flow into the gate when the transistor is held in its conducting state, and the resulting circuits have much lower power demands than do circuits made with bipolar transistors.

Another advantage is that the small, high-quality capacitor formed between the gate and the channel can be used to store information in the form of charge in the gate–channel capacitance, providing a built-in memory. These factors have resulted in the production of very large-scale integrated (VLSI) circuits, in particular the microprocessor chip and the memory chip.

Improvements in the speed of operation of FETs have been made by the use of ingenious materials technology, enabling higher electron mobilities to be obtained than for 'ordinary' doped silicon or gallium arsenide.

FETs made of polymers are under development, but their properties are not yet good enough to replace silicon transistors in practical applications.

15.14 Junction transistors

The junction transistor can be regarded as two p–n junctions connected back-to-back as shown in Fig. 15.18, where the current carriers, holes and electrons are also shown. We now have two potential barriers, one between the (left-hand) *emitter* and (centre) *base* regions and one between the base and the (right-hand) *collector* region. If we apply forward bias, reducing the barrier height between emitter and base regions, holes pass

Fig. 15.18

The junction transistor, with its biasing voltages.

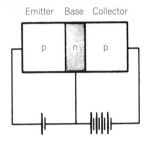

Emitter   Base   Collector

into the base region from the emitter – we say they are *injected* into the base. If the base region were thick, they would soon recombine with the electrons there. But the base is made especially thin so that, before this can happen, they find themselves swept into the collector, because the potential drop across the base–collector junction encourages this.

Now quite a small forward bias voltage across the emitter–base junction injects a large number of carriers into the base region from the emitter. At the collector junction a fairly large reverse-bias voltage is applied, which assists the flow of electrons into the collector from the base while opposing the flow of current in the opposite direction. Thus the current injected from the emitter enters at low voltage, and almost the same current (if few electrons are lost by recombination) is collected at a higher voltage, and hence a higher power. Like the FET, the junction transistor can be used as a power amplifier if the emitter–base voltage is varied in accordance with the magnitude of an alternating electrical signal. The much larger change in collector voltage results in amplified output power.

15.15 *Properties of amorphous silicon: α-Si*

Amorphous silicon has become commercially important because it permits the economic manufacture of arrays of thousands of transistor switches covering areas exceeding 30×30 cm, for display panels used in laptop computers and increasingly in desktop computers.

The electronic properties of silicon that were discussed earlier were dependent on the perfectly regular crystal structure, in which each atom is bonded to four neighbours via covalent bonds. In amorphous silicon there is no regular positioning of the atoms. Instead, they form a random network, some atoms having fewer than four neighbours with which to share their four valence electrons. If, for example, a given silicon atom has only three neighbours to which it can bond, its fourth valence electron forms what is called a *dangling* bond. A small amount of thermal energy can release this spare valence electron, causing it to migrate away, leaving behind an incomplete valence shell on the parent atom. The vacant energy level then becomes an 'electron trap', because any free electron migrating near it will be caught and localized within the atom. In a random amorphous structure there will be many incomplete bonds and consequently many electron traps. The electrical conductivity is thus determined by how far an electron may travel before being re-trapped. We can picture the electrons hopping from trap to trap under the influence of an electric field, giving rise to 'hopping conductivity'. In terms of the energy band diagram the traps are represented by 'localized' energy states for electrons situated between the valence and conduction bands, like the donor and acceptor states introduced earlier, but much larger in number. The frequent trapping of electrons in these states reduces their mobility enormously.

The important feature of pure α-Si is that its conductivity cannot be changed by doping with impurities using conventional methods, because the amorphous network is locally so flexible that it adjusts itself to complete any number of bonds to an impurity atom, and it is impossible to induce p- or n-type conductivity in the same way as in a single crystal. However, by incorporating hydrogen into the silicon, the conductivity can be made similar to that of intrinsic crystal silicon, although with a reduced electron mobility. Hydrogen-loaded silicon can be prepared from an electrical glow discharge in silane gas (SiH_4), by creating an ionized plasma in the gas with radio frequency energy. The SiH_4 dissociates into silicon and hydrogen, resulting in the deposition of a thin layer of solid silicon on any surface exposed to the discharge. Analysis of the resulting amorphous silicon showed that the dangling bonds are occupied by the hydrogen atoms, forming α-Si:H$_n$ groups, where n varies from 1 to 3. It is then possible to dope the material by including diborane gas (B_2H_6) with the silane, producing p-type films, while mixing phosphine (PH_3) in the silane produces n-type films. Thus it becomes possible to construct p–n junctions and FETs in the thin layer of α-Si:H$_n$, albeit with much lower mobilities of electrons and holes than are measured in single crystals. Despite the low mobility, amorphous Si devices find applications as cheap, low-power solar cells powering, for example, calculators and toys, and in flat-panel visual displays as mentioned earlier. The transistors act as switches, turning the individual picture elements of the display on and off. These would be impossibly expensive to lay down and connect using single crystal integrated circuits, but can be fabricated cheaply using similar deposition and etching methods to those described in the next chapter.

15.16 Metal-to-semiconductor contacts

It is necessary, when making amplifying or switching devices, that the metal contact to the semiconductor exhibits low resistance, and linear (i.e. non-rectifying) current–voltage characteristics over a wide range of current. Such contacts are referred to as 'ohmic'. Special precautions must be taken to ensure ohmic contacts, because without them the useful metals make rectifying contacts, as we shall explain using silicon as the practical example.

The basic characteristics of metal–semiconductor contacts can best be understood from a simple energy band diagram. It was mentioned above that, in an unbiased p–n junction, the Fermi energy must be constant throughout the material; the same will apply to a metal–semiconductor junction. In order to apply this principle, however, we must be able to define the Fermi energy level with respect to some fixed energy reference for both the metal and the semiconductor. This can be done by defining the *work function* for each, which is the energy that must be supplied to an electron at the Fermi level in order for it to escape from the solid altogether, i.e. for it to be emitted from the solid surface. In this definition it is assumed that the electron appears just outside the solid surface with zero kinetic energy; of course to detect the presence of the electron we would have to have the solid in vacuum. Now an electron emitted from the semiconductor would be indistinguishable from one emitted from the metal, so we take their energy just outside their respective surfaces to be the same and use this as our energy reference level. We can now interpret the diagrams shown in Fig. 15.19. In Fig. 15.19(a) the metal and the n-type semiconductor are not in contact, the reference energy (called the vacuum level) is shown, the work function energies are ϕ_m for the metal and ϕ_s for the semiconductor, and the diagram is drawn for the case $\phi_m > \phi_s$, which is true for all useful metals such as Al, Au, Pt, in contact with n-type Si.

When the metal and semiconductor are brought into contact, electrons flow from the semiconductor into the metal, because there are empty allowed

Fig. 15.19

Energy level diagrams for a
metal–semiconductor contact:
(a) metal and semiconductor
before contact; (b) equilibrium
state after contact.

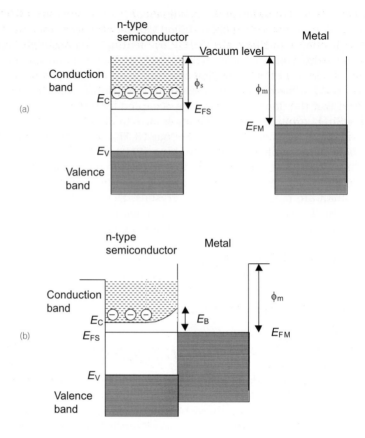

energy levels for electrons in the conduction band of the metal at energies
well above the Fermi level, E_{FM}, while in the semiconductor at the same
energy level there are more electrons, as it is nearer its Fermi level E_{FS}.

The effect of losing electrons from the semiconductor conduction band
will be to move the semiconductor Fermi level, E_{FS}, further from the edge
of the band and so the Fermi levels of metal and semiconductor are moved
towards each other. Figure 15.19(b) shows the equilibrium state, in which
there is a thin layer depleted of electrons at the semiconductor surface and
the Fermi level is the same throughout the two materials. The result is an
energy barrier E_B between the allowed electron levels in the metal and the
electrons in the conduction band of the semiconductor. It is called a
Schottky barrier and it causes the contact to behave electrically like a p–n
diode. Electrons may cross this barrier from metal to semiconductor only if
they have enough thermal energy to do so. The probability of an electron
crossing a barrier of height E_B is proportional to $\exp(-E_B/kT)$, and so is the
current that flows. Flow of electrons in the other direction, from the semiconductor to the metal, however, rises exponentially with the voltage applied
across the junction. The result is a rectifying current–voltage characteristic
given by Eq. (15.12) and illustrated in Fig. 15.15.

A similar sequence of argument shows that contact to a p-type semiconductor is rectifying when the work function of the metal is *less* than that
of the p-type silicon.

In order to make a contact non-rectifying and of low resistance, an n-type
semiconductor must be very heavily doped with n-type impurity at its

surface. The effect of this is to make the depleted region very thin indeed – so thin that it becomes transparent to electron waves near it, and they pass through almost unattenuated, a mechanism that we call 'quantum-mechanical tunnelling'.

While this theory predicts the general behaviour of metal–semiconductor contacts, the actual contact characteristics are variable. The reasons for this are that the work function of a material depends on the condition of its surface and is very sensitive to contamination by other materials. Furthermore, if interdiffusion between the metal and semiconductor occurs, their effective work functions may change drastically. Finally, differences in impurity content and structure mean that the work function of a metal in thin-film form does not have quite the same value as its bulk equivalent.

15.17 Advanced topic: Conduction processes in wide band gap materials

Electronic conduction occurs in non-metals having much larger energy gaps than silicon, but the details are more complex. We shall simply outline some of the factors involved, ignoring both conduction by ions and what happens in polycrystalline materials. What follows is relevant to, for example, the pyroelectric and ferroelectric crystals discussed in Chapter 18. We discuss four current flow mechanisms that can occur when a voltage is applied between two metal electrodes attached to opposite faces of a slab of a poorly conducting crystal.

Electrons (or holes) must flow first across an energy barrier that exists at the contacts, and then through the bulk of the material, before crossing a barrier into the second contact. The resistances of these three regions are electrically in series, and are shown in Fig. 15.20 as R_{C1}, R_B and R_{C2} respectively. Whichever is the largest of these resistances is the main limitation to current flow, and most (but not all) of the applied voltage appears across it. It is important to note that each of the three resistances varies in value with the voltage appearing across it, as described below.

Consider first the contact region. An energy barrier faces electrons trying to move from the electrode into the crystal, as described in the previous section. As a result, the conduction band edge E_C in Fig. 15.19 lies perhaps 0.5 eV or more above the Fermi level in the metal, and electrons must climb from E_F to E_C if they are to enter the semiconductor. The current changes exponentially with the voltage V_{C1} or V_{C2} across each barrier, just as explained in the previous section. If the energy barrier happens to be very thin, electrons can tunnel quantum-mechanically through it, giving a second mechanism for flow. Because of the exponential current–voltage relationship, the two contact resistances in Fig. 15.20 have very different values at any given value of the voltage applied across the slab of material.

Once electrons have surmounted the energy barrier in the contact region, they flow into the space between the electrodes, where, if their mobility is not large, their numbers accumulate. The space between the electrodes becomes filled with negative charge, which retards the flow of further

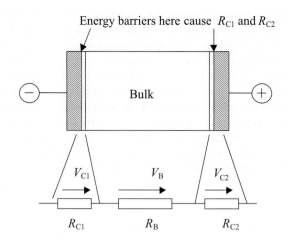

incoming electrons. Likewise, if more holes are present than electrons, this
space-charge is positive, deterring the entry of more holes. In either case the
resistance of the bulk material is high, and we refer to the current,
which is proportional to the square of the voltage V_B, as a 'space-charge-
limited' current.

Problems

15.1 Using the data in Table 15.2, calculate the
mobilities of electrons and holes (assuming
them to be equal) in grey tin, silicon and
germanium. Calculate also the mobility
that would be necessary for diamond
to have an *intrinsic* conductivity of
10^{-12} S m^{-1} if the hole and electron
mobilities were equal.

15.2 What concentration of n-type impurity is
necessary to explain the conductivity of
10^{-12} S m^{-1} in diamond if the mobility is
$0.18 \, m^2 V^{-1} s^{-1}$ for electrons, and intrinsic
conductivity can be neglected?

15.3 Calculate the carrier densities per m^3 in
intrinsic silicon and germanium at 100°C
using the value $N_C = 2.5 \times 10^{25} \, m^{-3}$, with
$E_g = 0.75$ eV in germanium, and $E_g = 1.1$ eV
in silicon. Assuming the hole and electron
mobilities to be inversely proportional
to temperature, estimate their values at
100°C (see Problem 15.1 for the values at
300 K) and hence calculate the resistivities
of silicon and germanium at this
temperature.

15.4 Calculate the total number of valence
electrons per cubic metre in germanium
(density $5.32 \times 10^3 \, kg \, m^{-3}$) and find what
fraction of them is available for conduction
at 300 K.

15.5 Calculate the intrinsic conductivity at 300 K
of the compound semiconductor gallium
antimonide (GaSb), for which the energy
gap is 0.70 eV. Take $N_C = 2.5 \times 10^{25} \, m^{-3}$ and
the mobilities of electrons and holes to be
$2.3 \, m^2 V^{-1} s^{-1}$ and $0.010 \, m^2 \, V^{-1} s^{-1}$
respectively. What concentration of n-type
impurity is required per m^3 to give GaSb a
conductivity of 100 S m^{-1}? Would the same
quantity of p-type impurity suffice?

15.6 Because Sb is a Group 15 element, an excess
of Sb in GaSb would make the compound
n-type. Using the data given in Problem 15.5,
calculate the excess atomic percentage of Sb
required to give a conductivity of 100 S m^{-1}.
The density of GaSb is $5.4 \times 10^3 \, kg \, m^{-3}$.

15.7 A piece of p-type silicon contains 10^{24}
acceptors per cubic metre. Calculate the

temperature at which intrinsic silicon has the same carrier concentration. Hence deduce qualitatively the manner in which the conductivity of the p-type material depends upon temperature.

15.8 What effect has an increase in temperature on the *I–V* characteristic of the diode shown in Fig. 15.15? Using Eqs (15.11a) and (15.12), calculate the current through the diode under 0.3 V *reverse* bias at 60°C, given that Fig. 15.15 is correct at 20°C.

15.9 How many holes and electrons would you expect to find in a piece of germanium containing an equal concentration N of donor and acceptor atoms? Calculate the conductivity of a piece of germanium containing 3×10^{22} donors and 8×10^{21} acceptors per cubic metre. The electron mobility in Ge is $0.39 \, \text{m}^2 \, \text{V}^{-1} \text{s}^{-1}$. (Assume all impurity atoms to be ionized and neglect intrinsic carriers.)

15.10 Assuming that in intrinsic silicon at room temperature 10% of the free electrons recombine with holes each second, calculate the approximate rate of release of energy that results. Does this liberation of energy produce a rise in the temperature of silicon?

15.11 In order for an electron to cross from the top of the valence band to the bottom of the conduction band energy must be supplied to it. If the source of energy is a photon, calculate the minimum frequency that the photon must have. What bearing does your result have on the optical properties of germanium?

15.12 The resistivity of many polymers such as polyethylene and PVC is very high. What does this tell you about the energy gap for electrons in a polymer?

15.13 As described in the text, the depletion layers at a p–n junction form a kind of capacitor. The capacity may be calculated approximately by assuming all the charge in each layer to be concentrated at the mid-plane of the layer. By writing expressions for the capacitance per unit area and the charge per unit area of the junction, deduce the expression

$$V_0 = eN_d W^2 / \varepsilon_r \varepsilon_0$$

for the voltage V_0 across the junction. Hence deduce the width W of the depletion layer in a silicon p–n junction, given $N_d = 10^{24} \, \text{m}^{-3}$, $V_0 = 1.0 \, \text{V}$ and $\varepsilon_r = 11.7$.

Self-assessment questions

1 A perfect covalently bonded crystal at the absolute zero of temperature will be a perfect insulator because

(a) the valence electrons are tied to the parent atoms

(b) electrons cannot move from bond to bond

(c) the electrons in the bonds form standing waves

2 If n electrons are liberated from bonds in a pure semiconductor the electrical conductivity is proportional to $2n$ because

(a) there are two electrons in each bond

(b) for each electron liberated a hole is created

(c) liberated electrons move twice as fast as those in bonds

3 The positive hole acquires a drift velocity in an electric field

(a) in the same direction as the field

(b) in the opposite direction to the field

(c) perpendicular to the field

4 The positive hole moves through the crystal by

(a) escaping from the bonds

(b) attaching itself to an electron in a bond

(c) reciprocal motion of electrons in the bonds

5 The total current through a semiconductor is given by

(a) the sum   (b) the difference

(c) the product

of the currents due to the holes and the electrons.

6 Free electrons and bonding electrons in a semiconductor have bands of permitted energy levels with an energy gap between them. The free electrons occupy the conduction band and the bonding electrons the valence band. Which of the following statements are correct?

(a) the conduction band is nearly full

(b) the valence band is nearly full

(c) holes are found mainly in the conduction band

(d) holes are found only in the valence band

(e) a hole reaches the conduction band by combining with an electron

(f) an electron reaching the conduction band leaves a hole in the valence band

(g) the bonding energy of an electron is equal to the energy gap

(h) when an electron recombines with a hole it falls from the conduction band to the valence band

(i) when an electron recombines with a hole, the hole rises to the conduction band

(j) when an electron and a hole recombine they must absorb energy

(k) when an electron and a hole recombine energy is liberated into the crystal

(l) electrons in the conduction band can acquire a net acceleration from a field because there are empty energy levels available

(m) an electron in the valence band cannot be accelerated by the field unless there are empty energy levels available

(n) holes cannot be accelerated by the field unless there are empty energy levels available

7 In a pure semiconductor having an energy gap of 0.65 eV and $N_C = 2.5 \times 10^{25}$ m^{-3}, the concentration of holes at a temperature of 300 K is

(a) 1.3×10^{14} m^{-3} (b) 8.6×10^{19} m^{-3}

(c) 9.8×10^{24} m^{-3}

8 A pure semiconductor has $E_g = 0.7$ eV, $N_C = 2.5 \times 10^{25}$ m^{-3}, $\mu_n = 2.3$ m^2 V^{-1} s^{-1}, $\mu_p = 0.01$ m^2 V^{-1} s^{-1}. Its conductivity at 300 K is

(a) 12 S m^{-1} (b) 0.13 S m^{-1}

(c) 6.4×10^{-6} S m^{-1}

9 The conductivity of a pure semiconductor

(a) is proportional to temperature

(b) rises exponentially with temperature

(c) decreases exponentially with increasing temperature

10 The slope of a graph of $\log_e$(conductivity) against reciprocal temperature for an intrinsic semiconductor with energy gap E_g is

(a) $-E_g/2k$ (b) $E_g/2k$ (c) E_g/k (d) $-E_g/k$

where k is Boltzmann's constant

11 The following questions concern impurity semiconductors and consist of an assertion and a reason. Give the answer (a), (b) or (c), as follows

(a) assertion correct, reason correct

(b) assertion correct, reason incorrect

(c) assertion wrong and therefore reason irrelevant

(i) A Group 15 impurity gives p-type conductivity *because* it has five valence electrons.

(ii) A Group 13 impurity gives p-type conductivity *because* it cannot take part in the covalent bonding system.

(iii) A Group 15 impurity is called a donor *because* it donates an extra electron to the covalent bonding system.

(iv) A Group 13 impurity is called an acceptor *because* it accepts an extra electron from the bonding system.

(v) Boron is a donor *because* it donates a hole to the valence band.

(vi) The energy difference between a donor level and the edge of the valence band is small *because* it easily accepts an electron.

(vii) In an n-type semiconductor the current carriers are almost exclusively electrons *because* the product of electron and hole densities is a constant for constant temperature.

(viii) In an n-type semiconductor the impurity atom concentration approximately equals the concentration of electrons in the conduction band *because* all the donor atoms are ionized.

12 Silicon contains 5×10^{28} atoms per m^3. If it is doped with 2 parts per million of arsenic the electron concentration is approximately

(a) 4×10^{23} m^{-3} (b) 10^{23} m^{-3}

(c) 2×10^6 m^{-3}

13 For the material of Question 12, using $N_C = 2.5 \times 10^{25}$ m^{-3} and $E_g = 1.15$ eV, the concentration of holes at 300 K will be

(a) 5×10^{28} m^{-3} (b) 10^{23} m^{-3} (c) 3.0×10^8

14 Each donor atom in an n-type semiconductor at normal temperatures

(a) carries a net positive charge

(b) carries a net negative charge

(c) is neutral

15 Each acceptor atom in a p-type semiconductor at normal temperatures

(a) carries a net positive charge

(b) carries a net negative charge

(c) is neutral

16 In a p–n junction there is a charge difference across the depletion layer, caused by the fixed impurity ions, such that

(a) the p side is positive and the n side negative

(b) the n side is positive and the p side negative

(c) both sides are positive but the n side has the bigger positive charge

17 The potential energy of an electron in a p–n junction is

(a) low on the n side and high on the p side

(b) high on the n side and low on the p side

(c) the same on each side but with a minimum at the junction

18 In a p–n junction that is in equilibrium and has zero bias voltage

(a) no holes or electrons cross the junction

(b) only electrons cross the junction

(c) equal numbers of electrons cross the junction in opposite directions

(d) the number of holes crossing in one direction equals the number of electrons crossing in the opposite direction

19 The height of the potential barrier at a p–n junction is determined solely by the densities of the impurity atoms on each side of the junction

(a) true   (b) false

20 A p–n junction is in forward bias when an external battery is connected with

(a) the positive terminal to the p-type side, negative to the n-type

(b) the positive terminal to the n-type side, negative to the p-type

21 In a silicon p–n junction the current at the temperature $T = 300$ K for a reverse bias voltage V such that $eV \gg kT$ is 10 μA. The current with 0.2 V forward bias will be

(a) 23 mA   (b) 3.35×10^{-9} A   (c) -10^{-5} A

22 The transistor is a power amplifier

(a) true   (b) false

23 The field effect transistor is basically a capacitor

(a) true   (b) false

24 The equation $np = n_i^2$ cannot apply in the source–drain channel of a MOSFET

(a) true   (b) false

25 When a negative gate voltage is applied to the n-channel MOSFET described in the text, the source–drain current will

(a) increase   (b) decrease

(c) reverse   (d) remain negligibly small

Answers

1 (c)	**2** (b)	**3** (a)	**4** (c)
5 (a)	**6*** (b), (d), (f), (g), (h), (k), (l), (n)	**7** (b)	**8** (a)
9 (b)	**10** (a)	**11** (i) (c); (ii) (b); (iii) (b); (iv) (a); (v) (c); (vi) (c); (vii) (a); (viii) (a)	**12** (b)
13 (c)	**14** (a)	**15** (b)	**16** (b)
17 (a)	**18** (c)	**19** (b)	**20** (a)
21 (a)	**22** (a)	**23** (a)	**24** (b)
25 (d)[†]			

* Note: (m) is untrue because an electron in a full band can accelerate by changing places with an electron at higher energy, the latter therefore being accelerated, (n) is true because holes actually *are* empty levels.

† Note: this channel becomes p-type, but the n source and drain contacts are isolated from the p-channel by the n–p diodes thus formed.

Semiconductor materials processing | 16

16.1 Introduction

The present age of microelectronics based on the silicon 'chip', properly called an integrated circuit, would not have been possible without the contributions of materials science, which underpin the whole microelectronics industry. This has been true at every stage in the development of integrated circuit technology and in this chapter we discuss the principles and methods involved, and introduce many relevant properties of the materials used.

We concentrate our discussion on silicon, which is the preferred material for about 95% of today's integrated circuits in a world-wide market that was worth almost US$100 billion in 2000. The other 5% is taken by materials with higher electron mobility than silicon (e.g. GaAs and Si-Ge) for use in faster transistors, or which have direct energy gaps and can emit light. The most important single reason why silicon is dominant is that its natural oxide, SiO_2, serves both as an excellent insulator and as a passivating (protecting) layer. No other material possesses such an easily grown, stable oxide[†]. Most of the techniques described below for silicon chips are nevertheless used in modified form to make chips from these other materials.

We begin by describing how highly perfect silicon crystals are grown as long cylinders with diameters currently between 10 and 30 cm, which are then sawn into circular wafers about 0.7 mm thick. Then we explain how transistors are made by impregnating one of the flat surfaces of the wafer with dopants, in predetermined patterns. Then an electrical wiring pattern is laid over the transistors by depositing a metallic layer over an insulating glassy SiO_2 layer. Contacts are simultaneously made to the underlying transistors through holes made previously in the SiO_2. As many as six differently patterned layers of metallic conductors are then added, separated by glassy insulators. A pattern of strategically placed holes called *vias* is made through each insulating glass layer to allow local contact between one metal layer and the next. The same pattern-defining process, called *lithography*, is used for each and every layer, and may be employed up to 20 times during the manufacture of an integrated circuit. Figure 16.1 illustrates a cross-section through the layers at the surface of a typical finished wafer, showing a small region about 10 μm in depth. Millions of transistors can be made simultaneously in each circuit, while many such circuits are made simultaneously, side by side on one face of the same silicon wafer.

[†] However, the discovery in 1990 at the University of Illinois of an insulation oxide of AlGaAs may enable increased commercial usage of this material.

Fig. 16.1

Many layers of materials are laid down on top of one another to form this small part of an integrated circuit, which is illustrated in cross-section.

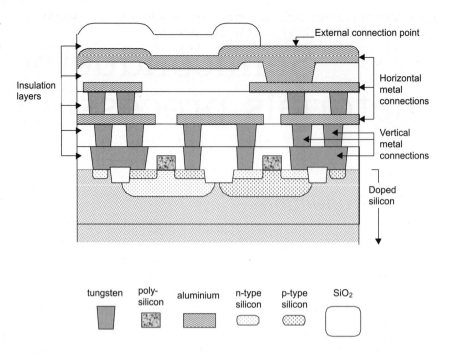

There are two main concerns throughout:

1. Lattice defects in the silicon wafer must be eliminated, and the concentration of dopant atoms must be closely controlled.
2. During the high-temperature processes needed to fabricate each overlayer, it is essential to prevent any deterioration in the properties of the underlying silicon transistors.

The layer-on-layer technique, using lithography to pattern each layer, is versatile enough to be used in a variety of other developing markets, because it is highly reproducible and allows the simultaneous manufacture of many identical components. It is used in making deceleration sensors for road vehicle airbags, liquid crystal display screens, microscopic mirror arrays for projection TV, and switching arrays for interconnecting optical fibres. Further applications are emerging rapidly as the 21st century advances. In addition, one or more of the individual processes described below is used in making surface coatings, ranging from metallization, for aluminized plastic bags, to ion implantation, for surface-hardened layers on mechanical parts for machine tools and vehicles.

16.2 Materials requirements

The discussion of p–n diodes and transistors in Chapter 15 leads us to expect that it is vital for their correct operation to control accurately the concentrations of holes and electrons. These are affected by (i) the concentrations of impurities introduced both deliberately and accidentally, and (b) the freedom of holes and electrons to move without becoming 'trapped' by a defect such as an unwanted impurity atom, or by a silicon atom that is not fully bonded to its neighbours. Such an atom may have a spare valence electron with no partner, and readily accepts a wandering free electron, making it unavailable for conduction.

What types of defect are damaging in this way? Box 16.1 lists some of them, in roughly decreasing order of importance, with the reasons why they are damaging. The one major exception to the rank order in the list is that an FET can be made to function adequately for restricted purposes in the presence of grain boundaries. For example, transistors made in *polycrystalline*, or even amorphous, silicon can be used as switches in liquid crystal display screens, where their rather slow speed of operation is not a major hindrance.

It is clear from the above that the level of purity with which silicon is made is critical. Because as little as one part per million of a dopant impurity in silicon has a profound effect on its electrical properties, it follows that

---

Box 16.1 Effects of lattice defects and impurities on electrons and holes in silicon

Grain boundaries – these can trap electrons or holes, and also they readily accommodate impurity atoms of all sizes. In each case, they normally deplete carrier concentrations in their vicinity and acquire charge in the process. On each side of such grain boundaries there is therefore a depletion layer, i.e. majority carriers are repelled by the trapped charge. Because electrons and holes have no way of avoiding a grain boundary in passing from one crystallite to the next, the effective mobilities of the remaining carriers are lowered. Grain boundaries are unwanted in transistors, because slow electrons make for slow switches; *requirement for integrated circuits*: zero in wafer.

Dislocations – edge dislocations must be removed from the neighbourhood of all p–n junctions because they act as precipitation sites for metallic impurities, which can cause a short circuit across the junction. They can also trap electrons or holes; *requirement* [*] *for integrated circuits*: < 1 per cm^2 surface.

Vacancies – an Si atom with a missing neighbour traps an electron or hole, which then needs more than kT of thermal energy to release it. Thus vacancies are found with either positive or negative charges ($+e$, $-e$ and $-2e$), which affect the position of the Fermi level and the carrier densities n and p; *requirement for integrated circuits*: concentration $< 1\%$ majority carrier density.

Heavy metal atoms may trap several electrons and holes with large binding energies (the energy needed to release the carriers). They therefore undo the effects of deliberate doping and must be held at very low concentrations; *requirement for integrated circuits*: < 1 ppba[†].

Carbon impurities readily become substitutional atoms, and can be tolerated except for their tendency to aid the precipitation of oxygen (see below) and other defects; *requirement for integrated circuits*: < 0.1 ppma[†].

Oxygen in concentrations above about 10 ppm atoms in silicon precipitates as an SiO_2 phase. The consequent change in volume puts the surrounding silicon into compression, generating dislocation loops. Some oxygen is unavoidable because silicon crystals are grown in an SiO_2 crucible (see Fig. 16.5), which is partly dissolved by the molten Si. Careful use of heating and cooling cycles during the fabrication of transistors ensures that oxygen remains in supersaturated solid solution at room temperature. Oxygen in interstitial sites at low concentrations is however beneficial, as it increases the yield strength of silicon by solution hardening (see Chapter 9) and scavenges metallic impurities. To limit its tendency to precipitate at high concentrations, the concentration must be uniform and not allowed to rise locally in any part of a silicon wafer; *requirement for integrated circuits*: uniform within range 19–30 ppma[†].

[*] Required for economic manufacture of circuits with up to 5 million transistors. From: S. M. Sze, *VLSI Technology*, 2nd edn, Wiley, 1988.
[†] ppma/ppba = parts per million/billion atoms.

a typical requirement for the starting material might be less than one impurity atom per 10^9 silicon atoms (1 ppba). Broadly similar considerations apply when devices are made with compound semiconductors such as GaAs, HgTe, etc.

The methods described below of processing purified silicon into thin slices (wafers) have been developed so as to eliminate such defects as far as possible from the surface region of each wafer in which the transistors are to be made.

16.3 Purification of silicon

Silicon, the commonest element on Earth, is mainly found in the form of its oxides, e.g. sand and quartz. These can be reduced in a furnace with carbon, at about 1450°C, to produce 98% pure metallurgical-grade silicon. Further purification then forms electronic-grade polycrystalline silicon in which all impurities, with the exception of carbon, are at concentrations below one part per billion (1 ppb). Concentrations of impurities are found by measuring the degree of absorption of infrared radiation, at a wavelength that is characteristic of each impurity. The accurate measurement of trace quantities of carbon is difficult because its characteristic wavelength, 16.5 μm, overlaps with the wavelengths characteristic of silicon, making the lower detection limit about 100 ppb; the carbon content is below this figure.

The standard purification process, used by industry over decades, is reduction with hydrogen of the gas trichlorosilane, SiHCl$_3$, condensed onto the heated surface of solid silicon. The first step is to produce SiHCl$_3$ by reacting pulverized metallurgical-grade silicon with anhydrous hydrogen chloride at 300°C using a catalyst. The main reaction that occurs is:

$$\text{Si (solid)} + 3\text{HCl (gas)} \rightarrow \text{SiHCl}_3 \text{ (gas)} + \text{H}_2 \text{ (gas)} + \text{heat}$$

together with an unwanted one:

$$\text{Si (solid)} + 4\text{HCl (gas)} \rightarrow \text{SiCl}_4 \text{ (gas)} + 2\text{H}_2 \text{ (gas)} + \text{heat}$$

Under the correct operating conditions approximately 90% of the product is trichlorosilane, the remainder being mainly silicon tetrachloride together with chlorides of the various impurities. Trichlorosilane is a liquid with a boiling point of 31.8°C. It can be separated from the unwanted chlorides by fractional distillation, reducing the electrically active impurity concentrations to less than 1 ppb.

Electronic-grade silicon is produced from the highly pure trichlorosilane, by depositing it on a solid surface from a chemical vapour. A similar method is used for *epitaxial growth* (explained in Section 16.7). The trichlorosilane is vaporized and diluted with high-purity hydrogen. It is passed into a reaction vessel (or *reactor*) in which a long thin rod of silicon is heated electrically to over 1000°C. The silicon is deposited on the heated rod by the following reaction:

$$\text{H}_2 + \text{SiHCl}_3 \rightarrow \text{Si} + 3\text{HCl}$$

The by-products of the reaction leave the reactor as a high-temperature gas stream containing H$_2$, HCl, SiHCl, SiCl$_4$ and a small amount of SiH$_2$Cl$_2$. The chlorosilanes are condensed and the trichlorosilane and hydrogen are recovered for recycling back into the reactor. Silicon tetrachloride is taken off as a useful by-product and the HCl can be removed in anhydrous form for recycling to the trichlorosilane production process.

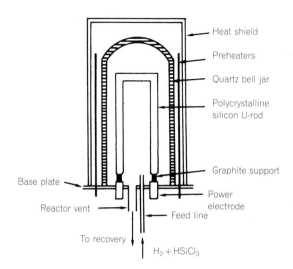

Fig. 16.2

Schematic cross-sectional diagram of a quartz deposition reactor.

A schematic diagram of the reactor is given in Fig. 16.2 and the process is depicted in Fig. 16.3. Commercial reactors are capable of producing cylinders (called *ingots*) up to 2 m long and 30 cm in diameter, which takes 200 or more hours of deposition.

The major shortcomings of the trichlorosilane technology are its relatively inefficient use of silicon and chlorine and the high energy consumption of the deposition reactors. As much as 70% of the silicon and 90% of the chlorine leaves the system in the form of silicon tetrachloride.

An alternative technology, which can reuse the gases with only minor by-products, is based on the pyrolysis (i.e. decomposition by heat) of silane, SiH_4. The process steps are (i) trichlorosilane production, (ii) silane production via several redistributions of the hydrogen/chlorine ratio in chlorosilane and (iii) silane pyrolysis to produce electronic-grade silicon. The principal reactions are:

Hydrogenation of $SiCl_4$

$$3SiCl_4 + Si + 2H_2 \rightarrow 4SiHCl_3$$

Catalysed redistribution of $SiHCl_3$ to SiH_2Cl_2

$$2SiHCl_3 \rightarrow SiH_2Cl_2 + SiCl_4$$

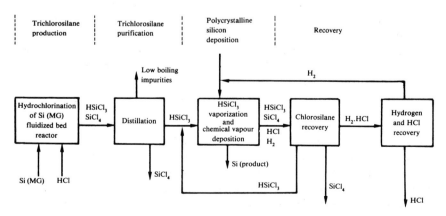

Fig. 16.3

Process flow block diagram for the production of polycrystalline silicon via the hydrogen reduction of trichlorosilane.

Fig. 16.4

Process flow block diagram for the
production of polycrystalline
silicon via the pyrolysis of silane.

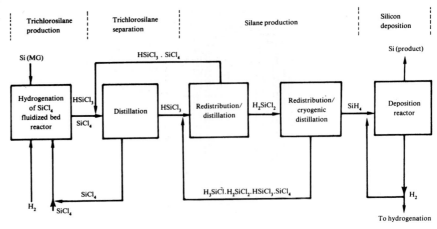

Catalysed redistribution of SiH_2Cl_2

$$3SiH_2Cl_2 \rightarrow SiH_4 + 2SiHCl_3$$

Silane pyrolysis

$$SiH_4 \ (heat) \rightarrow Si + 2H_2$$

All the chlorosilanes, silicon tetrachloride and hydrogen are reused by recycling them back through the preceding processes, resulting in conversion
of nearly 100% of the incoming silicon to electronic grade. The pyrolysis can
be carried out in the same type of reactor as for the trichlorosilane process,
depositing the silicon on a thin heated rod. A block diagram of the process is
given in Fig. 16.4.

The worldwide production capacity of electronic-grade silicon exceeds
10^7 kg per year.

16.4 Silicon crystal growth

The principle of growing crystals from a molten source is that of controlled
freezing as slowly as can be achieved. Of the many methods by which
crystals may be grown the technique almost universally chosen is that of
pulling the crystal slowly from the melt named after Czochralski, who
developed it as long ago as 1917.

A schematic diagram of a crystal puller is given in Fig. 16.5. An atmosphere of pure argon is used to prevent oxidation. To obtain a crystal with a
known lattice orientation, a small seed crystal is mounted with the desired
growth plane accurately parallel to the melt surface. It is then dipped into
the molten surface and the melt temperature is reduced until the molten
silicon begins to freeze on to the seed. Pulling is started and more material
freezes on to the crystal as it is withdrawn, the area of the solid–liquid
interface increasing while the rod grows in length. This forms a neck
whose diameter increases up to a constant value that is determined by the
temperature gradients and heat losses, and the rate of pulling. Silicon
expands considerably ($\sim$25%) in volume on crystallization and the rate of
pulling must allow for this expansion.

Fig. 16.5

A Czochralski crystal puller.

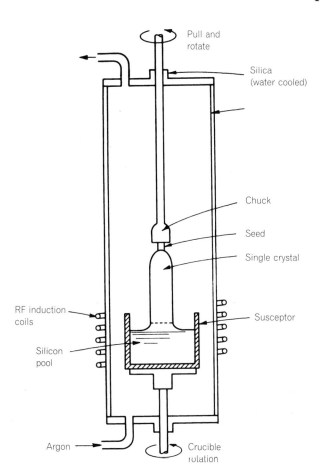

In the simplest view of the process the maximum rate of growth of the crystal is governed by the rate of heat loss, H_0 (watts) flowing from the crystal surface into the molten silicon. This must balance the latent heat of crystallization of the growing silicon. If a length dx and cross-sectional area A crystallizes in a time dt, then

$$H_0\, \mathrm{d}t = L\rho_{\mathrm{Si}}A\, \mathrm{d}x \qquad (16.1)$$

where ρ_{Si} is the density of solid silicon. Thus the maximum rate of pulling will be given by

$$\mathrm{d}x/\mathrm{d}t = H_0/L\rho_{\mathrm{Si}}A \qquad (16.2)$$

and we see that the area of crystal produced, i.e. the diameter of the crystal rod, can be controlled by the rate of pulling. Typically growth rates vary between 10^{-4} and 10^{-2} mm s^{-1}, being finally determined on an empirical basis for any particular crystal puller.

Because dislocations can become short circuits across p–n junctions (see Box 16.1), the usable diameter of crystal is determined by the area over which an acceptably low dislocation density can be obtained.

Briefly, the requisites for pulling a dislocation-free crystal are:

• The diameter of the starting seed crystal must be sufficiently small (1–2 mm) not to contain any dislocations itself.

- {111} glide planes should make a large angle with the growth direction; this is because dislocations generated by plastic deformation as the crystal grows concentrate in the {111} glide planes, and if the growth axis makes a large angle to the planes, the dislocations tend to grow out to the cylindrical surface, rather than up the rod.
- To favour climb of dislocations to the outer surface an adequate supply of vacancies should be available for their motion. This is favoured by rapid growth in a large temperature gradient.

As a result of improvements in silicon technology, the semiconductor industry moved from the use of 2 inch (51 mm) diameter wafers as standard during the 1960s, through several stages to mostly 8 inch (200 mm) wafers in the late 1990s. Wafers of 300 mm diameter were first used in chip production in Germany in 1998, and are becoming widespread.

The choice of crucible to contain the molten silicon has an important bearing on the purity of the crystal. Silicon reacts with graphite to form silicon carbide and it alloys with refractory metals, so these materials cannot be used. The best crucible is vitreous silica, which is used as a liner to a graphite crucible (a 'susceptor') heated by radio-frequency induction heating. Silicon slowly reacts with SiO_2 to form SiO, which is gaseous at the melting temperature. Thus the crucible is slowly attacked. The rate of attack is reduced by rotating the crucible and the crystal at the same rate and in the same direction during pulling. This has the effect of averaging out any thermal asymmetry in the radial directions (the reason for rotating the crystal in the first place) without introducing stirring of the melt, which would accelerate crucible erosion. A common contaminant of vitreous silica is boron, which dissolves in the melt as the crucible is eroded. This, of course, produces undesirable p-type doping.

16.5 Crystal doping

The slice of material in and on the surface of which the circuits are to be fabricated is commonly referred to as the *substrate*. A popular substrate is p-type with a resistivity in the range $0.01-0.5\,\Omega\,m$, depending on the application. This corresponds to acceptor densities in the vicinity of $10^{21}-10^{22}\,m^{-3}$, i.e. between one impurity atom in 10^7, and one silicon atom in 10^6. The introduction of this level of impurity requires successive stages in which a high-purity polycrystalline material is heavily doped to a resistivity of $\sim 10^{-4}\,\Omega\,m$, by the addition of a sufficient quantity of boron to be accurately weighed. This is then powdered and a small quantity added to the melt from which the crystal is pulled. A similar procedure is required to produce n-type substrate material, the dopant usually being phosphorus, arsenic or antimony. Because the impurity atoms can fit more readily into the disordered melt than into the solid, the number of impurity atoms per unit volume in solid silicon is less than that in liquid silicon held at the same temperature. The ratio of these concentrations is termed the *equilibrium distribution coefficient*. The resulting low take-up of impurity atoms in the crystal leads to a gradual enrichment of the dopant concentration in the molten pool as growth proceeds. As the crystal pulling continues, this produces a gradient of impurity concentration and hence of resistivity, along the length of the crystal.

Table 16.1

Distribution coefficients for typical
dopant impurities in silicon

Impurity	Distribution coefficient	Dopant type
B	0.8	p
Al	0.002	p
Ga	0.008	p
In	0.0004	p
P	0.35	n
As	0.30	n
Sb	0.023	n

Values of the distribution coefficient for typical dopants in silicon are given in Table 16.1. The nearer to unity is the value of the distribution coefficient, the less the enrichment effect in the molten silicon. From this point of view, boron is the best dopant. Where the distribution coefficient is small, the melt volume should greatly exceed the volume of crystal to be grown. It is also possible to feed fresh material of appropriate composition continuously (e.g. pure silicon) into the melt during growth to maintain its correct composition.

16.6 Wafer preparation

16.6.1 Cutting the ingot

The single-crystal ingot has to be cut up into thin circular wafers for further processing and these must have a perfectly flat, defect-free surface of an accurately oriented crystal plane.

After cutting off the seed and top ends of the ingot, its surface is ground by means of a rotating diamond wheel to reduce the diameter to the required value. Following this, a flat is ground along the length of the ingot, in a position that is located, using an X-ray technique, relative to a specific crystal direction. It is used to mechanically locate the wafer in automatic processing equipment, and permits orientation of integrated circuit devices with respect to particular crystallographic directions.

After grinding, the ingot is ready to be sliced into wafers using a diamond saw with a lubricant. The orientation of the wafer surface relative to the crystal axes is fixed by the angle of the saw. Usually trial wafers are cut, their orientation is determined by X-rays, and the angle of the saw adjusted until the correct orientation is obtained. For a $\langle 100 \rangle$ surface an accuracy of $1°$ is accepted; while $\langle 111 \rangle$ wafers are usually deliberately cut about $3°$ to the $\langle 111 \rangle$ plane, to aid epitaxial growth (see Section 16.7). The saw blade is annular, with the cutting edge on the inside perimeter of the annulus. This allows the blade to be mounted in a rigid frame and ensures that it does not bend during the cutting. The wafers are sliced with a thickness of 0.5–0.7 mm, depending on their diameter, up to one-third of the crystal becoming waste sawdust. Laser marking of each wafer identifies the doping type and surface orientation.

16.6.2 Lapping and chemical-mechanical polishing

The cutting process leaves the surface somewhat rough, so that it is followed by a series of mechanical and chemical processes, designed to create a wafer that is both flat enough for the later process of *lithography* (Section 16.9.1) to work effectively, and has a low density of defects at the surface.

The first step is mechanical polishing (called *lapping*) of both sides of the wafer, using a soft pad loaded with a mixture of alumina and glycerine as the polishing powder. Approximately $20\,\mu$m is taken off each side, giving a wafer with a surface roughness of only 1–2 μm in height. After lapping, the edges of the wafer are rounded to a curve by grinding. This prevents the edges being chipped during handling and also serves to limit the build-up of liquid photoresist at the edges during the photolithography process (Section 16.9).

These processes leave a high density of defects as a result of mechanical damage, extending to a depth of about $10\,\mu$m beneath the surface. To remove the damaged layer, the wafer is etched in a suitable solution.

After the first etching stage the surface is given a final polish to achieve a high degree of surface flatness, to an accuracy of between 5 and 10 μm across the whole face. A polishing pad of synthetic felt loaded with an aqueous solution of sodium hydroxide containing colloidal silica (particle size 10 nm) is pressed against the surface and continuously rotated. Typically, this removes about 25 μm of surface, by a combination of both chemical and mechanical processes, and is thus called *chemical-mechanical polishing*. The same type of process is used again at a later stage to flatten the surface before a subsequent layer is laid down (Section 16.15).

This is followed by a chemical clean with acid, base and/or solvent mixtures to remove the polishing residues. Chemical etching of silicon has been extensively studied. Because it is a covalent material it is very resistant to chemical attack, but it will readily react with oxygen. However, oxidation of the silicon surface results in the formation of a thin surface layer of silica, which is extremely resistant to the usual acids, but which can be dissolved in hydrofluoric acid. The simplest etch for silicon is a mixture of nitric and acetic acids, which oxidizes the silicon to SiO_2, together with hydrofluoric acid, which dissolves the oxide. Another factor to be considered is the different etch rate that applies to different orientations of the crystal surface. A (111) surface plane is close-packed in the diamond lattice and exposes fewer unattached bonds than other atomic planes. It is etched away more slowly than a (100) plane. But if there is extensive mechanical damage of the surface, parts of it will no longer present (111) planes, and the effect of the differential in etching rates will be to bring out the defects in the form of bumps or holes in the surface. It is for these reasons that many different etching methods have been developed for silicon.

The proper choice of etchant can either enhance or reduce the differential etching effect. In general an isotropic etchant will be used so that the silicon surface being etched is removed uniformly. This will be an etchant that reacts very quickly so that the material is dissolved away rapidly and differential effects are not able to develop. Sometimes, however, it is convenient to employ an *anisotropic etch* on (100) wafers. The (111) silicon planes, which lie at an angle of 57.7° to the surface, are attacked very much more slowly than other planes in a suitable etch (typically a potassium hydroxide–isopropanol mixture). By etching through a narrow window in a mask, a V-shaped groove is formed. This increases the available surface area in a given wafer, enabling either more or larger transistors, etc., to be packed into a given area of chip. This technique has been used mainly in the fabrication of high-power FETs and in making micromechanical devices from silicon.

16.7 Epitaxial growth

The first layer to be deposited on a wafer is often of silicon, made by growing on to the surface a thin, single-crystal layer by the process known as *epitaxy*, from a Greek word meaning 'arranging upon'. In this, atoms are deposited on to the crystal surface under conditions such that, under the influence of thermal agitation, each atom can 'run around' until it finds its correct crystal lattice position where it forms bonds to the atoms in the solid surface. In this way a new crystal is formed layer by layer so long as the supply of atoms continues. The source of silicon atoms will usually be a vapour, which may be produced by evaporating silicon from a molten pool in vacuum but, more usually, comes from a chemical reaction, as described below.

The use of an epitaxial layer has significant advantages in the fabrication of integrated circuits. Because the growing layer replicates lattice defects in the surface of the substrate, the degree of crystalline perfection cannot improve on that of the substrate. However, the so-called *epilayer* can have lower concentrations of impurities than the wafer. Moreover it can readily be uniformly doped p-type or n-type at any desired concentration by the inclusion of dopant atoms in the silicon vapour. Indeed, this is the only convenient means by which a lightly-doped surface layer can be created on a heavily-doped substrate. Typically an n-type layer $10 \, \mu m$ thick with a resistivity of $0.1 \, \Omega$-cm might be deposited on a p-type crystal wafer of resistivity $0.005 \, \Omega$-cm. The resulting p–n junction covering the whole surface area can act, if reverse-biased, as an effective insulating layer between the epilayer and the substrate. This prevents currents from flowing via the substrate between adjacent components formed in the epilayer. The uniformly doped epilayer, isolated from the substrate, is an ideal starting point for the fabrication of diodes, transistors, etc., by inserting further dopants as described in Sections 16.10–16.13.

Two principal methods of epitaxy are used – these are described in the next two sections.

16.7.1 Chemical vapour deposition

The commonest method for the preparation of epitaxial layers is *chemical vapour deposition* (CVD). In this method, the source of silicon atoms is silicon tetrachloride, $SiCl_4$, which is reduced by hydrogen on a hot silicon surface to deposited silicon atoms and gaseous hydrochloric acid. The reaction is

$$SiCl_4 + 2H_2O \rightarrow Si \text{ (solid)} + 4HCl \text{ (gas)}$$

A schematic diagram of a CVD reactor is given in Fig. 16.6. The cleaned and etched crystal wafers are placed on a graphite slab, which is heated by

Fig. 16.6

A vapour phase reactor for epitaxial growth.

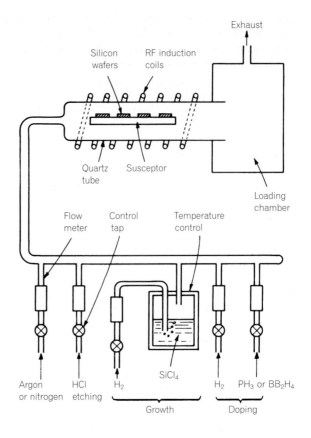

radio-frequency induction[†] to about 1200°C. Initially, hydrochloric acid vapour carried in a hydrogen stream is passed over the wafers, which etches away about $1\,\mu$m of their surface, cleaning them of all traces of contaminants. Growth is then started by switching off the acid gas and bubbling the hydrogen through a solution of silicon tetrachloride, which carries the silicon tetrachloride vapour to the hot substrates.

The rate of epitaxial growth is typically about $1\,\mu$m min^{-1}. This rate depends upon the vapour pressure of $SiCl_4$ in the reactor, determined by the temperature of the solution, which therefore has to be accurately controlled, usually in the range 0–30°C. Doping is achieved by mixing in with the $SiCl_4$ an appropriate gas containing dopant atoms, diluted in a stream of hydrogen. For n-type doping the most usual dopant gas is phosphine, PH_3, which decomposes to phosphorus and hydrogen on the hot substrate surface. If a p-type epilayer is required, diborane, B_2H_6, is used. These dopant gases are highly toxic and stringent safety precautions are necessary when they are used.

The conditions for epitaxial growth of good quality, single-crystal material are quite critical. The substrate temperature must be sufficiently high that the arriving atoms can diffuse rapidly enough across the crystal surface to find their required crystal lattice positions. Their rate of arrival must be slow enough to allow this process to occur for each atom in turn, and this limits the permissible growth rate to a maximum of about

---

[†] Electric currents are electromagnetically induced in the graphite by surrounding the apparatus with a coil carrying a radio-frequency current from a generator.

$2\,\mu\text{m min}^{-1}$. Above this rate, a polycrystalline layer is produced. For growth at a rate of $1\,\mu\text{m min}^{-1}$, the molecular fraction of $SiCl_4$ required in H_2 is about 0.01. If this fraction rises above about 0.25 a competing reaction sets in with the $SiCl_4$ reacting with solid silicon according to

$$SiCl_4 \text{ (gas)} + Si \text{ (solid)} \rightarrow 2SiCl_2 \text{ (gas)}$$

Under these conditions the crystal surface is actually etched away and the 'growth' rate becomes negative.

Like the seed crystal used in the Czochralski growth technique, the orientation of the substrate's crystal axes determines the orientation in the epitaxial layer. Thus a (100) face will grow a (100) epitaxial layer, and so on. One of the advantages of epitaxial technology is that it allows the use of a buried layer. For example, to make an n–p–n transistor we require a heavily doped n-type layer to make a low-resistance contact to its collector. This can be achieved by first diffusing n-type impurities into the p-type substrate to form the heavily doped layer (see next section), and then growing a more lightly doped epitaxial n-type layer over it. The transistor is subsequently fabricated in the epitaxial layer, by methods to be described in Section 16.8. To do that, it has to be possible to identify the position of the buried n-type region beneath the epitaxial layer. Fortunately the process of forming the buried n-type layer results in a shallow trough, about 50–100 nm deep, in the wafer surface, and this is replicated at the surface of the epitaxial layer, providing it is not made too thick.

16.7.2 Molecular beam epitaxy

When the epitaxial layer is grown from a vapour of atoms produced by evaporating molten silicon in a vacuum, the process is known as *molecular beam epitaxy (MBE)*. A long-established process, it has the disadvantage of an intrinsically slow rate of growth (0.01–$0.03\,\mu\text{m min}^{-1}$) and the need for very high-grade vacuum equipment. However, it offers a number of advantages over the CVD process, the most important of which is that substrate temperatures can be as low as 400°C during growth, as opposed to typically 900–1250°C for the CVD process. The significance of this in silicon technology is that, at the higher temperatures, there is out-diffusion from localized, doped areas such as buried layers. As a consequence, devices cannot be packed as close together on the substrate as is required in very large-scale integrated circuits (VLSI). The use of MBE overcomes this problem as well as allowing greater precision in doping the deposited layers.

MBE is a simpler process than CVD – atoms travel in straight lines from source to target, and no chemical reactions take place. Furthermore, there are no foreign atoms, such as chlorine and hydrogen, to become incorporated within the epitaxial layer. The basic process consists of evaporation of silicon and one or more dopant species in an ultra-high-vacuum (UHV) chamber, as illustrated in Fig. 16.7. The ultra-clean conditions in the stainless steel chamber minimize contamination of the silicon.

To evaporate silicon cleanly, a high-energy electron beam is focused to a spot on the surface of a silicon ingot, where it forms a small molten pool from which the silicon evaporates. The solid Si ingot is held in a water-cooled copper hearth so that its outer shell remains cool and acts as a crucible to contain the molten silicon, avoiding contamination of the molten pool. Dopants are evaporated from a small electrically heated furnace alongside, known as a Knudsen effusion cell. The rate of deposition of each

Fig. 16.7

Schematic of an MBE growth
system.

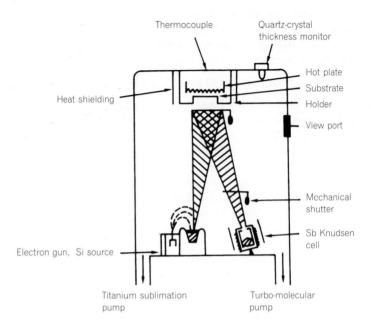

atom species on the substrate is determined by the vapour pressure above
the liquid, which is a strong function of temperature (see Section 16.15).

Thus the relative rates of evaporation of silicon and dopant can be
regulated by control of the source temperature, which controls the atom flux
reaching the substrate. Unfortunately, desirable dopants such as arsenic
and phosphorus evaporate too rapidly, and boron too slowly for precise control, and so antimony may be used for n-type doping and gallium
or aluminium for p-type. An advantage of this method is that the dopant
concentration can be varied during deposition by varying the Knudsen cell
temperature to produce a gradient in the dopant concentration, which is
desirable in some circuit components.

The method is, of course, not confined to the deposition of silicon. With
the increasing need for very fast digital electronic devices, ultra-high-
frequency amplifiers and semiconductor lasers, there is a high demand for
devices and circuits based on compound semiconductors such as GaAs,
GaAlAs, Si-Ge, InP and InSb, all of which can readily be fabricated with
MBE equipment.

Because deposition of the silicon epitaxial layer takes place in a
vacuum, it provides the opportunity of *in situ* cleaning of the substrate
surface before deposition. This is done by sputter cleaning (see Section
16.15.2), in which a beam of argon ions is directed at the substrate surface.
This removes any oxide and other contamination, leaving an atomically
clean, but partially disordered, surface. A brief anneal at 800–900°C is
sufficient to re-establish crystalline order in the perfectly clean surface on
which the epitaxial layer is to be grown.

16.8 Integrated bipolar transistor fabrication

We have by now covered the production of a single-crystal silicon wafer
with an epitaxially deposited layer on one surface. Let us say we wish to

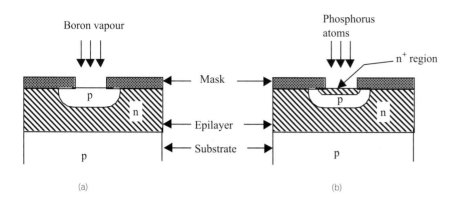

Fig. 16.8

The steps in bipolar transistor fabrication: (a) diffusion of boron for p-type conductivity; (b) introduction of phosphorus for n-type conductivity; (c) the finished transistor.

fabricate a bipolar n–p–n transistor in the n-type epilayer, which itself is to be used as the collector region for the transistor. It will be necessary to produce a p-type region for the base between two n-type regions which are the emitter and collector of the n–p–n transistor. The scientific principle involved is expressed by Eq. (15.1) in Section 15.6, i.e. the product np is independent of doping. Thus the introduction of more acceptor impurity atoms to increase hole concentration p must automatically reduce n, and so when sufficient boron is diffused into a small area of the n-type layer, it will be converted to p-type for the base region, as illustrated in Fig. 16.8(a). If this is followed by the insertion of, say, phosphorus, into a shallow portion of this p-layer [Fig.16.8(b)], the region is reconverted to n-type to complete the transistor. A schematic diagram of the sequence is shown in Fig. 16.8, in which (a) and (b) are the controlled diffusion stages and (c) is the finished transistor.

It is clear from Fig. 16.8 that a *masking layer* is required to define the areas into which the impurities are to be diffused. A convenient barrier to the impurity vapours is silicon dioxide (silica), through which the rate of diffusion of the impurity atoms is negligibly small at the diffusion temperatures. The minimum oxide thickness for this purpose depends on the nature of the impurity atom, but is generally in the region of 1 μm. The silica (SiO_2) layer can be produced by flowing a stream of oxygen over the hot silicon surface, producing the reaction

$$\text{Si (solid)} + O_2\text{(gas)} \xrightarrow[1000°C]{} SiO_2\text{(solid)}$$

The oxidation is more rapid in the presence of water vapour so that, when making a thick oxide layer, the oxygen may be bubbled through water on its passage into the furnace. The oxide growth rate is around 0.5 μm h^{-1}. The oxidation process is discussed in more detail in Section 16.14.

The transistor must also be isolated electrically from adjacent components, otherwise it would not operate correctly in a circuit. This can be achieved by using a further diffusion process to place a narrow border of p-type impurity in the wafer surface all around the transistor. This provides a p–n junction barrier around it, which operates just like the barrier between the epitaxial layer and the substrate.

Having grown an oxide masking layer over the whole of the silicon surface, we need to open 'windows' in the oxide, to enable impurity atoms to diffuse into the exposed area of silicon where each transistor is to be made. This can be relatively easily done by dissolving the silica in hydrofluoric acid (HF), which does not attack the silicon itself. Where higher precision is required, another method, *reactive ion etching* (see Section 16.9.2), is used instead. But first an etch pattern must be defined on the surface.

16.9 Patterning by photolithography and etching

16.9.1 Photolithography

During etching, it is necessary to protect the oxide from the etch process in places where windows are not required (see Fig. 16.9) and it is also important that the positions and sizes of the windows be accurately defined. For this purpose organic *photoresists* have been developed, which are the emulsions of an organic monomer in a suitably volatile carrier liquid. The sequence of processes used is illustrated in diagrammatic form in Fig. 16.9. Working in a yellow light, this material is sprayed onto the oxide surface, after which the horizontal wafer is set spinning at about 6000 rpm [Fig. 16.9(a)] to produce a uniform thickness of liquid on the surface by centrifugal force. The carrier liquid evaporates and the surface is gently baked to 150°C to solidify the layer of resist. The resist material is then in a condition such that it will polymerize if exposed to ultraviolet light. The polymerized material is not attacked by the hydrofluoric acid. The processing thus consists of preparing a mask, usually a film of chromium on glass, which masks the window areas so that they remain unexposed when the whole wafer is irradiated with ultraviolet light. After irradiation the unexposed resist is washed off in a suitable solvent, referred to as a *developer*, which will not remove the polymerized material. Thus the windows to the oxide are opened for etching it away. After the SiO_2 is etched, the polymerized resist is removed by exposing it to an oxygen-rich ionized gas plasma, which 'burns' it off. This is known as *plasma etching*. Because the exposed areas are not removed, the resist is called a *negative* emulsion, by analogy with photographic development. Because of the similarity it bears to processes used in producing etched plates for lithographic printing, the whole process is referred to as *photolithography*. There are many types of resist in use, including positive resists that on exposure become soluble in a solvent which does not dissolve the unexposed material.

The development of the photolithographic process has been crucial to the realization of integrated circuits. By using high-precision machines to locate the mask position, together with high-quality lenses and ultraviolet light, it was possible in 2000 to define and locate windows having minimum dimensions down to 0.1–0.2 μm, and there are plans to achieve

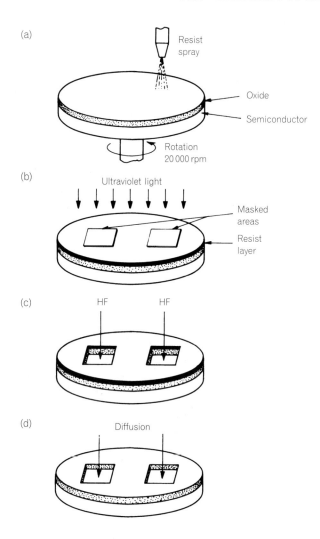

Fig. 16.9

The sequence of steps used in photolithography: (a) 'spinning-on' photoresist; (b) photoresist exposure with masking; (c) etching 'windows'; (d) diffusing in dopant.

0.07 μm dimensions in production in 2003. Achieving this level of accuracy is expensive and difficult, and all operations must be carried out in a totally dust-free environment.

Lithography can alternatively be undertaken with either X-rays or electron beams, in which case the mask-and-expose stage is replaced by 'writing' the pattern into a suitable resist material such as PMMA, by means of focused electrons or X-rays. Electron-beam lithography, as it is known, has for many years been used to pattern the chromium-on-glass master patterns that are required for photolithography, and, because of its superior accuracy, this technique, although slower in use, may at some future date replace photolithography in the manufacture of chips.

16.9.2 Etching windows

As mentioned above, two methods of etching the SiO_2 are available. The preferred method is one that etches down into the exposed layer below the window, without then etching sideways into the wall of the pit so created, i.e. etching should be *anisotropic*. Liquids such as HF – so-called wet etches – remove material equally in all directions (*isotropic* etching),

Fig. 16.10

Cross-section of a pit in silica isotropically etched by a liquid exhibits under-cutting of the resist.

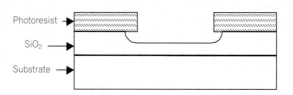

and hence produce a pit with a profile like that in Fig. 16.10. They are thus little used for finely detailed patterning of the layers such as the silica mask described above.

An alternative method that produces near-vertical sidewalls in the pit, called *reactive ion etching*, achieves directional etching by ensuring arrival of reactive gas ions from a single direction, perpendicular to the surface to be etched. The equipment and the process used for reactive ion etching are similar to those used for *sputter deposition*[†], which is discussed in Section 16.15.2. The etching ions are accelerated towards the target surface to be etched by a voltage difference applied between the target wafer and a counter-electrode placed nearby. An electric discharge ionizes the etch gas, which is fed into the intervening space. For etching silicon dioxide, fluorine-bearing gases such as CF_4 or the carbon-halide compounds known as *freons* are widely used. For successful results the products of the reaction between the etch gas and the silica must be volatile, for example CO_2 and SiF_4 result when the etch gas is CF_4. Obviously, too, we must not etch the masking photoresist, i.e. the etch process must be *selective*. The mechanisms leading to directional etching are not fully understood, but are connected with the energy brought in by incoming ions, which is largely lost to the wafer as soon as the ions are bonded to surface atoms. Arriving Cl^- or F^- ions carry enough energy to react chemically with the surface, forming a highly volatile product that vaporizes and is carried away in the surrounding gas. Because only horizontal surfaces, i.e. those normal to the incident ion direction are so bombarded, the etching is strongly directional. Little or no etching occurs on vertical surfaces, probably because etching atoms arrive there only by surface diffusion, and so have too little energy to initiate the reaction. Strong directionality makes this etching method particularly favourable when circuit components are packed very tightly together.

The same dry etching method, coupled with photolithography, can also be used for patterning layers of materials deposited at later stages in chip fabrication, such as silicon, aluminium, tungsten and polymers, for which suitable volatile reaction products are respectively SiF_4, $AlCl_3$, WF_6 and CO_2. Freons are again commonly used as the etching gases.

16.10 Impurity diffusion

The driving force for a diffusion process, as explained in Chapter 7, is a gradient in the concentration $N(x)$ of diffusing atoms. The diffusion flux, F, is given by Fick's Law [Eq. (7.32)]:

$$F = -D \frac{dN}{dx} \tag{16.3}$$

[†] Using this equipment the *target* is etched away, while in sputter deposition the removed material is deposited on a nearby *substrate*.

where dN/dx is the concentration gradient and D is the diffusion coefficient, measured in $m^2 s^{-1}$. This coefficient depends on the mechanism by which the diffusion occurs. In silicon the vacancy mechanism described in Section 8.3 results in a diffusion constant, D, with a strong temperature dependence, given by

$$D = D_0 \exp\left[-\frac{E_0}{kT}\right] \tag{16.4}$$

where D_0 and the activation energy E_0 are both constants. Group 13 and 15 impurities diffuse readily, with approximately the same average total activation energy $E_0 \simeq 3.5\,eV$ and with about the same value of $D_0 = 10^{-4}\,m^2\,s^{-1}$.

16.11 Prediction of depth profiles for diffused dopant concentrations

When fabricating integrated circuits, it is clearly important to be able to predetermine the depth of penetration and the profile of concentration versus depth of impurity atoms introduced into a semiconductor slice. To calculate this we can make use of the *transport equation* for diffusion of atoms, which was derived from Fick's Law in Section 8.3 and is repeated here:

$$\frac{\partial N}{\partial t} = -\frac{\partial^2 N}{\partial x^2} \tag{16.5}$$

where t as usual represents time.

The appropriate solution of this equation depends upon the initial conditions, i.e. on the values of $N(x,t)$ at particular depth x below the surface at the starting time. The following solution of the transport equation introduced in Section 8.3 is useful here:

$$N(x,\,t) = At^{-1/2}\exp(-x^2/4Dt) \tag{16.6}$$

Although this is not the most general solution, it is sufficient for our purposes.

Suppose now that a known quantity of dopant is placed in the surface, comprising a very thin, uniformly doped layer of thickness δ. If the initial impurity distribution is approximately a constant concentration N_0 per unit volume within a depth δ, then there is a finite number $Q = N_0\delta$ of atoms per unit area available for diffusion, and as diffusion proceeds this source will be depleted. Equation (16.6) then describes the evolution of diffusion, if we set $A = Q/\sqrt{\pi D}$, and if we assume that the process begins at a time t_0, which we set equal to $[\delta^2/\pi(d)]$. Equation (16.6) is graphed in Fig. 16.11 for three different times, and is called a *Gaussian* distribution. As the diffusion proceeds, the concentration $N(0,t)$ at the surface (i.e. where $x = 0$) becomes depleted with time t, as we can show by putting $x = 0$ in Eq. (16.6):

$$N(0,t) = \frac{Q}{\sqrt{\pi Dt}} \tag{16.7}$$

Equations (16.6) and (16.7) are valid only for times $t > t_0$, the time at which diffusion started.

Now suppose that, before diffusion began, the silicon was already doped p-type with a uniform concentration, P_b, of acceptors. Then, after diffusing

Fig. 16.11

The Gaussian distribution,
Eq. (16.6), for three successive
times, where $t_1 < t_2 < t_3$.

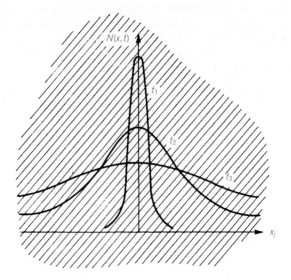

for a known time t, the depth x at which $N(x, t) = P_b$ will become, as explained in the next section, the depth of the interface between the p-type and n-type regions, i.e. the depth of the p–n junction. Thus we can find the junction depth x_j by putting $N(x_j, t) = P_b$ in Eq. (16.6) and solving for x_j, giving

$$x_j = \left[4Dt \ln\left(\frac{Q}{P_b \sqrt{\pi Dt}} \right) \right]^{1/2} \tag{16.8}$$

As time proceeds, the Gaussian distribution changes in accordance with Eq. (16.6), as illustrated in Fig. 16.11 for three successive times t_1, t_2 and t_3.

An alternative to supplying a fixed quantity Q of dopant per unit area is to expose the surface to a constant concentration of dopant, supplied in the form of a gas such as phosphine, PH_3 or diborane, B_2H_6 (both very toxic gases). The solution to Eq. (16.5) then has the shape of the error function plotted in Fig. 8.8, and the concentration at the surface no longer falls with time in the way shown above. The use of toxic gases naturally requires elaborate safety measures.

16.12 Formation of p–n junctions

Before the formation of p–n junctions or transistors in an integrated circuit chip, the silicon slice will already have been doped, typically by forming an n-type epitaxial layer as in Section 16.6. To produce a junction at some depth, x_j, below the surface, suppose that we diffuse in boron as a p-type impurity. The centre of the junction will be formed at the depth where the in-diffused boron concentration equals the original n-type impurity concentration. At this precise depth the doping results in *compensation* (see Section 16.8), that is, the two dopants effectively cancel each other out. The carrier concentrations n and p at this point are both equal to n_i, the intrinsic value. If the impurity is diffused from a finite source of boron in the way described above, the concentration profile will be Gaussian and so is given by Eq. (16.6). Calling the original n-type impurity concentration N_D, we can now calculate the depth x_j of the junction by replacing P_b by N_D in Eq. (16.8), i.e.

$$N_D = N(x_j, t) = (Q/\sqrt{\pi Dt}) \exp(-x_j^2/4Dt) \qquad (16.9)$$

Suppose, now, that we wish to continue processing to produce an n–p–n transistor. We will need a second stage in which n-type impurity atoms such as phosphorus are inserted in a layer even shallower than the depth x_j, to form the emitter of the transistor. To achieve a sharp drop in phosphorous concentration well before the depth x_j is reached, the doping technique normally used at this stage is ion implantation, described in the next section. The resulting individual doping profiles might be as shown in Fig. 16.12(a), giving the net majority carrier concentrations n or p shown in Fig. 16.12(b) as a function of depth.

A given area of a slice of silicon may be subjected to several heating cycles corresponding to different stages in the oxidation and diffusion processes. Following a diffusion process, further heating must not result in a major change in the existing distribution of dopants – in other words, subsequent processes must use lower temperatures. For this reason it is better that, if two or more diffusion processes are used in sequence, the first in-diffused species should be the one with the lowest diffusion coefficient, requiring the highest temperature. The next stage requires a lower temperature, and so the second dopant should be a faster diffuser. These requirements can be met by using first boron (p-type), for which the diffusion coefficient is $1.15 \times 10^{-13}\,\mathrm{cm^2\,s^{-1}}$ at 1200°C, followed by phosphorus (n-type), which has the *same* diffusion coefficient of $1.15 \times 10^{-13}\,\mathrm{cm^2\,s^{-1}}$ at the *lower* temperature of 1090°C.

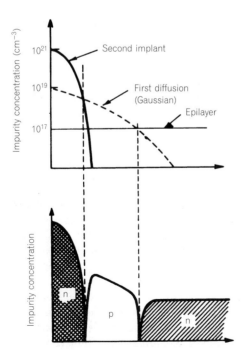

Fig. 16.12

Stages in the formation of an n–p–n transistor: (a) individual profiles for the successive stages; (b) resulting *net* impurity concentration distributions, i.e. the electron and hole concentrations as a function of depth below the surface.

16.13 Ion implantation

In the discussion of diffusion of dopant atoms into silicon in the previous sections, it was clear that abrupt changes in dopant concentration are difficult to achieve. A much more controllable method is *ion implantation*. In this, dopant atoms are vaporized, accelerated to a high velocity and directed at the silicon surface in a vacuum chamber. On reaching the crystal lattice they collide with silicon atoms, losing energy gradually as they bury themselves below the surface and eventually come to rest at an average depth below the surface determined by their initial energy. The actual number of dopant atoms delivered can be controlled quite closely by monitoring the ion current during implantation. The process causes disruption of the silicon lattice, due to ion collisions, which can be repaired by subsequent heat treatment. Ion implantation thus satisfies the conditions necessary for an effective doping process. Implantation energies for doping of silicon range from 10 to 200 keV, producing average penetration depths from 10 to 800 nm, depending on the type of atom implanted. Total doses range from 10^{12} to 10^{18} ions for each square cm of surface area.

In the energy range 10–200 keV, used for implantation into silicon, the energy loss mechanism that dominates is that of ion collisions with silicon atoms. The initial implanted ion energy of about 100 keV is much larger than the 10–20 eV energy binding a silicon atom in place, so that we can infer that the implanted atom will penetrate much closer to a silicon atom than the normal interatomic distance in the lattice. This allows us to ignore the relatively weak lattice forces and use classical dynamics to describe the collision between pairs of nuclei. Each collision is elastic, the energy lost by the incoming atom being entirely transferred to the target atom, which recoils away from its lattice site. For a 'head-on' collision the energy transferred, T_{max}, would be given by

$$T_{max} = E \left(\frac{4 M_1 M_2}{(M_1 + M_2)^2} \right)$$

where M_1 and M_2 are the atomic mass numbers of the ion and target atoms respectively and E is the initial energy of the implanted ion.

Each implanted ion follows a random path as it moves through the target and its average total path length is called its range R, which is a mixture of vertical and lateral motion. The average depth reached by the implanted ions is called the projected range R_p and the distribution of ions about this depth is approximately Gaussian in shape in both the vertical and lateral dimensions. A distribution shaped like one of those shown in Fig. 16.12 can be achieved by implanting the dopant through an SiO_2 surface layer, and by adjusting the energy to place the peak in concentration at the silicon surface.

Various theoretical models have been developed for the calculation of projected ranges and have resulted in reference tables from which the projected range, and spread in depth and width, can be found for a given atom type and implantation energy. A computer program called TRIM is widely used to model this process, and examples of its predictions for boron, phosphorus and arsenic implantations into silicon are given in Table 16.2.

The ranges quoted in Table 16.2 are calculated assuming implantation into an amorphous substrate, but these figures may be modified in a single crystal by the phenomenon of *channelling*. For ions moving in certain

Table 16.2

Ion implantation ranges and spreads in silicon for different dopant atoms and energies

Ion	Energy (keV)	Projected range (nm)	Longitudinal spread (nm)	Lateral spread (nm)
B (mass 11)	10	43	26	27
	50	208	82	98
	100	393	120	158
	200	700	159	231
P (mass 31)	10	17	8.5	8.7
	50	71	29	29
	100	142	51	52.5
	200	288	87	97
As (mass 75)	10	12.3	4.6	5.0
	50	38	12.2	13.4
	100	68	20	22
	200	128	34.5	38

crystallographic directions the atomic planes form a channel through which the ion may travel a relatively long distance, by glancing angle collisions with the atomic rows or planes. To avoid this, most implantations into silicon are carried out with the wafer tilted so that the ion beam enters at an angle of 7° from the normal to the surface plane.

As mentioned earlier, implantation damages the target by displacing many silicon atoms for each implanted ion. Indeed, high doses cause the silicon to lose its crystalline character completely and become amorphous. However, the crystalline structure can be recovered and the dopant atoms placed on substitutional lattice sites, as required to make them electrically active, by a suitable thermal annealing treatment. Raising the wafer temperature to 500°C during implantation induces instant annealing which maintains the crystal structure, but this may promote unwanted dopant diffusion elsewhere in the wafer.

Where a surface layer of the silicon has become completely amorphous, a moderately high temperature can induce regrowth of the undamaged crystal at the boundary with the damaged layer, by only small localized movements of atoms. This process, which occurs without any large-scale diffusion or melting, is called *solid-phase epitaxy*. In this way the boundary between the amorphous and crystalline regions moves towards the surface at a velocity that depends on the temperature, the doping and the crystal orientation. For example, in undoped silicon held at 600°C the boundary velocity is about $2\,\text{nm}\,\text{min}^{-1}$ for a $\langle 111 \rangle$ direction and about $30\,\text{nm}\,\text{min}^{-1}$ for a $\langle 100 \rangle$ direction.

When the implantation is not intense enough to create an amorphous layer, lattice repair occurs by the generation and diffusion of interstitials and vacancies. Because this process has a higher activation energy than solid-phase epitaxy, a temperature of about 900°C is required to remove all defects.

While these annealing treatments can be carried out in a furnace, heating the whole slice for up to 30 minutes may cause unwanted diffusion of dopants. For this reason *rapid thermal annealing* methods have been developed. These are generally based on heating by localized irradiation with high-energy light, from a laser, a xenon tube, or tungsten-halogen lamps. Rapid thermal annealing occurs because photons are absorbed by free electrons and holes in the silicon, which then transfer their energy to the lattice. Thus the heating rate depends on the doping level and temperature, which determine the number of carriers, as well as the photon flux and the surface emissivity of the silicon.

Commercial equipments for ion implantation are available from a number of manufacturers. In Fig. 16.13 the layout of a typical implanter is shown; Fig. 16.14 is a schematic diagram showing the features of the various parts. Beginning on the right-hand side of the diagram the

Fig. 16.13

The Imperial College ion beam facility (Whickham Ion Beam Systems Ltd).

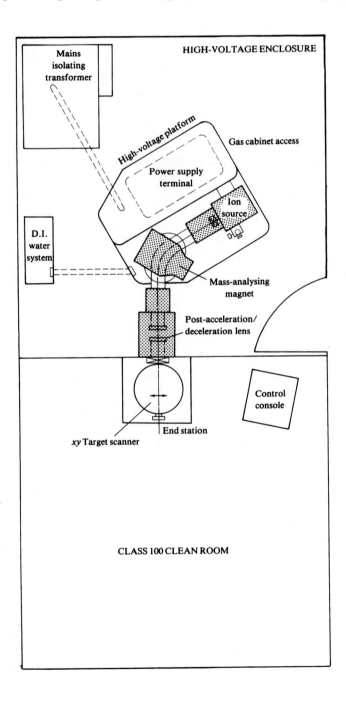

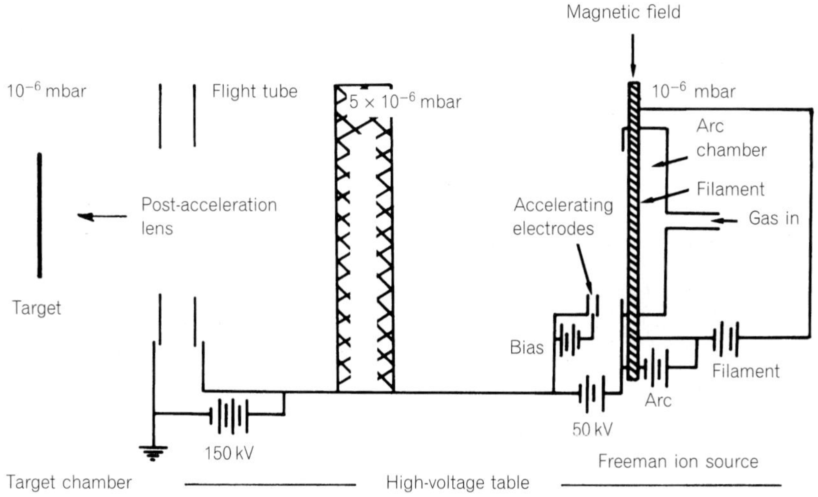

Fig. 16.14

Schematic diagram of the ion implantation equipment.

Freeman ion source converts a gas containing the desired implantation species to an ionized plasma. For example, for boron the gas may be BF_3, which is ionized to B^+ and F^- ions. The ionization is effected by electrons emitted from a hot tungsten filament. A magnetic field causes them to spiral about the filament, increasing their path length in the gas and hence their ionizing efficiency. The positive ions are extracted as a beam by means of suitable negatively biased extraction and accelerating electrodes. The ion beam travels down a curved flight tube under the combined influence of an accelerating electric field and a variable magnetic field. The strengths of these two fields are adjusted so that only ions of the desired mass are able to follow the curve down the centre line of the tube. An electrostatic lens system at the end of the tube directs the ion beam onto the target. The beam is scanned across the target by mechanical translation of the target holder in the x–y plane.

Implantation equipment is now a standard component of integrated circuit fabrication lines. Development of this technique has been so successful that it is used for almost all dopant stages in processing large chips. It is also being used in other industries, for example to surface harden metallic mechanical components by implanting foreign atoms that inhibit dislocation motion.

16.14 Thermal oxidation of silicon

As described earlier in this chapter, silicon is oxidized at various stages in the processing of an integrated circuit, to form diffusion barriers and electrically insulated areas. In MOS (metal–oxide–semiconductor) technology, in which the transistors formed are field-effect devices, the oxide gate insulator is a crucial element in the operation of the transistor itself. As a consequence, the thermal oxidation of silicon has been closely and extensively studied.

Experiments have shown that, in the course of thermal oxidation, oxygen diffuses through the oxide layer already grown, towards the Si/SiO_2 interface, where it reacts with the silicon surface to form more oxide.

Fig. 16.15

Schematic diagram of the
oxidation process.

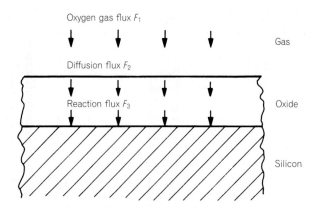

Fig. 16.15

Schematic diagram of the oxidation process.

The flow of oxygen molecules into the SiO_2 layer from the gas depresses the concentration of molecules in a thin boundary layer in the gas, close to the surface. Under steady-state conditions, the concentration of O_2 molecules in this surface layer drops until the diffusion rate through the oxide (which is proportional to the concentration gradient there) just equals the rate at which the reaction at the silicon surface consumes oxygen (which is proportional to the concentration there). The flows involved are illustrated schematically in Fig. 16.15.

For short times and thin oxide layers the surface reaction is the rate-limiting factor; the theoretical solution[†] predicts that the growth in the oxide thickness x_0 obeys the simple equation

$$x_0^2 + (B/C)x_0 = Bt \tag{16.10}$$

where B and C are constants that depend on (i) the diffusion coefficient D of oxygen molecules in SiO_2 (strongly dependent on temperature), (ii) the concentration of oxygen molecules in the gas, and (iii) the natural equilibrium concentration of oxygen in infinitely thick SiO_2 when in contact with the same gas.

For short oxidation times and hence thin oxide layers, $x_0^2 \ll (B/C)x_0$, so that Eq. (16.10) predicts a growth in thickness x_0 that is proportional to time. But when the layer thickness is greater than about a few hundredths of a micrometer, the term $(B/C)x_0$ becomes small compared with x_0^2. Then the thickness grows with time according to a parabolic law:

$$x_0^2 = Bt \tag{16.11}$$

The experimental results are found to follow closely the theoretical equations, so that timing the oxidation process allows good control of thickness.

The constants B and C both depend on temperature, following an activated law similar to that for the diffusion coefficient [Eq. (16.4)]. They are given by

$$C = C_1 \exp(-E_1/kT) \tag{16.12}$$

and

$$B = C_2 \exp(-E_2/kT) \tag{16.13}$$

[†] S. M. Sze (1998) *VLSI Technology*, pp. 100–103.

Gas	C_1 (μm h^{-1})	E_1 (eV)	C_2 (μm h^{-1})	E_2 (eV)
Dry O$_2$	6.23×10^6	2.00	7.72×10^2	1.23
H$_2$O + O$_2$	1.63×10^6	2.05	3.86×10^2	0.78

Table 16.3
Some values of quantities in
Eqs (16.12) and (16.13)

where the values of the constants C_1 and C_2 and the activation energies E_1 and E_2 are given in Table 16.3, both for oxidation in dry oxygen, and in a mixture of H$_2$O with O$_2$.

The presence of water vapour considerably increases the rate of oxidation, by a factor of about ten times at 1000°C. This is mainly a result of a much lower value of the activation energy, E_2, for wet oxidation.

The structure of the oxide formed on the silicon single-crystal surface is not itself crystalline, but rather is an amorphous network structure having a density close to that of silica glass. This is a more open structure than crystalline quartz, only 43% of the volume of oxide being occupied by atoms.

16.15 Metal conductors and interleaving insulation

Once the components have been formed on a chip by the methods described in this chapter, they must be connected to each other, to form a circuit, and to the outside world. This is done by depositing thin films of metal and patterning them to form the interconnection pattern on the chip and by bonding appropriate points in the pattern to very thin wires, to form the external connections. Three methods of depositing materials are described below. After deposition of each layer, it is patterned using photolithography followed by reactive ion etching (see Section 16.9.2). Chemical-mechanical polishing (Section 16.6.2) sometimes precedes the lithographic process, to flatten the surface and thus achieve greater pattern accuracy.

16.15.1 Vacuum evaporation

A common method of depositing metal on a chip is vacuum evaporation. In this technique the metal to be deposited is melted, usually by putting small lumps of it in an electrically heated 'boat' made of tungsten or molybdenum, in a vacuum chamber in which the pressure is 10^{-4} Pa or lower. Referring to Fig. 16.16, once the metal in the boat has formed a liquid pool, the shutter is opened and deposition starts. The rate at which atoms of metal leave a liquid surface in vacuum was first investigated, using mercury, by Hertz in 1882; his work was later extended by Knudsen. They found that the evaporation rate was proportional to the difference between the equilibrium vapour pressure of the liquid material and the vacuum pressure above it. This theory finally leads to an expression for the mass evaporation rate per unit area, F, which is

$$F = 77.77(M/T)^{1/2}P \quad (\text{kg m}^{-2}\,\text{s}^{-1}) \tag{16.14}$$

where M is the molecular weight of the evaporating species (in kg mol^{-1}), T is the temperature of the liquid (in K), P is the equilibrium vapour

Fig. 16.16

Schematic diagram of the vacuum evaporation process for metallization.

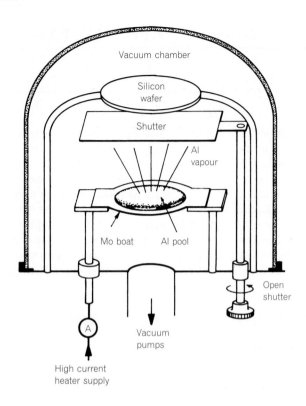

pressure (in Pa) of the liquid metal at the temperature T. It is assumed that P is so much greater than the pressure of other gases present that the latter can be neglected. Vapour pressures for most elements are published for practical ranges of temperature (see, for instance, *Handbook of Chemistry and Physics*, published annually by the Chemical Rubber Publishing Company in the USA). Taking aluminium as an example, its vapour pressure is 1.33×10^{-2} Pa at 972°C (1245 K) and its atomic weight is 26.98, giving an evaporation rate of $8.6 \times 10^{-7}\,\mathrm{g\,cm^{-2}\,s^{-1}}$. If we wished to deposit a square centimetre of layer with a thickness of 0.1 μm on a surface close to the molten aluminium, because the density of aluminium is $2.7\,\mathrm{g\,cm^{-3}}$, it would take 31 s.

16.15.2 Sputter deposition

It is sometimes necessary to deposit a thin film of metal, such as tungsten or platinum, that has such a high melting point that it cannot easily be melted in the vacuum chamber. This can be done by a process known as *sputtering*. If a gas, such as argon at reduced pressure, is subjected to a strong electric field it will ionize to form a *plasma*. The process by which the plasma is formed is that a relatively small number of free electrons, which are bound to exist in any gas because of the ionizing effect of stray cosmic ray particles and photons of high energy, are accelerated by the electric field. They acquire kinetic energy such that, when they collide with a gas atom they cause it to ionize, producing an electron and a positive ion. The new electron is, in turn, accelerated and causes ionization of another gas atom by colliding with it. The process continues to build up until the whole body of gas is ionized. The positive ions in the plasma will be

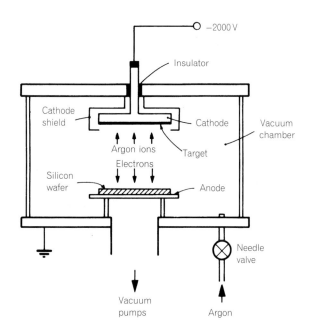

Fig. 16.17

Schematic cross-sectional diagram of the sputtering process.

attracted to the negative electrode (cathode) and the electrons to the positive electrode (anode) and give rise to a current in the external circuit. When an ion reaches the cathode it will collect an electron, supplied round the external circuit from the anode, and again become a neutral atom that returns to the gas and is then re-ionized. Thus the cathode surface is bombarded continuously by positive ions which, if the applied electric field is sufficiently high, arrive at the surface with considerable momentum. This is transferred to the atoms of the electrode itself by collision of the ion with the surface and, as a result, the atoms may be kicked out of the metal surface. This release of atoms from the metal, which is akin to splashing water out of a pool by throwing stones into it, is called sputtering. A substrate placed facing the cathode, called the 'target' electrode, will be coated with a film of the metal sputtered off the target surface. Because this process does not involve melting, thin films of refractory metals like platinum or tungsten can be produced by making the target of the appropriate metal. A schematic diagram of a sputtering system is shown in Fig. 16.17. The anode is the bottom electrode, which carries the silicon wafer and is electrically earthed. The top electrode carries the target material, which is to be sputtered, and is held at a high negative potential. A cathode shield, which is held at earth potential, surrounds the target and confines the ions to the surface to be sputtered. Compared with evaporation, sputtering is a relatively slow process, typically requiring 10 minutes to deposit $0.1\,\mu$m thickness of platinum.

Similar equipment is sometimes used for *sputter etching*, when a wafer that is to be etched replaces the target. More frequently this type of equipment is used for reactive ion etching, when a radio-frequency alternating electrical supply is used, instead of DC.

16.15.3 Chemical vapour deposition

The CVD process for depositing silicon, described in Section 16.7.1, can be adapted to the deposition of many materials, and is frequently used to

Table 16.4

Materials deposited by CVD: chemical reactions and typical temperatures used

Material deposited	Chemical reaction at wafer surface	Suitable wafer temperatures
Copper, Cu	$H_2 +$ organo-Cu compound $\rightarrow Cu +$ volatile product	200–450°C
Tungsten, W	$WF_6 + 3H_2 \rightarrow W + 6HF$ $2WF_6 + 3SiH_4 \rightarrow 2W + 3SiF_4 + 6H_2$	>300°C 400–500°C
Silicon dioxide, SiO_2	$Si(OC_2H_5)_4 \rightarrow SiO_2 + 2CO + H_2 + 3C_2H_6$ $SiCl_2H_2 + 2N_2O \rightarrow SiO_2 + 2N_2 + 2HCl$	650–750°C 900°C
Borophosphosilicate glass	$SiH_4 + O_2 \rightarrow SiO_2 + 2H_2$ $4PH_3 + 5O_2 \rightarrow 2P_2O_5 + 6H_2$ $\Big\}$ simultaneously $2B_2H_6 + 3O_2 \rightarrow 2B_2O_3 + 6H_2$	300–450°C
Silicon nitride, Si_3N_4	$3SiH_4 + 4NH_3 \rightarrow Si_3N_4 + 12H_2$ $3SiCl_2H_2 + 4NH_3 \rightarrow Si_3N_4 + 6HCl + 6H_2$	700–900°C 650–750°C

deposit uniform layers of metals such as tungsten and copper (but not aluminium), and also insulators such as SiO_2 and silicon nitride, which are needed between the various conducting layers.

Two or more vapours flow through a chamber containing the heated wafer, similar to that in Fig. 16.6. The deposited material is the result of a chemical reaction that takes place on the hot wafer surface, where the reacting gas molecules are temporarily 'trapped' for long enough for them to meet and react. The process is controlled by the wafer temperature, the rates of supply of the reacting gases into the chamber, and the concentration of a non-reactive carrier gas, which may be introduced to limit the rates of diffusion of reactants to the wafer surface. Some of the materials deposited in this way and some reactions that can be used to deposit them are listed in Table 16.4.

Silicon nitride and borophosphosilicate glasses are used as alternative insulators to SiO_2, because they can be deposited at lower temperatures, so avoiding diffusion of dopants already in place. Silicon nitride forms a particularly good protective coating for the completed chip, as it can be made very impervious to atmospheric pollution.

A very important metal missing in Table 16.4, namely aluminium, has no suitable volatile compound with which CVD could be undertaken. Copper, on the other hand, has a surface oxide that is permeable, allowing copper atoms to migrate and cause havoc in the behaviour of silicon transistors, unless prevented by adding a suitable barrier layer before the copper is deposited onto the chip.

16.16 Metallic interconnections

16.16.1 Electrical contact materials and interconnections

The choice of metals for making contacts to semiconductors was discussed in Chapter 15, where we remarked that it is all but impossible to predict just how good a contact a particular metal will make in practice. Aluminium has been chosen as the most widely used material in integrated circuits. It is used both in contact with silicon, and as general-purpose 'wiring'. Its big advantage over other low-resistivity metals is the possession of a stable oxide that *passivates* (i.e. renders unreactive) its surface. Aluminium is easily deposited by vacuum evaporation and the desired interconnection pattern is defined, using photolithography, on a mask over the aluminium, before etching away the unwanted metal.

To ensure low resistance contacts to the silicon, the circuits are heat treated following the deposition, typically at 550°C for 10 minutes. During heat treatment the aluminium reduces the thin native SiO_2, which occurs naturally on the silicon surface, and Si–Al contact is made. Dissolution of the silicon in the aluminium then occurs, up to about 1.5 atomic percent at 550°C. Unfortunately the dissolution is anisotropic, the (111) face of Si dissolving more slowly than the (110) and (100) faces. The result of this is that shallow, faceted pits tend to appear under the aluminium contact during annealing and, also, the dissolution may spread under the oxide adjacent to the contact area. When the heat treatment is over, and the structure returns to room temperature, some silicon precipitates out of the aluminium epitaxially on to the silicon surface and this precipitate is heavily doped with aluminium. Because aluminium is a Group 13 material, the precipitates are very p-type. The consequences of these processes are undesirable. A dissolution pit can short-circuit a p–n junction made a small depth below the silicon surface, and precipitation may produce spots of high electrical field between the p-type precipitate and a differently doped silicon surface. Nevertheless, for many purposes, aluminium contacts can be satisfactory. However, the present generation of integrated circuits uses a more complex metallization technology, in which successive layers of different metals are used, to enhance reliability.

Two examples are given in Fig. 16.18, both of which are based on the formation of metal silicides, which form good, non-rectifying, contacts to silicon but do not form pits or precipitates. A layer of platinum or palladium, merely 0.05 μm thick, is first sputtered on to the silicon and sintered at 600–700°C. This forms the silicides PtSi or Pd_2Si, and the surplus metal is etched away. Unfortunately, aluminium reacts with these silicides and, at 400°C, could penetrate and make direct contact with the silicon. To prevent this a 'diffusion barrier' of partially oxidized chromium is next laid down and finally, over that, an aluminium alloy containing 3% Cu is deposited. Figure 16.18(a) shows a cross-section through a contact to the n-type epitaxial layer in a wafer, while Fig. 16.18(b) shows the structure of a p-channel MOSFET. The inclusion of 3% Cu in the aluminium limits an undesirable phenomenon known as *electromigration*. Because the aluminium interconnections are both thin and narrow, quite small currents through them give rise to an enormous current per unit area of cross-section, up to about 10^4 A mm^{-2}. The very high flux of electrons flowing through the metal can carry enough momentum to literally 'blow' the metal atoms like a wind, along in the direction of electron flow. The aluminium atoms move most readily within grain boundaries, as a result of the lower density there than in the bulk. As a result the conductor forms a bulge where the downwind end of a grain boundary reaches the 'wire' surface, and correspondingly thins at the upwind (electrically negative) end, until eventually the resulting build-up of stress causes a mechanical rupture. Adding heavier copper atoms creates massive obstacles that reduce the electromigration effect, while its escape into the surrounding materials is prevented by the impermeable Al_2O_3 surface oxide.

Fig. 16.18

Cross-sectional diagrams of
metal–silicon contacts in
integrated circuits: (a) an ohmic
contact to an epilayer in a bipolar
transistor chip; (b) a p-channel
MOSFET.

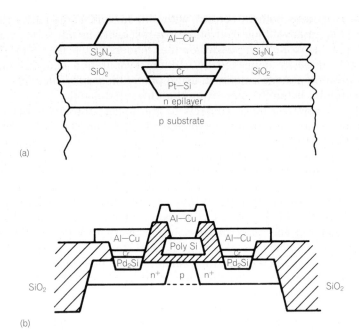

To connect one layer of aluminium 'wiring' to the next, tungsten is
frequently used, deposited so as to fill small holes etched in the SiO_2 insulation layer. When deposited by CVD, tungsten is found to fill these small
holes more reliably than does aluminium, which can only be deposited by
vacuum evaporation or sputtering.

At the time of writing much development activity is concentrated on
replacing aluminium entirely by copper, to take advantage of its lower
electrical resistance. This has raised a number of tricky materials problems,
largely because copper is very reactive, and because, unlike aluminium, it
possesses no stable, impermeable oxide. This has now been overcome by
the use of additional barrier layers, and the use of copper will become
widespread.

16.16.2 External connections

After all the metal layers and intervening insulating layers have been
completed, the silicon slice is mechanically divided into individual chips
by scribing lines on the surface between the chips and gently bending the
slice until it cracks along suitable crystallographic planes. Each chip is
then ready for testing and wire-bonding.

The designer of an integrated circuit must provide rectangular 'pads' of
metal at intervals around the edge of the chip, which are normally connected to more substantial metal pins by thin wires (Fig. 16.19). The wires
are bonded to these metal pads by 'ultrasonic welding', for aluminium wire
cannot be soldered. In this technique a wedge-shaped tool presses the wire
on to the aluminium pad and is vibrated up and down at an ultrasonic
frequency, literally welding the metals together. Aluminium wire may be
used to connect to aluminium pads whenever the chip package is to be
hermetically sealed against potential corrosion, but a preferred route for
permeable plastic packaging is to coat the pad with gold and use gold wire.
As gold does not adhere well to aluminium, an intermediate 'glue' layer of

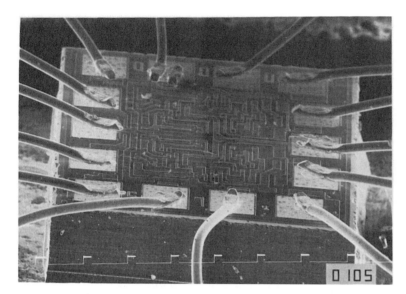

Fig. 16.19

An early, small-scale, integrated circuit, with aluminium wire bonded to its connection pads. The internal chip 'wiring' is just visible as a white outline.

chromium must be deposited first on the aluminium. To achieve thick pads, the gold is generally deposited by electroplating. When gold alone is used, *thermocompression bonding* becomes possible, in which the tool that presses the wire on to the pad is heated. The combination of heat and pressure produces the thermocompression bond, which is equivalent to a weld. The process is less violent than ultrasonic bonding and is less liable to failure by breakage of the wire.

Figure 16.19 shows an electron microphotograph of a chip wire-bonded to the pads.

Problems

16.1 In a crystal puller with a molten silicon charge, 99.9% of the heat input is lost by conduction and radiation. Calculate the heat input required to draw a 3-inch diameter crystal at a rate of $10^{-3}\,\mathrm{mm\,s^{-1}}$. The latent heat of crystallization of silicon is $49\,\mathrm{kJ\,mol^{-1}}$, its density is $2.33 \times 10^3\,\mathrm{kg\,m^{-3}}$ and its atomic weight is 28.

16.2 An impurity diffusing into silicon has diffusion coefficient values of $9.46 \times 10^{-16}\,\mathrm{cm^2\,s^{-1}}$ at 900°C and $1.07 \times 10^{-12}\,\mathrm{cm^2\,s^{-1}}$ at 1000°C. Using Eq. (16.4), determine the activation energy E_0 and the prefactor D_0.

16.3 A doped layer $0.01\,\mu\mathrm{m}$ thick on the surface of a p-type epitaxial layer is used as a source for a diffusion. If the p-type doping is 10^{23} boron atoms/m³ and the source contains 10^{25} phosphorus atoms/m³, determine the time taken to produce a p–n junction at a depth of $0.3\,\mu\mathrm{m}$ below the surface, given that the diffusion coefficient is $1.15 \times 10^{-13}\,\mathrm{cm^{-2}\,s^{-1}}$. What will be the surface concentration of phosphorus atoms after this time? [Note: Eq. (16.8) will yield the time, but an iterative calculation using a series of trial values of t is required.]

16.4 Using the values given in Table 16.3 determine the initial *linear* rate of growth of silicon dioxide at 1100°C, for both dry and wet oxygen.

16.5 Using the result of Problem 16.4 plot graphs of the growth of an oxide layer at 1100°C for dry and wet oxidation. Assume that when the growth process begins there is already an oxide layer of 5 nm thickness, and plot thickness against time up to a maximum thickness of $1\,\mu\mathrm{m}$. Hence show that changeover from linear to parabolic growth occurs at a thickness of about $0.05\,\mu\mathrm{m}$.

Self-assessment questions

1 Electronic-grade silicon requires all impurities to be at levels below 1 ppm atoms

(a) true   (b) false

2 On freezing, the volume of silicon

(a) increases   (b) decreases

3 In crystal pulling the diameter of the crystal depends upon

(a) temperature of melt

(b) temperature of solid

(c) rate of pulling

(d) volume change of silicon upon freezing

4 The dislocation content of a crystal may be reduced by increasing the rate of pulling

(a) true   (b) false

5 The concentration of dopant atoms during pulling of a doped crystal is greater in

(a) the solid   (b) the liquid

6 A single crystal slice is etched after polishing primarily to

(a) remove surface oxide

(b) remove mechanically damaged material

(c) make the slice thinner

7 Epitaxial growth is used to

(a) improve crystal perfection

(b) increase the doping level

(c) change the type of dopant

8 Producing a transistor by integrated circuit processing depends on the fact that, at a constant temperature

(a) $n + p =$ constant   (b) $np =$ constant

(c) $n = p =$ constant

9 In integrated circuit processing the silicon slice is oxidized to

(a) prevent the silicon evaporating

(b) act as a barrier to impurity diffusion

(c) prevent loss of dopant from the crystal

10 The driving force for impurity diffusion is

(a) concentration gradient

(b) temperature gradient

(c) chemical potential

11 The concentration profile for a dopant diffused into silicon is directly proportional to

(a) the surface concentration of impurity atoms

(b) the square root of diffusion coefficient

(c) the reciprocal of time

12 After diffusing a finite source of impurity atoms from the surface, the profile of their concentration versus depth is

(a) an error function (erfc)   (b) Gaussian

13 In forming an n–p–n transistor it is best to use, as the first n-type diffusion

(a) arsenic   (b) phosphorus

14 The rate at which an oxide layer grows on a silicon surface depends on the thickness of oxide already there

(a) true   (b) false

15 A *thin* oxide layer grows with time t at a rate that is

(a) constant   (b) proportional to $t^{-1/2}$

(c) proportional to $t^{1/2}$

16 The rate of evaporation of a molten metal in vacuum falls with increasing molecular weight

(a) true   (b) false

17 Sputter deposition is a physical rather than a chemical process

(a) true   (b) false

18 A single metal layer on a semiconductor will always make a satisfactory ohmic contact

(a) true   (b) false

19 Electromigration is the movement of metal ions due to

(a) high temperature   (b) high voltage

(c) high current density

20 Insulating layers of SiO_2 can be deposited by

(a) evaporation   (b) sputtering

(c) chemical vapour deposition

Answers

1	(b)	**2**	(a)	**3**	(a), (b), (c)	**4**	(a)
5	(b)	**6**	(b)	**7**	(a)	**8**	(b)
9	(b)	**10**	(a)	**11**	(a)	**12**	(b)
13	(a)	**14**	(a)	**15**	(a)	**16**	(b)
17	(a)	**18**	(b)	**19**	(c)	**20**	(c)

17 | Magnetic materials

17.1 Introduction

The ability of certain materials – notably iron, nickel, cobalt, and various of their alloys and compounds – to acquire a large, permanent magnetic moment is central to electrical engineering, electronics and information technology. The many applications of magnetic properties range from permanent magnets for loudspeakers or electrical actuators in cameras, vehicle doors, computer disk drives etc., to 'soft iron' and related materials for transformers or video recorders, and cover the use of almost every aspect of magnetic behaviour.

The many applications of magnetic properties include some that originated as far back as the nineteenth century, while others were only developed in the late twentieth century. For example, the computer disk drive shown in Fig. 1.4 contains at least five different magnetic materials, each with a different function. It has been standard practice for about forty years to use magnetic disks for long-term storage of digital information. To do this, binary data is represented as either '1' or '0' by magnetizing a small region of a disk of a suitable material in either of two opposite directions. A related method is used to store video information on magnetic tape. Thanks largely to new and improved magnetic materials, the rate at which information can be 'read' from tape or disk has increased by about a factor of 10^{10} in forty years. New magnetic materials are being discovered or invented with increasing frequency, and new uses are constantly being found for them.

As there is an extremely wide range of differing types of magnetic material, it is important to know, firstly, why only these materials should display strong magnetic properties and, secondly, to know what governs the differences between their behaviour. For example, why can some materials carry a permanent magnetic moment but have a low permeability, while others have a high magnetic permeability but do not retain their magnetization? These questions will be answered in the course of this chapter, but we begin in Section 17.2 by introducing the concept of the *magnetic dipole moment* of a body, and showing that its value is proportional to a body's volume. This enables us to relate its magnetic moment to both magnetic field strength and magnetic flux density, whose ratio is called the permeability of the material.

The magnetic moment of an individual atom is the fundamental unit to be understood, so in Section 17.3 we show how its value depends on the way the electronic states of an atom are filled up – in other words, it depends in a fairly straightforward way on the quantum numbers of the electrons in the atom.

Using this approach it becomes clear why the strongest magnetic properties amongst the elements are found in the *transition metals*, which are therefore found as constituents of most of the engineering materials in use today. The size of the magnetic moments of the commoner transition metal atoms are discussed in Section 17.4.

The next three sections explain the way in which atomic magnets align themselves when combined together in a solid. This leads first to the definition in Section 17.5 of two technically important classes of material: ferromagnetic and ferrimagnetic, and of four classes that are weakly magnetic: antiferromagnetic, paramagnetic, diamagnetic and spin glasses. We shall not discuss the last four in depth, but will concentrate on the properties of ferromagnetic and ferrimagnetic materials in the next three sections. The ways that their magnetic properties vary with applied magnetic field strength are covered in Sections 17.9–17.11, where they are divided into 'hard' and 'soft' types, depending on whether they retain their magnetized state or change it readily, when either the strength or direction of the applied field is altered. Finally, Section 17.12 discusses the magnetoresistance of some ferromagnetic materials, a property that is very useful for sensing and measuring magnetic fields.

17.2 *Magnetic moment of a body*

We know from experiments that if a magnetized solid is subdivided into smaller and smaller parts it is impossible to find a piece that carries a single magnetic charge, or monopole. Magnetic materials act always as dipoles, and for such materials it is possible to define a *magnetic dipole moment*.

In this respect the material behaves like an assembly of blocks (which we shall later identify as atoms) each carrying a circulating current, i (Fig. 17.1) and having the same magnetic dipole moment, ia, where a is the cross-sectional area of the block, i.e. the area enclosed by the current loop (see Problem 17.1). Thus if the solid is broken down into separate blocks, each block has a dipole moment.

Using this model we can calculate the magnetic moment of the whole body in Fig. 17.1 in terms of the dipole moment of each block. Now the net current at the boundary between interior blocks is zero, because currents in neighbouring blocks are equal and cancel one another. The only surface where no cancellation occurs is the outer surface of the body. There is thus

Fig. 17.1

The magnetic moment of a body is the sum of the magnetic moments of the elementary units making up the body.

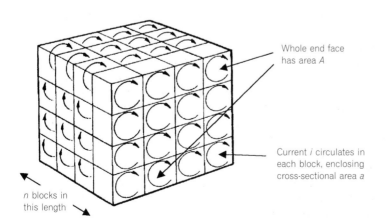

Whole end face has area A

Current i circulates in each block, enclosing cross-sectional area a

n blocks in this length

an outer surface current equal to the current circulating round n blocks, giving a total current equal to $i \times n$, if there are n blocks along that length of the solid lying perpendicular to the current direction. This current encloses the total cross-sectional area, A, of the body so that the total magnetic moment is inA, which equals $\Sigma(ia)$, the sum of the magnetic moments of all the blocks.

Clearly the magnetic moment is proportional to volume, so that the magnetic moment *per unit volume* is a characteristic of the material and not of the shape or size of the solid. The moment per unit volume is called the *intensity of magnetization*, often abbreviated as just *magnetization*, for which we use the symbol M. Because we could write the magnetic moment of a solid in terms of current $\times$ area, iA, its units are ampere-metres^2. Hence we find the units of the intensity of magnetization, M, to be A m^{-1}, amperes per metre.

Now, in the SI system of units the magnetic flux density, B, through a magnetized body is equal to the intensity of magnetization multiplied by the permeability of free space, μ_0, which is defined as exactly $4\pi \times 10^{-7}$ H m^{-1} (henrys per metre). This means that, when no external field is applied, we have $B = \mu_0 M$. When an externally applied magnetic field of strength H is also present, it adds an amount $\mu_0 H$ to the flux density, so we have the general result

$$B = \mu_0(H + M) \qquad (17.1)$$

Thus H and M share the same units, while B is measured in Teslas (T)[†]. However, μ_0 has the units of henrys per metre so that we have the dimensional relationship:

$$\text{Teslas} = (\text{henrys/m}) \times (\text{amperes/m})$$

Magnetic flux, for which the unit is the weber, is flux density B multiplied by area, thus webers $=$ Teslas $\times$ m^2 $=$ amperes $\times$ henrys.

We shall need the relationship in Eq. (17.1) between B, H and M later in this chapter, but in the meantime we return to the question of the origin of the magnetic moment.

17.3 The origin of atomic magnetic moments

The model of Fig. 17.1 suggests that a solid be divided into the smallest blocks possible (atoms) so that our first task is to consider the magnetic moment of single atoms and later to investigate the magnetic properties of a lattice of atoms.

We begin by considering an isolated atom. As we saw in Chapter 3, the magnetic quantum numbers, m_ℓ and m_s, of an electron are both related to the magnetic moment of that electron. To obtain the total magnetic moment of the atom we have to add up the magnetic moments of the electrons in it, taking into account the fact that the direction of each moment is given by the sign of m_ℓ and m_s. Thus, closed shells of electrons having equal numbers of electrons with positive and negative values of m_ℓ and m_s have no net magnetic moment because each electron cancels

---

[†] An alternative definition of the intensity of magnetization, M, which is used in some texts, gives it the units of Teslas. It is not introduced here, in order to minimize confusion.

Table 17.1

Directions of electron moments in a filled 2p shell

Electron	1	2	3	4	5	6
m_ℓ	-1	-1	0	0	$+1$	$+1$
Moment due to m_l	↓	↓	–	–	↑	↑
m_s	$+\frac{1}{2}$	$-\frac{1}{2}$	$+\frac{1}{2}$	$-\frac{1}{2}$	$+\frac{1}{2}$	$-\frac{1}{2}$
Moment due to m_s	↑	↓	↑	↓	↑	↓

another. This is shown in Table 17.1 where the dipole moment of each electron is represented by an arrow.

This then is the first important rule to note: the atomic magnetic moment comes only from incomplete electron shells.

Now the valence electrons together form an incomplete shell, so that we would expect all but the rare gases to exhibit magnetic properties. This is true for atoms in isolation, i.e. in the gaseous state, but when we come to consider solids we find only a few elements capable of permanently retaining a magnetic moment at room temperature.

If all the elements are studied at very low temperatures a lot more are found to be magnetic, while if we include compounds in our investigations yet more materials are found to be strongly magnetic.

However, not just any compound is found to be magnetic, as we show in Table 17.2. In this table the columns and rows represent the groups of the periodic table. At each intersection of a row and a column an example is given, where one exists, of a magnetic alloy or compound that is composed of an element from each of the two groups.

The table shows that, with one or two exceptions (which can be readily explained) all the magnetic materials contain *at least one transition element*.

The distinguishing feature of a transition element is that one of the inner electron shells is incomplete. This suggests immediately that the inner incomplete shell is the one that gives rise to the atomic magnetic moment, while the valence electrons are of no importance. This characteristic of the valence electrons is just a reflection of the rule we learnt in Chapter 5 about bonding: atoms bond together in such a way as to form closed shells of electrons. As we saw above, a closed shell carries no magnetic moment.

The exceptions in Table 17.2 may now be explained if we note that the bonding is unusual in these cases: it results in an unfilled inner shell. [For example in copper (II) oxide (CuO), the copper atom must lose two electrons to form an ionic bond with divalent oxygen and this leaves it with an unfilled 3d shell.]

17.3.1 Calculating the magnetic moments of atoms

Having established that the magnetic moment of an atom in the solid state is due only to an incomplete inner shell we now calculate what its magnitude should be. In Chapter 3 we noticed that the magnetic moment was related to angular momentum, and for each electron this takes two forms: the orbital and the spin angular momenta. For convenience we shall treat these separately. But first we have to discuss the values of m_l and m_s for the electrons in the unfilled shell.

When we considered the filling of energy levels in multi-electron atoms in Chapter 4 no indication was given of the sequence of filling the energy

Table 17.2 Magnetic materials

Legend:
* Antiferromagnetic
† Compound of two Group IIIB elements
? Magnetic properties uncertain
Dash denotes a disordered alloy

TRANSITION ELEMENTS (Groups 3–11)

GROUP →	1	2	3	4	5	6	7	8	9	10	11	12	13	14	15	16	17
	UH_3	—	—	—	—	—	—	—	—	—	—	—	—	—	—	—	—
	—	—	Gd–Sc	—	—	—	—	—	—	—	—	—	—	—	—	—	—
	TiH_2*	—	—	—	—	—	—	—	—	—	—	—	—	—	—	—	—
	—	—	—	—	CrV*	—	—	—	—	—	—	—	—	—	—	—	—
	Mn–H	—	YMn_5	—	—	$MnCr$*	—	—	—	—	—	—	—	—	—	—	—
	Fe–H	$FeBe_2$	PrRu	Fe_2Zr	VFe	Fe_3Cr	FeMn	—	—	—	—	—	—	—	—	—	—
	Co–H	Co–Be	$HoIr_2$	Co_4Zr	Nb–Co	$CrIr_3$	$MnRh$*	FeRh	—	—	—	—	—	—	—	—	—
	Ni–H	Ni_3Mg	$GdPd_2$	Ni–Zr	Ni–V	$CrPt_3$	$MnPt_3$	$FePd_3$	CoPt	—	—	—	—	—	—	—	—
	—	—	$AgDy$*	—	Au_4V	—	$MnAu_4$	FeAu	Co–Au	NiCu	—	—	—	—	—	—	—
	—	—	GdCd	$ZrZn_2$	—	—	$MnHg$*	Fe_2Cd	Co–Zn	$Ni–Hg_3$	—	—	—	—	—	—	—
	—	—	$CeAl_2$	—	—	$CrAl_2$*	MnB	FeB	Co_2B	Ni_3Al	—	—	$GaAl_2$†	—	—	—	—
	—	—	$NdGe_2$	—	—	$CrGe_2$	Mn_5Ge_2	FeC	Co–Si	Ni–Si	—	—	—	—	—	—	—
	—	—	TbN	—	—	CrAs	MnBi	Fe_2P	$Co_5–As_4$	Ni–Bi	Cu_2Sb*	—	—	—	—	—	—
	KO_2*	—	EuS	Ti_2O_3*	VSe*	CrTe	Mn_3O_4	Fe_3O_4	CoS_2	$Ni_2–Te$	CuO*	—	—	—	—	—	—
	—	—	$GdCl_3$	$TiCl_3$*	VCl_3*	CrI_3	$MnBr_2$*	$FeCl_2$*	CoF_2	NiF_2	AgF_2?	—	—	—	—	—	—
ELEMENTS	—	—	Gd	—	—	Cr*	Mn*	Fe	Co	Ni	—	—	—	—	—	—	—

levels in a subshell. The rule governing this was first deduced from the emission spectra of these elements, and is known as Hund's rule after its discoverer. Hund deduced from his spectroscopic measurements that levels of different m_l filled first with electrons having the same m_s value, that is, $m_s = +\frac{1}{2}$. Thus, the first level to fill is that with $m_\ell = -l$, $m_s = +\frac{1}{2}$; the second level to fill is that with $m_\ell = -(\ell - 1)$, $m_s = +\frac{1}{2}$, and so on. When, however, the shell is just half full then according to Pauli's principle no more electrons with $m_s = +\frac{1}{2}$ are allowed so that the second half-shell fills in the sequence $m_\ell = -l$, $m_s = -\frac{1}{2}$ then $m_\ell = -(l - 1)$, $m_s = -\frac{1}{2}$, and so on.

We may summarize this by saying that the electrons arrange themselves among the levels to give the maximum possible total spin angular momentum consistent with Pauli's principle. This is the usual form of Hund's rule and we may deduce from it that there must be an interaction between electrons when they have different wave functions, which tends to make their spin axes parallel. Moreover it seems that two electrons with different wave functions and with spins parallel have lower energy than two of the same wave functions with antiparallel spins. This cannot be explained without going deeply into the subject of quantum mechanics but we may note that this spin–spin interaction is a direct consequence of Pauli's principle combined with the coulomb repulsion of two like charges.

We can now return to the problem of calculating the atomic magnetic moment. Let us begin with the case of a single orbiting electron. The element scandium is an example of this situation, for there is but one electron in the 3d shell. According to the rules given above, the magnetic quantum numbers of this solitary electron are $m_\ell = -2$, and $m_s = \frac{1}{2}$. In Chapter 3 it is stated that the magnetic moment is proportional to the angular momentum and we must now derive the exact relationship between them.

We have seen that the magnetic moment, μ, of a current, i, flowing in a loop of radius r is given by

$$\mu = \pi r^2 i \tag{17.2}$$

In place of i we must put the equivalent current carried by the electron in its orbit. In the present case we wish to calculate the current flow around the circumference of the circular path of radius, $\bar{r}$, the mean radius of the wave function. The current carried around by the electron can be derived in the manner used earlier to obtain Eq. (13.10). Let us assume that all the electron's charge, $-e$, is moving around at a radius, $\bar{r}$, so that the amount of charge per unit length of the circumference is just $-e/2\pi\bar{r}$. The amount of charge passing a point on the circumference in unit time is then $-ev/2\pi\bar{r}$, where v is the velocity at which the charge is being transported. This velocity is just p/m, that is, the momentum divided by the mass. Thus we can write down the equivalent current, i, as follows

$$i = \frac{-e}{2\pi\bar{r}} \frac{p}{m} \tag{17.3}$$

The magnetic moment is obtained by combining Eq. (17.2) and Eq. (17.3):

$$\mu = -\pi\bar{r}^2 \frac{ep}{2\pi\bar{r}m} = -\frac{e}{2m} p\bar{r} \tag{17.4}$$

Now the product $p\bar{r}$ is equal to the angular momentum of the electron. Because the total orbital angular momentum is precessing about a fixed axis as described in Chapter 3, only its component along that axis contributes to the atomic magnetic moment. The perpendicular component is continuously rotating (see Fig. 3.7) and averages to zero.

Thus the angular momentum to be inserted in Eq. (17.4) to obtain the orbital magnetic moment μ_{orb} is equal to $m_\ell h/2\pi$, and the resultant expression for the magnetic moment is

$$\mu_{orb} = -m_\ell \frac{eh}{4\pi m} \tag{17.5}$$

Since m_1 is an integer, we can regard the quantity $eh/4\pi m$ as a natural unit of magnetic moment. It is called the Bohr magneton, symbol β, and the atom of scandium thus has an orbital magnetic moment of two Bohr magnetons, because its sole 3d electron has $m_\ell = -2$.

The spin magnetic moment is not obtained simply by replacing m_1 by m_s in Eq. (17.5) but the relationship is similar:

$$\mu_{spin} = -m_s \frac{eh}{2\pi m} \tag{17.6}$$

so that a single electron with $m_s = \frac{1}{2}$ has a spin magnetic moment equal to one Bohr magneton.

17.4 The 3d transition elements

The second element in the first transition series is titanium. It has two 3d electrons and according to Hund's rule their magnetic quantum numbers are respectively $m_\ell = -2$, $m_s = \frac{1}{2}$ and $m_\ell = -1$, $m_s = \frac{1}{2}$.

The total orbital magnetic moment of titanium is the sum of those of the two electrons, that is,

$$\mu_{orb} = \beta + 2\beta = 3\beta \qquad \text{(titanium)}$$

The spin magnetic moment is obtained in the same way but in this case the contributions of the two electrons are both one Bohr magneton:

$$\mu_{spin} = \beta + \beta = 2\beta \qquad \text{(titanium)}$$

In the same way the orbital and spin magnetic moments of vanadium, which has three 3d electrons, are both found to be 3β.

So far nothing has been said about the way in which the orbital and spin magnetic moments combine to give the total moment for the atom. This combination is just a little too complicated for consideration here and for our present purpose is unnecessary. For we are confining our attention to the 3d series of transition elements, which are different from the 4f series because in the former the orbital magnetic moment disappears in the solid state. This is because each atom in the crystal experiences an electric field due to the charge on all the neighbouring ions. This electric field alters the shape of the wave functions, and makes the total orbital angular momentum $m_\ell h/2\pi$ precess in a much more complicated way, with the result that, on the average, there is no component of orbital magnetic moment along any direction. The spin magnetic moment, on the other hand, is not affected by this electric field.

This 'crystal field', as it is called, does not affect the orbital moments of the magnetic electrons in the 4f transition metals, starting with lanthanum and cerium, because those electrons are deeper inside the atom and are shielded

Table 17.3

Spin distributions and magnetic moments in the first long period

Element	K	Ca	Sc	Ti	V	Cr	Mn	Fe	Co	Ni	Cu	Zn
Number of 3d electrons	0	0	1	2	3	5	5	6	7	8	10	10
Spin directions	—	—	↑	↑↑	↑↑↑	↑↑↑↑↑	↑↑↑↑↑	↑↑↑↑↑ ↓	↑↑↑↑↑ ↓↓	↑↑↑↑↑ ↓↓↓	↑↑↑↑↑ ↓↓↓↓↓	↑↑↑↑↑ ↓↓↓↓↓
Magnetic moment in Bohr magnetons	0	0	1	2	3	5	5	4	3	2	0	0

by the 5d and other electrons. Because these shells are full each acts like a charged sphere in shielding anything inside from external electric fields. The conclusion we draw from the above discussion is that the atomic moments of the 3d transition elements in the solid state are obtained by adding the spin magnetic moments of the electrons in the unfilled shell.

We have only to note that, at manganese, the 3d shell becomes half full, five electrons being present. Because of Pauli's principle the next electron to be added must have $m_s = -\frac{1}{2}$, so that its magnetic moment is opposite to, and subtracts from, those of the first five electrons. Thus iron has a spin moment of 4β, cobalt 3β, and so on. This is shown in Table 17.3.

17.5 Alignment of atomic magnetic moments in a solid

So far we have considered two effects arising from the close proximity of atoms to one another in a solid. These were (i) the forming of closed shells by the valence electrons to give zero resultant magnetic moment and (ii) the disappearance of the magnetic moment connected with the quantum number m_l, in the case of the unfilled inner 3d shell.

We now come to a third effect, which has to do with the directions adopted by the atomic 'magnets' relative to one another. This is governed by the interactions between the spins of electrons on neighbouring pairs of atoms, which are quantum mechanical in nature. These interactions, when they are large, generally manifest themselves as a tendency for adjacent atomic

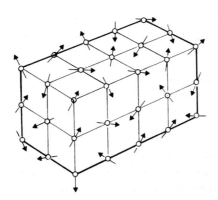

Fig. 17.2

In a paramagnet the atomic magnetic moments (represented by arrows) are all in different directions.

Fig. 17.3

Ferromagnetic iron, which has the b.c.c. structure.

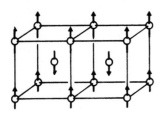

Fig. 17.4

Antiferromagnetic chromium, which has the b.c.c. structure.

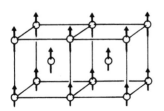

magnets to remain either parallel or antiparallel to one another. This leads to the classification of magnetic behaviour into the following groups:

- Materials in which the strength of the interactions is negligible compared to thermal energy. Thermal energy causes the directions of the atomic 'magnets' to be random, and fluctuating rapidly. Figure 17.2 shows a 'still' picture of this arrangement. An applied field can produce a degree of alignment, but this disappears on removal of the field. This behaviour is termed *paramagnetic*.
- Solids in which the atomic moments are all parallel (see Fig. 17.3 for iron). This is *ferromagnetic* behaviour.
- Materials in which the magnetic moments on alternate atoms point in opposite directions, and as a result just cancel one another in pairs, are termed *antiferromagnetic*. The example in Fig. 17.4 shows the arrangement in metallic chromium, the only element that is antiferromagnetic at room temperature.
- Materials in which the magnetic moments on alternate atoms point in opposite directions but are unequal in magnitude and therefore do not cancel one another, are called *ferrimagnetic*.
- For completeness, we mention the case of materials in which the electron shells are complete, and each atomic magnetic moment is zero when no magnetic field is applied. An applied magnetic field nevertheless induces a modified motion of the electrons, which results in a very weak magnetization, in the opposite direction to the magnetic field. This weak effect is termed *diamagnetism*, and is present in all materials, but is normally insignificant compared to ferromagnetic or ferrimagnetic effects.
- Finally, if atoms having a permanent magnetic moment are in dilute solution in a host material having closed shells (a diamagnetic host), then a randomly oriented arrangement of magnetic moments rather like Fig. 17.2 can become 'frozen in' at very low temperatures. This is a result of competing interactions between atomic magnets whose spacing is not regular, and the material is termed a *spin glass*.

Because of their practical importance, we shall concentrate mostly on the ferrimagnetic and ferromagnetic classes.

17.6 Parallel atomic moments – ferromagnetism

The most important of the ferromagnetic materials are all metallic, and iron, cobalt, nickel and gadolinium are the only elements that are ferromagnetic at room temperature. The origin of the force that aligns the moments in metals is beyond our scope, but it is known to arise from the wave-mechanical nature of electrons and it is closely related to the force that aligns the spins of electrons in a partly filled atomic shell.

There is a further peculiarity about metallic bonding. It is found that in the case of iron, cobalt and nickel the 3d electrons are able to wander through the lattice just like the 4s electrons. One effect of this is to reduce the effective magnetic moment of each electron so that instead of the expected values of the atomic magnetic moment calculated above, the measured values are as follows:

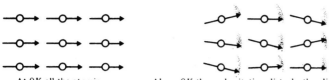

Fig. 17.5
The effect of thermal agitation on the alignment of atomic moments in a ferromagnet.

At 0 K all the atomic moments are aligned

Above 0 K thermal agitation disturbs the alignment Each moment fluctuates in direction about the average

Metal	No. of Bohr magnetons per atom (calculated)	No. of Bohr magnetons per atom (measured)
Fe	4	2.22
Co	3	1.72
Ni	2	0.61

If one of these elements is bonded ionically or covalently in a compound, however, the moment is not reduced because, in that case, none of the 3d electrons can move through the lattice.

17.7 Temperature dependence of the magnetization in a ferromagnetic material

The magnetic moment of a solid body is the sum of all the individual atomic moments. Thus in metallic nickel, with an atomic moment of 0.61β and an atomic weight of 58.7, the magnetic moment per kilogram is

$$\frac{6.02 \times 10^{26} \times 0.61 \times \beta}{58.7} \, \mathrm{A\,m^2\,kg^{-1}}$$

because by Avogadro's law there are 6.02×10^{23} atoms in a mole.

The magnetic moment per unit volume (called the *intensity of magnetization*, M) is the above value multiplied by the density, which for nickel is $8.8 \times 10^3 \, \mathrm{kg\,m^{-3}}$.

Thus, for nickel

$$M = \frac{6.02 \times 10^{26} \times 0.61 \times 9.27 \times 10^{-24} \times 8.8 \times 10^3}{58.7}$$

$$= 5.1 \times 10^5 \, \mathrm{A\,m^{-1}}$$

Now the measured value of the intensity of magnetization is exactly equal to this at a temperature close to 0 K. But at room temperature the measurement yields a value of $3.1 \times 10^5 \, \mathrm{A\,m^{-1}}$, while at 631 K and above nickel is not ferromagnetic at all! This is explained by the thermal agitation in the lattice, which disturbs the alignment of the atomic moments (Fig. 17.5), so reducing the component of each moment resolved along the direction of the net magnetization. Above the *Curie temperature* (631 K for nickel) the forces trying to align the moment are not strong enough even partially to overcome the thermal agitation and the atomic moments become randomly oriented – in fact the material behaves like a paramagnet. A graph relating M in nickel with temperature T is shown in Fig. 17.6.

Fig. 17.6
The intensity of magnetization of nickel plotted against the absolute temperature.

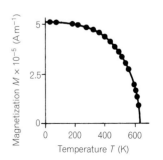

17.8 Antiparallel atomic moments – antiferromagnetic and ferrimagnetic materials

In chromium metal there are as many atomic moments pointing in one direction as in the opposite direction, resulting in exact cancellation, so that the material has no magnetic moment. This behaviour is called *anti-ferromagnetic* and it occurs in many compounds as well as some metals. Table 17.2 lists some examples. Above a temperature called the *Néel temperature*, the antiferromagnetic alignment is lost, and the material behaves like a paramagnet. For example the Néel temperature of chromium is 37°C.

When, in contrast to the antiferromagnet, alternate atoms carry *different* magnetic moments, as indicated in Fig. 17.7 by the sizes of the arrows, the moments do not cancel one another and there is a residual intensity of magnetization. These *ferrimagnetic* materials are of great engineering importance because, unlike all the usable ferromagnets they are insulators, being ionic–covalent compounds frequently incorporating oxygen. An example is the oxide of iron, Fe_3O_4, known to the ancients as Lodestone. Here, some of the iron atoms are divalent, while the rest are trivalent and hence have a different magnetic moment because one more electron is missing from the 3d shell. The net magnetization in this case is due solely to the Fe^{2+} ions, because the moments of the Fe^{3+} ions cancel one another.

Further ferrimagnets, known as *ferrites*, can be made by substituting other magnetic ions such as Mn or Cr, for the Fe^{2+} ions. These divalent transition elements have a different atomic magnetic moment to that of iron, and can be used to design for stronger magnetic properties. Ferrites with the properties of permanent magnets (see below) have the general formula $MO.6Fe_2O_3$, where M is typically Ba or Sr. Ferrites with 'soft' magnetic properties on the other hand generally have a cubic crystal structure and the formula $MO.Fe_2O_3$, where M is a transition metal such as Ni or Mn.

The insulating properties of ferrites are particularly important in applications that use alternating magnetic fields, which induce electric currents in a conducting material. The heat energy so generated comes from the power supply to the equipment, and is simply wasted.

17.9 Magnetization and magnetic domains

Section 17.5 explained that in the ferromagnetic elements the magnetic moments of all the atoms are parallel to one another at a temperature

Fig. 17.7

An imaginary ferrimagnet of cubic symmetry, in which alternate magnetic atoms have antiparallel moments of different magnitudes. In real ferrimagnets the crystal structure is usually more complex.

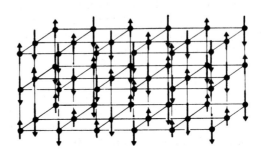

of 0 K. This would mean that any piece of iron, for example, would be spontaneously magnetized to saturation, that is, it would be a permanent magnet. On the contrary, in practice a piece of iron is generally unmagnetized in the natural state. On application of a magnetic field, H, it becomes magnetized to an extent determined by its susceptibility to the magnetic field. The intensity of magnetization, M, is related to H by the equation $M = \chi H$, where χ is the magnetic susceptibility. Since M and H share the same units, χ has no dimensions – it is just a number.

The magnetic flux density, B, in a magnetized solid has, as was explained earlier, two components and this is expressed in the relationship

$$B = \mu_0(H + M)$$

Using the relation $M = \chi H$ this can be transformed to read

$$B = \mu_0(H + \chi H) = \mu_0 H(1 + \chi)$$

The proportionality constant $(1 + \chi)$ is often lumped into one symbol: μ_r, the relative permeability; like χ it has no units.

Before going on to consider why M is not just a constant, independent of H, we shall study in more detail the observed behaviour of a piece of iron for example, when it is placed in a variable magnetic field.

Starting with the iron unmagnetized, we apply an increasing field and measure the flux density produced in the material. Referring to Fig. 17.8, the flux B rises along the line O–P–Q–R, reaching a saturation value, B_{sat}, beyond which any increase in the field has a negligible effect on B.

The intensity of magnetization at this point is equal to the value obtained from a graph like that in Fig. 17.6. If the experiment were performed at 0 K it would also be equal to the value given in Section 17.6 corresponding in iron to 2.22β per atom.

If now the field is reduced again, B does not follow the same curve but decreases more slowly along the line R–B_r; reversing the field and increasing it in the opposite direction causes the flux to follow the line B_r–$(-H_c)$–S, at which point the iron is again saturated but with the magnetization in the opposite direction. Reduction and re-reversal of the field carries the flux along the path S–$(-B_r)$–H_c–R and finally closes the loop. The closed path so described is called a *hysteresis loop* and it shows that when the field is removed after reaching the point R, the material retains a remanent flux density, B_r, and behaves as a 'permanent' magnet. In order to destroy the magnetization, it is then necessary to apply a reversed field equal to the coercive force H_c. The values of B_r and H_c may vary widely from material to material and some typical values are shown in Table 17.4

All the above features can be explained by the following theory.

It was suggested more than fifty years ago that the solid is indeed magnetized to saturation at all times but that it comprises a collection of small volumes called 'domains', each magnetized to saturation but each with its magnetization in a direction that is random with respect to all the others. The collection of randomly oriented domains gives the solid a zero overall magnetic moment. The boundaries between domains are called domain walls and subsequent research has led to ways of making these walls visible, confirming the theory. It has also been shown that in many cases the domains are not random but quite regular. As might be expected the domains in a single crystal are more regular than those in a polycrystalline solid. We show examples in Figs 17.9 and 17.10.

Fig. 17.8

The *B–H* curve of a ferromagnet or a ferrimagnet.

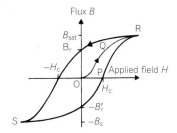

Table 17.4

Properties of some typical magnetic materials at room temperature

Material	Use	B_r (T)	B_{sat} (T)	H_s (A m^{-1})	μ_r	$(BH)_{max}$ (kJ m^{-3})
Fe (pure)		1.3	2.16	0.8	10^4	–
Fe (commercial)		1.3	2.16	7	150	–
Carbon steel (98Fe, 1Mn, 0.9C)	Permanent magnets	1.0	1.98	4×10^3	14	1.6
Silicon iron (Fe–3% Si)	Transformers	0.8	1.95	24	500 to 1500	–
Mu-metal (5Cu, 2Cr, 77Ni, 16Fe)	Magnetic shielding	0.6	0.65	4	2×10^4	–
MnZn ferrite (Mn$_{0.48}$Zn$_{0.52}$Fe$_2$O$_4$)	High-frequency transformers	0.14	0.36	50	1400	–
Alcomax (50.2Fe, 24Co, 13.5Ni, 8Al, 3Cu, 0.8Nb)	Permanent magnets	1.3	1.4	5×10^5	3	45
Barium ferrite (BaFe$_{12}$O$_{19}$)	Permanent magnets	0.40	0.41	1.5×10^5	1	30

Fig. 17.9

(a) Hypothetical domain arrangement in a demagnetized ferromagnet. (b) Domains in a polycrystalline sample of nickel.

(a)

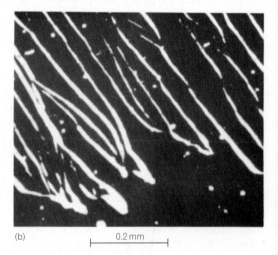

(b)

0.2 mm

The magnetization curve and the hysteresis loop can be interpreted in terms of domains. As an example, we take in Fig. 17.10 a single crystal of nickel, in which the process is clear. On first application of the field, in Fig. 17.10(b), the domain walls move so as to increase the volume of those domains whose direction of magnetization is nearest to the field direction. This is because, as in the examples sketched in Fig. 17.11, the lower domain has *lower magnetic energy*, and will therefore try to grow at the expense of regions with higher magnetic energy.

Referring back to Figs 17.10(c) and (d), it can be seen that, as the unfavourably orientated domains shrink, they become more resistant to removal, and rotation of the magnetization direction within whole domains begins to play a larger part in the process. In some soft magnetic materials

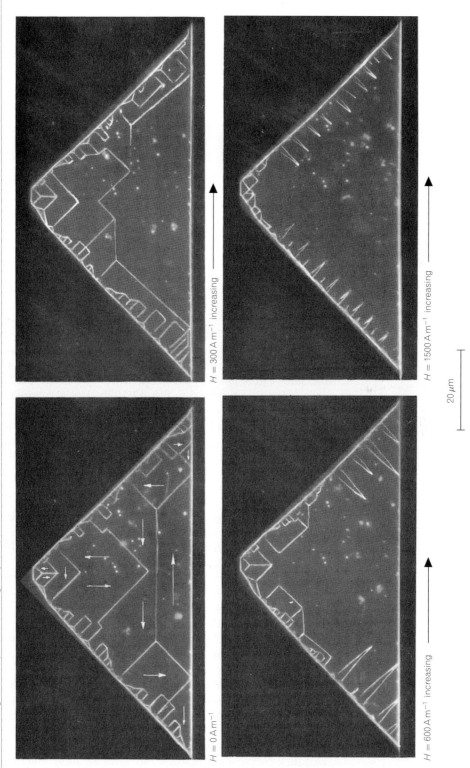

Fig. 17.10 The magnetization process in a single crystal specimen of nickel.

$H = 0\,\mathrm{A\,m^{-1}}$

$H = 300\,\mathrm{A\,m^{-1}}$ increasing

$H = 600\,\mathrm{A\,m^{-1}}$ increasing

$H = 1500\,\mathrm{A\,m^{-1}}$ increasing

$20\,\mu\mathrm{m}$

Fig. 17.11

Schematic diagram of a domain wall separating two domains, in the presence of a magnetic field *H*.

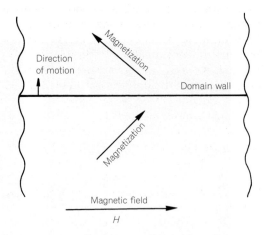

(i.e. those with high permeability), rotation of the magnetization is very easy, and occurs at much lower applied fields, simultaneously with domain wall motion – for example, in the amorphous Fe–Si–B alloys described in Section 17.10.

Saturation of the magnetization is only complete when all the unfavourable domains are removed – this occurs only at very high fields in the nickel sample of Fig. 17.10, fields several times the nominal saturation field as determined from the *B–H* loop.

This, then, explains qualitatively the shape of the initial magnetization curve (O–P–Q–R in Fig. 17.8), but it does not explain why in most cases the same curve is not followed in reverse when the field is reduced. In the case of the single crystal in Fig. 17.10 the initial magnetization curve is, in fact, followed quite closely in the reverse direction, but most practical materials contain large numbers of imperfections that act as obstacles to rotation of the magnetization and to domain wall motion. When the field is increased these obstacles are overcome by the energy supplied by the field but, when the field is removed, the defects prevent the domain walls from returning to their previous positions. The situation is rather like that of a vehicle pulling a trailer along a long hilly road – if the trailer comes unhitched it does not roll back right to its starting point, but merely to the bottom of the last hill. Thus, on removal of the magnetic field, the flux density does not return to zero but to a remanent value B_r that can be quite a large fraction of the flux density at saturation, as the *B–H* loop in Fig. 17.8 indicates. To return the domain structure to a random distribution with zero net moment it is necessary to supply more energy, either from a reversed applied field or in the form of heat.

This model enables us to understand the difference between hard and soft magnetic materials in a qualitative way. In a soft material there are few obstacles to domain wall motion or rotation of the magnetization; the coercive force is low and the magnetization changes follow the applied field as it increases or decreases, with little hysteresis. For example, in Table 17.4 it can be seen that careful purification of iron both increases the relative permeability and decreases the coercive force compared to iron of a commercial grade.

Conversely, if many obstacles to domain wall motion exist, the domains will lock in the fully magnetized position after magnetization has taken place, and both the coercive force and the remanent flux density will be large. Such a material would make a good permanent magnet, and some permanent magnets are indeed made from alloys in which a precipitate forms, the precipitated particles acting as obstacles.

The effectiveness of these obstacles to domain wall motion is dependent on (i) two basic material parameters, the anisotropy constant and the saturation magnetostriction constant (both defined below); and (ii) the changes in both of these parameters with position, which is determined by the type and density of lattice defects.

The *anisotropy constant* is a measure of the energy required to rotate the magnetization direction uniformly away from a structurally preferred direction, such as a crystal axis, along which it normally lies in the absence of any external influence – for example, any of the $\langle 111 \rangle$ directions in the nickel sample of Fig. 17.10. An increase in the anisotropy constant generally makes the motion of domain walls harder. This is because the effectiveness with which an obstacle opposes domain wall motion depends upon the ease with which the magnetization rotates from its original direction towards its final one, when a domain wall passes through the obstacle.

The second parameter, the *saturation magnetostriction constant*, is a measure of the sensitivity of the state of magnetization to stresses within the material. It is actually defined as the *strain* of the material when magnetized from zero flux density to saturation flux density, and it varies with the crystallographic direction of the magnetization. As the magnetization direction changes in a region through which a domain wall moves, there is an associated change in the lattice shape, i.e. there is a strain, which is proportional to the size of the magnetostriction constant. Domain wall motion thus involves less mechanical work, and hence is easier, when the magnetostriction constant is smaller.

These considerations lead to rather straightforward requirements for producing either a soft or a hard magnetic material.

Hard, or permanent, magnetic properties require:

- A large anisotropy constant (the effects of magnetostriction are small enough to be neglected in this case).
- Defects must usually be specifically introduced, in order to create variations of anisotropy with position and so oppose the motion of domain walls.

A soft material, with high permeability, requires

- A small anisotropy constant.
- A small magnetostriction constant.

If both of these constants can be simultaneously reduced to near zero magnitude, the permeability obtainable ought to be highest. This is exemplified in a variety of ternary (three-component) alloys of iron, one of which, Fe–Si–Al, is mentioned later.

It should be noted that these general rules cannot be allowed to overrule the requirement that a useful material must be able to sustain a high flux density, so that a high intensity of magnetization at saturation, M_s, is a prime necessity. Moreover, the intensity of magnetization in the absence of external influences should lie naturally along the direction required for the flux when the material is used. Thus a strip of material that is to be wound into the shape of a ring or toroid, for use as the core of a transformer, should normally have its magnetization direction aligned with the circumferential direction of the ring. In this way, the domain walls in the high permeability material are most readily moved by the magnetic field H that is created when a current is passed through a coil wound tightly around the core. For ease of assembly with the wire coils, the completed ring-shaped core is often cut into two C-shaped halves. In a similar way, a permanent magnet will have maximum value of its remanent flux density, B_r, if the natural axis of the material is aligned with the flux direction that is required in use.

We shall now discuss a selection of practical materials in the light of these criteria.

17.10 Some soft magnetic materials

17.10.1 Silicon–iron

Iron containing a few percent of silicon has been the dominant soft magnetic material in volume production for over 80 years. Silicon, like carbon, is very soluble in b.c.c. Fe – up to about 15% at temperatures below 800°C. As the percentage of Si added to Fe rises, the properties of the alloy change as follows.

1. The Si assists in the removal of interstitial oxygen by forming SiO_2 during melting. This is beneficial, because interstitial oxygen in α-iron increases both the coercive force H_c and the *hysteresis loss*, which is the energy expended in going once around the B–H loop.
2. The maximum permeability increases as the silicon content rises – only a little at first, then by a factor of about 2 between 3% and 6% Si. Simultaneously the coercive force H_c and the hysteresis loss decrease. The reasons underlying these improvements are that both the anisotropy constant and the appropriate saturation magnetostriction constant decrease, as indicated in Fig. 17.12. While the lowest values are found when the alloy contains more than 6% Si, it is not possible to reduce both anisotropy and magnetostriction simultaneously to negligible values in a binary alloy[†].

Fig. 17.12

Anisotropy constant K and magnetostriction constant λ_{100}, each plotted against the weight percentage of Si in Fe.

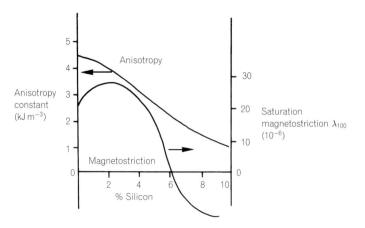

[†] The addition of aluminium to Fe–Si provides the extra degree of freedom needed to select a composition having zero values of both constants. The resulting alloy, trade name Sendust, has a maximum relative permeability between 80 000 and 110 000, but is very brittle, and hence is used only in small quantities for specialist purposes.

3. The Curie temperature decreases, and with it the intensity of magnetization at saturation. This is disadvantageous, because it reduces the maximum flux density at which the material may be used. Thus a transformer or motor, for example, would need to be larger, in proportion to the reduced flux density.

4. The electrical resistivity rises linearly from $1 \times 10^7 \, \Omega \, m$ to about $9 \times 10^{-7} \, \Omega \, m$ at 6% Si. This reduces the energy loss arising from the circulation of eddy currents when the magnetic flux changes with time. It is particularly important in machines and transformers operating with alternating currents.

5. The material becomes more brittle and less easily worked, especially above about 3% Si content. As a result, there are two distinct classes of Fe–Si transformer steels, produced using different techniques, one with a silicon content up to 3%, the other nearer 6%.

At Si concentrations up to 3% the alloy can be cold-rolled into thin sheets. Because ductility is greatest at low Si concentrations, the cheaper steels contain 1–2% Si. Cold rolling followed by annealing the 3% Si steel induces a marked texture, or preferred orientation, of the crystallites. Two different textures can be produced, illustrated in Figs 17.13(a) and (b). Because the magnetization lies naturally along the $\langle 100 \rangle$ directions, the rooftop or Goss texture [Fig. 17.14(a)] has high permeability along the rolling direction, while the four-square or cube texture of Fig. 17.14(b) gives similarly high permeability μ_r, both along and at right angles to the rolling direction. Unfortunately, the latter orientation is harder to produce so that, for the majority of applications, which require lower loss and higher permeability than is given by random orientation, the rooftop texture is used.

The techniques of hot and cold rolling that produce the four-square orientation have been perfected to the point where the sheet can be made very nearly monocrystalline, i.e. it has few large grains with their crystal axes nearly aligned. Simultaneously, the saturation flux density, B_s, is raised about 5% by this more elaborate process, enabling a significant reduction in the amount of steel needed in a transformer.

Fe–Si sheet can be produced with a higher silicon content by a further processing stage. The silicon content is raised, sometimes above 6 wt%, by in-diffusion of Si from a layer that is deposited onto the surface of the sheets by thermally decomposing $SiCl_4$ (a CVD process).

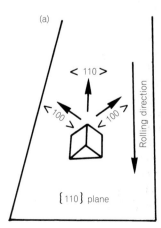

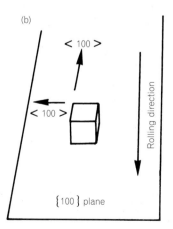

Fig. 17.13

Illustrating (a) rooftop or Goss texture, and (b) four-square texture, in a rolled Fe–Si sheet.

Fig. 17.14

The melt quenching process for producing fine-grained Fe–Si and amorphous metals.

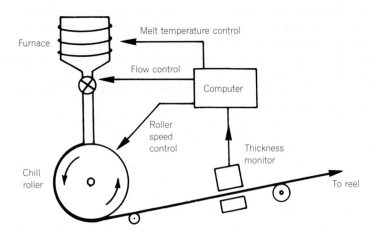

Silicon steel with 6% Si content can be made economically in narrow strips by an alternative technique of rapid cooling, called *melt casting*. A flowing stream of liquid Fe–Si is directed at the surface of a cooled rotating roller (Fig. 17.14), which rapidly removes the heat from the thin layer of metal in contact with it. The thin foil has a very small grain size (10–200 nm diameter), but is magnetically soft, because of the low anisotropy and magnetostriction. Total energy losses on reversal of the magnetic flux are higher than when the grain size is larger, but the cost of production is very much lower unless a very wide strip is required, for reasons to be given below.

17.10.2 Amorphous and nanocrystalline magnetic metals

Because anisotropy is related to the presence of crystalline order, the removal of that order should help to reduce the obstacles to domain wall motion, until only the effects of magnetostriction remain. This is indeed so; the anisotropy constant can be reduced to zero in iron that is made amorphous by rapid cooling, or quenching, at a rate of about $10^6 \, \mathrm{K \, s^{-1}}$. This inhibits the crystallization process, but because the glass transition temperature of most such metals is below room temperature, they are not useful materials. However, the introduction of sufficient boron, silicon, carbon or phosphorus into the alloy stabilizes the amorphous state at normal temperatures. These impurities, known as *metalloids*, readily form metallic alloys with the host metal without segregating, and, when present in sufficient quantity, they get in the way of the normal diffusion of host atoms, which is the process by which recrystallization occurs.

Amorphous metallic alloys are readily made by the same melt-casting technique as was described above for 6% Si in Fe. The ribbon so produced cannot be made more than about $50 \, \mu\mathrm{m}$ thick, otherwise its outer surface cools slowly enough to crystallize. Because the rate of production is fairly high (the foil speed is up to $50 \, \mathrm{m \, s^{-1}}$) the process is economic, although it is difficult to scale up for the production of wide sheet. This is because the nozzle's dimensions, and its distance from the roller, are critical in controlling the thickness. An alternative, more costly, method is to vaporize the metal in a vacuum and condense it onto a cooled surface.

Amorphous metals of suitable composition prepared by these methods can exhibit any of the types of magnetism observed in crystalline materials – diamagnetism, paramagnetism, ferromagnetism, etc. For example, amorphous $Pd_{80}Si_{20}$ is paramagnetic (the subscripts are *atomic* percentages), while $Mn_{75}P_{15}C_{10}$ is antiferromagnetic, and the amorphous alloy Tb–Fe is a ferrimagnet used as the magnetic digital storage layer in erasable optical computer memory disks.

Of most interest here, however, are the soft ferromagnetic glasses such as $Fe_{80}B_{20}$ or $Fe_{40}Ni_{40}P_{14}B_6$. Unfortunately, many of these show a slow variation in magnetic properties with time because of a stress relaxation process. This can sometimes be overcome by annealing at a temperature of 50–100°C below the crystallization temperature. The latter lies at about 400 ± 50°C in most cases. After much experimentation, the stable alloy $Fe_{78}B_{13}Si_9$ has become the basis of commercial materials, and is readily available as ribbon and bonded laminates, in widths from millimetres to tens of centimetres.

In these amorphous metals, the addition of the non-magnetic constituents, P, B or Si, adds electrons into the 3d electron energy band, so that the magnetic moment per atom at the temperature of 0 K is reduced below the values given in Section 17.6 for the pure metals. The average magnetic moment per atom falls approximately linearly with the addition of the metalloid, at a rate that depends on the electrons available in the metalloid for filling the 3d band. The rate of fall is about 3β for each P atom, 2β for each Si or C atom, and 1β for each B atom, when these are added to Co. In other magnetic hosts, notably Fe, the rate of fall is more variable from alloy to alloy – the B atom gives a fall of between 0.3β and 1.6β, but most commonly β. For example, the moment of $Fe_{80}B_{20}$ at 0 K can be calculated assuming about 1.1β per Fe atom alone (the B atoms themselves carry no magnetic moment). At room temperature, as expected, thermal agitation further lowers the average component of the atomic moment that lies along the direction of the magnetization.

Because the nearest magnetic neighbours are on average further apart than in a crystal, the strength of the interaction that holds their magnetic moments parallel is somewhat weakened. This is reflected in a slightly lower Curie temperature than in the crystalline counterpart.

Nevertheless, the ease with which the domain walls can be moved results in a much lower coercive force and consequently lower energy losses in an alternating magnetic field. Higher resistivity, in the range 0.2–$5.0\,\mu\Omega\,m$, also results from the random atomic arrangement, because electrons then collide more frequently with the lattice. This reduces the energy losses due to induced eddy-currents, compared with the equivalent crystalline material.

The attributes of low anisotropy and high electrical resistance give the commercial alloy 'Metglas', $Fe_{78}B_{13}Si_9$, a four-fold advantage over crystalline Fe–Si in respect of the total energy losses when used in a transformer at a frequency of 50 or 60 Hz. However, the brittle nature of the material and the problems of handling large quantities of such thin foil restrict its use to fairly small transformers and motors. Nevertheless, transformers capable of handling over 100 kW of power and showing improved losses have been made, so these materials have a promising future.

By annealing these amorphous materials at a temperature of several hundred °C, they can be made *nanocrystalline*, i.e. polycrystalline, but with a crystallite size in the range 10–500 nm. Like the amorphous parent material, the coercivity H_c remains low, and the permeability μ_r is correspondingly high. But the magnetostriction coefficient is much lower than in the amorphous phase, so that the properties are much less affected by mechanical strain. A similar method is used for making permanent magnetic materials from Nd–Fe–B alloys, as described in the next section.

17.11 Hard or 'permanent' magnetic materials

The requirement for a 'permanent' magnet is two-fold: it should support a high flux density, and it should not easily be demagnetized. These requirements can be combined in a single figure, the so-called energy product $(BH)_{max}$, which is the maximum value reached by the product of B and H at a point in the second quadrant of the B–H loop. This point is shown in Fig. 17.15 by the dashed line giving the locus of all such points. The energy product is a measure of the maximum magnetic energy stored per unit volume of the material, in units of J m^{-3}, and is inversely proportional to the volume needed for a given task.

It is possible to show theoretically that the energy product cannot exceed $\mu_0 M_s^2/2$, so that a high saturation intensity of magnetization, M_s, is the first requisite. In this respect the rare earth metals, with their large atomic moments, are very useful constituents. Because in some applications, very large fields are applied in the reverse direction to the flux, the coercive force H_c is another useful figure of merit. High coercive force goes along with a high anisotropy constant, which in turn derives from a very anisotropic lattice, i.e. a large difference in the dimensions of the unit cell.

Cobalt, with its hexagonal structure, is the one ferromagnetic element with a very anisotropic lattice, while among the simple ferrites, hexagonal barium ferrite, $BaO.6Fe_2O_3$ is the most notable example. Of these, barium ferrite is currently the most used, in part owing to the very short supply, and hence the high price, of cobalt (in spite of its relative abundance in the Earth's crust).

Fig. 17.15

B–H loops of some permanent magnet materials: (a) isotropic barium ferrite; (b) anisotropic Ba–Sr ferrite; (c) SmCo$_5$; (d) Nd–Fe–B; (e) Alcomax III (a cobalt-based alloy).

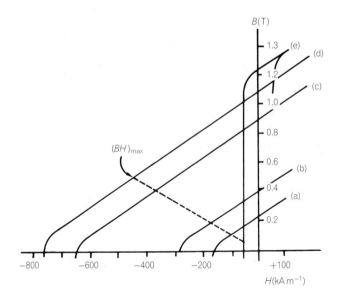

Barium ferrite, and its close relation strontium ferrite, have an unusually high coercive force, typically $250\,\text{kA}\,\text{m}^{-1}$, among the cheaper magnet materials: typical partial B–H loops in the second quadrant of some commercial magnets are compared in Fig. 17.15. Because the intensity of magnetization, M, is constant, all of these B–H loops have a constant slope of μ_0 until the field strength is close to $-H_c$, because the material remains fully magnetized until that point.

The very elongated unit cell of barium ferrite gives it a large anisotropy constant, which holds the magnetization direction fixed until very large reverse fields are reached. On the other hand, being a ferrite, it has a lower intensity of magnetization and lower saturation flux density compared to ferromagnetic alloys (see Table 17.4). For these reasons, barium ferrite magnets are commonly made shorter and fatter than their metallic counterparts. They are prepared by sintering at high temperature a compressed powder of the mixed oxides, and are not monocrystalline. A high density of grain boundaries provides the obstacles needed to hinder the motion of domain walls. The coercive force H_c is found to rise as the grain size is reduced, in an almost inverse relationship. The theoretically predicted limit is not reached in practice, because it is difficult to ensure uniformity in the height of the barriers to wall motion using powder processing techniques. The magnet is normally moulded in its required final shape, because its mechanical hardness makes subsequent cutting very slow.

If the powder grains are randomly oriented, their local c-axes are all randomly distributed. Hence, after removal of a magnetic field, their preferred magnetization directions lie within a solid angle of 2π centred on the average flux direction, as in Fig. 17.16(a). In this case, the net total magnetic flux along the average direction must be lower than is achieved if the grains are all aligned as in Fig. 17.16(b). This is apparent in a comparison of the B–H loops (a) and (b) in Fig. 17.15 for unaligned (hence isotropic) and aligned (anisotropic) materials respectively. To align the grains, a large magnetic field is applied while the milled grains are compacted together in a slurry, before sintering.

Ferrites are currently the most widely used permanent magnetic materials. This is because of the combination of good properties with a low cost of both the raw materials and the processing methods.

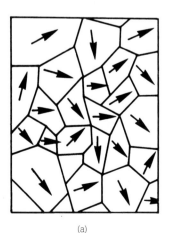 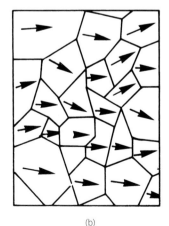

(a)   (b)

Fig. 17.16

The magnetization directions in a magnet of barium ferrite having (a) randomly oriented grains and (b) aligned grains.

In the search for materials with higher energy products $(BH)_{\text{max}}$, several intermetallic compounds containing rare earth metals were discovered in the 1960s and later. The earliest were $SmCo_5$ and Sm_2Co_{17}, with values of $(BH)_{\text{max}}$ close to $0.16\,\text{MJ}\,\text{m}^{-3}$ ($=\text{MT}\,\text{A}\,\text{m}^{-1}$), near the theoretical maximum of $0.19\,\text{MJ}\,\text{m}^{-3}$ for these materials. Like cobalt, they have hexagonal lattice symmetry, and must be carefully fabricated to form a microstructure that prevents the movement of domain walls. They can be made by sintering granules or powder in an inert atmosphere, and can only be cut subsequently with a saw made of silicon carbide or diamond.

The expense of separating samarium from other, more abundant, rare earth metals of almost identical chemistry, and the scarcity of cobalt, limited the application of these materials. The search for alternatives succeeded with the discovery in the 1980s of the ternary compound neodymium iron boride, $Nd_2Fe_{14}B$, having a theoretical maximum energy product of $0.5\,\text{MJ}\,\text{m}^{-3}$. The crystal lattice has tetragonal symmetry, although its structure has hexagonal features reminiscent of $SmCo_5$. Practical Nd–Fe–B magnets having energy products of more than $0.3\,\text{MJ}\,\text{m}^{-3}$ can be made both by traditional powder metallurgy techniques and by rapid solidification and annealing (to produce nanocrystalline material, as described for Fe–Si–B in the previous section). They are widely used in small electromechanical actuators operating at power levels up to about 1 kW. With their lower raw materials costs (Nd is more abundant than Sm), they have largely replaced samarium–cobalt.

17.12 Magnetoresistance

Magnetoresistance, as explained in Chapter 15, is a change of electrical resistance with magnetic field strength. The effects are largest in ferromagnetic materials, and such materials are routinely used to detect magnetically stored information in computer disk stores and video recorders, using an arrangement similar to that shown in Fig. 17.17 and described below.

The arrangement shown in the figure can also be used to demonstrate and measure the magnetoresistive effect. A thin strip of a magnetic material

Fig. 17.17

The magnetoresistance in a magnetic strip can be used to measure the strength of a magnetic field H applied in the direction shown.

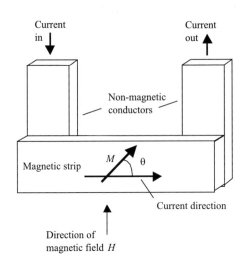

having high permeability, such as permalloy, $Ni_{80}Fe_{20}$, is initially magnetized along its preferred or 'easy' axis, which is horizontal in the diagram. An electrical contact made at either end of the strip of permalloy enables a current to flow parallel to that axis. If now a magnetic field H is applied in such a direction that it rotates the magnetization vector M into a new direction, so that it makes an angle θ with the direction of current flow, the electrical resistance of the strip changes. The change in resistance ΔR depends on the angle θ between the magnetization M and the current direction, according to the relation:

$$\Delta R = \left(\frac{\Delta \rho}{\rho}\right) R \cos^2 \theta \qquad (17.7)$$

where R is the resistance of the strip of magnetic material when $\theta = 0$, ρ is the resistivity, and $\Delta \rho/\rho$ is the *magnetoresistance coefficient* of the material, typically a few percent in value. Higher values can be engineered by alternating thin layers of magnetic and non-magnetic materials, but we shall concentrate here on the properties of uniform materials. The coefficient depends on temperature typically as shown in Fig. 17.18. It vanishes both at the magnetic Curie point T_c, and as the temperature approaches 0 K. It is a maximum in most materials at a temperature where the thermal fluctuations in the intensity of magnetization, M, are at their greatest, between $0.7T_c$ and $0.8T_c$. The magnetic resistance change ΔR owes its origin to these fluctuations, which cause local irregularities in the smooth average vector magnetization M, in just the same way as phonons change the regularity of the crystal lattice (see Chapter 7). The fluctuations are called *magnons*, for they are the magnetic equivalent of phonons, and conduction electrons that carry the current 'collide' with them as they do with phonons, so adding to the frequency of electron collisions, as discussed in Section 14.8. This results in an increase in resistivity, for resistivity is proportional to the collision frequency. The effect is to add a third term to the right-hand side of each of Eqs (14.18) and (14.19).

The sensitivity of the arrangement in Fig. 17.17 to an external magnetic field H depends upon the ease with which the field rotates the magnetization vector M from its preferred axis – in other words, it depends on the size of the anisotropy constant K that was introduced in Section 17.9.

The largest change in resistance ΔR is obtained with a combination of low anisotropy constant with large resistance R and high magnetoresistivity

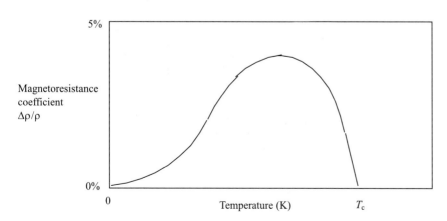

Fig. 17.18

The characteristic temperature variation for the magnetoresistance coefficient of metallic materials shows a peak near 3/4 of the Curie point T_c.

coefficient $\Delta\rho/\rho$ [see Eq. (17.7)]. The best properties require the operating temperature to be near the maximum in Fig. 17.18, and are found amongst the alloys of Ni, Fe and Co. Permalloy, $Ni_{80}Fe_{20}$, and the alloy $Ni_{82}Fe_{12}Co_6$, are the commonest choices for use in 'readout heads', used for detecting the tiny external magnetic field that results when information is stored as small magnetized regions in magnetic tape and disk recording systems. To do this, a current-carrying strip of permalloy as illustrated in Fig. 17.17 is moved across the surface of the magnetic disk or tape, at a distance above it of a few hundred nm. The voltage across the strip varies with the magnetic field produced by the stored information, and is measured electronically.

Problems

17.1 Calculate the couple on a circular current loop of area a carrying a current i when placed in a magnetic flux B that makes an angle θ with the axis of the loop. What is the corresponding couple on a short bar magnet whose magnetic moment (i.e. pole strength × distance between poles) is m? Hence deduce the magnetic moment of the current loop.

17.2 What are the magnetic moments, in Bohr magnetons, of the following ions: Ru^+, Nb^+, La^{2+}, Gd^{3+}, Pt^{2+}, Fe^{2+}, Fe^{3+}?

17.3 Calculate the intensity of magnetization at 0 K of iron, in which the atomic magnetic moment is 2.22 Bohr magnetons and the density is $7.87 \times 10^3\ kg\,m^{-3}$.

17.4 The chemical formula for Fe_3O_4 can be expressed in the form $(Fe^{2+}O)\,(Fe_2^{3+}O_3)$, so that there are twice as many Fe^{3+} ions as Fe^{2+} ions. Calculate the intensity of magnetization of Fe_3O_4 at 0 K, given that the density is $5.18 \times 10^3\ kg\,m^{-3}$.

17.5 In manganese ferrite, $MnFe_2O_4$, the Mn^{2+} ions take the place of the Fe^{2+} ions in Fe_3O_4, while the dimensions of the unit cell are almost identical. Calculate the intensity of magnetization at 0 K.

17.6 The magnetic moment of a paramagnetic material is directly proportional to the applied magnetic field. Aluminium is paramagnetic and has a susceptibility of 1.9

per cubic metre. Calculate the field required to give aluminium the same intensity of magnetization as nickel at room temperature (see Fig. 17.6). Is it possible to attain such a field in the laboratory?

17.7 A toroid of ferrite has the following dimensions: internal diameter 1.0 cm, outside diameter 2.0 cm, thickness 0.5 cm. Calculate the total flux through the specimen when it is magnetized circumferentially by a field of $100\ A\,m^{-1}$, if the intensity of magnetization is $12\,000\ A\,m^{-1}$.

17.8 Why does Mu-metal (see Table 17.4) make a good magnetic shield? What other material in the table would also serve? Why is it not used in practice?

17.9 What effect should cold working have on the properties of Mu-metal? How would you restore its original properties after fabrication of a shield?

17.10 The coercivity of a material is raised by incorporating microscopic voids which impede the motion of domain walls. What other magnetic property is influenced by the voids?

17.11 According to the discussion in Chapter 3 of the gyroscopic nature of magnetic moments, one would expect that, when a magnetic field is applied to a magnetized body, its magnetic moment should precess around the field. Why does this not occur?

Self-assessment questions

1 The magnetic dipole moment of a body is measured in units of

(a) $A\,m^{-1}$ (b) $A\,m^2$ (c) $Wb\,m^{-2}$ (d) $Wb\,m^2$

2 The intensity of magnetization of a body is defined as

(a) magnetic moment/volume

(b) magnetic moment $\times$ area

(c) magnetic moment $\times$ volume

3 The quantities B (magnetic flux density), H (magnetic field intensity) and M (intensity of magnetization) are related by the equation

(a) $B = \mu_0(H + M)$ (b) $M = \mu_0(H + B)$
(c) $H = \mu_0(M + B)$

4 The magnetic moment of a solid is the vector sum of the magnetic moments of its constituent atoms

(a) true   (b) false

5 The permanent magnetic moment of an atom is dependent on the quantum numbers

(a) n and ℓ (b) m_ℓ and m_s (c) ℓ and m_s

6 The resultant permanent magnetic moment of two electrons each with $m_\ell = 0$ and with opposed spins is

(a) twice the moment of one electron   (b) zero

7 A complete shell of electrons has

(a) the maximum possible permanent magnetic moment

(b) zero permanent magnetic moment

(c) a moment dependent on the particular values of m_1 and m_s

8 The resultant orbital magnetic moment of two electrons, one with $m_\ell = +2$ and one with $m_\ell = -2$ is

(a) zero   (b) twice that of each electron

(c) dependent on the values of m_s for the two electrons

9 An element can form a strongly magnetic solid only if its atoms have

(a) an incomplete valence shell

(b) an incomplete inner shell

(c) a vacant inner shell

10 To work out the size of the magnetic moment due to the electrons in a subshell, we use Hund's rule, which states that the electrons fill the states

(a) in the order of decreasing m_ℓ

(b) so that the maximum possible value of spin angular momentum is achieved consistent with Pauli's principle

(c) so that there is maximum cancellation of magnetic moments

11 The magnetic moment of a single electron due to its orbital motion around the nucleus is

(a) $-2m_\ell\beta$ (b) $-\dfrac{m_\ell}{\beta}$ (c) $-\dfrac{\beta}{2m_\ell}$ (d) $-m_\ell\beta$

12 The magnetic moment of a single electron due to its spin alone is

(a) $-2m_s\beta$ (b) $-\dfrac{m_s}{\beta}$ (c) $-\dfrac{\beta}{2m_s}$ (d) $-m_s\beta$

13 In the 3d transition elements the 'crystal field' due to the charges on neighbouring ions in the solid causes

(a) the spin magnetic moment to become negligible

(b) the spin magnetic moment to be a maximum

(c) the orbital magnetic moment to become negligible

(d) the orbital magnetic moment to be a maximum

14 In the table below, which of the elements are given the wrong value of magnetic moment in the solid state?

	(a)	(b)	(c)	(d)	(e)	(f)	(g)
Element	Ca	V	Cr	Mn	Fe	Ni	Cu
No. of 3d electrons	1	3	5	5	6	8	10
Atomic magnetic moment in Bohr magnetons	1	3	4	6	4	2	1

15 If the atomic magnetic moments are randomly oriented in a solid its magnetic behaviour is termed

(a) polycrystalline   (b) paramagnetic

(c) antiferromagnetic   (d) polymagnetic

16 A ferromagnetic material is one in which

(a) one constituent is iron

(b) the constituents are transition metal oxides

(c) the atomic magnetic moments are parallel

(d) the atomic magnetic moments are antiparallel and unequal

17 The intensity of magnetization M of a ferromagnetic solid

(a) increases with increasing temperature

(b) decreases with increasing temperature

(c) is independent of temperature

18 The dependence of M on temperature is caused by

(a) the presence of magnetic domains magnetized in varying directions

(b) a permanent misalignment of the atomic magnetic moments

(c) fluctuations in the directions of the atomic magnetic moments

19 The Curie temperature is the temperature at which

(a) the saturation intensity of magnetization becomes zero

(b) the domains become entirely randomly magnetized

(c) the atomic magnetic moment disappears

20 A material with unequal, antiparallel atomic magnetic moments is termed

(a) a ferrite   (b) a ferrimagnet

(c) an antiferromagnet

21 Within each magnetic domain in a ferromagnet all the atomic magnetic moments are

(a) antiparallel   (b) demagnetized

(c) parallel   (d) random

22 A piece of material has no net magnetic moment. It can only therefore be composed of domains magnetized in different directions

(a) true   (b) false

23 A piece of magnetic material has a net magnetic moment when no field is applied. It must therefore be ferromagnetic

(a) true   (b) false

24 If the domain walls in a magnetic material can be easily moved the material displays

(a) high permeability   (b) high flux density

(c) permanent magnetic behaviour

25 To increase the permeability of iron it is necessary to

(a) introduce carbon   (b) purify it

(c) alloy it with cobalt

26 Magnetic recording tape is most commonly made from

(a) small particles of iron   (b) silicon–iron

(c) ferric oxide

27 Permanent magnets are sometimes made by the aggregation of particles which are

(a) smaller than a magnetic domain width

(b) non-magnetic particles in a magnetic bonding medium

(c) smaller than a domain wall thickness

Each of the sentences in Questions 28–35 consists of an assertion followed by a reason. Answer:

(a) If both assertion and reason are true statements and the reason is a correct explanation of the assertion.

(b) If both assertion and reason are true statements but the reason is not a correct explanation of the assertion.

(c) If the assertion is true but the reason contains a false statement.

(d) If the assertion is false but the reason contains a true statement.

(e) If both the assertion and reason are false statements.

28 The magnetic moment of an iron atom in Fe_3O_4 is less than that in metallic iron *because* the orbital moment is reduced by the crystal field.

29 A piece of iron may have no magnetic moment *because* it is antiferromagnetic.

30 Copper cannot be antiferromagnetic *because* it possesses a full 3d shell of electrons.

31 Valence electrons cannot contribute to the magnetic moment of an isolated atom *because* their spins are always opposed in pairs.

32 In the 4f transition series of metals, the crystal field does not cause the orbital magnetic

moment to disappear in the solid *because* the 4f electrons are not deeply buried inside the atom.

33 The compound MnBi cannot be ferromagnetic *because* it contains no transition elements.

34 Ferrites are useful in high-frequency transformers *because* they have a lower saturation flux density than does silicon–iron.

35 Silicon–iron has a high remanent flux density *because* defects prevent the domain walls from returning to their original positions when the magnetizing field is removed.

Answers

1	(b)	2	(a)	3	(a)	4	(a)
5	(b)	6	(b)	7	(b)	8	(a)
9	(b)	10	(b)	11	(d)	12	(a)
13	(c)	14	(a), (c), (d), (g)	15	(b)	16	(c)
17	(b)	18	(c)	19	(a)	20	(b)
21	(c)	22	(b)	23	(b)	24	(a)
25	(b)	26	(c)	27	(c)	28	(d)
29	(c)	30	(a)	31	(e)	32	(c)
33	(e)	34	(b)	35	(c)		

18 | Dielectric, piezoelectric, ferroelectric and pyroelectric materials

18.1 Introduction

The broad subject of dielectric materials covers much more than simply a study of how materials respond to static electrical charges. Even the simplest materials exhibit strong changes in permittivity as we increase the frequency of an alternating voltage applied across them. Some materials, termed *piezoelectric*, respond electrically to pressure or heat, and conversely when they are electrically stimulated they respond mechanically or thermally. Most such materials are also ferroelectric, i.e. they have an electrical memory of a previous electrical stimulus. These materials are used in ultrasound scanners to generate and detect ultrasonic waves, to create audible sound waves, or to convert those waves into electrical signals. They may also be used in fine precision actuators, for example to position accurately the components inside a computer disk drive. All such devices, which convert energy between different forms, are called *transducers*.

Dielectric materials or insulators have the unique property of being able to store electrostatic charge. Very often the material is charged up by friction, as in the classic school experiment of rubbing a glass rod with dry silk. Virtually all of the modern plastic materials are good dielectrics: the charging up of nylon fabric through friction with other clothing gives rise to crackling sparks and sometimes to tangible arcs and shocks when the garment is removed in a dry atmosphere. One of the authors has seen a 3-inch spark from a nylon garment in the dry atmosphere of the South African high veldt. Bearing in mind that an arc struck in dry air at 5000 feet altitude needs an electric field of approximately $10^7 \, \mathrm{V \, m^{-1}}$, such a spark corresponds to a potential of around 100 000 V!

Before attempting to account for these phenomena in Section 18.2, we first consider how to describe a dielectric in terms of the ideas of energy bands discussed in connection with semiconductors in Chapter 15. In Sections 18.3 and 18.5 we summarize the basic electrical laws of electrostatics, and we use them in Section 18.4 to show how the concept of permittivity can be generalized when the electric field is alternating, to describe a material that is an imperfect insulator, and hence to define a figure of merit, the 'loss factor'. We then introduce in Sections 18.6–18.10 the definitions of polarization and polarizability, which measure the response of charges inside a dielectric to the externally applied field strength. In Section 18.11 this enables us to show how a study of the frequency dependence of the permittivity in alternating fields helps us to distinguish and identify four different physical mechanisms involving ions and electrons that contribute to the permittivity.

Section 18.12 introduces piezoelectric materials, which respond electrically to pressure and are widely used in electronic watches, computers and radio transmitters for accurate time measurement, as well as in the transducers mentioned above. Section 18.13 then describes a variety of different piezoelectric materials – crystalline, ceramic and polymeric. The topic of pyroelectricity (the electrical response to heat) and ferroelectricity are covered in Sections 18.14 and 18.15, and the chapter ends with an example of a method for measuring permittivity at low frequencies.

18.2 Energy bands in dielectrics

Dielectric materials are invariably substances in which the electrons are localized in the process of bonding the atoms together. Thus covalent or ionic bonds, or a mixture of both, or van der Waals bonding between closed-shell atoms all give rise to solids (or gases) exhibiting dielectric (insulating) properties. The energy band diagram for a crystal will be just like that of a semiconductor, with a valence band and a conduction band separated by an energy gap as was mentioned in Chapter 15. The gap is so large that, at ordinary temperatures, thermal energy is insufficient to raise electrons from the valence to the conduction band, which is, therefore, empty of electrons. Consequently, there are no free charge carriers and the application of an electric field will produce no current through the material.

This description applies to a perfect dielectric – in practice there will always be a few free electrons in the conduction band. These will be knocked there by stray high-energy radiation (such as cosmic rays) or by irradiation with visible or ultraviolet light and will have a relatively long lifetime before returning to the valence band because, once the electrons are free, the probability of their being recaptured by an empty bond is low because so few bonds are empty. However, when the energy gap exceeds about 3 eV the number of such electrons is so small that they are unable to give a significant current. In a good insulator the current, when a field of several tens of volts per mm is applied, will be of the order of 10^{-9} A or less.

It should be remembered that the concept of energy bands is strictly relevant only to a single crystal of material. This is because it is only in a single crystal that every atom and its bonding system is the same as every other, so that the energy that is required to liberate an electron from a bond is precisely the same everywhere in the solid. Many dielectrics used as insulators are highly disordered, so that the environment of each atom tends to be a little different from that of its neighbours, as in the glassy network structure described in Chapter 11. However, it is still possible to consider the energy band picture to apply but with the band edges smeared out somewhat. This allows for the fact that slightly different energies may be needed to energize an electron from a bond at different places in the material.

In the electrostatic charging phenomenon mentioned above charge is stored on the surface of the material, where it persists because there are no free carriers to neutralize it, and the charge itself becomes bound in the surface. Various mechanisms can be postulated whereby charge may be bound in the surface of the material, but none are easy to demonstrate experimentally. In a material with a repetitive although not necessarily regular structure, like glass, the surface must represent an interruption of the bonding system. There will be atoms that have been unable to complete a covalent or ionic bond with a neighbour because there is no neighbour with which to do so; there will be one or two valence electrons that are relatively loosely bound to their parent atoms and these may be transferred into the material used for the rubbing by mechanical work due to friction. The surface will then be left with a positive charge due to the loss of the negatively charged electrons. It is a demonstrable fact that a glass rod acquires positive charge on being rubbed. The positive charge will remain so long as it is unable to acquire replacement negative charge to compensate it. When there is water vapour in the air the H_2O molecules will readily ionize to OH^- and H^+ and the hydroxyl (OH^-) groups become attached to the surface to replace the lost negative charge. Thus the charge only persists in a dry atmosphere. In practice, in the process of fabrication of glass the surface usually becomes covered with hydroxyl ions and the frictional work tends to remove these rather than to remove electrons, as described in Section 18.1. This results in a positively charged surface just the same.

The case of polymer-based materials, such as resin, Bakelite, silk, nylon, etc., must be somewhat different. This is evidenced by the fact that on rubbing they acquire a negative charge. The structure of such materials comprises covalently bonded, long-chain molecules held together by van der Waals forces and there are no loosely bound electrons in the surface. However, all these molecules have side groups that are like electrostatic dipoles. The simplest case, that of the paraffins, is illustrated in Fig. 18.1.

Because the bonding electrons are concentrated mainly midway between the carbon and hydrogen ions, the positive hydrogen ion represents a positive charge spatially displaced from the negative charge in the bond, so forming a dipole. This occurs with other kinds of side groups so that water molecules, which are dipolar, can attach themselves by forming hydrogen bonds with the side groups. The action of the mechanical work due to friction may then be to ionize the water, i.e. to 'wipe off' a hydrogen ion, leaving the negatively charged hydroxyl group attached to the end of the polymer side chain. This would then give a negative surface charge. Again, in a damp atmosphere the hydroxyl would readily capture a hydrogen ion and so neutralize the surface charge.

We should emphasize that the above mechanisms are only tentative explanations and that the actual processes involved in the frictional surface charging of dielectrics are not yet well understood. However, the general

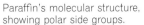

Fig. 18.1

Paraffin's molecular structure, showing polar side groups.

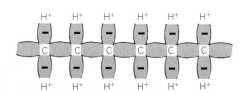

features of the explanations are well founded and lead to the conclusion that, because the sites for charge absorption are atoms or groups of atoms, it will be possible for a very large density of surface charge to be accumulated. For example, it can be shown that there are 10^{15} to 10^{16} 'dangling bonds' on $1\,m^2$ of the surface of a covalently bonded insulator. Thus the maximum charge density could be in the region of 10^{16} electrons per m^2, i.e. a charge of $1.6 \times 10^{-19} \times 10^{16} = 1.6 \times 10^{-3}\,C\,m^{-2}$. In order to estimate what this implies in terms of voltages we shall first recapitulate the basic laws of electrostatics.

18.3 Coulomb's law

Experiments on electrically charged bodies yield the following observations:

1. Like charges repel and opposite charges attract each other.
2. The force between charges is:
   (a) inversely proportional to the square of the distance between them;
   (b) dependent on the medium in which they are embedded;
   (c) acts along the line joining the charges;
   (d) proportional to the product of the charge magnitudes.

These facts are summarized in Coulomb's law of force, which gives the force, $\mathbf{F}$, on a charge q_2 due to a charge q_1 of like sign as

$$\mathbf{F} = \frac{q_1 q_2}{4\pi\varepsilon r^2}\,\mathbf{a}\quad\text{newtons} \tag{18.1}$$

where q_1, and q_2 are the charges in coulombs, r is the distance between them in metres, and ε is called the permittivity of the medium in which they are embedded. The vector $\mathbf{a}$ is a unit vector from q_2, pointing away from q_1 in the direction of the line joining the charges and reflects the fact that when the charges have the same sign the force is one of repulsion. The units of permittivity may be deduced from Eq. (18.1) thus:

$$\varepsilon = \frac{(\text{coulombs})^2}{\text{newtons}\,(\text{metres})^2}$$

The properties of a material as a dielectric enter into Coulomb's law only through this permittivity, which is also a measure of its ability to store charge. This follows from Coulomb's law of capacitance, which states that the capacity of a body to store charge is defined by the equation

$$Q = CV \tag{18.2}$$

where $+Q$ is the charge on one surface of the body and $-Q$ is the charge on the opposite surface, V is the potential drop between the surfaces, and C is the capacitance of the body. The dimensions of C are given by Eq. (18.2) as

$$\text{capacitance} = \frac{\text{coulombs}}{\text{volts}}$$

and the unit of capacitance is called the farad. The capacitance of a body is found experimentally to be an intrinsic property of the material that forms the body and of its geometry, that is, its physical shape. It is most conveniently defined in terms of a parallel plate capacitor. If we have two metal electrodes, each of area a square metres, separated by a distance,

Fig. 18.2

Parallel plate capacitor.

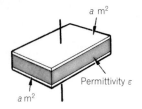

$a\ m^2$

$a\ m^2$

Permittivity ε

d metres and parallel to each other, filled with a material of permittivity, ε (Fig. 18.2), the capacitance of the system is given by

$$C = \frac{\varepsilon a}{d} \tag{18.3}$$

It can be shown that the potential energy stored by a capacitor is given by

$$E = \frac{1}{2}\frac{Q^2}{C} = \tfrac{1}{2}CV^2 \text{ joules} \tag{18.4}$$

when it has Q coulombs of charge stored. From this we see that

$$\text{joules} = \text{newton metres} = \frac{(\text{coulombs})^2}{\text{farads}}$$

Combining this with Eq. (18.2) shows that the dimensions of permittivity reduce to farads per metre. When the medium in the capacitor is empty space, the permittivity is written as ε_0 and has the value $\frac{1}{36\pi} \times 10^{-9}$ farads per metre.

The permittivity, ε, of a dielectric material may be related to the permittivity of vacuum by writing it as

$$\varepsilon = \varepsilon_r \varepsilon_0 \tag{18.5}$$

where ε_r is the relative permittivity and is simply a number. Values of relative permittivity for a range of materials are given in Table 18.1. It should be noted that dry air has a value approximately equal to one, the relative permittivity of a vacuum.

We may now return to the question of the potential of a friction-charged surface. We may take the surface as one plate of a capacitor with the other being the nearest earthed surface, that is, the nearest surface at zero potential.

Let us take the example of the 3-inch spark quoted in the introduction to the chapter. We can calculate the charge stored per unit area, knowing that

Table 18.1

Relative permittivity and loss factors of various materials

Material	Relative permittivity (ε_r) at 1 MHz	Loss factor (tan δ) at 1 MHz
Alumina	4.5–8.4	0.0002–0.01
Amber	2.65	0.015
Glass (Pyrex)	3.8–6.0	0.008–0.019
Mica	2.5–7.0	0.0001
Neoprene (rubber)	4.1	0.04
Nylon	3.4–3.5	0.03–0.04
Paraffin wax	2.1–2.5	0.003 ($\approx$ 900 Hz)
Polyethylene	2.25–2.3	0.0002–0.0005
Polystyrene	2.4–2.75	0.0001–0.001
Porcelain	6.0–8.0	0.003–0.02
PTFE (Teflon)	2.0	0.0002
PVC (rigid)	3.0–3.1	0.018–0.019
PVC (flexible)	4.02	0.1
Rubber (natural)	2.0–3.5	0.003–0.008
Titanium dioxide	14–110	0.0002–0.005
Titanates (Ba, Sr, Ca, Mg and Pb)	15–12 000	0.0001–0.02

the field strength E at the surface is $10^7\,\mathrm{V\,m^{-1}}$ at the point when the air breaks down and conducts the spark. Using Gauss' law, which relates charge to the product $\varepsilon \times E \times$ area, it is straightforward to show that the charge per unit area of the surface exactly equals the value of εE in the air immediately adjacent to the surface. Since the permittivity of air is close to ε_0, the charge q per unit area just equals $(10^{-9}/36\pi) \times 10^7 = 8.84 \times 10^{-5}\,\mathrm{C\,m^{-2}}$.

This charge would correspond to q/e single electronic charges, that is to $(8.8 \times 10^{-5})/(1.6 \times 10^{-19}) = 5.52 \times 10^{14}$ electronic charges in an area of one square metre. Now, in a nylon fabric there are about 10^{21} dipolar side chains per square metre of surface so that this rough calculation suggests that on average one side chain in 10^6 acquires a single electronic charge as a result of friction.

18.4 A.C. permittivity

Permittivity, $\varepsilon = \varepsilon_r\varepsilon_0$, has been defined as a property of the medium by means of Coulomb's law. If a parallel plate capacitor has a capacitance, C_0, in air, and the space between the plates is then filled by a medium of relative permittivity ε_r then, neglecting fringing effects, the capacitance becomes $C = \varepsilon_r C_0$.

When an alternating electromotive force, V, is applied across an ideal capacitor an alternating current, i, will flow, its value being determined by the reactance of the capacitor, the value of which is given by $1/\omega C$, where $\omega = 2\pi f$, f being the frequency. Now a fundamental property of a capacitor is that the current that flows to and from it due to its charging and discharging successively is $90°$ out of phase with the alternating voltage applied across it. This means that the current and voltage are related as shown in Fig. 18.3. In real capacitors, containing a dielectric, the phase angle is found not to be exactly $90°$ but is less than $90°$ by some small angle, δ. This is due to the flow of a small component of current which is in phase with the applied voltage as the current would be in a resistor. This resistive current, shown in Fig. 18.3, combines with the capacitive current to give a total current which is $(90 - \delta)°$ out of phase with the applied voltage. The resistive current is due to the dielectric actually acting as a very poor conductor and it is often referred to as leakage current. The magnitude of the leakage current is a property of the dielectric and can be represented in an equivalent circuit diagram by showing the capacitor to have a resistance in parallel with it, as in Fig. 18.4. The value of the resistance is determined by the size of the leakage current.

To express this phenomenon mathematically we represent the relative permittivity by a complex number so that

$$\varepsilon_r = \varepsilon' - j\varepsilon'' \tag{18.6}$$

Here ε' represents the part of relative permittivity that increases capacitance and ε'' represents the 'leakage' or 'loss'. The derivation of this

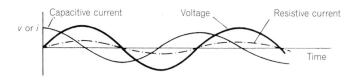

Fig. 18.3

Voltage–current phase relationships in a capacitor.

Fig. 18.4

Equivalent circuit of lossy
dielectric.

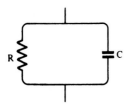

expression is given, for those familiar with j-notation in a.c. theory, in Section 18.16 at the end of the chapter. It is shown there that the 'loss angle', δ, is given by

$$\tan \delta = \frac{\varepsilon''}{\varepsilon'} \tag{18.7}$$

and $\tan \delta$ is called the *loss factor* of the dielectric. This can range in value from 10^{-5} for the very best insulators to 0.1 for rather poor dielectrics. Some typical loss factors are included in Table 18.1.

18.5 Electric flux density

When a dielectric material is placed in an electric field already existing in a homogeneous medium such as air, it has the effect of changing the distribution of the field to a degree depending upon its relative permittivity, that is, the electric field intensity is a function of the medium in which it exists. We represent this situation mathematically by defining an electric flux density **D** in the dielectric by the equation

$$\mathbf{D} = \varepsilon \mathbf{E} \tag{18.8}$$

where electric field strength **E** is a vector, because it has magnitude and direction, and so therefore is the flux density. The units of flux density may be deduced from Eq. (18.3) and are coulombs per square metre. Now the electric field strength at a point is defined as the force per unit charge on a positive test charge placed at that point. Thus, from Eq. (18.1),

$$\mathbf{E} = \frac{q}{4\pi \varepsilon r^2} \mathbf{a} \tag{18.9}$$

will be the field at a point distant r from a charge q.

Combining Eqns (18.9) and (18.8) we have

$$\mathbf{D} = \frac{q}{4\pi r^2} \mathbf{a} \tag{18.10}$$

and we see that, since ε does not appear in the equation, the flux density arising from a point charge, q, is independent of the medium and is a function of the charge and its position only. If we take q to be at the centre of a sphere of radius, r, the flux density will be the same at all points on the surface. The total flux, Ψ, crossing the surface will be the flux density multiplied by the area of the sphere, hence, using Eq. (18.10),

$$\Psi = 4\pi r^2 \frac{q}{4\pi r^2} = q \text{ coulombs}$$

Thus we see that the total flux crossing the surface of a sphere with a charge at its centre is equal to the value of the charge.

This statement is generalized in Gauss' law, which states that, for any closed surface containing a system of charges, the flux out of the surface is equal to the charge enclosed.

18.5.1 Electrostatic potential

In Section 18.3 we introduced potential drop V in connection with Coulomb's law of capacitance. A potential drop between two points A and B is the difference in the electrostatic potentials at the points A and B

Fig. 18.5

Illustration of calculation of potential.

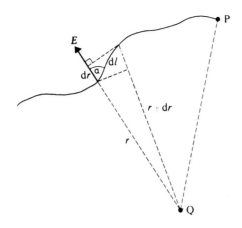

respectively. Formally this is defined as the work done on a unit charge in moving it from A to B when an electric field exists between the two points (Fig. 18.5). Now the force on a unit positive charge in a field E is, by definition, equal to the field, i.e. $F = E$. Work done is force times distance so that, if the charge moves a distance dl, the work done is E dl. However, the path between points A and B may not necessarily be parallel to the field E, as in the figure. If there is an angle, say α, between the direction of E and dl, then the work done is $E \cos \alpha$ dl.

For a unit positive charge in a field, E, we define potential difference mathematically as

$$V_A - V_B = -\int_B^A E \cos \alpha \, \mathrm{d}l \qquad (18.11)$$

where V_A is the potential at A, V_B is the potential at B, and dl is an element of the path of length l between A and B.

Because only difference in potential has been defined, the potential at a given point can only be stated with respect to some other point of known, or arbitrarily fixed, potential. We usually meet this difficulty by defining as zero the potential of a point an infinite distance from the region in which we are interested. Thus the potential at some point P is given by

$$V_P = -\int_\infty^P E \cos \alpha \, \mathrm{d}l \qquad (18.12)$$

Suppose that the field at P is due to a charge Q and we bring a unit charge up from infinity to the point P by any path. From the diagram, it is clear that, treating dl as a straight line, d$l \cos \alpha = \mathrm{d}r$, and the field due to a point charge Q is given by Eq. (18.9). Because it is the magnitude only of the field that we require, the unit vector in Eq. (18.9) is dropped. Thus

$$V_P = -\int_\infty^{r_P} \frac{Q}{4\pi \varepsilon r^2} \, \mathrm{d}r$$

i.e.

$$V_P = \frac{Q}{4\pi \varepsilon r_P}$$

This will be recognized as the expression used for the potential energy of an electron in Eq. (2.21), in which, instead of unit charge, we are considering an electron of charge $-e$ in the field due to a charge Q equal to $+e$. Thus the potential energy of the electron is

$$V_P = \frac{-e^2}{4\pi\varepsilon r_P}$$

Since potential energy is equal to work, which is given by force times distance, the units are newton-metres, i.e. joules. Thus potential energy, as defined above, has the dimensions of energy.

The total energy of a unit positive charge at a point P in the field due to a system of charges at various distances from it will be given by

$$V_P = \sum_{i=1}^{n} = \frac{Q_i}{4\pi\varepsilon r_i}$$

where we have charges $Q_1, Q_2, Q_3 \ldots Q_n$ at distances $r_1, r_2, r_3 \ldots r_n$ from the point P.

In general, differentiating Eq. (18.12), we may write the relation between field and potential. However, because of the vector nature of field we can only write the relation for a specific direction. Thus for, say, a field E_x, in the x-direction

$$E_x = -\frac{\partial V}{\partial x}$$

Putting this equation in words: the field in a given direction is equal to the negative value of the gradient of the potential in that direction.

18.6 Polarization

When connected to a battery, a capacitor having air as its dielectric will charge until the free charges on each plate produce a potential difference equal to the battery voltage, as illustrated in Fig. 18.6(a). A dielectric increases the charge storage capacity of a capacitor by neutralizing some of the free charges that would otherwise contribute to the potential difference opposing the battery voltage. More charge can, as a result, flow into the capacitor, which then has an increased storage capacitance given by $\varepsilon_r C_0$, where C_0 is the original capacitance in air. We visualize this effect as arising from alignment of electrostatic dipoles in the dielectric under the influence of the field between the capacitor plates, as shown in Fig. 18.6(b). The dipoles lie head-to-tail in long, electrically neutral chains, but with a positive charge at one end and a negative charge at the other. The uncompensated positive charge lying adjacent to the negative capacitor plate will neutralize some, but not all, of the charge on the plate. Similarly the negative end of the dipole chain will neutralize some of the charge on the positive capacitor plate.

For an applied battery voltage V, the charge carried by the air-filled capacitor will be $Q_0 = C_0 V$ and, on insertion of the dielectric, $Q_0 = \varepsilon_r C_0 V$ where ε_r is defined by Eq. (18.6).

Because we are not, in this argument, considering energy loss mechanisms, ε'' is assumed to be zero so that ε may be replaced by ε' and we can write

$$\frac{Q}{\varepsilon'} = C_0 V \quad \text{i.e. } V \propto \frac{Q}{\varepsilon'}$$

The implication is that, of the total charge Q, only a fraction, Q/ε', contributes to neutralization of the applied voltage, the remainder $Q[1 - (1/\varepsilon')]$

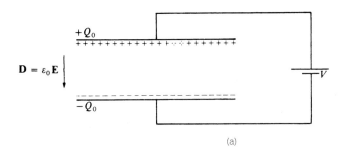

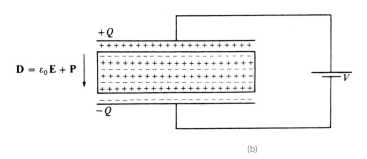

Fig. 18.6

(a) Charges on the plates of an air-filled capacitor. (b) Modification of charges on capacitor plates caused by insertion of a solid dielectric.

being bound charge that is neutralized by the polarization of the dielectric. We define the polarization of the dielectric in terms of this bound charge. The polarization, **P**, is equal to the bound charge per unit area of the dielectric surface and is measured in coulombs per square metre – the same units as the flux density, **D**; like **D**, the polarization is a vector quantity.

Thus we may imagine the electric flux density in a dielectric to be a result of two causes: first, the flux density that would be set up in the space occupied by the dielectric by an applied electric field, and, second, the polarization of the dielectric that results from the electric field. Thus we may write

$$\mathbf{D} = \varepsilon_0\mathbf{E} + \mathbf{P} \tag{18.13}$$

In the example of the parallel plate capacitor discussed above, **D**, **E** and **P** are all parallel to each other so that their magnitudes D, E and P may be used in Eq. (18.13), i.e. $D = \varepsilon_0 E + P$.

In Eq. (18.8) we defined the flux density **D** such that

$$\mathbf{D} = \varepsilon_r\varepsilon_0\mathbf{E} \tag{18.14}$$

and using this with Eq. (18.13) we have

$$\mathbf{P} = \mathbf{D} - \varepsilon_0\mathbf{E} = \varepsilon_0\mathbf{E}(\varepsilon' - 1) \quad \text{and} \quad \mathbf{P} = \chi\mathbf{E} \tag{18.15}$$

where $\chi = \varepsilon_0(\varepsilon' - 1)$ is called the *dielectric susceptibility* of the medium and is given by

$$\chi = \frac{\text{bound charge density}}{\text{free charge density}}$$

The measurement of polarization (which is the analogue of magnetization of a magnetic material) is based on Eq. (18.15) and consists in the measurement of χ or, in practice, of ε'. This is dealt with at the end of the chapter.

18.7 Mechanisms of polarization

Permittivity is essentially a macroscopic, or averaged out, description of the properties of a dielectric. To understand exacty what is happening in the material when an electric field is applied, we have to link the permittivity to atomic or molecular mechanisms which describe the processes of polarization of the material.

On the macroscopic scale we have defined the polarization, **P**, to represent the bound charges at the surface of the material.

Two point electric charges, of opposite polarity, $+Q$ and $-Q$, separated by a distance, d, represent a dipole of moment μ, given by

$$\mu = Qd \qquad (18.16)$$

The moment is a vector whose direction is taken to be from the negative to the positive charge.

We now have, in the polarized dielectric, P bound charges per unit area and if we take unit areas on opposite faces of a cube separated by a distance l, the moment due to unit area will be

$$\mu = Pl \qquad (18.17)$$

For unit distance between the unit areas $l = 1$ and we have $\mu = P$ per unit volume. Thus the polarization, P, is identical to the electric moment per unit volume of the material. This moment may be thought of as resulting from the additive action of N elementary dipoles per unit volume, each of average moment $\bar{\mu}$, therefore

$$P = N\bar{\mu} \qquad (18.18)$$

Furthermore, μ may be assumed to be proportional to a local electric field inside the dielectric that is not necessarily the same as the applied field E. If this is denoted by E_{int}, being the value of the field acting on the dipole, we define

$$\bar{\mu} = \alpha E_{\text{int}} \qquad (18.19)$$

where α is called the *polarizability* of the dipole, that is, the average dipole moment per unit field strength. The dimensions of α are

$$\frac{\text{coulombs} \times \text{metres}}{(\text{volts/metre})} = \varepsilon \times (\text{metres})^3$$

Thus we have

$$\mathbf{P} = (\varepsilon - 1)\varepsilon_0 \mathbf{E} = N\alpha \mathbf{E}_{\text{int}} \qquad (18.20)$$

This is referred to as the Clausius equation.

Since α is defined in terms of dipole moment, its magnitude will clearly be a measure of the extent to which electric dipoles are formed by the atoms and molecules. These may arise through a variety of mechanisms, any or all of which contribute to the value of α. Thus, for convenience, we regard the total polarizability to be the sum of four individual polarizabilities, each arising from one particular mechanism, i.e.

$$\alpha = \alpha_e + \alpha_a + \alpha_d + \alpha_i$$

where the terms on the right-hand side are the individual polarizabilities illustrated in schematic form in Fig. 18.7. We will now discuss each of these in turn.

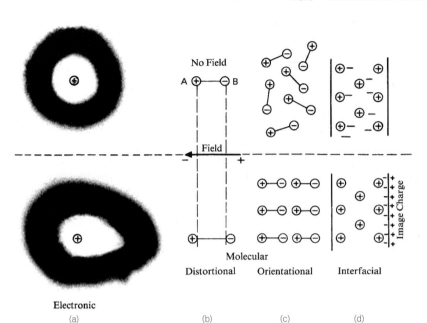

Fig. 18.7

Polarization mechanisms.

18.8 Optical polarizability (α_e)

An atom comprises a positively charged inner shell surrounded by electron clouds having symmetries determined by their quantum states. When a field is applied the electron clouds are displaced slightly with respect to the positive charges because the force, due to an electric field, on a negative electron is in the opposite direction to the force on a positive ion core. Thus there is, on average, a displacement between the ion core and the centre of gravity of the orbiting electrons, as illustrated in Fig. 18.7(a). We now have positive and negative charges separated by a small distance, which constitute an electric dipole having a dipole moment μ_e. The strength of this induced moment is proportional to the intensity of the local electric field in the region of the atom in accordance with Eq. (18.7), i.e. $\mu_e = \alpha_e \mathbf{E}_{int}$, where α_e is called the optical polarizability; it is sometimes also referred to as the electronic polarizability.

18.9 Molecular polarizability (α_a and α_d)

Consider a diatomic molecule made up of atoms A and B as shown in Fig. 18.7(b). Because of the interaction between them there is a redistribution of electrons between the constituent atoms which should, generally, be axially symmetrical along AB. It may be expected that the diatomic molecule will possess a dipole moment in the direction AB, except where the atoms A and B are identical, when the dipole moment should vanish for reasons of symmetry. Molecules having a large dipole moment are described as 'polar', an example being hydrochloric acid, in which there is a displacement of charge in the bonding between H^+ and the Cl^- ions. This gives rise to a configuration in which a positive charge is separated from a negative charge by a small distance, thus forming a true dipole.

Under the influence of an applied field the polarization of a polar substance will change by virtue of two possible mechanisms. Firstly, the field may cause the atoms to be displaced, altering the distance between them and hence changing the dipole moment of the molecule. This mechanism is called *atomic polarizability*, α_a. Secondly, the molecule as a whole may rotate about its axis of symmetry, so that the dipole aligns itself with the field. This is *orientational polarizability* (α_d) [Fig. 18.7(c)].

HCl is, of course, a gas or liquid in which each molecule is complete in itself and is bound to the others only by van der Waals forces. In an ionically bonded solid, such as NaCl, the ions Na^+ and Cl^- are arranged in a regular extended lattice. The unit cell of the lattice is the face-centred cubic structure pictured in Fig. 6.9(b) and, referring to that diagram, we may take the dark-coloured spheres as Na^+ and the light-coloured ones as Cl^-. Considering each set separately, a moment's thought will show that the centre of gravity of each set lies at the centre of the unit cell. The crystal is centro-symmetric and thus the centres of gravity of the positive and of the negative sets of charges coincide at the centre and the whole array will have no resultant electric dipole moment. When an electric field is applied, all the positive charges will tend to move in one direction and all the negative ones in the opposite direction. Because of the powerful bonding forces these movements will be small, but they result in the centres of gravity of the positive and negative sets of charges being slightly displaced from each other. As a result the array acquires a small dipole moment and the material exhibits atomic polarizability. Because of the fixed positions of the atoms with respect to each other, however, the orientational mechanism described above will be impossible.

18.10 Interfacial polarizability (α_i)

In a real insulating material there inevitably exists a number of defects such as lattice vacancies, impurity ions, and so on, together with some free electrons. Under the influence of an applied field some or all of these may migrate through the material towards the electrode that has opposite polarity to their charge. If they reach the electrode and are able to discharge there, the ions by acquiring electrons from the electrode and electrons by escaping into the electrode, the result is a loss current through the dielectric. However, if not all of them can discharge there results a pile-up of charge, of opposite types, in a thin layer close against the surface of each electrode. This gives the dielectric a dipole moment and constitutes a separate mechanism of polarization, known as *interfacial polarizability*, as illustrated in Fig. 18.7(d).

Any or all of the above mechanisms may contribute to the behaviour in an applied field. As has been described earlier, they are lumped together in a phenomenological constant, α, the polarizability, defined by Eq. (18.19).

18.11 Classification of dielectrics

In general any or all of the above mechanisms of polarization may be operative in any material. The question is how can we tell which are the important ones in a given dielectric? We can do this by studying the frequency dependence of permittivity.

Imagine, first of all, a single electric dipole in an electric field. It will, given time, line itself up with the field so that its axis lies parallel with the field; if the field is reversed, the dipole will turn itself round through 180° so that it again lies parallel with the field. When the electric field is an alternating one the dipole will be continually switching its position in sympathy with the field. For an assembly of dipoles in a dielectric the same will apply, the polarization alternating in sympathy with the applied field. If the frequency of the field increases a point will be reached when, because of their inertia, the dipoles cannot keep up with the field and the alternation of the polarization will lag behind the field. This corresponds to a reduction in the apparent polarization produced by the field, which appears in measurements as an apparent reduction in the permittivity of the material. Ultimately, as the field frequency increases, the dipoles will barely have started to move before the field reverses, and they try to move the other way. At this stage the field is producing virtually no polarization of the dielectric. This process is generally called *relaxation* and the frequency beyond which the polarization no longer follows the field is called the relaxation frequency.

Considering now the various mechanisms of polarization we can predict, in a general way, what the relaxation frequency for each one might be.

Electronic polarizability relies on the position of electrons relative to the core of an atom. Since the electrons have extremely small mass they have little inertia and can follow alternations of the electric field up to very high frequencies. In fact, relaxation of electronic polarizability is not observed until the visible or ultraviolet light range of the frequency spectrum.

In atomic polarizability we require individual ions to change their relative positions. Now we know that these atoms vibrate with thermal energy and the frequencies of the vibrations correspond to those of the infrared wavelengths of light. Thus the relaxation frequencies are in the infrared range.

Orientational polarizability refers to actual reorientation of groups of ions forming dipoles. The inertia of these groups may be considerable so that relaxation frequencies may be expected to occur in the radio frequency spectrum.

In the case of interfacial polarizability a whole body of charge has to be moved through a highly resistive material and this can be a very slow process. The relaxation frequencies for this mechanism can be as low as fractions of a Hertz.

In Fig. 18.8 we show a curve of the variation of relative permittivity, ε', with the logarithm of frequency over the entire spectrum, as it might be expected to occur for a hypothetical material showing all these effects. A slight extension of the above description is necessary in connection with this diagram.

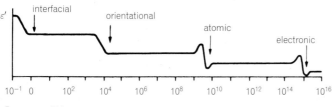

Frequency (Hz)

Fig. 18.8

Frequency spectrum of relative permittivity for hypothetical material.

Relaxation arises from the *inertia* of the system of charges. In many instances there will also be a *restoring force*, acting on the charges, which opposes the force due to the applied electric field. For example, in electronic polarization the attraction between the outer electrons and the inner ionic core of the atom will resist any attempt to pull the electron further away from the core. Similarly, in an ionic solid the positions of the ions are determined by a balance of attractive and repulsive forces and any attempt to displace the ion brings these forces into play to restore it to its original position. Thus these systems combine inertia and restoring force and in any mechanical system where this is true, resonance is possible. The classical example is a weight suspended on a coil spring.

In the presence of heavy damping, such as would arise if the weight were immersed in a pot of oil, the resonance effect is damped out and the small force is insufficient to produce significant movement of the weight. In this case the system exhibits relaxation since, if the frequency of alternation of the force is raised it will produce progressively less movement. The same considerations apply in the dielectrics; in the electronic and atomic mechanisms the damping is generally small and we observe resonance, rather than relaxation, at the characteristic frequencies. Resonance causes an increase in the displacements of the charges so that the alternating polarization, and therefore the permittivity, at first rises in value. As the applied field frequency rises above the resonant frequency it passes through an antiphase condition, where it opposes the vibrations of the charges, and the polarization falls to a very low value. Finally, as the frequency is raised still further, complete relaxation occurs and the polarization resumes a steady value in which the particular mechanism is no longer operative. This process gives the characteristic resonance shape to the permittivity curves shown in Fig. 17.8 for electronic and atomic polarization mechanisms. In most systems the damping is sufficiently large that orientational and interfacial polarization mechanisms do not show resonance behaviour.

The basic electronic and atomic polarizabilities, α_e, α_s and α_d, lead to a general classification of dielectrics. All dielectrics fall into one of three groups:

(a) Non-polar materials, which show variations of permittivity in the optical range of frequencies only; this includes all those dielectrics having a single type of atom, whether they be solids, liquids or gases.
(b) Polar materials having variation of permittivity in the infrared as well as the optical region; the most important members of this group are the ionic solids, such as rock salt and alkali halide crystals in general.
(c) Dipolar materials that also show orientational polarization; this embraces all materials having dipolar molecules of which one important common one is water. The chemical groups O—H and C=O are dipolar and may impart dipolar properties to any material in which they occur.

18.12 Piezoelectricity

'Piezo' is derived from the Greek word meaning 'to press', and the piezoelectric effect is the production of electricity by pressure. It occurs only in insulating materials and is manifested by the appearance of charges on the surfaces of a single crystal which is being mechanically deformed. It is easy to see the nature of the basic molecular mechanism involved. The application of stress has the effect of separating the centre of gravity of the positive charges from the centre of gravity of the negative charges, producing a dipole moment. Clearly, whether or not the effect occurs depends upon the symmetry of the distributions of the positive and negative ions. This restricts the effect so that it can occur only in those crystals not having a centre of symmetry since, for a centro-symmetric crystal, no combination of uniform stresses will produce the necessary separation of the centres of gravity of the charges. Crystals may be divided into 32 classes on the basis of their symmetry. Of these, 20 show the property of piezoelectricity because of their low symmetry. This description makes it clear that the converse piezoelectric effect must exist. When an electric field is applied to a piezoelectric crystal it will strain mechanically. There is a one-to-one correspondence between the piezoelectric effect and its converse, in that crystals for which strain produces an electric field, will strain when an electric field is applied.

In a piezoelectric crystal the polarization, $\mathbf{P}$, is related to the mechanical stress, $\mathbf{T}$, or, conversely, the electric stress, $\mathbf{E}$, is related to the mechanical strain, $\mathbf{S}$. We define a *piezoelectric coefficient d* relating polarization to stress and strain to field by

$$d = \left(\frac{\partial P}{\partial T}\right)_E = \left(\frac{\partial S}{\partial E}\right)_T \qquad (18.21)$$

where the suffix E indicates that the field is held constant and the suffix T that the stress is held constant. In words, the piezoelectric coefficient is given by the rate of change of polarization with stress at constant field, or the rate of change of strain with field at constant stress. The units of d will be coulombs per newton or metres per volt.

Because the polarization, field, stress and strain are all vector quantities, the value of d will depend on the relative directions of the quantity involved as well as their magnitudes. There are two types of stress, linear and shear, along each of three axes, giving six possibilities. Thus, in general,

$$P_i = \sum_j d_{ij} T_j \quad (i = 1, 2, 3; j = 1, \ldots, 6) \qquad (18.22)$$

and

$$S_j = \sum_i d_{ij} E_i \quad (i = 1, 2, 3; j = 1, \ldots, 6) \qquad (18.23)$$

where d_{ij} are the piezoelectric coefficients. There will be 18 of them (three possible values of i times six possible values of j) but, with $d_{ij} = d_{ji}$ there are 15 independent ones. How many of these are non-zero depends on the symmetry of the crystal.

An alternative piezoelectric coefficient, g, may be defined as

$$g = \left(\frac{-\partial E}{\partial T}\right)_P = \left(\frac{\partial S}{\partial T}\right)_T \qquad (18.24)$$

where the suffix P indicates constant polarization and T constant stress. The dimensions of g will be $m^2\,C^{-1}$.

The relationship between g and d can be seen, by inspection of Eqs (18.21) and (18.24), to be

$$d = \varepsilon g \qquad (18.25)$$

In practical applications the important property of a piezoelectric is its effectiveness in converting electrical to mechanical energy or vice versa. This is given by its *coupling coefficient k*, which is defined by

$$k^2 = \frac{\text{Electrical energy converted to mechanical energy}}{\text{Input electrical eneregy}}$$

or (18.26)

$$k^2 = \frac{\text{Mechanical energy converted to electrical energy}}{\text{Input mechanical energy}}$$

The magnitude of k is proportional to the geometric mean of the piezoelectric coefficients d and g and is a measure of the ability of the material both to detect and to generate mechanical vibrations.

Piezoelectric crystals are widely used to control the frequency of electronic oscillators. If a crystal is cut in the form of a thin plate it will have a sharp mechanical resonance frequency determined by the dimensions of the plate. In a suitable circuit, this resonance can be excited by an applied alternating voltage the frequency of which it then controls, giving a very stable electronic oscillator working a fixed frequency. Such circuits are universally used to provide the fixed frequency 'clock' pulses in computers and watches and to control the frequencies of radio transmitters.

18.13 *Piezoelectric materials*

Historically, quartz was the first piezoelectric material to be used in practical devices. This was because large single crystals occur in nature

Table 18.2

Material	Formula	Piezoelectric coefficients (C N^{-1})	Relative permittivity
Quartz	SiO_2	$d_{11} = -2.25 \times 10^{-12}$ $d_{14} = 0.85 \times 10^{-12}$	$e_{11} = e_{12} = 4.58$
Ammonium dihydrogen phosphate (ADP)	$NH_4H_2PO_4$	$d_{36} = 5 \times 10^{-11}(0°C)$	$e_{11} = 44.3$ $e_{33} = 20.7$
Potassium dihydrogen phosphate (KDP)	KH_2PO_4	Similar to ADP	
Lithium niobate	$LiNbO_3$	$d_{33} = 1.6 \times 10^{-11}$ $d_{15} = 7.4 \times 10^{-11}$	$e_{11} = 85.2$ $e_{33} = 28.7$
Lithium tantalate	$LiTaO_3$	$d_{33} = 8 \times 10^{-12}$ $d_{15} = 2.6 \times 10^{-11}$	$e_{11} = 53.5$ $e_{33} = 43.4$
Rochelle salt	$NaKC_4H_4O_6 \cdot 4H_2O$	$d_{14} = 2.33 \times 10^{-9}$ $d_{25} = -5.6 \times 10^{-11}$ $d_{36} = 1.16 \times 10^{-11}$	$e_{11} = 3000$ $e_{33} = 11$
Lead zirconate titanate (PZT)	$PbTi_{0.48}Zr_{0.52}O_3$	$d_{31} = -9.4 \times 10^{-11}$ $d_{33} = 2.23 \times 10^{-10}$	$e = 730$
Polyvinylidene fluoride (PVDF)	$(CH_2-CF_2)_n$	$d_{31} = 1.82 \times 10^{-11}$ $d_{32} = \sim 3 \times 10^{-12}$	$e = 160-200$

and they are relatively cheap. It is still universally used in quartz crystal electronic oscillators because its piezoelectric coefficients vary only slowly with temperature. Other materials have higher piezoelectric activities but are also pyroelectric (see next section) and are therefore very temperature-sensitive. Table 18.2 lists some practical piezoelectric materials together with their piezoelectric coefficients and permittivities. Two modern developments have been a key feature in the rapid growth of piezoelectric devices and applications, namely ceramic and plastic piezoelectrics.

18.13.1 Ceramics

Ceramic oxide compositions, based on the ferroelectric oxides listed in Table 18.3, can generally be produced by thoroughly mixing the constituents as oxides and calcining the mixture at a temperature that gives substantial interdiffusion of cations. The calcine is then finely ground and hot-pressed into the desired shape. After application of electrodes the material is poled (see later) to give it a uniaxial polarization. The most successful ceramic piezoelectric to date is lead zirconate titanate (PZT) having the general formula $PbTi_{1-z}Zr_zO_3$, z having a value around 0.52. This gives a saturation polarization of 47 microcoulombs per square metre and a d coefficient in the region of 10^{-10}C/N, i.e. 50 times that of quartz. To get an idea of what these figures mean, consider a rod of $1\,mm^2$ cross-sectional area and length 1 cm; its capacitance between electrodes on each end, using the value of relative permittivity in Table 18.2, will be 6.45×10^{-13}F. Suppose we apply an impulse stress of 10 N (hit it with a hammer!) – the charge on the electrodes would be 10^{-9}C and the voltage between its ends would be 1550 V. This sensitivity has been widely exploited in record player pick-ups, microphones, force transducers for measuring pressure and weights, and even so-called electronic cigarette or gas lighters, which use the voltage across the crystal to generate a spark. Unfortunately the piezoelectric activity has a rather large temperature dependence, which limits its use in some applications.

18.13.2 Plastics

In 1969 Kawai (*Jpn J. Appl. Phys.*, **8**, 975) reported a strong piezoelectric effect in polyvinylidene fluoride (PVDF). This is a polymer having the basic monomeric unit CH_2—CF_2 and is similar to PTFE in that it is chemically very inert. In 1972 Bell Laboratories discovered that it could also be pyroelectric (see Section 18.14).

Piezoelectricity in polymers arises because many of them have regions where the polymer chains are ordered and form localized crystalline phases surrounded by amorphous regions. At least four different crystalline phases have been identified in PVDF and in the untreated form 50–90% of the volume is crystalline, mainly in the α form. In this phase the polymer chains have a relatively high electric dipole moment but the crystal structure is such that in adjacent polymer chains the dipoles are aligned antiparallel and overall the material is non-polar. The β phase has all the fluorine atoms facing in one direction relative to the backbone and is piezoelectric. The β phase is typically produced by mechanically stretching a film of the material at temperatures between 50 and 100°C, producing extensions in length of 400–500%. The stretching can be performed biaxially or uniaxially and produces regions of the polar β phase which are randomly oriented throughout the material.

In order to exhibit piezoelectricity the material must now be poled. This can be done by depositing aluminium electrodes on the top and bottom of the sheet, heating to about 100°C and applying a large field of $8 \times 10^9 \, \mathrm{V \, m^{-1}}$. This has the effect of reorienting the dipoles of the β phase in the direction of the applied field. Finally the specimen is brought down to room temperature with the electric field still applied, locking in the dipole orientation.

The importance of PVDF in practical applications is that it can be produced in large areas, relatively cheaply. It has been used for large-area ultrasonic receivers, particularly in underwater sonar systems. One novel application has been to use it as the analogue of human skin, because its response to pressure emulates the sense of touch.

18.14 Pyroelectricity and ferroelectricity

In accounting for the piezoelectric effect we found it necessary to consider the symmetry of the crystal. When under stress, the centres of gravity of the positive and negative charges separated, forming an electrostatic dipole and hence a polarization of the crystal. There are many materials in which the symmetry is such that the centres of gravity of the positive and negative charges are separated even without a stress being applied. These will exhibit *spontaneous polarization*, which means that there must be permanent electrostatic charge on the surfaces of the crystal, with one face positive and another negative, depending on the direction of the polarization vector. However, the observer would not, in general, be aware of these charges because the atmosphere normally contains sufficient free positive and negative ions to neutralize the free surface charge by being attracted to, and adsorbed on, the surface.

The spontaneous polarization will be a strong function of temperature, because the atomic dipole moments vary as the crystal expands or contracts. Heating the crystal will tend to desorb the surface neutralizing ions, as well as changing the polarization, so that a surface charge may then be detected. Thus the crystal appears to have been charged by heating. This is called the *pyroelectric effect*. It was observed in the natural crystal tourmaline as early as the 17th century. In the 18th and 19th centuries many experiments were made to characterize the phenomenon and it was these that led to the discovery of piezoelectricity by the Curies in 1880.

In 1920 it was discovered, by Valasek, that the polarization of Rochelle salt, which was known to be pyroelectric, could be reversed by the application of an electric field. By analogy with ferromagnetic materials, which have a spontaneous magnetic polarization that can be reversed by applying a magnetic field, materials of this type were described as *ferroelectric*.

Of the 20 piezoelectric crystal classes, 10 are characterized by the fact that they have a unique polar axis and possess spontaneous polarization; these are the pyroelectric crystals. A crystal is said to be ferroelectric when it has two or more orientational states of polarization, in the absence of an applied electric field, and can be switched from one to another of these states by the application of an electric field. Most pyroelectrics are also ferroelectric.

For the majority of ferroelectrics there exists a prototype crystal phase at high temperature which is sufficiently symmetrical not to exhibit

Table 18.3

Name	Formula	Curie temperature T_c (°C)	Spontaneous polarization P_s (C cm^{-2}) (at T °C)	
Barium titanate	$BaTiO_3$	135	26.0	(23)
Lead titanate	$PbTiO_3$	490	>50	(23)
Potassium niobate	$KNbO_3$	435	30.0	(25)
Lithium niobate	$LiNbO_3$	1210	71.0	(23)
Lithium tantalate	$LiTaO_3$	665	50	(25)
Potassium dihydrogen phosphate (KDP)	KH_2PO_4	−150	4.75	(−177)
Guanidinium aluminium sulphate hexahydrate (GASH)	$C(NH_2)_3Al(SO_4)_2 \cdot 6H_2O$	None	0.35	(23)
Triglycine sulphate (TGS)	$(NH_2CH_2COOH)_3H_2SO_4$	49	2.8	(20)
Rochelle salt	$NaKC_4H_4O_6 \cdot 4H_2O$ upper	24	0.25	(5)
	lower	−18		
PLZT (ceramic)	$Pb_{0.88}La_{0.08}Zr_{0.35}Ti_{0.65}O_3$	∼97	∼47	(20)

spontaneous polarization in the absence of a field. As the crystal is cooled it undergoes a phase transition to a ferroelectric phase at a temperature, T_c, called the *Curie temperature*. Below T_c, in the absence of an applied field, there are at least two directions along which spontaneous polarization can develop. To minimize the depolarizing fields, different regions of the crystal polarize in each of these directions, each volume of uniform polarization being called a domain. The result is a domain structure which reduces the net macroscopic polarization nearly to zero and the crystals consequently exhibit very small, if any, pyroelectric effects until they are 'poled'. Poling of a crystal is done by applying a strong electric field across it as it cools through the Curie temperature. This has the effect of forcing the polarization into the direction of the field throughout the crystal, eliminating the domain structure. On removing the field at room temperature the single direction of polarization will remain in many crystals, because the energy contained in the depolarizing field is less than the energy required to switch the polarization direction to form domains. Crystals are usually poled before being used in practical applications.

There are hundreds of ferroelectric materials and a comprehensive list is given in the book by Lines and Glass[†]. Table 18.3 lists some of the more commonly used materials.

18.14.1 Molecular mechanisms

A large number of possible mechanisms of ferroelectricity have been discovered. It is clear from careful X-ray analysis, however, that the dipole moment is associated with distortion of molecular groups, while the

---

[†] *Principles and Applications of Ferroelectrics and Related Materials*, M.E. Lines and A.E. Glass, Oxford University Press (1977).

Fig. 18.9

(a) The perovskite crystal structure
of barium titanate; (b) sectional
view of the structure through a
(100) face without and with
applied field.

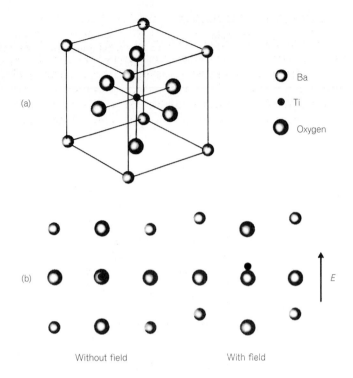

(a)

Ba

Ti

Oxygen

(b)

Without field   With field

E

cooperative alignment of the resulting dipoles, to give spontaneous polarization, is dependent on the interatomic bonding in the crystal.

In the case of the uniaxial crystals, and of some sulphates, the dipole
moment is due to the deformation of atomic groups such as SO_4, SeO_4,
AsO_4, etc., in which the undeformed state is a symmetrical arrangement of
the oxygen ions around the central sulphur (or other) ion. As the crystal
cools past the Curie temperature the crystal strains spontaneously, the
central ion is slightly displaced, and the atomic groups acquire a dipole
moment. The spontaneous strain is due to ordering of the hydrogen bonds
(present in all the uniaxial ferroelectrics), resulting in an order–disorder
phase change in the crystal. The ordered bonds act to align the induced
dipoles causing spontaneous polarization. The determining factor in the
order–disorder phase change is the (thermodynamic) free energy associated
with the crystal structure, and is outside the scope of the present text.

The case of the multiaxial ferroelectrics of perovskite and pyrochlore
structure differs somewhat. Here the basic molecular structure is an
octahedral arrangement of oxygen ions around a central ion such as Ti in
$BaTiO_3$, as shown in Fig. 18.9(a). Figure 18.9(b) shows a sectional view of
the unit cell as seen through one face and it is easy to see that the centres
of gravity of the positive and negative charges coincide. Below the Curie
temperature the crystal strains spontaneously with the Ti^{4+} and Ba^{2+} ions
moving upwards with respect to the O^{2-} ions, separating the centres of
gravity of the charges and producing a dipole moment. The alignment
of the moments, however, is not due to hydrogen bonds but to coupling
between the oxygen ions. This coupling is partially ionic and partially
covalent, and the directional property of the covalent bond is responsible
for the alignment.

18.14.2 Dielectric behaviour

The total permittivity of normal dielectrics decreases with decreasing temperature. In ferroelectric materials the permittivity and susceptibility increase with decreasing temperature, going through a sharp maximum at the Curie temperature and thereafter falling further, as temperature decreases. Above the Curie temperature the dielectric susceptibility follows a Curie–Weiss law of a type also encountered in magnetism, that is,

$$\chi \simeq \frac{C}{T - T_c} \tag{18.27}$$

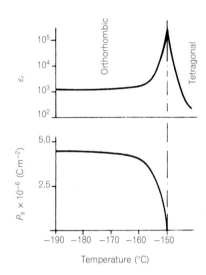

Fig. 18.10

Permittivity and polarization of KDP as a function of temperature.

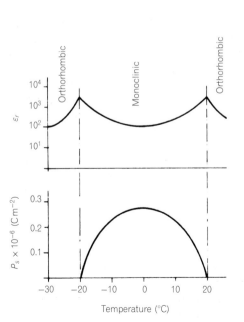

Fig. 18.11

Permittivity and polarization of Rochelle salt as a function of temperature.

where C is called the Curie constant. For ferroelectrics containing hydrogen bonding, C is of the order of 100 K. The temperature dependence of permittivity and of spontaneous polarization of potassium dihydrogen phosphate (KDP) are shown in Figs 18.10(a) and (b) and for Rochelle salt in Figs 18.11(a) and (b) over the region of their Curie temperatures.

The high permittivity, typically of the order of 10 000 at the Curie temperature, is readily understandable. At this temperature the ions are on the point of moving into or out of the position corresponding to spontaneous polarization; consequently an applied field will be able to produce relatively large shifts, with big changes in dipole moment, corresponding to a high permittivity. The falling susceptibility below T_c corresponds to an increasing degree of spontaneous saturation of polarization in the material.

The high relative permittivities of ferroelectric materials are potentially of considerable practical importance, and much effort has been put into taking advantage of them in capacitors. Piezoelectric behaviour and chemical instability as well as the rapid variation of permittivity with temperature have, however, limited their application.

18.15 Pyroelectric devices

The electric field developed across a pyroelectric crystal can be remarkably large when it is subjected to a small change in temperature. We define a *pyroelectric coefficient p* as the change in flux density in the crystal due to a change in temperature, i.e.

$$p = \frac{\partial D}{\partial T}$$

the units of which are coulombs per square centimetre per degree. For example, a crystal with a typical pyroelectric coefficient of $10^{-8}\,\mathrm{C\,cm^{-2}\,K^{-1}}$ and a relative permittivity of 50 develops a field of $2000\,\mathrm{V\,cm^{-1}}$ for a 1 K temperature change.

At equilibrium, the depolarization field due to the polarization discontinuity at the surfaces of the pyroelectric crystal is neutralized by free charge. This usually arises from free (mobile) electrons and holes in the crystal itself, where it can be regarded as a wide-band-gap semiconductor, or from an external circuit connected to the electrodes, or both. When the crystal temperature changes so that an excess of charge appears on one of the polar faces, a current will flow in the external circuit, the sense of the current flow depending on the direction of the polarization change. After the initial surge, the current dies away exponentially with time and eventually falls to zero until another temperature change comes along.

Pyroelectric devices can be used to detect any radiation that results in a change in temperature of the crystal, but are generally used for infrared detection. Because of its extreme sensitivity a temperature rise of less than one-thousandth of a degree can be detected. The detector must be designed so that the heat generated in the crystal by the radiation does not flow away too quickly and a typical design is shown in Fig. 18.12. Such detectors are widely used in burglar alarms, which detect the thermal radiation from a human body. By using a pyroelectric as the sensitive screen in a television camera tube, infrared images can be formed from the differing heat radiation from the scene being viewed, so that the operator can 'see' in the dark. These are used in a wide variety of satellite and military applications.

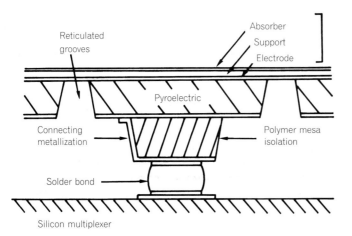

Fig. 18.12
Thermal isolation using a
mesastructure.

18.16 Advanced topic: Complex permittivity

We assume a capacitor of value C_0 in air to be filled with a medium of relative permittivity ε_r. When an alternating voltage V is applied across it the current, i, will be 90° out of phase with the voltage and will be given by

$$i = \mathrm{j}\omega\varepsilon_\mathrm{r}C_0 V \tag{18.28}$$

provided that the dielectric is a 'perfect' one. This current is just the normal capacitor charging and discharging current, which is 90° out of phase with the applied voltage and consumes no power. In general, however, an in-phase component of current will appear, corresponding to a resistive current between the capacitor plates. Such current is due entirely to the dielectric medium and is a property of it. We therefore characterize it as a component of permittivity by defining relative permittivity as

$$\varepsilon_\mathrm{r} = \varepsilon' - \mathrm{j}\varepsilon'' \tag{18.29}$$

Combining this with Eq. (17.28) we have

$$i = \mathrm{j}\omega(\varepsilon' - \mathrm{j}\varepsilon'')C_0 V \qquad \text{or} \tag{18.30}$$
$$i = \omega\varepsilon''C_0 V + \mathrm{j}\omega\varepsilon'$$

Thus the current has an in-phase component $\omega\varepsilon''C_0 V$ corresponding to the observed resistive current flow. The magnitude of ε'' will be defined by the magnitude of this current.

It is conventional to describe the performance of a dielectric in a capacitor in terms of its *loss angle*, δ, which is the phase angle between the total current, i, and the purely quadrature component i_c.

If the in-phase component is i_L, then

$$|i| = (|i_\mathrm{L}|^2 + |i_\mathrm{c}|^2)^{1/2} \tag{18.31}$$

and

$$\tan\delta = \frac{|i_\mathrm{L}|}{|i_\mathrm{c}|} = \frac{\omega\varepsilon''C_0 V}{\omega\varepsilon'C_0 V} = \frac{\varepsilon''}{\varepsilon'} \tag{18.32}$$

This is the result quoted as Eq. (18.7).

Fig. 18.13

A Schering bridge for the measurement of complex permittivity.

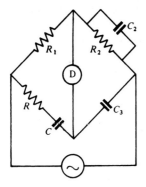

18.17 Measurement of permittivity

A wide variety of methods exists for the measurement of permittivity, the particular method adopted being determined by the nature of the specimen and the frequency range in which the measurement is to be made.

The measurement of the real part of the relative permittivity, ε', is generally done by measuring the change in capacitance of a capacitor, brought about by the introduction of the dielectric between its electrodes. The imaginary part, ε'', is found by measurement of $\tan \delta$, the loss factor resulting from the introduction of the dielectric.

The most usual type of measurement at low frequencies employs a bridge circuit working in the range 10^2–10^7 Hz. The commonest circuit is the Schering bridge illustrated in Fig. 18.13. The dielectric-filled capacitor is represented by the component C and the dielectric loss by R. It should be noted that, as shown in Fig. 18.4, the loss is properly represented by a resistance *in parallel with* the capacitor. However, this can always be simulated by a resistor R_s in series with a capacitor C_s, which will produce the same relationship between current and voltage as a parallel resistor R_p in parallel with a capacitor C_p. The relationships between these components are then

$$R_p = \frac{1}{\omega^2 C_s^2 R_s} \quad \text{and} \quad R_s = \frac{1}{\omega^2 C_p^2 R_p}$$

In the bridge circuit, C_2 and C_3 are variable, calibrated capacitors and the resistors R_1 and R_2 are usually made equal. In use, C_2 and C_3 are adjusted until there is zero current through the detector D.

Initially, suppose that the value of C without the dielectric present is C_0, and $R = 0$. If, in addition, $R_1 = R_2$, the bridge is balanced when

$$C_0 = C_3 \quad \text{and} \quad C_2 = 0$$

With the dielectric inserted, these equations become

$$C = C_3 \quad \text{and} \quad R_2 = \frac{R_1 C_2}{C_3} \tag{18.33}$$

Because $\tan \delta = \omega C R$ for a series R–C circuit, we have $\tan \delta = \omega C_2 R_1$ and, as R_1 is fixed, the dial of C may be calibrated directly in values of $\tan \delta$. Because $C/C_0 = \varepsilon'$, using Eq. (18.7) the value of the complex permittivity for the dielectric is readily found.

Problems

18.1 State Coulomb's law of force between electric charges. A positive sodium ion and a negative chlorine ion are a distance 0.5 nm apart in vacuum. Calculate the attractive force between them if each has a charge of magnitude equal to the charge on the electron. What will be the value of this force if the ions are embedded in water that has a relative permittivity of 80? What is the relevance of your calculation to the solubility of rock salt in water?

18.2 A large flat plate of an insulating material is rubbed so that its surface is charged electrostatically and placed, in air, 10 mm from a metal plate of the same area that is connected electrically to ground. If the charge on the insulator is equivalent to 10^{14} electrons per m^2, what will be the potential difference between the insulator and the metal? Would you expect a spark to jump from the insulator to the plate if the breakdown strength of the air is $3 \times 10^7 \, V \, m^{-1}$?

18.3 Explain why a complex permittivity is attributed to a lossy dielectric material. Show that the flux density due to a point charge is independent of the medium in which it is embedded. In a parallel plate capacitor the distance between the plates, each of area $1\,cm^2$, is $1\,mm$. The dielectric has a relative permittivity of 10 and it holds $10^{-9}\,C$ of charge. Find the flux density in the dielectric, and hence deduce the potential difference between the plates.

18.4 Summarize the atomic and molecular mechanisms of polarization indicating, with reasons, in what range of frequency you would expect relaxation for each one.

18.5 Explain what is meant by the term ferroelectric when applied to a dielectric material. How could you determine whether or not a material was (a) piezoelectric, (b) pyroelectric, (c) ferroelectric?

18.6 Silicon has a relative permittivity of 11.7 at frequencies high enough to ignore all but its optical polarizability. It contains 4.82×10^{28} atoms per m^3. Calculate the dipole moment of each atom in a field of $10^4\,V\,m^{-1}$, and so find the effective distance at this field strength between the centre of the electron cloud in each atom and the nucleus. How does this compare with the Si–Si bond length?

18.7 A coaxial cable comprises an inner conducting cylinder of radius a and an outer conductor of radius b. If the inner conductor carries a charge $+q$ per unit length, the outer will have a charge $-q$ per unit length induced on its inner surface. Use Gauss' law to determine an expression for the electric field strength in the dielectric medium, of permittivity ε, between the conductors.

18.8 For the coaxial cable of Problem 18.7, obtain an expression for the potential difference between the conductors. Hence find the capacitance per unit length of a cable having an insulator of permittivity 2.56 and an outer conductor whose diameter is 10 times that of the inner.

18.9 The relaxation frequency for interfacial polarization can be estimated using the model in Fig. 18.4, a capacitance C connected in parallel with the resistance R of the dielectric slab filling the capacitor. Show that the alternating current through the pair is largely independent of the resistance R (and hence independent of interfacial charge flowing through the slab) once the frequency is much larger than $1/2\pi RC$ (the *relaxation frequency*). Hence find the relaxation frequency, and the loss factor of the slab at a frequency of $1\,kHz$, when $R = 10^8\,\Omega$ and $C = 100\,pF$.

Self-assessment questions

1 In a dielectric material nearly all the electrons are localized on atoms or in bonds

(a) true   (b) false

2 A dielectric can be regarded as a semiconductor with an energy gap that is

(a) small   (b) large

3 A dielectric surface positively charged by friction retains its charge in a dry atmosphere because there are no free electrons in the solid to neutralize it

(a) true   (b) false

4 A dielectric surface can only be charged negatively by friction if there are no free electrons in the solid

(a) true   (b) false

5 The force between charges is independent of the medium in which they are embedded

(a) true   (b) false

6 The permittivity of a medium has the units

(a) Farad metre   (b) Farad $(metre)^{-1}$

(c) $(Farad\ metre)^{-1}$ (d) is a number

7   The capacitance of a parallel plate capacitor is inversely proportional to the voltage between the plates

(a) true   (b) false

8   If the distance between the plates of a parallel plate capacitor is initially small and then is doubled the capacitance is

(a) doubled   (b) halved

(c) increased by a factor of 4

9   The charge held by a parallel plate capacitor is proportional to the area of its plates for a fixed applied voltage

(a) true   (b) false

10 Permittivity is represented as a complex number to take account of losses in the dielectric

(a) true   (b) false

11 The loss factor of a capacitor in which the dielectric has permittivity given by $\varepsilon' - j\varepsilon''$ is equal to

(a) $\tan^{-1}(\varepsilon''/\varepsilon')$ (b) $\varepsilon'/\varepsilon''$ (c) $\varepsilon''/\varepsilon'$

12 An equivalent circuit for a lossy capacitor comprises a resistor and capacitor in parallel. If the loss were large the resistor would be small

(a) true   (b) false

13 List the true statements in the following question. The loss angle of a capacitor represents

(a) the phase angle, θ, between total current and total voltage

(b) $(90° - \theta)$

(c) the phase angle between the resistive component of current and the applied voltage

(d) the phase angle between the total current and its capacitive component

14 List the true statements in the following question. The electric flux density at any point in a medium distant r from a point charge q is

(a) a vector

(b) proportional to the electric field at the same point

(c) dependent on the permittivity of the medium

(d) proportional to the distance from the charge

(e) has units of coulombs per square metre

15 The electrostatic potential at any point P in an electric field E due to a point charge Q is the work done on a unit positive charge in bringing it to the point P from infinity. The potential is

(a) a vector quantity

(b) positive if the charge Q is negative

(c) positive if the charge Q is positive

(d) independent of the path followed

(e) given in joules per metre

(f) proportional to the permittivity of the medium

(g) equal to the gradient of the field

16 The polarization in a dielectric is

(a) the free charge per unit volume of the dielectric

(b) the bound charge per unit volume of the dielectric

(c) the bound charge per unit area of the dielectric

17 The dielectric susceptibility is the polarization per unit electric field

(a) true   (b) false

18 The polarizability of a dipole is the average dipole moment per unit field strength

(a) true   (b) false

Each of the sentences in Questions 19–25 consists of an assertion followed by a reason. Answer:

(a) If both assertion and reason are true statements and the reason is a correct explanation of the assertion.

(b) If both assertion and reason are true statements but the reason is *not* a correct explanation of the assertion.

(c) If the assertion is true but the reason contains a false statement.

(d) If the assertion is false but the reason contains a true statement.

e) If both the assertion and reason contain false statements.

19 Displacement of the electron cloud round a nucleus by an electric field is called optical polarizability *because* its relaxation frequency is in the optical range.

20 Optical polarizability arises in all substances *because* the electron cloud round a nucleus experiences a force in an electric field in the same direction as does the nucleus.

21 Molecular polarizability occurs in all substances *because* diatomic molecules can possess a dipole moment.

22 HCl is a polar molecule *because* it comprises a positive charge separated from a negative charge by a small distance.

23 Rock salt cannot acquire a dipole moment in an electric field *because* the crystal is centro-symmetric.

24 Orientational polarization generally occurs in solids *because* the atoms can move freely.

25 Interfacial polarization in a solid has a very low relaxation frequency *because* it requires the movement of a whole body of charge through a resistive medium.

26 A piezoelectric material is one that generates bound electrostatic charge on its surface when it is mechanically deformed

(a) true   (b) false

27 A pyroelectric material must also be piezoelectric

(a) true   (b) false

28 Pyroelectricity is only found in centro-symmetric crystals

(a) true   (b) false

29 All ferroelectric materials exhibit pyroelectricity

(a) true   (b) false

30 Poling of a ferroelectric material creates a domain structure

(a) true   (b) false

31 The Curie temperature for a ferroelectric is the temperature above which its spontaneous polarization disappears

(a) true   (b) false

32 Barium titanate is a ferroelectric because its lattice strains spontaneously above the Curie temperature

(a) true   (b) false

Answers

1	(a)	2	(b)	3	(a)	4	(a)
5	(b)	6	(b)	7	(b)	8	(b)
9	(a)	10	(a)	11	(c)	12	(a)
13	(b), (d)	14	(a), (b), (e)	15	(c), (d)	16	(c)
17	(a)	18	(a)	19	(a)	20	(a)
21	(a)	22	(b)	23	(a)	24	(b)
25	(a)	26	(b)	27	(a)	28	(b)
29	(a)	30	(b)	31	(a)	32	(b)

19 | Optical materials

19.1 Introduction

The evidence that materials selectively absorb or reflect different parts of the visible spectrum of light is all around us, in the natural and artificial colours of everyday objects. This selectivity naturally extends beyond the range of wavelengths we can see, into those invisible parts of the electromagnetic spectrum, the ultraviolet (shorter wavelengths) and the infrared (longer wavelengths). In the course of this chapter we will cover the whole of this broader spectral range, from a wavelength of about $15\,\mu$m in the infrared down to about 150 nm in the ultraviolet.

The optical behaviour displayed by materials can be thought of merely as a special case of the response of a solid to the application of an electric field. This is because light is an electromagnetic wave that consists of electric and magnetic fields travelling together through space. The difference from behaviour discussed in the preceding chapter is that, in the wavelength range mentioned above, the electric field strength is alternating in sign and amplitude (e.g. sinusoidally) at a very high frequency indeed: between about 2×10^{14} Hz and 2×10^{15} Hz.

The electric field in an electromagnetic wave can exert forces on, and its energy can be absorbed by, either the electrons or the atomic nuclei in a material, because both carry electrostatic charges. At the far infrared end of the spectrum, where the frequency is lowest, the rate of change of the electric field is small enough that the nuclei in a crystal lattice can respond to the alternating force exerted on them.

The effects of this are described in Section 19.3. As the frequency of the incoming wave increases and approaches the natural frequency of vibration of the atomic lattice, the amplitude of the induced nuclear motion increases and the energy absorbed increases correspondingly, reaching a maximum at the natural, or resonant, frequency. Beyond that frequency, the induced motion and the energy absorption decrease, due to the inertia of the nuclei. An optical wave having a frequency f substantially above the resonant frequency is transmitted with very little absorption by the lattice of ions if the quantum of energy hf is below the threshold for exciting transitions of the electrons between allowed energy levels. The absorption and scattering of ultraviolet and visible radiation, on the other hand, is primarily a result of the excitation of electrons into higher energy levels, although nuclear motion, as so often, can influence the effects we observe.

Thus in insulators and semiconductors that have an 'energy gap' in the electronic energy bands which is greater than the photon energy, we can neglect the free electrons when discussing infrared spectra, and can explain their optical properties in terms of the crystal lattice absorption alone. Insulators that have electronic energy levels separated by an energy difference equal to the photon energy hf, however, show strong electronic absorption, as explained in Section 19.4. In metals, however, and also in very narrow band gap semiconductors, the electrons dominate the behaviour even at infrared frequencies, as seen in Section 19.5. The connection between absorption and the refractive index is explored in Section 19.6, and scattering of light is discussed in Section 19.8.

All of the above ideas are used in Section 19.9 to explain the properties of optical fibre materials used in communications. In Sections 19.10–19.12 we introduce the principles underlying the emission of light from materials, and the amplification of light in lasers. These have important applications ranging from communications, in sending fast optical pulses down optical fibres, to welding and cutting of materials with kilowatts of focused light available from a carbon dioxide laser. The chapter ends with a explanation of liquid crystal properties.

The study of absorbed or emitted radiation as a function of wavelength is called spectroscopy, and is a powerful tool for the materials scientist to use in characterizing solids, liquids and gases. The electronic energy levels that can be studied in this way are not only characteristic of the lattice and even the ionic species present, but also contain information about the immediate surroundings in which the atoms are situated. This last point can be understood if we recognize that (i) the number of occupied energy levels depends on the state of ionization of an atom and that (ii) the solutions to Schrödinger's equation must be slightly altered if the potential distribution in and around the nucleus is modified by the presence of neighbouring ions. We begin, therefore, with a discussion of the spectroscopy of absorbed radiation.

19.2 Absorption spectroscopy

The degree of absorption of light by a medium can be determined from the ratio of the transmitted and incident intensities for a slab of material, if allowance is made for the energy reflected at the front and back surfaces. The intensity decreases with distance inside the absorbing medium according to an exponential law. This is simply due to the fact that in any small thickness δx, a fraction $\alpha \delta x$ of the intensity of the light is absorbed, where α is called the absorption coefficient, and has the dimensions (length)$^{-1}$.

Putting this mathematically, the fall δI in intensity I over the distance δx is given by

$$\delta I = -\alpha \delta x I$$

the minus sign being necessary because I decreases as x increases. In the limit $\delta x \rightarrow 0$, the intensity I will obey the differential equation

$$dI/dx = -\alpha I$$

The solution of this equation gives the intensity I at position x as

$$I = I_0 \exp(-\alpha x)$$

where I_0 is the intensity at the position where x = 0. By taking natural logarithms,

$$\alpha x = -\ln(I/I_0) \qquad (19.1)$$

This result is known as Lambert's law, from which it can be seen that the absorption coefficient α is the slope of a graph of $-\ln I$ against x.

When the material being studied is in solution, or is a compound or mixture, the concentration of the absorbing species also influences the strength of absorption. In many cases, the absorption coefficient α is directly proportional to the concentration C, i.e. $\alpha = KC$, where K is a constant. Combining this with Eq. (19.1) leads to the Lambert–Beer law

$$\ln(I/I_0) = -KCx$$

It is usual practice to use base-10 logarithms in place of the natural logarithm, i.e.

$$\log(I/I_0) = -aCx$$

where the coefficient a is called the absorptivity of the species whose concentration is C. When C is expressed in moles per litre, and x is in centimetres, this coefficient is known as the molar absorptivity, and is given the symbol ε, so that

$$\log(I/I_0) = -\varepsilon Cx$$

Thus ε is expressed in units of litre mole^{-1} cm^{-1}.

Beer's law strictly applies only at low concentrations, though deviations at high concentration are often small enough to be ignored.

If a light source with a broad spectral output (a 'white' light) is concentrated on a semi-transparent material, the absorptivity can be measured for each wavelength. The simplest method of measuring the wavelength dependence of absorptivity is to pass the transmitted light through a prism which separates the various wavelengths, and the resulting spectrum can be recorded on a photographic film, as in Fig. 3.15, or can be scanned by a suitable small detector which produces an output current or voltage proportional to intensity. Commercial spectrometers for these measurements are readily available.

The radiant energy absorbed by the material may be

(a) re-radiated without change in wavelength, i.e. scattered into another direction of propagation;
(b) re-radiated at a different wavelength, if the absorbing charge subsequently makes a transition to a different energy level to that from which it was first excited;
(c) lost to kinetic energy of atomic motion (i.e. to thermal energy) if the atom(s) concerned 'collide' with their neighbours.

Which of these processes occurs depends on their relative probabilities: we often refer to *transition probabilities* between energy levels.

All the above processes, including (a), can give rise to a reduction in the intensity transmitted through a material, and hence show up in an absorption spectrum.

Before discussing some of the features to be found in absorption spectra, let us study the absorption due to induced nuclear motion alone. We shall find that it displays features which are universally found even in electronic spectra.

19.3 A model for crystal lattice absorption

Crystal lattice absorption dominates at infrared frequencies in ionic insulators and semiconductors, for example LiF, MgO, InSb and ZnS.

The motion of ions in a crystal lattice was discussed in Chapter 7 and illustrated in Fig. 7.7. When alternate ions have equal and opposite charges, we have a simple two-dimensional model of a CsCl lattice. Forces on neighbouring ions due to the electric field are opposite in direction, and in the normal mode of vibration thus excited they oscillate in antiphase motion, as illustrated in Fig. 7.7(b). The natural frequency of this mode lies in the infrared, and incident radiation at this same frequency will be most strongly absorbed because the amplitude of the motion it can create is a maximum at the natural frequency. The degree of absorption at this frequency is limited by the damping of the oscillatory motion.

To calculate how the absorption of energy is expected to vary with the frequency of the incoming radiation, we can use the above model in which we assume for simplicity that the negative ions (say) are fixed, while the positive ions each of mass M move under the constraints of the 'springs' attached to them. (This approximates to the actual motion if the negative ions are very much more massive than the positive ions.) These exert a restoring force proportional to the displacement x. Then each positive ion with charge q obeys the equation of damped simple harmonic motion, under the influence of an oscillatory driving force $F = qE_0 \cos \omega t$, where E_0 is the amplitude of the alternating electric field strength. The equation of motion, which simply states that the mass × acceleration equals the sum of the forces acting, is

$$M\ddot{x} = -kx - \beta\dot{x} + qE_0 \cos \omega t \qquad (19.2)$$

where $-kx$ is the restoring force, and $-\beta\dot{x}$ is the damping force, which is assumed to be proportional to the velocity of the ions.

The resulting alternating displacement of the ions follows the driving force, but with a phase difference we shall call ϕ, i.e. the solution of Eq. (19.2) has the form

$$x = A \cos(\omega t + \phi) \qquad (19.3)$$

By substitution of this equation into Eq. (19.2) and solving for A and ϕ we obtain the results

$$A = \frac{qE_0}{M} \left[\frac{1}{(\omega_0^2 - \omega^2)^2 + \gamma^2\omega^2} \right]^{1/2} \qquad (19.4)$$

and

$$\tan \phi = -\gamma\omega/(\omega_0^2 - \omega^2) \qquad (19.5)$$

where

$$\omega_0 = \sqrt{k/M} \quad \text{and} \quad \gamma = \beta/M$$

Here ω_0 is the natural frequency of oscillation, and γ depends on the strength of damping. The rate P (J s^{-1}) at which energy is absorbed from the driving force is the damping force multiplied by dx/dt, the velocity

$$P = (\beta\dot{x}) \, dx/dt = \beta(\dot{x})^2$$

The mean of P over many cycles thus depends on $\langle \dot{x}^2 \rangle$, the mean of the square of the velocity, thus using Eq. (19.3) the mean power $\langle P \rangle$ is given by

$$\langle P \rangle = \beta \langle \dot{x}^2 \rangle = \frac{\beta}{2} (\omega A)^2 \qquad (19.6)$$

A plot of the mean absorbed power $\langle P \rangle$ against angular frequency ω is shown in Fig. 19.1 and it represents the shape of the absorption spectrum line of this model 'oscillator'. The similarity to the measured shape of the transmission spectrum of lithium fluoride shown in Fig. 19.2 is obvious.

However, as shown in Chapter 7, in a crystalline solid there is no single natural frequency ω_0, but a band of frequencies, the 'optical' phonon band. Each frequency in the band corresponds to a mode of oscillation of the lattice with a different periodicity, i.e. wavelength. Only one of the modes absorbs strongly – the one for which the mode wavelength coincides with the *optical* wavelength at the same frequency. In all other modes, the ionic motion cannot remain in phase with the force exerted by the optical wave as it travels through the material. The phase angle ϕ between the two changes continuously from 0 to π and on to 2π as the waves proceed. The work done by the optical wave on the ions is the product of a force $F = F_0 \cos \omega t$ with displacement $x = x_0 \cos(\omega t + \phi)$. The work done, $W = Fx_0 \cos \omega t \cos(\omega t + \phi)$ varies with ϕ and is even negative if $\frac{1}{2}\pi < \phi < \frac{3}{2}\pi$! In fact, when averaged over all phase angles 0 to 2π the net energy absorbed is just zero. For absorption to occur both the frequencies *and* the wavelengths of the optical

Fig. 19.1

Absorption versus frequency predicted from Eq. (19.4).

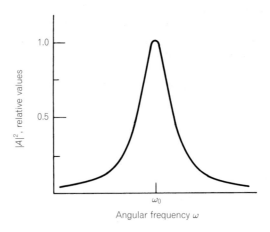

Fig. 19.2

Transmission spectrum of a thin slice of lithium fluoride. Relative transmitted intensity is plotted against wavelength.

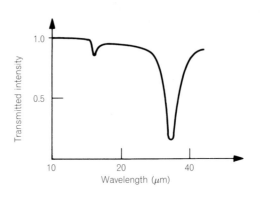

and lattice waves must coincide. Because only one frequency f_0 is absorbed, the spectrum of light transmitted through the crystal and recorded on a photographic plate shows a narrow dark line at the wavelength $\lambda = c/f_0$ corresponding to that frequency, where c is the velocity of light. We call this a *line spectrum*, for obvious reasons. The width at half-intensity of the absorption line, shown in profile in Fig. 19.1, can be found from the condition that the amplitude A in Eq. (19.4) must be $1/\sqrt{2}$ of its peak value at half-intensity. This occurs when $(\omega_0^2 - \omega^2) = \gamma\omega$, i.e. $\Delta\omega \simeq \frac{1}{2}\gamma$ since $\gamma \ll \omega_0$, so that the *total* width of the line is $2\Delta\omega = \gamma = \beta/M$. Thus the damping coefficient β in the equation of motion alone determines the absorption line width. In practice this damping, or energy loss, from the vibrational motion (a phonon) is governed by 'collisions' with other phonons and/or with defects.

In molecular solids (which includes the vast range of organic solids), vibrations of one molecule couple only weakly to the next, owing to the relative weakness of intermolecular bonds. The concept of lattice vibrations or phonons must then be replaced by treating each molecule, or even each group within a molecule, as a separate vibrator.

For example, the methylene group as illustrated in Fig. 19.3 can vibrate in several characteristic ways, each with a characteristic natural frequency which causes an absorption at the corresponding optical wavelength. A list of wavelengths for some modes of this and other groups is given in Table 19.1. Note that bending vibrations of the group have lower characteristic frequencies than stretching modes of the same group, because a bond displays a lower stiffness when bent than when stretched or compressed. The natural frequency is proportional to the square root of the stiffness, i.e. the force constant, k, as seen in Eq. (19.5).

The absorptivity due to a fundamental vibrational mode is proportional to the square of the rate of change of the dipole moment with displacements of the atoms. Thus symmetric motions, as in Fig. 19.3(a), which produce almost no change in the dipole moment, do not show up in infrared absorption

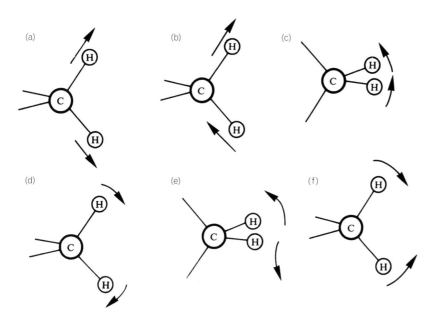

Fig. 19.3

Vibrational modes of a CH_2 group. (a) Symmetric stretching; (b) asymmetric stretching; (c) wagging; (d) rocking; (e) twisting; (f) scissor.

Group and mode		Wavelength (μm)	Strength
—C—H	Stretch	3.03	Strong
$=C\begin{smallmatrix}H\\H\end{smallmatrix}$	Stretch	3.23–3.25	Medium
$>CH_2$	Rocking	14	Weak
—OH	Stretch	2.7–3.1	Variable
—O—H	Bend	7.09–7.94	Strong
N—H	Stretch	2.8–3.03	Medium
N—H	Bend	6.3–6.7	Weak

spectra. But asymmetric motions such as in Fig. 19.3(b) are said to be
infrared 'active', and produce significant absorption.

19.4 Electronic absorption in insulators

The absorption spectra of different materials, while at first sight being very
dissimilar, are made up of features that broadly fall into two categories:
absorption *lines* such as we have already studied, and absorption *edges*,
which must wait until Section 19.6.

Frequent mention is also made in textbooks on spectroscopy of *absorption bands*, which are no more than a series of overlapping lines.

Absorption of visible and ultraviolet energy by electrons also frequently
gives line spectra. That this must be so can be understood from what we
have already learned about electronic energy levels in materials. If energy
levels are discrete and narrow, like those of the hydrogen gas whose
emission spectrum is shown in Fig. 3.15, absorption of energy occurs over
a narrow band of frequencies centred on the one given by Planck's law

$$hf = E_1 - E_0 \qquad (19.7)$$

where E_1 and E_0 are the energies of the final and initial levels respectively.
Thus line spectra are observed in the visible and ultraviolet regions of
the spectrum, as well as in the infrared. This can be seen in Fig. 19.4, which
shows transmission spectra of some organic dye molecules. The wavelength of the principal absorption maximum determines the colour, and its
width the 'purity' of the colour: narrow absorption lines give bright, 'pure'
colours. It can be seen from Fig. 19.4 that the absorption peaks in these
molecules, which have alternate single and double bonds ('conjugated'
bonds), shifts towards longer wavelengths (lower frequencies) as the molecule lengthens. The positions of these peaks and those of many other
conjugated molecules can be predicted remarkably well with the aid of a
very simple model. The electrons in the conjugated bonds are assumed to
be free to move right along the molecule, whose length L along the zig-zag
bonds is calculable from known bond lengths. The 'electron in a box'
model discussed in Section 2.6 can then be employed to predict energy

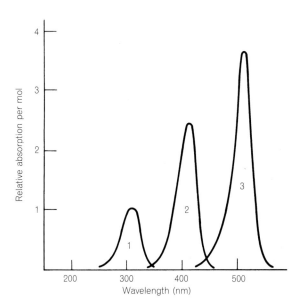

Fig. 19.4

Ultraviolet spectra of the dyes
$Me_2N=(CH-CH=CH)_n-NMe_2$,
where Me is a methyl group; and
n is given in the figure.

differences between the highest filled level E_0 and the lowest unfilled level E_1. The resulting equation for the absorption peak λ_{max}, is

$$\lambda_{max} = \frac{8mcL^2}{h(N+1)} \qquad (19.8)$$

This equation predicts the values of 170 nm, 270 nm and 350 nm for the three dyes in Fig. 19.4. While agreement with the observed, values is only approximate, it is surprisingly good for such a simple model, especially as λ_{max} is sensitive to the molecular surroundings, a fact which the equation above does not predict. However the model can predict the relative strengths of absorption fairly well. These and other data convince us that electrons are indeed rather free to move within organic molecules having conjugated bonds: the molecules behave rather like minute metallic conductors. This knowledge is useful for predicting other properties of organic materials. Naturally the λ_{max} of dye molecules can be predicted much more accurately by other, better, models, and nowadays the absorption peak of a dye can be routinely predicted to within 10 nm or less before it has been synthesized. It is notable that the shapes of the absorption lines in Fig. 19.4 are not dissimilar to the one predicted for lattice absorption in Fig. 19.1. A classical model for the absorption of electrons can indeed be quite useful.

In more complex organic molecules and in inorganic insulators there may be many more energy levels, producing many absorption lines that may overlap with one another. Furthermore, the width of an absorption line often increases substantially when atoms or molecules are condensed to form solids or liquids. The effect is analogous to that which broadens electron energy levels into bands in a solid, and the resulting absorption bands can be quite broad. Atomic vibrations in molecules and solids can also broaden electronic absorption lines.

The identification and separation of these lines and bands is a major task of spectroscopists, and the detailed shape of an absorption spectrum can often be used as a 'fingerprint' for the identification of unknown substances.

19.5 Electronic absorption in metals

The 'free electrons' in organic molecules display narrow absorption lines because their electronic energy levels are separated by energies comparable to those of photons. In contrast the electron energy bands in metals allow the conduction electrons freedom to respond to electromagnetic waves of almost any frequency. It is this strong response of the conduction electrons which prevents a metal from transmitting light except when thinned to less than 100 nm.

The seeming paradox, that metals both absorb strongly and reflect strongly, is no contradiction. The strong reflecting power results from the high conductivity – a perfect conductor would be a perfect reflector. This is because an accelerating charge radiates an electromagnetic energy, and the electrons are being rapidly accelerated backwards and forwards. Most of the incident energy is re-radiated backwards, while little penetrates the surface. Having penetrated, that little is absorbed, just because the conductivity is finite, so that kinetic energy acquired by electrons is readily transferred to the lattice by collision, as described in Chapter 13 in the discussion of conduction. The strength of the absorption at a frequency f is a consequence of the large concentration of electrons that lie within an energy hf (typically 1–2 eV) below the Fermi level. Strong absorption and reflection persist as the frequency is raised throughout the infrared, visible and near ultraviolet parts of the spectrum, until in the far ultraviolet, around a wavelength of about 100 nm, the metal becomes significantly more transparent. It is only at these high frequencies, above 10^{16} Hz, that the inertia of the electrons makes itself felt and they fail to move sufficiently to either absorb or reflect the incoming wave.

It is obvious from the differences in the colours of metals that the strength of absorption varies across the optical part of the spectrum. This is due to variations in the absorption by electrons lying well below the Fermi level, and is due to the fact that most common metals have partially empty d shells. Absorption is usually strongest at higher photon energies, i.e. at the blue end of the spectrum. The metal therefore reflects the complementary colour – yellow if the absorption is weak, but reddish, as in copper, when the absorption is rather stronger.

19.6 Electronic absorption in semiconductors

Because semiconductors have many fewer free carriers, free carrier absorption such as occurs in metals is very weak. Lattice absorption can occur in the infrared, as we have seen in Section 18.3. As long as the photon energy hf is less than the energy gap E_g between the valence band and the conduction band, little or no absorption by electrons can occur. The only vacant levels at an energy which the valence electrons can reach are the few holes at the top of the valence band.

Hence significant absorption only occurs at frequencies f such that $hf > E_g$, the energy gap. Figure 19.5 shows such a case: the spectrum of germanium. This shows features that can be related to the details of the energy bands in germanium, which are not nearly as simple as was

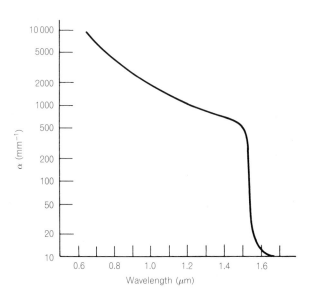

Fig. 19.5

The absorption coefficient α of germanium plotted against wavelength.

described in Chapter 14. The rise in absorption at higher frequencies seen in Fig. 19.5 is called an absorption edge. This is because, on the low frequency (longer wavelength) side of the main 'edge' the absorption rises exponentially with wavelength over a long range. Thus the shape differs noticeably from the $1/(\omega_0^2 - \omega^2)^2$ dependence given by Eq. (19.4) when $\gamma\omega$ is small. From Fig. 19.5 it is clear that germanium absorbs strongly like a metal at visible wavelengths, and as might be expected it looks metallic to the eye.

From the energy gaps of other semiconductors given in Table 14.1 the wavelengths at which the absorption edges occur can be calculated using $hc/\lambda = hf = E_g$. Thus silicon, like germanium, appears metallic with an absorption edge at the infrared end of the spectrum, whilst cadmium sulphide is yellow (absorption edge at 540 nm) and diamond is of course visibly transparent because it only absorbs ultraviolet light at wavelengths shorter than 230 nm.

19.7 The refractive index

From the absorption of light, we now turn to another property of transparent solids which is of interest to us: the refractive index. The theory of electromagnetic waves tells us that the refractive index is equal to the square root of the relative dielectric permittivity of a material, determined at the frequency of oscillation of the electromagnetic wave. This, of course, is controlled by the motion of the charges in a material in response to the electric field of a light wave. It can therefore be predicted from the mathematical model developed in Section 19.3. The change in the refractive index with frequency of wavelength is usually termed *dispersion* of the index.

The equations of Section 19.3 enable us to calculate the polarizability (see Chapter 17) of the moving charges, which is just $qA\cos\phi/E_0$ for each charge. A and $\cos\phi$ are to be found from Eqs (19.4) and (19.5) respectively.

Equation (18.20) gives us a relation between permittivity ε and polarizability α which, however, needs modification to relate the internal

Fig. 19.6

Refractive index and absorption variations across an absorption line, predicted by Eq. (19.10).

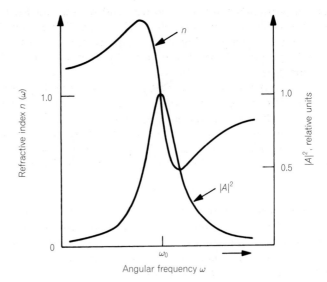

electric field experienced by moving charge to the applied field $E_0 \cos \omega t$. When this is done, and n^2 is substituted for ε, Eq. (18.20) leads to the result

$$\frac{n^2 - 1}{n^2 + 2} = \frac{N\alpha}{\varepsilon_0} = \frac{NqA\cos\phi}{3\varepsilon_0 E_0} \tag{19.9}$$

Inserting A and $\cos\phi$ from Eqs (19.4) and (19.5) gives the result

$$\frac{n^2 - 1}{n^2 + 2} = \frac{Nq^2}{3M\varepsilon_0}\left[\frac{(\omega_0^2 - \omega^2)}{(\omega_0^2 - \omega^2) + \gamma^2\omega^2}\right] \tag{19.10}$$

The resulting shape for $n(\omega)$ is shown in Fig. 19.6, together with the associated power absorption, for an arbitrary density N of moving charges. This reproduces fairly closely the observed behaviour in at least some materials but, as with the absorption spectrum, broadening is common in solids and liquids. The refractive index changes with frequency in a characteristic way as the absorption line is crossed. Note that, except in a narrow range near the absorption peak, the index increases with the frequency. The exceptional behaviour close to the absorption peak is often called anomalous dispersion – a rather outdated description now that the theory is so well established!

19.8 Scattering of optical radiation

Besides direct absorption, there are various mechanisms by which a beam of radiation is scattered within a material, which contribute to the loss of energy in the beam.

The simplest of these, Rayleigh scattering, is an important contributor to the energy loss in highly pure, transparent materials such as are used for optical fibres. Rayleigh scattering is essentially due to small, random fluctuations in the refractive index from point to point which in an amorphous substance arise from the density and composition fluctuations consequent upon a random structure. Such variations also occur in crystalline materials, due to thermal motion of the lattice, and other lattice defects. Variations of index can be particularly large in liquids and gases, and Rayleigh scattering is responsible for the blue colour of the sky. The reason for this is that the scattered energy is inversely proportional to the fourth power of the wavelength and so is much stronger – by about 2^4 – at the blue end of the visible spectrum.

We now explain how this comes about, by considering the scattered radiation to be due to induced dipoles located at the randomly sited points where the index differs from the average values.

The dipole moments of these induced dipoles are proportional to the incident light amplitude, just as in the previous section. So the dipole moment p varies cosinusoidally in sympathy with the driving force, i.e.

$$p = p_0 \cos \omega t \qquad (19.11)$$

where $\omega = 2\pi f = 2\pi c/\lambda$ is the angular frequency of the incident radiation of wavelength λ. The total scattered amplitude is the sum of many wavelets originating at random distances from a point of observation, so that their relative phases are random. But the total scattered intensity will be proportional to the intensity of each scattered wavelet.

Now according to electromagnetic theory, a moving charge radiates electromagnetic energy with an amplitude proportional to its acceleration (a steady current will not radiate) so that the wavelet has amplitude proportional to d^2p/dt^2. From Eq. (19.11)

$$\frac{d^2p}{dt^2} = p_0\omega^2 \cos \omega t = \frac{p_0 c^2}{\lambda^2} \cos \omega t$$

Hence the scattered *intensity*, being proportional to the square of this amplitude, is proportional to $1/\lambda^4$. To compare this result with experiment, we shall use the example of attenuation of light in optical fibres designed for long-distance communication. Because of their current importance, they deserve more than a brief mention.

19.9 Optical fibres

The scattering and absorption of light in glasses, and particularly in silica, has received much attention because of the cost reductions obtainable in telephone systems by the use of optical fibres. These fibres, which consist primarily of silica, have a cylindrical core of 5–50 μm diameter doped with GeO_2 to give a slightly higher refractive index. The light is guided in the core by total internal reflection at the interface with the pure silica cladding.

In 1970 the best fibres then made attenuated light of 850 nm wavelength by a factor of 100 over a 1 km length. Careful study and control of the sources of absorption had by 1978 reduced the attenuation by 1 km of fibre to a factor of less than 1.2 at a wavelength of 1.3 μm. The wavelength dependence of the measured attenuation typical in recent fibres is shown in Fig. 19.7.

Rayleigh scattering is clearly the main limiting factor for wavelengths up to about 1 μm. At longer wavelengths the curve is reminiscent of spectral absorption 'lines'. Indeed, it has been shown that the first few absorption lines are 'overtones', or harmonics, of an absorption line at about 2.8 μm wavelength which results from the vibration of O—H bonds inadvertently incorporated in the silica. The exclusion of water from the fibre is thus vital to achieve the lowest possible attenuation of the guided light. Exclusion

Fig. 19.7

Ratio of transmitted intensity I to incident intensity I_0 typical of 1 km of silica fibre, plotted against wavelength. The lower curve shows the estimated effect of Rayleigh scattering alone.

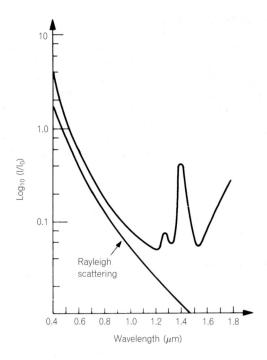

Fig. 19.7

Ratio of transmitted intensity I to incident intensity I_0 typical of 1 km of silica fibre, plotted against wavelength. The lower curve shows the estimated effect of Rayleigh scattering alone.

of other absorbing impurities such as Na, Fe, Cu is even more important, and indeed the allowed impurity levels rival those achieved in semi-conductors for transistor manufacture.

The magnitude of the Rayleigh scattering in any given material must depend upon the amplitude of both density and composition fluctuations. The variations in density of a glass can be attributed to thermally driven fluctuations arising from the Brownian motion in the liquid glass before it freezes. The amplitude of the fluctuations must therefore be dependent on the freezing temperature, being greater for a higher freezing point. A high melting point material such as silica might, therefore, be expected to show a higher Rayleigh scattering loss than a lead glass of much lower melting temperature. But the reverse is the case, since in multi-component lead glasses local composition fluctuations cause bigger refractive index changes than does the non-uniform density. Hence silica is the preferred material for long-distance optical fibre communication. At still longer wavelengths than those in Fig. 19.7, the attenuation is predominantly due to absorption by induced vibration of the constituent atoms in the glass. Each interatomic bond oscillates at a different characteristic frequency and hence absorbs most strongly at the corresponding wavelength, as indicated in Table 19.2.

Table 19.2

Characteristic stretching frequencies of bonds in glasses for optical fibres

Bond	Wavelength (μm)
Si—O	9.0
B—O	7.3
P—O	8.0
Ge—O	11.0

19.10 Luminescence

Luminescence is the generic term used to describe the emission of electro-magnetic radiation after the prior absorption of energy, often in the form of radiation. Thus photoluminescence, which includes fluorescence and phosphorescence, is the emission of light after photons have been absorbed. Fluorescence occurs via energy transitions of electrons that do not involve a change of electron spin, while phosphoresence involves a change of spin, and so is much slower than fluorescence. Typically, fluorescence is over within about 10^{-5} to 10^{-6} seconds of the absorption process. Phosphores-cence decays more slowly, in 10^{-4} to 10 seconds. In both cases the lifetime is defined as the time required for population of the excited states to decay to 1/e of its original value after the source of excitation is switched off. In this section we shall concentrate on fluorescence.

A typical energy level diagram showing fluorescence transitions is given in Fig. 19.8. The states S_0 and S_1 both have zero spin, and are split into several vibrational energy levels. Having zero spin, they are called singlet states, while the energy levels labelled T_1 are associated with an electronic state having a spin quantum number of one, called a triplet state.

In a fluorescence process, a molecule or solid in the lowest vibrational level of the singlet ground state S_0 is excited by an incoming photon into a higher vibrational level of the first excited singlet state S_1. The vibrational energy of the excited state disappears rapidly: it is communicated to neigh-bouring atoms or molecules in about 10^{-12} s, a time much shorter than the excited state lifetime of 10^{-8} to 10^{-6} s quoted above. The excited state may then decay with emission of a photon of fluorescent radiation, usually to an upper vibrational level of the ground state. Because of the similarity between the vibrational levels of the ground and excited states, the fluor-escent spectrum is often somewhat like a mirror image of the absorption spectrum of the same material, as in the example in Fig. 19.9.

Phosphorescence occurs following an alternative decay process, via the triplet state T_1, which is also shown in Fig. 19.8. This process has lower probability as it requires a flip (a reversal of direction) of the spin of the electron involved. Decay to the ground state must then occur with another spin flip, either by phosphorescent emission, or by emission of vibrational energy.

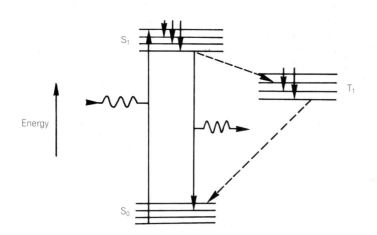

Fig. 19.8

Schematic energy level diagram showing fluorescence via singlet levels S_0 and S_1, and phosphorescence via a triplet state T_1.

Fig. 19.9

Emission and absorption spectra for benzo(ghi)perylene in benzene.

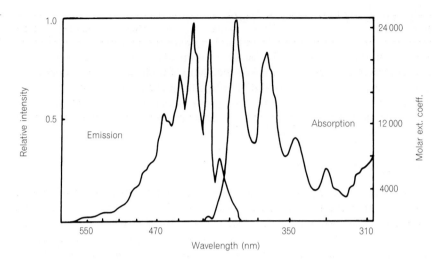

Fig. 19.9

Emission and absorption spectra for benzo(ghi)perylene in benzene.

Note that the luminescent spectrum always has longer wavelength (lower energy) than the exciting radiation. If both fluorescence and phosphorescence occur in the same material, the latter has the longer wavelength for a similar reason.

19.10.1 Applications of luminescence

Fluorescence is important for two quite different reasons. Inorganic 'phosphors', as they are confusingly called, are used for producing light in fluorescent tubes and in lasers, and will be covered in Sections 19.10.2 and 19.11. Fluorescence in organic materials, too, is used in lasers, but in addition it is most useful as an analytical tool for the scientist and engineer. This is because it is more sensitive to structural changes than is absorption spectroscopy, and can often be used in the analysis of trace contaminants in both industrial and biological environments, down to levels of parts per trillion in some cases. It is relatively easy to filter the fluorescent light from the incident excitation of a shorter wavelength, so that the few photons produced by a trace impurity can be detected against a 'dark' background. Thus the drug LSD (lysergic acid diethylamide) can be detected at levels of $1 \, \text{ng ml}^{-1}$ in a 5 ml sample of blood plasma or urine.

Fluorescence occurs in aromatic molecules, and in those containing stable arrangements of conjugated double bonds. Such molecules have delocalized electrons with excited singlet states at energies of 2–3 eV which fluoresce readily. Substituents for hydrogen that withdraw electrons from the conjugated system (e.g. $-Cl$, $-Br$, $-I$, $-NO_2$, $-COOH$) will tend to inhibit the fluorescence, while substituents that help to delocalize the electrons enhance fluorescence (e.g. $-NH_2$, $-OH$, $-F$, etc.). Thus aniline fluoresces, while nitrobenzene does not. The importance of such molecules in industrial chemistry and as drugs and carcinogens is what makes fluorescence spectroscopy of such value.

The reader is advised to consult a textbook of analytical chemistry for further details on this topic.

Table 19.3

Some important lamp phosphors

Host	Host formula	Activator	Emission colour
Zinc silicate	Zn_2SiO_4	Mn	Green
Calcium halophosphate	$Ca_5(PO_4)_3(F, Cl)$	Sb, Mn	Blue to pink
Calcium tungstate	$CaWO_4$	None	Deep blue
Yttrium oxide	Y_2O_3	Eu	Red

(Source: *Lamps and Lighting*, Thorn-EMI Lighting Ltd, 1983)

19.10.2 Phosphors

Phosphors are inorganic photoluminescent solids used for the generation of light. They consist of an inactive host (often an insulating oxide or wide band gap semiconductor, e.g. $CaWO_4$, ZnS), which is doped with a small percentage of luminescent atoms. The latter are commonly transition metal ions, e.g. Ag^+, Cr^{2+}, Nd^{3+}, because these have suitable excited states between 2 and 3 eV above the ground state.

The fluorescent light tube is perhaps the best known application of photoluminescence. In the tube, a mercury vapour discharge produces ultraviolet light efficiently, predominantly at the two wavelengths of 185 and 254 nm. This irradiates a phosphor coated on the tube walls, consisting of one or more of the combinations of host and active dopant listed in Table 19.3. These materials are chosen for their efficiency in converting ultraviolet light into visible light, and because their performance deteriorates little with rising temperature and age. Their lack of toxicity and ease of preparation in fine-particle form are also important to the manufacturer.

Other means of 'pumping' luminescent atoms into their excited state will also cause luminescence, including exposure to an energetic electron beam. This, called cathodoluminescence, is the mechanism at work in the cathode ray tube. For colour tubes the three phosphors normally used are Cu^+-doped ZnS (green), Ag^+-doped ZnS (blue) and Eu^{3+}-doped YVO_4 (red).

19.11 Principles of lasers: optical amplification

This chapter would hardly be complete without a discussion of **L**ight **A**mplification by **S**timulated **E**mission of **R**adiation – the fundamental mechanism underlying the LASER. Stimulated emission is, in fact, just another form of luminescent emission.

To understand stimulated emission, let us recall that spontaneous emission of a photon occurs when an electron falls from a higher energy level to a lower one, while absorption of a photon causes a quantum jump from a lower to a higher energy level. But what if an electron in a high energy level collides with a photon? Clearly the electron makes a quantum jump, perhaps to a higher level by absorbing the photon. But it is equally probable that the jump will be downward with the emission of a photon of energy equal to that of the photon which stimulated the jump. This is the process called stimulated emission. The three processes are compared diagrammatically in Fig. 19.10.

Fig. 19.10

A representation on an energy
level diagram of (a) spontaneous
emission, (b) absorption,
(c) stimulated emission.

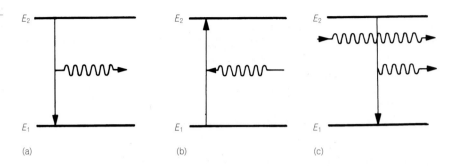

Fig. 19.10

A representation on an energy
level diagram of (a) spontaneous
emission, (b) absorption,
(c) stimulated emission.

The proof that the stimulated emission and absorption processes are equally likely lies in the symmetry of the equations which describe the interaction between photon and electron.

One more feature of stimulated emission which is most important is this: the electromagnetic wave which is the emitted photon has the same phase everywhere as the stimulating photon wave. In fact the two photons can be represented as a single wave with twice the intensity of a single photon: the two photons are, of course, indistinguishable. We sometimes say that the two photons are *coherent*, meaning simply that they have the same phase.

Now, here is the possibility of an optical amplifier, for we have two photons where there was one. What is required to observe it is a material containing plenty of atoms in excited states, and rather few in a lower energy level (not necessarily the ground state) into which the excited electrons may go on receipt of a stimulus from a suitable photon.

Now a material that has more electrons in higher energy levels than low can hardly be in thermal equilibrium – the ratio of the population of the higher level to that in the lower in equilibrium is given by the exponential Boltzmann factor $\exp -\Delta E/kT$. To make this ratio greater than unity is impossible unless energy is continuously put into the material to 'pump' electrons into the higher levels. It is not surprising that energy is needed, for we cannot expect to amplify the intensity of an optical beam by breaking the first law of thermodynamics. We can relate the populations of the upper and lower energy levels E_2 and E_1 to the amplification of the intensity, in the following way. Suppose a plane wave of a frequency f and intensity I is passing through the laser, in which there are N_2 atoms per unit volume in the upper level and N_1 in the lower energy level. If the probability of a stimulated transition of one atom from level 2 to level 1 is W_s, then the probability of absorption causing an atom to move from level 1 to level 2 is also equal to W_s. The net flow of atoms from 2 to 1 is thus

$$(N_2 - N_1)W_s$$

But W_s must also be proportional to the intensity I of the plane wave. The net power P generated per unit volume is then

$$P = (N_2 - N_1)W_s hf = (N_2 - N_1)w_s I$$

where w_s is just a proportionality constant. If there were no energy loss by other means (never the case) the intensity *increase* per unit length is equal to P so that we can write

$$\frac{\mathrm{d}I}{\mathrm{d}z} = (N_2 - N_1)w_s I$$

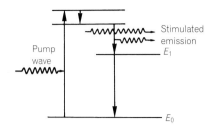

Fig. 19.11

Energy level diagram of a four-level laser.

where z is the distance, and the solution of this equation is

$$I(z) = I(0) \exp \gamma z$$

where $I(0)$ is the intensity at $z = 0$ and $\gamma = (N_2 - N_1)$. w_s is called the *gain* of the laser medium. (Note that it has the same dimensions as the absorption coefficient of the medium.) We see that gain exists only as long as N_2 is greater than N_1. If the atoms are pumped directly from the lower to the higher level, the pump power needed would be very high, since the lower level contains few atoms available to absorb the pump energy. Less pump power is needed if level 1 is not the ground state (as in Fig. 19.11), for two reasons. One is that the density of atoms N_0 in the ground state can be much higher than either N_1 or N_2, so that the pump power is readily absorbed. But in addition the lower lasing level population N_1 is held low since atoms can decay rapidly to the ground state. The energy levels and transitions in such a laser are indicated in Fig. 19.11. This scheme is called a *four-level* laser, since the pump more often than not excites the electron into a fourth level, higher than E_2 (the upper lasing level) just as in fluorescence.

The conditions in which the four-level laser will produce gain can be determined from the equations describing the rate at which the populations N_2 and N_1 change with time.

If the atoms are pumped up to the level E_2, at rate R per unit volume per second, then the rate of increase of N_2 is less than R because

(a) atoms decay to level 1 by spontaneous emission at a rate N_2/τ_2 where τ_2 is the mean lifetime of an atom in level 2 before this decay process occurs, and because
(b) the net rate of stimulated emission given earlier as $(N_2 - N_1)W_s$ also reduces N_2.

The rate of increase of N_2 is thus

$$\frac{dN_2}{dt} = R - \frac{N_2}{\tau_2} - (N_2 - N_1)W_s$$

In similar fashion the rate of increase of N_1 is given by

$$\frac{dN_1}{dt} = \frac{N_2}{\tau_2} - \frac{N_1}{\tau_1} + W_s(N_2 - N_1)$$

where τ_1 is the natural lifetime of level 1, by decay to the ground state of the atom.

Since in the steady state both dN_2/dt and dN_1/dt are zero, the above two simultaneous equations give the result:

$$N_2 - N_1 = \frac{R(1 - \tau_1/\tau_2)}{W_s + 1/\tau_2}$$

This equation tells us the vital condition for achieving gain: $N_2 - N_1$ is only positive if $\tau_2 > \tau_1$, i.e. the natural lifetime of level 2 must be greater than that of level 1 in the four-level laser.

The pump energy for a laser may be provided in several ways, though the obvious way (employed in the first lasers) is to use an intense source of light, such as a flash lamp. Naturally this method only produces pulses of amplified light, but it is still used today in some types of pulsed lasers.

Alternatively, an electrical discharge in a gas will excite atoms to high energy levels – after all, such a discharge radiates spontaneously. However, a neon or a sodium lamp is not called a laser, although stimulated emission probably occurs to a small extent in the discharge.

Oddly, we reserve the acronym LASER for what in electronic parlance is actually an *oscillator*. The difference is that, in an oscillator, a portion of its output is fed back into the input to make it continue oscillating at a fixed frequency. In electronic terms, the neon or sodium lamp produces a very noisy output – that is, one which is continually changing phase, and hence cannot be at a single, fixed, frequency. The constancy of the output phase and frequency is, perhaps, the most important characteristic of the laser, and is what gives it its unique qualities as a light source.

The feedback of output to input is done by mirrors, as shown in Fig. 19.12. One of these is made partially transmitting to enable a fraction of the light to be used externally. Since oscillation can only be sustained if the fed-back optical wave arrives back in phase with itself after a complete round-trip between the mirrors, the only wavelengths λ which will oscillate must satisfy the equation

$$N\lambda = 2L \qquad (19.12)$$

where N is an integer. Thus the mirror separation L also plays a role in determining the spectral distribution of the output light.

Fig. 19.12

Mirrors MI and M2 reflect the amplified beam in a laser back upon itself to sustain the oscillation. The pump is a source either of light or electrical power.

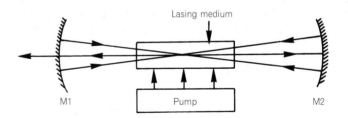

Fig. 19.13

Typical output intensity versus wavelength for a laser.

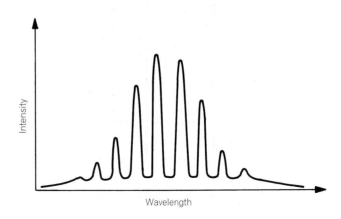

A typical distribution is given in Fig. 19.13. Each narrow emission line is centred on a wavelength that satisfies Eq. (19.12), whilst the envelope of the peaks is determined by (though not equal to) the natural spectral width of the gain γ of the lasing material defined earlier.

19.12 Laser materials

It is clear that a material which fluoresces is likely to be a candidate as a lasing material. We shall select three examples to illustrate good laser materials: the solid-state neodymium laser, the liquid dye laser and the semiconductor laser.

19.12.1 The neodymium laser

The neodymium laser is a fairly straightforward example of a four-level laser. The low-lying energy levels of the Nd^{3+} ion are shown in Fig. 19.14. In the electronic ground state the Nd^{3+} ion has three electrons in its 4f shell having parallel spins (total spin $S = 3/2$) and with orbital angular momentum $L = 3 + 2 + 1 = 6$ (see Chapter 5). The letter I in the figure, the sixth in the series PDFGHIJ ..., signifies that $L = 6$. There are four possible combinations of L and S, giving the following values of J: $6 - 3/2$, $6 - 1/2$, $6 + 1/2$ and $6 + 3/2$, corresponding to the values of the subscripts in the figure.

In the first excited electronic state an electron is promoted to the next 4f level where $l = 0$ giving $L = 2 + 1 = 3$, i.e. an F state, but its spin is unchanged, so that the total spin is still $S = 3/2$. Only the lowest two levels are illustrated.

Three possible lasing transitions are shown, of which one $(0.914\,\mu m)$ ends on the ground state, and therefore has a highly populated lower level. It consequently provides less gain than the other two. The laser may be pumped at $0.914\,\mu m$, $0.805\,\mu m$, or at one of several shorter wavelengths at which the ion absorbs strongly. The excited ion relaxes rapidly into the lowest level of the first excited state, just as any fluorescent species.

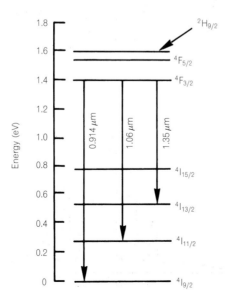

Fig. 19.14

Energy levels of the Nd^{3+} ion involved in lasing.

When the ion is in a solid, these energy levels are little affected by the surrounding anions, because the 4f electrons are shielded from their fields by the filled 5s, 5p, 5d shells, as described in Chapter 17. Hence by putting a small quantity (about 5% by weight) of Nd_2O_5 into a suitable rod of silicate glass, an excellent lasing rod can be made. Large rods, pumped by a powerful xenon flashlamp, can produce brief (less than 1 ns) but 'giant' pulses of light at $1.314\,\mu$m or $1.064\,\mu$m, with peak powers of a few gigawatts. Alternatively, the glass can be made in optical fibre form, pumped down the core by a few tens of milliwatts of infrared light of wavelength $0.914\,\mu$m from a cheap semiconductor laser to give an output at the more useful wavelength of $1.3\,\mu$m (see Section 19.12).

Nd^{3+} ions in crystals can, however, produce about ten times as much optical gain as the same ions in a glass. This is because the surroundings experienced by the Nd ions in a glass vary from site to site in the random structure, so detuning the energy levels just far enough from their average positions that many of them provide little or no gain at the frequency of an incoming photon. However, Nd^{3+} ions in a crystal of yttrium aluminium garnet (YAG), where each is surrounded uniformly by eight oxygen ions, have energy levels which all coincide, and all will contribute to the gain. Nd-YAG laser rods are readily obtainable, and are capable of producing higher power pulses than Nd-glass.

19.12.2 Dye lasers

Fig. 19.15

Vibrational splitting of the electronic levels in an organic dye may allow stimulated emission over a range of wavelengths.

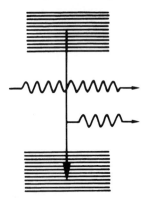

Organic dyes are also four-level lasing materials, having a closely spaced set of vibrational levels associated with each electronic level, as shown in Fig. 19.15. Because there are so many lower levels, lasing is possible over an almost continuous set of frequencies, making the laser *tunable*. Tuning is usually achieved by putting a diffraction grating into the optical path, in such a way that only one diffracted wavelength reaches the laser mirror, and no spontaneously emitted photons at neighbouring frequencies can be amplified. By rotating the grating, the wavelength selected can be varied. The dye is usually pumped by another laser operating at a shorter wavelength. Fairly high pump powers are needed.

19.12.3 Semiconductor lasers and LEDs

The semiconductor laser currently is of great economic significance, owing to its high efficiency, small size, ease of pumping, and its range of applications in communications, data recording and data retrieval from, for example, compact discs. It is also frequently used to optically pump other solid lasing materials.

It makes use of the radiation that results from the recombination of an electron in an energy level close to the bottom of the conduction band with a hole (an empty level – see Chapter 15) close to the top of the valence band. The transition is illustrated in Fig. 19.16. The energy released in this process equals the energy gap E_g of the semiconductor, so that the wavelength is given approximately by $hc/\lambda = E_g$.

Only in certain semiconductors is this process efficient. The commonest such semiconductor is gallium arsenide, with $E_g = 1.4\,$eV, for which $\lambda = 0.85\,\mu$m. By substituting either the gallium or arsenic with Al, In or P, the wavelength can be 'tuned' over the range 0.6–3.0 μm, and lasers are commercially made over this whole range. Silicon and germanium are

Fig. 19.16

Stimulated emission across the energy gap of a semiconductor.

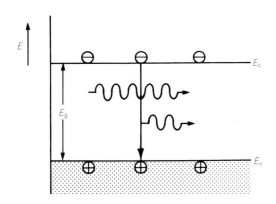

not normally used, because their energy bands are such that electron-hole recombination can only produce radiation with the help of a phonon of suitable momentum. The lack of such phonons makes this a highly improbable process, although the probability increases when *two* phonons are involved.

The energy level scheme is like that of the dye laser, but the pumping mechanism is quite different. The laser is made in the form of a p–n junction diode, through which a high current is passed. As a result, a large fraction of the high concentration of electrons present on the n-type side of the junction is injected across the depletion region into the p-type side. There they recombine with the large population of holes, which is created and maintained by the p-type doping.

Thus far, this description applies equally to an LED and to a laser. The LED has no *optical feedback* from a pair of mirrors, as does a laser. To make the device into a laser, mirrors are made by cleaving a GaAs crystal along a [110] plane at either end. As the refractive index of GaAs is high (near 3.5), the natural reflectance of the surface is also relatively high, being close to 30%.

In order to raise the population of conduction electrons in the p-type region to the level required for producing gain (about $2 \times 10^{24}\,\mathrm{m}^{-3}$), both sides of the junction must be very heavily doped, and the forward bias voltage applied to the diode must be equal to the energy gap voltage E_g/e, which is nearly 1.4 V in GaAs.

The efficiency of energy conversion is good. Each electron-hole pair produces a photon of energy E_g. The electrical energy used to produce the photon is also E_g because it equals the electron's charge e, multiplied by the voltage E_g/e across the junction. Thus almost no energy is wasted in pumping electrons up into energy levels higher than the conduction band edge E_C. However, the overall efficiency is not 100%, as many electrons recombine outside the region between the mirrors, while others recombine at lattice defects without radiating photons.

Efficiencies of 20–40% are typical within the junction. Efficiency is further reduced because the high current density needed (typically $4000\,\mathrm{A\,cm}^{-2}$) generates considerable heat. This must be removed by a heat sink making direct contact with the junction. The heat sink is usually much larger than the laser itself, which is only perhaps $0.1 \times 0.25 \times 0.05$ mm. Individual lasers with light output powers of a few watts are available commercially at the time of writing, limited mainly because higher output power causes optical damage at the mirror faces.

19.13 Liquid crystals

A liquid crystal might be better described as a liquid with some crystalline features. It is a molecular liquid in which there is a substantial degree of regularity in the orientations of the molecules, and often in their positions, too. Liquid crystals are formed by organic molecules with rod-like shapes, and many thousands of such organic molecular liquids exhibit this *mesophase*, a phase of matter that is intermediate, as the name implies, between liquid and crystal.

High viscosity is a common signature of a liquid crystal, and so also are strongly *anisotropic* properties, i.e. properties depending on direction relative to a reference axis. Thus most liquid crystals have a refractive index that varies with the orientation of the molecules, giving them a unique appearance under a microscope when illuminated with polarized light.

We look first at a typical example of a liquid crystal, the organic substance para-azoxyanisole, abbreviated as PAA (more correctly 4,4'-dimethoxy-azoxybenzene), whose molecular structure is given in Table 19.4. PAA is a liquid crystal in the temperature range between its melting point, 118°C,

Table 19.4

Transition temperatures of some liquid crystals

Molecular name	Type	Transition temperatures (°C)		Latent heat (kJ mol^{-1})
Para-azoxyanisole, PAA	Nematic	Crystal ↔ nematic	118	29.6
		Nematic ↔ isotropic	135.5	0.574
4'-heptyl-4-cyanobiphenyl	Nematic	Crystal ↔ nematic	28.5	
		Nematic ↔ isotropic	42	
n-(4-methoxybenzylidene)-4'-butylaniline, MBBA	Nematic	Crystal ↔ nematic	22	18.06
		Nematic ↔ isotropic	47	0.582
Cholesteryl benzoate	Cholesteric	Crystal ↔ cholesteric	147	
		Cholesteric ↔ isotropic	186	
4,4'-dinonyl azobenzene	Smectic	Crystal ↔ smectic A	38	
		Smectic A ↔ smectic B	41	
		Smectic B ↔ isotropic	54.1	

and its *clearing point* of 135.5°C, the temperature at which it transforms into a clear liquid with low viscosity. Within that limited temperature range it has unusually high viscosity and a milky appearance. The transformation at 135.5°C to a clear liquid, like that at 118°C, is accompanied by a density change and the absorption of latent heat. This well-defined transition thus displays characteristics like those of a melting crystal, and the need for an input of heat energy is strongly suggestive of an internal structural change. Structure in liquid crystals is readily confirmed using X-ray diffraction, as described earlier for crystals.

Liquid crystals like PAA are termed *thermotropic* because temperature determines the appearance and disappearance of the phase. High pressure also changes the state of a liquid crystal, shifting its melting and clearing points to higher temperatures. Other substances, like soaps, form *lyotropic* liquid crystals when dissolved in a solvent (often water). These are discussed later. Materials whose thermotropic temperature range brackets room temperature are widely used in display screens for laptop computers, car dashboards, telephones, etc. This is one reason why we concentrate on thermotropic liquid crystals here.

Three principal types of structural regularity are normally distinguished among liquid crystals, although on geometrical grounds many more are possible. The three types of molecular ordering are illustrated schematically in Fig. 19.17, and Table 19.4 gives examples of substances of the three types.

(a)

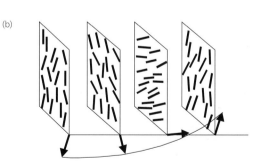

(b)

Fig. 19.17

Three schemes of molecular ordering in thermotropic liquid crystals: (a) nematic, (b) cholesteric (or chiral nematic), (c) smectic. Each short line represents one molecule.

(c)

The simplest are *nematic* liquid crystals – PAA is one of these. The long molecules align themselves as shown in Fig. 19.17(a), approximately parallel to one another, so their average direction defines a single axis of symmetry that we call the *director*. Unlike a crystal, the molecules in a nematic liquid crystal are not ordered in *position*. Thermal agitation causes individual molecules to move their positions randomly relative to other molecules, as happens in normal liquids, but also to wobble continually in their orientation by up to a few tens of degrees. The degree of alignment of the molecules decreases as the temperature rises and is characterized by an *order parameter* that is rather similar to the intensity of magnetization, M, in a ferromagnet.

In *cholesteric* liquid crystals, often called *chiral nematic*, the average molecular direction (called the *director*) spirals from place to place about an axis as in Fig. 19.17(b), with a characteristic periodicity that varies widely from substance to substance.

Smectic liquid crystals show alignment of molecules relative to an axis, like a nematic liquid crystal, but in addition the molecules arrange themselves in layers [Fig. 19.17(c)]. Six different smectic types occur: the director may be parallel in neighbouring layers, as in the figure, or it may change direction from layer to layer, either randomly or in some regular way. Positional order within the layers is also found in other smectic types. As you can see from the last row in Table 19.4, some liquid crystals change their structure between the melting and clearing points.

In its natural state, the milky appearance of a liquid crystal is caused by light scattered by many microscopic *domains* of aligned molecules, typically of micrometer dimensions. These are exactly like the grains of a polycrystalline material, the domains being separated by *domain boundaries*, across which the director changes its direction within the length of a few molecules. Just as crystalline sugar or common salt grains scatter light strongly, so do these domains. The director within each domain has no preferred orientation, so that its direction will change abruptly from one domain to the next. Now the refractive index and the permittivity both differ with different orientations of the director relative to the electric field of the light. The reason for this lies in the molecular structure. Both the structure and the conjugated bonding evident in Table 19.4 are similar to those of the dye molecules discussed in Section 19.4. Thus when the electric field is parallel to the length of the molecule the refractive index has one value, and a different one when the electric field lies at right angles to the molecule. As with dye molecules, this behaviour is a result of the ease with which electrons can be driven from one end of the long molecule to the other. The milky appearance is thus explained by the scattering mechanism described in Section 19.8, caused by differences in the refractive index from point to point.

Now the directors within all domains of a liquid crystal can be aligned by the use of a strong static magnetic or electric field – the long axis of each molecule rotates under the field's influence until it lies, as close as thermal agitation will allow, either parallel to or perpendicular to the field axis, depending on which direction corresponds to the highest permittivity. As a result, the milkiness disappears, and the liquid crystal becomes transparent, just like a large, perfect crystal of salt or sugar. At the same time, the uniformly aligned liquid crystal has a refractive index that depends on the polarization of a beam of light passing through it. Suppose that a beam of light travels in a direction *perpendicular* to the axis along which the molecules are aligned. The refractive index, and hence also the speed at which the light travels, differs according to whether this light beam is polarized parallel to the alignment axis or at right angles to it. Moreover, if the director in the LC is made to spiral around the axis of the light beam, as in Fig. 19.17(b), the axis of polarization of the light follows the same spiral as the director! It is these properties that are used in display panels.

A display panel is a thin cell of glass containing a nematic liquid crystal. The cell is equipped with electrodes designed to rotate the director, usually through 90°, into the direction of travel of the light beam. When the director is so aligned, a beam of polarized light passes through with unchanged polarization. But the director of the LC is arranged to rotate as in Fig. 19.17(b) through 90° between front and back of the cell when no electric field is applied from the electrodes. The beam of light will then have its polarization rotated through the same angle. This cell is effectively an electrically rotatable polarizer of light. By placing the cell between a pair of crossed polarizers consisting of polymer sheets, it can be made into an on/off light valve. The design and successful construction of a workable cell also requires an understanding of a rather complicated interaction of the liquid crystal director with the walls of the cell.

It is because the long organic molecules can change the polarization of light that, in a microscope using polarized light, these liquids display complex patterns, differing according to whether they are nematic, spiral nematic or smectic. For example, when no field is present, nematic liquid crystals are characterized by patterns like intertwined threads (*nematic*, meaning thread-like, is derived from a Greek word).

Lyotropic liquid crystals also exhibit many different molecular arrangements. Composed of a solvent and solute, the solute molecules consist of long flexible chains with a polar group attached at one end. The polar group is strongly attracted by dipolar forces to the solvent molecules, while the flexible chain does not mix easily with the solvent. These properties determine the structures that are observed. Each is stable only over a limited range of solvent concentrations.

As an example, the molecules of common soap – sodium lauryl sulphate – when dissolved in water go through a series of different structural arrangements as the concentration of water is increased. What these arrangements have in common is that the polar groups form layers that

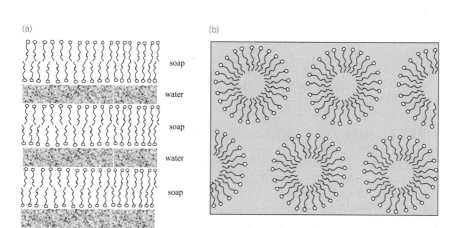

(a)

soap
water
soap
water
soap

(b)

Fig. 19.18

Two arrangements of soap molecules in water, each forming a lyotropic liquid crystal.

surround the hydrophobic chains, 'protecting' the latter from contact with water molecules. Two such arrangements are shown in Figs 19.18(a) and (b), where small circles represent the polar groups terminating the soap molecules. Diagram (b) could represent the internal structure of either cylinders or spheres, each of which occurs at a different water concentration. These and many other arrangements are found in liquid crystals in the natural world, indeed an enormous range of naturally occurring lyotropic liquid crystals are of interest as biological materials, and hence lie largely outside the scope of this book.

Problems

19.1 Estimate the absorption coefficients of (a) a metal foil of thickness 50 nm which reflects 40% and transmits 20% of the light incident upon it, and (b) window glass, which absorbs 90% of the incident light in a thickness of 0.2 m.

19.2 By substitution of Eq. (19.3) into Eq. (19.2), obtain the results quoted in Eqs (19.4) and (19.5).

19.3 With the aid of Eq. (19.5) give a dimensioned sketch of the variation with frequency of the phase angle between atomic displacement and electric field of the incoming wave.

19.4 Remembering that an oscillating charge radiates energy in phase with its *acceleration*, show that the oscillator described by Eqs (19.3)–(19.5) re-radiates a wave lagging 90° behind an incident wave of the frequency ω_0.

19.5 Use the 'electron in a box' model of Section 2.6 to find an expression for the energy difference between two adjacent energy levels having quantum numbers N and $N + 1$. Hence prove the result quoted in Eq. (19.8).

19.6 Describe some of the processes which give rise to attenuation of light in optical fibres.

19.7 Explain what is meant by the statement that the output of a laser is *coherent*. Why is the light from an incandescent source not coherent?

19.8 Find the separation between the emission lines in the spectrum of the output from

(a) a helium–neon laser with mirror separation 0.5 m and wavelength 632.8 nm;

(b) a gallium arsenide semiconductor laser operating at 850 nm wavelength and with mirror separation 0.3 mm.

Self-assessment questions

1 The frequencies of infrared radiation are higher than those of visible light

(a) true   (b) false

2 Crystal lattice absorption occurs predominantly in which part of the spectrum?

(a) visible   (b) infrared   (c) ultraviolet

3 Electronic absorption occurs at infrared frequencies in

(a) insulators   (b) metals   (c) semiconductors

(d) organic molecules   (e) glasses

4 The following processes contribute to the intensity loss on transmission through a medium

(a) electronic absorption

(b) electron–phonon collisions

(c) photon–phonon collisions

(d) stimulated emission

(e) Rayleigh scattering

5 Crystal lattice absorption in insulators gives rise to a spectrum of

(a) narrow lines

(b) a broad band of wavelengths

(c) an absorption edge

6 Metals become fairly transparent beyond wavelengths

(a) just longer than visible

(b) just shorter than visible

7 On the long wavelength side of an absorption edge the absorption varies with wavelength

(a) exponentially   (b) logarithmically

(c) parabolically

8 The energy gap in a semiconductor determines the wavelength below which the absorption rises strongly

(a) true   (b) false

9 The refractive index is which function of the permittivity ε measured at optical frequencies?

(a) square   (b) square root   (c) cube

10 Refractive index dispersion is the change in refractive index with

(a) photon energy   (b) wavelength

(c) energy gap   (d) absorption

11 Anomalous dispersion occurs

(a) far from an absorption line

(b) at the edges of an absorption line

(c) at the centre of an absorption line

12 Rayleigh scattering is due to

(a) fluctuations in absorption

(b) stimulated emission

(c) vibrations of O–H bonds

(d) fluctuations in refractive index

13 Rayleigh scattering varies with wavelength λ as

(a) λ (b) λ^{-1} (c) λ^2 (d) λ^{-2}

(e) λ^3 (f) λ^{-3} (g) λ^4 (h) λ^{-4}

14 Density fluctuations in glasses cause stronger Rayleigh scattering than do composition fluctuations

(a) true   (b) false

15 Stimulated emission occurs whenever

(a) an electron makes a transition to a lower energy level

(b) a photon is absorbed by an electron

(c) both (a) and (b) occur simultaneously

(d) a photon causes an electron to emit radiation

16 Two photons are coherent when

(a) they travel at the same speed

(b) their wavelengths are the same

(c) their phases are the same

(d) they obey Planck's equation

17 A laser requires mirrors because

(a) they provide optical feedback

(b) they invert the electron populations

(c) they determine the wavelength at which lasing occurs

18 The number of electrons in the upper level of a laser must be

(a) higher than the equilibrium population

(b) higher than the population in the lower level

(c) lower than the population in the lower level

Each of the sentences in Questions 19–25 consists of an assertion followed by a reason. Answer:

(a) If both assertion and reason are true statements and the reason is a correct explanation of the assertion.

(b) If both assertion and reason are true statements but the reason is not a correct explanation of the assertion.

(c) If the assertion is true but the reason contains a false statement.

(d) If the assertion is false but the reason contains a true statement.

(e) If both assertion and reason contain false statements.

19 Crystal lattice absorption cannot readily be observed in metals *because* it is masked by electronic absorption

20 Pure silica shows a lower Rayleigh scattering loss than lead glass *because* it contains no elements other than Si and O

21 A laser requires a pump energy source *because* stimulated emission is impossible without it

22 Crystal lattice absorption occurs at a single frequency *because* only one phonon frequency travels at the same speed as the absorbed radiation

23 Organic dye molecules are coloured *because* electrons in them are confined to the molecule

24 Less pump power is needed in a two-level laser than in a three-level laser *because* it is easier to increase the population in the upper level

25 Metals reflect strongly *because* they also absorb strongly

Answers

1 (b)	2 (b)	3 (b)	4 (a), (c), (e)
5 (a)	6 (b)	7 (a)	8 (a)
9 (b)	10 (b), (a)	11 (c)	12 (d)
13 (h)	14 (b)	15 (d)	16 (c) + (b)
17 (a)	18 (b)	19 (a)	20 (a)
21 (c)	22 (a)	23 (b)	24 (e)
25 (b)			

Further reading

Chapter numbers are given in square brackets.

General

Chalmers, B. (1982) *The Structure and Properties of Solids*. Heyden, London.

Cotterill, R. (1985) *The Cambridge Guide to the Material World*. CUP, Cambridge.

Crane, F. A. A. and Charles, J. A. (1984) *Selection and Use of Engineering Materials*. Butterworth Heinemann, Oxford.

Easterling, K. (1988) *Tomorrow's Materials*. Institute of Metals, London.

Van Vlack, L. H. (1982) *Materials for Engineering*. Addison-Wesley, London.

Van Vlack, L. H. (1985) *Elements of Materials Science and Engineering*, 5th edn. Addison-Wesley, London.

Basic physics and chemistry

Shriver, D. F. and Atkins, P. W. (1999) *Inorganic Chemistry*, 3rd edn. OUP, Oxford.

Structure of materials

Ashby, M. F. and Jones, D. R. H. (1998) *Engineering Materials 2: An Introduction to Microstructure, Processing and Design*, 2nd edn. Butterworth Heinemann, Oxford [9, 10, 11, 13].

Campbell, I. M. (2000) *Introduction to Synthetic Polymers*, 2nd edn. OUP, Oxford [13].

Porter, D. A. and Easterling, K. E. (1992) *Phase Transformations in Metals and Alloys*. Nelson Thornes, Cheltenham [9].

Verhoeven, J. D. (1975) *Fundamentals of Physical Metallurgy*. Wiley, Chichester [8, 9].

Wright, J. D. (1995) *Molecular Solids*, 2nd edn. CUP, Cambridge [6].

Thermal behaviour

Bett, K. E., Rowlinson, J. S. and Saville, G. (1975) *Thermodynamics for Chemical Engineers*. Athlone Press, London [7].

Swalin, R. A. (1972) *Thermodynamics of Solids*, 2nd edn. John Wiley, New York [7].

Tabor, D. (1991) *Gases, Liquids and Solids*, 3rd edn. London, Penguin [7].

Warn, J. W. R. and Peters, A. P. H. (1996) *Concise Chemical Thermo-dynamics*. Stanley Thornes, Cheltenham [7].

Mechanical properties

Gordon, J. E. (1991) *The New Science of Strong Materials*. London, Penguin [13].

Harris, B. (1986) *Engineering Composite Materials*. Institute of Metals, London [11].

Hull, D. (1981) *An Introduction to Composite Materials*. CUP, Cambridge [11].

Le May, I. (1981) *Principles of Mechanical Metallurgy*. Arnold, London [9, 10].

Matthews, F. L. and Rawlings, R. D. (1994) *Composite Materials: Engineering and Science*, London, Chapman & Hall [13].

McCrum, N. G., Buckley, C. P. and Bucknall, C. B. (1997) *Principles of Polymer Engineering*, 2nd edn. Oxford, Oxford Science Publications.

Electrical and magnetic properties

Jiles, D. C. (1998) *Introduction to Magnetism and Magnetic Materials*. Nelson Thornes, Cheltenham [17].

Jiles, D. C. (2001) *Introduction to Electronic Properties of Materials*, 2nd edn. Nelson Thornes, Cheltenham [14–16].

Leaver, K. D. (1997) *Microelectronic Devices*, 2nd edn. Imperial College Press, London [14, 15].

Lines, M. E. and Glass, A. E. (1977) *Principles and Applications of Ferroelectrics and Related Materials*. OUP, Oxford [18].

O'Handley, R. C. (1999) *Modern Magnetic Materials*. John Wiley and Sons, Chichester [17].

Solymar, L. and Walsh, D. (1998) *Electrical Properties of Materials*, 6th edn. OUP, Oxford [13–19].

Sze, S. M. (1985) *Semiconductor Devices – Physics and Technology*. Wiley, New York [14–16].

Data books

Banccio M. (ed.) (1993) *American Society of Metals Reference Book*, 3rd edn. ASM International, Ohio.

Kaye, G. W. C. and Laby, T. H. (1995) *Tables of Physical and Chemical Constants*, 16th edn. Longman, London.

Lide, D. R. (ed.) *Handbook of Chemistry and Physics*. Chemical Rubber Publishing Company, New York (published annually).

Rice, R. W. (ed.) (1993) *Electrical Engineering Materials Reference Guide*. Marcel Dekker, New York.

Appendix 1
Units and
conversion factors

SI units

(1) Basic units

quantity	unity	unit symbol
length	metre	m
mass	kilogramme	kg
time	second	s
electric current	ampere	A
thermodynamic temperature	Kelvin	K
luminous intensity	candela	cd
amount of substance	mole	mol

(2) Derived units

quantity	unit	unit symbol
force	newton	$N = kg\,m\,s^{-2}$
work, energy, heat	joule	$J = N\,m$
power	watt	$W = J\,s^{-1}$
pressure	pascal	$Pa = N\,m^{-3}$
	electrical units	
potential	volt	$V = W\,A^{-1}$
resistance	ohm	$\Omega = V\,A^{-1}$
charge	coulomb	$C = A\,s$
capacitance	farad	$F = A\,s\,V^{-1}$
electric field strength	–	$V\,m^{-1}$
electric flux density	–	$C\,m^{-2}$
	magnetic units	
magnetic flux	weber	$Wb = V\,s$
inductance	henry	$H = V\,s\,A^{-1}$
magnetic field strength	–	$A\,m^{-1}$
intensity of magnetization	–	$A\,m^{-1}$
magnetic flux density	tesla	$T = Wb\,m^{-2}$

(3) Unit conversion factors

Length, volume

$$1 \text{ mil} = 2.54 \times 10^{-5} \text{ m}$$
$$1 \text{ in} = 2.54 \text{ cm} \qquad 1 \text{ in}^3 = 16.39 \text{ cm}^3$$
$$1 \text{ ft} = 0.3048 \text{ m} \qquad 1 \text{ ft}^3 = 0.02832 \text{ m}^3$$
$$1 \text{ mile} = 5280 \text{ ft} = 1.609 \text{ km}$$
$$1 \, \mu\text{m} = 10^{-6} \text{ m} = 39.37 \, \mu\text{in}$$
$$1 \, \text{Å} = 10^{-10} \text{ m}$$
$$1 \text{ gal} = 0.1605 \text{ ft}^3 = 4546 \text{ cm}^3$$
$$1 \text{ US gal} = 0.1337 \text{ ft}^3 = 3785 \text{ cm}^3$$
$$1 \, \text{l} = 10^{-3} \text{ m}^3 = 1000 \text{ cm}^3$$

Mass

$$1 \text{ lb} = 0.4536 \text{ kg}$$
$$1 \text{ ton} = 2240 \text{ lb} = 1016 \text{ kg}$$
$$1 \text{ tonne} = 1000 \text{ kg}$$

Density

$$1 \text{ lb/in}^3 = 27.68 \text{ g/cm}^3$$
$$1 \text{ lb/ft}^3 = 16.02 \text{ kg/m}^3$$
$$1 \text{ g/cm}^3 = 10^3 \text{ kg/m}^3$$

Force

$$1 \text{ pdl} = 0.1383 \text{ N}$$
$$1 \text{ lbf} = 32.17 \text{ pdl} = 4.448 \text{ N}$$
$$1 \text{ tonf} = 9964 \text{ N}$$
$$1 \text{ kgf} = 2.205 \text{ lbf} = 9.807 \text{ N}$$
$$1 \text{ dyn} = 10^{-5} \text{ N}$$

Power

$$1 \text{ hp} = 550 \text{ ft lbf/s} = 0.7457 \text{ kW}$$
$$1 \text{ ft lbf/s} = 1.356 \text{ W}$$

Torque

$$1 \text{ lbf ft} = 1.356 \text{ N m}$$
$$1 \text{ tonf ft} = 3037 \text{ N m}$$

Energy, work, heat

$$1 \text{ ft lbf} = 1.356 \text{ J}$$
$$1 \text{ kWh} = 3.6 \text{ MJ}$$
$$1 \text{ Btu} = 1055 \text{ J} = 252 \text{ cal} = 778.2 \text{ ft lbf}$$
$$1 \text{ cal} = 4.187 \text{ J}$$
$$1 \text{ hp h} = 2.685 \text{ MJ}$$
$$1 \text{ erg} = 10^{-7} \text{ J}$$

Pressure stress

$$1 \text{ lbf/in}^2 = 0.07031 \text{ kgf/cm}^2 = 6895 \text{ Pa}$$
$$1 \text{ tonf/in}^2 = 157.5 \text{ kgf/cm}^2 = 15.44 \text{ MN/m}^2$$
$$1 \text{ kgf/cm}^2 = 0.09807 \text{ MN/m}^2$$
$$1 \text{ lbf/ft}^2 = 47.88 \text{ N/m}^2 = 47.88 \text{ Pa}$$
$$1 \text{ ft H}_2\text{O} = 62.43 \text{ lbf/ft}^2 = 2989 \text{ Pa}$$
$$1 \text{ in Hg} = 70.73 \text{ lbf/ft}^2 = 33\,896 \text{ Pa}$$
$$1 \text{ mm Hg} = 1 \text{ torr} = 133.3 \text{ Pa}$$
$$1 \text{ Int atm} = 1.013 \times 10^5 \text{ Pa} = 14.70 \text{ lbf/in}^2$$
$$1 \text{ bar} = 10^5 \text{ Pa} = 14.50 \text{ lbf/in}^2$$

Temperature
$$1°C = 1.8°F$$
$$T\,K = T°C + 273.15°C$$

Dynamic viscosity
$$1 \text{ poise (g cm s)} = 0.1 \text{ kg/m s} = 0.1 \text{ Ns/m}^2$$
$$1 \text{ kgf s/m}^2 \qquad = 0.9807 \text{ kg/m s}$$
$$1 \text{ lbf s/in}^2 \qquad = 6895 \text{ kg/m s}$$

Kinematic viscosity
$$1 \text{ ft}^2/\text{s} = 0.09290 \text{ m}^2/\text{s}$$
$$1 \text{ in}^2/\text{s} = 6.452 \text{ cm}^2/\text{s}$$

Thermal conductivity
$$1 \text{ cal/cm s K} = 418.7 \text{ J/m s K}$$

Electrical units
The following conversion factors are from the CGS system to the SI system.
(Note: in the CGS system $1 \text{ e.m.u.} = 10^{10} \text{ e.s.u. of charge.}$)

capacitance	1 e.s.u.	$= \frac{1}{9} \times 10^{-11} \text{ F}$
charge	1 e.m.u.	$= 10 \text{ C}$
current	1 e.m.u.	$= 10 \text{ A}$
electric field strength	1 e.s.u.	$= 3 \times 10^4 \text{ V/m}$
electric flux density	1 e.s.u.	$= (1/12\pi) \times 10^{-5} \text{ C/m}^2$
electric polarization	1 e.s.u.	$= \frac{1}{3} \times 10^{-5} \text{ C/m}^2$
inductance	1 e.m.u.	$= 10^{-9} \text{ H}$
intensity of magnetization	1 e.m.u.	$= 10^3 \text{ A/m}$
magnetic field strength	1 e.m.u.	$= (1/4\pi) \times 10^3 \text{ A/m}$
magnetic flux	1 e.m.u.	$= 10^{-8} \text{ Wb}$
magnetic flux density	1 e.m.u.	$= 10^{-4} \text{ T}$
magnetic moment	1 e.m.u.	$= 10^{-3} \text{ Am}^2$
magnetomotive force	1 e.m.u.	$= (10/4\pi) \text{ A}$
mass susceptibility	1 e.m.u./g	$= 4\pi \times 10^{-3}/\text{kg}$
potential	1 e.m.u	$= 10^{-8} \text{ V}$
resistance	1 e.m.u.	$= 10^{-9} \,\Omega$

Appendix 2
Physical constants

Avogadro's number	N	$= 6.023 \times 10^{23}/\text{mol}$
Bohr magneton	β	$= 9.27 \times 10^{-24}\,\text{A}\,\text{m}^2$
Boltzmann's constant	k	$= 1.380 \times 10^{-23}\,\text{J/K}$
electron volt	eV	$= 1.602 \times 10^{-19}\,\text{J}$
electron charge	e	$= 1.602 \times 10^{-19}\,\text{C}$
electron rest mass	m_e	$= 9.109 \times 10^{-31}\,\text{kg}$
electronic charge to mass ratio	e/m_e	$= 1.759 \times 10^{11}\,\text{C/kg}$
energy for $T = 290\text{K}$	kT	$= 4 \times 10^{-21}\,\text{J}$
energy of ground state H atom (Rydberg energy)		$= 13.60\,\text{eV}$
Faraday constant	F	$= 9.65 \times 10^7\,\text{C/mol}$
permeability of free space	μ_0	$= 4\pi \times 10^{-7}\,\text{H/m}$
premittivity of free space	ε_0	$= (1/36\pi) \times 10^{-9}\,\text{F/m}$
Planck's constant	h	$= 6.626 \times 10^{-34}\,\text{Js}$
proton mass	m_p	$= 1.672 \times 10^{-27}\,\text{kg}$
proton to electron mass ratio	m_p/m_e	$= 1836.1$
Average radius of wave function of H atom in ground state		$= 0.529 \times 10^{-10}\,\text{m}$
		$= 0.529\,\text{Å}$
standard gravitational acceleration	g	$= 9.807\,\text{m/s}^2$
		$= 32.17\,\text{ft/s}^2$
Stefan–Boltzmann constant	σ	$= 5.67 \times 10^{-8}\,\text{J/m}^2\,\text{s}\,\text{K}^4$
universal constant of gravitation	G	$= 6.67 \times 10^{-11}\,\text{N}\,\text{m}^2/\text{kg}^2$
		$= 3.32 \times 10^{-11}\,\text{lbf}\,\text{ft}^2/\text{lb}^2$
universal gas constant	R	$= 8.314\,\text{J/mol}\,\text{K}$
velocity of light in vacuo	c	$= 2.9979 \times 10^8\,\text{m/s}$
volume of 1 mol of ideal gas at N.T.P.		$= 22.42\,\text{l}$

Appendix 3
Physical properties of elements

Atomic number	Element	Density at 20°C (10^3 kg m^{-3})	Melting point (°C)	Boiling point (°C)	Young's modulus (10^{10} N m^{-2})	Shear modulus (10^{10} N m^{-2})	Poisson's ratio
1	Hydrogen (H)	0.00009	−259.2	−252.7			
2	Helium (He)	0.00018		−268.9			
3	Lithium (Li)	0.534	180.5	1330	1.15		
4	Beryllium (Be)	1.848	1277	2770	29.65	14.48	0.08
5	Boron (B)	2.34	(2100)	(2550)			
6	Carbon (C)				0.69		
	Graphite	2.25	3700[a]	4830			
					(Polycrystalline)		
	Diamond	3.52			82.74	34.48	0.25
7	Nitrogen (N)	0.00125	−210	−196			
8	Oxygen (O)	0.00143	−219	−183			
9	Fluorine (F)	0.0017	−219.6	−188			
10	Neon (Ne)	0.0009	−249	−246			
11	Sodium (Na)	0.971	97.8	892	0.9		
12	Magnesium (Mg)	1.74	650	1105	1.72		0.3
13	Aluminium (Al)	2.699	660	2450	6.9	2.62	0.34
14	Silicon (Si)	2.33	1410	2680	11.0		
15	Phosphorus (P) (white)	1.83	44.2	280			
16	Sulphur (S) (yellow)	2.07	119	445			
17	Chlorine (Cl)	0.0032	−101	−34.7			
18	Argon (A)	0.0018	−189.4	−186			
19	Potassium (K)	0.86	63.7	760	0.34		
20	Calcium (Ca)	1.55	838	1440	2.07	0.69	0.31
21	Scandium (Sc)	2.99	1540	2730			
22	Titanium (Ti)	4.507	1670	3260	11.58	4.14	0.34
23	Vanadium (V)	6.1	1860	3400	13.79	5.03	0.36
24	Chromium (Cr)	7.19	1875	2665	24.82		
25	Manganese (Mn)	7.43	1245	2150	15.86		
26	Iron (Fe)	7.87	1536	3000	19.65	7.93	0.28
27	Cobalt (Co)	8.85	1495	2900	20.68		0.31
28	Nickel (Ni)	8.90	1453	2730	21.37	7.93	0.31
29	Copper (Cu)	8.96	1083	2600	12.41	4.62	0.35
30	Zinc (Zn)	7.13	419.5	906	9.65	3.45	0.35
31	Gallium (Ga)	5.91	29.8	2240	0.97		

Atomic number	Element	Density at 20°C (10^3 kg m^{-3})	Melting point (°C)	Boiling point (°C)	Young's modulus (10^{10} N m^{-2})	Shear modulus (10^{10} N m^{-2})	Poisson's ratio
32	Germanium (Ge)	5.32	937	2830	7.58		
33	Arsenic (As)	5.72		613[a]			
34	Selenium (Se)	4.79	217	685			
35	Bromine (Br)	3.12	−7.2	58			
36	Krypton (Kr)	0.0037	−157	−152			
37	Rubidium (Rb)	1.53	38.9	688	0.023		
38	Strontium (Sr)	2.60	768	1380	1.72	0.69	0.28
39	Yttrium (Y)	4.47	1510	3030			
40	Zirconium (Zr)	6.49	1852	3580	9.65	3.45	0.34
41	Niobium (Nb) Columbium (Cb)	8.57	2470	4900	10.34	3.72	0.38
42	Molybdenum (Mo)	10.22	2610	5550	34.48		
43	Technetium (Tc)		(2100)	(3900)	40.68		
44	Ruthenium (Ru)	12.2	(2500)	(4900)	41.37	18.61	0.25
45	Rhodium (Rh)	12.44	1965	4500	28.96		
46	Palladium (Pd)	12.02	1552	4000	11.72	4.83	0.39
47	Silver (Ag)	10.49	960.8	1761	7.58	2.76	0.38
48	Cadmium (Cd)	8.65	320.9	765	5.52	2.76	0.29
49	Indium (In)	7.31	156.2	2000	1.10		
50	Tin (Sn)	7.298	231.9	2270	4.69	1.72	0.36
51	Antimony (Sb)	6.62	630.5	1380	7.58		
52	Tellurium (Te)	6.24	449.5	990	4.14		
53	Iodine (I)	4.94	113.7	183			
54	Xenon (Xe)	0.0059	−112	−108			
55	Caesium (Cs)	1.903	28.7	690		0.17	
56	Barium (Ba)	3.5	714	1640	1.24		
57	Lanthanum (La)	6.19	930	3470	6.90		
58–71	(Rare earth elements)						
72	Hafnium (Hf)	13.09	2250	5400	13.79	3.45	0.37
73	Tantalum (Ta)	16.6	2980	5400	18.62	18.62	
74	Tungsten (W)	19.3	3410	5900	34.48	15.17	0.17
75	Rhenium (Re)	21.04	3170	5900	48.27	20.69	0.26
76	Osmium (Os)	22.57	(3000)	5500	55.16	23.44	0.25
77	Iridium (Ir)	22.5	2455	5300	51.71	22.06	0.26
78	Platinum (Pt)	21.45	1769	4530	14.48	5.52	0.39
79	Gold (Au)	19.32	1063	2970	8.27	2.76	0.42
80	Mercury (Hg)	13.55	−38.6	357			
81	Thallium (Tl)	11.85	303	1457	0.69	0.28	0.45
82	Lead (Pb)	11.36	327.4	1725	1.79	0.55	0.45
83	Bismuth (Bi)	9.80	271.3	1560	3.17		
84	Polonium (Po)		250				
85	Astatine (At)		(300)				
86	Radon (Rn)	0.01	(−70)	−61.8			
87	Francium (Fr)		(27)				
88	Radium (Ra)	5.0	700				
89	Actinium (Ac)		(1000)				
90	Thorium (Th)	11.66	1750	(3850)	7.58	2.76	0.30
91	Protactinium (Pa)	15.4	(1200)				
92	Uranium (U)	19.07	1132	3820	17.24	7.58	0.24

(Numbers in parentheses are uncertain)   [a] Sublimes

Appendix 4
Fourier analysis:
an introduction

In Chapter 2 we referred to the fact that combining sinusoidal waveforms of different frequencies produces a wave whose amplitude, as a function of position, differs from the amplitudes of the two constituent waves. It is possible to demonstrate this using simple trigonometrical formulae. Mathematically, a plane wave whose amplitude varies with time, t, can be represented by the expression

$$y = A \sin \omega t \qquad\qquad \text{(A4.1)}$$

where y is the amplitude at any instant of time, A is the maximum amplitude, and ω is 2π times the frequency.

If the wave is travelling in a given direction at a velocity $c\,\mathrm{m\,s^{-1}}$, then it will take a time $x/c\,\mathrm{s}$ to reach a point x m from its point of origin, i.e. the variation of amplitude at the point x will lag, in time, behind the variation at the origin. Thus the wave will then be represented either by

$$y = A \sin[2\pi f\,(t - x/c)]$$

or, because $c = f\lambda$ and $\omega = 2\pi f$, by the equivalent expression

$$y = A \sin(\omega t - 2\pi x/\lambda) \qquad\qquad \text{(A4.2)}$$

Using the formula for $\sin(A - B)$ we may write this as

$$y = A[\sin \omega t \cos(2\pi x/\lambda) - \cos \omega t \sin(2\pi x/\lambda)]$$

If we take $\omega t = \pi/2$, then $y = \cos 2\pi x/\lambda$, while, at some subsequent time such that $\omega t = \pi$, we have $y = \sin(2\pi x/\lambda)$. These are illustrated in Fig. A4.1 and it is easy to see that the equation describes a wave travelling steadily through space.

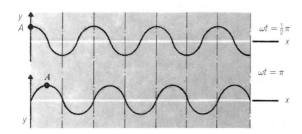

Fig. A.4.1

A travelling wave.

If we were to combine two waves of the same frequency and amplitude travelling in opposite directions, one in the $+x$ and the other in the $-x$ direction, the individual waves would be given by

$$y_1 = A \sin(\omega t - 2\pi x/\lambda) \quad \text{and} \quad y_2 = A \sin(\omega t + 2\pi x/\lambda)$$

so that using the trigonometrical formula for $\sin(A+B)$

$$y = y_1 + y_2 = 2A \sin \omega t \cos(2\pi x/\lambda) \qquad \text{(A4.3)}$$

In this case the variation of the wave with distance, x, is the same for all times, t, being represented by $\cos 2\pi x/\lambda$ in every case. This is shown in Fig. A4.2 and is called a standing wave. In it there are fixed positions in space called nodes, distance $\lambda/2$ apart, where the amplitude of the wave is always zero and positions called antinodes where the wave varies with time between zero and a maximum value $2A$.

When we wish to deduce the shape of a combination of travelling waves of different frequencies we may do so by representing them by trigonometric functions having differing values of ω, λ and amplitude A and adding them up. In principle this can be done by using the trigonometric formulae to simplify the expressions and plotting a graph of the resulting equation. However, this rapidly becomes clumsy algebraically, and the simplest method for the more straightforward cases is to draw graphs representing each of the component waves and add them up, point by point. This is, in effect, what has been done in Fig. 2.1.

The mathematical technique known as Fourier analysis simplifies the algebraic complexities. This is based on Fourier's theorem, which states that any single-valued periodic function, $y = f(t)$, having a period 2π, may be expressed in the form

$$y = c + A_1 \sin(\omega t + \phi_1) + A_2 \sin[2(\omega t + \phi_2)] + A_3 \sin[3(\omega t + \phi_3)] + \cdots \quad \text{(A4.4)}$$

Because we can write

$$A \sin(\omega t + \phi) = a \sin \omega t + b \cos \omega t$$

where we have put $a = A \cos \phi$ and $b = A \sin \phi$, the right-hand side of Eq. (A4.4) may be rewritten as

$$y = c + a_1 \sin \omega t + a_2 \sin 2\omega t + a_3 \sin 3\omega t + \cdots$$

$$+ b_1 \cos \omega t + b_2 \cos 2\omega t + b_3 \cos 3\omega t + \cdots$$

that is

$$y = c + \Sigma a_n \sin n\omega t + \Sigma b_n \cos n\omega t \qquad \text{(A4.5)}$$

In order to use Eq. (A1.5) to analyse a complex wave it is necessary to determine the coefficients $c, a_1, a_2, a_3, \ldots$ and $b_1, b_2, b_3, \ldots$ This is done by multiplying both sides of the equation by a suitable factor and integrating between limits. If the multiplying factor is correctly chosen, all terms vanish except those that will give the required coefficient.

The details of this are more properly the province of a mathematical textbook. However, we give here the results, together with an example of their application.

$$c = \frac{1}{2\pi} \int_0^{2\pi} y\, d(\omega t) \tag{A4.6}$$

$$a_n = \frac{1}{2\pi} \int_0^{2\pi} y \sin n\omega t\, d(\omega t) \tag{A4.7}$$

$$b_n = \frac{1}{2\pi} \int_0^{2\pi} y \cos n\omega t\, d(\omega t) \tag{A4.8}$$

where n is a positive integer.

It should be realized that for a travelling wave a graph of amplitude against time for a fixed distance will be the same as that for amplitude against distance for a fixed time. In the first instance we are standing at a fixed position and plotting the wave as it goes past and in the second case we take a look at the wave over a length of space at a fixed time; in each case the pattern is the same.

As an example we consider the series of pulses shown in Fig. A4.3 in which amplitude is plotted as a function of ωt. From $\omega t = 0$ to $\omega t = \pi$ the equation for the curve is $y = d$, from $\omega t = \pi$ to $\omega t = 2\pi$ the equation is $y = 0$ and the wave has a period 2π. Using Eq. (A1.6) and these values for y, we can write

$$c = \frac{1}{2\pi} \int_0^{\pi} y\, d(\omega t) + \frac{1}{2\pi} \int_\pi^{2\pi} y\, d(\omega t)$$

$$= d/2$$

This is the mean level of the wave.

Equation (A1.7) gives in a similar way the result

$$a_n = \frac{1}{2\pi} \int_0^{\pi} y \sin n\omega t\, d(\omega t) + \frac{1}{2\pi} \int_\pi^{2\pi} y \sin n\omega t\, d(\omega t)$$

$$= \frac{d}{n\pi} (1 - \cos n\pi)$$

Where n is odd, the bracket $(1 - \cos n\pi) = 2$, and when n is even, then $(1 - \cos n\pi) = 0$. Thus

$$a_1 = \frac{2\pi}{d}, \quad a_2 = \frac{2\pi}{3d}, \text{etc.}$$

and

$$a_1 = a_2 = a_3 = \ldots = 0$$

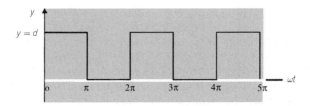

Fig. A4.3

A periodic square wave.

From Eq. (A4.8) we obtain in a similar way

$$b_n = \frac{1}{2\pi} \int_0^{\pi} y \cos n\omega t \, \mathrm{d}(\omega t) + \frac{1}{2\pi} \int_{\pi}^{2\pi} y \cos n\omega t \, \mathrm{d}(\omega t) = 0$$

It turns out that all the cosine terms are zero and the wave is represented by the expression

$$y = \frac{d}{2} + \frac{2d}{\pi} \left(\sin \omega t + \tfrac{1}{3} \sin 3\omega t + \tfrac{1}{5} \sin 5\omega t + \ldots \right) \qquad \text{(A4.9)}$$

Appendix 5
Wave mechanics

A5.1 Elementary mathematics of wave mechanics

In Section 2.6 we introduced the idea of representing an electron by a mathematical wave function, symbol Ψ, and explained that Ψ is a *complex* quantity of the form $A + jB$.

It remains to choose a mathematical expression that will represent a complex wave as a function of time t and position x. Putting ω for the angular frequency and λ for the wavelength, we can combine a suitable real function

$$\Psi' = A\cos\left(\omega t + \frac{2\pi x}{\lambda}\right)$$

with an imaginary one

$$\Psi'' = jA\sin\left(\omega t + \frac{2\pi x}{\lambda}\right)$$

to form the wave function, thus

$$\Psi = \Psi' + \Psi'' = A\exp[j(\omega t + kx)] \qquad (A5.1)$$

where we have combined both terms using de Moivre's theorem, and for brevity we have put $k = 2\pi/\lambda$.

We can separate the space-dependent and time-dependent parts of Eq. (A5.1) by writing it as the product of two exponentials, that is:

$$\Psi = A\exp(j\omega t)\exp(jkx)$$

or alternatively, we could write this in the form

$$\Psi = \psi\exp(j\omega t) \qquad (A5.2)$$

where

$$\psi = A\exp(jkx) \qquad (A5.3)$$

and ψ represents *just the space variation* of the wave. Equation (A5.2) is a *general* expression for a complex wave having a fixed frequency, in which ψ may be any function of x, not necessarily of the form $\exp(jkx)$.

Considering now the probability defined in Section 2.6, Eq. (2.18), we have

$$\Psi\Psi^* = \psi\psi^*\exp(j\omega t)\exp(-j\omega t)$$

$$= \psi\psi^*\exp(0)$$

$$= \psi\psi^*$$

Thus if, in a given situation, we can find an expression for ψ and hence for ψ^*, we can determine the most probable position for the electron in a given region of space. Because the electron carries a charge, the product $-e\psi\psi^*$, when plotted against position, gives a measure of the distribution of charge over the same region. These are the most useful applications of the theory and in the majority of cases we can forget about the time-variation term.

A5.2 *Schrödinger's wave equation*

All these ideas are brought together in a general equation describing the wave function, ψ, which is known as Schrödinger's wave equation.

We saw that Eq. (A5.3) represents the space variation of the wave function. Differentiating that expression for ψ with respect to x yields

$$\frac{d\psi}{dx} = jkA\exp(jkx)$$

and differentiating again we have

$$\frac{d^2\psi}{dx^2} = -k^2 A\exp(jkx)$$

Using Eq. (A5.3) gives just

$$\frac{d^2\psi}{dx^2} = -k^2\psi \tag{A5.4}$$

Substituting for k^2 using Eq. (2.20) we now express the right-hand side in terms of the electron's total energy E and its potential energy E_p, giving the result:

$$\frac{d^2\psi}{dx^2} + \frac{8\pi^2 m}{h^2}(E - E_p)\psi = 0 \tag{A5.5}$$

and this is the differential wave equation for a wave moving in the x-direction. Its solution yields information about the position of the electron in an environment described by the potential energy, E_p, which itself may be a function of position, x.

Equation (A5.5) has been developed by starting with the equation for a 'plane' wave, Eq. (A5.3), but it does not follow that all wave shapes will satisfy the same equation. In reality, Eq. (A5.5) cannot be derived – it was put forward by Schrödinger as an *assumption,* to be tested by its ability to predict the behaviour of real electrons. A simple example involving solution of the equation for specific conditions appears later in this appendix. Other examples are included in the problems at the end of Chapter 2.

In three dimensions described by the Cartesian coordinates x, y and z, Schrödinger's equation becomes a *partial differential equation,* as follows:

$$\frac{\partial^2\psi}{\partial x^2} + \frac{\partial^2\psi}{\partial y^2} + \frac{\partial^2\psi}{\partial z^2} + \frac{8\pi^2 m}{h^2}(E - E_p)\psi = 0 \tag{A5.6}$$

The potential energy, E_p, can be dependent on x, y and z, and the solution for ψ obtained from this equation will be a function of x, y and z.

Alternatively, the *polar coordinates* (r, θ, ϕ) may be used instead for problems where this is more convenient, such as for the hydrogen atom in

Chapter 3. θ and ϕ are angular coordinates denoting the many points around the nucleus at the *same* radius r. This transforms Schrödinger's equation into the form

$$\frac{1}{r^2}\frac{\partial}{\partial r}\left(r^2\frac{\partial\psi}{\partial r}\right)+\frac{1}{r^2\sin^2\theta}\left[\frac{\partial}{\partial\theta}\left(\sin\theta\frac{\partial\psi}{\partial\theta}\right)+\frac{\partial^2\psi}{\partial\phi^2}\right]$$

$$+\frac{8\pi^2m}{h^2}(E-E_\mathrm{p})\psi=0 \qquad\qquad (A5.7)$$

This looks a formidable equation to solve for the function $\psi(r,\theta,\phi)$, so in Chapter 3 we concentrate on the solutions that are independent of both θ and ϕ. If we set $\partial\psi/\partial\theta=0$ and $\partial\psi/\partial\phi=0$, Schrödinger's equation becomes

$$\frac{1}{r^2}\frac{\partial}{\partial r}\left(r^2\frac{\partial\psi}{\partial r}\right)+\frac{8\pi^2m}{h^2}(E-E_\mathrm{p})\psi=0 \qquad\qquad (A5.8)$$

In any of the above forms, the equation is called the *time-independent* Schrödinger equation. Its solution gives the space-dependent part, ψ, of the wave function, Ψ, and it is to be interpreted that the value of $|\psi|^2$ at any point is a measure of the probability of observing an electron at that point.

A5.3 Electron confined in a one-dimensional 'box'

By way of an example of solving Schrödinger's equation for a given environment, we consider the case discussed in Chapter 2, Section 2.9, in which the potential energy, E_p, of the electron is zero within a length L along the x-axis, and infinite outside it, as illustrated in Fig. 2.15. Because the electron can never acquire infinite energy it should not be able to exist where $E_\mathrm{p}=\infty$, so we expect to find the solution $\psi=0$ to the wave equation for this case. Mathematically, we put $E_\mathrm{p}=\infty$ for values of $x<0$ and for $x>L$, while $E_\mathrm{p}=0$ wherever $0<x<L$.

The one-dimensional wave equation is just Eq. (A5.5):

$$\frac{\mathrm{d}^2\psi}{\mathrm{d}x^2}+\frac{8\pi^2m}{h^2}(E-E_\mathrm{p})\psi=0 \qquad\qquad (A5.5)$$

At all points where E_p is infinity, the term $-E_\mathrm{p}\psi$ in this equation is so much greater than all other terms that they can be neglected, and we can therefore straight away write

$$-E_\mathrm{p}\psi=0$$

which, because $E_\mathrm{p}=\infty$, can only be true if $\psi=0$. Thus the condition $E_\mathrm{p}=\infty$ at all points outside the length L automatically ensures that $\psi=0$ there, as expected, and the probability of the electron being there is exactly zero.

Inside the length L we do not expect ψ to be zero. But exactly at the points $x=0$ and $x=L$, a finite value for ψ is not possible, for ψ cannot be a discontinuous function of x at these points. A discontinuous function has an infinite gradient, so the value of $\mathrm{d}^2\psi/\mathrm{d}x^2$ would be indeterminate at $x=0$ and $x=L$, so that Schrödinger's equation could not be satisfied there. We therefore conclude that ψ is zero at $x=0$ and $x=L$, and these values give the appropriate *boundary conditions* for the solution of Schrödinger's equation within the distance L.

We turn now to the problem of finding a solution to Schrödinger's equation inside this 'box' on the x-axis between 0 and L. Because the potential energy E_p is zero inside this region, Schrödinger's equation is simply

$$\frac{d^2\psi}{dx^2} + \frac{8\pi^2 m}{h^2} E\psi = 0 \qquad (A5.9)$$

Because the potential energy of the electron is zero, the energy E in this equation is just the kinetic energy, E_k, which is given by Eq. (2.20) in terms of the length k of the wave-vector. Rewriting that equation we have

$$k^2 = \frac{8\pi^2 mE}{h^2} \qquad (A5.10)$$

and substituting this into Eq. (A5.9) gives Eq. (A5.4) again:

$$\frac{d^2\psi}{dx^2} = -k^2\psi$$

We already know one solution of this standard type of differential equation: it is given in Eq. (A5.3) as $A\exp(jkx)$. Another is $B\exp(-jkx)$. The most general solution, as described in standard mathematical textbooks, is the sum of these two expressions:

$$\psi = A\exp(jkx) + B\exp(-jkx) \qquad (A5.11)$$

This may be tested by differentiating Eq. (A5.11) and substituting into Eq. (A5.4).

To evaluate the constants A and B we must consider the *boundary conditions* described above. So, by substituting $\psi=0$ and $x=0$ in Eq. (A5.11) we find that

$$0 = A\exp(0) + B\exp(-0)$$
$$= A + B$$

Thus $A = -B$ and Eq. (A5.11) may be written as

$$\psi = A[\exp(jkx) - \exp(-jkx)]$$

But it is known that

$$\sin\theta = \frac{e^{j\theta} - e^{-j\theta}}{2j}$$

so that

$$\psi = 2jA\sin kx \qquad (A5.12)$$

Now at the second boundary, where $x=L$, the condition $\psi=0$ again must hold so that

$$0 = 2jA\sin kL$$

which can only be true if $kL = n\pi$, where n is an integer, a *quantum number*. This implies that the only valid values of k are given by the equation

$$k = \frac{n\pi}{L} \qquad (A5.13)$$

So with this value of k Eq. (A5.12) becomes

$$\psi = 2jA\sin\frac{n\pi x}{L} \qquad (A5.14)$$

Note that negative values of n have no significance, because the oscillation of the wave function in time means that $\sin(-kx)$ is indistinguishable from $\sin kx$.

The question now arises of finding a value for the constant A, and this is done by what is called *normalization* of the solution. Because $|\psi|^2$ is to be interpreted as a probability, by the very definition of the term, the *probability* of finding the electron in the whole of space must be 1.0. Hence we must arrange that the total sum of $|\psi|^2$ over all space is unity. Expressed mathematically, this means that the integral of $|\psi|^2$ over the volume of all space equals 1, that is

$$\int_0^\infty |\psi|^2 \, dU = 1 \qquad\qquad (A5.15)$$

where dU is an infinitesimal element of volume.

In the present problem, 'all space' is confined simply to the whole x-axis, so that Eq. (A5.15) becomes

$$\int_{-\infty}^\infty |\psi|^2 \, dx = 1 \qquad\qquad (A5.13)$$

Now any region where $\psi = 0$ gives no contribution to the integral, so that we need only integrate this equation from $x = 0$ to $x = L$. Inserting into it the value of ψ from Eq. (A5.14) gives the following condition that A must satisfy, namely:

$$\int_0^L 4A^2 \sin^2\left(\frac{n\pi x}{L}\right) dx = 1$$

To evaluate this integral we first transform it with a well-known trigonometrical identity to give

$$4A^2 \int_0^L \frac{1}{2}\left[1 - \cos\left(\frac{2n\pi x}{L}\right)\right] dx = 1$$

After performing the integration we have

$$2A^2 \left[x - \frac{L}{2n\pi}\sin\left(\frac{2n\pi x}{L}\right)\right]_0^L = 1$$

which, on inserting the limits of integration, becomes

$$4A^2 \left[\frac{L}{2} - \frac{L}{4n\pi}\sin(2n\pi)\right] = 1$$

Because $\sin 2n\pi = 0$, we conclude that $2A^2 L = 1$, leading to the following result for A:

$$A = \left(\frac{1}{2L}\right)^{1/2}$$

This can now be used to write Eq. (A5.14) without any unknown quantities, as follows

$$\psi = 2j\, A \sin\frac{n\pi x}{L} \qquad\qquad (A5.17)$$

This is the equation quoted in Chapter 2.

The energy E of the electron may now be found by combining Eqs (A5.10) and (A5.13):

$$E = \frac{h^2 n^2}{8mL^2}$$

Answers to problems

Chapter 2

2.3 820 eV

2.4 5.92×10^{-29} m

2.5 0.35 nm

2.6 0.0078 nm

2.7 1.25×10^{14} m^{-2}

2.8 2.5×10^{18} m^{-2} s^{-1}

2.9 8.64×10^{9}

2.10 2.3×10^{5} m^{-1}, 2.0×10^{9} m^{-1}

Chapter 3

3.1 $3.36 \times 10^{-24} n^{2/3}$ J $(n = 1, 2, 3 \ldots)$

3.2 0.106 nm; 0.06 nm

3.3 -2.17×10^{-18} J; -5.42×10^{-19} J; -2.41×10^{-19} J; 2.46×10^{15} Hz; 2.91×10^{15} Hz; 4.55×10^{14} Hz; $n = 3$ to $n = 2$

3.4 5, 5, 7

3.5 $-e^{4}m/2h^{2}\varepsilon_{0}^{2}n^{2}$

3.9 0.1 eV

Chapter 4

4.5 Ne, Mg, S

4.6 As L shell, plus: $\ell = 2$, $m_{\ell} = 0$, ± 1, ± 2, $m_{s} = \pm \frac{1}{2}$

4.8 See Table 4.1

4.9 $1s^{2}2s^{2}2p^{6}3s^{2}3p^{6}3d^{10}$

Chapter 5

5.1 10.6 eV

5.2 $109^{\circ}\,28'$

5.3 3.51×10^{-14} N, 4.39×10^{-9} N

5.6 1.33 eV, 4.8 eV

Chapter 6

6.1 $(\sqrt{3} - 1)$, 0.104 nm
6.4 9° 36', 28° 6'
6.5 2, 4
6.7 2.25×10^3 kgm^{-3}
6.8 8.50×10^{28} m^{-3}, 8.50×10^{28} m^{-3}
6.10 0.183 nm, 0.195 nm, 0.127 nm, 0.144 nm
6.11 {111}

Chapter 7

7.2 1.5 km s^{-1}
7.3 7×10^{-6}, 2.1×10^{-5}
7.6 0.921; Pb: 2×10^{12} Hz; Au: 3.5×10^{12} Hz; NaCl: 5.8×10^{12} Hz; Fe: 7.5×10^{12} Hz; Si: 1.4×10^{13} Hz; Diamond: 3.9×10^{13} Hz.
7.7 3 translational, 3 rotational, 2 vibrational, $4R$ per mole.
7.10 591 K, 4.49×10^{-9} kg m^{-2} s^{-1}.

Chapter 8

8.1 9.2×10^6; 7.2×10^{-21}
8.2 2×10^{-19} J/atom
8.4 8.75×10^{-8} N/m
8.5 5.7×10^4 J/m^3

Chapter 9

9.1 22.86 cm^2
9.2 4.78%
9.3 40 mm^3
9.5 0.527
9.6 0.4998, 1 GN m^{-2}
9.7 206 GN m^{-2}, 80.5 GN m^{-2}
9.8 3.0 kN s m^{-2}
9.9 2.0 kN m^{-2}
9.13 116 MN m^{-2}

Chapter 10

10.1 (a) 62 wt% Sn
   (b) $\alpha_e = 18$ wt% Sn, $\beta_e = 97$ wt% Sn
   (c) none
   (d) $\approx 20\%$
   (e) $\alpha \approx 95\%$, $\beta \approx 5\%$ (90 wt% Sn alloy)
       $\alpha \approx 42\%$, $\beta \approx 58\%$ (eutectic alloy)
   (f) ≈ 20
10.3 (b) 100% γ, 0.4 wt% C
   (c) $\alpha \approx 88\%$, cementite $\approx 12\%$
   (d) eutectic mixture of $\gamma \approx 49\%$ and cementite $\approx 51\%$

Chapter 11

11.2 (a) ~63%;

(b) Liquid 5.5 wt%, cristobalite ~ 0%, mullite ~ 72 wt% Al_2O_3

11.5 ΔT_c for A and B are 719°C and 63°C respectively. Select A as higher ΔT_c and higher K

Chapter 12

12.1 (a) No: e.g. POM backbone contains —O—;

(b) yes; (c) no: e.g. PMMA is amorphous; (d) yes; (e) no: e.g. PB.

12.2 160 kg mol^{-1}, 250 kg mol^{-1}

12.4 (a) 10.5, 105 nm; (b) 3.3, 10.5 nm;

(c) 0.22, about 1×10^{-45}: should both be zero

12.5 (a) 2.415×10^{26}; (b) 6.48 links per section;

(c) 25.8 repeat units per section

12.6 128. (a) 5.18; (b) 24.7

12.7 (a) 0.9 s; (b) 0.69 s

12.8 P: crystallinity 48.4%, density 920 kg m^{-3}

Q: crystallinity 69.8%, density 952 kg m^{-3}

12.10 (a) 49 MN m^{-2}; (b) 14 GN m^{-2}; (c) 90.1 MN m^{-2}

Chapter 13

13.2 1.5 mm, 1500 MN m^{-2}

Chapter 14

14.2 8.5×10^{28} electrons/m^3, 7.4×10^{-7} m/s

14.4 4.09, 2.48, 8.54, 6.76 eV

14.5 4.1×10^{-3}, 6.7×10^{-3}, 4.4×10^{-3}, 9.0×10^{-4}, 7.9×10^{-4} m^2 V^{-1} s^{-1}

14.6 2.3×10^{-14}, 3.8×10^{-14}, 2.5×10^{-14} s

14.7 1.58×10^6 m s^{-1}, 5.28×10^{-8} m

1.39×10^6 m s^{-1}, 3.68×10^{-8} m

1.39×10^6 m s^{-1}, 3.48×10^{-8} m

14.8 2.6×10^{-8} Ω m, 79 nm, 1.06 nm

Chapter 15

15.1 0.58, 0.26, 0.074, 2.84×10^{26} m^2 V^{-1} s^{-1}

15.2 3.48×10^7 m^{-3}

15.3 9.47×10^{17} m^{-3}, 76 Ω m

2.19×10^{20} m^{-3}, 0.24 Ω m

15.4 1.77×10^{29} m^{-3}; 1 in 2×10^{10}

15.5 12.3 Ω^{-1} m^{-1}, 21.7×10^{20} m^{-3}

15.6 1.59×10^{-6} at%

15.7 2060 K

15.8 0.51 mA

15.9 1.37×10^3 Ω^{-1} m^{-1}

15.10 1.06×10^{-4} W

15.11 1.57×10^{14} Hz
15.12 2.54×10^{-8} m

Chapter 16

16.1   18.6 kW
16.2 $E_D = 3.5$ eV, $D_0 = 1$ cm^2/s
16.3 $t = 1271$ s, $n(O, t) = 4.67 \times 10^{23}$ m^{-3}
16.4   Linear: Dry 0.288 μm/h, wet 4.94 μm/h
           Parabolic: Dry 0.024 μm^2/h, wet 0.532 μm^2/h

Chapter 17

17.2   3, 4, 1, 7, 2, 4, 3
17.3 1.75×10^6 A m^{-1}
17.4 4.97×10^5 A m^{-1}
17.5 6.21×10^5 A m^{-1}
17.6 1.63×10^5 A m^{-1}
17.7 7.6×10^{-5} Wb

Chapter 18

18.1 9.25×10^{-10} N, 1.16×10^{-11} N
18.2   18.1 kV
18.3 10^{-5} C m^{-2}, 113 V
18.6 1.96×10^{-35} C m, 8.75×10^{-18} m
18.7 $E = q/2\pi\varepsilon_r\varepsilon_0 r$
18.8 $V = (q/2\pi\varepsilon) \ln(b/a)$, 12.5 pF m^{-1}
18.9   16 Hz, 0.016

Chapter 19

19.1   (a) 22 μm^{-1}
           (b) 11.5 m^{-1}
19.8   (a) 0.8 pm
           (b) 2.4 nm

Index